Tectonics of Strike-Slip Restraining and Releasing Bends

Geological Society books refereeing procedures

The Society makes every effort to ensure that the scientific and production quality of its books matches that of its journals. Since 1997, all book proposals have been refereed by specialist reviewers as well as by the Society's Books Editorial Committee. If the referees identify weaknesses in the proposal, these must be addressed before the proposal is accepted.

Once the book is accepted, the Society Book Editors ensure that the volume editors follow strict guidelines on refereeing and quality control. We insist that individual papers can only be accepted after satisfactory review by two independent referees. The questions on the review forms are similar to those for *Journal of the Geological Society*. The referees' forms and comments must be available to the Society's Book Editors on request.

Although many of the books result from meetings, the editors are expected to commission papers that were not presented at the meeting to ensure that the book provides a balanced coverage of the subject. Being accepted for presentation at the meeting does not guarantee inclusion in the book.

More information about submitting a proposal and producing a book for the Society can be found on its web site: www.geolsoc.org.uk.

It is recommended that reference to all or part of this book should be made in one of the following ways:

Cunningham, W. D. & Mann, P. (eds) 2007. *Tectonics of Strike-Slip Restraining and Releasing Bends.* Geological Society, London, Special Publications, **290**.

Waldron, J. W. F, Roselli, C. & Johnston, S. K. 2007. Transpressional structures on a Late Palaeozoic intracontinental transform fault, Canadian Appalachians. *In*: Cunningham, W. D. & Mann, P. (eds) *Tectonics of Strike-Slip Restraining and Releasing Bends.* Geological Society, London, Special Publications, **290**, 367–385.

GEOLOGICAL SOCIETY SPECIAL PUBLICATION NO. 290

Tectonics of Strike-Slip Restraining and Releasing Bends

EDITED BY

W. D. CUNNINGHAM
University of Leicester, UK

and

P. MANN
University of Texas at Austin, USA

2007
Published by
The Geological Society
London

THE GEOLOGICAL SOCIETY

The Geological Society of London (GSL) was founded in 1807. It is the oldest national geological society in the world and the largest in Europe. It was incorporated under Royal Charter in 1825 and is Registered Charity 210161.

The Society is the UK national learned and professional society for geology with a worldwide Fellowship (FGS) of over 9000. The Society has the power to confer Chartered status on suitably qualified Fellows, and about 2000 of the Fellowship carry the title (CGeol). Chartered Geologists may also obtain the equivalent European title, European Geologist (EurGeol). One fifth of the Society's fellowship resides outside the UK. To find out more about the Society, log on to www.geolsoc.org.uk.

The Geological Society Publishing House (Bath, UK) produces the Society's international journals and books, and acts as European distributor for selected publications of the American Association of Petroleum Geologists (AAPG), the Indonesian Petroleum Association (IPA), the Geological Society of America (GSA), the Society for Sedimentary Geology (SEPM) and the Geologists' Association (GA). Joint marketing agreements ensure that GSL Fellows may purchase these societies' publications at a discount. The Society's online bookshop (accessible from www.geolsoc.org.uk) offers secure book purchasing with your credit or debit card.

To find out about joining the Society and benefiting from substantial discounts on publications of GSL and other societies worldwide, consult www.geolsoc.org.uk, or contact the Fellowship Department at: The Geological Society, Burlington House, Piccadilly, London W1J 0BG: Tel. +44 (0)20 7434 9944; Fax +44 (0)20 7439 8975; E-mail: enquiries@geolsoc.org.uk.

For information about the Society's meetings, consult *Events* on www.geolsoc.org.uk. To find out more about the Society's Corporate Affiliates Scheme, write to enquiries@geolsoc.org.uk.

Published by The Geological Society from:
The Geological Society Publishing House, Unit 7, Brassmill Enterprise Centre, Brassmill Lane, Bath BA1 3JN, UK

(*Orders*: Tel. +44 (0)1225 445046, Fax +44 (0)1225 442836)
Online bookshop: www.geolsoc.org.uk/bookshop

British Library Cataloguing in Publication Data

A catalogue record for this book is available from the British Library.

ISBN 978-1-86239-238-0

Typeset by Techset Composition Ltd, Salisbury, UK

Printed by MPG Books Ltd, Bodmin, UK

Distributors

North America
For trade and institutional orders:
The Geological Society, c/o AIDC, 82 Winter Sport Lane, Williston, VT 05495, USA
Orders: Tel +1 800-972-9892
Fax +1 802-864-7626
E-mail gsl.orders@aidcvt.com

For individual and corporate orders:
AAPG Bookstore, PO Box 979, Tulsa, OK 74101-0979, USA
Orders: Tel +1 918-584-2555
Fax +1 918-560-2652
E-mail bookstore@aapg.org
Website http://bookstore.aapg.org

India
Affiliated East-West Press Private Ltd, Marketing Division, G-1/16 Ansari Road, Darya Ganj, New Delhi 110 002, India
Orders: Tel +91 11 2327-9113/2326-4180
Fax +91 11 2326-0538
E-mail affiliat@vsnl.com

Contents

Tectonics of strike-slip restraining and releasing bends

W. D. CUNNINGHAM[1] & P. MANN[2]

[1]*Department of Geology, University of Leicester, Leicester LEI 7RH, UK (e-mail: wdc2@le.ac.uk)*

[2]*Institute of Geophysics, Jackson School of Geosciences, 10100 Burnet Road, R2200, Austin, Texas 78758, USA (e-mail: paulm@ig.utexas.edu)*

One of the remarkable tectonic features of the Earth's crust is the widespread presence of long, approximately straight and geomorphically prominent strike-slip faults which are a kinematic consequence of large-scale motion of plates on a sphere (Wilson 1965). Strike-slip faults form in continental and oceanic transform plate boundaries; in intraplate settings as a continental interior response to a plate collision; and can occur as transfer zones connecting normal faults in rift systems and thrust faults in fold–thrust belts (Woodcock 1986; Sylvester 1988; Yeats *et al.* 1997; Marshak *et al.* 2003). Strike-slip faults also are common in obliquely convergent subduction settings where interplate strain is partitioned into arc-parallel strike-slip zones within the fore-arc, arc or back-arc region (Beck 1983; Jarrard 1986; Sieh & Natawidjaja 2000).

When strike-slip faults initiate in natural and experimental settings, they commonly consist of en échelon fault and fold segments (Cloos 1928; Riedel 1929; Tchalenko 1970; Wilcox *et al.* 1973). With increased strike-slip displacement, and independent of fault scale (Tchalenko 1970), fault segments link, and the linked areas along the 'principal displacement zone' may define alternating areas of localized convergence and divergence along the length of the strike-slip fault system (Fig. 1; Crowell 1974; Christie-Blick & Biddle 1985; Gamond 1987). Typically, divergent and convergent *bends* are defined as offset areas where bounding strike-slip faults are continuously linked and continuously curved across the offset, whereas more rhomboidally shaped *stepovers* are defined as zones of slip transfer between overstepping, but distinctly separate and subparallel strike-slip faults (Wilcox *et al.* 1973; Crowell 1974; Aydin & Nur 1982, 1985). However, fault stepovers may evolve into continuous fault bends as the bounding faults and connected splays propagate and link across the stepover (e.g. Zhang *et al.* 1989; McClay & Bonora 2001). Thus, the two terms 'stepover' and 'fault bend' are often used interchangeably.

Bends that accommodate local contraction are referred to as restraining bends, and those that accommodate extension are referred to as releasing bends (Fig. 1; Crowell 1974; Christie-Blick & Biddle 1985). Double bends have bounding strike-slip faults which enter and link across them, whereas single bends are essentially strike-slip fault-termination zones. Restraining and releasing bends are widespread on the Earth's surface, from the scale of major mountain ranges and rift basins to sub-outcrop-scale examples (Swanson 2005; **Mann** this volume). Releasing bends have also been documented along oceanic transforms connecting spreading ridges (Garfunkel 1986; Pockalny 1997), and extra-terrestrial restraining bends have been interpreted to occur on Europa and Venus (Koenig & Aydin 1998; Sarid *et al.* 2002).

Strike-slip restraining and releasing bends are sites of localized transpressional and transtensional deformation, respectively. Thus, bends are characterized by oblique deformation that is ultimately controlled by larger-scale relative plate motions either acting on relatively straight, long interplate boundaries (Garfunkel 1981; Mann *et al.* 1983; Bilham & Williams 1985; Bilham & King 1989) or acting across more complex zones of intraplate deformation where faults tend to be shorter, less continuous and more arcuate (**Cunningham** this volume). Within the bend, oblique deformation may be accommodated by oblique-slip faulting or partitioned into variable components of strike-slip and dip-slip fault displacements (Jones & Tanner 1995; Dewey *et al.* 1998; Cowgill *et al.* 2004*b*; **Gomez *et al.*** this volume). As seen in deeply eroded outcrop exposures or from subsurface geophysical surveys, double restraining bends and releasing bends commonly define positive and negative flower structures respectively, and strike-slip bends or 'duplexes' in plan view (Fig. 1; Lowell 1972; Sylvester & Smith 1976; Christie-Blick & Biddle 1985; Harding 1985; Woodcock & Fisher 1986; Dooley *et al.* 1999), although considerable structural variation and complexity occurs (Barka & Gulen 1989; May *et al.* 1993; Wood *et al.* 1994; Waldron 2004; Barnes *et al.* 2005; Decker *et al.* 2005; Parsons *et al.* 2005). Single bends commonly have horsetail splay fault

From: CUNNINGHAM, W. D. & MANN, P. (eds) *Tectonics of Strike-Slip Restraining and Releasing Bends.* Geological Society, London, Special Publications, **290**, 1–12.
DOI: 10.1144/SP290.1 0305-8719/07/$15.00

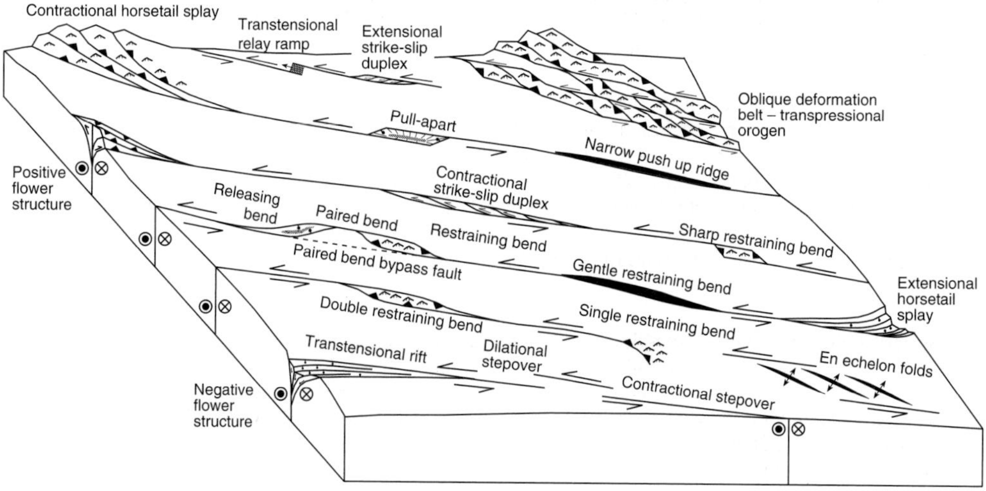

Fig. 1. Tectonic features associated with strike-slip restraining and releasing bends.

geometries in plan view, with strike-slip displacements terminally accommodated by oblique-slip and dip-slip faulting (McClay & Bonora 1997). Adjacent restraining and releasing bends called 'paired bends' by **Mann** (this volume) are commonly described from strike-slip systems in all tectonic settings and may reflect a volumetric balancing between crustal thickening and uplift at restraining bends, and crustal thinning and basin formation at releasing bends (Woodcock & Fischer 1986).

Restraining bends are sites of topographic uplift, crustal shortening and exhumation of crystalline basement (Segall & Pollard 1980; Mann & Gordon 1996; McClay & Bonora 2001), whereas releasing bends are sites of subsidence, crustal extension, significant basin sedimentation, high fluid flow, and possible volcanism (Aydin & Nur 1982; Mann *et al.* 1983; Hempton & Dunne 1984; Dooley & McClay 1997). Restraining bends and releasing bends are commonly elongate, lazy-S- or Z-shaped features in plan view, and they may form the dominant topographic and structural feature within a deforming region. With increased strike-slip offset, S- and Z-shaped pull-apart basins may evolve into more rhomboidally shaped features (Mann *et al.* 1983).

Restraining bends produce elongate, individual massifs with anomalously high topographic elevations such as the Denali Range in Alaska (Fitzgerald *et al.* 1993), the Lebanon and Anti-Lebanon ranges of the Middle East (**Gomez *et al.*** this volume), or the Cordillera Septentrional on the island of Hispaniola (Mann *et al.* 1984, 2002). Releasing bends produce pull-apart basins and fault-bounded troughs that comprise some of Earth's lowest topographic depressions, such as the Dead Sea (ten Brink *et al.* 1999), Death Valley (Christie-Blick & Biddle 1985) and submarine basins underlying the Gulf of Aqaba (Elat; Ben-Avraham 1985), the Cayman trough (Leroy *et al.* 1996, AAPG) and the Gulf of California (Persaud *et al.* 2003).

Restraining and releasing bends along both continental and oceanic strike-slip faults may act as barriers to earthquake propagation (King & Nabelek 1985; Sibson 1985; Barka & Kadinsky-Cade 1988) or conversely, they may provide nucleation sites for major earthquakes (e.g. Shaw 2006). There are also documented cases of large fault bend earthquakes ($M > 7$) having complex rupture mechanisms with multiple faults being activated within the bend, as well as major faults rupturing through the bend (Bayarsayhan *et al.* 1996; Harris *et al.* 2002). Because the length of fault segment rupture is proportional to earthquake magnitude (Scholz 1982), identification of fault bends between parallel strike-slip fault segments that may act as seismic propagation barriers is important in assessing the potential severity of future earthquakes in areas of active strike-slip faulting. Documenting three-dimensional fault connectivity and kinematics within an individual fault bend is important for assessing whether the bend may act as a future earthquake propagation barrier (**Graymer *et al.*** this volume).

In addition to earthquake hazards, tectonically active fault bends have other societal relevance. Restraining bends may:

1. exhume crystalline basement rocks that contain important mineral deposits (e.g. Pinheiro & Holdsworth 1997);

2. host hydrocarbons in their interiors and flanking basins (Christie-Blick & Biddle 1985; Escalona & Mann 2003; Decker *et al.* 2005); and
3. form major topographic uplifts that provide a locally significant rain catchment area and potential groundwater resources (Gobi Altai and Altai restraining bends, Cunningham *et al.* 1996; Cunningham 2005, this volume).

The societal significance of releasing bends includes the following:

1. pull-apart basins form depressions containing significant sedimentary accumulations that may host hydrocarbons, metalliferous deposits, evaporites and other industrial minerals (e.g. the Vienna Basin, Hamilton & Johnson 1999; Hinsch *et al.* 2005);
2. releasing bends may be zones of high heat flow and crustal dilation that can be exploited as sources of geothermal energy, such as the Coso geothermal area of California (Lees 2002) and the Cerro Prieto geothermal area of Mexico (Glowacka *et al.* 1999); and
3. releasing bends may create large valleys that provide fertile agricultural land and flat-lying urbanized areas, such as the Imperial Valley of southern California, the Silicon Valley of northern California, the Vienna Basin of Austria, and the Dead Sea–Sea of Galilee Valley in the Middle East.

Origin and evolution of strike-slip fault bends

Factors that influence and control the origin and progressive development of restraining and releasing bends are complex and numerous, but can be grouped into several major research themes.

Fault geometry and reactivation

The shape, topography and internal architecture of a fault bend is fundamentally controlled by several factors, including the orientation of the plate motion vector relative to the master strike-slip fault; the original width of the stepover; and whether the bend is a strike-slip fault termination; a double bend along a single continuously linked strike-slip fault; or a stepover where parallel strike-slip fault segments are offset and may or may not overlap. For example, wide stepovers may contain fewer faults that bridge the gap between master strike-slip faults, whereas narrow stepovers may have greater linkage between major faults within the bend (Dooley & McClay 1997; McClay & Bonora 2001). Because fault bends typically form in mechanically heterogeneous crust, pre-existing faults and basement fabrics may be reactivated instead of new faults generated. The orientations of reactivated older structures are unlikely to be ideal for either pure strike-slip or pure dip-slip motions, thus oblique-slip displacements on reactivated faults are typically important within fault bends, and workers should therefore be aware of field criteria that indicate fault reactivation (Holdsworth *et al.* 1997).

Strain magnitude and distribution

Strike-slip displacements along master faults that enter a fault bend will be partially or wholly accommodated by deformation within the bend (Segall & Pollard 1980). Thus, large displacement strike-slip faults are capable of producing the largest restraining and releasing bends. However, small restraining and releasing bends may also exist along major strike-slip faults, especially when early formed bends are bypassed as the system evolves (Bennett *et al.* 2004; **Mann** *et al.* this volume), or when fault bends nucleate late in the history of a strike-slip fault system (Sieh & Natawidjaja 2000), or when the releasing stepover and basin depocentre has progressively migrated along the master strike-slip system, instead of maintaining a fixed position relative to the adjacent sliding blocks (**Wakabayashi *et al.*** 2004; this volume; Lazar *et al.* 2006). Depending on the angle between the master strike-slip fault and the far-field displacement direction, the degree of strain partitioning of oblique deformation within the bend into separate thrust, normal and strike-slip displacements will control bend evolution. Kinematic partitioning of non-coaxial strike-slip and coaxial strains is common when the far-field displacement direction is strongly oblique ($<20°$) to the deformation zone boundary (Dewey *et al.* 1998). In addition, three-dimensional strain in strike-slip settings typically involves vertical-axis rotations (e.g. Jackson & Molnar 1990). Thus, the progressive evolution of a fault bend may involve local vertical axis rotations within the bend, and vertical axis rotations in the larger region that the bend occurs within (Luyendyk *et al.* 1980; Westaway 1995; Cowgill *et al.* 2004*b*). This may be indicated by changes in strike trends, and can be proven palaeomagnetically (Luyendyk *et al.* 1985). Progressive vertical-axis rotations within a fault bend will result in changing fault kinematics as the faults rotate relative to the external stress field. Vertical-axis rotations may thus lead to fault abandonment and propagation of new faults. In addition, strain hardening processes may operate locally within the bend and may influence whether old faults remain active or lock up (Cowgill *et al.* 2004*a*).

Stress field considerations

The orientation of the maximum horizontal stress ($S_{H_{max}}$) relative to the deformation zone boundary will strongly influence the degree of transpression or transtension within the fault bend region (Tikoff & Teyssier 1994). Fault bends that form where $S_{H_{max}}$ is at a high angle to the deforming zone will tend to have large dip-slip displacements, thus forming large restraining bend mountains (e.g. Karlik Tagh Range, China, Cunningham *et al.* 2003) or wide and deep releasing-bend basins (e.g. Sea of Japan, Jolivet *et al.* 1994) When regional plate-motion changes lead to stress-field changes in transform boundary settings, ratios of strike-slip to dip-slip displacements within fault bends will change and the fundamental architecture and topographic development of the bend will reflect that change. In addition, fault bends may switch from transtensional to transpressional systems or vice versa, if the original fault bend was at a low angle relative to $S_{H_{max}}$ (Tikoff & Teyssier 1994). Thus, transtensional basins may become inverted and restraining bends may be cross-cut by overprinting transtensional faults (**Legg *et al.*** this volume). Stress fields within individual fault bends may also evolve with progressive faulting, structural compartmentalization and increased mechanical interaction between intersecting faults, resulting in fault motions that are internally guided (Muller & Aydin 2004; **Waldron *et al.*** this volume; **Fodor** this volume).

Feedback between climate, topography, faulting and thermal history

Long-term climate patterns and mountain erosion rates compete with mountain uplift and influence the extent of topography generation or destruction for all mountain ranges, including restraining bends (Anderson 1994; Willett *et al.* 2001). If a restraining bend achieves a steady state between uplift and erosion, then its dimensions will stabilize, and thus individual faults will tend to remain active and new faults may not form (Beaumont *et al.* 1991; Norris & Cooper 1997; Willett 1999). In addition, larger releasing bends that evolve into marine basins may also influence local climate, driving changes in precipitation, erosion and rates of sediment deposition – thus influencing the overall dimensions of the releasing bend basin (e.g. Sea of Marmara, Turkey).

Progressive exhumation of deeper crustal rocks in restraining bends by uplift and erosion, and in releasing bends through normal faulting and erosion, will lead to changes in the thermal evolution of the fault bend. This can be demonstrated by fission-track and other geothermometric data which reveal the timing and rates of exhumation (Fitzgerald *et al.* 1995; Blythe *et al.* 2000; Batt *et al.* 2004). If heat flow increases within a restraining bend region, then it may lead to increased buoyancy and topographic uplift. This may lead to positive feedback between uplift and erosion and progressive exhumation of mid-crustal rocks similar to the crustal aneurysm model proposed for structural culminations in the Himalayan syntaxes (Zeitler *et al.* 2001). In releasing bend settings, high extensional strains in pull-apart basins may lead to increased heat flow and possibly volcanism; extrusive rocks may then constitute volumetrically significant basin fill (Dhont *et al.* 1998).

Because all of these factors will be different for every fault bend, it follows that restraining and releasing bends should be diverse in nature – with each one having unique topographic, geomorphological, architectural and evolutionary characteristics. Analogue models of fault bends have recreated some of the fault patterns and topographic characteristics of natural examples, and have documented progressive stages of evolution (Hempton & Neher 1986; Dooley & McClay 1997; McClay & Bonora 2001); however, they have not included many of the factors considered here, and so they must be regarded as somewhat generic.

With this previous work in mind, and in order to bring together workers from around the world who are actively investigating strike-slip fault bends, an international meeting on the tectonics of strike-slip restraining and releasing bends in continental and oceanic settings was convened in London on 28–30 September 2005, under the auspices of the Geological Society of London. This volume includes contributions that were presented at the conference, and new results by others whose research connects with the conference theme.

This volume

The 17 papers included in this volume cover a variety of topics and regions related to tectonics, geology and geophysics of restraining and releasing bends. The papers are organized into three major themes: (1) bends, sedimentary basins and earthquake hazards; (2) restraining bends, transpressional deformation and basement controls on development; and (3) releasing bends, transtensional deformation and fluid flow. Many papers have multiple emphases, and these subdivisions are general guides to subject content only. The brief discussion below is meant to summarize key results and conclusions of each study, without being exhaustive. In addition to this volume, the reader is referred to several other important volumes which cover related topics of strike-slip fault tectonics (Sylvester 1988); continental transpressional and transtensional tectonics (Holdsworth *et al.* 1998); intraplate strike-slip deformation belts (Storti *et al.* 2003); strike-slip deformation, basin formation and sedimentation (Biddle & Christie-Blick 1985); and continental wrench tectonics and hydrocarbon habitat (Harding 1985; Zolnai 1991).

The volume begins with a review paper by **Mann**, which contains a global compilation of active and ancient releasing and restraining bends, with the aim of defining common modes of origin and tectonic development. He identifies five main tectonic settings for strike-slip faults and related bends:

1. oceanic transforms separating oceanic crust and offsetting mid-oceanic spreading ridges;
2. long and linear plate-boundary strike-slip fault systems separating two continental plates whose plate-boundary kinematics can be quantified for long distances along strike by a single pole of rotation (e.g. the San Andreas fault system of western North America);
3. relatively shorter, more arcuate, indent-linked, strike-slip fault systems bounding escaping continental fragments in zones of continent–continent or arc–continent collision (e.g. the Anatolian plate);
4. straight to arcuate trench-linked strike-slip fault systems bounding elongate fore-arc slivers generated in active and ancient fore-arc settings by oblique subduction (e.g. Sumatra); and
5. continental interior, intraplate strike-slip fault systems removed from active plate boundaries, formed on older crustal faults, but acting as 'concentrators' of intraplate stresses.

By far the most common, predictable and best-studied setting for restraining and releasing bends occurs in continental plate boundary strike-slip fault systems, where arrays of two to eight en echelon pull-apart basins mark transtensional fault segments, and single and sometimes multiple large restraining bends mark transpressional segments; fault areas of transtension versus transpression are determined by the intersection angles between small circles about the interplate pole of rotation and the trend of the strike-slip fault system. These longer and more continuous boundary strike-slip systems also exhibit a widespread pattern of 'paired bends' or 'sidewall ripouts', or adjacent zones of pull-aparts and restraining bends that range in along-strike-scale from kilometres to hundreds of kilometres. Four evolutionary models to explain the origin and evolution of bends are presented:

1. progressive linkage of en echelon shears within a young evolving shear zone;
2. formation of lenticular 'sidewall ripout' structures at scales ranging from outcrop to regional;
3. interaction of propagating strike-slip faults with pre-existing crustal structures such as ancient rift basins; and
4. concentration of regional maximum compressive stress on pre-existing, basement structures in intraplate continental regions.

With over 225 modern bends in this global compilation, it follows from uniformitarian principles that restraining bends and releasing bends must also have been widespread in the geological past. Whilst most modern restraining and releasing bends have obvious topographic expression, ancient examples may occur in regions that have been eroded flat, or are buried beneath sedimentary cover, or are overprinted by younger orogenic events and thus may be difficult to discern.

Bends, sedimentary basins, and earthquake hazards

Many of the original ideas regarding continental transforms and restraining and releasing bends were developed by workers in California investigating the San Andreas system (e.g. Crowell 1974). Early work focused on the on-land expression of the diffuse plate boundary, whereas it is now recognized that some of the interplate strain is also accommodated offshore in the southern California borderland region. This subject is addressed by **Legg *et al.***, who provide a review of restraining bends and releasing bends formed during the last 20 Ma along the diffuse Pacific–North American transform boundary in the southern California borderlands. By combining multi-beam swath bathymetry, high-resolution seismic imaging, earthquake and geological data, they describe two major strike-slip restraining bends in the largely submarine setting – structures that are beautifully imaged because of the diminished effects of submarine erosion as compared to on-land examples. Pre-existing rift trends have influenced the stepping geometry of the major strike-slip faults, and there is a common association of double restraining bends bounded by releasing bends, an association which is also seen in on-land examples in southern California. The largest restraining bends approach 100 km in length, and could be sites of major (*M*7 or greater) earthquakes. Because of their offshore setting and rugged topography, the active faults within large restraining bends pose a potential tsunami hazard which is only now being appreciated.

An important consideration for understanding the evolution of restraining and releasing bends is whether their location remains fixed with respect to adjacent laterally moving blocks or whether they migrate along strike with increased fault displacements. **Wakabayashi** addresses this question by looking at numerous examples from the San Andreas system, where he documents the sedimentary and structural record of stepover migration, including 'wakes' of deposits trailing behind migrating stepovers as well as the sedimentary and structural expression of migrating inversion of former releasing bends. Importantly, he compares the sedimentary wake length to the overall displacement of the master fault system, in order to quantify the magnitude of stepover migration. Two end members are possible: stepovers that migrate for the entire duration of strike-slip displacement, and those that remain fixed. Fixed restraining bends will tend to have the largest structural relief and greatest erosional exhumation, and will form regionally significant topographic and structural culminations. However, other smaller restraining bends with less relief may have existed for just as long, but their migration limits their topographic and structural development, because uplifted/downdropped areas are soon abandoned as deformation moves along strike. The implication is that size of bend may not reflect longevity.

Although fault bends are widely regarded as earthquake propagation barriers for most major active strike-slip fault systems (Sibson 1985; Barka & Kadinsky-Cade 1988), surface fault complexity may mask relatively simple patterns at depth (>5 km deep). This is expected if the faults within the stepover define a flower structure and surface faults root into a singular master fault. In a paper by **Graymer *et al.*** the authors demonstrate that carefully located hypocentres beneath both restraining and releasing bends in California, where large earthquakes have occurred, define singular or simple fault patterns. They conclude that stepover zones provide less of an impediment to through-going rupture than previously assumed. Exceptions may be those large bends which have complex multilayered fault patterns reflecting both strike-slip and thrusting displacements and which have master strike-slip displacements migrating through the bend and eventually bypassing the bend. Their conclusions complement results from Lettis *et al.*'s (2002) compilation of 30 historical strike-slip earthquake ruptures involving 59 stepover basins, which indicated that strike-slip events with small to large displacements usually propagate through stepovers less than 1–2 km wide. With increasing displacements, 2–4-km-wide stepovers may be through-ruptured. However, stepovers of 4–5 km width always arrest fault rupture, regardless of the amount of displacement.

Although all major continental transform boundaries tend to have restraining and releasing bends along them, the Scotia–Antarctic transform boundary is particularly interesting, because it has both oceanic and continental crustal elements along its length, and stepover nucleation and development are directly related to the distribution of the two different types of crust. **Bohoyo *et al.*** present new geophysical data that are used to image and map the distribution of restraining and releasing bends along this remote submarine plate boundary. Their most important conclusion is that the distribution of releasing bend basins, and other bathymetric troughs formed by transtension, is directly controlled by the distribution and shape of rheologically weak continental fragments which rift more easily than surrounding, stronger oceanic crust. In contrast, restraining bends and areas of transpression occur at the interface between crust types, where oceanic crust underthrusts continental blocks.

Restraining bends, transpressional deformation and basement controls on development

One of the most interesting distant effects of the Indo–Eurasia collision is the active intraplate transpressional mountain building in western China and, in particular, Mongolia. In a paper by **Cunningham**, the geological and structural characteristics of 12 separate restraining-bend mountain ranges from the Altai, Gobi Altai and easternmost Tien Shan are reviewed and compared. All bends have individual structural, topographic and dimensional characteristics, due to multiple factors which influence their characteristics, including: stepover width; pre-existing basement structures, angular relation between main fault trace and $S_{H_{max}}$; and local tectonic setting. The significant structural and topographic diversity challenges simplistic models of restraining-bend evolution. In addition, there is an orogenic continuum: from isolated restraining bends along major strike-slip faults, to thrust-dominated transpressional ridges with only minor strike-slip transfer zones between thrust ridges. The key point is that as restraining bends grow along-strike and across-strike they may coalesce with neighbouring ranges and lose their individual identities. Mature oblique deformation belts may obscure the original restraining bends uplifts which were the initial nucleation sites for mountain building.

The island of Jamaica is essentially the morphological expression of a large Late Miocene to Recent restraining bend along the North

America–Caribbean plate boundary, and its origin and progressive development are described by **Mann *et al.*** By analysing geodetic, geological and seismic data they document the continued uplift and evolution of the bend in eastern Jamaica (Blue Mountains), along with less well-expressed bends in central and western Jamaica. However, seismicity along the south coast of the island suggests that a more linear short-cut fault is developing, which will bypass the range, and that the interplate strain may progressively transfer to that fault system. Another important conclusion of their study is that the restraining bends of Jamaica were initially localized by older basement faults related to Palaeogene rifts which trend northwesterwards and oblique to the evolving plate boundary.

Seyrek *et al.* provide a careful correlation of Pleistocene basalts across the northern Dead Sea transform boundary, coupled with new age data to calculate slip rates along the plate boundary since the Pliocene (*c.* 3.73 Ma). An important implication of their work is that the calculated displacement vectors and slip rates require that the northern continuation of the transform in southern Turkey must be convergent, and that the entire Amanos Mountain and Karasu Valley region constitutes a gentle restraining bend with active uplift in the Amanos Mountains. The overall geometry and kinematics are very similar to the Lebanon stepover, where ***Gomez et al.***, using geological and geomorphological fieldwork, cosmogenic dating, seismicity data, GPS results, and the analysis of relative plate motions, propose a two-stage history to the bend: an early wrench-dominated stage, followed by the modern strain-partitioned transpression-dominated stage. The switch was probably driven externally by changes in relative plate motions. The recognition of strain-partitioned deformation within the modern bend has implications for the regional seismic hazard; multiple strike-slip faults are active within the bend region and growing anticlines may hide seismically active blind thrusts.

A pair of papers by **Smith *et al.*** and **Morley *et al.*** addresses the structural evolution of transpressional zones along the active, left-lateral Mae Ping Fault Zone in central Thailand. By using satellite images, geological maps and magnetic data, they document a regional-scale strike-slip duplex within and adjacent to the Khlong Lan restraining bend. Importantly, Morley *et al.* review published cooling-age data and provide new apatite and zircon fission-track data to document the spatial and temporal evolution of uplift and exhumation within the restraining bend. Deformation and exhumation appear to have been focused at the corners of the restraining bend, as blocks migrate out of the bend during progressive strike-slip displacements in a manner similar to that described by Cowgill *et al.* (2004*a*, *b*) for the Akato Tagh bend on the Altyn Tagh Fault of Tibet. Smith *et al.* and Morley *et al.* also link their results from Thailand to an evolving deformation regime initially driven by terrane collision, but later driven by escape tectonics due to the Indo–Eurasia collision.

An unusual example of a restraining bend formed within a fold and thrust belt is presented by **Zampieri *et al.*** who document a north–south polydeformed relay zone cutting across the Italian Alps, a zone that has localized transpressional deformation at a prominent restraining stepover. Liassic and Palaeogene north–south extensional structures were reactivated during Alpine compression as a strike-slip relay zone within the thrust belt. Reactivation of normal faults and inversion of the older graben fill produced a complex restraining bend with different degrees of shortening on either side of the stepover, due to juxtaposed sequences with strongly contrasting rheological properties. There are very few studies of restraining bends formed locally along transfer faults within major fold-and-thrust belts, especially where older extensional structures have been reactivated; however, their study suggests that other examples await discovery.

In another example of polyphase deformation in an intraplate restraining bend, but in an older Palaeozoic setting, **Waldron *et al.*** document the detailed and complex internal structures within a portion of a flower structure that formed during right-lateral Carboniferous strike-slip movement along the Minas fault zone in Nova Scotia. Their study underscores the importance of detailed structural analysis to unravel different phases of folding, thrusting and oblique deformation within a zone of localized transpression. The structures that they observe are consistent with a progressive change in the local angle of convergence, increased strain partitioning, and pure-shear dominated transpression. An important implication of their work is that as the restraining bend developed, significant topography was generated, and the deformation migrated laterally from inner, dominantly steep transpressional zones, to outer, low-angle zones enhanced by the presence of Late Palaeozoic evaporite layers that promoted low-angle thrust faulting and gravitational spreading.

Releasing bends, transtensional deformation and fluid flow

Single restraining and releasing bends that form at strike-slip fault terminations are usually sites of oblique deformation where strike-slip displacements are progressively transformed into dip-slip displacements. The architecture and fault kinematics within such transitional zones can be very complex. **Mouslopoulou *et al*.** address this problem in a detailed study of the termination zone of the North Island strike-slip fault system against the Taupo back-arc rift of northern New Zealand. They find that the strike-slip fault system splays into five main strands as it approaches the rift. These strike-slip faults link with the rift margin normal faults, but do not displace the normal faults. This geometry requires that the faults bend and that slip vectors on the strike-slip faults must progressively steepen to accommodate increasing dip-slip components near the active Taupo Rift. Data from displaced landforms, fault trenching, gravity and seismic profiles are used to quantify displacements, and to evaluate rotational and non-rotational mechanisms of displacement transfer.

In a similar study investigating the kinematic linkage between strike-slip faults and extensional faults, but on a local outcrop scale, **Fodor** documents the role of transtensional relay ramps in accommodating displacement transfer within releasing bends along Upper Tertiary strike-slip faults in the Vertes Hills of the Pannonian basin in Hungary. His study comes from a mined area with exceptional exposures. Normal and oblique-normal faults change their strike, dip and slip vectors systematically to accommodate extension across the relay ramp. Fault slip inversion for different groups of faults demonstrates that inclusion of transfer-zone faults modifies the results of palaeostress calculations, because displacements on the transfer-zone faults are not governed by the regional stress field, but by their bounding strike-slip faults (i.e. guided slip). This influence of bounding-strike-slip faults on local stress fields should be considered by anyone attempting palaeostress calculations using fault stepover data.

Releasing bends and dilational stepovers are typically complex sites of fracturing, veining and fluid flow. In a theoretical and field-based study using outcrop data from the Carboniferous Northumberland basin in England, **DePaola *et al.*** document how deformation within dilational stepovers with low angles of oblique divergence ($<30^\circ$) may evolve from wrench- to extension-dominated transtension as strain increases. Veins, dykes, fracture meshes and faults record progressive transtensional deformation, including reactivation of earlier-formed structures. The complex pattern of structures within the stepover may inhibit development of a through-going single fault. Therefore, the stepover may be long-lived and persist as a site of subsidence, and provide long-term enhanced structural permeability favourable to fluid migration and mineral precipitation.

It is well known that many world-class mineral deposits have formed where fluid flow is focused in dilational sites along fault bends (Sibson 2001; Cox 2005). **Berger** presents structural, lithological and geochronological evidence which reveals the importance of fault bends in controlling the locations of volcanism and associated epithermal volcanic centre-related hydrothermal gold and silver systems in Nevada. Specifically, by analysing the temporal and spatial evolution of volcanism and the relationship between high-grade gold deposits and faults that formed at the stepover, he concludes that migrating corner zones where strike-slip faults link to oblique-slip and dip-slip faults are important sites of hydrothermal veining, and that high-grade bonanza ores were deposited along abandoned normal-fault systems *following* stepover migration. Another complicating factor is that abandoned stepovers have locally experienced contractional deformation and inversion, further affecting the permeability structure within the stepover. The author also documents a rarer case of hydrothermal mineralization in extensional veins within a restraining bend in the Excelsior Mountains of Nevada.

Summary

Because restraining and releasing bends often occur as singular self-contained domains of complex deformation, they provide appealing natural laboratories for Earth scientists to study fault processes; earthquake seismology; active faulting and sedimentation; fault and fluid-flow relationships; links between tectonics and topography; tectonic and erosional controls on exhumation; and tectonic geomorphology. A major challenge for future workers will be to untangle the deformational history of those regions where multiple fault bends have grown large enough to interact, coalesce and structurally interfere. Finally, the deep expression of fault bends and the manner in which they are coupled to ductilely deforming lower crust and lithospheric mantle remain poorly understood (Tikoff *et al.* 2004).

This special publication stems from an international conference hosted by the Geological Society of London during September 2005, on the tectonics of strike-slip restraining and releasing bends in continental and oceanic settings. We are grateful to all authors for contributing to the volume, and are indebted to the many conscientious referees who provided professional reviews ensuring the high quality of the volume.

The following colleagues and friends kindly helped with the reviewing of the papers submitted for this volume: P. Barnes, Wellington, New Zealand; G. Batt, London, UK; C. Burchfiel, California, USA; R. Burgmann, California, USA; C. Childs, Dublin, Republic of Ireland; E. Cowgill, California, USA; S. Cox, Canberra, Australia; N. de Paola, Perugia, Italy; T. Dooley, Texas, USA; M. Gordon, Texas, USA; C. Henry,

Nevada, USA; J.-C. Hippolyte, Savoie, France; P. Klipfel, Nevada, USA; L. Lavier, Texas, USA; M. Legg, California, USA; K. McIntosh, Texas, USA; S. Mazzoli, Naples, Italy; M. Menichetti, Urbino, Italy; C. Morley, Bangkok, Thailand; M. Murphy, Texas, USA; D. Peacock, Llandudno, Wales, UK; A. Robertson, Edinburgh, Scotland, UK; J. Rowland, Auckland, New Zealand; G. Scheurs, Berne, Switzerland; L. Seeber, New York, USA; R. Sibson, Dunedin, New Zealand; M. Swanson, Maine, USA; A. Sylvester, California, USA; F. Taylor, Texas, USA; D. van Hinsbergen, Utrecht, Netherlands; J. Waldron, Alberta, Canada; R. Walker, Oxford, UK; L. Webb, New York, USA; N. Woodcock, Cambridge, UK.

References

ANDERSON, R. 1994. Evolution of the Santa Cruz Mountains, California, through tectonic growth and geomorphic decay. *Journal of Geophysical Research*, **99**, 20 161–20 179.

AYDIN, A. & NUR, A. 1982. Evolution of stepover basins and their scale independence. *Tectonics*, **1**, 91–105.

AYDIN, A. & NUR, A. 1985. The types and role of stepovers in strike-slip tectonics. *In*: BIDDLE, K. T. & CHRISTIE-BLICK, N. (eds) *Strike-Slip Deformation, Basin Formation, and Sedimentation*. SEPM Special Publications, **37**, 35–44.

BARKA, A. A. & GULEN, L. 1989. Complex evolution of the Erzincan basin (western Turkey). *Journal of Structural Geology*, **11**, 275–283.

BARKA, A. A. & KADINSKY-CADE, K. 1988. Strike-slip fault geometry in Turkey and its influence on earthquake activity. *Tectonics*, **7**, 663–684.

BARNES, P., SUTHERLAND, R. & DELTEIL, J. 2005. Strike-slip structure and sedimentary basins of the southern Alpine fault, Fiordland, New Zealand. *Geological Society of America Bulletin*, **117**, 411–435.

BATT, G., BALDWIN, S. L., COTTAM, M. A., FITZGERALD, P. G., BRANDON, M. T. & SPELL, T. L. 2004. Cenozoic plate boundary evolution in the South Island of New Zealand: new thermochronological constraints. *Tectonics*, **23**, doi: 10.1029/2003TC001527 TC4001.

BAYARSAYHAN, C., BAYASGALAN, A., ENHTUVSHIN, B., HUDNUT, K., KURUSHIN, R. A., MOLNAR, P. & OLZIYBAT, M. 1996. 1957 Gobi-Altay, Mongolia, earthquake as a prototype for southern California's most devastating earthquake. *Geology*, **24**, 579–582.

BEAUMONT, C., FULSACK, P. & HAMILTON, J. 1991. Erosional control of active compressional orogens. *In*: MCCLAY, K. R. (ed.) *Thrust Tectonics*. Chapman and Hall, New York, 1–18.

BECK, M. E., JR, 1983. On the mechanism of tectonic transport in zones of oblique subduction. *Tectonophysics*, **93**, 1–11.

BEN-AVRAHAM, Z. 1985. Structural framework of the Gulf of Elat (Aqaba), northern Red Sea. *Journal of Geophysical Research*, **90**, B1, 703–725.

BENNETT, R., FRIEDRICH, A. & FURLONG, K. 2004. Codependent histories of the San Andreas and San Jacinto fault zones from inversion of fault displacement rates. *Geology*, **32**, 961–964.

BIDDLE, K. T. & CHRISTIE-BLICK, N. 1985. Glossary – strike-slip deformation, basin formation and sedimentation. *In*: BIDDLE, K. T. & CHRISTIE-BLICK, N. (eds) *Strike-Slip Deformation, Basin Formation and Sedimentation*. SEPM Special Publications, **37**, 375–384.

BILHAM, R. & KING, G. 1989. The morphology of strike-slip faults: examples from the San Andreas fault, California. *Journal of Geophysical Research*, **94**, 10 204–10 216.

BILHAM, R. & WILLIAMS, P. 1985. Sawtooth segmentation and deformation processes on the southern San Andreas fault, California. *Geophysical Research Letters*, **12**, 557–560.

BLYTHE, A., BURBANK, D., FARLEY, K. & FIELDING, E. 2000. Structural and topographic evolution of the central Transverse Ranges, California, from apatite fission track, U–Th/He, and digital elevation model analysis. *Basin Research*, **12**, 97–114.

CHRISTIE-BLICK, N. & BIDDLE, K. T. 1985. Deformation and basin formation along strike-slip faults. *In*: BIDDLE, K. T. & CHRISTIE-BLICK, N. (eds) *Strike-Slip Deformation, Basin Formation, and Sedimentation*. SEPM Special Publications, **37**, 1–34.

CLOOS, H. 1928. Experimente zur inneren Tektonic. *Centralblatt fur Mineralogie*, **1928B**, 609–621.

COWGILL, E., ARROWSMITH, R., YIN, A., XIAOFENG, W. & ZHENGLE, C. 2004*a*. The Akato Tagh bend along the Altyn Tagh fault, northwest Tibet, 2: active deformation and the importance of transpression and strain hardening within the Altyn Tagh system. *Geological Society of America Bulletin*, **116**, 1443–1464.

COWGILL, E., YIN, A., ARROWSMITH, R., FENG, W. & SHUANHONG, Z. 2004*b*. The Akato Tagh bend along the Altyn Tagh fault, northwest Tibet, 1: smoothing by vertical-axis rotation and the effect of topographic stresses on bend-flanking faults. *Geological Society of America Bulletin*, **116**, 1423–1442.

COX, S. F. 2005. Coupling between deformation, fluid pressures, and fluid flow in ore-producing hydrothermal systems at depth in the crust. *Economic Geology*, **100th Anniversary Volume**, 39–76.

CROWELL, J. C. 1974. Origin of late Cenozoic basins of southern California. *In*: DICKINSON, W. R. (ed.) *Tectonics and Sedimentation*. SPEM Special Publications, **22**, 190–204.

CUNNINGHAM, W. D. 2005. Active intracontinental transpressional mountain building in the Mongolian Altai: defining a new class of orogen. *Earth and Planetary Science Letters*, **240**, 436–444.

CUNNINGHAM, W. D., OWEN, L. A., SNEE, L. & LI, JILIANG. 2003. Structural framework of a major intracontinental orogenic termination zone: the easternmost Tien Shan, China. *Journal of the Geological Society, London*, **160**, 575–590.

CUNNINGHAM, W. D., WINDLEY, B. F., DORJNAMJAA, D., BADAMGAROV, G. & SAANDAR, M. 1996. Late Cenozoic transpression in southwestern Mongolia and the Gobi Altai–Tien Shan connection. *Earth and Planetary Sciences*, **140**, 67–82.

DECKER, K., PERESSON, H. & HINSCH, R. 2005. Active tectonics and Quaternary basin formation along the Vienna basin transform fault. *Quaternary Science Reviews*, **24**, 307–322.

DEWEY, J. F., HOLDSWORTH, R. E. & STRACHAN, R. A. 1998. Transpression and transtension zones. *In*: HOLDSWORTH, R. E., STRACHAN, R. A. & DEWEY, J. F. (eds) *Continental Transpressional and Transtensional Tectonics*. Geological Society, London, Special Publications, **135**, 1–14.

DHONT, D., CHOROWICZ, J., YURUR, T., FROGER, J.-L., KOSE, O. & GUNDOGDU, N. 1998. Emplacement of volcanic vents and geodynamics of Central Anatolia, Turkey, *Journal of Volcanology and Geothermal Research*, **85**, 33–54.

DOOLEY, T. & MCCLAY, K. 1997. Analog modeling of strike-slip pull-apart basins. *AAPG Bulletin*, **81**, 804–826.

DOOLEY, T., MCCLAY, K. & BONORA, M. 1999. 4D evolution of segmented strike-slip fault systems: applications to NW Europe. *In*: FLEET, A. & BOLDY, S. (eds) *Petroleum Geology of Northwest Europe: Proceedings of the 5th Conference, Petroleum Geology '86*, Geological Society, London, 215–225.

ESCALONA, A. & MANN, P. 2003. Three-dimensional structural architecture and evolution of the Eocene pull-apart basin, central Maracaibo basin, Venezuela. *Marine and Petroleum Geology*, **20**, 141–161.

FITZGERALD, P., SORKHABI, R., REDFIELD, T. & STUMP, E. 1993. Uplift and denudation of the central Alaska range: a case study in the use of apatite fission-track thermochronology to determine absolute uplift parameters. *Journal of Geophysical Research*, **100**, 20 175–20 192.

FITZGERALD, P., STUMP, E. & REDFIELD, T. 1993. Late Cenozoic uplift of Denali, and its relation to relative plate motion and fault morphology. *Science*, **259**, 497–500.

GAMOND, J. 1987. Bridge structures as sense of displacement criteria in brittle fault zones. *Journal of Structural Geology*, **9**, 609–620.

GARFUNKEL, Z. 1981. Internal structure of the Dead Sea leaky transform (rift) in relation to plate kinematics. *Tectonophysics*, **80**, 81–108.

GARFUNKEL, Z. 1986. Review of oceanic transform activity and development. *Journal of the Geological Society*, London, **143**, 775–784.

GLOWACKA, E., GONZALEZ, J. & FABRIOL, H. 1999. Recent vertical deformation in Mexicali Valley and its relationship with tectonics, seismicity, and the exploitation of the Cerro Prieto geothermal field, Mexico. *Pure and Applied Geophysics*, **156**, 591–614.

HAMILTON, W. & JOHNSON, N. 1999. The Matzen project – rejuvenation of a mature field. *Petroleum Geoscience* **5**, 119–125.

HARDING, T. P. 1985. Seismic characteristics and identification of negative flower structures, positive flower structures, and positive structural inversion. *AAPG Bulletin*, **69**, 582–600.

HARRIS, R., DOLAN, J. F., HARTLEB, R. & DAY, S. M. 2002. The 1999 Izmit, Turkey, earthquake: a 3D dynamic stress transfer model of intra-earthquake triggering. *Bulletin of the Seismological Society of America*, **92**, 245–255.

HEMPTON, M. & DUNNE, L. 1984. Sedimentation in pull-apart basins: active examples in eastern Turkey. *Journal of Geology*, **92**, 513–530.

HEMPTON, M. & NEHER, K. 1986. Experimental fracture, strain, and subsidence patterns over en echelon strike-slip faults: implications for the structural evolution of pull-apart basins. *Journal of Structural Geology*, **8**, 597–605.

HINSCH, R., DECKER, K. & WAGREICH, M. 2005. 3-D mapping of segmented active faults in the southern Vienna basin. *Quaternary Science Reviews*, **24**, 321–336.

HOLDSWORTH, R. E., BUTLER, C. A. & ROBERTS, A. M. 1997. The recognition of reactivation during continental deformation. *Journal of the Geological Society*, London, **154**, 73–78.

HOLDSWORTH, R. E., STRACHAN, R. A. & DEWEY, J. F. (eds) 1998. *Continental Transpressional and Transtensional Tectonics*. Geological Society, London, Special Publications, **135**.

JACKSON, J. A. & MOLNAR, P. 1990. Active faulting and block rotations in the western Transverse ranges, California. *Journal of Geophysical Research*, **95**, 22 073–22 087.

JARRARD, R. D. 1986. Terrane motion by strike-slip faulting of fore-arc slivers. *Geology*, **14**, 780–783.

JOLIVET, L., TAMAKI, K. & FOURNIER, M. 1994. Japan Sea, opening history and mechanism: a synthesis. *Journal of Geophysical Research*, **99**, 22 237–22 260.

JONES, R. R. & TANNER, P. W. G. 1995. Strain partitioning in transpression zones. *Journal of Structural Geology*, **17**, 793–802.

KADINSKY-CADE, K. & BARKA, A. A. 1989. Effects of restraining bends on the rupture of strike-slip earthquakes. *In*: SCHWARTZ, D. P. & SIBSON, R. (eds) *Proceedings of Conference XLV on Fault Segmentation and Controls of Rupture Initiation and Termination*. USGS Open-File Report, **89-315**, 181–192.

KING, G. & NABELEK, J. 1985. Role of fault bends in the initiation and termination of earthquake rupture. *Science*, **228**, 984–987.

KOENIG, E. & AYDIN, A. 1998. Evidence for large-scale strike-slip faulting on Venus. *Geology*, **26**, 551–544.

LAZAR, M., BEN-AVRAHAM, Z. & SCHATTNER, U. 2006. Formation of sequential basins along a strike-slip fault: geophysical observations from the Dead Sea basin. *Tectonophysics*, **421**, 53–69.

LEES, J. 2002. Three-dimensional anatomy of a geothermal field, Coso, southeast-central California. *In*: GLAZNER, A., WALKER, J. & BARTLEY, J. (eds) *Geologic Evolution of the Mojave Desert and Southwestern Basin and Range, Boulder, Colorado*, Geological Society of America Memoirs, **195**, 259–276.

LEROY, S., LEPINAY, B., MAUFFRET, A. & PUBELLIER, M. 1996. Structure and tectonic evolution of the Eastern Cayman Trough (Caribbean Sea) from multichannel seismic data. *AAPG Bulletin*, **141**, 222–247.

LETTIS, W., BACHHUBER, J., WITTER, R., BRANKMAN, C., RANDOLPH, C., BARKA, A., PAGE, W. & KAYA, A. 2002. Influence of releasing step-overs on surface fault rupture and fault segmentation: examples from the 17 August, 1999, Izmit earthquake on the North Anatolian fault, Turkey. *Bulletin of the Seismological Society of America*, **92**, 19–42.

LOWELL, J. D. 1972. Spitsbergen Tertiary orogenic belt and the Spitsbergen Fracture Zone. *Geological Society of America Bulletin*, **83**, 3091–3102.

LUYENDYK, B. P., KAMMERLING, M. J. & TERRES, R. 1980. Geometrical model for Neogene crustal rotations in southern California. *Bulletin of the Geological Society of America*, **91**, 211–217.

LUYENDYK, B. P., KAMMERLING, M. J., TERRES, R. & HORNAFIUS, J. S. 1985. Simple shear of southern California during Neogene time suggested by paleomagnetic declinations. *Journal of Geophysical Research*, **90**, 12 454–12 466.

MCCLAY, K. & BONORA, M. 2001. Analog models of restraining stepovers in strike-slip fault systems. *AAPG Bulletin*, **85**, 233–260.

MANN, P. & GORDON, M. 1996. Tectonic uplift and exhumation of blueschist belts along transpressional strike-slip fault zones. *In*: BEBOUT, G., SCHOLL, D., KIRBY, S. & PLATT, J. (eds) *Dynamics of Subduction Zones*, Geophysical Monographs, **96**, American Geophysical Union, Washington, DC, 143–154.

MANN, P., MATUMOTO, T. & BURKE, K. 1984. Neotectonics of Hispaniola: plate motion, sedimentation, and seismicity at a restraining bend. *Earth and Planetary Science Letters*, **70**, 311–324.

MANN, P., CALAIS, E., RUEGG, J. C., DEMETS, C., DIXON, T., JANSMA, P. & MATTIOLI, G. 2002. Oblique collision in the northeastern Caribbean from GPS measurements and geological observations. *Tectonics*, **21**, 1057, doi:10.1029?2001TC001304.

MARSHAK, S., NELSON, W. & MCBRIDE, J. 2003. Phanerozoic strike-slip faulting in the continental interior platform of the United States: examples from the Laramide orogen, mid-continent, and Ancestral Rocky Mountains. *In*: STORTI, F., HOLDSWORTH, R. & SALVINI, F. (eds) *Intraplate Strike-slip Deformation Belts*, Geological Society, London, Special Publications, **210**, 159–184.

MAY, S., EHMAN, K., GRAY, G. & CROWELL, J. 1993. A new angle on the tectonic evolution of the Ridge basin, a 'strike-slip' basin in southern California. *Geological Society of America Bulletin*, **105**, 1357–1372.

MULLER, J. R. & AYDIN, A. 2004. Rupture progression along discontinuous oblique fault sets: implications for the Karadere rupture segment of the 1999 Izmit earthquake, and future rupture in the Sea of Marmara. *Tectonophysics*, **391**, 283–302.

NORRIS, R. & COOPER, A. 1997. Erosional control on the structural evolution of a transpressional thrust complex on the Alpine fault, New Zealand. *Journal of Structural Geology*, **19**, 1323–1342.

PARSONS, T., BRUNS, T. & SLITER, R. 2005. Structure and mechanics of the San Andreas–San Gregorio fault junction, San Francisco, California. *G^3, Geochemistry Geophysics Geosystems*, **6**, Q01009, doi: 10.1029/2004GC000838.

PERSAUD, P. ET AL. 2003. Active deformation and shallow structure of the Wagner, Consag and Delfin basins, northern Gulf of California, Mexico. *Journal of Geophysical Research*, **108**, doi:10.1029/2002JB001937, 2003.

PINHEIRO, R. V. L. & HOLDSWORTH, R. E. 1997. The structure of the Carajas N-4 ironstone deposit and associated rocks: relationship to Archaean strike-slip tectonics and basement reactivation in the Amazon region, Brazil. *Journal of South American Earth Sciences*, **10**, 305–319.

POCKALNY, R. 1997. Evidence of transpression along the Clipperton Transform: implications for processes of plate boundary organization. *Earth and Planetary Science Letters*, **146**, 449–464.

RIEDEL, W. 1929. Zur Mechanik geologischer Brucherscheinungen. *Zentralblatt für Mineralogie, Geologie und Paleontologie*, **1929B**, 354–368.

SARID, A. R., GREENBERG, R., HOPPA, G. V., HURFORD, T. A., TUFTS, B. R. & GEISSLER, P. 2002. Polar wander and surface convergence of Europa's ice shell: evidence from a survey of strike-slip displacement. *Icarus*, **158**, 24–41.

SCHOLZ, C. H. 1982. Scaling laws for large earthquakes: consequences for physical models. *Bulletin of the Seismological Society of America*, **72**, 1–14.

SEGALL, P. & POLLARD, D. 1980. Mechanics of discontinuous faults. *Journal of Geophysical Research*, **85**, 4337–4350.

SIBSON, R. H. 1985. Stopping of earthquake ruptures at dilational fault jogs. *Nature*, **316**, 248–251.

SIBSON, R. H. 2001. Seismogenic framework for hydrothermal transport and ore deposition. *Reviews in Economic Geology*, **14**, 25–50.

SIEH, K. & NATAWIDJAJA, D. 2000. Neotectonics of the Sumatran fault, Indonesia. *Journal of Geophysical Research*, **105**, 28 295–28 326.

SHAW, B. 2006. Initiation propagation and termination of elastodynamic ruptures associated with segmentation of faults and shaking hazard. *Journal of Geophysical Research*, **111**, B08302, doi:1029/2005JB004093.

STORTI, F., HOLDSWORTH, R. E. & SALVINI, F. (eds) 2003. *Intraplate Strike-Slip Deformation Belts*. Geological Society, London, Special Publications, **210**.

SWANSON, M. 2005. Geometry and kinematics of adhesive wear in brittle strike-slip fault zones. *Journal of Structural Geology*, **27**, 871–887.

SYLVESTER, A. G. 1988. Strike-slip faults. *Geological Society of America Bulletin*, **100**, 1666–1703.

SYLVESTER, A. & SMITH, R. 1976. Tectonic transpression and basement controlled deformation in the San Andreas fault zone, Salton trough, California. *AAPG Bulletin*, **60**, 2081–2102.

TCHALENKO, J. S. 1970. Similarities between shear zones of different magnitudes. *Geological Society America Bulletin*, **81**, 1625–1640.

TEN BRINK, U. ET AL. 1999. Anatomy of the Dead Sea transform: does it reflect continuous changes in plate motion? *Geology*, **27**, 887–890.

TIKOFF, B. & TEYSSIER, C. 1994. Strain modelling of displacement-field partitioning in transpressional orogens. *Journal of Structural Geology*, **16**, 1575–1588.

TIKOFF, B., RUSSO, R., TEYSSIER, C. & TOMASSI, A. 2004. Mantle-driven deformation of orogenic zones and clutch tectonics. *In*: GROCOTT, J., MCCAFFREY, K. J. W., YAYLOR, G. & TIKOFF, B. (eds) *Vertical Coupling and Decoupling in the Lithosphere*. Geological Society, London, Special Publications, **227**, 41–64.

WAKABAYASHI, J., HENGESH, J. V. & SAWYER, T. L. 2004. Four-dimensional transform fault processes: progressive evolution of step-overs and bends. *Tectonophysics*, **392**, 279–301.

WALDRON, J. 2004. Anatomy and evolution of a pull-apart basin, Stellarton, Nova Scotia. *Geological Society of America Bulletin*, **116**, 109–127.

WESTAWAY, R. 1995. Deformation around stepovers in strike-slip fault zones. *Journal of Structural Geology*, **17**, 831–846.

WILCOX, R. E., HARDING, T. P. & SEELY, D. R. 1973. Basic wrench tectonics. *AAPG*, **57**, 74–96.

WILLETT, S. D. 1999. Orogeny and orography: the effects of erosion on the structure of mountain belts. *Journal of Geophysical Research*, **104**, 28 957–28 981.

WILLETT, S. D., SLINGERLAND, R. & HOVIUS, N. 2001. Uplift, shortening, and steady state topography in active mountain belts. *American Journal of Science*, **301**, 455–485.

WILSON, J. T. 1965. A new class of faults and their bearing on continental drift. *Nature*, **207**, 343–347.

WOOD, R., PETTINGA, J., BANNISTER, S., LAMARCHE, G. & MCMORRAN, T. 1994. Structure of the Hanmer strike-slip basin, Hope fault, New Zealand. *Geological Society of America Bulletin*, **106**, 1459–1473.

WOODCOCK, N. 1986. The role of strike-slip fault systems at plate boundaries. *Philosophical Transactions of the Royal Society of London A*, **317**, 13–29.

WOODCOCK, N. & FISCHER, M. 1986. Strike-slip duplexes. *Journal of Structural Geology*, **8**, 725–735.

YEATS, R., SIEH, K. & ALLEN, C. 1997. *The Geology of Earthquakes*, Oxford University Press, New York and Oxford, 568 pp.

ZEITLER, P. K., MELTZER, A. S. *ET AL.* 2001. Erosion, Himalayan geodynamics, and the geomorphology of metamorphism. *GSA Today*, **11**, 4–8.

ZHANG, P., BURCHFIEL, C., CHEN, S. & DENG, Q. 1989. Extinction of pull-apart basins. *Geology*, **17**, 814–817.

ZOLNAI, G. 1991. Continental wrench-tectonics and hydrocarbon habitat. *AAPG Continuing Education Course Note Series,* **30**, American Association of Petroleum Geologists Education Department, Tulsa, Oklahoma.

Global catalogue, classification and tectonic origins of restraining- and releasing bends on active and ancient strike-slip fault systems

P. MANN

Institute for Geophysics, Jackson School of Geosciences, University of Texas at Austin, 10100 Burnet Road, R2200, Austin, Texas 78758, USA (e-mail: paulm@ig.utexas.edu)

Abstract: Restraining- and releasing bends with similar morphology and structure have been described by many previous studies of strike-slip faults in a variety of active and ancient tectonic settings. Despite the documentation of at least 49 restraining and 144 releasing bends along active and ancient strike-slip faults in the continents and oceans, there is no consensus on how these structural features are named and classified, or how their wide range of structures and morphologies are controlled by the distinctive strike-slip tectonic settings in which they form. In this overview, I have compiled published information on the strike-slip tectonic setting, size, basin and bend type, age, and models for active and ancient releasing and restraining bends. Examples of bends on strike-slip faults are compiled and illustrated from five distinctive active strike-slip settings:

1. oceanic transforms separating oceanic crust and offsetting mid-oceanic spreading ridges;
2. long and linear plate-boundary strike-slip fault systems separating two continental plates whose plate-boundary kinematics can be quantified for long distances along strike by a single pole of rotation (e.g. the San Andreas fault system of western North America);
3. relatively shorter, more arcuate indent-linked strike-slip fault systems bounding escaping continental fragments in zones of continent–continent or arc–continent collision (e.g. the Anatolian plate);
4. straight to arcuate trench-linked strike-slip fault systems bounding elongate fore-arc slivers generated in active and ancient fore-arc settings by oblique subduction (e.g. Sumatra); and
5. cratonic strike-slip fault systems removed from active plate boundaries, formed on older crustal faults, but acting as 'concentrators' of intraplate stresses.

By far the most common, predictable and best-studied settings for restraining and releasing bends occur in continental-boundary strike-slip fault systems, where arrays of two to eight en échelon pull-apart basins mark transtensional fault segments and single and sometimes multiple large restraining bends mark transpressional segments; fault areas of transtension versus transpression are determined by the intersection angles between small circles about the interplate pole of rotation and the trend of the strike-slip fault system. These longer and more continuous boundary strike-slip systems also exhibit a widespread pattern of 'paired bends' or 'sidewall ripouts', or adjacent zones of pull-aparts and restraining bends—that range in along-strike-scale from kilometres to hundreds of kilometres. En échelon arrays of pull-apart basins are also observed on active 'leaky' or transtensional oceanic transforms, but restraining bends are rarely observed. In indent-linked strike-slip settings, strike-slip fault traces bounding escaping continental fragments tend to be more arcuate, less-continuous, and more splayed – but paired bends are common. Trench-linked strike-slip fault patterns closely mimic the trends of the subduction zone; these strike-slip faults can vary from long and continuous to short and arcuate, depending on the trace of the adjacent subduction zone. Paired bends are also observed in this setting. Bends on active, cratonic strike-slip fault form isolated, seismically active structures that act as 'stress concentrators' for intraplate stress. Cratonic strike-slip faults are generally not associated with pull-apart basins, and therefore paired bends are not observed in this setting.

The most likely geological models for the formation of releasing, restraining bends, and paired bends along boundary and trench-linked strike-slip faults include:

1. progressive linkage of en échelon shears within a young evolving shear zone; this model is not applicable to older strike-slip fault traces that have accumulated significant, lateral fault offsets;

From: CUNNINGHAM, W. D. & MANN, P. (eds) *Tectonics of Strike-Slip Restraining and Releasing Bends.* Geological Society, London, Special Publications, **290**, 13–142.
DOI: 10.1144/SP290.2 0305-8719/07/$15.00

2. formation of lenticular 'sidewall ripout' structures at scales ranging from outcrop to regional; ripouts are thought to form as a response to adherence or sticking along an adjacent and relatively straight strike-slip fault zone; this structural concept may help to explain the large number of paired bends embedded within strike-slip systems, sinusoidal curvature along the traces of many strike-slip faults, and the episodic nature of lateral shifts in the main strike-slip fault zone;
3. interaction of propagating strike-slip faults with pre-existing crustal structures such as ancient rift basins. Propagation of new strike-slip faults and interaction with older structures may occur on plate boundary, indent-linked, and trench-linked strike-slip faults; and
4. concentration of regional maximum compressive stress on pre-existing, basement fault trends in stable cratonic areas can produce active restraining-bend structures; periodic release of these bend-related stress concentrations is one of the leading causes of intraplate earthquakes within otherwise stable cratons.

The bulges [restraining bends] and gaps [releasing bends] became apparent when the [tectonic] restoration was made. The bulges appeared as areas of discrepancy from which sections of the blocks have been eliminated by faulting, folding, or warping, while the gaps are now occupied by bodies of water, or by sediments and lava. From: A. M. Quennell, 1956, *Tectonics of the Dead Sea Rift, 20th Session of the International Geological Congress*, Mexico City, Mexico.

Significance of strike-slip bends and unresolved questions

Previous work

The description and tectonic origin of releasing and restraining bends on strike-slip faults have fascinated researchers since their first recognition on active strike-slip faults by A. M. Quennell (1956) along the Dead Sea fault in 1956 (cf. Girdler 1989); by Kingma (1958) and Lensen (1958) along the Alpine–Macquarie Ridge fault system of New Zealand; and by Crowell (1974*a*) along the San Andreas fault zone of California (Fig. 1). Kingma (1958) postulated that the formation of basins at fault gaps, and their eventual deformation at fault overlaps, ultimately leads to the difficulty of recognizing and reconstructing ancient strike-slip zones in the rock record. Crowell (1974*a*, *b*, 1976) applied the early three-dimensional concepts of strike-slip fault deformation of Quennell, Kingma and Lensen, and larger-scale, early plate-tectonic concepts outlined by Carey (1958), to Late Cenozoic tectonics and sedimentation along the San Andreas fault zone in California (Fig. 1a). Using examples from the active San Andreas fault system, Crowell (1974*a*, *b*) named areas of extensional deformation at fault bends 'releasing bends' (Fig. 1b) and areas of convergent deformation at fault bends 'restraining bends' (Fig. 1c). Later workers, like Cowgill *et al.* (2004*a*, *b*) also use Crowell's (1974*a*) terms 'double bend' for a bend developed at an en échelon offset in a fault, and 'single bend' for a fault that terminates within the bend.

Objectives of this paper

This overview paper to the special volume on strike-slip bends synthesizes 50 years of bend-related research, spanning many fields of the geosciences, and focuses attention on important but poorly answered questions about the tectonics of large-offset oceanic and continental strike-slip faults that include:

1. what tectonic processes are responsible for producing prominent or subtle releasing and restraining bends along mainly straight strike-slip faults? (Fig. 1);
2. are these tectonic processes the same or do they differ in the five main tectonic settings for active strike-slip faults (Woodcock 1986; Fig. 2 & Table 1): for example, does a strike-slip fault formed in a collisional setting show the same type of bends as those seen on an oceanic transform fault or a strike-slip fault formed in a fore-arc setting by oblique subduction?
3. why are adjacent restraining and releasing or 'paired bends' produced along the length of many active and ancient strike-slip faults? (Fig. 1); and
4. how do single and paired bends evolve morphologically and structurally with the progressive offset of the strike-slip fault?

Scope of this paper

The central topic of this paper is the tectonic setting, progressive geological evolution, and possible tectonic controls of strike-slip releasing- and restraining bends occurring in natural settings (Fig. 1a & Table 2). Many of these descriptions of natural

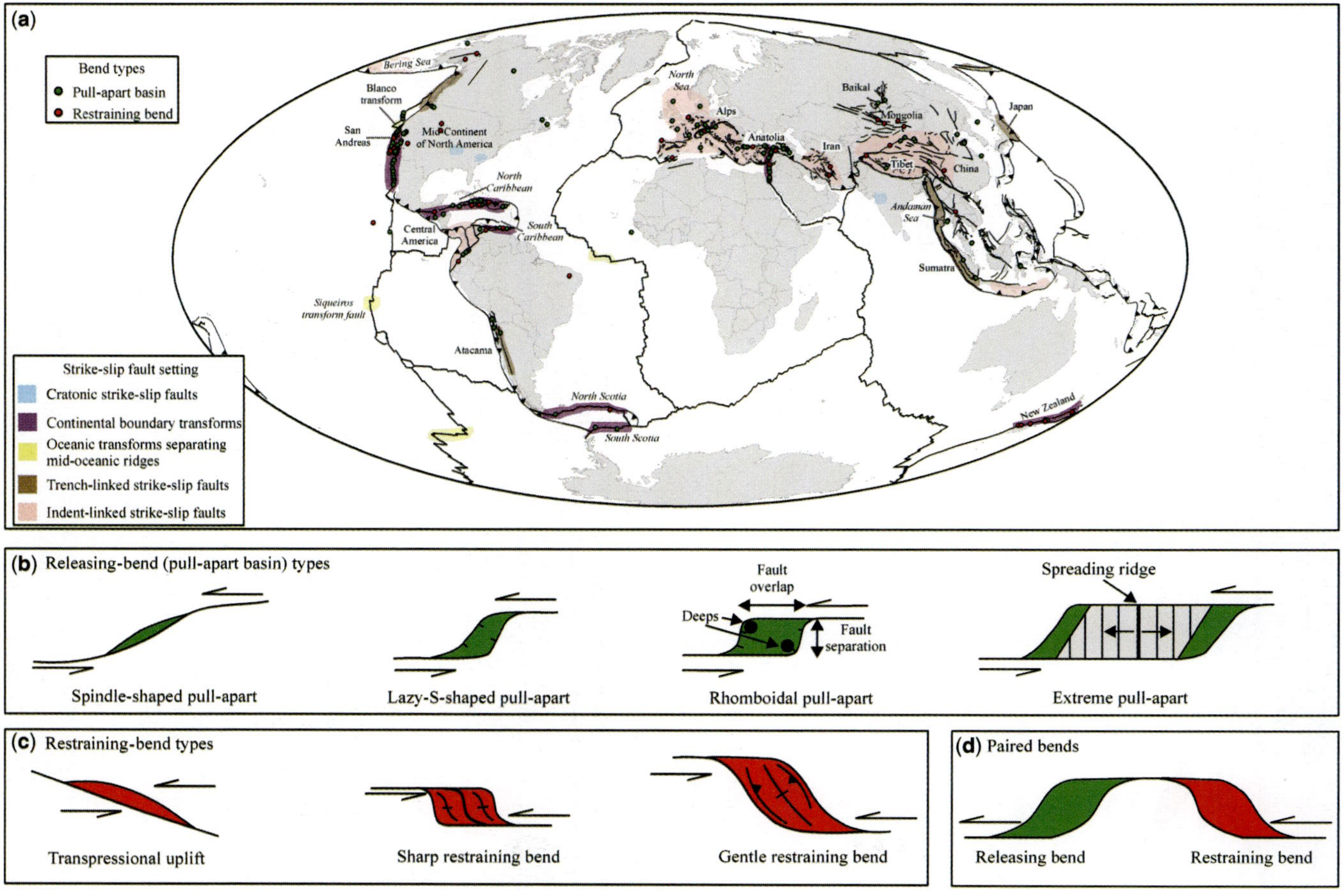

Fig. 1. (**a**) Global map showing distribution of pull-apart basins (green dots) and restraining bends (red dots) on active and ancient faults: (1) continental boundary transforms in purple; (2) oceanic transforms separating mid-oceanic ridges in yellow; (3) trench-linked strike-slip faults in brown; (4) indent-linked strike-slip faults in pink; and (5) cratonic strike-slip faults in blue. Most of the restraining and releasing bends that are shown on this map are active – although also included are well-studied ancient inactive examples. The main strike-slip fault systems discussed in this paper are labelled. (**b**) Releasing bend (pull-apart basin) types. Spindle-shaped or nucleating pull-apart, lazy-S-shaped pull-apart, rhomboidal pull-apart and extreme pull-apart. Refer to text for discussion. (**c**) Restraining-bend types: transpressional uplift, sharp restraining bend, gentle releasing bend. Refer to text for discussion. (**d**) Paired bend, or releasing and restaining bend. Refer to text for discussion.

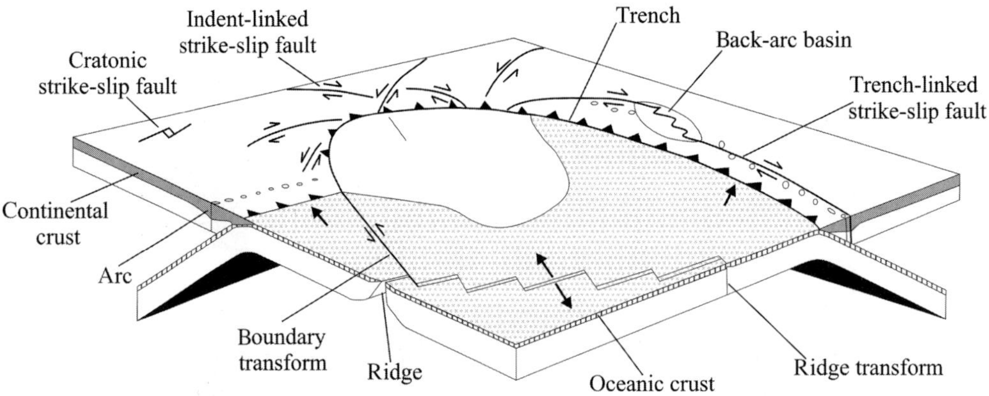

Fig. 2. Strike-slip fault tectonic settings, modified from Woodcock (1986).

bends (e.g. Segall & Pollard 1980; Aydin & Nur 1982; Ben-Avraham & Zoback 1992; Dooley & McClay 1997; McClay & Bonora 2001) have focused on the geology and mechanics of individual and geographically isolated bend structures and do not attempt to relate bends to their tectonic setting, their relationship to regional plate vectors, or their regional geology (Fig. 2). For this reason, this review uses a large number of maps of regional strike-slip fault systems to emphasize the occurrences of releasing and restraining bends that do not occur singly but instead occur in arrays of multiple, en échelon basins or bends. Such en échelon bend arrays indicate that regional tectonic controls and regional plate vectors can be important factors of bend development for along-fault strike distances of tens to hundreds of kilometres (Le Pichon & Francheteau 1978; Garfunkel 1981; Mann *et al.* 1983; Woodcock 1986; Fig. 2).

In the maps and descriptions in this paper, I will emphasize bend structures that have been described from strike-slip systems other than the well-studied San Andreas fault zone of the western USA and Mexico (Fig. 1), which is the most commonly used example of a continental strike-slip system in various chapters on strike-slip faulting in structural and tectonic undergraduate and graduate textbooks: Twiss & Moores (1992); Hatcher (1995); Ingersoll & Busby (1995); Moores & Twiss (1995); Davis & Reynolds (1996); Yeats *et al.* (1997); and Allen & Allen (2005).

The non-US areas of recent bend observations discussed in this paper include the Caribbean, the submarine Macquarie Ridge south of New Zealand, the Scotia Sea, the Tibetan Plateau, the Mongolian Altai, the Lake Baikal region, SE Asia, Sumatra, and the Philippines (Fig. 1 & Table 2). It is especially important to include these non-San Andreas examples in a review of this type, since the San Andreas fault system represents only one of the five types of strike-slip faults that are described here, and therefore its bends may not be representative for all types of strike-slip fault (Table 1).

Because this paper focuses on strike-slip bends, it is not intended to be a comprehensive review of all aspects of strike-slip faulting, seismology, geodesy, deformation and sedimentation. The interested reader requiring a broader review of strike-slip faulting is referred to the above-mentioned textbooks, along with specialized volumes and review papers on strike-slip faulting by: Ballance & Reading (1980), Sylvester (1984), Lowell (1985), Biddle & Christie-Blick (1985), Harding (1985), Woodcock (1986), Sylvester (1988), Woodcock & Schubert (1989), Harding (1990), Holdsworth *et al.* (1998), and Storti *et al.* (2003). A concise and well-illustrated summary of the morphology, seismicity and active tectonics of many of the world's best-studied strike-slip faults is given in a special issue of *Annalae Tectonicae* (1992, edited by Bucknam & Hancock 1992).

Strike-slip fault terminology

Fault terms used in this paper

Any description and discussion of strike-slip bends requires the use of a specialized vocabulary that has evolved over the past 50 years, and classification schemes for strike-slip faults that have been proposed and modified since the late 1980s. For a paper describing so many different strike-slip fault zones, I will adopt a specific set of terms that are illustrated on Figure 1 and used consistently throughout this paper.

Table 1. *Characteristics and examples of four major classes of strike-slip faults discussed in this paper (modified from Woodcock 1986) (cf. Figure 2)*

	Oceanic transforms separating mid-oceanic spreading ridges	Continental boundary strike-slip faults	Indent-linked strike-slip faults	Trench-linked strike-slip faults	Cratonic strike-slip faults
Crustal type	Ocean–ocean	Continent–continent, continent–ocean	Continent–continent, arc–continent	Continent–continent, arc–continent	Continent–continent
Active duration	Usually <5 Ma	Tens of millions of years; thousands of kilometres	Tens of millions of years; 10–200 km	Tens of millions of years	Tens of millions of years
Strike-slip offset	Typically <100 km but some much larger; later dip-slip	New or reactivated	Reactivate old structures where possible, but some new strands	Hundreds of kilometres	Kilometres
New or reactivated?	New, but may nucleate at old rift margin features	Very variable, non-marine to deep marine clastics or non-deposition	Non-marine clastic and playa sedimentation	New, but including old segments: may nucleate on or localize arc	Reactivated faults commonly inherited from rifts or plutons
Coeval sedimentation in fault zone	Pelagic carbonates and cherts, metaliferous sediments, ophiolitic screes	Rare in transpressive zones, variable but often alkaline basalt in transtension	Rare silicic lavas	Arc-derived clastics, non-marine to marine or non-depositional	Not observed
Coeval volcanism in fault zone	Cut MORB tholeiitic basalts; coeval more alkaline basalts	Not observed	Not observed	Arc tholeiites or calc-alkaline associations; high-K shoshonites may be common	Not observed
Coeval plutonism in fault zone	Ophiolitic plutonics; serpentinite diapirs on faults	Rare in transpression	Probably S-type granites	Gabbros and I-type granites	Not observed
Review papers	DeLong *et al.* (1979); Bonatti *et al.* (1979); Fox & Gallo (1984); Garfunkel (1986); Searle (1986); Fornari *et al.* (1989)	Mann *et al.* (1983); Sylvester (1988); Yeats *et al.* (1997)	Sylvester (1988); Yeats *et al.* (1997)	Fitch (1972); Karig (1979); Jarrard (1986); Yeats *et al.* (1997)	Marshak *et al.* (2003); Gangopadhyay & Talwani (2003); Schulte & Mooney (2005)

Table 2. *Locations of select basins and bends shown in regional maps*

Map ID	Bend or basin type (cf. Fig 1)	Ancient or active	Age of activity	Bend name	Country	Associated master faults	Selected references
				Active releasing bends			
1	Spindle-shaped pull-apart	Active	L. Quaternary	Algeciras Basin	Colombia	Algeciras	Velandia *et al.* (2005)
2	Spindle-shaped pull-apart	Active	L. Quaternary	Ararat Basin	Turkey; Armenia; Iran; Azerbaijan	Sardarapat–Nakhichevan; Dogubayazit; Maku	Karakhanian *et'al.* (2004)
3	Spindle-shaped pull-apart	Active	Quaternary	Balsillas Basin	Colombia	Algeciras	Velandia *et al.* (2005)
4	Spindle-shaped pull-apart	Active	Neogene	Bichuan	China	Nanding River–Xiaojin	Deng *et al.* (1986)
5	Spindle-shaped pull-apart	Active	Quaternary	Cantil Valley composite pull-apart basin; Koehn Lake Basin	USA (California)	Garlock	Aydin & Nur (1982); Aydin & Nur (1985)
6	Spindle-shaped pull-apart	Active	Quaternary	Clonard Basin	Haiti	Enriquillo–Plantain Garden	Mann *et al.* (1983); Mann *et al.* (1995)
7	Spindle-shaped pull-apart	Active	Miocene–L. Quaternary	Eastern Alps	Austria	Mur–Murz	Sachsenhofer *et al.* (2003)
8	Spindle-shaped pull-apart	Active	Quaternary	El Paraiso Basin	Colombia	Algeciras	Velandia *et al.* (2005)
9	Spindle-shaped pull-apart	Active	Quaternary	E'shan Basin	China	Qujiang	Swanson (2005)
10	Spindle-shaped pull-apart	Active	Quaternary	Lago Fagnano Basin	Chile	Magallanes–Fagnano	Lodolo *et al.* (2003)
11	Spindle-shaped pull-apart	Active	Late Quaternary	Lake Glynn Wye Basin	New Zealand	Hope	Freund (1971); Cowan (1990); Cowan (1994)
12	Spindle-shaped pull-apart	Active	L. Quaternary	Mesquite Basin	USA (California)	Brawley–Imperial	Fuis *et al.* (1982); Mann *et al.* (1983); Dooley & McClay (1997)
13	Spindle-shaped pull-apart	Active	L. Quaternary	Neiva Basin	Colombia	Garzon	Chorowicz *et al.* (1996)
14	Spindle-shaped pull-apart	Active	L. Quaternary	North Aegean Trough	Aegean Sea	Athos	Dooley & McClay (1997)

15	Spindle-shaped pull-apart	Active	L. Quaternary	South Golbaf	Iran	Gowk	Walker & Jackson (2004)
16	Spindle-shaped pull-apart	Active	L. Quaternary	Various unnamed basins	Iceland	Selsund	Angelier *et al.* (2004)
17	Lazy-S pull-apart	Active	Oligocene–L. Quaternary	Lake Baikal	Russia	Baikal	Petit *et al.* (1996); Lesne *et al.* (1998)
18	Lazy-S pull-apart	Active	Pliocene–L. Quaternary	Salina del Fraile	Argentina	El Fraile	Dooley & McClay (1997); Reijs & McClay (2003)
19	Lazy-S pull-apart	Active	Quaternary	San Juan de Villalobos–Yunguillo Basin	Colombia	Yunguillo	Velandia *et al.* (2005)
20	Lazy-S pull-apart	Active	Neogene	Siabu Basin	Indonesia	Sumatra	Lelgemann *et al.* (2000)
21	Lazy-S pull-apart	Active	Quaternary	Siqueiros (four basins)	Pacific Ocean	Siqueiros	Fornari *et al.* (1989)
22	Lazy-S pull-apart	Active	L. Quaternary	Unnamed basins (South Scotia Ridge)	Antarctica	South Scotia Ridge	Galindo-Zaldivar *et al.* (1996); Galindo-Zaldivar *et al.* (2002)
23	Lazy-S pull-apart	Active	Eocene–L. Quaternary	Virgin Islands Basin	Virgin Islands	Unnamed	Jany *et al.* (1990)
24	Lazy-Z pull-apart	Active	Oligocene–L. Quaternary	Bohai Basin	China	Tanlu	Allen *et al.* (1997)
25	Lazy-Z pull-apart	Active	Neogene–L. Quaternary	Central Basin	Turkey	North Anatolian	Barka & Gulen (1989); Armijo *et al.* (1989); Le Pichon *et al.* (2001); Armijo *et al.* (2002); Rangin *et al.* (2004)
26	Lazy-Z pull-apart	Active	Cenozoic	Cerro de la Mica	Chile	Atacama	McClay & Bonora (2001)
27	Lazy-Z pull-apart	Active	Quaternary	Clear Lake	USA (California)	Mountain	McLaughlin & Nilsen (1982); Hearne *et al.* (1988)
28	Lazy-Z pull-apart	Active	L. Plio–Quaternary	Dagg Basin	New Zealand	Alpine	Barnes *et al.* (2001); Barnes *et al.* (2005)
29	Lazy-Z pull-apart	Active	Miocene–Quaternary	Death Valley Basin	USA (California)	Southern Death Valley; Fish Lake Valley–Northern Death Valley–Furnace Creek	Burchfiel Stewart (1966); Christie-Blick & Biddle (1985)
30	Lazy-Z pull-apart	Active	L. Plio–Quaternary	Five Fingers Basin	New Zealand	Alpine	Barnes *et al.* (2005)
31	Lazy-Z pull-apart	Active	L. Plio–Quaternary	George Basin	New Zealand	Alpine	Barnes *et al.* (2005)

(*Continued*)

Table 2. Continued

Map ID	Bend or basin type (cf. Fig 1)	Ancient or active	Age of activity	Bend name	Country	Associated master faults	Selected references
32	Lazy-Z pull-apart	Active	L. Oligocene to M.'Miocene	Gulf of Paria Basin	Venezuela	El Pilar; Central Range	Babb & Mann (1999); Flinch *et al.* (1999)
33	Lazy-Z pull-apart	Active	L. Quaternary	Karakoram Basin	China	Karakoram	Murphy & Burgess (2006)
34	Lazy-Z pull-apart	Active	Pliocene (?)–Quaternary	La Gonzalez Basin	Venezuela	Bocono	Christie-Blick and Biddle (1985)
35	Lazy-Z pull-apart	Active	L. Quaternary	Lake Tahoe	USA (California)	Walker Lane	Unruh *et al.* (2003)
36	Lazy-Z pull-apart	Active	L. Plio–Quaternary	Looking Glass Basin	New Zealand	Alpine	Barnes *et al.* (2005)
37	Lazy-Z pull-apart	Active	Neogene–L. Quaternary	Mor	Israel	Dead Sea	Bartov & Sagy (2004)
38	Lazy-Z pull-apart	Active	Late Quaternary	Mount Lyford duplex	New Zealand	Hope	Eusden *et al.* (2000)
39	Lazy-Z pull-apart	Active	Early Pliocene–L. Quaternary	Niksar Basin	Turkey	North Anatolian	Mann *et al.* (1983); Tatar *et al.* (1995); Barka *et al.* (2000); Kocyigit & Erol (2001)
40	Lazy-Z pull-apart	Active	L. Quaternary	Offshore Santa Catalina Island	USA (California)	San Diego Trough–Catalina	Legg *et al.* (2007)
41	Lazy-Z pull-apart	Active	Neogene–L. Quaternary	Panamint	USA (California)	Walker Lane	Burchfiel *et al.* (1987)
42	Lazy-Z pull-apart	Active	L. Miocene–L. Quaternary	Ridge Basin	USA	San Andreas; San Gabriel	May *et al.* (1993)
43	Lazy-Z pull-apart	Active	L. Quaternary	Round Valley	USA (California)	Bartlett Springs	McLaughlin & Nilsen (1982); Hearne *et al.* (1988)
44	Lazy-Z pull-apart	Active	Neogene–L. Quaternary	Saline valleys	USA (California)	Walker Lane	Burchfiel *et al.* (1987)
45	Lazy-Z pull-apart	Active	L. Plio–Quaternary	Secretary Basin	New Zealand	Alpine	Barnes *et al.* (2005)
46	Lazy-Z pull-apart	Active	Quaternary	Sunda Strait	Sumatra	Sumatran	Lelgemann *et al.* (2000)
47	Lazy-Z pull-apart	Active	Oligocene–L. Quaternary	Tunka	Russia	Tunka	Larroque *et al.* (2001)
48	Rhomboidal pull-apart	Active	Quaternary	Alarcon Basin (Gulf of California)	USA (California)	San Andreas	Mann *et al.* (1983)
49	Rhomboidal pull-apart	Active	L. Quaternary	Altamira–Pitalito Basin	Colombia	Granadillo; El Cedro; Pitalito	Velandia *et al.* (2005)

50	Rhomboidal pull-apart	Active	Oligocene–L. Quaternary	Andamen Sea	Thailand	Seuliman; Sumatra	Curray (2005)
51	Rhomboidal pull-apart	Active	L. Quaternary	Aragonese Basin	Gulf of Aqaba	Arava	Ben-Avraham (1985); Erhardt *et al.* (2005); Garfunkel (2005)
52	Rhomboidal pull-apart	Active	E. Miocene–L. Quaternary	Arnona Basin	Gulf of Aqaba	Arava	Ben-Avraham (1985)
53	Rhomboidal pull-apart	Active	Neogene–L. Quaternary	Baitiquiri	Caribbean Sea	Oriente	Calais & Mercier de Lepinay (1991)
54	Rhomboidal pull-apart	Active	Quaternary	Barahta	Israel	Rachaya; Si'on; Shab'a	Heimann *et al.* (1990)
55	Rhomboidal pull-apart	Active	Not given in paper	Bayburt Basin	Turkey	Kelkit–Coruh	Kocyigit & Erol (2001)
56	Rhomboidal pull-apart	Active	Quaternary	Bend region	USA (California)	San Clemente	Legg *et al.* (2007)
57	Rhomboidal pull-apart	Active	L. Quaternary	Beng Co	China		Armijo *et al.* (1989)
58	Rhomboidal pull-apart	Active	Neogene–L. Quaternary	Cabo Cruz	Caribbean Sea	Oriente	Calais & Mercier de Lepinay (1991)
59	Rhomboidal pull-apart	Active	L. Oligocene or E. Miocene to L. Quaternary	Cariaco Basin	Venezuela	Moron; El Pilar; San Sebastian	Schubert (1982); Mann *et al.* (1983); Jaimes (2003)
60	Rhomboidal pull-apart	Active	Not given in paper	Carmen Basin (Gulf of California)	USA (California)	San Andreas	Mann *et al.* (1983)
61	Rhomboidal pull-apart	Active	Pliocene	Cascadia	Offshore USA (Oregon)	Blanco transform	Embley & Wilson (1992)
62	Rhomboidal pull-apart	Active	Neogene–L. Quaternary	Chivirico	Caribbean Sea	Oriente	Calais Mercier de Lepinay (1991)
63	Rhomboidal pull-apart	Active	L. Quaternary	Cholame Valley Basin	USA (California)	San Andreas	R. Brown & Vedder (1967); Shedlock *et al.* (1990); Graymer *et al.* (2007)
64	Rhomboidal pull-apart	Active	Miocene–Pliocene	Cinarcik Basin	Turkey	North Anatolian	Okay *et al.* (2000); Le Pichon *et al.* (2001); Seeber *et al.* (2006)
65	Rhomboidal pull-apart	Active	Neogene	Coyote Ridge	USA (California)	Coyote Ridge (San Jacinto)	Sharp (1975)
66	Rhomboidal pull-apart	Active	Neogene–L. Quaternary?	Daisy Bank	USA	Daisy Bank	Goldfinger *et al.* (1996)
67	Rhomboidal pull-apart	Active	E. Miocene–L. Quaternary	Dakar Basin	Gulf of Aqaba	Dead Sea	Ben-Avraham (1985)

(Continued)

Table 2. Continued

Map ID	Bend or basin type (cf. Fig 1)	Ancient or active	Age of activity	Bend name	Country	Associated master faults	Selected references
68	Rhomboidal pull-apart	Active	Late Miocene–L. Quaternary	Damxung Corridor Basin	China	Damxung–Jiali	Armijo *et al.* (1986); Edwards & Ratschbacher (2005)
69	Rhomboidal pull-apart	Active	E. Miocene–E. Pliocene (L. Quaternary)	Dead Sea Basin	Jordan	Dead Sea; Jordan; Arava	Ten-Brink & Ben-Avraham (1989); Lazar *et al.* (2006)
70	Rhomboidal pull-apart	Active	Not given in paper	Delfin Basin (Gulf of California)	USA (California)	San Andreas	Mann *et al.* (1983)
71	Rhomboidal pull-apart	Active	Tertiary	Dungun Basin	Malaysia	Dungun	Dooley & McClay (1997)
72	Rhomboidal pull-apart	Active	Quaternary	East Blanco	Offshore USA (Oregon)	Blanco transform	Embley & Wilson (1992)
73	Rhomboidal pull-apart	Active	Pliocene–Quaternary	Erciyes Basin	Turkey	Central Anatolian	Kocyigit & Erol (2001)
74	Rhomboidal pull-apart	Active	Quaternary	Erzincan Basin	Turkey	Tercan–Askale	Aydin & Nur (1982); Barka & Gulen (1989); Kocyigit & Erol (2001)
75	Rhomboidal pull-apart	Active	Not given in paper	Farallon Basin (Gulf of California)	USA (California)	San Andreas	Mann *et al.* (1983)
76	Rhomboidal pull-apart	Active	Not given in paper	Ghab Basin	Syria	Dead Sea	Brew *et al.* (2001); Adiyaman & Chorowicz (2002); Westaway (2004)
77	Rhomboidal pull-apart	Active	L. Quaternary	Golden Gate	USA (California)	San Andreas	Zoback *et al.* (1999); Wakabayashi (2007)
78	Rhomboidal pull-apart	Active	Quaternary	Gorda	Offshore USA (Oregon)	Blanco transform	Embley & Wilson (1992)
79	Rhomboidal pull-apart	Active	Not given in paper	Guaymas Basin (Gulf of California)	USA (California)	San Andreas	Mann *et al.* (1983)
80	Rhomboidal pull-apart	Active	Neogene–L. Quaternary	Gulu	China	Gulu	Edwards & Ratschbacher (2005)
81	Rhomboidal pull-apart	Active	Quaternary	Hamner Basin	New Zealand	Hope	Freund (1971); Wood *et al.* (1994)
82	Rhomboidal pull-apart	Active	L. Quaternary	Homestead Valley Basin	USA (California)	Homestead Valley; Johnson Valley	Sowers *et al.* (1994)
83	Rhomboidal pull-apart	Active	Not given in paper	Hula Basin	Israel; Lebanon	Dead Sea	Heimann & Ron (1987); Zilberman *et al.* (2000)

84	Rhomboidal pull-apart	Active	Neogene-L. Quaternary	Jianchuan–Hequin	China	Xiaojinhe	Deng *et al.* (1986)
85	Rhomboidal pull-apart	Active	Not given in paper	Karasu Basin	Turkey	Tercan–Askale	Kocyigit & Erol (2001)
86	Rhomboidal pull-apart	Active	Not given in paper	Karliova Basin	Turkey	East Anatolian	Kocyigit & Erol (2001)
87	Rhomboidal pull-apart	Active	Not given in paper	Kinneret Zemah Basin	Israel	Dead Sea	Rotstein & Bartov (1989)
88	Rhomboidal pull-apart	Active	Not given in paper	Lake Elsinore Basin	USA (California)	Elsinore	Mann *et al.* (1983)
89	Rhomboidal pull-apart	Active	Not given in paper	Lake Hazar Basin	Turkey	East Anatolian	Mann *et al.* (1983); Hempton & Dunne (1983)
90	Rhomboidal pull-apart	Active	L. Miocene	Lake Matano Basin	Indonesia	Matano	Parkinson & Dooley (1996)
91	Rhomboidal pull-apart	Active	L. Miocene	Lake Poso Basin	Indonesia	Poso	Parkinson & Dooley (1996)
92	Rhomboidal pull-apart	Active	Quaternary	Laohuyiaoxian	China	Xiaokou	Zhang *et al.* (1989)
93	Rhomboidal pull-apart	Active	L. Quaternary	Little Lake Valley	USA (California)	Maacama	McLaughlin & Nilsen (1982); Hearne *et al.* (1988)
94	Rhomboidal pull-apart	Active	L. Quaternary	Little Sulfur Creek	USA (California)	Maacama	McLaughlin & Nilsen (1982); Hearne *et al.* (1988)
95	Rhomboidal pull-apart	Active	Miocene–L. Quaternary	Macquarie Ridge	Australia	Australian–Antarctic plate boundary	Massell *et al.* (2000); Daczko *et al.* (2003)
96	Rhomboidal pull-apart	Active	Not given in paper	Malatya Basin	Turkey	Malatya–Ovacik	Kocyigit & Erol (2001)
97	Rhomboidal pull-apart	Active	Not given in paper	Mazatlan Basin	USA (California)	San Andreas	Mann *et al.* (1983)
98	Rhomboidal pull-apart	Active	Pliocene–L. Quaternary	Merced	USA (California)	San Andreas	Wakabayashi (2007)
99	Rhomboidal pull-apart	Active	Quaternary	Mirogoane Lakes Basin	Haiti	Enriquillo–Plantain Garden	Mann *et al.* (1983); Mann *et al.* (1995)
100	Rhomboidal pull-apart	Active	L. Plio–Quaternary	Nancy Basin	New Zealand	Alpine	Barnes *et al.* (2005)
101	Rhomboidal pull-apart	Active	Quaternary	Northern Gulf of Aqaba	Israel	Dead Sea	Ehrhardt *et al.* (2005)

(Continued)

Table 2. Continued

Map ID	Bend or basin type (cf. Fig 1)	Ancient or active	Age of activity	Bend name	Country	Associated master faults	Selected references
102	Rhomboidal pull-apart	Active	Quaternary	Not given in paper	New Zealand	Alpine; Boo Boo; Marlborough	Barnes (2005)
103	Rhomboidal pull-apart	Active	L. Quaternary	Olema	USA (California)	San Andreas	Wakabayashi (2007)
104	Rhomboidal pull-apart	Active	Not given in paper	Pescadero	USA (California)	San Andreas	Mann *et al.* (1983)
105	Rhomboidal pull-apart	Active	Quaternary	Pitalito Basin	Colombia	San Francisco	Velandia *et al.* (2005)
106	Rhomboidal pull-apart	Active	Quaternary–L. Quaternary	Rio Lempa Basin	El Salvador	El Salvador (San Vicente segment and Berlin segment)	Corti *et al.* (2005)
107	Rhomboidal pull-apart	Active	L. Quaternary	San Jacinto Basin	USA (California)	Casa Loma; Claremont	Langenheim *et al.* (2004)
108	Rhomboidal pull-apart	Active	Late Quaternary	San Pablo Bay	USA (California)	Hayward; Rodgers Creek	Parsons *et al.* (2003)
109	Rhomboidal pull-apart	Active	L. Quaternary	San Pedro Martir (Gulf of California)	Mexico	San Pedro	Mann *et al.* (1983); Legg *et al.* (2007)
110	Rhomboidal pull-apart	Active	Pliocene-Quaternary	Sea of Galilee	Israel	Dead Sea	Hurwitz *et al.* (2002)
111	Rhomboidal pull-apart	Active	Quaternary	Sibundoy Basin	Colombia	Sibundoy; San Francisco	Velandia *et al.* (2005)
112	Rhomboidal pull-apart	Active	Quaternary	Singkarak	Indonesia	Sumatran; Metawai	Sieh & Natawidjaja (2000)
113	Rhomboidal pull-apart	Active	Eocene–L. Quaternary	St Croix Basin	St Croix	Right-lateral strike-slip	Jany *et al.* (1990)
114	Rhomboidal pull-apart	Active	Quaternary	Surveyor Basin	Offshore USA (Oregon)	Blanco transform	Embley & Wilson (1992)
115	Rhomboidal pull-apart	Active	Neogene–L. Quaternary	Swan Islands	Honduras	Swan Islands	Mann *et al.* (1991)
116	Rhomboidal pull-apart	Active	L. Quaternary	Tabriz Basin	Iran	North Tabriz	Karakhanian *et al.* (2004)
117	Rhomboidal pull-apart	Active	Pliocene–Early Quaternary	Tasova–Erbaa	Turkey	North Anatolian	Barka *et al.* (2000)

118	Rhomboidal pull-apart	Active	Miocene-Pliocene	Tekirdag Basin	Turkey	North Anatolian	Okay *et al.* (1999); Le Pichon *et al.* (2001); Seeber *et al.* (2006)
119	Rhomboidal pull-apart	Active	L. Oligocene to E.'Miocene–L. Quaternary	Tesac Basin	Atlantic offshore	Magallanes–Fagnano	Lodolo *et al.* (2003)
120	Rhomboidal pull-apart	Active	Quaternary	Tiburon river valley	Haiti	Enriquillo–Plantain Garden	Mann *et al.* (1983); Mann *et al.* (1995)
121	Rhomboidal pull-apart	Active	E. Miocene–L. Quaternary	Tiran Basin	Gulf of Aqaba	Dead Sea	Ben-Avraham (1985)
122	Rhomboidal pull-apart	Active	Neogene–L. Quaternary	Unnamed	China	Xianshuihe; Ganzi–Yushu	Allen *et al.* (1991)
123	Rhomboidal pull-apart	Active	Quaternary–L. Quaternary	Vanadzor	Armenia	Pambak-Sevan–Sunik	Karakhanian *et al.* (2004)
124	Rhomboidal pull-apart	Active?	E. Quaternary–L. Quaternary ?	Various unnamed pull-apart basins	USA (Nevada)	Olinghouse	Aydin & Nur (1982)
125	Rhomboidal pull-apart	Active	Not given in paper	Wagner (Gulf of California)	Mexico	San Andreas	Mann *et al.* (1983)
126	Rhomboidal pull-apart	Active	E. Quaternary–L. Quaternary	Walker Lake Basin	USA (Nevada)	Walker Lane	Link *et al.* (1985)
127	Rhomboidal pull-apart	Active	Quaternary	West Blanco Basin	Offshore USA (Oregon)	Blanco transform	Embley & Wilson (1992)
128	Rhomboidal pull-apart	Active	Late Neogene–L. Quaternary	West Jamaica (Hendrix)	Jamaica	Duanvale; Swan Islands	Mann *et al.* (2007*a*)
129	Rhomboidal pull-apart	Active	Neogene–L. Quaternary	Windward Passage	Caribbean Sea	Oriente	Calais & Mercier de Lepinay (1991)
130	Spindle-shaped pull-apart	Active	Late Quaternary	Poplars basin	New Zealand	Hope	Cowan (1990); Cowan (1994)
131	Extreme pull-apart	Active	Neogene–L. Quaternary	Andaman Sea	India; Thailand	Sumatra–Sagaing	Curray (2005)
132	Extreme rhomboidal pull-apart	Active	Eocene–L. Quaternary	Cayman Trough	Northern Caribbean Sea	Oriente; Swan	Rosencrantz *et al.* (1988); Leroy *et al.* (2000)
				Ancient releasing bends			
133	Spindle-shaped pull-apart	Ancient	Eocene	Chiwaukum Basin	USA (Washington)	Entiat; Leavenworth	Johnson (1985)
134	Spindle-shaped pull-apart	Ancient	Eocene	Chuckanut–Puget–Naches Basin	USA (Washington)	Puget; Straight Creek	Johnson (1985)

(*Continued*)

Table 2. Continued

Map ID	Bend or basin type (cf. Fig 1)	Ancient or active	Age of activity	Bend name	Country	Associated master faults	Selected references
135	Spindle-shaped pull-apart	Ancient	L. Palaeozoic–Quaternary	Danish Basin	Sweden; Bornholm; Baltic Sea	Colonus Shale Trough; Romeneasen; Kullen Rjngsjon Andrawum; Christianso	Erlstrom *et al.* (1997)
136	Spindle-shaped pull-apart	Ancient	?–Quaternary	Dongchuan Basin	China	East Xiaojian	Swanson (2005)
137	Spindle-shaped pull-apart	Ancient	Ordovician–Devonian	Gaspe Basin	Canada	Lac Cascapedia; Mont Albert; Shickshock Sud	Sacks *et al.* (2004)
138	Lazy-S pull-apart	Ancient	Mesozoic	Caleta Coloso Basin	Chile	Bolfin; Jorgillo; Coloso	Cembrano *et al.* (2005)
139	Lazy-S pull-apart	Ancient	Oligocene–E. Miocene	Salar Grande Basin	Chile	Atacama	Reijs & McClay (1998)
140	Lazy-S pull-apart	Ancient	Archaean	Whim Creek Basin	Australia	Loudens	Krapez & Eisenlohr (1998)
141	Lazy-Z pull-apart	Ancient	Jurassic–E. Cretaceous	Chaoshui Basin	China	North Qilian Shan	Vincent & Allen (1999)
142	Lazy-Z pull-apart	Ancient	Jurassic–E. Cretaceous	Huahai-Jinta Basin	China	North Qilian Shan	Vincent & Allen (1999)
143	Lazy-Z pull-apart	Ancient	Cretaceous	Jiaolai Basin	China	Tan–Lu; Jimo	Zhang *et al.* (2003)
144	Lazy-Z pull-apart	Ancient	Jurassic–E. Cretaceous	Minle Basin	China	North Qilian Shan	Vincent Allen (1999)
145	Lazy-Z pull-apart	Ancient	Cretaceous	Zhucheng Basin	China	Tan–Lu; Baichihe; Wulian	Zhang *et al.* (2003)
146	Rhomboidal pull-apart	Ancient	Miocene–Pliocene	Bir Zreir graben	Egypt	Southern Diagonal; Northern Diagonal	Eyal *et al.* (1986)
147	Rhomboidal pull-apart	Ancient	Cenozoic	Bovey Basin	England	Stickepath–Lustleigh	Holloway & Chadwick (1986)
148	Rhomboidal pull-apart	Ancient	Late Cretaceous–Cenozoic	Catacamas	Honduras	Guayape	Gordon & Muehlberger (1994)
149	Rhomboidal pull-apart	Ancient	Permian	Collio Basin	Italy	Valcamonica; Giudicarie	Bertoluzza & Perotti (1997)

150	Rhomboidal pull-apart	Ancient	Mesozoic–Cenozoic	Coso Basin	USA (Nevada)	Airport Lake; Owens Valley	Feng & Lees (1998); Unruh *et al.* (2003);
151	Rhomboidal pull-apart	Ancient	Ordovician–Devonian	Culm Basin	France	Nort-sur-Erdre	Shelley & Bossiere (2001)
152	Rhomboidal pull-apart	Ancient	Quaternary	Dayinshui	China	Haiyuan	Zhang *et al.* (1989)
153	Rhomboidal pull-apart	Ancient	Triassic	Donaustauf Basin	Germany	Pfahl; Danube	Mattern (2001)
154	Rhomboidal pull-apart	Ancient	Cretaceous	Eumsung Basin	Korea	Kongiu	Ryang & Chough (1999)
155	Rhomboidal pull-apart	Ancient	Jurassic–Cretaceous	Guinea marginal plateau	Guinea	Unnamed	Benkhelil *et al.* (1995)
156	Rhomboidal pull-apart	Ancient	Eocene	Icotea	Venezuela	Icotea	Escalona & Mann (2003)
157	Rhomboidal pull-apart	Ancient	Jurassic–Cretaceous	Inner Moray Firth Basin	North Sea	Great Glen	Dooley *et al.* (1999)
158	Rhomboidal pull-apart	Ancient	Jurassic	Kuzigongsu pull-apart	Kazakostan; Kyrgyzstan; China	Karatau/Talas–Fergana	Allen *et al.* (2001)
159	Rhomboidal pull-apart	Ancient	Carboniferous–Miocene (?)	Magdalena Basin	Canada	Long Range; Belleisle; Kennebacasis; Caldonia	Mann *et al.* (1983)
160	Rhomboidal pull-apart	Ancient	Triassic–E. Cretaceous	Mazhan Basin	China	Tancheng–Lujian (Tan–Lu)	Hong & Miyata (1999)
161	Rhomboidal pull-apart	Ancient	E. Proterozoic	Nonacho Basin	Canada	Noman; King; Macinnis; Salkeld	Aspler & Donaldson (1985)
162	Rhomboidal pull-apart	Ancient	Palaeogene	Norton Basin	USA (Alaska)	Kaltag	Fisher *et al.* (1979)
163	Rhomboidal pull-apart	Ancient	Oligocene–E. Miocene	Oschiri Basin	Italy	Olbia	Funedda & Oggiano (2005)
164	Rhomboidal pull-apart	Ancient	Pre-Tertiary	Overton Arm Basin	USA	Lake Mead	Aydin *et al.* (1990)
165	Rhomboidal pull-apart	Ancient	Tertiary	Petrockstow Basin	England	Stickepath–Lustleigh	Holloway & Chadwick (1986)
166	Rhomboidal pull-apart	Ancient	Not given in paper	Salt Lake Basin	China	Haiyuan	Zhang *et al.* (1989)
167	Rhomboidal pull-apart	Ancient	Quaternary	Shaoshui Basin	China	Haiyuan	Zhang *et al.* (1989)
168	Rhomboidal pull-apart	Ancient	L. Jurassic–E. Cretaceous	Soria strike-slip Basin	Spain	Unnamed	Guiraud & Seguret (1985)

(*Continued*)

Table 2. Continued

Map ID	Bend or basin type (cf. Fig 1)	Ancient or active	Age of activity	Bend name	Country	Associated master faults	Selected references
169	Rhomboidal pull-apart	Ancient	Carboniferous–Permian	St Etienne Basin	France	France Massif Central	Mattauer & Matte (1998)
170	Rhomboidal pull-apart	Ancient	Palaeogene	St George	USA (Alaska)	Beringian	Worrall (1991)
171	Rhomboidal pull-apart	Ancient	Late Palaeozoic	Stellarton Basin	Canada	Late Palaeozoic	Waldron (2004)
172	Rhomboidal pull-apart	Ancient	Plio–Quaternary	Upper Rhine Graben	Germany	Left-lateral strike-slip	Schumacher (2002)
173	Rhomboidal pull-apart	Ancient	Tertiary (Miocene)	Vienna Basin	Austria; Czech Republic; Slovakia	Waschberg	Mann *et al.* (1983); Christie-Blick & Biddle (1985); Royden (1985); Hinsch *et al.* (2005)
174	Rhomboidal pull-apart	Ancient	E. Miocene–M. Miocene	Virginia City	USA (Nevada)	Comstock; Occidental	Berger (2007)
175	Extreme pull-apart	Ancient	Neogene	Sea of Japan	Japan	Unnamed	Lallemand & Jolivet (1986); Burgmann *et al.* (1994)
				Active restraining bends			
1	Transpressional uplift	Active	Neogene–L. Quaternary	Betic Range	Spain	Trans-Alboran	Keller *et al.* (1995)
2	Transpressional uplift	Active	Cenozoic-Late Neogene	Border Ranges	USA (Alaska)		Little (1990); Roeske *et al.* (2003)
3	Transpressional uplift	Active	Late Quaternary	Charwell 'fault wedge'	New Zealand	Hope	Eusden *et al.* (2005)
4	Transpressional uplift	Active	Quaternary	Clipperton	Pacific Ocean	Clipperton	Pockalny (1997)
5	Transpressional uplift	Active	Late Neogene–L. Quaternary	Mecca Hills	USA (California)	San Andreas	Sylvester & Smith (1976)
6	Transpressional uplift	Active	Late Neogene–L. Quaternary	Southern Alps	New Zealand	Central Alpine	Little *et al.* (2005)
7	Sharp restraining bend	Active	Neogene–L. Quaternary	Almacik Mountains	Turkey	North Anatolian	Saribudak *et al.* (1990)
8	Sharp restraining bend	Active	Neogene–L. Quaternary	Araya–Paria	Venezuela	El Pilar	Audemard *et al.* (2006)

9	Sharp restraining bend	Active	Oligocene–L. Quaternary	Chainat Ridge	Thailand	Mae Ping	Smith *et al.* (2007)
10	Sharp restraining bend	Active	L. Quaternary	Middleton Place Summerville	USA (South Carolina)	Woodstock; Ashley River	Gangopadhyay & Talwani (2003); Marshak *et al.* (2003)
11	Sharp restraining bend	Active	?–L. Quaternary	Ocotillo Badlands	USA (California)	Coyote Creek	Sharp & Clark (1972); Brown & Sibson (1989); Brown *et al.* (1991); Lutz *et al.* (2006)
12	Sharp restraining bend	Active	L. Quaternary	Rokko Range	Japan	Unnamed	Sangawa (1986)
13	Sharp restraining bend	Active	Late Neogene–L. Quaternary	Unnamed	Mexico	Polochic; unnamed	Guzmán-Speziale (2001)
14	Sharp restraining bend	Active	Late Neogene	Villa Vasquez	Hispaniola	Septentrional	Mann *et al.* (1999)
15	Gentle restraining bend	Active	Pliocene (?)–L. Quaternary	Akato Tagh	China	Altyn Tagh	Cowgill *et al.* (2004*a*); Cowgill *et al.* (2004*b*)
16	Gentle restraining bend	Active	Pliocene (?)–L. Quaternary	Alpine Fault 'Big Bend'	New Zealand	Alpine	Campbell (1992); Cowan (1994)
17	Gentle restraining bend	Active	Cenozoic–L. Quaternary	Atacama	Chile	Atacama	Swanson (2005)
18	Gentle restraining bend	Active	Neogene–L. Quaternary	Bingol	Turkey	East Anatolian	Hempton & Dunne (1983); Saroglu *et al.* (1992)
19	Gentle restraining bend	Active	post-M. Miocene–L. Quaternary	Blue Mountains	Jamaica	Plantain Garden; Blue Mountain; Yallahs; Wagwater	Mann *et al.* (1985); Mann & Gordon (1996); Mann *et al.* (2007*a*)
20	Gentle restraining bend	Active	Quaternary	Chicibu–Sanbagawa belts	Japan	Median; Butsuzo	Mann & Gordon (1996); Yeats *et al.* (1997)
21	Gentle restraining bend	Active?	Quaternary	Crespi Knoll	USA (California)	Palos Verdes	Legg *et al.* (2007)
22	Gentle restraining bend	Active	Quaternary	Dagg Ridge	New Zealand	Alpine	Barnes *et al.* (2005)
23	Gentle restraining bend	Active	Neogene–L. Quaternary	Denali	USA (Alaska)	Denali	Fitzgerald *et al.* (1993)
24	Gentle restraining bend	Active	Cenozoic–L. Quaternary	Domeyko	Chile	Domeyko	Swanson (2005)

(*Continued*)

Table 2. Continued

Map ID	Bend or basin type (cf. Fig 1)	Ancient or active	Age of activity	Bend name	Country	Associated master faults	Selected references
25	Gentle restraining bend	Active	L. Quaternary	Dumbe	USA	Santa Monica–Dume	Sorlien *et al.* (2006)
26	Gentle restraining bend	Active	Pliocene–Quaternary	East Bay Hills	USA (California)	Hayward	Aydin & Page (1984)
27	Gentle restraining bend	Active	Not given in paper	Echo Hills	USA (Nevada)	Bitter Spring Valley	McClay & Bonora (2001)
28	Gentle restraining bend	Active	Late Neogene–L. Quaternary	Ecuadorian Andes	Ecuador; Colombia	Pallatanga; Rio Chingual–La Sofia	Ego *et al.* (1996)
29	Gentle restraining bend	Active	L. Quaternary	Ganos	Turkey	North Antolian	Armijo *et al.* (1999); Seeber *et al.* (2004)
30	Gentle restraining bend	Active	Quaternary	George Ridge	New Zealand	Alpine	Barnes *et al.* (2005)
31	Gentle restraining bend	Active	L. Quaternary	Guarapiche	Venezuela	El Pilar	Audemard (2006)
32	Gentle restraining bend	Active	Cenozoic	Gurvan Bogd mountains	Mongolia	Gurvan Bogd	Cunningham *et al.* (1996*a*); Bayasgalan *et al.* (1999)
33	Gentle restraining bend	Active	Late Quaternary	Hayward–Calaveras	USA (California)	Hayward–Calaveras	Graymer *et al.* (2007)
34	Gentle restraining bend	Active	Late Neogene–L. Quaternary	Hispaniola	Hispaniola	Septentrional	Mann *et al.* (1984); Mann *et al.* (1999); Prentice *et al.* (2003)
35	Gentle restraining bend	Active	L. Quaternary	Kaikoura Ranges	New Zealand	Hope; Jordan; Kekerengu	Van Dissen & Yeats (1991)
36	Gentle restraining bend	Active	L. Quaternary	Khlong Lan	Thailand	Inthanon; Sukhothai	Smith *et al.* (2007)
37	Gentle restraining bend	Active	L. Quaternary	Lasuen Knoll	USA (California)	Palos Verdes	Fisher *et al.* (2004); Legg *et al.* (2007)
38	Gentle restraining bend	Active	L. Quaternary	Lebanon and Anti-Lebanon ranges	Lebanon; Syria	Dead Sea; Levant; Yammouneh	Gomez *et al.* (2007)
39	Gentle restraining bend	Active	Quaternary	Lome Prieta	USA (California)	San Andreas; Sargent-Berrocal	Schwartz *et al.* (1990)
40	Gentle restraining bend	Active	Neogene–L. Quaternary	Massif du Macaya	Haiti	Enriquillo–Plantain Garden	Mann *et al.* (1983); Mann *et al.* (1995)
41	Gentle restraining bend	Active	L. Quaternary	Mission Hills	USA (California)	Hayward–Calaveras	Andrews *et al.* (1993); Burgmann *et al.* (2006); Manaker *et al.* (2005)

42	Gentle restraining bend	Active	L. Quaternary	Mt Oxford	New Zealand	Porters Pass–Amberley	Cowan (1992); Cowan *et al.* (1996)
43	Gentle restraining bend	Active	Pliocene–L. Quaternary	Mt. Diablo	USA (California)	Greenville; Concord	Burgmann *et al.* (2006); Wakabayashi (2007)
44	Gentle restraining bend	Active	Cenozoic	Nemegt Uul	Mongolia	Gobi–Tien Shan	Cunningham *et al.* (1995); Owen *et al.* (1999)
45	Gentle restraining bend	Active	L. Quaternary	Palos Verdes Hills	USA (California)	Palos Verdes	Ward (1994); Fisher *et al.* (2004); Legg *et al.* (2007)
46	Gentle restraining bend	Active	Miocene	Pangong Range	India	Karakoram	Searle & Phillips (2005)
47	Gentle restraining bend	Active	Tertiary	Pijnacker	Netherlands	Not given in paper	McClay & Bonora (2001)
48	Gentle restraining bend	Active	Late Quaternary	Porters Pass	New Zealand	Porters Pass	Paterson Campbell (1998)
49	Gentle restraining bend	Active	Late Neogene–L. Quaternary	Santa Catalina Island	USA (California)	Santa Catalina	Ward (1994); Mann & Gordon (1996); Fisher *et al.* (2004); Legg *et al.* (2007)
50	Gentle restraining bend	Active	Neogene–L. Quaternary	Santa Cruz Mountains	USA (California)	San Andreas	Anderson (1990); Schwartz *et al.* (1990); Burgmann *et al.* (1994); Anderson (1994)
51	Gentle restraining bend	Active	L. Cretaceous–L. Cenozoic	Sierra de las Minas	Guatemala	Motagua; Polochic	Aydin & Nur (1982); Mann & Gordon (1996)
52	Gentle restraining bend	Active	Late Neogene–L. Quaternary	South Georgia Island	Antarctica	North Scotia Ridge	Mann *et al.* (2003)
53	Gentle restraining bend	Active	L. Cenozoic	Sutai Range	Mongolia	Tonhil	Cunningham (1996*b*); Cunningham *et al.* (2003*c*)
54	Gentle restraining bend	Active	Quaternary	Tararua Ranges	New Zealand	Wellington	Beanland (1995)
55	Gentle restraining bend	Active	Eocene–L. Quaternary	Tortola ridge	Caribbean Sea	Anegada passage	Jany *et al.* (1990)
56	Gentle restraining bend	Active	Late Neogene–L. Quaternary	Transverse Ranges (San Bernardino Mountains)	USA (California)	San Andreas	Spotila & Anderson (2004)
57	Gentle restraining bend	Active	Miocene	Transverse Ranges (San Gabriel Mountains)	USA (California)	San Andreas	Blythe *et al.* (2000)

(Continued)

Table 2. Continued

Map ID	Bend or basin type (cf. Fig 1)	Ancient or active	Age of activity	Bend name	Country	Associated master faults	Selected references
58	Gentle restraining bend	Active	L. Quaternary	Transverse Ranges (San Gorgonio Pass)	USA (California)	Eastern California	Spotila & Anderson (2004)
59	Gentle restraining bend	Active	Late Neogene–L. Quaternary	Unnamed	Venezuela	Bocono	Schubert (1980)
60	Gentle restraining bend	Active	L. Quaternary	Unnamed, near Bam 2004 epicentre	Iran	Gowk–Sabzevaran	Talebian *et al.* (2004)
61	Gentle restraining bend	Active	L. Quaternary	Various restraining bends	Iceland	Selsund	Angelier *et al.* (2004)
62	Gentle restraining bend	Active	Late Neogene–L. Quaternary	Wairau 'big bend'	New Zealand	Wairau Fault	Yeats & Berryman (1987); Berryman *et al.* (1992)
63	Gentle restraining bend	Active	L. Quaternary	Zahret el-Qurein (Zahret Qurein)	Jordan	Jordan Valley	Garfunkel (1981)
				Ancient restraining bends			
64	Sharp restraining bend	Ancient	Cenozoic–Palaeocene	Cascade Mountains	USA (Washington)	Ross Lake	Miller (1994)
65	Sharp restraining bend	Ancient	Early Cenozoic	Owl Creek	USA (Wyoming)	North Owl Creek; Shotgun Butte	McClay & Bonora (2001)
66	Sharp restraining bend	Ancient	Tertiary	Unnamed	Netherlands	Unnamed	Dooley *et al.* (1999)
67	Gentle restraining bend	Ancient	Not given in paper	Anninghe	China	Xianshuihe; Zemuhe	Swanson (2005)
68	Gentle restraining bend	Ancient	Tertiary; Palaeogene	As Pontes Basin (Ebro Basin)	Spain	Penedes Valles; Gandesa–Ulldemolins; El Camp; Pont d'Armentera	Anadon *et al.* (1985)
69	Gentle restraining bend	Ancient	Palaeogene	Black Hills	USA (Alaska)	Beringian	Worrall (1991)
70	Gentle restraining bend	Ancient	Late Precambrian	Brazilion craton	Brazil	Pabos	Corsini *et al.* (1996)
71	Gentle restraining bend	Ancient	Late Palaeozoic	Casco Bay	USA (Maine)	Norumbega; Flying Point	Swanson (2005)

72	Gentle restraining bend	Ancient	Devonian–Carboniferous	Cobequid Highlands	Canada	Rockland Brook; Cobequid	Miller *et al.* (1995)
73	Gentle restraining bend	Ancient	Eocene?	Gasherbrum Range	Pakistan	Karakoram	Searle & Phillips (2005)
74	Gentle restraining bend	Ancient	Ordovician–Devonian	Gaspé Belt	Canada	Lac Cascapedia; Mont Albert; Shickshock Sud	Sacks *et al.* (2004)
75	Gentle restraining bend	Ancient	Mesozoic–Cenozoic	Jargaland	Mongolia	Haar Us Naar	Cunningham *et al.* (1996*a*); Howard *et al.* (2003); Cunningham (2007)
76	Gentle restraining bend	Ancient	Cretaceous–Tertiary	Magallanes	Argentina	North Scotia Ridge	Cunningham (1995); Lodolo *et al.* (2003)
77	Gentle restraining bend	Ancient	Mesozoic–Cenozoic	Mt Tormeno	Italy	Tormeno; Gamonda	Zampieri & Massironi (2005)
78	Gentle restraining bend	Ancient	Permian–Tertiary	Quealy Dome	USA (Wyoming)	South Quealy	Stone (1995); McClay & Bonora (2001)
79	Gentle restraining bend	Ancient	Cenozoic	Unnamed	Iran	Purang	Walker & Jackson (2004)

Green, pull-apart basins or releasing bends. Red, restraining bends.

In this paper, I reserve *transform fault* only for strike-slip faults in oceanic crust that offset and abruptly transform motion into orthogonal oceanic spreading ridges as originally defined by Wilson (1965; Figs 1 & 2; Table 1). Yeats *et al.* (1997) discuss some of the problems in the wider application of the term *transform fault* to continental strike-slip faults like the San Andreas and Sumatra faults, where strike-slip motion is not abruptly transformed at their ends on to other orthogonal structures, and where the strike-slip fault may only accommodate only a fraction of the total plate motion. Because of the problems of the application of the term *transform* to continental settings, I use the general term *strike-slip fault* for sub-aerial strike-slip faults in continental settings at any scale, whose slip vectors are roughly parallel to their strike (Sylvester 1988; Fig. 1). A *strike-slip fault system* has a number of kinematically related and subparallel strike-slip faults typically within a broad plate-boundary zone up to several hundred kilometres in width (e.g. the San Andreas strike-slip system). Two commonly used terms also used in this review are *transpression* and *transtension*, broadly defined as steep, strike-slip-influenced deformation zones that deviate from simple shear by a component of shortening (transpression) or extension (transtension) across the zone (Sanderson & Marchini 1984; Fossen & Tikoff 1998).

Following Sylvester (1988) and Yeats *et al.* (1997), I avoid using the words *wrench* or *transcurrent*, which were widely used in pre-plate tectonic literature and in articles and books about petroleum exploration in strike-slip fault settings (cf. Wellman 1955; Moody & Hill 1956; Wilcox *et al.* 1973; Freund 1974; Sylvester 1984; Harding *et al.* 1985; Lowell 1985; Mount & Suppe 1987; Jamison 1991; Zolnai 1991; Milnes 1994). Both *wrench* and *transcurrent* were originally defined in the early 20th century with genetic and/or dynamic connotations (cf. Sylvester 1988, for a historical review and the modern usage problems of both terms).

Releasing-bend terms used in this paper

For strike-slip bends of an extensional character (right-stepping offset for a right-lateral fault, left-stepping offset for a left-lateral fault, Fig. 1b), I will use the terms that Crowell (1974*a*, *b*) developed for structures along the San Andreas fault zone. These terms include *releasing bend* and *releasing double bend* (Crowell 1974*a*) and *pull-apart basin* (Burchfiel & Stewart, 1966; Mann *et al.* 1983) that are used interchangeably in this paper (cf. fig. 1 of Cunningham & Mann 2007; Figs 1b, c & d). The two approximately parallel strike-slip faults bounding pull-aparts or push-ups are simply referred to as *master faults* (Rodgers 1980). The distance between master faults is their *fault separation* or pull-apart basin width; the distance of overlapping master faults is their *fault overlap* and roughly defines the length of the pull-apart basin (Fig. 1b). *Pull-apart* is sometimes used in petroleum geology literature to refer to rift basins produced by continental breakup (cf. Lowell 1985), but the strike-slip usage that I will follow has precedence.

Active pull-apart basins are easily recognizable along active strike-slip faults, because the basins form topographically low, fault-bounded depressions that are commonly the sites of marine embayments, inland lakes and closed topographic depressions, which in the case of the Death Valley pull-apart (−85 m b.s.l.), the Dead Sea pull-apart (−411 m b.s.l.), and multiple pull-aparts in the southern Imperial Valley of California and northern Mexico (−40–65 m b.s.l.). Therefore pull-aparts have some of the lowest, sub-aerial topographic elevations in continental areas (Figs 3–4; Table 2). Crustal thinning produced by localized extension within the stepover region localizes rapid deposition of clastic sediments eroded from surrounding highlands (Hempton & Dunne 1983) as well as intrusive, volcanic and geothermal activity at the surface (Aydin *et al.* 1990) or intercalated within the basin sediments (Einsele 1986). Pull-apart depressions of a variety of sizes form major discontinuities along otherwise straight segments of the main ‘through-going’ strike-slip fault, or ‘principal displacement zone’ (Tchalenko & Ambraseys 1970; Figs 3 & 4; Table 2).

Other synonyms for pull-apart basins are *gaps* (Quennell 1956; 1958) *tectonic depressions* (Clayton 1966), *wrench grabens* (Belt 1968), *rhomb grabens* (Freund 1971; Aydin & Nur 1982; Bahat 1983; Heimann *et al.* 1990), *dilational fault jogs* (Sibson 1985, 1986*a*,*b*, 1987), *extensional duplexes* (Woodcock & Fischer 1986; Swanson 2005); *sidewall basins* (Gibbs 1989); *tensile bridges* (Gamond 1987); *stepover basins* (Wakabayashi *et al.* 2004, 2007), and *dilational stepovers* (Oglesby 2005). T. Dooley (pers. comm., 2006) uses the term *in-line depressions* as basins along strike-slip faults that are formed between Riedel shears that amplify as an anastomosing fault develops by R- and P-shear linkage (cf. the *tensile bridges* of Gamond 1987).

Rhombochasms were defined by Carey (1958), in his early outline of global plate tectonics, as a special case of parallel-sided pull-apart basins that are floored by oceanic crust. *Intra-transform spreading centres* were proposed by Fornari *et al.* (1989) as a special case of pull-aparts on oceanic transform faults that have developed short spreading centres and are floored in part by oceanic crust. *Sag ponds*, which are common along continental strike-slip faults may be related to the

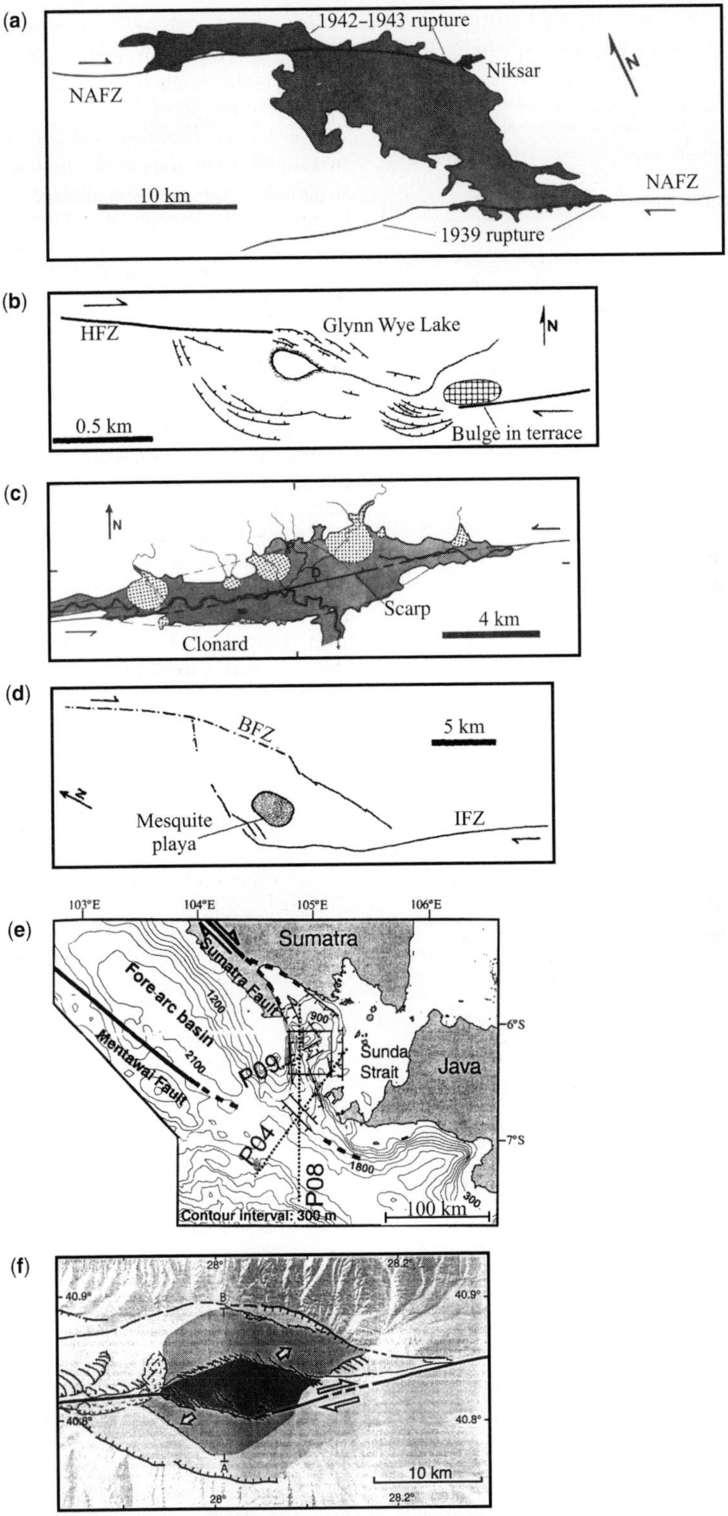

Fig. 3. See caption on p. 36.

formation of small, underlying pull-apart basins by subtle, small shifts in the trace of the fault by metres to hundreds of metres (Sylvester 1988).

Pull-apart basins are one type in a family of small but potentially thick, strike-slip-controlled basins that include *fault-wedge basins* (Crowell 1976; Biddle & Christie-Blick 1985), *fault-angle depressions* (Ballance & Reading, 1980), *ramp basins* (P. Mann *et al.* 1991) or *push-down basins* (Cobbold *et al.* 1993), and *fault-flank depressions* (Crowell 1976). The characteristics and tectonic controls on these other types of strike-slip basins are reviewed by Biddle & Christie-Blick (1985), Cobbold *et al.* (1993), Ingersoll and Busby-Spera (1995), and Allen & Allen (2001).

Fault termination basins

Fault termination basins or *single releasing bends* in the terminology of Crowell (1974*a*, *b*) are an important class of strike-slip basin that warrants illustration in Figure 5 because of their similarity to and possible confusion with classical pull-aparts as illustrated in Figures 3–4. These basins terminate strike-slip displacement along a single strike-slip fault or fault zone by diffuse and splayed extensional structures in a characteristic 'horsetail' pattern that ranges in width from about 30 km for the sub-aerial, Neogene La Tet Fault in the Pyrenees Mountains of Spain (Cabrera *et al.* 1988; Fig. 5a), to larger examples along the strike-slip Median Tectonic Line in Japan (Itoh *et al.* 1998); along the Motagua strike-slip fault of northern Central America (Guzmán-Speziale 2001); along the North Island strike-slip faults of New Zealand (Mouslopoulou *et al.* 2007); and to one extreme example over 1000 km in width that Briais *et al.* (1993) proposed for the offshore termination of the Red River strike-slip fault zone in the South China Sea (Fig. 5b). These extensional features are confined to one side of a single strike-slip fault, and therefore differ in character from a localized extension in a pull-apart basin confined by two, subparallel, master strike-slip faults (Fig. 1b). Fault termination features are not discussed further in this paper and are not included as part of the bend compilation shown on Figure 2.

Duplex terminology

In this paper, I do not use the *duplex terminology* for strike-slip bends that was adapted from thrust fault terminology by Woodcock & Fischer (1986). In this terminology, *ramps* are synonyms for strike-slip bends; *straights* are synonyms for strike-slip faults subparallel to the regional slip vector; *extensional duplex* is a synonym for pull-apart basin; *contractional duplex* is a synonym for restraining bend; and *trailing contractional* or *extensional imbricate fans* are synonyms for horsetail-type fault splays at the ends of strike-slip faults.

The duplex terminology is used for outcrop studies of strike-slip fault bends (Swanson 1989*a*, 2005; Cruikshank *et al.* 1991; Laney & Gates 1996), but is also widely used for descriptions of larger-scale bends (Swanson 2005; Cunningham 2007; Gomez *et al.* 2007). The Crowell (1974*a*, *b*) and Woodcock & Fischer (1986) terminology for bend structures are both illustrated and compared on Figure 1 of Cunningham & Mann (2007).

Restraining-bend terms used in this paper

Frequently used terms include *gentle restraining bend* and *sharp restraining bend* (Crowell 1974*a*; Fig. 1c). Active restraining bends are also easily recognizable, because they localize topographically elevated ridges or folded mountain belts and expose fault-bounded, elongate belts of deeper crustal rocks (Karig 1979; Cunningham 1995; Roeske *et al.* 2003; Cunningham *et al.* 2007) that include blueschist-facies metamorphic rocks in some localities (Mann & Gordon 1996; Fig. 1c; Table 2).

Sharp restraining bends are rhomboidal in shape, are aligned along the general trend of major strike-slip faults, and range in scale up to several kilometres in length and up to hundreds of metres in width (Fig. 6b & c; Table 2). The areas of sharp restraining bends are characterized by miniature fold-and-thrust belts and marked by a

Fig. 3. Active examples of recently nucleated and relatively young spindle-shaped pull-aparts, as interpreted in the basin evolutionary model of Mann *et al.* (1983). Examples are taken from several different tectonic settings or strike-slip faults that are discussed in this paper. (**a**) Niksar pull-apart basin along the North Anatolian fault zone, Turkey (from Seymen 1975; Hempton & Dunne 1983). (**b**) Glynn Wye Lake pull-apart along the Hope Fault, New Zealand (from Freund 1971; Cowan 1990). (**c**) Clonard pull-apart basin along the Enriquillo–Plantain Garden fault zone, southern peninsula of Haiti (from Mann *et al.* 1983, 1995). (**d**) Mesquite playa pull-apart basin along the Brawley–Imperial faults, Imperial Valley, California (modified from Johnson & Hadley 1976). (**e**) Sunda Strait pull-apart basin at the SE termination of the Sumatra fault zone between the islands of Sumatra and Java (from Lelgemann *et al.* 2000). (**f**) Central pull-apart basin along the North Anatolian fault zone in the Sea of Marmara, from Armijo *et al.* (2002). Arrows show the direction of extension inferred from en échelon normal faults.

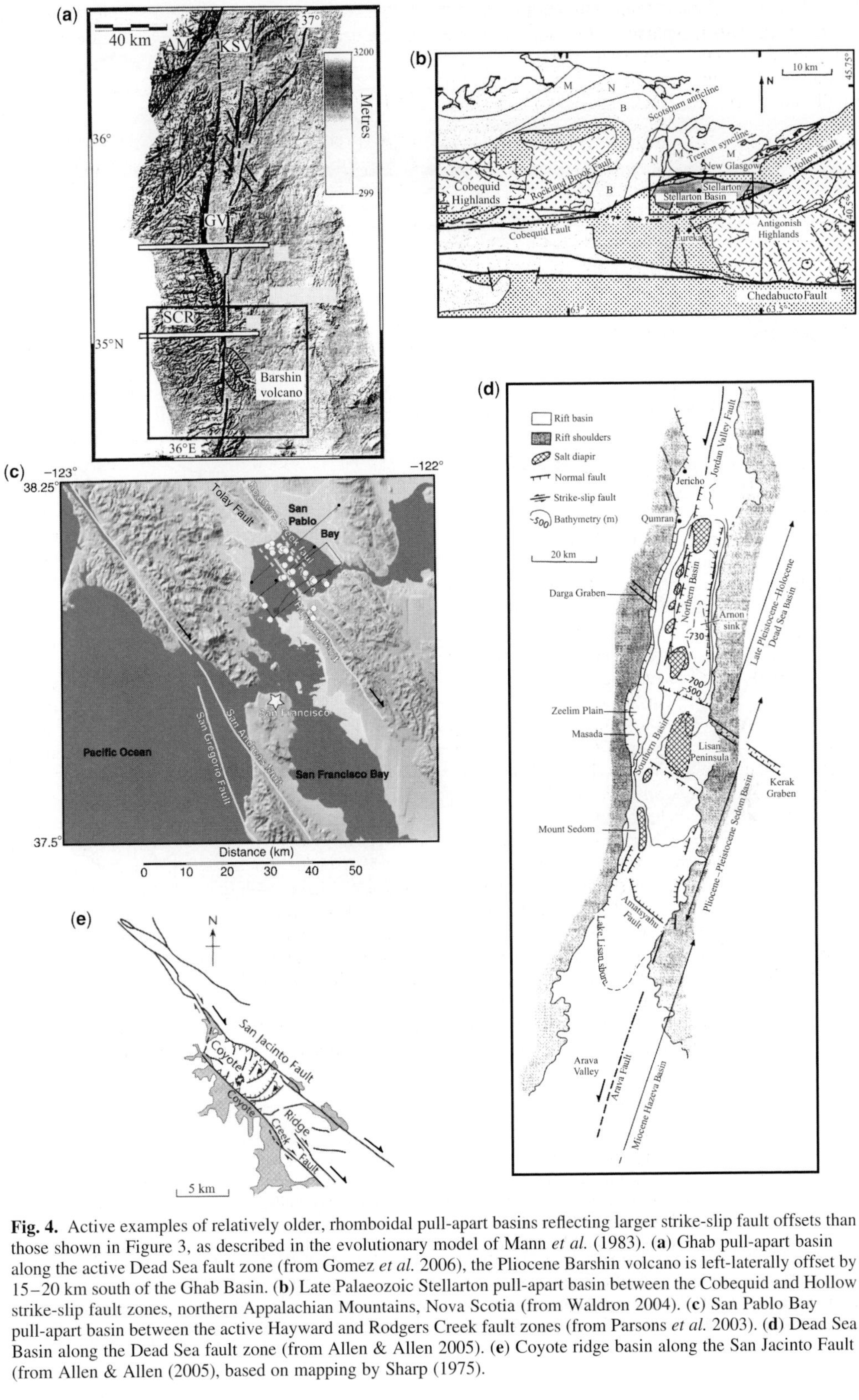

Fig. 4. Active examples of relatively older, rhomboidal pull-apart basins reflecting larger strike-slip fault offsets than those shown in Figure 3, as described in the evolutionary model of Mann *et al.* (1983). (**a**) Ghab pull-apart basin along the active Dead Sea fault zone (from Gomez *et al.* 2006), the Pliocene Barshin volcano is left-laterally offset by 15–20 km south of the Ghab Basin. (**b**) Late Palaeozoic Stellarton pull-apart basin between the Cobequid and Hollow strike-slip fault zones, northern Appalachian Mountains, Nova Scotia (from Waldron 2004). (**c**) San Pablo Bay pull-apart basin between the active Hayward and Rodgers Creek fault zones (from Parsons *et al.* 2003). (**d**) Dead Sea Basin along the Dead Sea fault zone (from Allen & Allen 2005). (**e**) Coyote ridge basin along the San Jacinto Fault (from Allen & Allen (2005), based on mapping by Sharp (1975).

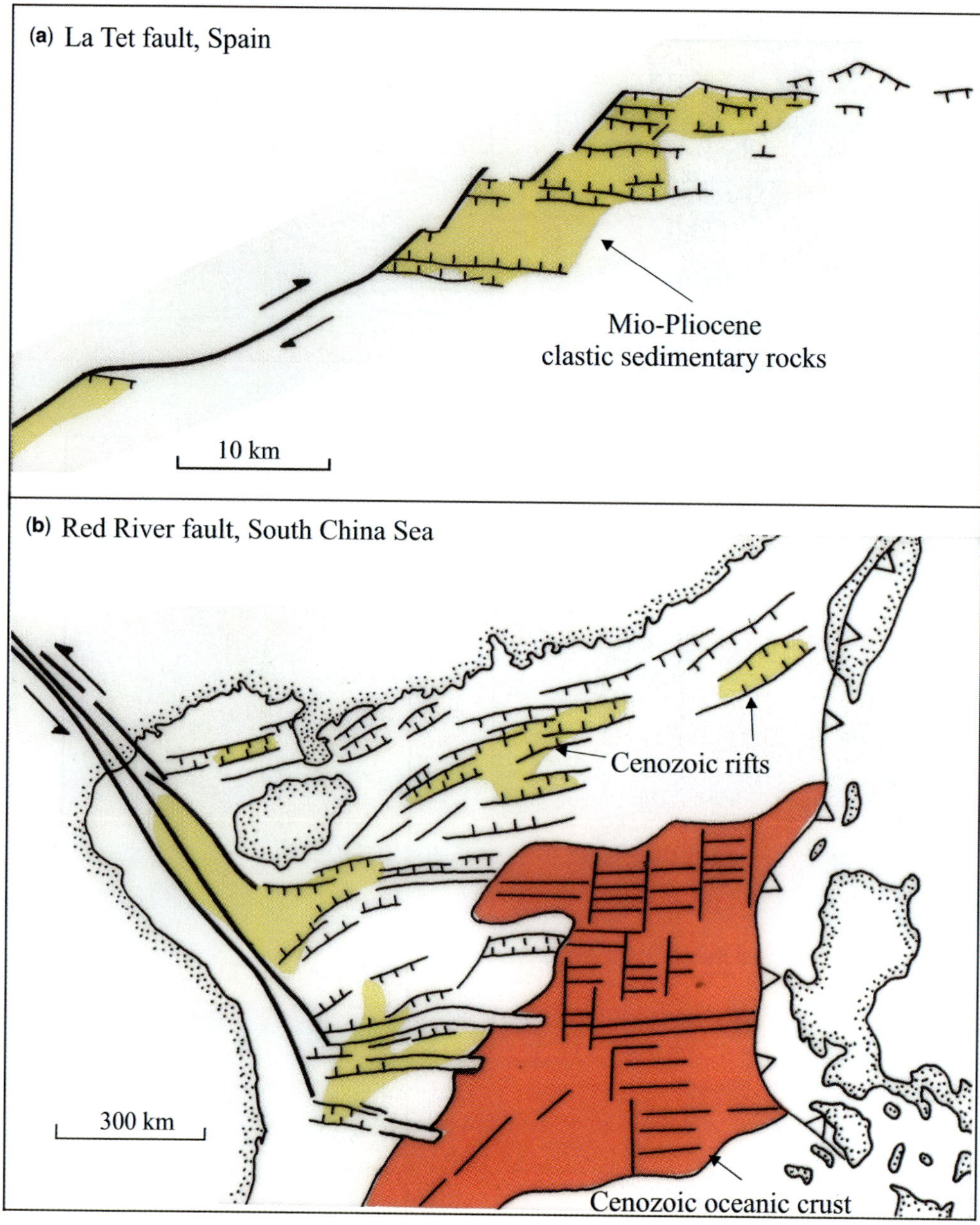

Fig. 5. Examples of strike-slip fault termination basins that are relatively common and share some characteristics with the stepover type of pull-aparts shown in Figures 3 & 4. (**a**) Cerdanya clastic basin (shaded yellow) formed by Late Miocene diffuse normal faulting at the termination of the La Tet strike-slip fault, Spain (modified from Cabrera *et al.* 1988). (**b**) Basins (shaded yellow) and area of oceanic spreading (shaded red) proposed as fault termination effects by Briais *et al.* (1993) for the Red River Fault of the South China Sea.

localized topographic uplift centred on the fold–thrust belt (Brown *et al.* 1991; Fig. 6c). *Gentle restraining bends* are more common, exhibit the characteristic lazy S and Z shapes (Fig. 6d & e) and range in size from small features a few kilometres in width to the world's largest bends up to 200 km wide and several kilometres in elevation (Mann & Gordon 1996; Fig. 6f & g). *Paired*

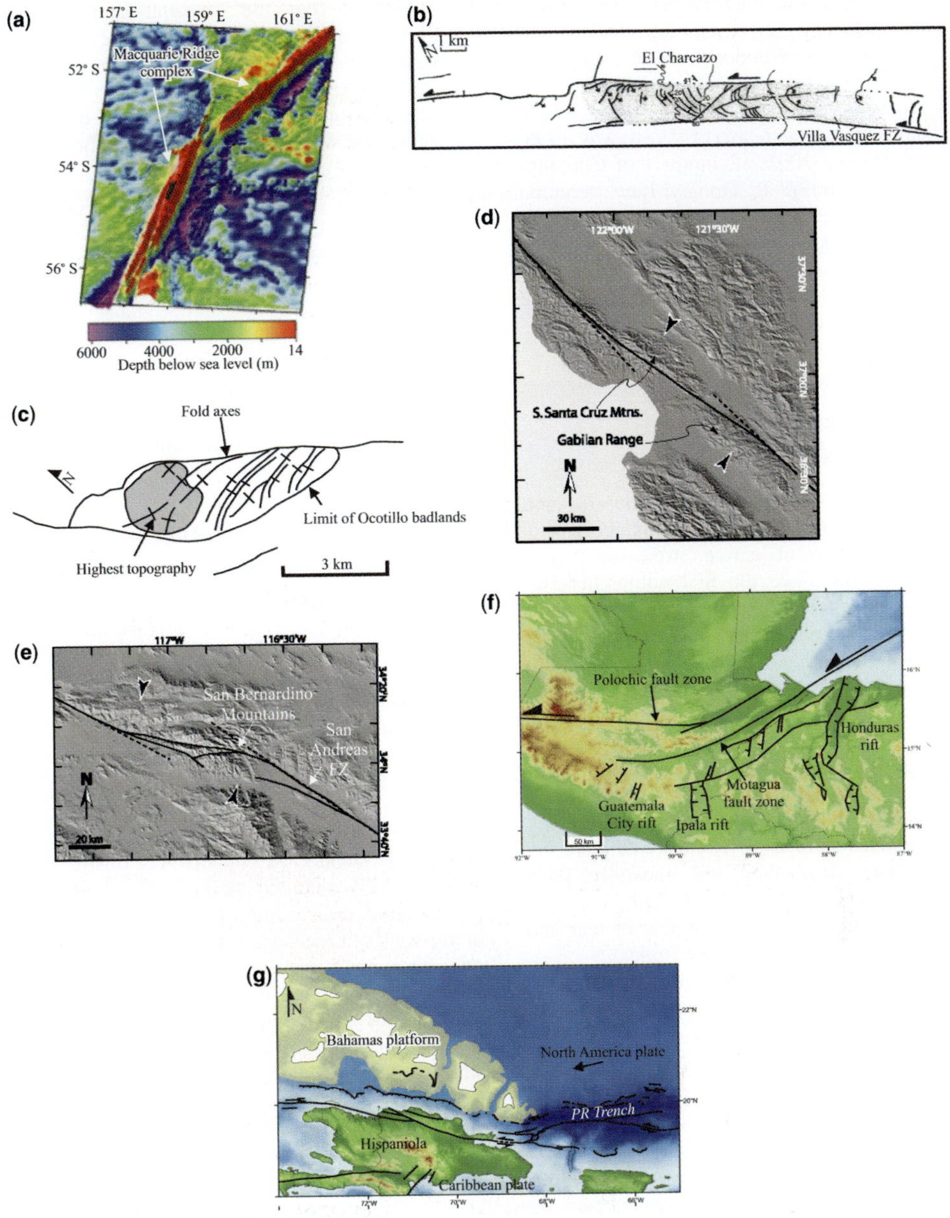

Fig. 6. (**a**) Macquarie Ridge from Dazcko *et al.* (2003). (**b**) Villa Vasquez fault zone and basin, from Mann *et al.* (1999). (**c**) Sketch of Ocotillo badlands, based on Segall & Pollard (1980). (**d**) Santa Cruz Mountains restraining bend, from Cowgill *et al.* (2004*a*). (**e**) San Bernardino Mountains restraining bend, from Cowgill *et al.* (2004*a*). (**f**) Polochic fault zone and Motagua fault zone, from Mann & Gordon (1996). (**g**) Bahama Bank and Hispaniola and corresponding restraining bends, from Mann *et al.* (2003).

bends are restraining and releasing bends that form adjacent to one another (Fig. 1d).

Synonyms for sharp and gentle restraining bends include *rhomb horsts* (Freund 1971; Aydin & Nur 1982; Bahat 1983); *pressure ridges* (Garfunkel *et al.* 1981); *compressional stepovers* (Segall & Pollard 1980; Aydin & Nur 1985; Westaway 1995; Oglesby 2005); *push-up blocks* (Mann *et al.* 1983); *anti-dilational fault jogs* (Sibson 1985, 1986*a*, *b*, 1987; Brown *et al.* 1991); *push-up* *swells* (Heimann & Ron 1987; Reches 1987); *structural knots* (Muehlberger & Gordon 1987); *compressional duplexes* (Woodcock & Fischer 1986; Gomez *et al.* 2007; Cunningham *et al.* 2007), *compressile bridges* (Gamond 1987); *pop-ups* (Stone 1995), and *restraining stepovers* (Westaway 1995; McClay & Bonora 2001; cf. figure 1 of Cunningham & Mann 2007). T. Dooley (pers. comm., 2006) uses the term *in-line uplifts* to refer to uplifts along strike-slip faults that are formed between Riedel shears that amplify as an anastomosing fault develops by R- and P-shear linkage (cf. *compressile bridges* of Gamond 1987).

Global classification of active strike-slip faults

Discussion of the settings of active releasing and restraining bends on strike-slip faults requires a global classification of active strike-slip faults. Woodcock (1986) made the first comprehensive, global classification of active strike-slip faults based on their tectonic origin and setting, which was significantly modified in the subsequent classifications by Sylvester (1988), Yeats *et al.* (1997), and Ingersoll & Busby (1995). In this paper, I will adopt Woodcock's (1986) original scheme that is summarized in Figure 2 & Table 1, but with some modifications that take into account more recent work. Sylvester (1988) adds two categories of strike-slip faults, *tear faults* and *transfer faults*, which are not discussed here since they are generally short (<50 km) and devoid of significant fault bends (cf. Escalona & Mann 2006, for a review of tear and transfer terminology and global examples).

Using Woodcock's (1986) classification, four major categories of strike-slip faults are shown schematically in Figure 2:

1. *ridge transform faults in oceanic crust* that offset spreading ridges (e.g. Siqueiros, Blanco, Clipperton transforms);
2. *boundary transform faults* that separate different plates and strike parallel to the direction of plate motion (e.g. the San Andreas Fault of California, the Alpine Fault of New Zealand, the Dead Sea Fault of the Middle East); this class includes propagating, slab-bounding transforms bounding plates in subduction to strike-slip transition areas like the Scotia and Caribbean (Bilich *et al.* 2001); this type of strike-slip fault has recently been compiled and named *STEP faults*, or *subduction–transform edge propagators* (Govers & Wortel 2005);
3. *trench-linked strike-slip faults* that accommodate the horizontal component of oblique subduction (e.g. the Atacama fault zone of Chile and the Median Tectonic Line of Japan); and
4. *indent-linked strike-slip faults* related to continent–continent collision (Red River fault zone, Altyn Tagh fault zone, North Anatolian fault zone).

In addition to these four categories from Woodcock (1986) I have added a fifth category that I will call *cratonic strike-slip fault systems* that are removed from active plate boundaries, form sharp restraining bends based on older crustal faults, and act as 'concentrators' of intraplate stresses (Talwani 1999; Gangopadhyay & Talwani 2003). In this paper, I will organize my descriptions of strike-slip fault systems into these four categories with emphasis on faults of each category that exhibit either releasing, restraining or paired bends.

Observations on the evolution of releasing bends on strike-slip faults from natural examples and models

Most previous workers have focused their investigations of pull-apart basins on some observable or measurable aspect of single natural basins, at scales ranging from 'outcrop basins' whose 'basin fill' is typically calcite or quartz, to large-scale sedimentary basins up to several hundred square kilometres in size and containing complex sedimentary facies totalling over 10 km in thickness (Hempton & Dunne 1983; Ingersoll & Busby 1995; Figs 3–4). The detailed geological mapping and description of these natural pull-apart examples at outcrop to basinal scales have provided the observational basis and parameters for continuum models such as the geological model proposed by Mann *et al.* (1983) for natural crustal-scale pull-apart basins (Fig. 7), as well as many modelling studies using laboratory analogue models (e.g. Dooley & McClay 1997) or theoretical models (e.g. Golke *et al.* 1994). In this section, I will review the main results of laboratory and theoretical studies and summarize how each type of study has furthered our current understanding of the continuum evolution of natural pull-aparts (Fig. 7).

For reasons of space I will place particular emphasis on the representative discoveries and insights made during the last two decades. Because of the numerous papers on pull-aparts, dating from the mid-1980s, this compilation (Fig. 1 & Table 2) has probably omitted many relevant papers and therefore should not be considered as a complete listing. Moreover, the compilation in Table 2 only contains crustal-scale bends, and does not include outcrop-scale pull-apart features as discussed by previous workers like Aydin & Nur

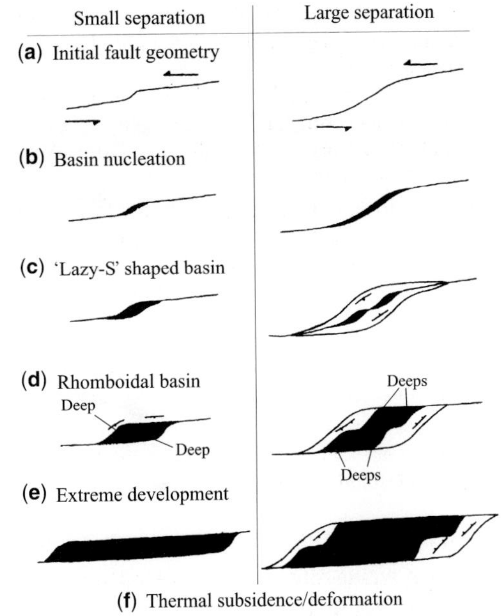

Fig. 7. Classification of pull-apart basins, as interpreted in the evolutionary model proposed by Mann *et al.* (1983). In both evolutionary sequences, a starting point of a single continuous fault with a bend is assumed, but to the left the fault bend exhibits a relatively smaller separation, or distance between the two straight fault segments. For the smaller separation case, the evolution proceeds from a nucleating basin, to a 'lazy-Z' shaped basin, to a rhomboidal basin, to an extremely elongate basin possibly floored by oceanic crust formed at a short spreading centre. For the larger separation case, two individual basins nucleate, follow the same evolutionary stages, and eventually merge to form a wider composite basin.

(1982), Bahat (1983), and Gamond (1987). In this section, I will not take into account the tectonic setting of the basins described, partly to make the fundamental point that the morphology and evolution of pull-aparts is remarkably similar, despite their four major tectonic settings proposed by Woodcock (1986) that are summarized in Figure 2 & Table 1.

Continuum of evolution in natural pull-apart basins

What causes the extensional bends on strike-slip faults?

Previously proposed models for the occurrence of solitary or arrays of pull-aparts along strike-slip faults include:

1. a nearby or changing pole of rotation produces fault curvature in the strike-slip fault, or a greater degree of transtension along the trace of the fault (Menard & Atwater 1968; Garfunkel 1981; Embley & Wilson 1992; Fig. 8a);
2. a strike-slip fault remains oblique to steady divergent motion (Fig. 8b);
3. interaction of the strike-slip fault with a pre-existing crustal heterogeneity results in a fault bend at or near their point of intersection (Mann *et al.* 1985; Mann *et al.* 2007*a*; Fig. 8c);
4. incompatible slip at a junction with another strike-slip fault or fault-bounded block bends the fault (Freund 1974; Bohannon & Howell 1982; Fig. 8d);
5. variable rock properties along the length of the fault will cause the fault to bend during its progressive offset and assume a sinusoidal fault trace in map view (Bridwell 1975); alternatively, fault-zone 'adherence' or frictional locking will lock the principal fault and form a new subparallel but curving splay fault (the 'sidewall ripout' of Swanson 1989*a*, 2005); and
6. intersection of a strike-slip fault with a zone of greater extensional or convergent strain – particularly in zones of continental collision–bends the fault at or near the point of intersection (Sengör *et al.* 1985; Fig. 8d). These mechanisms could also be responsible for forming restraining bends.

Controls of relative plate motion on pull-apart opening

For transtensional opening along the length of a fault related to a changing pole of rotation, Garfunkel (1981) proposes three possible scenarios for large-scale plate-tectonic control on the formation of pull-apart basins along a continental-boundary strike-slip fault like the San Andreas or Dead Sea strike-slip fault systems (Fig. 9).

In the simplest scenario, pure strike-slip motion occurs along some segments of the boundary that are segmented by either releasing- or restraining bends when the fault strike deviates slightly from the local strike direction (Fig. 9b). A paired bend (Fig. 1d) forms when releasing- and restraining segments form adjacent to one another and are similar in areal extent and fault separation (Fig. 9a). Fault segments of pure strike-slip motion are parallel to small circles of plate rotation about the interplate pole of rotation (thin arrows on Figure 9a point in the direction of the pole).

In a second scenario, shown in Figure 9b, the presence of many releasing- and restraining-bend segments causes the average trend of the overall

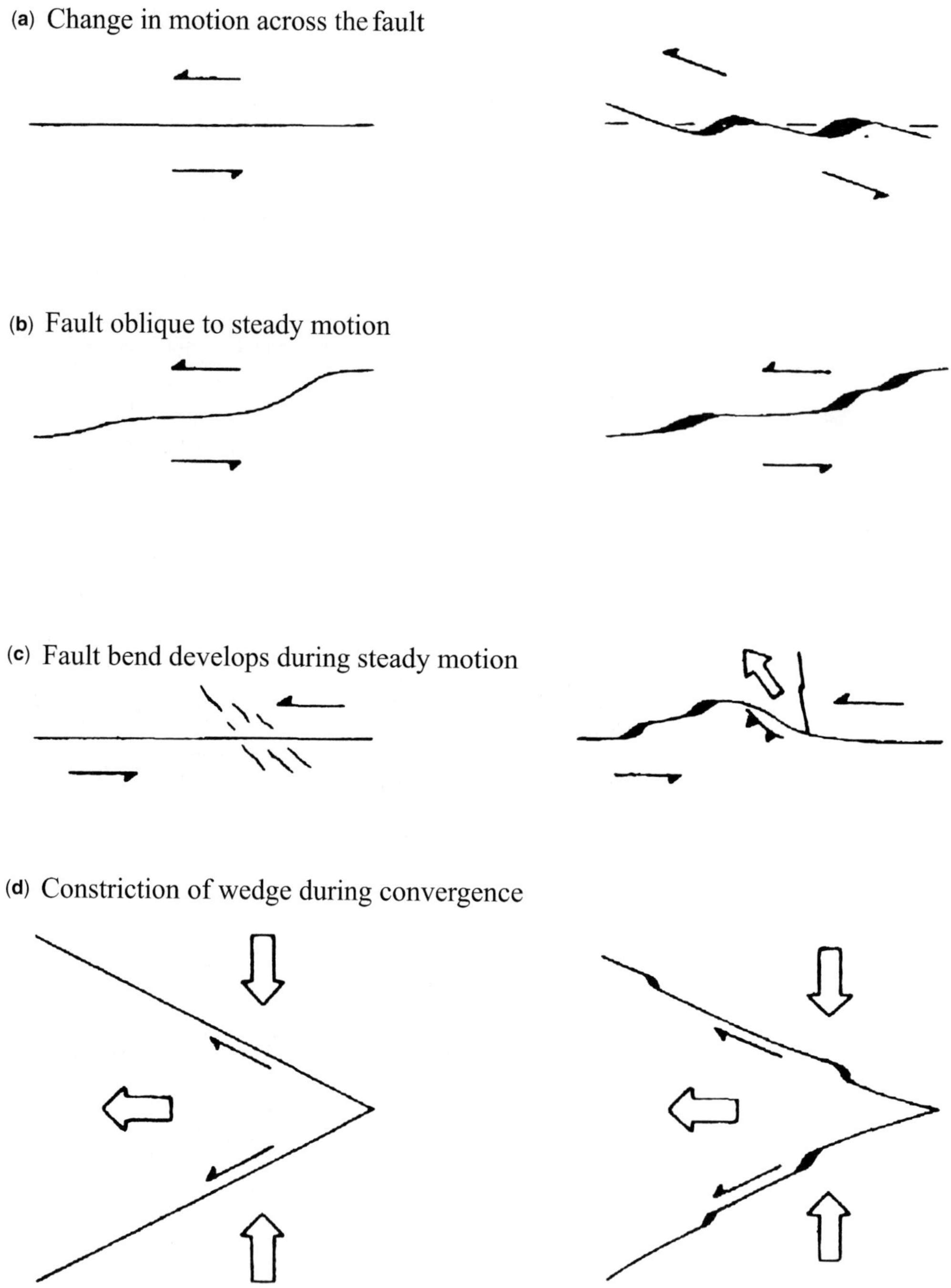

Fig. 8. Possible tectonic scenarios to explain the initiation of bends along strike-slip faults and the formation of pull-apart basins (modified from Mann *et al.* 1983). (**a**) Change in motion across fault. (**b**) Fault oblique to steady motion. (**c**) Fault bend develops during steady motion to create a double or paired bend consisting of adjacent restraining- and releasing-bend segments. (**d**) Constriction of a triangular-shaped wedge or 'escape block' during continental convergence. See text for discussion.

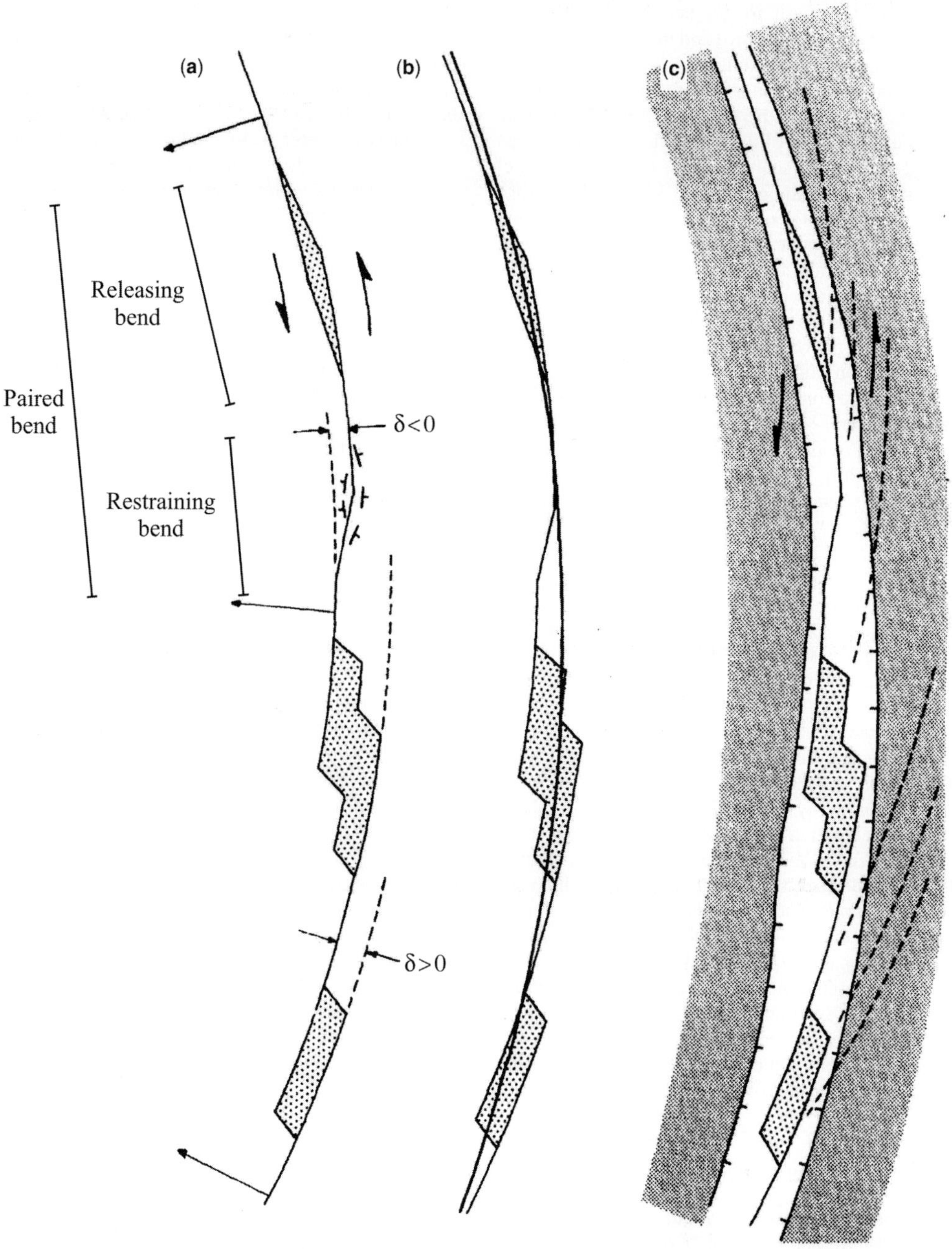

Fig. 9. Creation of pull-apart basins in a continental strike-slip fault tectonic setting, modified from Garfunkel (1981). (**a**) In the simplest tectonic interpretation, thin arrows point to the direction of the pole of relative motion that controls the kinematics of the strike-slip fault system. (**b**) Same tectonic setting as (a), with the plate boundary approximated by a circular arc that does not define either the pole of rotation or control the local fault kinematics. (**c**) Same situation, but with a small component of transverse motion added; the overall relative plate motions are given by the broken lines. See text for discussion.

plate-boundary fault to not accurately define the pole of rotation. In the third and most complex scenario, a component of additional transtension across the overall plate boundary results in the production of new surface area by the stretching of the plate edges and creation of a prominent fault valley that would correspond with the plate-boundary-parallel Imperial Valley along the San Andreas fault zone (Yeats *et al.* 1997) or the Dead Sea Valley along the Dead Sea fault zone (ten Brink *et al.* 1999; Fig. 9c). Overall, fault slip is resolved or 'partitioned' into a component of pure strike-slip aligned along the trend of the fault valley, and a component of orthogonal extension aligned at right angles to the trend of the fault valley.

This more complex scenario results in a complex structure and morphology: pull-apart basins embedded within a wider strike-slip valley (Fig. 9c). Pure strike-slip faults embedded in the fault valley are not necessarily parallel to the small circles of rotation (indicated by broken lines in Fig. 9c). In any or all of these scenarios, there are many potential geological complexities if the pole of rotation and plate kinematics were to suddenly change over large areas of the fault zone and the pre-existing fault traces and basins would have to adapt to the change in kinematics (ten Brink *et al.* 1999).

Parkinson & Dooley (1996) have presented a model similar to that of Garfunkel (1981) and Mann *et al.* (1983) for the predicted locations of en échelon arrays of pull-apart basins at places where the small circles of rotation obliquely intersect the trend of the plate boundary. They postulate that straight strike-slip fault segments, which are more efficient at accommodating slip than curved or irregular traces (Aydin & Nur 1985), will bidirectionally propagate until the strike of the segment is no longer compatible with the changing plate trajectory. To achieve maximum efficiency, most of the bends formed as accommodation structures are releasing bends, although some restraining bends will form as a result of non-parallel fault segments (Fig. 9a).

Nucleation of pull-apart basins

Early maps of pull-aparts by Quennell (1956, 1958), Carey (1958), and Clayton (1966) in New Zealand and along the Dead Sea fault zone were schematically shown as 'rhomb grabens' or 'sharp pull-aparts' (Crowell 1974*a*), with orthogonal angles between the master strike-slip faults and the normal faults accommodating basin extension. Careful field mapping by Freund (1971), and more recently by Cowan (1990), of active fault traces of the right-lateral Hope fault zone of New Zealand, shows that the basin-bounding master faults can be non-parallel in strike and appear as 'spindle' (Mann *et al.* 1983) or 'almond' (Dooley & McClay 1997) shapes (Fig. 3b–d & f) on the map rather than the classical, rectilinear rhomboidal shapes (Fig. 4a–e).

If the two master faults 'underlap', then they are connected by a short, oblique fault segment that makes a 10–15° angle with the master faults as seen in pull-aparts at Glynn Wye Lake on the Hope Fault (Clayton 1966; Freund 1971; Cowan 2000; Fig. 3b), the Clonard Basin on the Enriquillo–Plantain Garden fault zone in the northern Caribbean (Mann *et al.* 1983, 1995; Fig. 3c), Mesquite Playa on the Brawley and Imperial fault zones in southern California (Weaver & Hill 1978; Fuis *et al.* 1982; Fig. 3d); and the Central Basin on the North Anatolian fault zone in the Sea of Marmara (Fig. 3f) (Armijo *et al.* 2002).

This initial master-fault geometry observed for young basins is similar to the *gentle restraining-bend* geometry proposed by Crowell (1974*a*; Fig. 1b). The nucleating basins may be marked by obliquely striking faults at a discontinuity along the master faults that are not associated with subsiding basins (e.g. en échelon faults in the Tiburon area of the Enriquillo–Plantain Garden fault zone in SW Haiti–Mann *et al.* 1983, 1995; en échelon faults at the Laoqianning–Kangding fault junction on the Xianshuihe fault zone, Allen *et al.* 1991). When the width of the basin (or spacing between master faults) is relatively small (<5 km), the young basins have a characteristic 'spindle shape' morphology, as seen, for example, at Glynn Wye Lake, New Zealand; Clonard Basin, Haiti; Mesquite Playa in southern California; and the Central Basin in the Sea of Marmara, Turkey (Fig. 3a). Non-parallelism of master faults leads to space problems at the edge of the pull-apart basin, problems that are expressed morphologically by bulging, folding and uplift of Quaternary terraces as observed at Glynn Wye Lake on the Hope Fault (Clayton 1966; Freund 1971; Cowan 1990; Fig. 3b).

Mann *et al.* (1983) have interpreted these spindle-shaped pull-aparts to represent the initial stage in a continuum model for pull-apart development, as shown schematically in Figure 1b & Figure 7a. Because of the limited amount of opening, the structural depression overlying nucleating pull-aparts illustrated on Figure 7a–f is generally not as prominent as the more evolved rhomboidal pull-apart basin types illustrated on Figure 3b–d. Many nucleating pull-aparts, including those shown in Figure 3a–d & f are above sea-level, sub-aerial, or covered by small, shallow lakes, and exhibit low-relief fault scarps that localize only gradual facies changes across the depression (Hempton 1983; Hempton & Dunne 1983; Link *et al.* 1985).

Lazy S- and Z-shaped pull-apart basins

Following nucleation of pull-aparts at releasing bends with little or no fault overlap (Fig. 7a), continued strike-slip offset produces basin shapes that Mann *et al.* (1983) colloquially called *lazy-S shape* for pull-aparts between left-lateral faults (and *lazy-Z shape* for those between right-lateral faults (cf. the shapes of the Niksar and Sunda basins in Fig. 3a & e). These basin shapes may be modified by local deformation associated with lengthening non-parallel master faults, as seen in the Bichuan Basin of China (Deng *et al.* 1986).

Mann *et al.* (1983) proposed that the lazy S- and Z-shapes of pull-apart basins represent a transitional stage between incipient spindle-shaped basins between master faults with no overlap and rhomboidal basins or 'rhomb grabens' between overlapping master faults (Fig. 7a). Active examples of lazy S- and Z-shaped pull-aparts have been identified in many diverse tectonic and sedimentary settings, including the mountainous region of the northern Andes of Colombia and Venezuela (Schubert 1980; Velandia *et al.* 2005); the deeply sedimented Guinea strike-slip margin of the Equatorial Atlantic (Benkhelil *et al.* 1995); and the sparsely sedimented Blanco oceanic transform fault in the Pacific Ocean (Embley *et al.* 1987).

Larger fault separations, as shown in Figure 7, lead to more prominent S- and Z-shaped basins, along with the possibility of dual, coalescing basins formed in the stepover area, as proposed by Aydin & Nur (1982) and observed from subsurface studies in the Salton Trough of southern California (Fuis *et al.* 1982); the Mor Basin within the Dead Sea Basin (Bartov & Sagy 2004); the Bohai composite pull-apart basin of eastern China (Allen *et al.* 1997; Allen 1998); and the Bichuan Basin in western China (Deng *et al.* 1986).

Increased extension across S- and Z-shaped basins along normal to oblique-slip normal faults at 35–50° to the master faults results in a prominent topographic and sedimentary basin bounded by prominent normal faults at the flanks of the basin. This stage of pull-apart is typically a deep depression surrounded by uplifted and deeply eroded highlands of basement rocks. For example, Death Valley, a prominent Z-shaped pull-apart basin of Miocene–Recent age, is bounded by low-angle normal faults (Burchfiel & Stewart 1966; Biddle & Christie-Blick 1985; Serpa 1988; Cowan *et al.* 2003). The floor of the basin is 85 metres below sea-level, and about 3 km lower than the peaks in the surrounding mountains. The neighbouring Panamint and Saline valleys are geomorphically similar pull-aparts also bounded by low-angle normal faults (Burchfiel *et al.* 1987). The world's largest S-shaped pull-apart basin is Lake Baikal in Russia, covering over 31 494 km^2 and exhibiting 4477 m of structural relief between the deepest part of the lake (−1637 m) and the adjacent peaks (2840 m) (Petit *et al.* 1996; Lesne *et al.* 1998). Because of the large amount of relief, sedimentation in large S- and Z-shaped basins like Death Valley and Lake Baikal is highly asymmetrical and consists of alluvial-fan deposits localized along fault scarps forming the sides of the basin (Hempton 1983; Hempton & Dunne 1983). Proximal fan deposits grade into distal sand and mud facies on the flat basin floor.

Rhomboidal pull-apart basins

Pull-aparts with increased opening and offset along the strike-slip master faults evolve from lazy Z and S shapes to assume the more familiar rhomboidal or 'rhomb graben' appearance (Fig. 7d) and exemplified in the representative rhomb-shaped pull-aparts compiled on Figure 4a–e. In the terminology of Crowell (1974*a*, *b*), rhomboidal pull-aparts would correspond with his *sharp pull-aparts* (Fig. 1b). In most cases, basin lengthening to rhomboidal shapes does not involve basin widening; however, coalescence of adjacent basins may form larger composite pull-aparts with greater fault separations, as proposed by Aydin & Nur (1982) and shown schematically in Figure 7d & e.

Rhomboidal basins are longer (greater master fault overlap), wider (greater fault separation), and form deeper topographic basins than less-evolved spindle-shaped and S- and Z-shaped basins (Fig. 7). A good example of a rhomboidal basin is the Dead Sea pull-apart basin, whose 8–10-km-wide basin floor is −411 m b.s.l. and is filled by more than 10 km of clastic and evaporitic sedimentary rocks (Garfunkel & Ben-Avraham 1996; Fig. 4d).

As in spindle-shaped basins, non-parallelism of master faults leads to bulging of terraces, topographic uplift, and localized inversion of basinal sediments along the misaligned basin edge in the rhomboidal Hanmer pull-apart of New Zealand (Wood *et al.* 1994). Hempton & Dunne (1983) compiled stratigraphic thickness information from 20 modern and ancient basins of spindle, S and Z, and rhomboidal shapes, to conclude that the stratigraphic thickness and asymmetry increases with fault overlap (i.e. rhomboidal basins contain thicker basinal fills than spindle or S- and Z-shaped basins). In general, the master faults and transverse normal faults that create the subsidence in pull-aparts are not exposed at the surface, and must be imaged using the seismic-reflection techniques that are discussed below. Notable exceptions to this rule are the Coyote Ridge rhomboidal pull-apart basin along the San Jacinto fault zone in southern California (Sharp 1975) and an unnamed rhomboidal pull-apart between the Xianshuihe and Ganzi–Yushu fault zones of SW China (Allen *et al.* 1991). For both examples, orthogonal normal faults within the overlap area of the master faults are well exposed.

An interesting geological characteristic observed in several rhomboidal pull-apart basins is the presence of basinal deeps at the distal ends or 'inside corners' of the basin separated by a shallow sill, as shown schematically in Figure 7d. The deeps are roughly circular and may or may not be fault-bounded; they are generally distributed diagonally across the floor of the basin. Examples of rhomboidal pull-aparts with twin, diagonal deeps include the Mirogoane Basin of Haiti (Mann *et al.* 1983); the Vienna Basin of Austria (Royden *et al.* 1982; Hinsch *et al.* 2005); the Elat, Aragonese and Dakar basins of the Gulf of Aqaba (Elat) (Ben-Avraham *et al.* 1979); the Cariaco Basin of Venezuela (Schubert 1982; Jaimes 2003); the Ghab Basin of Syria (Brew *et al.* 2001) and the Jianchuan–Hequin Basin on the Xiaojinhe fault zone in western China (Deng *et al.* 1986).

Inversion of rhomboidal basins

Examples of partially inverted rhomboidal pull-apart basins commonly deformed as a result of non-parallel master faults include the Hanmer pull-apart along the Hope fault zone of New Zealand (Wood *et al.* 1994); the offshore Dagg Basin along the Alpine fault zone of New Zealand (Barnes *et al.* 2001, 2005); and the Late Palaeozoic Stellarton Basin of Nova Scotia (Waldron 2004). Reversal of slip along master faults and subsequent rhomboidal pull-apart deformation has been proposed in the Cenozoic Catacamas pull-apart basin of Honduras (Gordon & Muehlberger 1994) and the Cenozoic Bovey Basin of SW England (Holloway & Chadwick 1986).

Migration of depocentres within rhomboidal pull-apart basins

Outcrop-based mapping by Zak & Freund (1981), Wakabayashi *et al.* (2004), and Wakabayashi *et al.* (2007) and subsurface mapping using wells, seismic-reflection and gravity methods by ten Brink *et al.* (1999) and Lazar *et al.* (2006) all suggest that pull-apart depocentres do not remain stationary for periods of time from hundreds of thousands to millions of years, but instead migrate along the strike of the master faults that control them. In the Dead Sea pull-apart, Zak & Freund (1981) first noted that the depression is occupied by three distinct Miocene to Recent sedimentary basins whose depocentres are displaced northward over a lateral distance of 50 km with time. They postulated that the thinned, basinal area of the Dead Sea was fixed to the Levant plate and was detached from and lagging behind the relative northward motion of the thicker Arabian plate. Ten Brink & Ben-Avraham (1989) postulated that migration of the depocentre over a master-fault-parallel distance of 50 km reflects the directional propagation of the tips of both master faults; this propagation results in the northward migration of depocentres recognized by Zak *et al.* (1981). Lazar *et al.* (2006) have revised these previous models in their sequential basins model that includes a number of active sub-basins formed between en échelon strands of the main plate boundary fault. These basins become larger and deeper as motion progresses, and are bound by transverse normal faults.

Along the strands of the northern San Andreas fault zone, Wakabayashi *et al.* (2004; 2007) relates a 'wake' of deformed and inverted pull-apart or fault valley sediments to localized inversion effects similar to those described for slightly non-parallel master faults of the Hope fault zone of New Zealand (Freund 1971; Wood *et al.* 1994; Cowan *et al.* 2000). The net effect of this migrating depocentre model is for the active pull-apart to show less structural relief and stratigraphic thickness than predicted for the amount of total offset on the master faults.

Extreme pull-apart basin development

With continued strike-slip offset and minimal inversion effects, rhomboidal pull-apart basins may lengthen indefinitely into long, narrow oceanic basins, whose basin length, or master-fault overlap, can exceed basin width, or fault separation, by a factor of 10. The Cayman Trough, a pull-apart basin dominating the North America–Caribbean strike-slip system, is a unique example of an active pull-apart basin which has evolved until its master-fault overlap of approximately 1400 km greatly exceeds its fault separation of 100–150 km. Studies of magnetic anomalies in the floor of the Cayman Trough suggest that its spreading ridge initiated between the Early Eocene (Leroy *et al.* 1996; 2000) and the Mid-Eocene (Rosencrantz *et al.* 1988).

Other extreme pull-aparts in continental-margin settings occupied by short spreading ridges and underlain in part by oceanic crust include one of the five pull-aparts along the Blanco transform in the Pacific (Embley & Wilson 1992); an array of five pull-aparts in the Andaman Sea (Curray 2005); and four pull-aparts in the Gulf of California (Einsele 1986). Lallemand & Jolivet (1986) proposed that the oceanic crust of the Sea of Japan originated as an extreme, oceanic-floored pull-apart formed by oblique motion in a back-arc setting in Miocene to Recent times.

Models for pull-aparts based on heat flow and thermal modelling

Because of the small size of most pull-aparts, localized heat formed during crustal extension is quickly dissipated laterally through the colder basin walls by conduction (Bradley 1983) and by convection through fluid migration along dilatant fault planes (Sibson 1985; 1986*a*, *b*; 1987; Fig. 7f). Narrower basins formed by limited fault separation between master faults would also cool quickly. Because of their rapid cooling, pull-aparts would exhibit more rapid subsidence than predicted by the uniform extension model applicable to stretched continental margins (Pitman & Andrews 1985; Allen *et al.* 2001).

More rapid subsidence of pull-aparts related to lateral heat loss could contribute to deep, sediment-starved basins like the Cayman Trough of the Caribbean (Rosencrantz *et al.* 1988) and the Ridge pull-apart basin of southern California (Crowell & Link 1982; May *et al.* 1993) which are deepening faster than the rate of sedimentary infilling (Rosencrantz *et al.* 1988). Bradley (1983) and Mann *et al.* (1983) interpreted a two-stage sedimentary history for the 250 000 km^2 Magdalena Basin that conforms well to predictions based on crustal stretching models. Stretching of the lithosphere within the rhomboidal pull-apart was accompanied by crustal thinning and increased heat flow. Following the rift phase, recovery of isotherms caused slower subsidence of the enlarged basin that buried the pull-apart.

Complexly faulted pull-apart basins

The Z-shaped Niksar (Tatar *et al.* 1995); the rhomboidal Erzincan (Barka & Gulen 1989), pull-apart basins along the North Anatolian fault zone, and the rhomboidal Hula Basin on the Dead Sea transform (Heimann & Ron 1987; Zilberman *et al.* 2000), show that these basins are intersected by other active faults, some of which are not parallel to master faults assumed by previous workers such as Aydin & Nur (1982), Mann *et al.* (1983) and Hempton & Dunne (1983). Therefore, their morphology, continued opening and deformation are not simply the result of the simple pull-apart opening model shown in Figure 7 but are likely to reflect more complex tectonic origins.

Extinction of rhomboidal pull-apart basins

Zhang *et al.* (1989) described an array of four active rhomboidal pull-apart basins along a 100-km-long segment of the Haiyuan fault zone in NW China. Several of these basins exhibit the development of a new strike-slip fault along one of the basin-margin master faults (with a normal component of throw) or diagonally crossing the centre of the Quaternary basin fill. These new diagonal faults on the basin floor differ from those oblique basinal faults described from spindle-shaped basins as shown in Figure 3b–d & f, because in the case of the Haiyuan basins they occur in rhomboidal basins with 5–10 km of master fault overlap. Zhang *et al.* (1989) concluded that the migration of the master strike-slip faults toward the center of the basin would lead to the end of extension across the basin; a decrease in the areal size of the pull-apart basin; and thereby would result in the process of 'pull-apart extinction'. Zhang *et al.* (1989) speculated that the likely mechanism for this process was the tendency of the arcuate Haiyuan fault zone to straighten itself.

Examples of rhomboidal pull-aparts with documented, diagonal strike-slip faults that ruptured seismogenically across the Quaternary floor of the basin in the manner described by Zhang *et al.* (1989) for the Haiyuan fault zone, include a 1951 event that ruptured the rhomboidal Beng Co pull-apart basin in Tibet (Armijo *et al.* 1986); a 1966 event that ruptured the rhomboidal Cholame pull-apart along the San Andreas fault zone (Brown *et al.* 1967); and the 1992 Landers event that ruptured the Homestead Valley basin between two parallel master faults (Sowers *et al.* 1994).

Hurwitz *et al.* (2002) interpreted a grid of seismic-reflection data from the rhomboidal Sea of Galilee pull-apart, and concluded that motion on one master fault on the western side of the basin has ceased, with all motion now concentrated on a single master fault. This implies that stretching of the Sea of Galilee basin floor has ceased and that the pull-apart phase of the basin history has ended.

Preservation potential of active and ancient pull-apart basins

Aydin & Nur (1982) tabulated information on 64 active and ancient pull-aparts and concluded that most were rhomboidal, with length to width ratios of about 3:1. Subsequent work has shown that some of the measured populations were made from other types of strike-slip basins and not pull-aparts as defined above (i.e. compare their table 1 with the pull-aparts listed in my Table 2). Nevertheless, their observation supports the larger number of rhomboidal pull-aparts (Fig. 4a–e) than either spindle-shaped basins (Fig. 3 b–d & f), S- or Z-shaped basins (Fig. 3a & e), or extreme rhomboidal basins (the Cayman Trough), as compiled in my Table 2.

One explanation for the larger number of rhomboidal basins is the tendency for older, longer basins with higher ratios to either become extinct by the inversion processes described above or to be destroyed by a sudden change in the strike-slip or overall tectonic environment (Mann *et al.* 1983). In a global study of the longevity of sedimentary basin types, Woodcock (2004) concluded that strike-slip basins are relatively small and short-lived, typically in existence for 3–10 Ma. In contrast, basins formed along divergent margins can persist for 100–200 Ma.

Because rhomboidal basins are larger and contain greater sedimentary thicknesses, they are more likely to be preserved in pre-Cenozoic, eroded mountain belts. The now inverted, rhomboidal Stellarton pull-apart basin in the Appalachian Mountains of eastern Canada formed by *c.* 10 km of right-lateral shear during the late Palaeozoic and localized a depocentre now containing up to 3 km of alluvial-fan, coal, and lacustrine sediments (Waldron 2004).

Other areas that contain well-preserved, ancient pull-apart basins include the sheared margins of the Equatorial Atlantic (Benkhelil *et al.* 1995; Miskelly 2002), the southern (Escalona & Mann 2003) and northern (Meneses-Rocha 2001) Caribbean margin; southern California (Crowell & Link 1982; May *et al.* 1993); Mesozoic belts in China (Vincent and Allen 1999; Chen *et al.* 2001; Zhang *et al.* 2003) and Korea (Ryang & Chough 1999): the Mesozoic Iberic range of Spain (Guiraud & Seguret 1985); the Neogene foreland basin of the Betic Range of Spain (Van der Straaten 1993); the Cenozoic Pyrenean Orogeny (Santanach *et al.* 2005); the Mesozoic and Cenozoic basins of the North Sea (Dooley *et al.* 1999); and the Carboniferous orogenic belts of Europe (Mattauer & Matte 1998; Mattern 2001); Shelley & Bossiere 2001; Titus *et al.* 2002).

Reinterpreted pull-apart basins

Subsequent structural and fault kinematic studies by Hossack (1984) have shown that the spectacularly exposed Palaeozoic Norwegian basins are not pull-aparts bounded by steeply-dipping master strike-slip faults as originally proposed on the basis of the original mapping and stratigraphic studies by Steel & Gloppen (1980). Instead, structural studies have shown that the basins formed as rifts above low-angle normal faults related to orogenic collapse along pre-existing, low-angle Caledonian thrust faults. Over 25 km of sediment accumulated in the largest basin over a period of 14 Ma.

The 13.5-km-thick Ridge Basin of southern California was long thought to be a Late Miocene–Pliocene transpressional basin (Crowell 1974*a*; Crowell & Link 1982), but is now known to be an inverted pull-apart basin formed along a low-angle normal fault imaged by seismic-reflection methods (May *et al.* 1993). The circular Silver Pit of the southern North Sea, that was once interpreted as a pull-apart basin (Smith 2004) is now thought to be a meteorite impact crater (Stewart & Allen 2005).

Pull-aparts in magmatic settings

Pull-aparts have been proposed as important mechanisms for controlling the emplacement of large plutons along trench-linked strike-slip faults in volcanic arc settings such as in the active Central American arc (Girard & van Wyk de Vries 2005), and in ancient orogenic belts (Lacroix *et al.* 1998; Talbot *et al.* 2005).

Complexities and alternative geological models for pull-apart basins

Transform normal extension: an alternative model for pull-apart opening

Ben-Avraham (1992) and Ben-Avraham and Zoback (1992) proposed that there may be a significant component of transform or master fault-normal extension across major strike-slip plate boundary faults like the Dead Sea, San Andreas and Caribbean faults. The observational support for this idea is the large-scale asymmetry seen on 2D seismic-reflection profiles across the basinal fill of the plate-boundary strike-slip faults that, in their view, would require a relatively larger extensional vs. strike-slip component to form. They attributed this basinal asymmetry to fault-normal extension induced by far-field transtensional stress acting on a weak fault. Conversely, far-field transpressional stresses across weak strike-slip faults may lead to a large component of shortening across major strike-slip plate-boundary faults, as observed in well-bore breakouts along the San Andreas fault zone (Mount & Suppe 1987).

Block rotations: an alternative model for pull-apart opening

Detailed seismological studies have shown the existence of rotating fault blocks around vertical axes whose block edges are defined by transverse faults crossing the intervening areas between sub-parallel zones of strike-slip faults. Active examples include the restraining-bend area of the eastern Transverse Ranges (Nicholson *et al.* 1985; Nicholson *et al.* 1986) and the pull-apart area of the Imperial Valley in California (Nicholson *et al.* 1986).

Block rotations may be important to accommodate the progressive passage of one fault block around a releasing bend, as proposed by Neugebauer (1994) for pull-aparts along the North Anatolian fault zone in Turkey. Heimann *et al.* (1990) proposed that large-scale rotation of the basin block of the rhomboidal Barahta pull-apart basin on the northern Dead Sea fault zone provides an alternative model for basin opening (rather than extension along transverse normal faults in the pull-apart model as shown in Fig. 7). Wesnousky (2005) illustrates large-scale block rotations in a 50-km-wide pull-apart segment of the central Walker Lane fault zone in the western USA. Mann (1997) compiled several examples where large-scale plate rotations could lead to transtensional opening and pull-apart formation along major strike-slip faults that bound rotating blocks or plates.

Estimates of rotational strain in pull-apart basins

Waldron (2004) uses the Palaeozoic Stellarton pull-apart basin in Nova Scotia, Canada, to illustrate the significance of rotational strains within the basin. This rotational component of strain is commonly neglected in most structural studies of mature pull-apart basins which accommodate significant offset. For example, at higher shear strains at more advanced stages of pull-apart evolution, previously formed normal faults of the type shown on a simple shear-strain ellipse (e.g. Wilcox *et al.* 1973; Biddle & Christie-Blick 1985) would begin to show inversion along reverse faults and large rotations.

Other approaches and observations on pull-apart basins

Models for pull-aparts based on subsurface geophysical imaging

Due to their thick sedimentary fill and lack of basement exposure, geophysical surveys of pull-aparts, including gravity, refraction and seismic-reflection methods, have improved our understanding of how pull-aparts evolve on now deeply buried master and normal faults. A particular focus of subsurface geophysical surveys in rhomboidal basins has been to determine the exact depth of fill of pull-apart basins and how master faults and transverse normal faults have controlled the observed basement subsidence.

Regional Bouguer gravity mapping by ten Brink *et al.* (1993) and ten Brink *et al.* (1999) of the Dead Sea strike-slip fault has revealed a string of small subsurface basins and strike-slip related depocentres along the trace of the plate-boundary fault that vary in size, shape, and depth, and bound relatively short (25–50 km) and discontinuous fault segments. These authors suggest that the basins have formed in response to continuous adjustment of the plate boundary to small changes in the relative motion between the African and Arabia plates. Regional aeromagnetic mapping by Hatcher *et al.* (1981) shows how large east–west magnetic anomalies are truncated and offset more than 100 km by the Dead Sea Fault. This is the same amount of offset concluded from several studies of surface bedrock geology.

The depth-to-basement and the dip of basin-controlling normal faults in pull-apart basins are key factors for relating total basin subsidence to offset on master faults (Arbenz 1984; Escalona & Mann 2003; Lazar *et al.* 2006). These studies are also useful for determining whether the sedimentary fill of ancient basins extends deep enough into the crust to be preserved in deeply eroded, ancient mountain belts affected by episodes of widespread strike-slip faulting produced during the final stages of continental convergence.

The depth of penetration of strike-slip faults and associated basins into the crust and upper mantle is controversial because existing geophysical images taken to upper-mantle depths are neither clearly diagnostic nor available for most active and ancient strike-slip fault zones (cf. Lemiszki & Brown 1988; Serpa *et al.* 1988; Rotstein & Bartov 1989; Beaudoin 1994; Garfunkel & Ben-Avraham 1996; Hurwitz *et al.* 2002).

Whether transverse normal faults form low-angle listric or normal faults (Arbenz 1984) or are steeply dipping normal faults as shown in most idealized opening models, awaits improved deep-seismic imaging of the basement surface buried deep beneath pull-apart basins. Using geological mapping and shallow seismicity in the upper crust, previous workers have proposed that overlapping faults bounding releasing and restraining bends merge before reaching mid-crustal levels (Clayton 1966; Sharp 1975; Little 1990; Graymer *et al.* 2007). Others propose that some ancient, exhumed restraining-bend structures can persist to mid-crustal depths (Miller 1994). Deep, wide-angle geophysical profiles of major strike-slip faults indicate that active plate-boundary strike-slip faults may persist as vertical faults through the upper mantle rather than soling out at the base of the crust as predicted by some geological and palaeomagnetic studies (Hole *et al.* 1996; Henstock *et al.* 1997). A more revealing approach has been to use 3D seismic reflection from the petroleum industry to map the internal structure of an ancient Eocene pull-apart in Venezuela (Escalona & Mann 2003).

Confirming the existence of these faults in rhomboidal and extreme basins is key for determining the value of the pull-apart opening model of Crowell (1974*a*, *b*) and Mann *et al.* (1983), where normal faults striking transverse to the master faults are predicted as the primary mechanism for basin opening (Fig. 7). The transform-normal extension of Ben-Avraham (1992) and Ben-Avraham & Zoback (1992) would not require transverse normal faults, and would instead predict basement normal faults parallel to the master faults. Block-rotation models would also not predict regular patterns of transverse normal faults terminating on parallel master faults (Neugebauer 1994). Another issue is the dip of the transverse normal faults: are they listric and low-angle normal faults at right angles to the master faults as proposed by Arbenz (1984) for the Dead Sea pull-apart, or are they more steeply dipping? Geophysical surveys have confirmed the presence of steeply dipping transverse faults consistent with the pull-apart model in the southern basin of the Dead Sea pull-apart (Lazar *et al.* 2006).

Denser grids of seismic-reflection profiles are revealing, in that some basins considered to be rhomboidal pull-aparts on the basis of their surface are in fact underlain by a lazy S- or Z-shaped subsurface basement structure. For example, the northern Gulf of Aqaba (Ehrhardt *et al.* 2005), the Cariaco Basin of Venezuela (Jaimes 2003), and the Sea of Galilee (Hurwitz *et al.* 2002) are underlain by a gentle releasing bend, rather than a rhomboidal pattern of parallel master faults and transverse normal faults as postulated on the basis of earlier seismic-reflection surveys with less resolution and penetration (Schubert 1982; Ehrhardt *et al.* 2005).

Models for pull-aparts at depth, based on high-resolution seismological data

High-resolution recording of earthquakes on creeping strike-slip faults and during large strike-slip events, particularly along the well-monitored San Andreas fault zone, has allowed precise mapping of pull-apart basins down to the base of the brittle crust. Recent monitoring of activity beneath the Cholame pull-apart near Parkfield, California, indicates that the 1-km-wide stepover observed by surficial and shallow fault traces (Shedlock *et al.* 1990) does not extend deeper than 5 km (Bakun *et al.* 2005).

Models for pull-aparts as earthquake rupture boundaries

Pull-apart basins have received considerable attention from seismologists, because of the possibility that they act as boundaries or end-points for the seismogenic propagation of fault slip (Sibson 1985, 1986*a*, *b*, 1987). Sibson's hypothesis, based on observed clusterings of aftershocks in the Cholame, Coyote Lake and Mesquite basins on the San Andreas fault system states that ruptures are arrested because of suction forces in the fluid-saturated sediments of the releasing bends. Aftershock clusters following the termination of the mainshock are caused by re-equilibration of fluid pressures in complex dilatational fracture networks. This same process, at deeper levels of the crust, may account for enhanced mineralization in releasing-bend segments (Sibson 1987; Berger 2007; De Paola *et al.* 2007).

Models for pull-aparts based on mesoscopic outcrop studies

Pull-apart structures filled by calcite and quartz are common in brittle faults deforming sedimentary and crystalline rocks. The structural evolution of these outcrop-scale 'pull-aparts' have been described in detail by many workers in different areas where there are surface breaks associated with seismogenic strike-slip faulting (Tchalenko & Ambraseys 1970) and within ancient, deformed rocks (cf. Kim *et al.* 2004, for a recent review).

Despite material property differences and scaling issues, there are many intriguing similarities between these active outcrop-scale structures and sediment-filled pull-apart and fault-termination basins, as described and interpreted by Gamond (1987) and Kim *et al.* (2004). Bahat (1983) and Gamond (1987) propose that sediment-filled pull-apart basins could evolve by the opening along *en échelon* tensile fractures in a manner similar to that seen in outcrop-scale fractures. By the same argument, connecting, compressive 'bridges' are inferred as the manner in which restraining bends form on larger, natural fault systems (Cowgill *et al.* 2004*a*, *b*).

Laboratory analogue modelling of pull-apart basins

The simplicity of the natural pattern of pull-apart opening between parallel master faults (Fig. 7) lends itself well to laboratory-scaled sandbox models. Hempton & Neher (1986), Dooley & McClay (1997), Rahe *et al.* (1998), and Basile & Brun (1999) produced analogue models very similar to the main characteristics of pull-aparts known from geological and geophysical studies, including such details as S- and Z-shaped incipient basins, the twin depocentres in rhomboidal basins, and diagonal strike-slip faults marking the extinction of the normal mode of pull-apart basin opening. Sims *et al.* (1999) produced an analogue pull-apart model in which a weak and strong décollement horizon was introduced at depth. Weak décollements resulted in scattered and asymmetrical sub-basins bounded by oblique-slip faults similar to pull-apart environments like the Gulf of Paria in Trinidad (Babb & Mann 1999; Flinch *et al.* 1999) whilst strong décollements produced more focused basins bounded by normal faults similar to the Death Valley pull-apart.

Despite the close similarities between the geometries of the analogue models and those of the natural examples, all authors modelling pull-aparts state the limitations of applying analogue models to natural systems. For example, these models cannot:

1. accurately simulate the thermal, flexural and isostatic effects generated by or associated with faulting in the upper crust;
2. take into account the effects of pore-fluid pressures and compaction which are to be expected in natural systems (Dooley & McClay 1997); and
3. account for the fact that the modelled fault stepover cannot be bypassed or cut-off at a basement level (T. Dooley, pers. comm. 2006).

However, the modelling by Sims *et al.* (1999) did attempt to address the second problem by inserting a ductile layer designed to simulate buried shale or salt lithologies. Dooley (pers. comm., 2006) notes that in many pull-apart analogue models the presence of a cross-basin strike-slip fault zone indicates that the sidewall fault systems are not 100% efficient and thus some component of differential motion must be accommodated across the basin floor. Despite the presence and activity of the cross-basin strike-slip fault, models show that the stepover continues to expand on its normal faults, no matter how much progressive offset is used for the model.

Numerical models of pull-aparts

Numerical models have been successful in simulating some of the natural features of pull-apart basins, including features of pull-aparts such as the twin depocentres in rhomboidal basins (Rodgers 1980); distribution and orientation of secondary fractures within the master fault overlap area (Segall & Pollard 1980); the symmetry of the basin between the master faults (ten Brink *et al.* 1996); and basin deepening towards the more active master fault (Golke *et al.* 1994; Bertoluzza & Perotti 1997)

Evolution of restraining bends based on geological observations

Introduction

Because active restraining bends form positive topographic features, sometimes reaching several kilometres in elevation, sites of bends expose older and deeper crustal rocks, with much of their uplift history possibly characterized by an erosional unconformity separating the older rocks in the topographic centre of the bend from surrounding sedimentary rocks at lower elevations that have been eroded off the bend area (Fig. 6a–g). For this reason, techniques used to study restraining bends differ from those described above for pull-apart basins. For example, surface geomorphology and thermochronology of basement rocks are more relevant to restraining-bend uplifts than they are for understanding pull-apart topographic depressions. Conversely, seismic-reflection imaging has had little impact on understanding restraining bend uplifts–particularly if the uplifts are composed of non-layered and difficult-to-image crystalline rocks.

Because of these limitations and the fact that pull-apart basins are far more numerous than restraining bends (cf. compilations by Aydin & Nur 1982; Bahat 1983; Table 2 of this paper), there has been significantly less of a global effort in the classification, mapping and modelling of restraining bends than there has been for pull-apart basins. Unlike pull-apart basins, the morphological classes of restraining bends shown on Figure 1b occur on very different scales, and in some cases, in different tectonic settings. For this reason, there is presently no observationally based, geological continuum model – as in the case of pull-apart basins (Fig. 7).

In this review of restraining bends, I will review existing classification schemes and propose a new restraining-bend classification (Fig. 1c) based on the global compilation of restraining bends on Table 2. It is hoped that this preliminary descriptive scheme for classifying restraining bends will lead future workers to derive a more comprehensive and genetic classification.

Morphological classification of strike-slip restraining bends

Classification

Using previous terminology proposed by Crowell (1974*a*, *b*) and Mann & Gordon (1996), I propose a threefold morphological classification of active and ancient restraining bends, shown in Figure 1c. The three bend classes include:

Transpressional uplifts. These are adjacent to straight segments of strike-slip faults that are oblique to the direction of plate or block motion (Fig. 1c). Technically, this class is not a fault 'bend', since the main strike-slip fault remains straight or very gently curving. Nevertheless, I include this type in the classification because a narrow, elongate topographic uplift and exposure of older rocks is produced as the result of the oblique intersection of the fault with the direction of plate/block motion (Fig. 1c). These uplifts can persist for along-strike distances of tens to hundreds of kilometres, and they combine elements of strike-slip (simple shear) with an element of shortening (pure shear). Examples of transpressional uplifts include the uplift of narrow strips of oceanic crust, including ophiolites along the Macquarie Ridge south of New Zealand (Massell *et al.* 2000; Fig. 6a); narrow strips of mid- to deep levels of continental crust along the Alpine Fault of New Zealand (Little *et al.* 2005); the Border Ranges fault zone of SE Alaska (Little 1990; Roeske *et al.* 2003); and the Betic Range of SE Spain (Keller *et al.* 1995).

Much of the 1000-km length of the two continuous faults forming straight segments of the Macquarie Ridge transpressional uplift are slightly oblique (*c.* 5–10°) to the direction of Australia–Pacific plate motion, and for that reason control the uplift and exhumation of narrow (10–40 km) transpressional uplifts of sheared oceanic basement rocks adjacent to the active strike-slip fault zones (Massell *et al.* 2000; Meckel *et al.* 2005; Fig. 6a). Increasing obliquity of plate motion in a southward direction leads to widening of the elongate transpressional uplift as seen on the southern segment of the Macquarie Ridge that exposes a partial ophiolite sequence on Macquarie Island (Daczko *et al.* 2003; Fig. 6a). Within this broad transpressional system, smaller pull-aparts coexist along the main trend of the strike-slip fault that bisects the transpressional ridge (Daczko *et al.* 2003; Fig. 6a).

Sharp restraining bends. These form at discrete fault stepovers and produce rhomboidal topographic uplift and exposure of older rocks (Fig. 6b & c). Sharp restraining bends are generally much smaller in size (fault separations generally <15 km) than the third and most prevalent type of bend, the *gentle restraining bend* (fault separations <200 km; Fig. 6d–g; Table 2). Two examples of mapped, sharp restraining bends deforming Late Neogene, clastic sedimentary rocks along active strike-slip faults are the Villa Vasquez bend (length 14 km, width 2 km) along the Septentrional fault zone in Hispaniola (Mann *et al.* 1999; Fig. 6b) and the Ocotillo badlands bend along the Coyote Creek fault segment of the San Jacinto fault zone in southern California (width 1.5 km, length 8 km) (Sharp & Clark 1972; Brown *et al.* 1991; Lutz *et al.* 2006; Fig. 6c).

The relatively small sizes and rarity of active sharp restraining bends probably indicate that they are easily bypassed, become extinct, and are eroded during progressive offset of the strike-slip fault zone. For example, dating and palaeomagnetic work by Brown *et al.* (1989) has shown that the sharp restraining bend forming the uplifted topography of the Ocotillo badlands formed less than 1 Ma; only reflects a small fraction of the total offset on its master faults; and is accompanied by a limited amount of vertical-axis rotation (Fig. 6c). As in the case of pull-apart basins, formation of sharp bends may be linked to small lateral shifts or 'switchyard behaviour' in the strike-slip fault trace (Brown *et al.* 1991) – perhaps in response to a changing pole of rotation (Fig. 9).

Wesnousky (1988) has compiled the fault stepover distances for both pull-apart and restraining bends on the San Andreas and North Anatolian fault zones, and concluded that faults with greater total offsets exhibit fewer restraining and releasing bends and overall straighter fault traces. This observation confirms the idea that smaller irregularities like sharp bends with fault separations less than a few kilometres are bypassed or 'smoothed' by faults with greater offsets, and therefore do not evolve into longer sharp restraining bends. Earthquake rupture data compiled by Harris *et al.* (2002) and Graymer *et al.* (2007) show that bends less than 5 km wide generally do not act as rupture boundaries for ground breaks for strike-slip earthquakes like the 1999 Izmit earthquake in Turkey.

Gentle restraining bends. These form the familiar shapes colloquially called *lazy-S shape* for gentle bends between right-lateral faults, as shown in Figure 1C, and *lazy-Z shape* for those between left-lateral faults. As with evolving pull-apart basins (Fig. 7), the S and Z shapes of gentle restraining bends in strike-slip plate boundary settings (Fig. 2) become more prominent as the master-fault separation ranges upward from values of 1.3 km (Echo Hills restraining bend, Nevada – Campagna & Aydin 1991) to values of 6 km (Montejunto restraining bend, Portugal – Curtis 1999), 20 km (Santa Cruz restraining bend, California, Burgmann *et al.* 1994, Fig. 6d) to greater than 20 km (San Bernardino restraining bend, eastern Transverse Ranges, California, Spotila *et al.* 2001, Fig. 6e).

These gentle, bend fault geometries produce elongate to domal topographic uplifts mini-fold–thrust belts; localized exposure of older deeper crustal igneous and metamorphic rocks in the core of the uplift; and a radiating pattern of alluvial fans and fan deltas that accommodate the erosional stripping of the bend uplift (Mann & Gordon 1996; Fig. 6d–g). The lineaments marking the master strike-slip fault traces in the bend area itself are sometimes obscured by the rugged topography of the core of the bend, but are well expressed by valleys of the master faults on either side of the

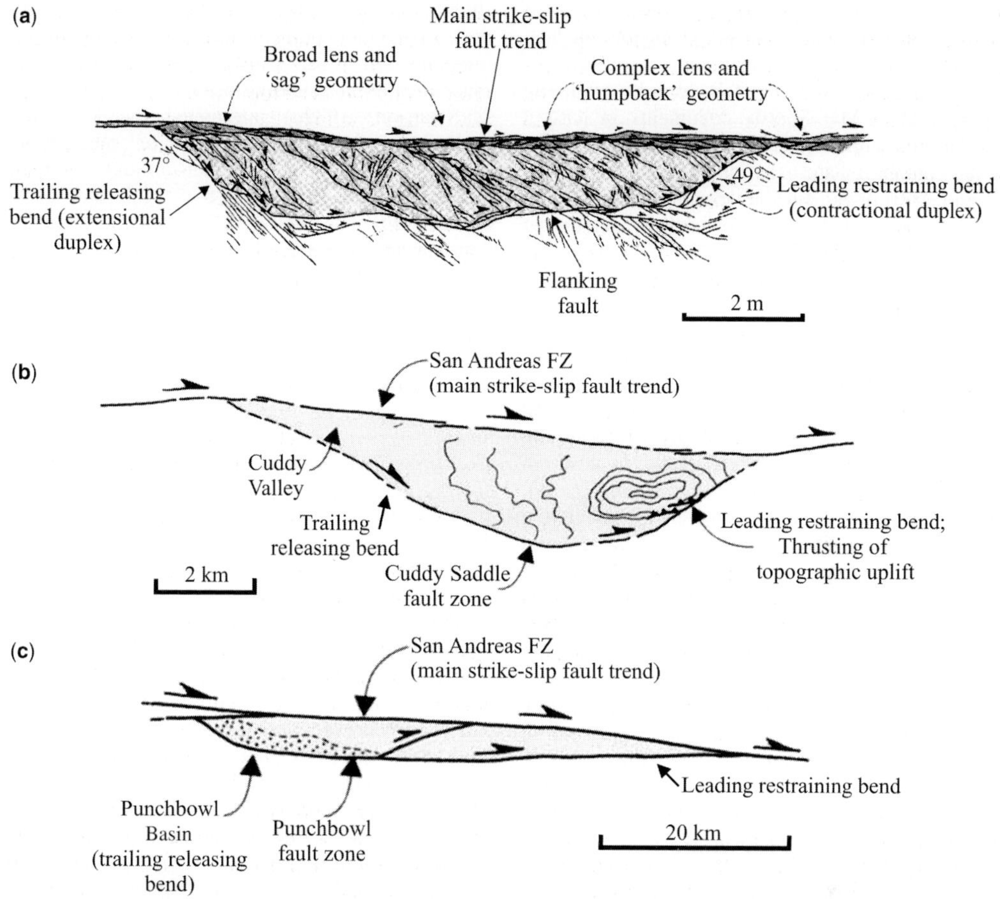

Fig. 10. Swanson (2005) proposes that 'sidewall ripouts' are scale-independent, based on his compilation of fault geometries for active and ancient strike-slip faults. (**a**) Outcrop-scale (12-metre-long) example of a 'sidewall ripout' (light grey) mapped from a Palaeozoic outcrop at Cape Elizabeth, Maine, by Swanson (2005). This structure exhibits imbrication and thrusting in a single obliquely thrusting 'leading ramp', fault and extension along three normal, oblique-slip faults in a 'trailing ramp'. Darker grey indicates the complexly deformed and linear, main strike-slip fault zone. (**b**) Regional-scale (10-km-long) example of a Neogene 'sidewall ripout', compiled by Swanson (2005) from mapping of the San Andreas and Cuddy Saddle fault zones at the western end of the Transverse Ranges by T. Davis and Duebendorfer (1987). Area of thrusting and topographic uplift is interpreted as a 'leading ramp'; the low-lying Cuddy Valley is interpreted as a 'trailing ramp'. (**c**) Regional-scale (55-km-long) example of a Neogene 'sidewall ripout', compiled by Swanson (2005) from mapping of the San Andreas and Punchbowl fault zones by Dibblee (1968). The Punchbowl Basin was filled by Neogene clastic sedimentary rocks, and is interpreted as a 'trailing ramp'.

bend (e.g. Massif du Macaya restraining bend in Haiti, Mann *et al.* 1995).

Gentle restraining bends are by far the most common, as compiled for this paper, and are the largest in their areal extent. Two examples of lazy-S gentle restraining bends along the San Andreas fault system are shown on Figure 6d & e. The Santa Cruz restraining bend on the San Andreas fault zone in northern California formed by a fairly subtle 5–10° change in the strike of the San Andreas fault zone (Anderson 1990, 1994; Schwartz *et al.* 1990; Burgmann *et al.* 1994; Fig. 6d). The San Bernardino Mountains bend on the San Andreas fault zone of southern California forms a more prominent lazy-S shape created by a larger 15–20° change in strike of the southern San Andreas fault zone (Spotila *et al.* 2001; Fig. 6e).

Both gentle bends exhibit fault separations seven to ten times the length of fault separations for the smaller sharp restraining bends shown in Figure 6b & c. Since these gentle bends have larger fault separations, bypassing with increasing fault offset (Wesnousky 1988) would be less likely, and therefore the bends might be expected to persist longer in the rock record. Using thermochronological methods, Burgmann *et al.* (1994) proposed that the Santa Cruz bend was originally uplifted at 12 Ma; Spotila documents at least a 1.8 Ma uplift history for the San Bernardino bend using mapping and thermochronology; Blythe *et al.* (2000) document the beginning of the bend-related uplift of the San Gabriel Mountains to have started at 7 Ma.

The gentle restraining bends with the largest areal coverage are found in a particular active tectonic environment: subduction to strike-slip transition areas (Bilich *et al.* 2001). These areas are the tectonic transitions between strike-slip plate boundaries and subduction boundaries, and include the gently curving boundaries between the North America and Caribbean plates in northern Central America (Mann & Gordon 1996; Fig. 6f); the North America and Caribbean plates in the Greater Antilles (Mann *et al.* 2003; Fig. 6g); and the South America and Scotia plates in southernmost South America (Cunningham 1995). The curvature of the faults near their intersection is related to either trenchward propagation of the strike-slip faults as described by Govers & Wortel (2005) or oblique collisional indentation as produced by the Bahama carbonate platform on the NE Caribbean plate (Mann *et al.* 2003; Fig. 6g).

Pattern of adjacent restraining and releasing bends at outcrop to regional scales

Sidewall ripouts on mesoscopic strike-slip faults

Swanson (1989*a*) reported asymmetrical sidewall ripout geometries from coastal exposures of mesoscopic, Late Palaeozoic strike-slip fault zones in southern Maine and western Greenland (Fig. 10a). *Sidewall ripouts* in plan view are plano-convex fault lenses and slabs where a splay fault diverges away from a section of the main fault to create an oblique-slip *flanking fault*, which rejoins the main trace on another lateral ramp further along strike (Fig. 10a).

The overall fault pattern formed in plan view is an asymmetrical fault lens flanking the main strike-slip fault trace. The asymmetry of the fault lens leads to the development of a gentle restraining bend or *contractional duplex* (Woodcock & Fischer 1986) at the leading edge of the fault lens, and the development of a gentle releasing bend or *extensional duplex* at the trailing edge of the fault lens (Swanson 1989*b*; Fig. 10a). Slight bulging or 'humpback' geometry deforms the straight fault trace to accommodate displacement at the restraining bend, whilst 'sag' geometry accommodates displacement at the releasing bend.

Swanson (1989*a*) interpreted the kinematic mechanism for the formation of mesoscopic ripouts to be the result of adhesion and increased friction on the main fault trace, followed by a lateral jump of the active fault surface to the adjacent curving fault trace and activation of the adjacent restraining and releasing bends. Revived movement on the locked, straight fault trace may lead to deactivation of the flanking fault and the shearing off and abandonment of the entire fault lens.

Significance of sidewall ripouts at regional scales

Swanson (2005) recognized similar sidewall ripout fault patterns on many active crustal-scale, strike-slip faults in all three types of strike-slip faults worldwide (Fig. 2). Proposed examples of regional-scale ripouts on *boundary strike-slip faults* included the San Andreas and San Gabriel fault zones of southern California (Fig. 10b & c). His proposed examples of *trench-linked strike-slip faults* with sidewall ripouts with adjacent releasing and restraining bends included the Atacama and Domeyko fault zones in Chile and the Yalakom fault zone of British Columbia. His proposed examples of *indent-linked strike-slip faults* included the Anninghe, East Xiaojiang, and Qujiang fault zones of the SE Tibetan Plateau in China.

For active crustal-scale faults, the surface expression of ripout translation includes a *leading gentle restraining bend* characterized by high topography, erosion, folding and oblique thrusting, and a *trailing gentle releasing bend* characterized by low topography, normal faulting and clastic sedimentation (Swanson 2005) (Fig. 10). Active, crustal-scale ripout slabs identified by Swanson (2005) range from 5 to 500 km in length. For example, his proposed 15-km-long ripout defined by the straight San Andreas Fault and the arcuate Cuddy Saddle Fault at the western end of the Transverse Ranges in southern California, shows a low valley topography at its trailing end of the proposed ripout structure and an uplifted topography with minor thrusting at its leading edge (Davis & Duebendorfer 1987; Fig. 10b). A larger, 65-km-long, 0.5-km-wide ripout structure is defined by the straight San Andreas fault zone and the arcuate Punchbowl fault zone (Dibblee 1968; Fig. 10c). The trailing edge of the ripout is defined by the Punchbowl Basin, filled by Neogene clastic sedimentary rocks (Dibblee 1968).

Paired strike-slip fault bends

Several previous workers have noticed the tendency for restraining and releasing bends to occur in pairs adjacent to one another and to exhibit almost the identical amount of master-fault separation (Figs 1d & 9a). In other words, the net effect of a paired bend is to locally deviate the main strike-slip fault trace, but to return the trace to the original and regional strike of the fault once the bend is passed (Fig. 1d). In map view this is exactly the same fault pattern as shown for regional ripout structures proposed by Swanson (2005; Fig. 10a–c). However, an alternative term for this pattern of adjacent releasing and restraining bends that I will use in this discussion is *paired bends*, which, unlike the term *sidewall ripouts*, is purely descriptive and has no genetic connotations of tectonic origin.

In Figure 11, I show four examples of paired bends that are described in greater detail in the latter part of this paper, they range in master-fault separations from 3 km in length for the Coyote Creek fault zone in southern California (Fig. 11a), to 50 km for the North Anatolian fault zone in the Sea of Marmara (Fig. 11b), to 175 km for the San Andreas fault system of southern California (Fig. 11c), to 50 km along the Dead Sea fault system (Fig. 11d). Each of these examples shows a similar pattern of an array of *en échelon* pull-apart basins adjacent to a large gentle restraining bend. Following the work of Swanson (2005), it became apparent to me that the paired bend, or crustal-scale 'sidewall ripout', is present at many scales within the plate boundary and trench-linked settings for strike-slip faults (Woodcock 1986; Fig. 2 & Table 1).

Comparison of tectonic models for restraining bends

Regional tectonic controls to bend fault traces

The same regional tectonic mechanisms summarized in Figure 8 for bending strike-slip fault traces to form pull-apart basins may also apply to natural examples of individual restraining bends (Figs 6a–g) and to paired bends (Fig. 11a–d). However, in this section, I will compare the predictions of three additional geological mechanisms with those shown in Figure 6 that have been proposed specifically for individual restraining bends and paired bends.

Shear-zone model

Gamond (1987) and Cowgill *et al.* (2004*a*) inferred that the origin of single and paired restraining bends at outcrop to regional scales was a natural consequence of progressive evolution and offset in the strike-slip fault zone (Fig. 12a). In this model, two sets of *en échelon* features form as an initial configuration in a developing shear zone: first a diffuse set of R-shears that eventually link up via bridging P-shears in an evolutionary sequence that has been observed and well documented in analogue models using:

1. various laboratory media (Tchalenko & Ambraseys 1970; Hempton & Neher 1986);
2. surface ruptures of natural strike-slip faults deforming relatively isotropic Quaternary sediments (Tchalenko & Ambraseys 1970); and
3. and in mesoscopic outcrops of rocks that have experienced brittle deformation (Gamond 1987).

The result of the shear-zone evolution is the present configuration consisting of a through-going fault trace with a sinusoidally curving trace that would form the sites of paired bends during progressive shear. Another model for the formation of a large single restraining bend mapped by Cowgill *et al.* (2004*a*, *b*) along the Altyn Tagh Fault includes rotation of large crustal blocks that deformed the fault zone and formed stepped bends in the fault as the fault propagated from west to east.

Sidewall ripouts

Swanson (1989*a*) interpreted the kinematic mechanism for the formation of mesoscopic ripouts at mesoscopic (Fig. 10a) and crustal (Fig. 10b & c) scales to be the result of adhesion or increased friction on the main strike-slip fault trace, followed by a lateral jump of the active fault surface to the adjacent curving fault trace and activation of the adjacent restraining and releasing bends (Fig. 12b). This model is similar to the finite-element model of Bridwell (1975) for forming sinusoidal strike-slip fault traces.

Revived movement on the locked, straight fault trace may lead to deactivation of the flanking fault and the shearing off of the entire asymmetrically shaped fault lens or 'sidewall ripout' (Fig. 12b). In his crustal-scale examples from the San Andreas and other fault systems, Swanson (2005) used dating of faulting and basinal deposits to show the history of abandonment and reactivation of the main fault trace and the splay fault defining the sidewall ripout (Fig. 10b & c).

Sidewall ripouts along strike-slip faults are analogous to lenses or 'horses' in contractional duplex structures on thrust faults. Progressive thrust offset leads to 'shunting' of the horse from the hanging wall to the footwall and then back to the hanging wall (Woodcock & Fischer 1986) (Fig. 12b).

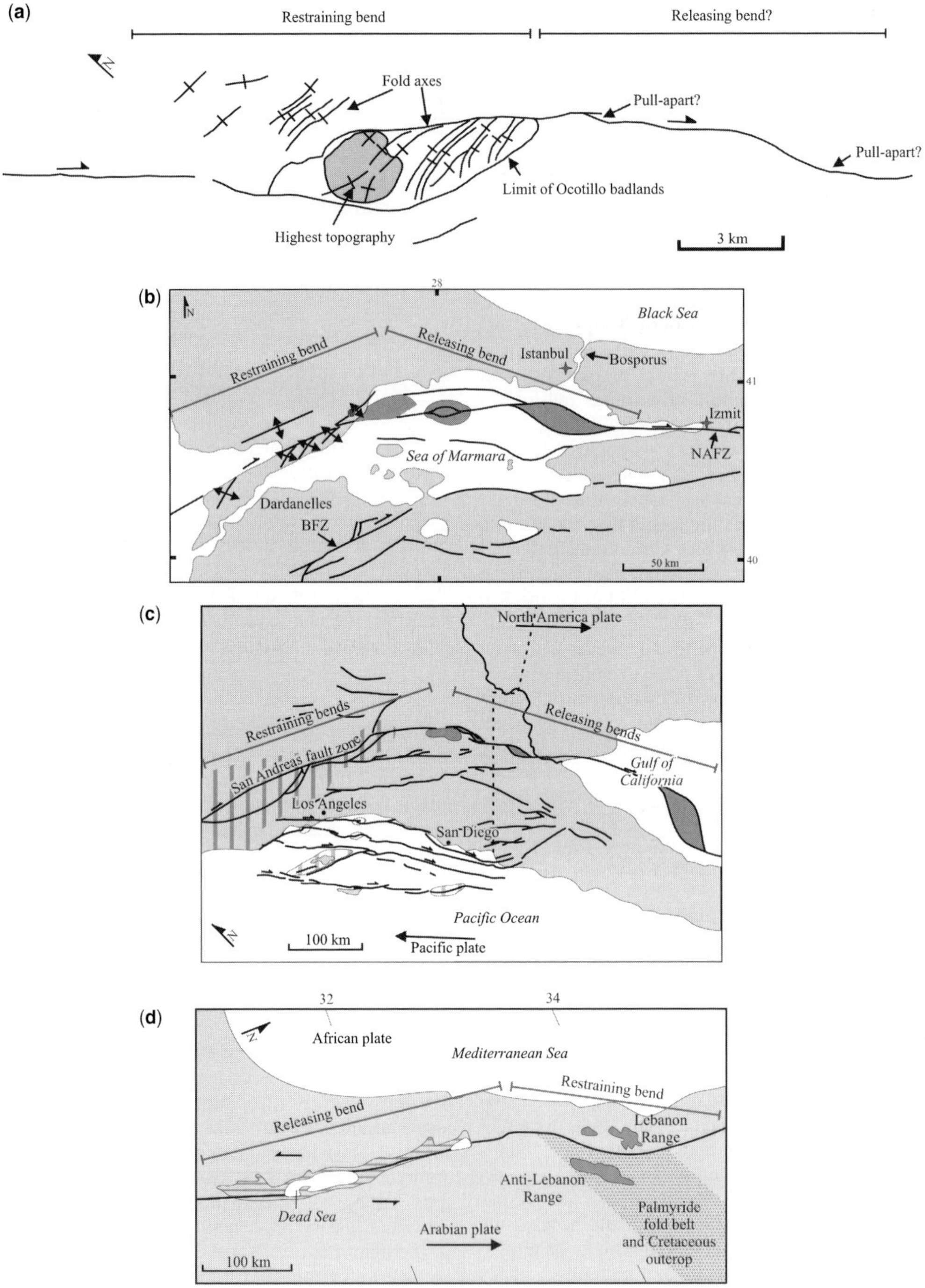

Fig. 11. Examples of large-scale paired-bend structures. The main strike-slip fault appears to contain releasing bends in one direction of the fault and restraining bends in the opposite direction. (**a**) Schematic map of the Ocotillo restraining bend along the Coyote Creek fault zone in southern California, modified from Segall & Pollard (1980), based on the original mapping of Sharp & Clark (1972). An inferred paired restraining- and releasing bend is indicated. (**b**) Sea of Marmara paired bend. (**c**) Southern California paired bend. (**d**) Dead Sea–Lebanon paired bend.

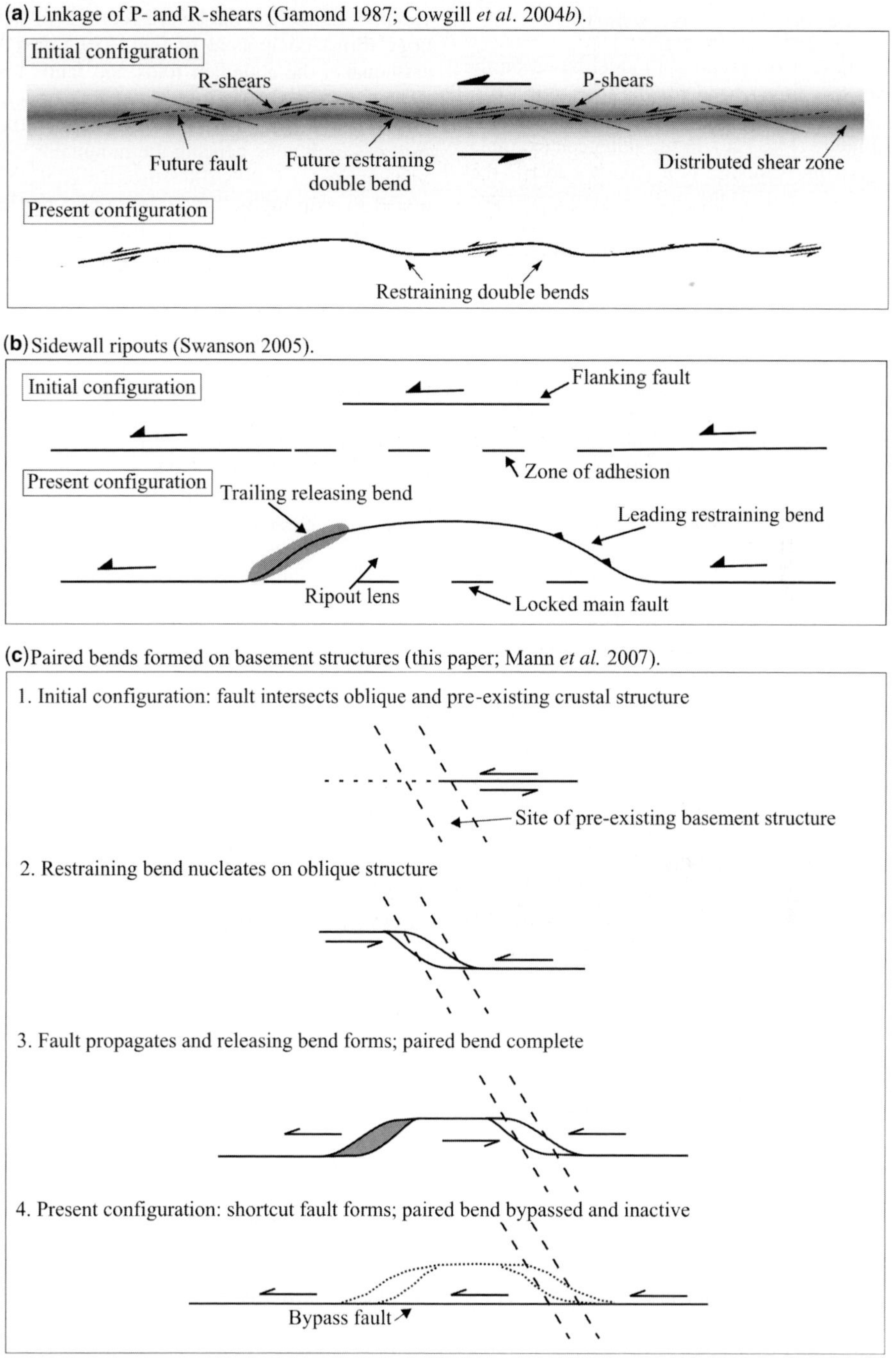

Fig. 12. (**a**) Model for the formation of adjacent releasing and restraining bends (i.e. double bends or paired bends) along strike-slip faults, based on the linkage of P- and R-shears (modified from fig. 12 of Cowgill *et al.* 2004*b*; see also Gamond 1987). (**b**) Model for the formation of adjacent extensional and contractional 'ramps' at the edges of 'sidewall ripouts' based on mesoscopic mapping and compilations of regional strike-slip fault traces, basins, and bends by Swanson (2005). (**c**) Model for the formation of adjacent releasing and restraining 'paired bends', based on Mann *et al.* (2007*a*) and the compilation done for this paper.

Basement control on paired bends

Mann *et al.* (2007*a*) propose a model based on the Jamaican restraining bend for the interaction of a propagating strike-slip fault and a pre-existing basement structure (Fig. 12c). In the Jamaican example, the pre-existing structure is a Palaeogene rift that trends at a 45° angle to the strike-slip zone and precedes the onset of strike-slip faulting. One puzzling question is: why does the strike-slip fault return to its original trend by forming an adjacent pull-apart basin as shown in the third drawing in Figure 12c? An identical map-view fault pattern is seen along other paired bends, as shown on the four paired-bend examples compiled in Figure 11.

One structural explanation is that the pull-apart basin forms as a geometric accommodation to shortening at the restraining bend localized on the pre-existing rift structure (Woodcock & Fischer 1986). To maintain zero bulk-volume change in a plane-strain strike-slip system, the excess volume of deformed crust 'piling up' at the restraining bend during progressive strike-slip displacement must be balanced by a volume deficit by thinning the crust in the adjacent, along-strike area of the releasing bend. Since mass balance of the crust is preserved, the area of thinned crust is equivalent to the thickening in the restraining bend, and therefore the approximate shapes and master-fault separations would remain similar. Whilst some of these crustal excesses or deficiencies will be accommodated by localized uplift and subsidence of the ground surface near both bends, most of the crustal deformation at the bend will be propagated laterally through the crust – hence the need for a releasing bend to develop adjacent to the restraining bend, and for the master fault to shift laterally to its original trend, as shown schematically in Figure 1d and in the natural examples of sidewall ripouts/paired bends in Figures 10 & 11.

Different approaches and observations on natural restraining bends

Tectonic geomorphology

A prominent feature of strike-slip restraining bends is their localized steep topography that commonly forms anomalously high peaks compared with the surrounding region (Fig. 6a–g). The steep drainage slopes and high local relief of the elongate Santa Cruz Mountains are strongly symmetrical about a lazy-S-shaped left-stepping restraining bend in the San Andreas fault zone (Fig. 6d), and this geomorphology has been modelled to support its tectonic origin by advection of crust and shortening around the bend (Anderson 1990, 1994; Burgmann *et al.* 1994). Schwartz *et al.* (1990) note that horizontal compression across the bend is consistent with the large thrust component of the 1989 Loma Prieta earthquake; the abundant folds and faults mapped in the southern Santa Cruz Mountains by Aydin & Page (1984); and the steep SW dip of the San Andreas fault zone beneath the mountains.

Gomez *et al.* (2006, 2007) used a digital elevation model to map high-altitude, low-relief surfaces in the 200-km-long Lebanese restraining bend. These surfaces are used as reference points to assess net post-Miocene uplift and strain partitioning effects.

Cowgill *et al.* (2004*a*) use a digital elevation model to document the tectonic geomorphology of the Akato Tagh gentle bend along the Altyn Tagh Fault in Tibet. Topographic profiles show that the bend uplift is asymmetrical, with the highest and widest areas of the range expressed as two topographic maxima located within the inside corners of the double bend. Structural studies show that the topographic elevated inside corners of the bend are regions of strong fault-perpendicular shortening, and that this shortening may cause the master fault to rotate and reduce and 'smooth' the restraining-bend angle through time (Wesnousky 1988).

Cunningham (2007) reviews the tectonic geomorphology of 12 individual restraining bends and fault-termination zones in the Altai, Gobi Altai and Tien Shan of central Asia. Cretaceous–Palaeocene summit peneplains provide a useful topographic datum for determining relative amounts of topographic uplift and the structural asymmetry of the ranges.

Comparative structure and tectonics of major restraining bends

The regional structure and seismicity of the larger and better-studied gentle restraining bends, including the San Gabriel and San Bernardino Mountains of the Transverse Ranges of California; the Lebanese bend of the Dead Sea Fault; the Hispaniola bend of the northern Caribbean; and the Southern Alps of New Zealand (reclassified in this paper as a transpressional ridge – Fig. 1b) have been compared at regional scales by Scholz (1977), Mann *et al.* (1984), Mann & Burke (1984), Yeats & Berryman (1987), and Cowgill *et al.* (2004*a*). On the basis of their mapped structure, Cowgill *et al.* (2004*a*) classify the Santa Cruz gentle bend (or double bend in their terminology) as a *thrust-dominated bend* with the highest elevations, steepest slopes, and largest local relief, all within 10 km of the San Andreas Fault and bounded by inside topographic corners of the bend in the Santa Cruz Mountains and Gabilan Range (Burgmann *et al.* 1994) (inside corner highs are indicated by arrows in Fig. 13a).

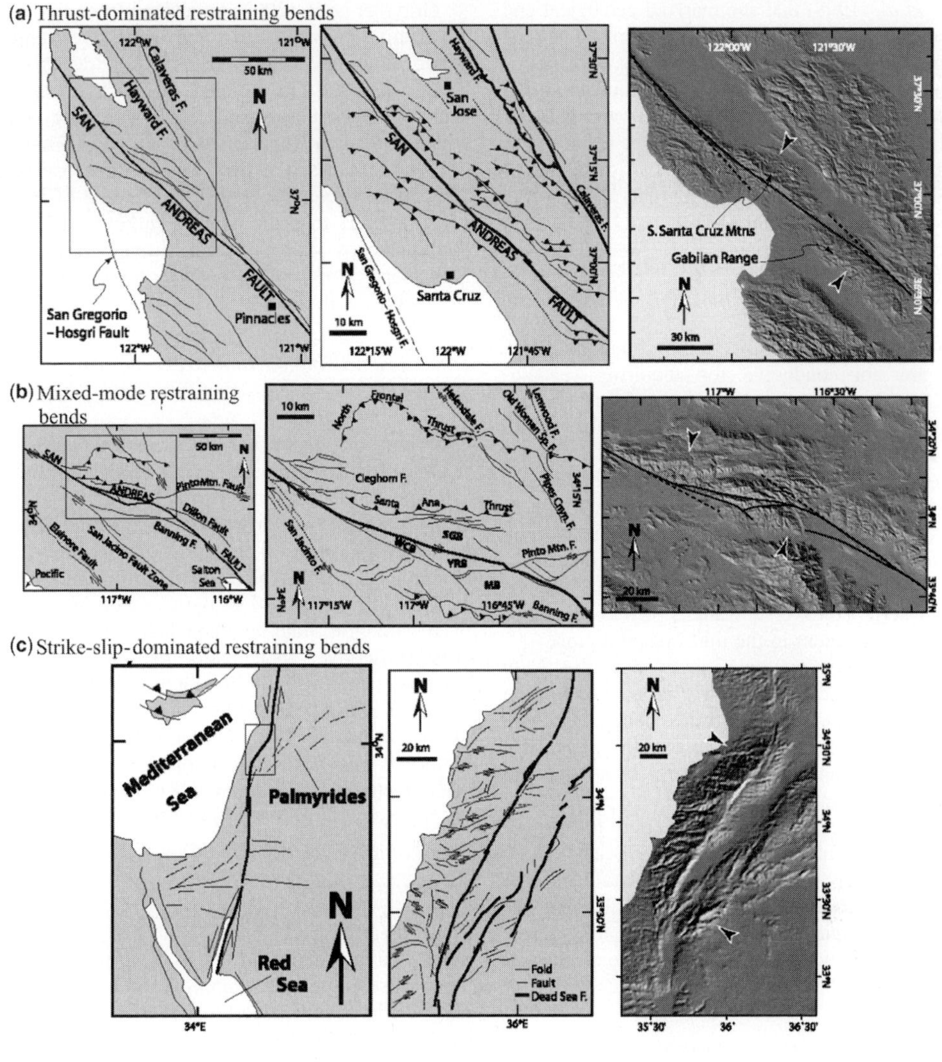

Fig. 13. Three types of restraining bends based on Cowgill *et al.* (2004*b*). (**a**) Thrust-dominated Santa Cruz Mountains. (**b**) Mixed-mode Big Bend on the San Andreas Fault in southern California. (**c**) Strike-slip-dominated Lebanon Ranges. Black triangles indicate corner topographic highs.

The San Bernardino Mountains exhibit similar nodes of topographic uplift in the inside corners, but their structure is classified by Cowgill *et al.* (2004*b*) as a *mixed-mode bend* by combining thrusting in the outside of the bend and strike-slip faulting in its internal areas (Fig. 13b). The Lebanese bend, which exhibits similar inside corner topographic uplifts, is interpreted as a *strike-slip dominated bend*, although Gomez *et al.* (2007) describe a mixed-mode, or strain-partitioned, style of deformation with internal strike-slip faulting and an external zone of folding and thrusting (Fig. 13c). Calais *et al.* (2002) and Mann *et al.* (2003) proposed a similar mixed-mode, or strain-partitioned, style of deformation for the Hispaniola bend, using GPS results (Fig. 6g).

McKenzie (1972) first noted that triangular-shaped crustal wedges in the vicinity of the Transverse Ranges bend of southern California may indicate a process of 'tectonic escape' or lateral migration of continental blocks away from the maximum area of bend-related shortening in the central Transverse Ranges. This same idea of tectonic escape was proposed in a more recent paper by

Walls *et al.* (1998) that summarized geological and earthquake evidence for east–west crustal escape along strike-slip and oblique-slip faults and north–south shortening by thrusting and conjugate strike-slip faulting in the region of the Transverse Ranges. This escape style of tectonics could be classified as 'mixed mode' in the terminology of Cowgill *et al.* (2004*a*), since the kinematics involve a shortening and strike-slip escape component.

Geophysical imaging and the deep structure of restraining bends

Because bends form positive, mountainous areas, they are not conducive for acquiring seismic-reflection data. For this reason, the deep structure of bends remains poorly known and controversial. For example, R. S. Yeats has argued both for shallow detachment of the Transverse Ranges as a 'crustal flake' (Yeats 1981) and for its deeply rooted structure in an undetached thick crust (Yeats *et al.* 1994). Balanced structural cross-sections drawn by Namson and Davis (1988) for the western Transverse Ranges invoke a major inferred detachment in the mid-crustal region.

Seismic-reflection data show that the New Zealand transpressional uplift is a pop-up-type structure on 2D seismic profiles that extends to depths of 20–30 km (Little *et al.* 2005). Industry seismic-reflection data from Syria show that the Lebanese restraining bend formed on a pre-existing Cretaceous rift system now inverted to form a 100-km-wide fold–thrust belt (McBride *et al.* 1990). Industry seismic-reflection data from the Hispaniola restraining bend show a 'mixed-mode' style of strain partitioning, with outer fold–thrust belts and subduction zones flanking an interior zone of strike-slip faults (Mann *et al.* 2003).

Seismological data to map downward continuation of restraining bends

High-resolution recording of earthquakes on creeping strike-slip faults and during large strike-slip events – particularly along the well-instrumented and monitored San Andreas fault zone has allowed precise mapping of restraining bends down to the base of the brittle crust. For example, Schwartz *et al.* (1990) have shown that aftershock hypocentres of the 1989 Loma Prieta earthquake define a 5-km-wide, 10-km-deep, SW-dipping restraining bend between the San Andreas and Sargent–Berrocal fault zones. The Mission Hills restraining bend connects the Hayward and Calaveras faults and forms a topographic uplift and one of the fastest uplifting areas in this region (based on radar interferometry, Burgmann *et al.* 2006). Manaker *et al.* (2005) have used relocated earthquakes to define the bend down to a depth of 10 km.

Graymer *et al.* (2007) review evidence for other down-dip extensions of other small restraining bends along the San Andreas Fault, determined using high-resolution earthquake hypocentres. They conclude that restraining or releasing bends with master-fault separations of less than 5 km are often not expressed as two faults in the subsurface. This result is consistent with the finding of Harris *et al.* (2002) that the Izmit (Turkey) strike-slip earthquake of 1999 ruptured through bends with fault separations of less than 5 km and fault angles of less than 30°.

Deep crustal bends and preservation potential

The depth of penetration of restraining bends into the mid- and lower crust is a key question for understanding their preservation potential in deeply eroded ancient mountain belts (Little 1990). Miller (1994) mapped an example of a mid-crustal example of a rhomboidal restraining bend with a 10 km fault separation in the Cascade Mountains of Washington State, and concluded that an upper crustal stepover penetrated to the mid-crust. Corsini *et al.* (1996) describe a ductile, mid-crustal restraining bend formed in the Brazilian craton during the Late Precambrian. The deformation pattern is very similar to ramp–flat transitions in thrust tectonics (Woodcock & Fischer 1986).

Palaeomagnetic data from restraining bends

Palaeomagnetism has been used to study large-scale block rotations about vertical axes within the gentle Transverse Ranges restraining bend by Luyendyk (1991; Fig. 13a); widespread vertical-axis block rotations within the gentle Lebanese restraining bend (Ron *et al.* 1990; Fig. 13c); and vertical axis rotations within smaller rhomboidal-type bends including the Ocotillo badlands (Brown *et al.* 1991; Fig. 11a) and the rhomboidal Almacik restraining bend or inferred 'crustal flake' actively being displaced and rotated along the North Anatolian fault zone (Saribudak *et al.* 1990; Table 2).

Uplift history and thermochronology of restraining bends

Fission-track dating has been used to study the localized uplift history of the Denali restraining bend in Alaska (Fitzgerald *et al.* 1993), the Transverse Ranges bend (Blythe *et al.* 2000; Spotila *et al.* 2001) and the Santa Cruz mountains bend (Burgmann *et al.* 1994). Spotila *et al.* (2001) note that the uplift mechanism of deep-crustal rocks is most likely a response to the gentle bend fault geometry rather than larger-scale kinematic models calling for fault-normal shortening along the entire length of the San Andreas fault zone (Mount & Suppe 1987). Cowgill *et al.* (2004*a*) noted that the anomalous topographic elevations marking one inside corner of the Santa Cruz gentle bend (Fig. 13a) correspond with one area of rapid uplift as measured by fission-track dating (Burgmann *et al.* 1994).

Modelling approaches to understanding restraining bends

Block and kinematic models

Several areas of large-scale (30°), counterclockwise block rotations have been identified in the Lebanese restraining bend (Ron *et al.* 1990). These rotations have been modelled either as neotectonic block rotations associated with the present-day bend (Ron 1987; Westaway 1995) or an earlier deformation related to the inversion of a Cretaceous rift system and the formation of the Palmyride fold–thrust belt (Walley 1998; Westaway 2004). Gomez *et al.* (2007) attempts to reconcile both interpretations by proposing a poorly dated two-stage history:

1. an earlier deformation involving less-oblique plate motion at the bend and controlled by a more distant pole of rotation; this deformation phase was characterized by distributed strike-slip faulting, less crustal shortening and large-scale block rotations (Ron 1987); and
2. a later phase of deformation involving increased oblique plate motion and increased strain partitioning into an inner zone of strike-slip faulting and an outer zone of folding and thrusting.

The palaeomagnetic recognition of significant vertical axis rotations in the region of the Transverse Ranges of southern California has led to block models by Luyendyk (1991), Dickinson (1996), and Ingersoll & Rumelhart (1999). All these models propose large rigid blocks that experience greatest rotations in the restraining-bend area in the Miocene and Pliocene, and that basinal areas were created at their block edges. GPS surveys by Feigl *et al.* (1993) show only smoothly varying motion of crustal material around the Big Bend area of southern California, and therefore indicate that all large-scale block rotations have ended.

Analogue models of restraining bends

McClay & Bonora (2001) have produced laboratory analogue models of restraining bends that simulate many of the observed features of natural bends. Three experimental conditions are produced: underlapping master faults which produce 'lozenges' or gentle bends; master faults with no overlap that produce rhomboidal basins; and master faults with overlap that also produce rhomboidal basins. In cross-sections, the bends show pronounced structural asymmetry and doubly plunging anticlines. As in analogue models of releasing bends, the models cannot accurately simulate the anisotropy of the upper crust or the effects of pore pressure and compaction.

Numerical models of restraining bends

Numerical modelling to simulate restraining-bend deformation has been carried out by Rodgers (1979), Segall & Pollard (1980), Du & Aydin (1995) and Brankman & Aydin (2004). Rodgers (1979) approximated the Big Bend in the San Andreas fault zone by a series of straight dislocation surfaces, and calculated the displacement and stress fields for each fault segment. The vertical displacement fields resemble the general topographic features of the Transverse Ranges, suggesting that the range topography is bend-related. The north–south principal stress orientations also correspond with earthquake and other observations. Segall & Pollard (1980) and Du & Aydin (1995) simulated fracture patterns within restraining bends using different fault geometries.

Tectonic setting and description of natural bends on transform faults separating oceanic plates

Objectives

In this and following sections, I will describe the formation of both releasing and restraining bends in the five types of natural, plate-boundary-zone settings summarized in Figure 1 & Table 1. There are two reasons for including these maps and descriptions. First, regional maps are necessary to illustrate the presence of paired bends, which, as previously noted by Swanson (2005) are commonly not recognized as genetically linked pairs of releasing- and restraining-bend structures (Figs 10 & 11). The tendency in previous discussions of either releasing (Aydin & Nur 1982; Mann *et al.* 1983) or restraining bends (Segall & Pollard 1980; McClay & Bonora 2001) has been to focus on individual structures rather than considering a longer segment of the fault that may be controlling an array of fault bends. Second, regional maps are more valuable to illustrate the close relationship between fault patterns within the strike-slip system and large-scale plate motions, a relationship which is particularly well known from plate modelling studies and

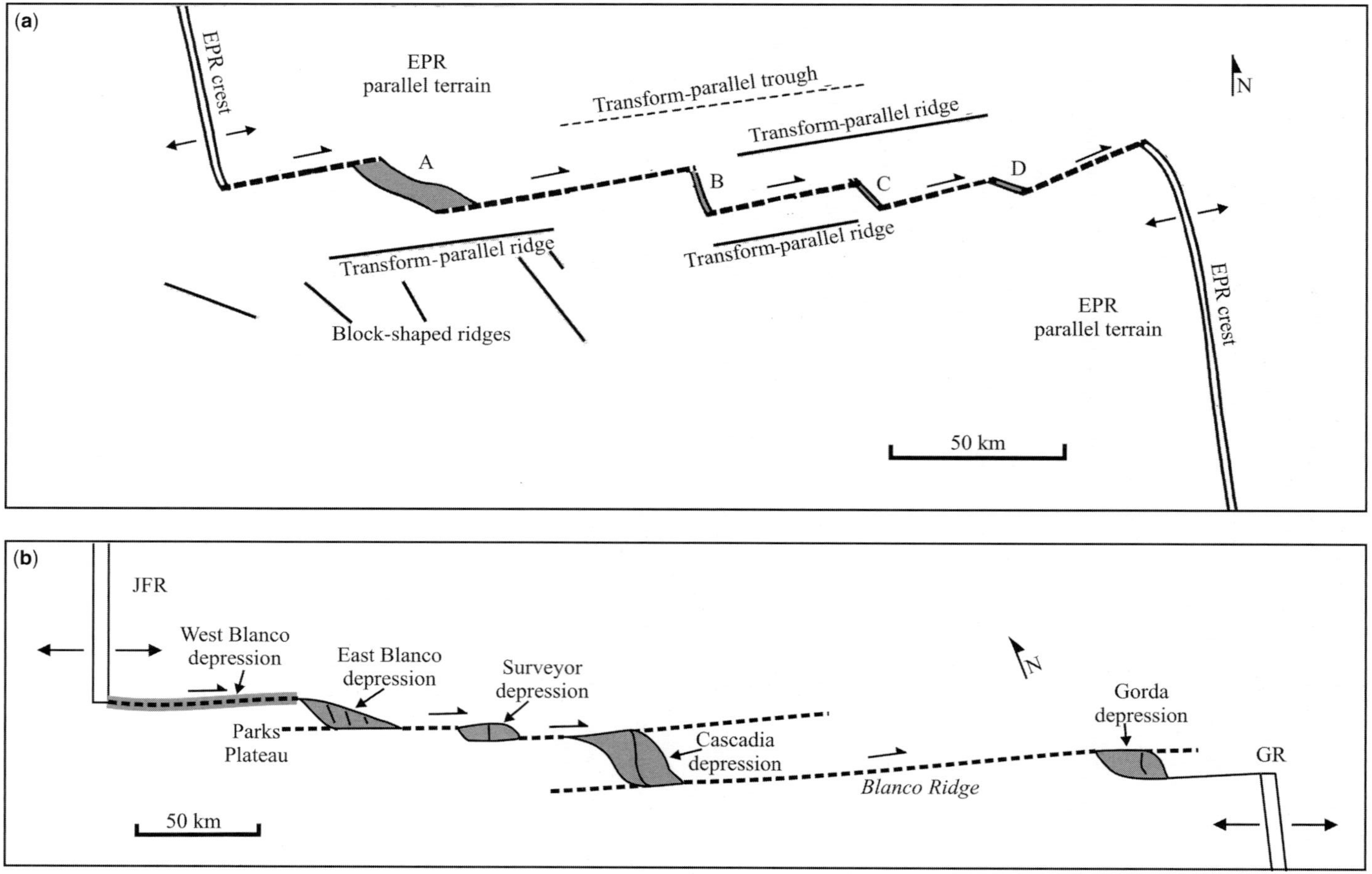

Fig. 14. (**a**) Array of four pull-apart basins found along the Siqueiros transform on the fast-spreading (63 mm/a) East Pacific Rise (EPR), modified from Fornari *et al.* (1989). (**b**) Array of four pull-apart basins found along the Blanco transform offsetting the fast-spreading (60 mm/a) Juan de Fuca (JFR) and Gorda (GR) ridges, modified from Embley *et al.* (1987).

GPS-based geodesy. As illustrated in Figures 8 and 9, the angles defined between the small circle about the pole of rotation and the strike of the fault are critical for determining the locations of individual or multiple arrays of pull-apart basins located along boundary transforms (Le Pichon & Francheteau 1978; Garfunkel 1981; Mann *et al.* 1983; Bilham & Williams 1985; Parkinson & Dooley 1996). Regional maps of strike-slip faults are also important for making the point that paired bends are not common for indent-linked strike-slip faults within interior areas of continental crust (Cunningham 2007) but do occur along trench-linked strike-slip faults.

Oceanic transforms

This class of faults connecting the offset ends of the mid-oceanic spreading ridges (Fig. 2) has been studied in detail using a variety of methods, including earthquake seismology, high-resolution bathymetry, studies of magnetic anomalies, and direct bottom sampling (Fox & Gallo 1984; Garfunkel 1986; Searle 1986; Fig. 14). Oceanic transforms typically form linear, steep-sided valleys, with the actively deforming zone marked by a single, narrow belt of deformation, or 'principal displacement zone' (Fox & Gallo 1984; Searle 1986). This fault morphology differs from wider, more arcuate and multi-branched patterns of faults seen in strike-slip faults affecting continental crust (Figs 4–6 & 10–11). The simpler and more focused patterns of deformation along oceanic transform faults probably reflect the thinner and more isotropic nature of oceanic crusts of varying ages that is lacking in the amount of inherited fabrics and cross-cutting features present in continental crust (Fox & Gallo 1984).

Another interesting contrast between oceanic transforms and continental strike-slip faults is the predominance of releasing bends, including those that have formed short spreading ridges (Garfunkel 1986; Searle 1986; Fornari *et al.* 1989). Only a single transpressional ridge (Fig. 1c) has been identified on the Clipperton transform fault in the Pacific Ocean (Pockalny 1997; Table 2). The lack of discrete restraining bends means that paired bends characteristic of most continental faults (Figs 10–11) have also not been identified on oceanic transform faults.

Pull-apart array on the Siqueiros transform, Pacific Ocean

Fornari *et al.* (1989) identified an array of four pull-aparts, or 'intra-transform spreading centres' on the Siqueiros transform (Fig. 14a). The pull-apart shapes are similar to those seen in continental settings: two of the basins exhibit lazy-Z morphology between non-overlapping master faults; the other two basins are more orthogonal to the fault trace and rhomboidal in morphology. Geological data indicate that the basins formed from west (older basins) to east (younger basins) over the past 1 Ma. The stability of the East Pacific Rise for the past 1.5 Ma argues against a small change in plate motion to form the basins, as shown schematically in Figure 7a. Fornari *et al.* (1989) propose that the basins originated by the interaction of small melt anomalies in the mantle with lithospheric fractures within the transform fault.

Pull-apart array on the Blanco transform, Pacific Ocean

Detailed seismological, bathymetric and side-scan sonar mapping of the Blanco transform revealed an array of five, rhomboidal pull-apart basins arranged along the 350-km length of the transform that offsets the Gorda and Juan de Fuca spreading ridges in the NE Pacific Ocean near Washington State (Dziak *et al.* 1991; Embley & Wilson 1992; Fig. 14a). The geometry of the basins (Fig. 14b) is strikingly similar to rhomboidal pull-aparts formed in continental settings (Fig. 4). The centrally located and largest rhomboidal basin, the Cascadia depression, exhibits inwardly facing back-tilted fault blocks and is inferred to contain a short spreading centre active over the past 5 Ma. The basins have formed over the past 5 Ma as the result of the continual reorientation of the Blanco transform fault in response to changing plate motion. Plate reconstructions by Embley & Wilson (1992) to explain the formation of the pull-apart array favour a complex model involving propagating ridges instead of a simpler change in motion across a single, pre-existing, linear transform fault (cf. Figure 8a).

Transpressional ridge on the Clipperton transform, Pacific Ocean

Pockalny (1997) describes a currently active transpressional ridge using high-resolution bathymetry and earthquake data. He attributes the tectonic origin of transpression to a small change in the spreading direction along the East Pacific Rise, starting about 0.4–0.5 Ma.

Tectonic setting and description of natural bends on long, strike-slip faults separating continental plates

San Andreas fault system

Tectonic setting. The San Andreas fault system, extending 2100 km from the mouth of the Gulf of California to Cape Mendocino in northernmost

California, is undoubtedly the world's best-studied continental-plate-boundary strike-slip fault system (Woodcock 1986; Page 1990; Brown *et al.* 1991; Fig. 15 & Table 1). In this section, I summarize the current state of knowledge of restraining and releasing bends compiled from literature on the San Andreas fault system, with particular emphasis on their tectonic settings relative to the small circles of rotation between the North America and Pacific plates (Fig. 15). I also review evidence for the existence of adjacent, 'paired bends' (Fig. 11) or regional 'sidewall ripouts' (Fig. 12) within the San Andreas fault system, whose occurrence and significance may be underestimated (Swanson 2005). For reviews of the tectonic origin of the San Andreas fault system, its offset history, and

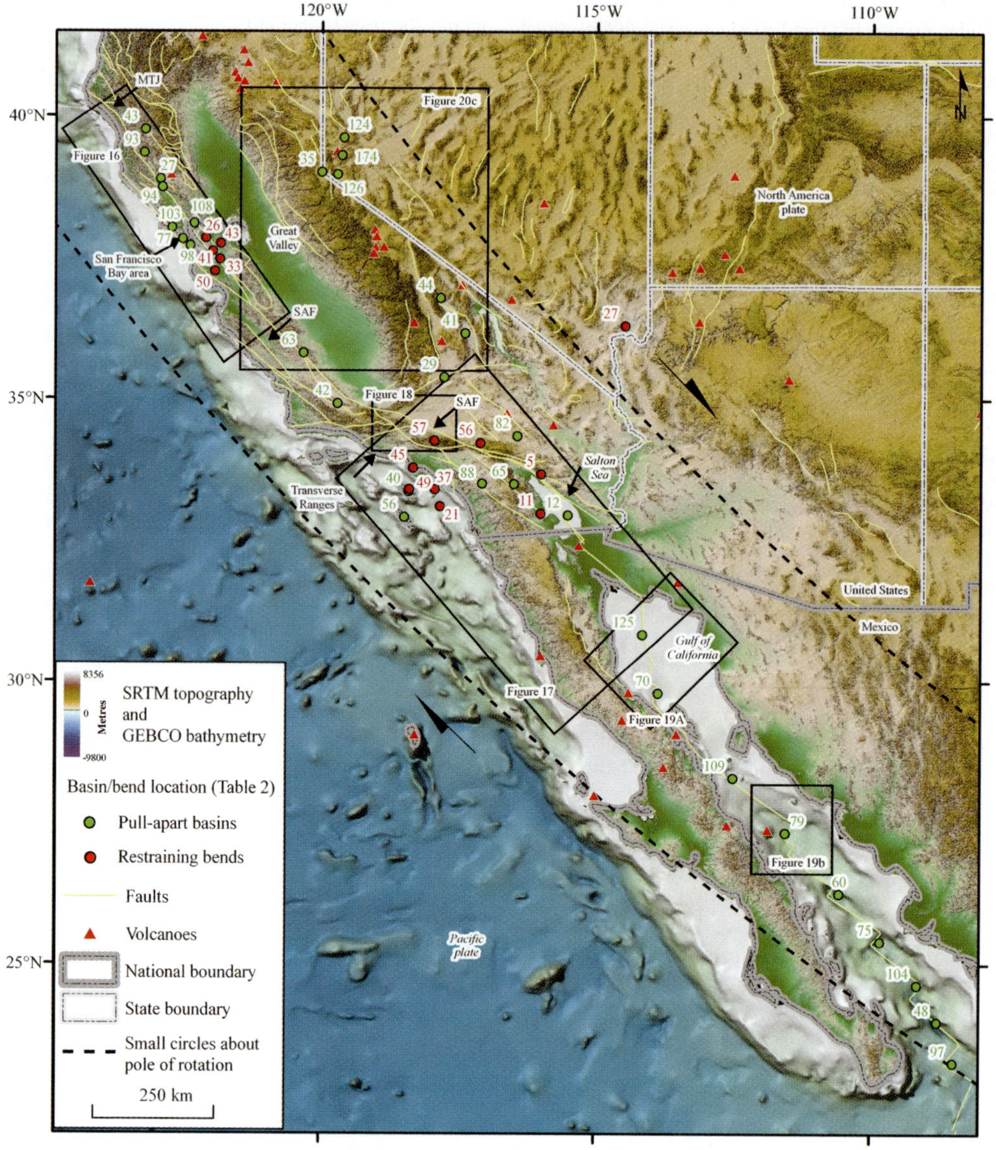

Fig. 15. Regional tectonic map of the San Andreas strike-slip fault system in California and the Gulf of California, using SRTM topography and GEBCO bathymetry as a base map. Yellow lines are active faults taken from the compilation by Levine (1995). Black lines represent small circles of rotation, about the pole of rotation, proposed by Argus & Gordon (2001). Boxes indicate areas of more detailed maps in Figures 16–20.

its palaeogeography, the reader is referred to Page (1990) and Powell & Weldon (1992).

The pole of North America–Pacific plate motion is distant and defines small circles roughly parallel to the concave-eastward traces of the San Andreas fault system (Argus & Gordon 2001; Fig. 15). The traces of the San Andreas fault system are smooth and anastomosing, and can be followed as continuous features for hundreds of kilometres. Analysis of GPS and plate kinematic data indicates that the overall character of the San Andreas Fault is slightly transpressional in character, with the fault strands accommodating about 3 mm/a of orthogonal shortening and about 37 mm/a of right-lateral strike-slip faulting (Wesnousky 2005). This slight transpressional character of the fault can account for fault-parallel faults and folds that parallel straight traces of the fault in central and northern California that are not characterized by restraining bends (Mount & Suppe 1987). Most of the 300–450 km of total right-lateral strike-slip displacement has occurred on the main San Andreas fault zone; parallel zones of strike-slip faults have offsets ranging from a few kilometres to 40 km (Powell & Weldon 1992; Wesnousky 2005).

The San Andreas fault zone can be more precisely described as the boundary separating the Pacific plate and the Sierran microplate, the latter of which includes the Great Valley and Sierra Nevada of eastern California and moves about 12 mm/a relative to the North American plate. This motion is occurring mainly in a zone of transtensional, right-lateral deformation separating the eastern edge of the Sierra microplate from the North American plate in the Great Basin (Unruh *et al.* 2003; Wesnousky 2005).

Bends on the San Andreas Fault. Three tectonic settings for the formation of San Andreas fault bends compiled on Table 2 can be defined:

1. a region in northern California that extends from north of the Transverse Ranges to Cape Mendocino where the traces of the strike-slip system are close to parallel or slightly convergent (<5°) with respect to the small circles of rotation;
2. a region of southern California where the traces of the strike-slip fault system are more convergent (5–15°) with respect to the small circles of rotation; this region contains major restraining bends at the topographically elevated Transverse Ranges and San Bernardino Mountains; and
3. a region of southernmost California, northern Mexico and the Gulf of California where the traces of the strike-slip fault system are divergent (5–10°) with respect to the small circles of rotation; this region of the broad, low-lying Imperial Valley of California and northern Mexico and the Gulf of California is characterized by *en échelon* pull-apart basins that lack associated restraining bends (Fig. 15).

Strike-slip bends of the slip-parallel region of northern California

Regional fault pattern of bends. The dominant bend structure in northern California is the lazy-S-shaped, Santa Cruz Mountains gentle restraining bend defined by a left-step (fault separation of 20 km) in the San Andreas fault zone south of San Francisco Bay (Fig. 16). Burgmann *et al.* (2006) have used GPS and the interferometric synthetic aperture radar (INSAR) method in the San Francisco Bay area to isolate vertical tectonic uplift rates at sub-mm per year precision. This survey indicates tectonic uplift at about 1 mm/a over a broad area of the Santa Cruz restraining bend (Fig. 16).

North of the bend, the San Andreas Fault follows the coastline and is marked by a prominent but narrow (*c.* 2–3 km) fault valley that is occupied by several isolated releasing rhomboidal bends at the Golden Gate and in Olema Bay with fault separations less than 2 km and poorly known transverse normal faults (Zoback *et al.* 1999; Wakabayashi *et al.* 2004, 2007). Parsons *et al.* (2005) propose a more complex fault-wedge basin model (Biddle & Christie-Blick 1985) for the simple pull-apart model shown by Zoback *et al.* (1999) and Wakabayashi *et al.* (2004, 2007).

This strand of the San Andreas Fault ruptured for several metres over a minimum onshore distance of 220 km in 1906, from the central part of the gentle Santa Cruz bend to the offshore region of the fault in northern California, to produce an *M*7 earthquake. This strand accommodates 40 mm/a of Pacific–North America motion, or about 90% of the total motion between the Pacific plate and the Sierran/North America plate (Argus & Gordon 2001).

As the fault passes seaward at Point Arena and extends northward to the most recently developed part of the fault near Punta Gorda and beyond to the Mendocino triple junction, a significant deviation occurs in the trace of the San Andreas Fault as the fault bends to the NE and defines a fault stepover distance of 50 km. This part of the fault is characterized by an unusually large component of vertical slip (0.5 cm/a, NE side up) (Brown 1991). Merritts & Bull (1989) have attributed this enhanced vertical slip to the recent passage of the Mendocino triple junction, but an alternative explanation is that part of the fault forms a gentle releasing bend that is largely located on fault splays offshore (McCulloch 1987) and defines the

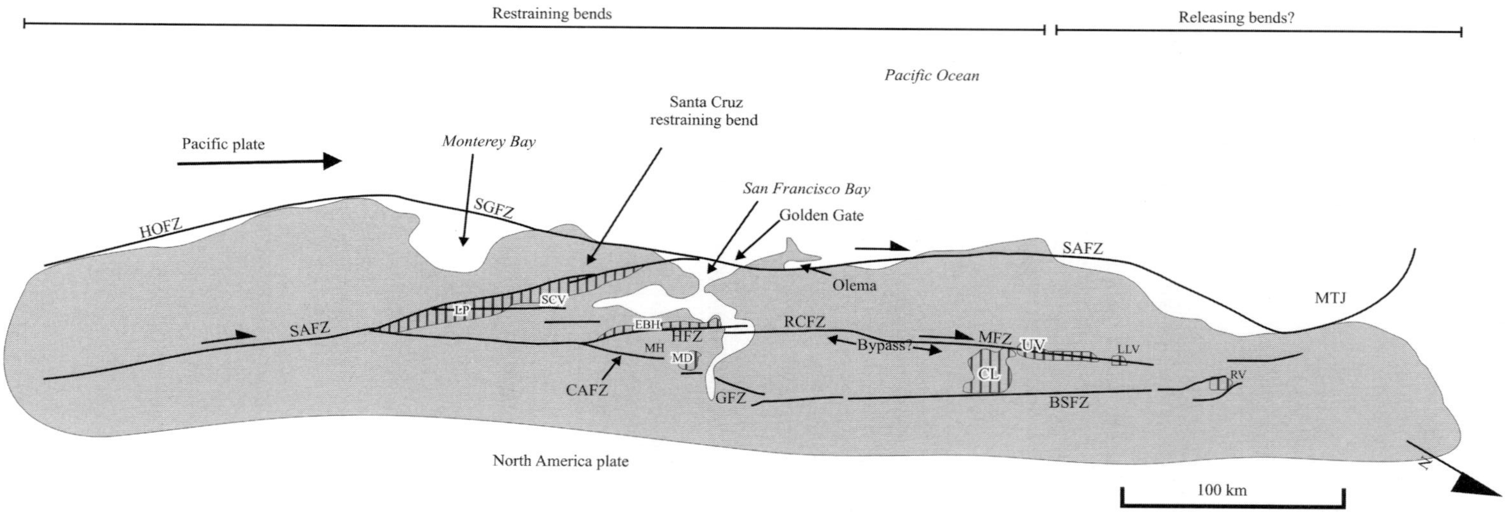

Fig. 16. Schematic map of the active traces of the San Andreas fault system in northern California, modified from Wesnousky (2005), McLaughlin & Nilsen (1982), Burgmann *et al.* (2006) and Aydin & Page (1984), and their relation to North America–Sierra Nevada–Great Valley microplate–Pacific plate motions, from Argus & Gordon (2001). See the regional tectonic map in Figure 15 for the location. An inferred paired of restraining and releasing bends is indicated on the San Andreas fault zone, along with a possible bypass fault system across the base of the paired bend (Calaveras–Hayward–Rogers Creek–Maacama fault zones). Note how the coastline of northern California mirrors the coastal and offshore paired-bend structure. HOFZ, Hosgri fault zone; SGFZ, San Gregorio fault zone; SAFZ, San Andreas fault zone; HFZ, Hayward fault zone; CAFZ, Calaveras fault zone; GFZ, Green Valley fault zone; RCFZ, Rodgers Creek fault zone; BSFZ, Bartlett Springs fault zone; LP, Loma Prieta; SCV, Santa Clara Valley; EBH, East Bay Hills; MH, Mission Hills; MD, Mount Diablo; CL, Clear Lake; MFZ, Maacama fault zone; UV, Ukiah Valley; LLV, Little Lake Valley; RV, Round Valley; MTJ, Mendocino triple junction.

coastline where Holocene uplift rates of 0.5 cm/a have been measured (Brown 1991; Fig. 16). I propose that this bend may represent the releasing-bend (cf. Fig. 1d) component of the Santa Cruz restraining bend to the south, since both share approximately the same amount of fault separation (Fig. 16).

East of the San Andreas Fault is a slightly segmented series of faults that diverges from the San Andreas at the Calaveras fault zone near the SW end of the Santa Cruz gentle restraining bend (Fig. 16). This second strand of the fault system, which is less continuous than the San Andreas strand 75 km to the west includes the Calaveras, Hayward, Rodgers Creek and Maacama faults (McLaughlin & Nilsen 1982; Wakabayashi 2007). In northern California, the northern strands of both this fault and the parallel San Andreas Fault represent the youngest part of the system that has propagated northward in the wake of the Mendocino triple junction (Henstock *et al.* 1997; Wakabayashi *et al.* 2004, 2007; Fig. 16). A third parallel strand representing the easternmost active strands of the San Andreas fault zone is even more discontinuous, and consists of the Green Valley and Bartlett Springs fault zone (McLaughlin & Nilsen 1982; Fig. 16).

Argus & Gordon (2001) note that the Calaveras–Maacama trend and the parallel San Andreas trend are very close to parallel to the small circle of motion between the Pacific and Sierran microplates (Fig. 15). An interesting regional structural correlation is that restraining bends are mainly confined to the southern part of the strike-slip system in a 120-km-long zone adjacent to the large Santa Cruz gentle restraining bend (Fig. 16). These much smaller bends include the Mission Hills gentle restraining bend connecting the Hayward and Calaveras faults (Manaker *et al.* 2005); the Mount Diablo gentle restraining bend between the Concord and Greenville fault zones (Wakabayashi 2007); and the East Bay Hills gentle restraining bend along strands of the Hayward fault zone (Aydin & Page 1984). Burgmann *et al.* (2006) has identified recent topographic uplifts with all of these bends using the INSAR method.

The northern part of this strike-slip system from San Pablo Bay (the easternmost extension of San Francisco Bay) northward to Mendocino appears to be transtensional and characterized by a lack of restraining bends and the presence of either pull-apart basins or fault wedge basins (McLaughlin & Nilsen 1982; Fig. 16). This inland region contains small pull-aparts or fault wedge basins ranging in fault separation from 1.5 to 2 km, and are either occupied by lakes or shallow-marine areas or have been mapped as outcrop areas of Late Neogene continental sedimentation (McLaughlin & Nilsen 1982; Wakabayashi *et al.* 2004, 2007).

These basins occurring at *en échelon* stepovers or aligned along the fault traces or at complex fault traces include:

1. the rhomboidal San Pablo pull-apart underlying San Pablo Bay (Parsons *et al.* 2003);
2. the Little Sulfur Creek, Ukiah Valley, and Little Lake Valley basins, pull-aparts or fault wedge basins aligned along the Maacama fault zone; and
3. the Clear Lake, Potter Valley, and Round Valley basins along the Lake Mountain, Bartlett Springs and Round Valley fault zones (McLaughlin & Nilsen 1982; Fig. 16).

Consistent with its transtensional setting, the Clear Lake Basin is also characterized by volcanic and geothermal activity, beginning at about 0.6 Ma, and subsidence rates of 1.7 mm/a for the past 0.6 Ma (Hearn *et al.* 1988).

Paired-bend interpretation. A simple interpretation of the regional fault pattern in Figure 16 is that there are paired bends on the San Andreas fault strand to the west, the Calaveras–Hayward–Rodgers Creek–Maacama fault zone in the centre and the Concord–Greenville–Lake Mountain–Bartlett Springs and Round Valley fault zones to the east. All three faults exhibit a convex westward shape that is most pronounced in the case of the San Andreas fault and is a result of the Santa Cruz gentle restraining bend and perhaps its paired bend located off the coast of northern California. Given the direction of interplate slip, the northern part of this arcuate belt of three parallel strike-slip faults north of San Pablo Bay is characterized by pull-apart or fault wedge basins along their traces, along with volcanic and geothermal activity centred on those basins.

The largest estimate of right-lateral displacement on the eastern strands of the fault system is 24 km on the Calaveras fault zone in contrast to the San Andreas strand that has a right-lateral offset on basement terranes of 450 km (Wesnousky 2005). The low offsets on the eastern strands are consistent with the idea that these faults initiated fairly recently – about 8 Ma according to McLaughlin & Nilsen (1982) – and have propagated northward in the wake of the Mendocino triple junction since that time (Henstock *et al.* 1997). Geological and fission-track studies of the Santa Cruz Mountains bend indicate that its earliest uplift occurred about 12 Ma (Middle Miocene), and since that time about 2 km of exhumation has occurred at the bend (Anderson 1990; Burgmann *et al.* 1994).

In the framework of paired bends, the eastern fault strands may represent a 'bypass fault' that is reducing the paired-bend asperity, or 'sidewall ripout' (Swanson 2005, Fig. 12b), developed along the San Andreas fault zone over at least the past 12 Ma and given the large offset, perhaps much longer. Because the inferred eastern bypass fault system is straighter and parallel along its length to the small circles of plate rotation, its activation would increase the efficiency of the strike-slip system as described by Wesnousky (1988). It is puzzling why the two parallel strands of the San Andreas fault system inherit the same westward convexity that in turn localizes aligned zones of restraining bends to the south of San Francisco Bay on the San Andreas fault zone and to the east of the bay on the Hayward fault zone (Fig. 16). Two possibilities are that these trends are controlled by inherited basement structures (Fig. 12c), or that they reflect the sidewall ripout process proposed by Swanson (2005; Fig. 12b). It seems unlikely and coincidental that two parallel strike-slip fault strands would align themselves by linkage of P- and R-shears as shown in Figure 12a.

Strike-slip bends of the convergent region of southern California

Regional fault pattern of bends along the Big Bend of the San Andreas fault zone and northern Gulf of California. The dominant bend structure in southern California is the lazy-S-shaped, Transverse Ranges gentle restraining bend, defined by a left-step (fault separation of 20 km) in the San Andreas fault zone north and east of Los Angeles (Fig. 15). In this region, the small circles of plate rotation are oblique by 15–20° to the fault traces and thereby provide a tectonic setting for the formation of the largest restraining-bend complex and high topography (3502 m in the San Bernardino Mountains) in California (Argus & Gordon 2001; Fig. 17a). As the San Andreas fault trace approaches the Salton Sea, it curves, straightens, and assumes a more slip-parallel orientation, and it is characterized by structures that are either slip-parallel or slightly transpressional (Bilham & Williams 1985; Fig. 17a). Because of excellent outcrops in a dry desert climate, this segment of the fault is the setting for the classic outcrop-based mapping study of strike-slip transpressional deformation by Sylvester & Smith (1976) in the Mecca Hills north of the Salton Sea (Fig 17b).

These areas of localized fault transpressional are best characterized as gentle bends connected by zones of pure slip faulting (Bilham & Williams 1985; Fig. 17b). The kinematic and morphological properties of each segment shown in Figure 17b depend entirely on the fault strike, despite very subtle changes in fault strike that are as small as 3°. Note that these fault strike changes are so subtle that the distinction between a gentle fault bend and a linear transpressional uplift (Fig. 1c) may not be clear.

At a point west of the Salton Sea, the San Andreas Fault abruptly curves to the SE and assumes a SE trend oblique in the extensional sense to the small circles of plate rotation (Fig. 17a). Along this segment of the fault that extends from the Salton Sea to the mouth of the Gulf of California, the strike-slip system is transtensional and therefore provides the tectonic setting for the formation of an *en échelon* array of 11 active pull-apart basins exhibiting the complete range of developmental morphologies that are summarized on Figures 1b & 7 (Mann *et al.* 1983).

Pull-apart basin morphologies include: spindle-shaped types (Upper Wagner Basin on Fig. 18a; Henyey & Bischoff 1973); rhomboidal types (Delfin Basin in Fig. 18a; Henyey & Bischoff 1973; Persaud *et al.* 2003) and rhomboidal type with oceanic spreading centres (Guaymas Basin on Figure 18b; Bischoff & Henyey 1974). The spindle-shaped basins have small fault overlaps and incipient coalescence with adjacent basins, reflecting their recent nucleation in the sub-sea-level areas SE of the Salton Sea area near the top of the convex eastward arc in the San Andreas Fault (Fig. 17). Larger, more isolated rhomboidal basins indicative of older basins formed by larger strike-slip offsets are found to the SE in the submarine Gulf of California (Fig. 15).

Transtensional structures and sub-sea-level topography within the Imperial Valley and Gulf of California reflect both the presence of pull-apart basins along the trend of the San Andreas Fault and also the activity of large normal faults bounding the walls of the valley. Umhoefer & Stone (1996) have pointed out that some of the transtensional motion is accommodated or 'partitioned' on the valley-parallel faults which form large escarpments bounding both the Imperial Valley and Gulf of California, similar to that discussed by Garfunkel (1981) in the Dead Sea, which is shown schematically in Figure 9c.

Regional pattern of bends in the California borderlands. The pattern of bends in the offshore California borderlands mirrors the pattern of bends along the onland San Andreas fault zone (Legg *et al.* 2007; Fig. 17). Most restraining bends are concentrated in the NW area of the California borderlands near the city of Los Angeles and the Transverse Ranges, whilst most releasing bends and straight strike-slip fault segments are concentrated in the SE areas near the city of San Diego

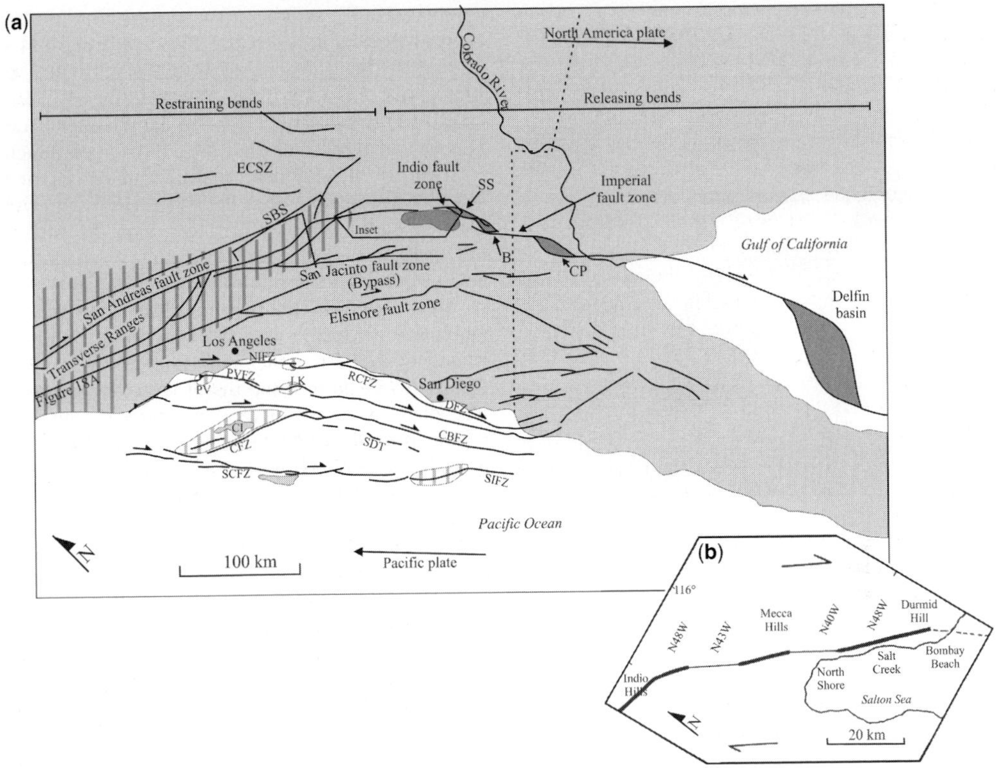

Fig. 17. (**a**) Schematic map of the active traces of the San Andreas fault system in southern California, from Wesnousky (2005), Persaud *et al.* (2003) and Dixon *et al.* (2002) (see the regional map in Fig. 15 for location). Schematic map of the active traces of the San Andreas fault system in the California borderlands and coastal southern California, modified from Legg *et al.* (2004), and their relation to plate motions, from Argus & Gordon (2001). Inferred paired restraining- and releasing bends are indicated for the three fault systems in the NE part of the map. Note how the coastline of southern California mirrors the paired-bend structure of the coastal and offshore fault systems. The San Clemente–San Isidro fault system in the SW part of the map and furthest offshore does not display the paired-bend pattern. (**b**) Map of the trace of the San Andreas fault zone to the NW of the Salton Sea, showing the close relationship between the strike of the fault and localized restraining bends (heavy lines) characterized by topographic uplift (modified from Bilham & Williams 1985).

and the Mexican border (Fig. 17). The strike-slip faults define a broad arc with restraining bends in the NW area, including the Palos Verdes, Lasuen Knolls and Santa Catalina Island gentle restraining bends (Ward 1994; Mann & Gordon 1996; Fisher *et al.* 2004; Legg *et al.* 2007). Pull-aparts are relatively rare and small in size in the California borderlands relative to restraining bends (Legg *et al.* 2007).

Paired-bend interpretation. The arcuate and sub-parallel strike-slip faults in southern California exhibit a convex eastward shape that is most pronounced for the Transverse Ranges restraining bend and its paired releasing bend, starting in the Salton trough area and extending into the northern Gulf of California (Fig. 17). The convex NE curvature of faults in the California borderlands is less than that along the San Andreas fault zone, but still results in the Santa Catalina Island restraining bend parallel to the Transverse Ranges bend (Fig. 17). Southward-curving strike-slip fault trends produce transtensional pull-apart that are most pronounced along the San Andreas fault zone in the Salton Trough and northern Gulf of Mexico. As in the case of the bends of northern California, it is puzzling why parallel sets of strike-slip faults along the San Andreas and California Borderlands trends exhibit the same sense of curvature across such a wide area of the plate-boundary zone. Two possibilities are that these trends are reactivating inherited structures in the basement (Fig. 12c), or alternatively that they reflect the sidewall ripout process proposed by Swanson (2005; Fig. 12b).

(a)

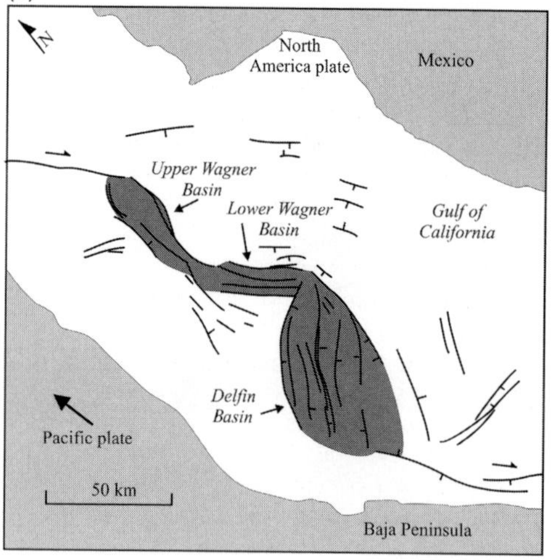

(b)

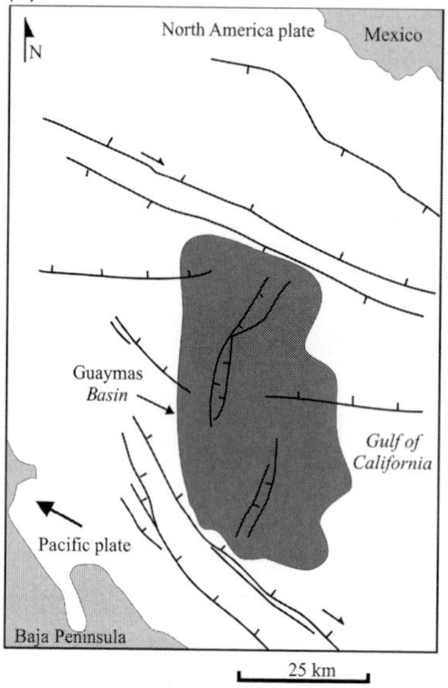

Fig. 18. (**a**) Schematic map of the Upper Wagner, Lower Wagner and Delfin pull-apart basins in the northern Gulf of California, modified from Henyey & Bischoff (1973) (see the regional map in Figure 15 for the location). The direction of plate motion is from Argus & Gordon (2001). (**b**) Schematic map of the Guaymas pull-apart basin in the central Gulf of California, modified from Bischoff & Henyey (1974).

Development of the Transverse Ranges restraining bend. Powell & Weldon (1992) proposed that the earliest Transverse Ranges bend was active from the Early to Mid-Miocene (18–13 Ma), and that this early bend accumulated up to 110 km of right-lateral displacement. This early phase of bend formation led to the cessation of slip on the San Andreas Fault, and transfer of slip to the San Gabriel fault zone from the Early Miocene (14–12 Ma) to Early Pliocene (4–5 Ma) (Crowell & Link 1982; Fig. 19a). Seismic profiling by May *et al.* (1993) has shown that the San Gabriel Fault is a low-angle normal fault (Fig. 19b) upon which 13.5 km of Mio-Pliocene marine and non-marine sedimentary fill has been deposited (Fig. 19c). As the San Gabriel Fault shifts to a more SE strike, the basin disappears and the fault localizes the uplift in the San Gabriel Mountains (Fig. 19a).

Swanson (2005) has interpreted this fault pattern and kinematic history as a sidewall ripout: an Early Miocene shift from the San Andreas to the San Gabriel Fault defined the ripout with its characteristic plan-view fault pattern. When the ripout was active from the Early Miocene to Early Pliocene, a basin formed behind the ripout as a releasing bend at the trailing edge of the block, and the San Gabriel mountain range formed at the leading edge. As motion shifted back to the San Andreas Fault, the basin and restraining bend were abandoned to allow a straighter fault trace (Fig. 19a).

The latest phase of Transverse Ranges uplift occurred as motion was re-established on the San Andreas fault trace in Late Pliocene to Recent times, and is recorded by Late Pliocene and Quaternary folds, thrust faults and reverse faults (Namson & Davis 1988). Although an earlier late Mid-Eocene to Late Oligocene folding and uplift event affected the Transverse Ranges, the latest event is responsible for most of the present structural relief and topography in the area. In a study of fission-track ages across the central Transverse Ranges, Blythe *et al.* (2000) found that the most recent phase of uplift in the San Gabriel Mountains thought to be linked to its restraining-bend history started at about 7 Ma, with an accelerated phase beginning at 3 Ma. The topographically higher San Bernardino Mountains to the east have experienced less dissection than the San Gabriel mountains, suggesting that the most recent phase of cooling and bedrock uplift began about 3 Ma ago. Spotila *et al.* (2001) used apatite ages to constrain the data indicating that a small fault sliver in the San Bernardino restraining bend was exhumed 3–6 km at rates of 5 mm/a over the past 1.8 Ma.

Rhomboidal pull-aparts like the Wagner, Delfin, and Guaymas basins in the northern and central Gulf of California formed at about 3 Ma ago (Henyey & Bischoff 1973; Bischoff & Henyey 1974; Boehm 1984) and are therefore roughly contemporaneous with the main phase of uplift and shortening on the Transverse Ranges gentle restraining-bend uplift (Fig. 17). The younging trend in pull-apart basin opening from the northern Gulf of California to the Salton Trough is also supported by the morphology of the spindle-shaped basins beneath the

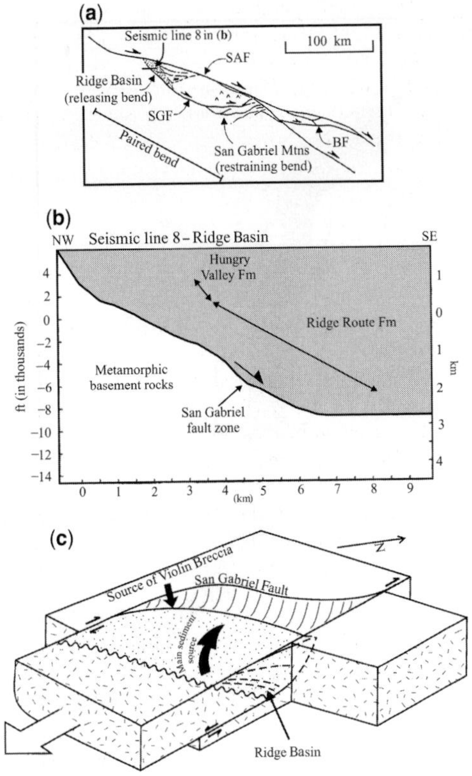

Fig. 19. (**a**) Schematic map of the active traces of the San Andreas fault system in southern California, from Swanson (2005) (see the regional map in Figure 15 for the location). The location of the onland, multi-channel seismic line crossing the Ridge Basin and the bounding San Gabriel fault low-angle normal fault zone, is shown in (**b**) Prior to this geophysical information, the San Gabriel fault zone was interpreted as a high-angle strike-slip fault zone (Crowell 1974*a*, *b*; Crowell & Link 1982). (**b**) Half-graben structure formed by mainly low-angle, normal slip on the San Gabriel Fault during Late Miocene times (modified from May *et al.* 1993). Wedge-shaped sedimentary deposits of the Ridge Basin above the San Gabriel normal fault are 5 km thick and consist of alluvial-fan, lacustrine and shallow marine sedimentary rocks. (**c**) Block diagram illustrating the low-angle normal fault formed at a releasing bend along the Mio-Pliocene San Gabriel fault zone, from May *et al.* (1993). This releasing bend was inferred by Swanson (2005) to have formed a trailing extensional ramp system or releasing bend (i.e. Ridge Basin) adjacent to the San Gabriel leading thrust system or restraining bend (cf. map in (**1d**)). In the terminology of this paper, the adjacent bends would be called a paired bend, as shown in Figure 14a. The modern San Andreas fault zone bypassed this ancient paired bend about 4–5 Ma ago, and subsequent transpressional deformation has exhumed the Ridge Basin in its present-day outcrop area.

Salton Trough, which started to form less than 1 Ma ago (Fuis *et al.* 1982) suggested started to form less than 1 Ma ago. Therefore, the youngest area of restraining-bend deformation in the San Bernardino Mountains uplifted from 1.8 Ma ago (Spotila *et al.* 2001) is paired or adjacent to the youngest area of spindle-shaped pull-apart basin formation in the Salton Trough, active from about 1 Ma ago. Older areas of restraining-bend formation in the San Gabriel Mountains (Blythe *et al.* 2000) are paired with older areas of rhomboidal pull-apart opening in the Gulf of California (Persaud *et al.* 2003; Fig. 17).

One structural explanation to explain the symmetry and synchronous development of the paired bend is that the excess volume of deformed crust 'piling up' at the Transverse Ranges must be balanced by a volume deficit by thinning the crust in the adjacent, along-strike area of the Salton Sea pull-aparts (Fig. 17).Since mass balance of crust is preserved, the area of thinned crust is equivalent to the thickening in the restraining bend, and therefore the approximate shapes and master-fault separations would remain similar.

Strike-slip fault bends in the Walker Lane transtensional zone

A 100-km-wide diffuse transtensional deformation along Walker Lane fault zone and Eastern California shear zone occurs at the eastern edge of the Sierra microplate, and has accommodated a maximum of 150 km of right-lateral displacement since its inception (Oldow 2003; Wesnousky 2005; Fig. 20a). Unruh *et al.* (2003) and Wesnousky (2005) have both noted that the fault trends obliquely with respect to the small circles of rotation between the Sierran microplate and North America, and that this misalignment has produced a highly transtensional fault that accommodates about 11 mm/a of right-lateral shear (Fig. 20a).

Because of this misalignment, Unruh *et al.* (2003) view the 500-km-long central segment of the Walker Lane fault zone to be a releasing stepover composed of many individual rift basins and transtensional faults (Fig. 20b). Discrete pull-aparts within this larger pull-apart may include several small and steep-sided basins, including the classic Z-shaped Death Valley pull-apart basin (Burchfiel & Stewart 1966; Serpa *et al.* 1988; Cowan *et al.* 2003); Lake Tahoe Basin (Unruh *et al.* 2003); the Coso Basin and geothermal field (Lees 2002); and the Walker Lake pull-apart basin (Link *et al.* 1985). Wesnousky (2005) notes the presence of a 100-km-wide, rhomboidal stepover in the central Walker Lane that is deforming by rotation of three rectangular blocks bounded by antithetic

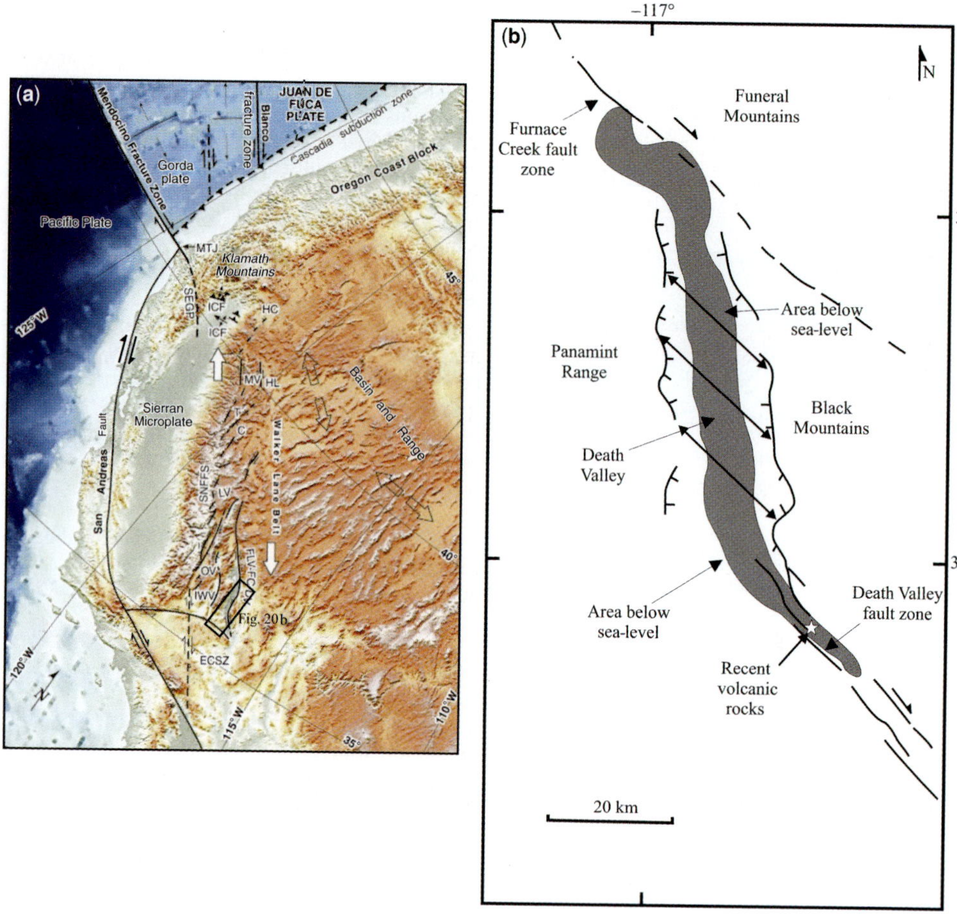

Fig. 20. (**a**) Oblique Mercator projection of the western Cordillera about the preferred Sierra Nevada–North America pole of rotation of Argus & Gordon (2001) (from Unruh *et al.* 2003). Direction of instantaneous Sierra Nevada–North America motion is vertical everywhere in the projection. Faults of the Walker Lane shear zone are transtensional. (**b**) Schematic map of the active, S-shaped Death Valley pull-apart basin along the Furnace Creek and Death Valley fault zones in southern California, modified from Christie-Blick & Biddle (1985). The grey area indicates part of the basin, from sea-level to a minimum elevation of 86 m below sea-level.

(left-lateral) strike-slip faults. He notes that the overall diffuse and transtensional character of Walker Lane may be analogous to an earlier stage in the structural development of the San Andreas fault system before sufficient slip accumulated to yield the now more concentrated San Andreas strike-slip fault zone.

Tectonic setting and geological development of bends along the Dead Sea strike-slip fault system

Tectonic setting. The Dead Sea fault system extends from its intersection with the East Anatolian strike-slip fault of SE Turkey in the north to the Gulf of Elat and the Red Sea oceanic spreading centre in the south (Fig. 21). Like the San Andreas Fault, the Dead Sea Fault ranks as one of the world's best-studied strike-slip faults (Quennell 1956, 1958; Freund *et al.* 1970; Daeron *et al.* 2005; Gomez *et al.* 2007). In this section, I will focus on the current state of knowledge of its major restraining and releasing bends, with particular emphasis on their tectonic setting relative to the small circles of rotation between the Africa and Arabia plates. More detailed descriptions of the Dead Sea fault, including its origin, offset history and palaeogeography, are given by Hempton (1987), Garfunkel *et al.* (1981) and Gomez (2007).

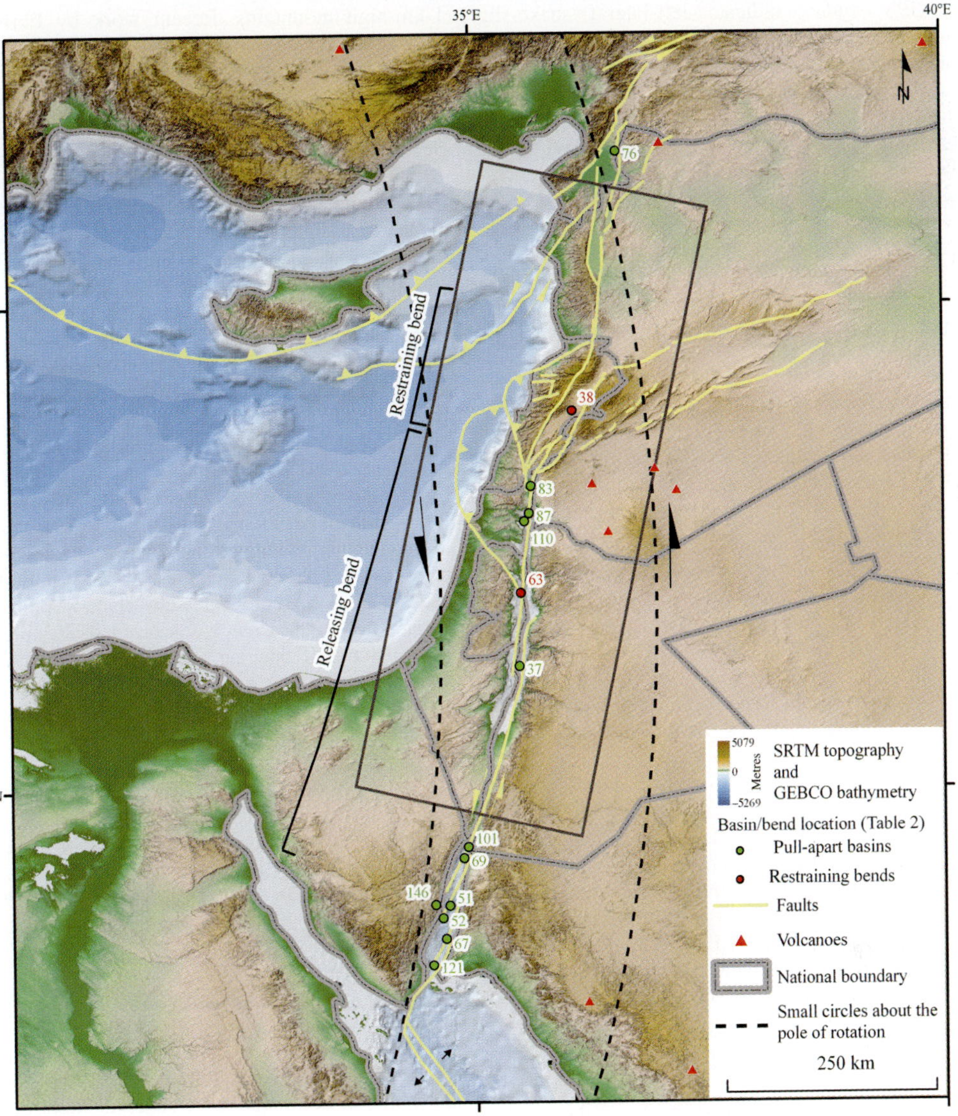

Fig. 21. Regional tectonic map of the Dead Sea strike-slip fault system in the Middle East, using SRTM topography and GEBCO bathymetry as a base map. Yellow lines are active faults, modified from Gomez *et al.* (2006, 2007). Black lines represent small circles of rotation about the pole of rotation proposed by McClusky *et al.* (2003) using GPS observations. An inferred paired restraining- and releasing bend is indicated.

Unlike the multi-branched San Andreas and Walker Lane fault systems, the Dead Sea Fault is remarkably linear and focused on a single major strand along much of its length. The total amount of left-lateral offset (110 km) is known from geological studies (Freund *et al.* 1970) and aeromagnetic mapping (Hatcher *et al.* 1981) because the fault cuts at a high angle across Precambrian and younger rocks. Geological studies indicate that offset occurred in two main phases. The first phase of *c.* 60 km of left-lateral displacement occurred during the Mid- and Late Miocene. The second phase began in the Early Pliocene and continues to the present-day. It may have been activated by an episode of 'ridge push' related to the onset of rifting and seafloor spreading in the Red Sea (Hempton 1987).

GPS studies indicate left-lateral strike-slip movement at rates of *c.* 5–8 mm/a with an increasing convergent component along the northern part of the fault (McClusky *et al.* 2003). Rates of motion also increase from 5.6 mm/a in the south to 7.5 mm/a along the northern segment of the fault (Fig. 21). These predicted, GPS-based small circles of rotation differ significantly from more fault-parallel small circles, based on a pole derived from plate modelling by Le Pichon and Francheteau (1978) and used by Garfunkel (1981) and Mann *et al.* (1983) in predicting zones of releasing and restraining bends along the fault.

Bends on the Dead Sea Fault. Gomez *et al.* (2006) has discussed the geological implications of the more precise GPS-based plate motion model on the observed bend structures along the fault. The Dead Sea fault system can be divided into three tectonic settings for restraining and releasing bends based on small circles:

1. a *c.* 400-km-long section of *en échelon* pull-apart basins from the Gulf of Elat through the Dead Sea pull-apart, Jordan River Valley and Sea of Galilee pull-apart; this section is roughly parallel to small circles in the Gulf of Elat, but becomes oblique and convergent by angles of 10–15° in the northern part of the segment;
2. a *c.* 200-km-long NE-striking, lazy-Z-shaped gentle restraining bend through the Mount Lebanon and Anti-Lebanon Ranges; this section is characterized by the maximum obliquity (*c.* 15–20°) of the small circles; and
3. a *c.* 250-km-long, north–south striking section in NW Syria and SE Turkey that contains one large, rhomboidal pull-apart basin (Brew *et al.* 2001).

Gomez *et al.* (2005) suggests that partitioning can explain the coexistence of pull-apart basins like the Ghab Basin within a predicted tectonic environment of tectonic transpression. A similar explanation was offered to explain the apparent existence of pull-apart basins on the transpressional Macquarie Ridge complex (Daczko *et al.* 2003).

Strike-slip bends of the southern Dead Sea fault system

The dominant bend structures in this slip-parallel southern segment of the fault are an array of three large rhomboidal pull-apart basins in the Gulf of Aqaba (Elat) (Ben-Avraham *et al.* 1979; Garfunkel & Ben-Avraham 1986; Ben-Avraham & Tibor 1993; Fig. 21). These basins underlie the 2-km-deep Gulf of Aqaba (Elat) that is flanked by 1-km-high mountains. Recent work by Ehrhardt *et al.* (2005) at the northern end of the northernmost basin show a gentle lazy-Z-type fault geometry rather than the more rectilinear fault geometry proposed by previous workers. To the north, the Dead Sea fault trace forms a linear scarp in the rift-like Arava Valley and intersects the Dead Sea rhomboidal pull-apart basin in the north.

With a 150 km length and an 8–10 km width, the Dead Sea Basin is the largest pull-apart on the Dead Sea fault system (ten Brink *et al.* 1999; Fig. 21). The basin started to form about 15 Ma or earlier, during the first phase of strike-slip motion, and reached about half its present length before the end of the Miocene (Garfunkel & Ben-Avraham 1996). Its sediment fill, including a massive salt layer, is about 5 km over half its length and reaches a maximum of about 10 km. According to mapping by Ben-Avraham *et al.* (1990) and Garfunkel & Ben-Avraham (1996), the pull-apart grew by becoming longer in the direction parallel to strike-slip motion, as predicted in the standard pull-apart model (Fig. 7). More recent subsurface mapping by Lazar *et al.* (2006) has emphasized basin opening by rotation and opening along a series of blocks underlying the basin. The lack of high-quality imaging of the extended basement beneath the Dead Sea pull-apart basin precludes a conclusive statement about the precise mechanism of basin opening.

North of the Dead Sea pull-apart, two other smaller but prominent pull-aparts include the Sea of Galilee, where motion is now thought to occur only on one master fault (Hurwitz *et al.* 2002) and the Hula pull-apart basin (Heimann & Ron 1987; Zilberman *et al.* 2000) that is adjacent to the Lebanese restraining bend.

Lebanese gentle restraining bend

The lazy-Z gentle restraining bend is cross-cut by the northern continuation of splayed faults of the Dead Sea fault system including a thrust splay extending along the Mediterranean coast of Lebanon (Daeron *et al.* 2005; Gomez *et al.* 2007) (Fig. 22b). Of these splays, only one fault, the Yammouneh, appears be through-going and interconnects the northern and southern Dead Sea fault zone. Gomez *et al.* (2006, 2007) have proposed that transpressive deformation is partitioned into 4.0 mm/a of strike-slip movement along the Yammouneh strike-slip fault, and 2.8 mm/a of orthogonal shortening on folds affecting the coastal area of Lebanon. Schattner *et al.* (2006) has mapped offshore convergent structures which are continuous with the onland shortening structures. Evidence for large-scale block rotations from

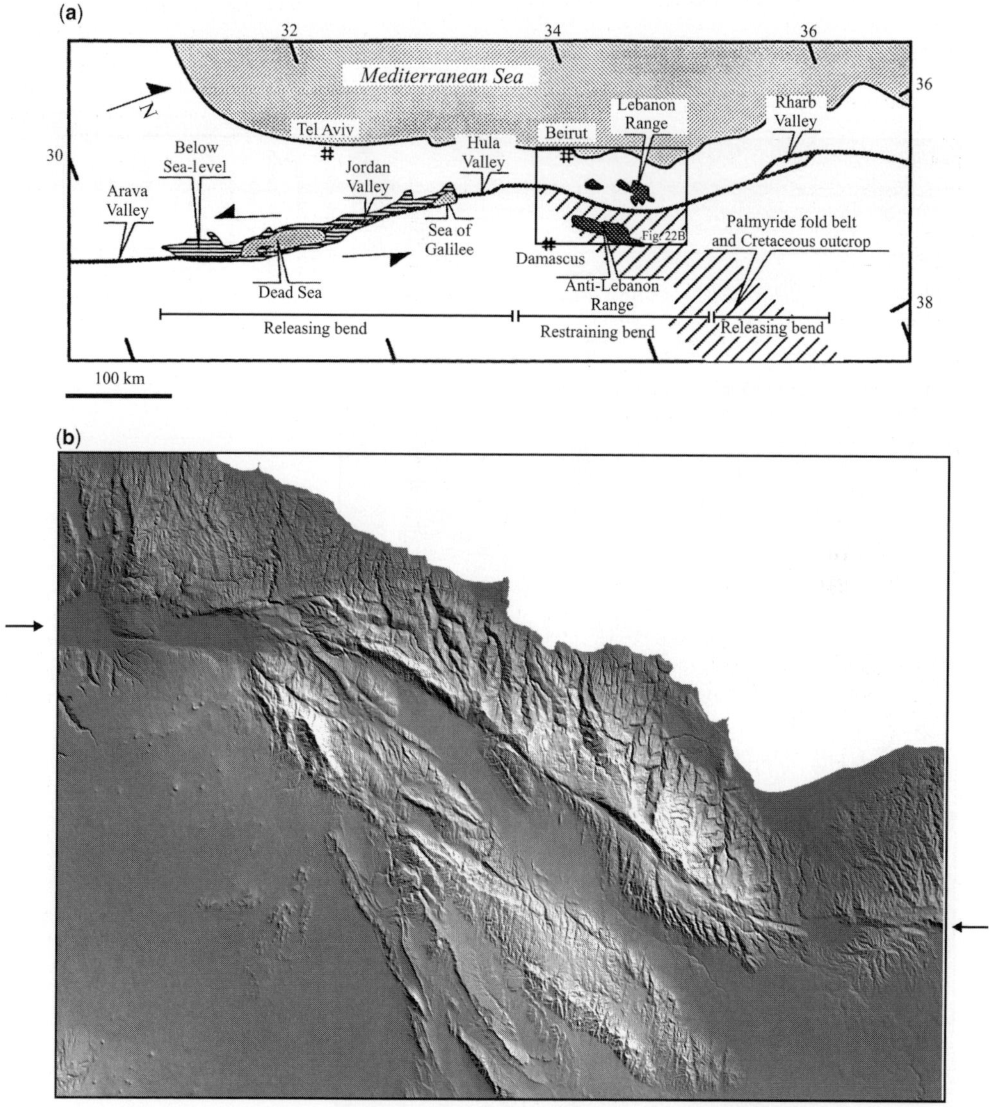

Fig. 22. (**a**) Schematic map of the active traces of the Dead Sea fault system in the Middle East, from a more detailed map in Figure 21. An inferred paired restraining- and releasing bend is indicated. Unlike the restraining bends of the San Andreas fault system, shown in Figures 16–20, the Lebanon restraining bend probably formed by inversion of an obliquely intersecting rift system of Cretaceous age (the present-day inverted Palmyride fold belt). See the regional tectonic map in Figure 21 for location. (**b**) SRTM topographic image of the Lebanon Range and Anti-Lebanon Range from Daeron (2005). Arrows indicate the main trace of the fault.

palaeomagnetic data is interpreted by Gomez *et al.* (2007) as an earlier phase of motion prior to the formation of a single throughgoing strike-slip fault zone in a manner similar to that described by Wesnousky for the Walker Lane fault zone in the western US.

Strike-slip bends of the northern Dead Sea fault system

The Ghab rhomboidal pull-apart basin is interpreted by Brew *et al.* (2001) as having formed during the second phase of Dead Sea fault zone evolution,

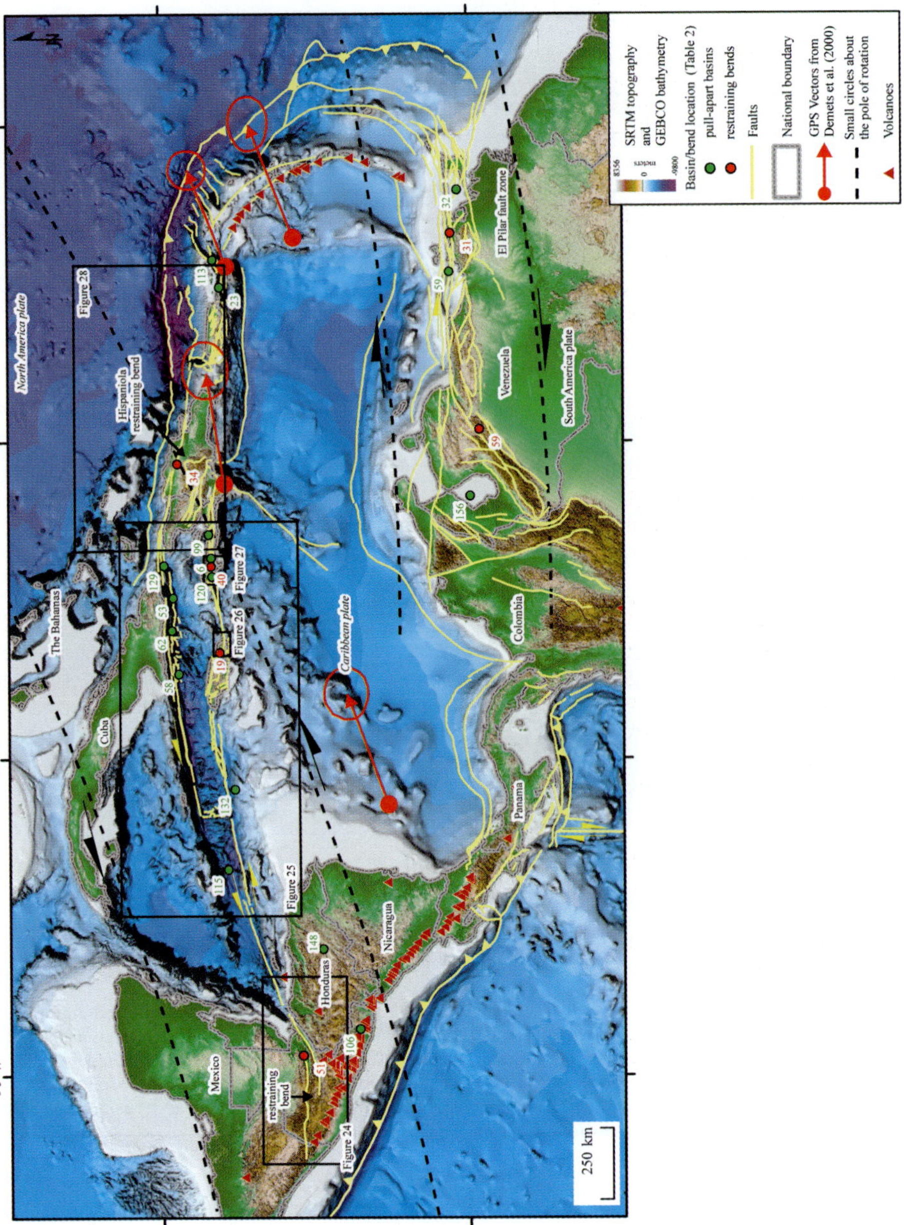

Fig. 23. For caption see p.77.

starting in the latest Miocene and earliest Pliocene. Some extension and subsidence has occurred on transverse normal faults, but most subsidence has occurred along the eastern master fault. Gomez *et al.* (2005) has pointed out that the GPS-based poles of rotation predict that the northern Dead Sea fault segment containing the Ghab pull-apart is highly transpressive, with the angle of intersection between the fault trace and the small circles being as much as 25°. This transpression can explain the uplifted margins of the northern Dead Sea fault that extend well beyond the extensional influence of the Ghab pull-apart basin.

Paired-bend interpretation. The geometry of the Lebanese gentle restraining bend with topographic elevations over 2 km and the Dead Sea–Jordan Valley with minimum elevations of −400 m along the Dead Sea fault system suggests that the two structures constitute a large paired bend (Fig 22a). As in the cases of the Californian paired bends, there is a similarity in the width and dimensions of the adjacent bends that may be explained by a mass balance between the excess crustal mass in the restraining bend with a thinned crustal mass in the Dead Sea pull-apart area.

Early paired-bend history. Seismic reflection and drill-hole data from Syria suggest that the Lebanese restraining bend formed at the site of a Permian–Triassic (or even older) failed rift of the Levantine margin that was inverted to form the Lebanon and Anti Lebanon Mountains and Palmyrides fold–thrust belt by Late Mesozoic and Cenozoic compression (McBride *et al.* 1990; Fig. 22a). Some of the inversion of the Palmyrides may be a far-field effect of closure or 'zippering' between Arabia and the Zagros collisional belt (Hempton 1987; Brew *et al.* 2001). The ancient rift trend subtends an angle of about 45° with the Dead Sea fault system. Initial formation of a paired bend is inferred to predate the formation of the Dead Sea–Jordan Valley releasing bend as a result of the fault re-adjusting to its bent configuration (Fig. 12c) (Mann *et al.* 2007*a*). Unfortunately, the initial age of the Lebanese restraining bend is not well known.

The timing of the Dead Sea pull-apart is well constrained. Ten Brink and Ben-Avraham (1989) interpret the localized occurrence of evaporites to indicate that the Dead Sea pull-apart formed in Middle to Late Pliocene times. Farther south, Ben-Avraham (1985) has inferred that the Gulf of Elat pull-aparts are also Pliocene in age (Fig. 21).

Tectonic setting and geological development of bends along the northern and southern Caribbean strike-slip fault systems

Caribbean strike-slip systems. The Caribbean plate is moving eastward relative to North and South America along two strike-slip fault systems along its northern and southern edges (Fig. 23). As the Caribbean plate moves eastward, the tips of these strike-slip faults also propagate eastward as the edge of the Lesser Antilles subduction zone. For this reason, these and tectonically similar faults in other parts of the world have been recently classified as *STEP faults*, or *subduction–transform edge propagators* (Govers & Wortel 2005).

The North America–Caribbean strike-slip system is a broad zone of deformation, extending 3200 km from northern Central America in the west to the Lesser Antilles Arc in the east. Most of the plate-boundary zone faults are submarine, with sub-aerial exposures of faults in northern Central America (Pflaker 1976; Fig. 24), in Jamaica and the southern peninsula of Hispaniola (Haiti) (Fig. 27) and in north-central Hispaniola (Fig. 28). The dominant structural element in the central plate boundary zone is the Cayman Trough, a submarine pull-apart basin formed by at least 1100 km of oceanic spreading at the Mid-Cayman spreading centre (Rosencrantz *et al.* 1988; Leroy *et al.* 2000; Fig. 25b). The spreading centre has been active since the Mid-Eocene, and is currently spreading at a rate of 15 mm/a (Rosencrantz *et al.* 1988).

The direction and rate of relative motion of the Caribbean plate has been quantified by GPS measurements (DeMets *et al.* 2000; Jansma *et al.* 2000; Mann *et al.* 2002), and these data can be used to predict the style of deformation along the length of both the northern and southern Caribbean

Fig. 23. Regional tectonic map of the northern and southern Caribbean strike-slip fault systems, using SRTM topography and GEBCO bathymetry as a base map. Yellow lines are active faults taken from compilations by Mann *et al.* (2002), Audemard & Serrano (2001) and Cowan *et al.* (2000). Black lines in the northern Caribbean represent small circles of rotation about the North America–Caribbean pole, based on GPS results from DeMets (2001). Dashed black lines in the southern Caribbean represent small circles of rotation about the South America–Caribbean pole, based on GPS results from Perez *et al.* (2001) and Weber *et al.* (2001).

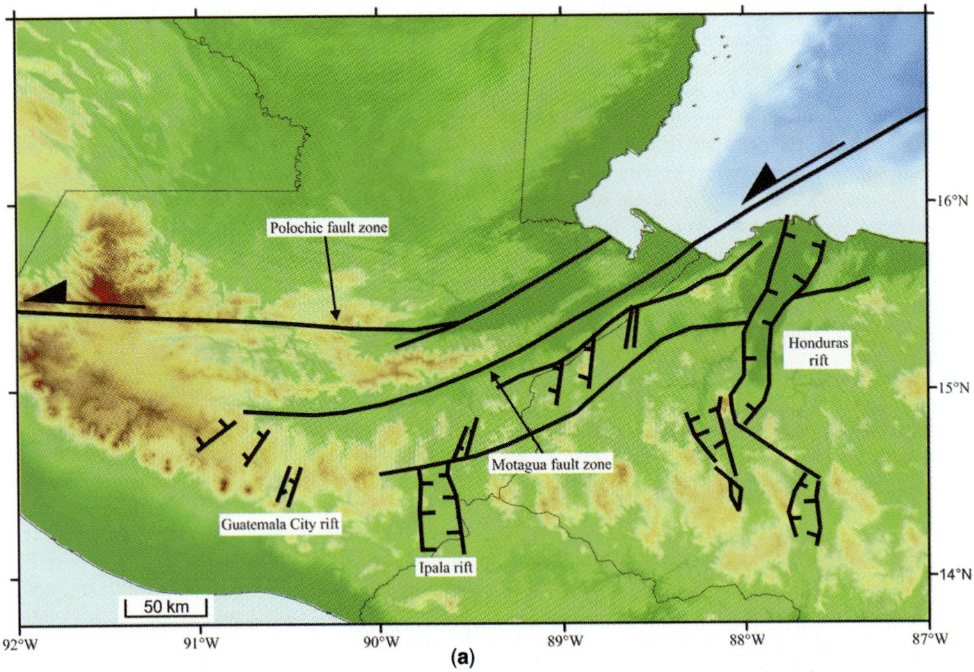

(a)

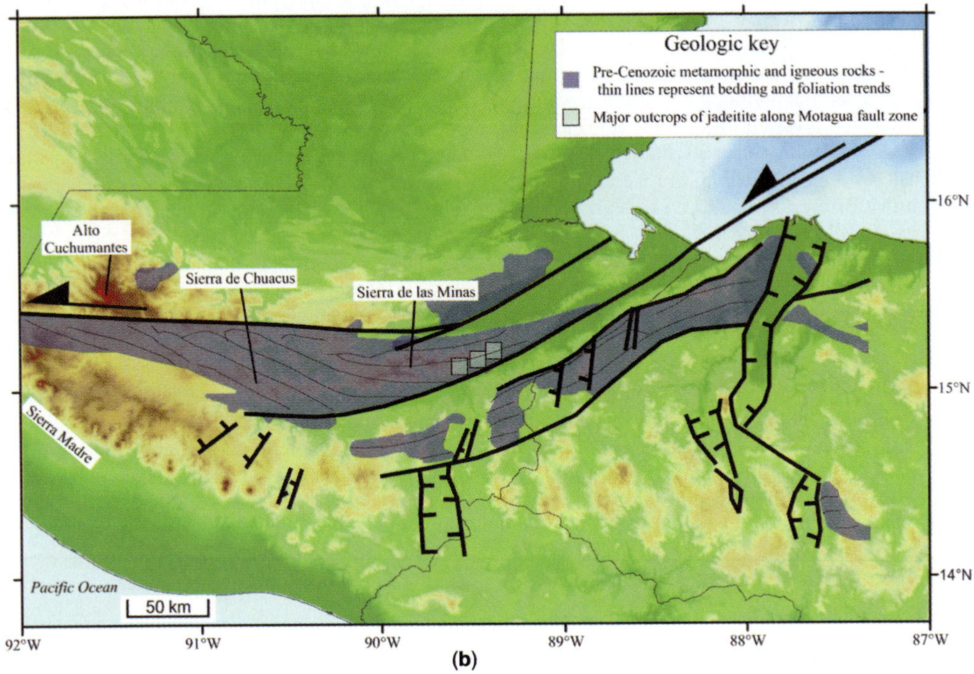

(b)

Fig. 24. (**a**) Topography of northern Central America and its relation to the left-lateral strike-slip faults separating the North America and Caribbean plates, modified from Mann & Gordon (1996). Note that the direction of relative plate motion of North America and the Caribbean is approximately east–west, and that the areas of highest and lowest topography are adjacent to the segment of the fault most oblique to that direction. The structure is interpreted as a paired bend – with a restraining bend in the mountainous area to the west, and a releasing bend in the deep, fault-controlled valleys and rifts to the east. (**b**) Map showing the ages of exposed rock units in northern Central America. Mann *et al.* (1996) proposed that the topographic highlands and exposures of deeper crustal rocks in the Sierra de las Minas, Sierra de Chuacus, and the eastern part of the Sierra Madré may be related to the combined effects of restraining bends in that area. Blueschist (jadeitite) outcrops are restricted to the area of the three squares shown in the Sierra de las Minas.

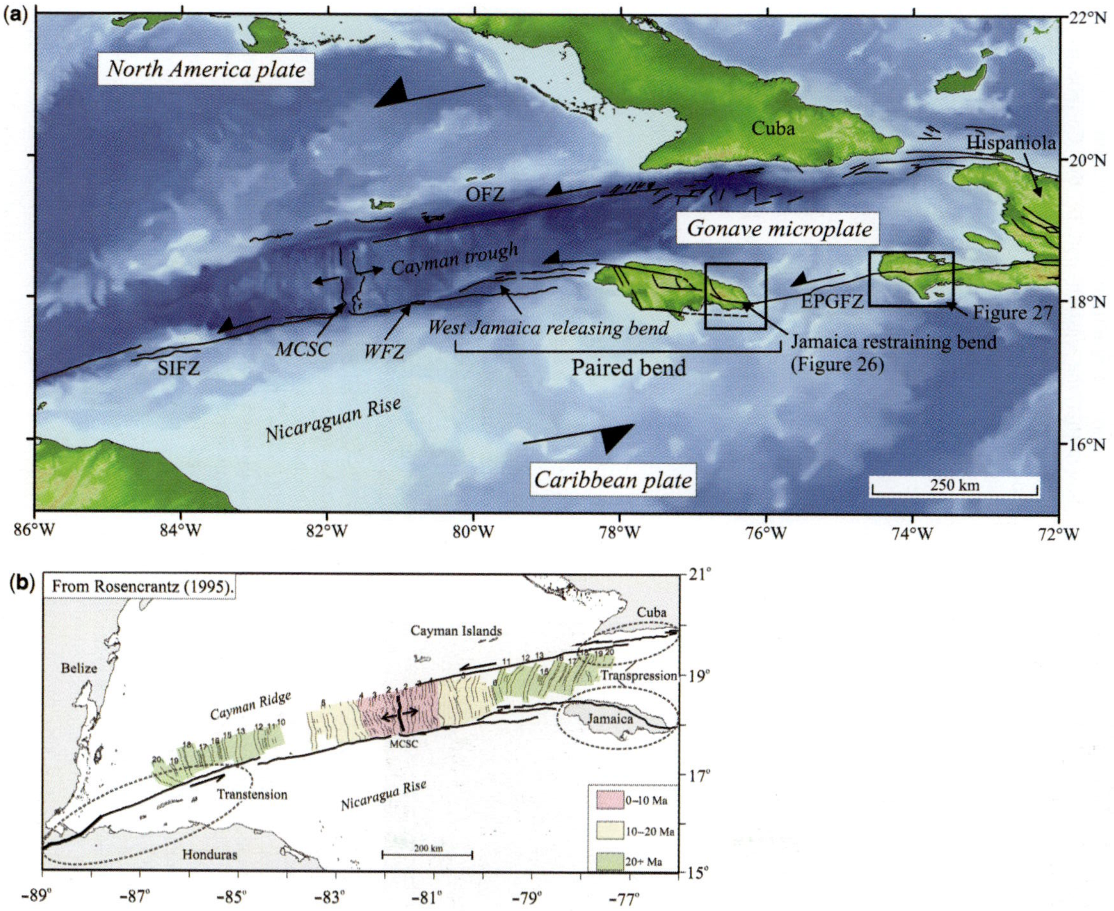

Fig. 25. (**a**) Regional tectonic map of the North America–Caribbean strike-slip fault system, using SRTM topography and GEBCO bathymetry as a base map. Black lines are active faults. An inferred paired restraining- and releasing bend is indicated. (**b**) Map of oceanic crust in the Cayman Trough, with black lines representing magnetic anomalies formed at the Mid-Cayman spreading centre (MCSC), as identified by Rosencrantz (1995). The oldest magnetic anomalies of Early Eocene age (*c.* 43 Ma) indicate that the Cayman Trough has accumulated about 1000 km of slip during its long life.

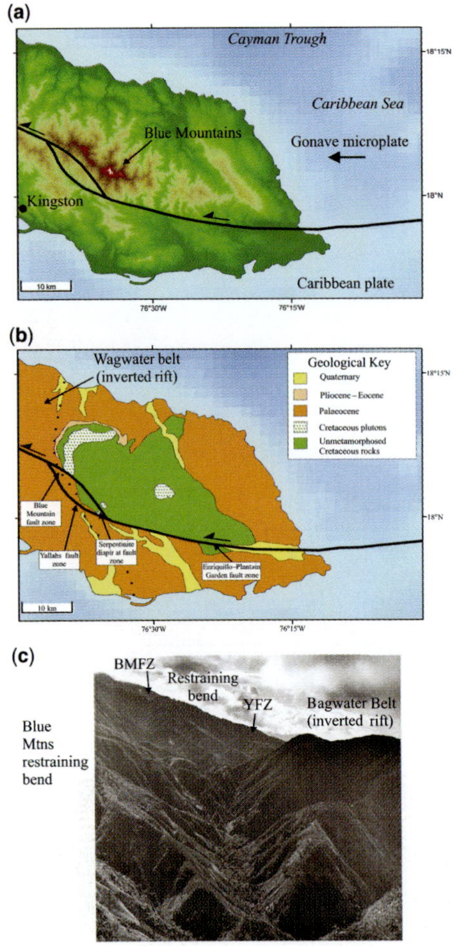

Fig. 26. (**a**) SRTM topography of eastern Jamaica, and its relation to active traces of the left-lateral strike-slip faults separating the Gonave microplate in the north (see Fig. 25a for location) from the Caribbean plate to the south. Note that the direction of North America–Caribbean relative motion is east–west, and that the area of higher topography is adjacent to the segment of the fault most oblique to this direction. (**b**) Map showing the ages of exposed rock units in the restraining-bend area. Note the presence of Cretaceous amphibolites, greenschists and blueschists along the upthrown side of the restraining bend, and fan deltas composed of Quaternary gravels that form a radial pattern around the restraining-bend uplift. The Palaeogene Wagwater rift (inverted during the Late Miocene to Recent) is inferred to be the basement structure responsible for the formation of the restraining bend. (**c**) Aerial view looking SE along the Blue Mountain (BMFZ) and Yallahs (YFZ) fault zones that form reverse faults bounding the SE edge of the Jamaica restraining bend and Blue Mountains. The Yallahs Fault originated as a Palaeogene, synsedimentary normal fault bounding the now-inverted Wagwater Rift.

strike-slip fault zones (Fig. 23). Small circles of rotation between the North America and Caribbean plates predict transtension along the western part of the plate boundary in northern Central America and the Cayman Trough; pure strike-slip near the Cayman spreading centre; and transpression in a belt extending from southern Cuba to Puerto Rico at rates of about 20 mm/a (Mann *et al.* 2003; Fig. 23).

In northern South America, GPS surveys have shown that the eastern part of the Caribbean–South America plate boundary is relatively simple, with small circles parallel to the trend of east-west strike-slip faults (Perez *et al.* 2001; Weber *et al.* 2001; Fig. 23). Rates are about 20 mm/a, and transpression is predicted only in the easternmost area near Trinidad (Weber *et al.* 2001). In this section I will focus on the current state of knowledge of its major restraining and releasing bends along the northern and southern boundaries within the tectonic framework established by these plate-motion studies. More detailed descriptions of the Caribbean plate and its strike-slip fault systems, including their origin, offset history and palaeogeography, are given by Pindell & Barrett (1990) and Mann *et al.* (1995).

Strike-slip bends of the North America–Caribbean strike-slip system

Northern Central America restraining bend. The western end of the main strike-slip fault (Motagua fault zone) of the North America–Caribbean strike-slip plate boundary curves to the NW in northern Central America to form a large gentle bend with high topographic elevations up to 4000 m and elongate, fault-aligned exposures of metamorphic rocks including the high-pressure mineral jadeite (Mann & Gordon 1996; Harlow *et al.* 2004; Fig. 24). This curvature of the fault intersects small circles of plate rotation to predict a strong component of tectonic transpression along the curved part of the fault trace, which ruptured along a 200-km distance to produce the *M*7.2 earthquake of 1976 (Pflaker 1976).

Interactions between the Motagua Fault and other neighbouring strike-slip faults may also play a role in the regional deformation and topographic uplift in this highly curved area of the fault. In addition to the gentle bend along the Motagua fault zone, Mann & Gordon (1996) infer a 300-km-long and 100-km-wide sharp restraining bend or stepover between the Motagua and Polochic faults (Fig. 24a). This stepover localizes the extremely steep-sided Sierra de Minas, bounded by the two master faults and reaching elevations of over 3 km (Fig. 24a). Guzman-Speziale & Meneses-Rocha (2000) propose a tectonic model, whereby

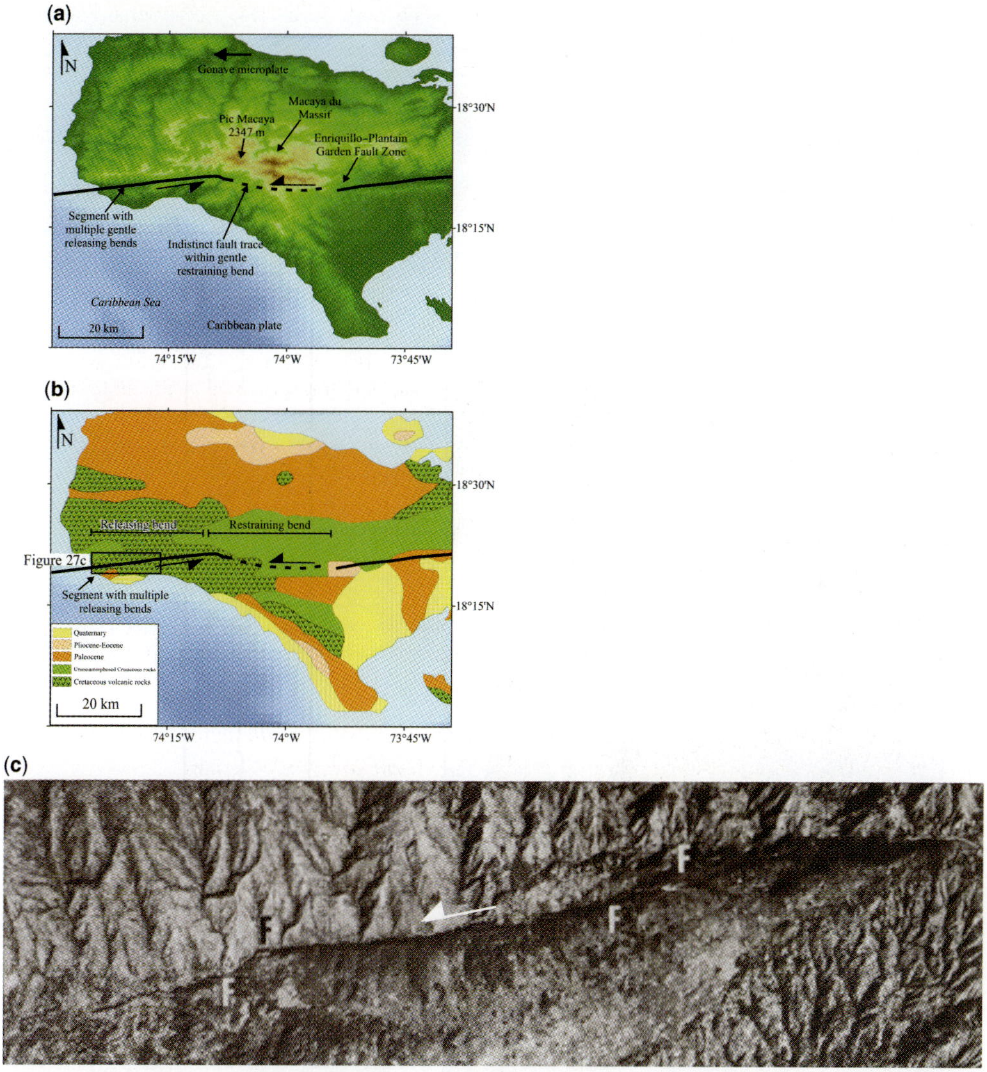

Fig. 27. (**a**) SRTM topography of the western part of the southern peninsula of Haiti, and its relation to active traces of the left-lateral strike-slip faults separating the Gonave microplate in the north (see Fig. 25a for location) from the Caribbean plate to the south. Note that the direction of North America–Caribbean relative motion is east–west, and that the area of higher topography (Massif du Macaya) is adjacent to the segment of the fault most oblique to this direction. The structure is interpreted as a paired bend, with the Massif du Macaya restraining bend to the east, and a releasing bend in the deep, fault-controlled valleys to the west. (**b**) Map showing the ages of exposed rock units in the restraining-bend area. Cretaceous volcanic and sedimentary rocks are exposed in a belt parallel to the fault; a domal structure is not as prominent as in eastern Jamaica (as seen in Fig. 26b) probably because the width of the restraining-bend stepover is only 15 km (as opposed to the 55-km distance in Jamaica). (**c**) Aerial photograph of incipient pull-aparts forming in the releasing-bend part of the paired-bend structure interpreted in Figure 27a and b (from Mann *et al.* 1983; original photo scale: 1:40 000). Letter F marks the approximate termination of dominant oblique-slip fault segments at releasing bends. Overlapping of strike-slip fault segments appears to be occurring at the eastern bend. Basin shapes suggest young pull-aparts similar to those compiled in Figure 3a–f.

motion from the Polochic Fault is transferred to another large, sharp restraining bend that they postulate on the basis of Late Neogene folding and faulting in southern Mexico.

The eastern part of the curved Motagua Fault is characterized by active rifts trending at right angles to the fault zone (Pflaker 1976; Fig. 24a). These features either represent fault-termination structures

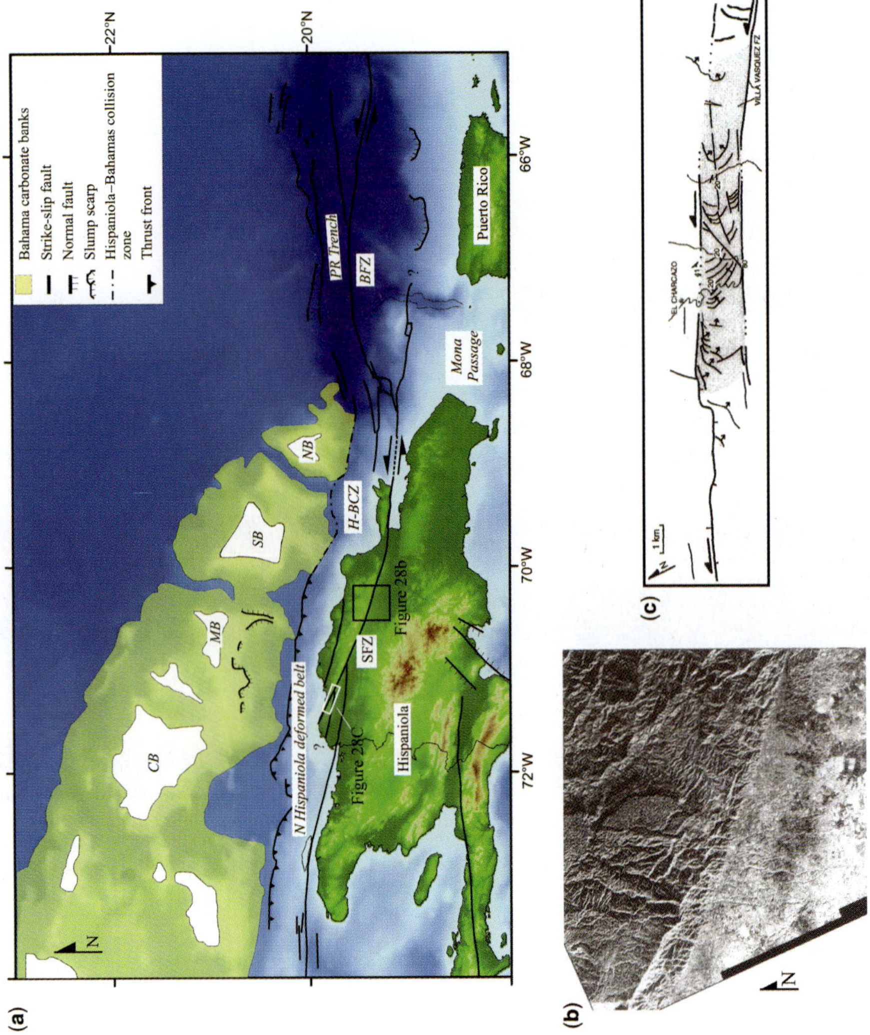

Fig. 28. For caption see p. 83.

(Guzmán-Speziale 2001) similar to those shown in Figure 5, or may represent an extreme example of partitioning of extension along the highly curved and transtensional eastern segment of the Motagua fault zone (Mann *et al.* 2007*b*; Rogers & Mann 2007).

Cayman Trough extreme pull-apart basin. The Cayman Trough is an extreme rhomboidal pull-apart basin that forms a 100 by 1400 km elongate depression with maximum water depths greater than 6000 m (Fig. 25b). North–south trending ridges in the centre of the trough are produced by slow spreading along the Mid-Cayman spreading centre. The northern and NW margins of the trough are bounded by elongate submarine ridges that extend near or above sea-level. The SE margin of the basin is marked by the island of Jamaica and a prominent bathymetric depression west of Jamaica: the West Jamaica pull-apart basin (Fig. 25a). Studies of marine magnetic anomalies and aeromagnetic anomalies within the Cayman Trough by Rosencrantz (1995) show similar patterns of basin opening that began quickly from about 49 to 26 Ma then slowed down considerably between 26 and 20 Ma, and maintained a slow but steady spreading rate from 20 Ma to the present-day (Fig. 25b). These anomalies show that the basin maintained its original master-fault separation distance of about 100 km through its 49 Ma of evolution. From present-day plate motions, the western and central part of the trough is transtensional to pure strike-slip, whilst the eastern part near Cuba and Jamaica is transpressional (Fig. 23). Transpression in this area has been documented in the form of narrow fold–thrust belts by Calais & Mercier de Lepinay (1991) and Leroy *et al.* (1996).

Paired bends in Jamaica and western Haiti. Jamaica is uplifted along a gentle restraining bend in the eastern part of the island as discussed in detail by Mann *et al.* (2007*a*; Fig 26a). Strike-slip faulting occurs along the Enriquillo–Plantain Garden fault zone which bounds the southern edge of the Gonave microplate (DeMets & Wiggins-Grandison 2007; Fig. 25a). Mann *et al.* (1995, 2003) proposed that the Gonave microplate detached from the North America plate about 5 Ma in response to oblique collision between the Caribbean plate and the Bahama carbonate platform. The direction of Gonave–Caribbean plate motion is roughly east–west at rates of *c.* 10 mm/a (DeMets & Wiggins-Grandison 2007).

In eastern Jamaica the Enriquillo–Plantain Garden Fault forms a prominent topographic limit separating older, exhumed Cretaceous–Palaeocene igneous, metamorphic and sedimentary rocks of the Blue Mountains from Eocene–Late Quaternary clastic and carbonate rocks fringing the domal uplift (Mann *et al.* 1985) (Fig. 26b). Elongate outcrops of blueschists and serpentinites are locally emplaced along the faulted edge of the restraining bend. As in many gentle bends, the topographic relief of the Jamaican bend is greatest near the most curved segments of the strike-slip faults bounding the bend. Other restraining bends are present in the central and western parts of the island (Mann *et al.* 2007*a*).

As in the case of the gentle Lebanese restraining bend, the Jamaica bend appears to have nucleated on a pre-existing Palaeogene rift structure that trends about 45° to the strike-slip fault (Mann *et al.* 1985, 2007*a*). This rift is now inverted and forms high mountains adjacent to the bend (Fig. 26c).

The West Jamaica pull-apart basin is inferred to represent the rhomboidal pull-apart basin paired to the restraining bends found in Jamaica (Fig. 25a). The southward step in the fault to form the 2-km-deep depression shifts the fault trace to the original trend of the Enriquillo–Plantain Garden fault zone east of Jamaica in a manner similar to that seen in the California and Lebanese bends.

Southern Haiti paired bend. A smaller paired bend structure is present along the Enriquillo–Plantain Garden fault system on the western end of the southern peninsula of Haiti (Fig. 27a). A 10-km wide right-step in the fault trace produces the 2-km-high Massif de la Macaya which exposes Late Cretaceous limestone and basalt at its core (Fig. 27b). Just to the east of the restraining bend, the fault steps southward along a fault trace

Fig. 28. (**a**) SRTM topography and GEBCO bathymetry of the islands of Hispaniola and Puerto Rico in the NE Caribbean, and their relationship to active traces of the left-lateral strike-slip faults separating the North America plate in the north from the Caribbean plate and Gonave microplate to the south. The north step in the North America–Caribbean strike-slip fault system across northern Hispaniola is related to the oblique collision and partial subduction of the Bahama carbonate platform on the North America plate with the island of Hispaniola on the Caribbean plate and Gonave microplate. Key to abbreviations: CB, Caicos Bank; MB, Mouchoir Bank; SB, Silver Bank; NB, Navidad Bank; SFZ, Septentrional left-lateral strike-slip fault zone; PR Trench, Puerto Rico Trench. (**b**) Seasat radar image, showing uplifted topography and exposure of older rock units along the central curved part of the Septentrional fault zone, modified from Mann *et al.* (1998). (**c**) Sharp restraining bend or push-up structure along the western Septentrional fault zone, modified from Mann *et al.* (1998).

containing four left-stepping gentle releasing bends interpreted to be incipient pull-apart basins (Mann *et al.* 1983; Fig. 27c).

Hispaniola gentle restraining bend. Hispaniola forms a prominent restraining bend in the main trace of the North America–Caribbean strike-slip plate-boundary fault, the Septentrional fault zone (Mann *et al.* 1984; Mann *et al.* 1998; Prentice *et al.* 2003). The bend is gentle, with a maximum curvature of 10° in the Septentrional fault zone (Fig. 28a). The northern range of Hispaniola is on the northern upthrown side of the Septentrional strike-slip fault, and it forms a prominent topographic limit separating older, Cretaceous–Early Pliocene igneous metamorphic (including blueschists), and sedimentary rocks from a basinal section in the valley to the south (Fig. 28b). The direction of Caribbean–North America plate motion in this area reaches its maximum amount of transpression in this area, with rates of about 20 mm/a (Mann *et al.* 2002). The correspondence of structural and topographic doming suggests that transpression is active.

The formation of the curvature and gentle bend in the Septentrional fault zone is probably related to the oblique collision of the Bahama carbonate platform on the North America plate (Dolan *et al.* 1998; Mann *et al.* 2003). Based on tectonic reconstructions, Mann *et al.* (1999, 2003) estimate that this oblique collision between the Bahama Platform and the NE Caribbean plate in Hispaniola began about 5 Ma ago and continues to the present day.

The Septentrional fault zone is the site of a single, sharp restraining bend (Fig. 1c) along one of its fault splays. The Villa Vasquez bend measures 15 km long by 3 km and exposes a folded belt of Miocene and Pliocene rocks with a badlands-type topography. The location of this sharp bend within an overall gentle bend suggests that tectonic transpression may have led to its formation.

Strike-slip bends of the South America–Caribbean strike-slip system

Introduction. There are two zones of strike-slip faulting in northern South America that accommodate relative motion between different plates or blocks in different tectonic settings (Fig. 1, Table 1). The simpler system is the El Pilar strike-slip fault system from central Venezuela to Trinidad that accommodates eastward motion of the Caribbean plate relative to South America (Fig. 23). The El Pilar fault zone can be classified as a continental-plate-boundary strike-slip fault system that can be described by small circles of plate motion, as shown on Figure 23 (Woodcock 1986; Table 1). Two rhomboidal pull-aparts are present along the El Pilar fault system: the Cariaco Basin in Venezuela (Schubert 1982) and the Gulf of Paria Basin in easternmost Venezuela and Trinidad (Flinch *et al.* 1999; Babb & Mann 1999) along with one intervening restraining bend on the Araya–Paria Peninsula (Audemard *et al.* 2006).

Tectonic setting and geological development of bends along the Scotia Sea strike-slip fault systems

Scotian strike-slip systems. The Scotia Sea region – bounded by an arc to the east and strike-slip fault systems to the north and south – has long invited tectonic comparisons to the Caribbean plate (cf. Royden 1993) (cf. Figs 23 & 29). However, the plate structure, active plate kinematics and tectonic history of the two plates differ in many respects. The Scotia region consists of two small plates: the larger Scotia plate to the west largely composed of oceanic crust, and the small South Scotia plate to the east formed in a back-arc position west of the South Sandwich Arc (Fig. 29). Both plates have formed over the past 30 Ma in response to changing relative motion between the much larger South America and Antarctica plates (Cunningham *et al.* 1995, 1998) and in the case of the smaller South Sandwich plate, to slab rollback of the subducted South Atlantic slab beneath it (Royden 1993).

GPS studies (Smalley *et al.* 2003) and earthquake studies (Pelayo & Wiens 1989; Giner-Robles *et al.* 2003) and plate models (Thomas *et al.* 2003) show that the Scotia plate is bounded to the north and south by left-lateral strike-slip fault systems which deform much of the plate within a left-lateral shear couple. Unlike the Caribbean plate, which experiences a component of north–south shortening between North and South America, the Scotia plate is bounded by two larger plates that are moving in a direction close to east–west, and therefore lack a significant component of north–south shortening (Thomas *et al.* 2003).

Bends along the North Scotia strike-slip fault system. The North Scotia Ridge is a series of islands and submarine banks that forms the northern edge of the Scotia plate and extends 2000 km from the southern tip of South America (Tierra del Fuego) to South Georgia Island in the western South Atlantic Ocean. Like the northern Caribbean, most of the strike-slip system is covered by water, and must be studied using GPS instruments places on islands, earthquake seismology, or marine geophysical techniques. Earthquake focal mechanisms indicate that motion is east–west strike-slip and that the component of convergence increases in an

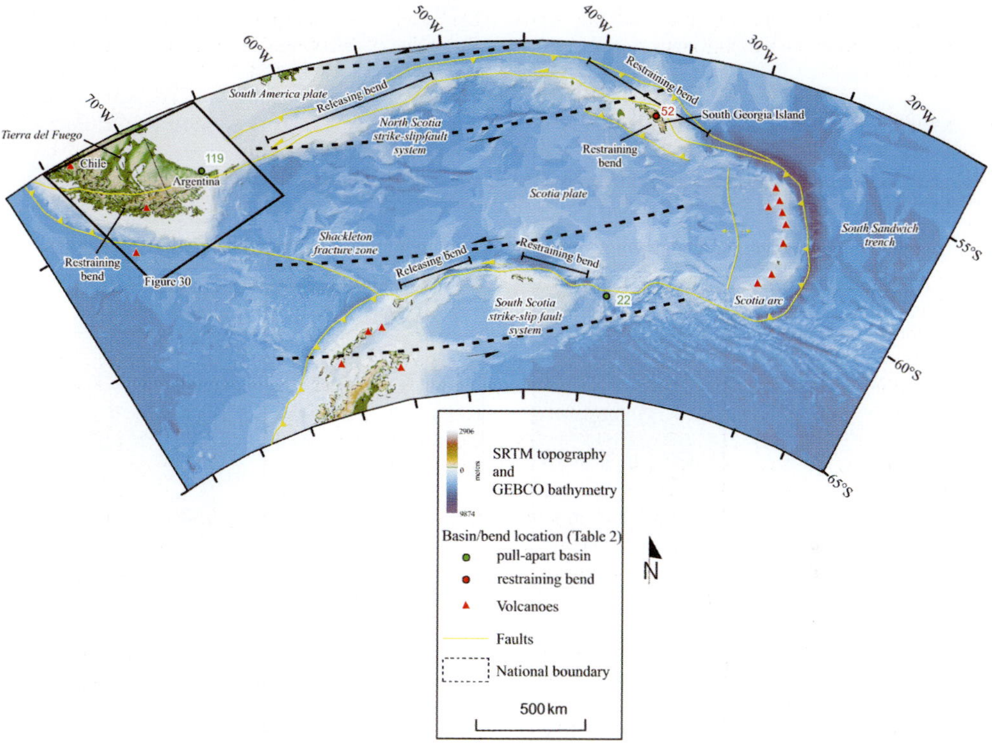

Fig. 29. Regional tectonic map of the north Scotia strike-slip fault system in the area of southernmost South America (Tierra del Fuego), using SRTM topography and GEBCO bathymetry as a base map. Yellow lines are active faults taken from compilations by Cunningham (1995), Lodolo *et al.* (2003), Mann *et al.* (2003) and Smalley *et al.* (2003). Dashed black lines are small circles about the pole of rotation, from Thomas *et al.* (2003). Note that the direction of Scotia–South America relative motion is east–west, and that the area of higher topography in Tierra del Fuego and South Georgia Island is adjacent to the segment of the fault most oblique to this direction. This large restraining bend is similar to those found in other strike-slip to subduction transition areas, including northern Central America (Fig. 24a & b) and Hispaniola (Fig. 28a).

eastward direction, with a maximum of convergence near South Georgia Island (Pelayo & Wiens 1989; Cunningham *et al.* 1998; Fig. 29). The North Scotia Ridge is flanked to the north by a large accretionary prism believed to have been formed by an oblique component of convergence across the left-lateral strike-slip plate boundary (Cunningham *et al.* 1998). The exact trace of the main strike-slip fault or faults parallel to the North Scotia Ridge is not well mapped, but is inferred on the basis of isolated structural features and bathymetry by Cunningham *et al.* (1998) and Mann *et al.* (2002). The fault makes landfall on Tierra del Fuego and has been mapped as a curving trace by Lodolo *et al.* (2003) (Fig. 30).

Bends have been best studied in Tierra del Fuego where Cunningham (1993) has interpreted the large curving trace of the fault as a gentle restraining bend (Fig. 1c), similar to that described above at the western end of the North America–Caribbean strike-slip boundary in northern Central America (Fig. 24). Smalley *et al.* (2003) has used GPS to show that plate motion in the bend area of Tierra del Fuego is almost entirely east–west and therefore consistent with large amounts of shortening in the NW-trending part of the bend. This gentle bend explains the formation of elevated topography along the trace of the fault, exposure of high-grade metamorphic rocks, and the deformation of younger sedimentary rocks (Cunningham 1995). Near the SE edge of the bend, Lodolo *et al.* (2003) have interpreted a rhomboidal pull-apart basin at a left-step in the fault trace at Lago Fagnano.

To the west, Vogt *et al.* (1976) and Mann *et al.* (2003) have interpreted South Georgia Island as a major gentle restraining bend formed at a major stepover in the strike-slip fault trace along the North Scotia Ridge. The island trends at right angles to the small circles of plate rotation; is associated with thrust type focal mechanisms;

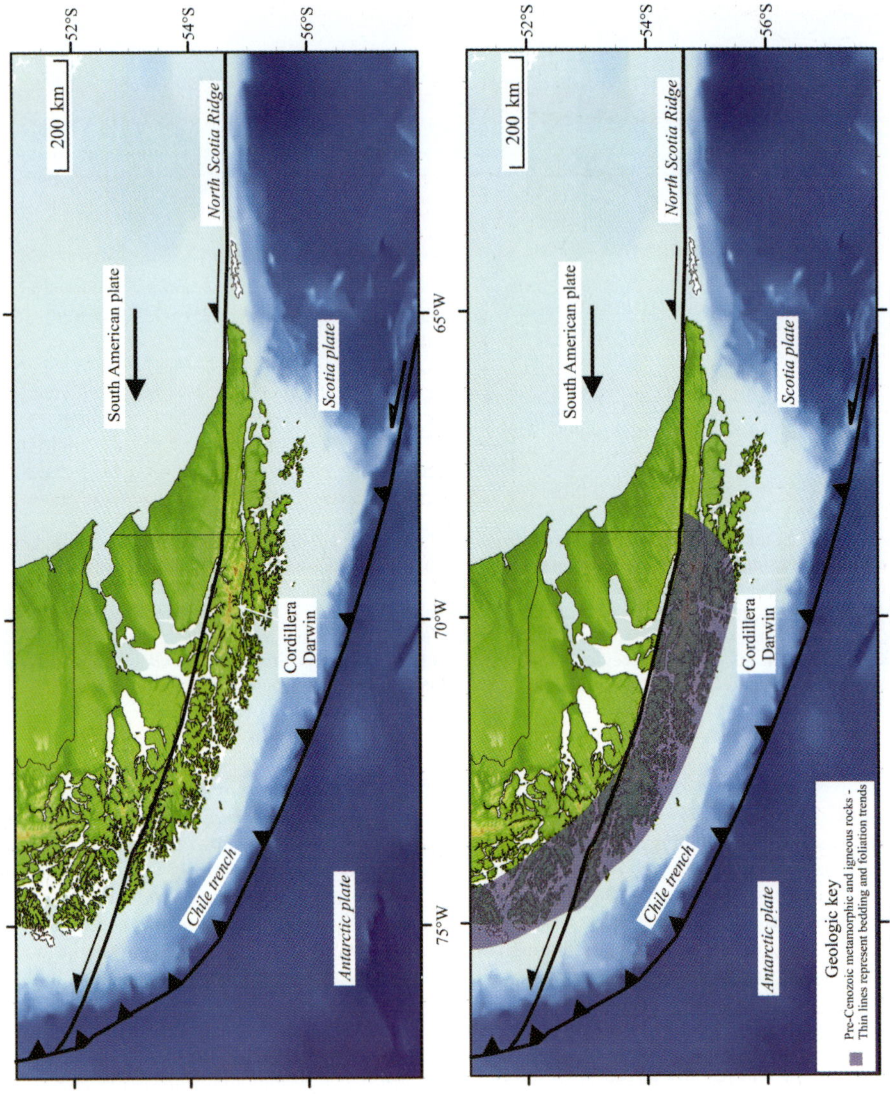

Fig. 30. For caption see p. 87.

mainly exposes highly deformed, older Mesozoic rocks; and is remarkably small (3500 km^2) for its maximum elevations over 3 km. Mann *et al.* (2003) have interpreted the South Georgia bend as a result of the indentation of the North Scotia Ridge by the Northeast Georgia Rise, a Mesozoic large igneous province formed during the early opening of the South Atlantic.

Given the curvature of the North Scotia Ridge, it is possible that the South Georgia restraining bend is adjacent to a releasing bend along the poorly mapped western part of the ridge. This hypothesis could be tested by more detailed mapping of the fault trace along the western part of the North Scotia Ridge.

A much larger gentle bend on the Scotia-South America plate boundary is formed by the northward curvature of the plate boundary fault adjacent to the Cordillera Darwin in Tierra del Fuego (Fig. 30). Cunningham *et al.* (1995) attribute the uplift and high topography of there pre-Cenozoic metamorphic and igneous rocks exposed in the Cordillera Darwin to shortening at the bend.

Bends along the South Scotia strike-slip fault system. Left-lateral shear along the South Scotia Ridge occurs on a much more complex and discontinuous morphological boundary than the North Scotia Ridge (Fig. 29). Marine surveys by Acosta & Uchupi (1996) and Galindo-Zaldivar *et al.* (1996) showed a rotating block structure with local deeps that were inferred to be collapsed block edges or pull-apart basins. Earthquakes indicate that much of the margin is transtensional in nature (Thomas *et al.* 2003).

Tectonic setting and geological development of bends along the Alpine–Macquarie Ridge strike-slip fault systems

The 650-km-long right-lateral Alpine fault zone of New Zealand links the subduction zone at the southern end of the Kermadec Trench off the NE coast of the North Island of New Zealand with a smaller subduction zone along the SW end (Fiordland) of the South Island and on the northern Macquarie Ridge complex (Puysegur Ridge) south of New Zealand (Figs 6a, 31). Oblique subduction beneath the North Island gives way to an increasing component of strike-slip along the Alpine fault zone. The rate of motion between the Pacific and Australia plates varies from 40 mm/a on the North Island to 35 mm/a on the South Island (Fig. 31). For complete reviews of the geology of the Alpine Fault–Macquarie Ridge, the reader is referred to Berryman *et al.* (1992), Lebrun *et al.* (2000) and Meckel *et al.* (2005).

Bends along the northern Alpine fault zone

The Alpine Fault is remarkably linear over much of its length. At its northern end, the fault splays into five major, parallel splays that are collectively called the Marlborough faults (Fig. 32a). The northernmost splay, the Wairau Fault, shows a gentle restraining bend that has been compared with the Transverse Ranges (Yeats & Berryman 1987) but is not associated with marked topographic uplift (Berryman *et al.* 1992).

More striking are the fault splays which curve in a more easterly direction that is transtensional with respect to the small circles of plate rotation (Fig. 32a). This small transtensional environment related to the Marlborough fault splays is the only tectonic environment for pull-aparts on the onland Alpine fault zone.

These faults provide the active tectonic setting for the spindle-shaped Poplars Grove and Glynnwye Lake pull-apart basins (Freund 1971; Cowan 1990) and the much larger, rhomboidal Hanmer Basin (Wood *et al.* 1994). The misalignment of the southern edge of the Hanmer Basin with the regional slip vector has resulted in a local belt of transpression along the SE edge of the basin.

Central Alpine fault. The Central Alpine fault onshore is a straight trace that lacks major bends (Berrryman *et al.* 1992; Fig. 31). The small circles of rotation intersect the fault trace at a small angle that are responsible for some small, shallow-crustal transpressional structures not visible at the scale of Figure 31 (Norris & Cooper 1995). This segment of the fault is characterized by a 50-km-long area of higher uplift rate, more relief, deeper exhumation,

Fig. 30. (**a**) Regional tectonic map of the North and South Scotia strike-slip fault systems, using SRTM topography and GEBCO bathymetry as a base map. GPS studies show that the direction of plate motion is east–west and left-lateral for both fault systems. Laterally extensive paired bends are interpreted along both boundaries. (**b**) Regional tectonic map of the North and South Scotia strike-slip fault systems, using SRTM topography and GEBCO bathymetry as a base map. Yellow lines are active faults taken from compilations by Mann *et al.* (2003). GPS studies show that the direction of plate motion is east–west and left-lateral for both fault systems. Laterally extensive paired bends are interpreted along both boundaries. Metamorphic outcrop (from Cunningham 1991) is shown in purple.

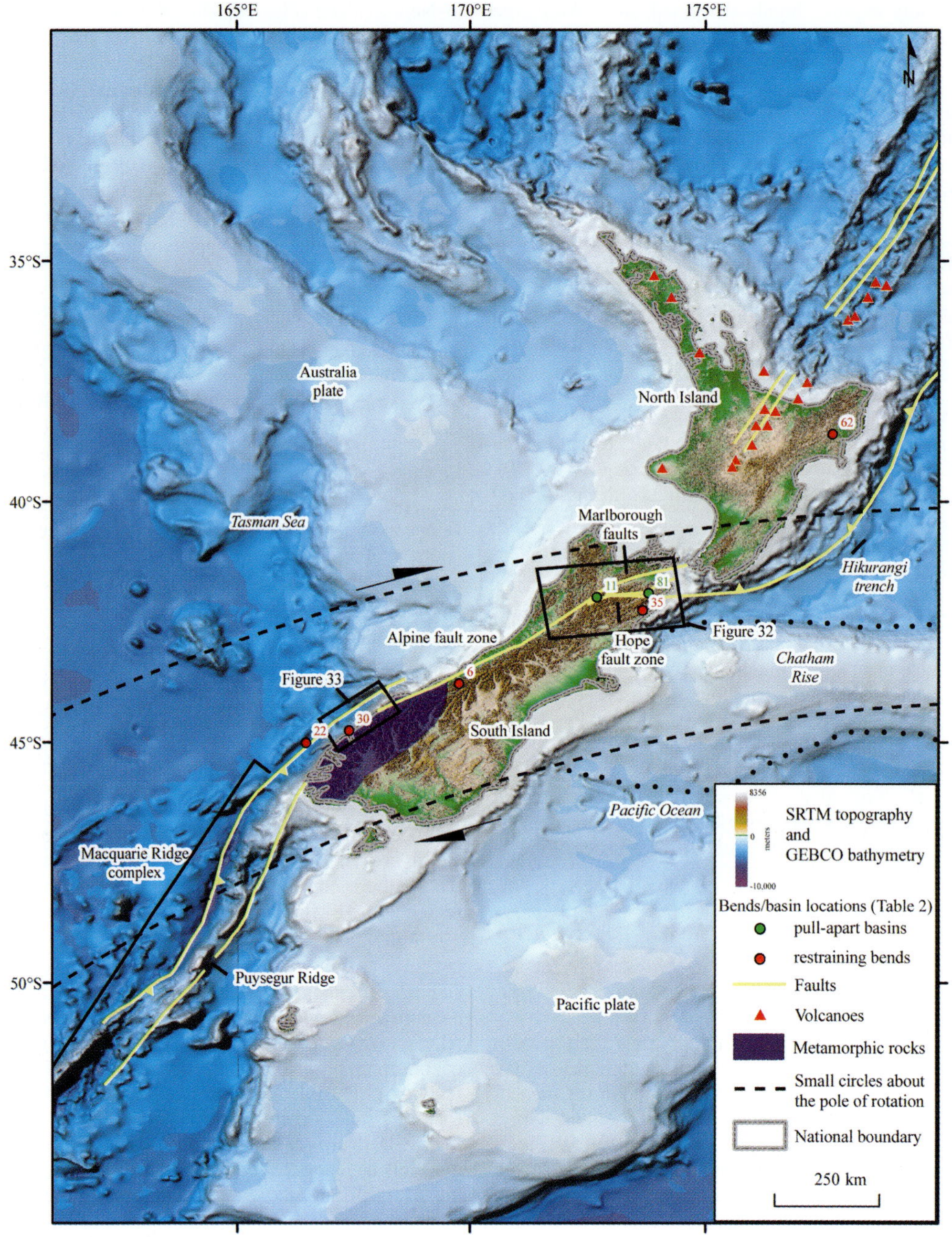

Fig. 31. Regional tectonic map of the Alpine strike-slip fault system in the area of New Zealand and the northern Macquarie Ridge complex, using SRTM topography and GEBCO bathymetry as a base map. Yellow lines are active faults, taken from a compilation by Barnes *et al.* (2005). Dashed black lines represent small circles of rotation.

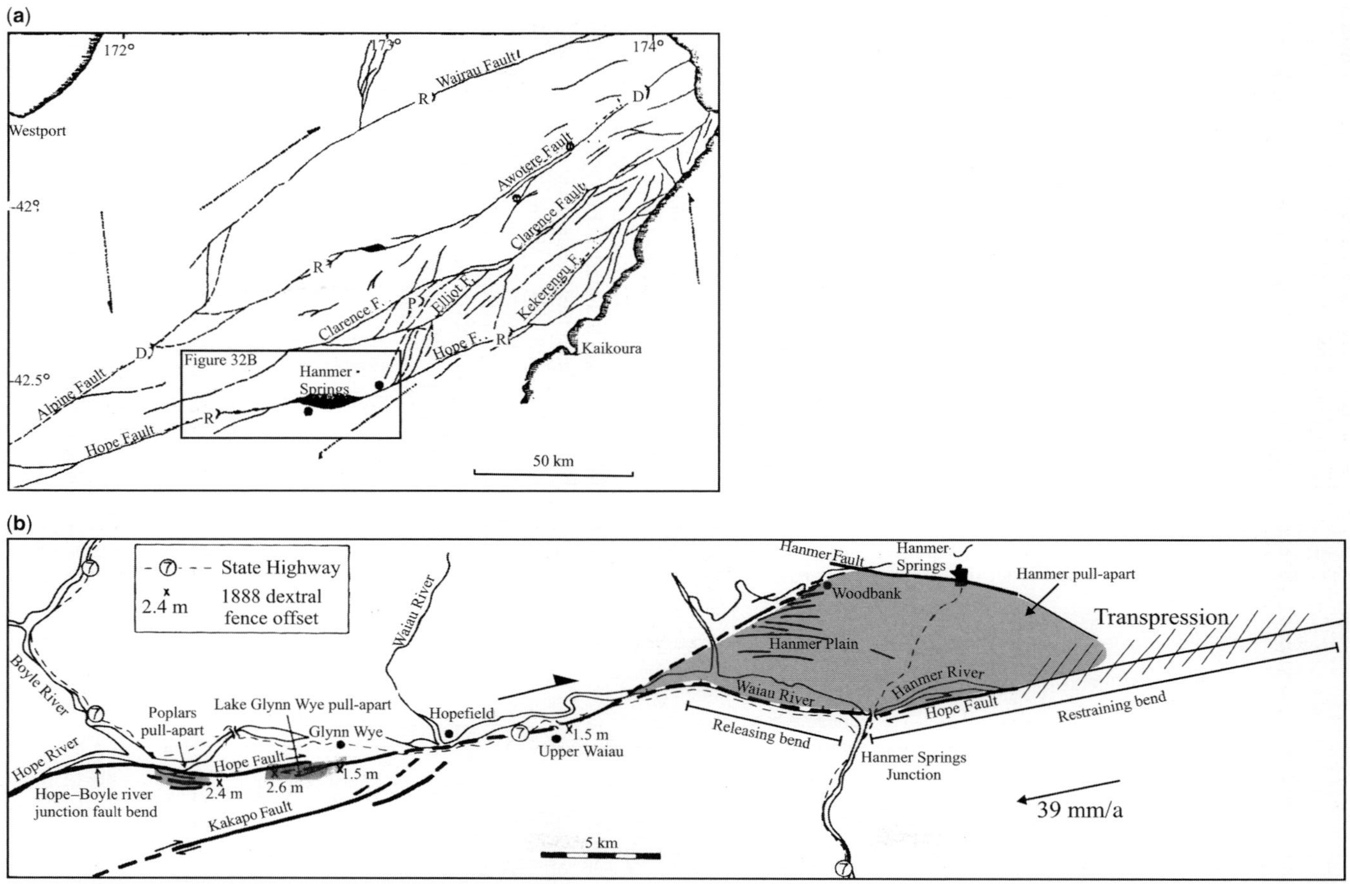

Fig. 32. (**a**) Map of the Marlborough strike-slip fault system at the northern end of the South Island, from Freund (1974). Black dots show measured offsets on the faults. Black areas are the larger pull-aparts developed on the Hope and Atwater fault zones. Note that at this scale the faults show no obvious paired-bend geometries. Instead, the subparallel strike-slip faults are uniformly right-stepping in a NE direction. (**b**) More detailed map of the Hope fault zone, showing the array of three pull-apart basins: Poplars, Lake Glynn Wye and Hanmer (modified from Wood *et al.* 1994). At this basin scale, these authors inferred one paired zone of a releasing bend and a restraining bend along the southern edge of the Hanmer pull-apart basin.

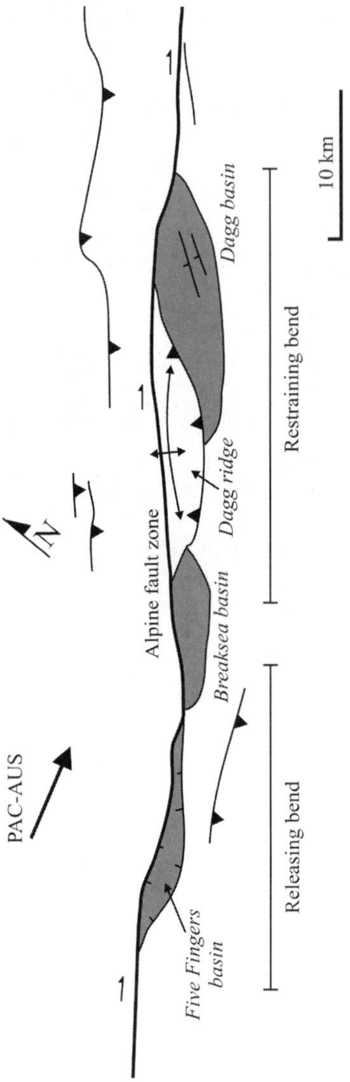

Fig. 33. Schematic map of strike-slip faulting along the west coast of the South Island, by Barnes *et al.* (2005) (cf. Fig. 31 for location), where Pacific–Australia plate motion is slightly oblique to the fault and partitioned into strike-slip faulting and a zone of offshore thrust faulting similar to northern Hispaniola (cf. Fig. 28a–d). A paired bend consisting of a releasing bend (Five Fingers Basin) and a restraining bend (Dagg Ridge) is inferred from the fault pattern.

and a narrower fault width than surrounding areas (Little *et al.* 2005). The hanging-wall uplift of deep-crustal rocks from beneath the Southern Alps in this area occurs as a localized and narrow transpressional uplift (Fig. 1c) that is not associated with a distinct bend in the fault trace.

Southern Alpine fault. The Southern Alpine fault passes offshore and has been surveyed by Barnes *et al.* (2001, 2005) using geophysical methods (Fig. 33). This offshore segment of the fault is characterized by an adjacent releasing bend (Five Fingers Basin) and restraining bend (Dagg Ridge) that together constitute a paired bend (Fig. 33). Barnes *et al.* (2001, 2005) discuss the ephemeral character of the ridges and basins during progressive strike-slip deformation.

Bends along the Puysegur Ridge. The Puysegur Ridge and trench have been shown by marine surveys to exhibit a small subduction feature at the base of its slope associated with a short subducted slab. A zone of strike-slip faults on the trench slope and top of the ridge also accommodate oblique plate motions (Lebrun *et al.* 2000; Fig. 33). The overall morphology of the Puysegur Ridge is that of a lazy-S gentle restraining bend (Fig. 31).

Bends along the Macquarie and Hjort trenches. Transpressional motion persists along the Macquarie and Hjort ridges south of the Puysegur Ridge. The two ridges are separated at a change in their trends by a bathymetric break and deep area in the ridge that is underlain by a single small rhomboidal pull-apart basin (Massell *et al.* 2000; Daczko *et al.* 2003; Fig. 34). Like the central Alpine fault zone, the two ridges exhibit straight morphologies suggestive of transpressional uplifts (Fig. 1c). Daczko *et al.* (2003) postulate other smaller pull-apart basins along the length of the ridges.

The Hjort Trench assumes a more transpressional setting as it curves to the southeast (Fig. 34a). Meckel *et al.* (2005) propose that it is evolving into a subduction zone similar to that observed for the Puysegur Trench.

Tectonic setting and description of natural bends on indent-linked strike-slip faults in active zones of continental collision and tectonic escape

Regions of continental collision and tectonic escape

As compiled by Woodcock (1986), an important tectonic setting for strike-slip faults are areas of continental collision (Fig. 1 & Table 1). Colliding continental

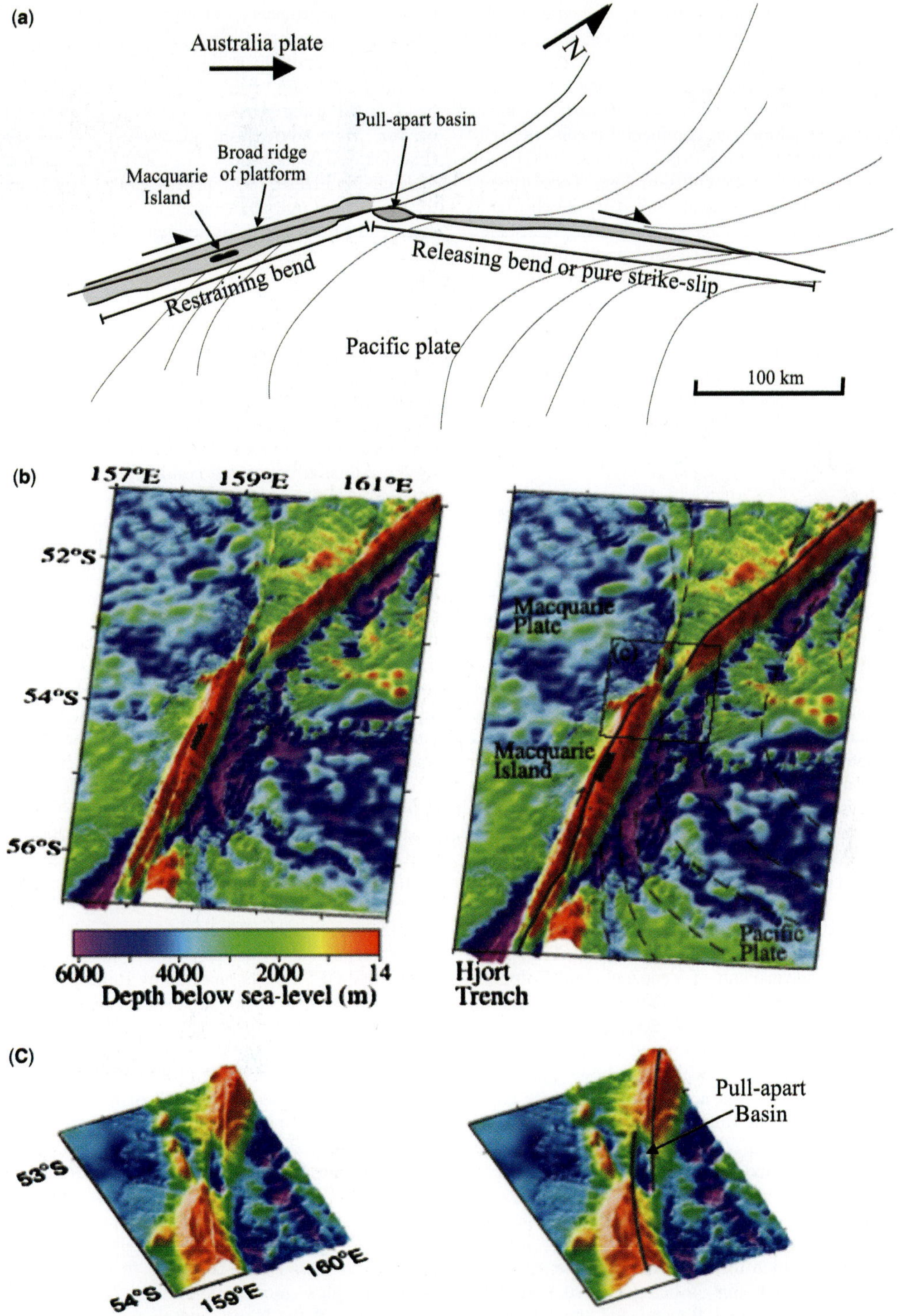

Fig. 34. (**a**) Schematic map of the Macquarie strike-slip fault system on the Macquarie Ridge south of New Zealand, modified from Daczko *et al.* (2003). The wider platform on which Macquarie Island is located (Macquarie Ridge) is oblique to plate motion and is interpreted to be a restraining bend. Adjacent to the restraining bend is a small but 1.2-km-deep pull-apart basin. The narrower MacDougall Ridge to the north is more parallel to plate motion, and is interpreted as a strike-slip-controlled ridge with less of a transpressional component. Curving fracture zone trends on both sides of the fault are consistent with right-lateral shearing. (**b**) Multi-beam bathymetric maps of the transition between the two fault segments separated by the deep pull-apart basin, from Dazcko *et al.* (2003). (**c**) Close-up of pull-apart basin, from Dazcko *et al.* (2003).

edges commonly have irregularly shaped edges that are mismatched at points of collision and therefore lead to indentation of one continent by the promontory of another. Continental shortening can be taken up thrust faulting and thickening but also by conjugate strike-slip faulting and continental escape of continental fragments along strike-slip faults into uncollided segments of the collision zone. These type of collision-related or 'indent linked' strike-slip faults commonly are associated with releasing fault bends and other types of strike-slip, partly because the escaping blocks undergo significant rotations (Mann & Gordon 1996) and because of reactivation of previous structural grains in the crust (Cunningham 2007). In this section I will review restraining bends in seven areas involving indent-linked strike-slip faults: NW South America, Anatolia, Northern Europe, Iran, Tibet, Mongolia and the Lake Baikal region of Russia.

Tectonic setting and geological development of bends along the active NW South America strike-slip fault systems

The NW corner of South America is a complex, indent-linked zone that accommodates the northward and NE motion or 'escape' of blocks of NW South America into southern South America (Trenkamp *et al.* 2002; Fig. 35a). These indent-linked strike-slip faults are less-continuous, more arcuate and it is more difficult to relate their motion to poles of rotation than the El Pilar strike-slip system to the east that is controlled by a single continental-plate pair (Perez *et al.* 2001; Weber *et al.* 2001; Fig. 23).

Specific strike-slip faults within this system include the Bocono Fault of western Venezuela (Schubert 1980; Fig. 35b), the Santa Marta fault zone of Colombia (Mann & Burke 1984), the Eastern Andean Fault of Venezuela and Colombia (Pennington *et al.* 1979), the Algeciras Fault of Colombia (Velandia *et al.* 2005) and faults of a large restraining bend proposed in the Ecuadorian Andes (Ego *et al.* 1996). GPS shows that all of these faults act across a broad zone to move continental and oceanic fragments in a northward direction, perhaps as a consequence of the collision of the Panama island arc in Late Miocene to Recent times (Mann *et al.* 2006).

One of the best-studied faults in this zone with prominent bends is the Bocono Fault of western Venezuela, which forms one edge of the triangular Maracaibo Block that is being ejected northward along the right-lateral Bocono and the left-lateral Santa Marta Fault in Colombia to the east (Fig. 35a). The Bocono fault zone exhibits a paired bend defined by the La Gonzalez gentle releasing bend, and an adjacent, unnamed restraining bend with high topography (Schubert 1980; Fig. 35b).

An ancient pull-apart, the Icotea Basin, related to the early evolution of this strike-slip boundary, has been studied using 3D seismic data in the subsurface of the Maracaibo Basin in western Venezuela (Escalona & Mann 2003; Fig. 36). These studies have clearly imaged the transverse faults responsible for an Eocene pull-apart basin, which have eluded researchers in more thickly sedimented pull-aparts.

Tectonic setting and geological development of bends along the active Anatolian strike-slip fault systems

Introduction. The Anatolian region (roughly corresponding with the country of Turkey) is a well-known example of 'tectonic escape' triggered by the collision between a continental promontory on the Arabian plate and the Eurasian plate (Sengör *et al.* 1985; Hempton 1987; Fig. 37). Two major strike-slip faults, the North and East Anatolian faults, were formed about 5 Ma ago, in response to the collision, and propagated from east to west (Sarolgu *et al.* 1992; Armijo *et al.* 2002). Because the faults have opposed shear senses, they expel the Anatolian continental fragment westward into the Mediterranean Sea, where oceanic crust of the eastern Mediterranean Sea can be subducted beneath the fragment. In addition to these two major strike-slip faults, other smaller strike-slip faults were formed within the Anatolian block and exhibit fault bends of restraining and releasing types (Sengör *et al.* 1985; Kocyigit & Erol 2001).

On the basis of geological information, Sengör *et al.* (1985) inferred that large areas of Anatolia and the Caucasus were internally deformed during the collision process. Whilst this may be true prior to the formation of the large indent-linked strike-slip faults, GPS results reported by Reilinger *et al.* (2006) show that blocks adjacent to the presently active large strike-slip faults are behaving rigidly, with little evidence for the type of internal strain envisioned by Sengör *et al.* (1985). GPS measurements show that the escape path of the Anatolian Block is highly arcuate along a counterclockwise path and proceeds at very rapid rates in the range of 20–30 mm/a (McClusky *et al.* 2000; Reilinger *et al.* 2006).

One possible mechanism for pull-apart and other basin formation related to north–south constriction of the Anatolian Block is shown in Figure 8d; a more complex model involving the westward propagation of the North Anatolian Fault is given by Chorowicz *et al.* (1999). The only place along the fault where the North Anatolian fault trace is out of alignment with the plate vector is in the

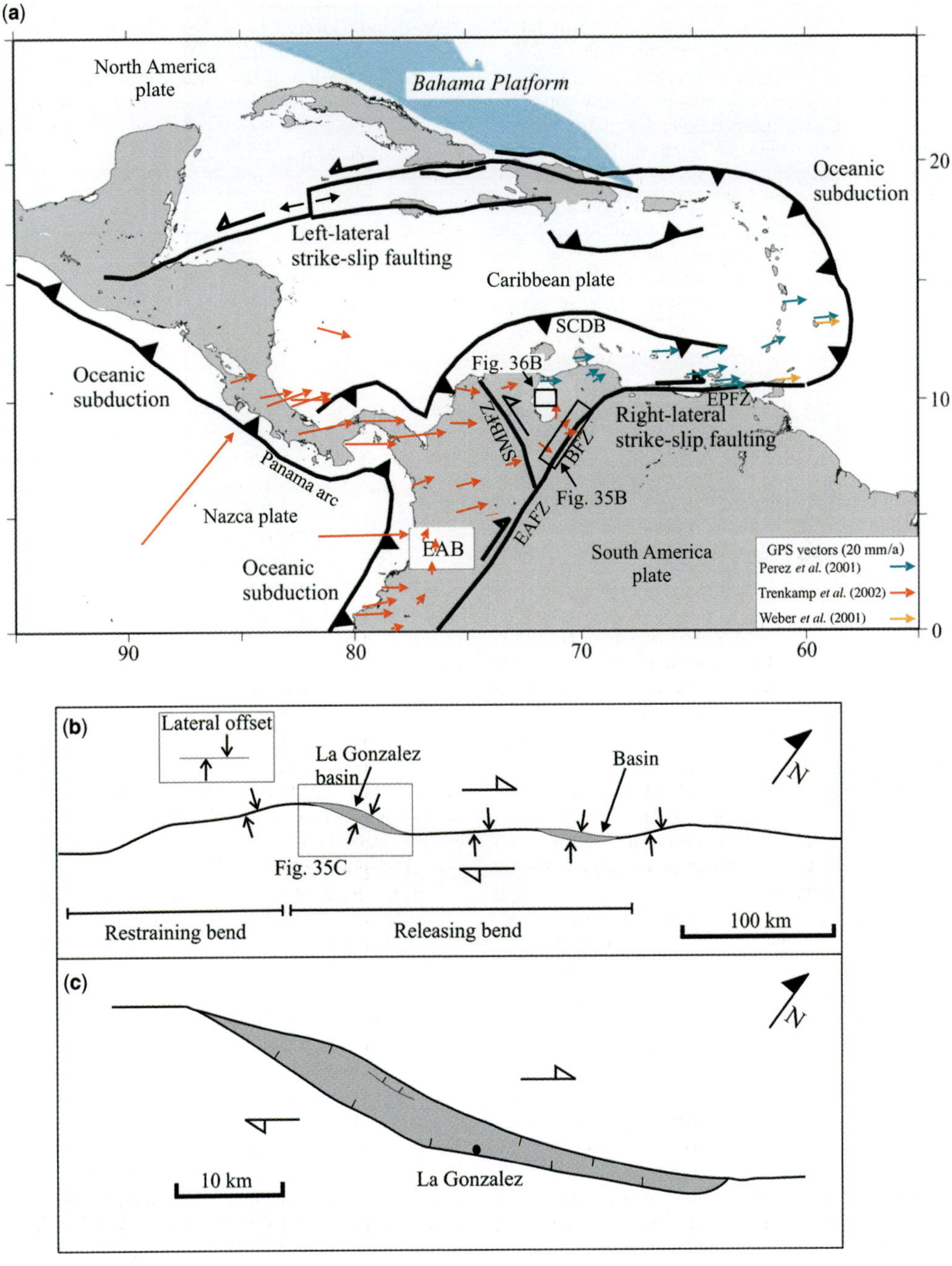

Fig. 35. (**a**) Tectonic map and GPS velocity field (red arrows) of northern South America, modified from Mann *et al.* (2006). Sources of the GPS data are Perez *et al.* (2001), Trenkamp *et al.* (2002), and Weber *et al.* (2001), and are colour-coded, according to author, in the inset box. The Maracaibo Block of NW South America is a large escape block activated by convergence between the Panama Arc and NW South America and moving to the north and NE along the left-lateral SMBFZ (Santa Marta–Bucaramanga fault zone) and the right-lateral Bocono fault zone (BFZ). (**b**) Map of the Bocono fault zone of Venezuela, modified from Schubert (1980), showing possible paired-bend structure. (**c**) Map of the La Gonzalez pull-apart basin similar to young pull-apart basins with a lazy-Z shape (cf. examples in Fig. 3a–f).

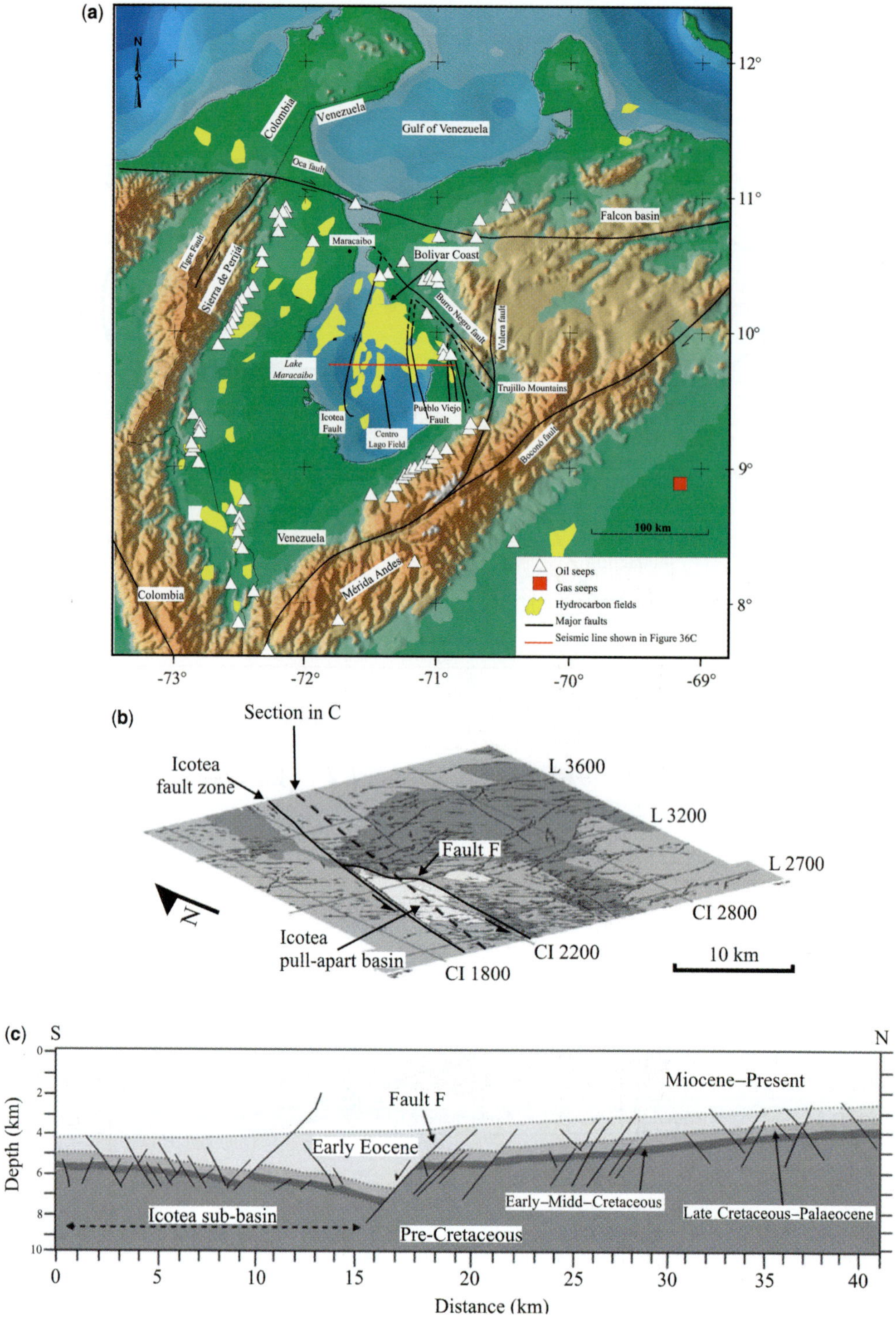

Fig. 36. (**a**) Tectonic setting of the Eocene Icotea pull-apart basin in the Maracaibo basin, Venezuela (from Escalona & Mann 2003). (**b**) Two-way time slice from a 3D seismic survey, showing a pull-apart structure preserving younger Eocene sedimentary rocks in the left-stepping pull-apart. (**c**) Cross-section through the Eocene pull-apart basin, showing basin formation along normal fault F, and the inactivity of faults in post-Eocene times. Compare the structure of the pull-apart on the normal fault with that of the Ridge Basin of southern California in Figure 19.

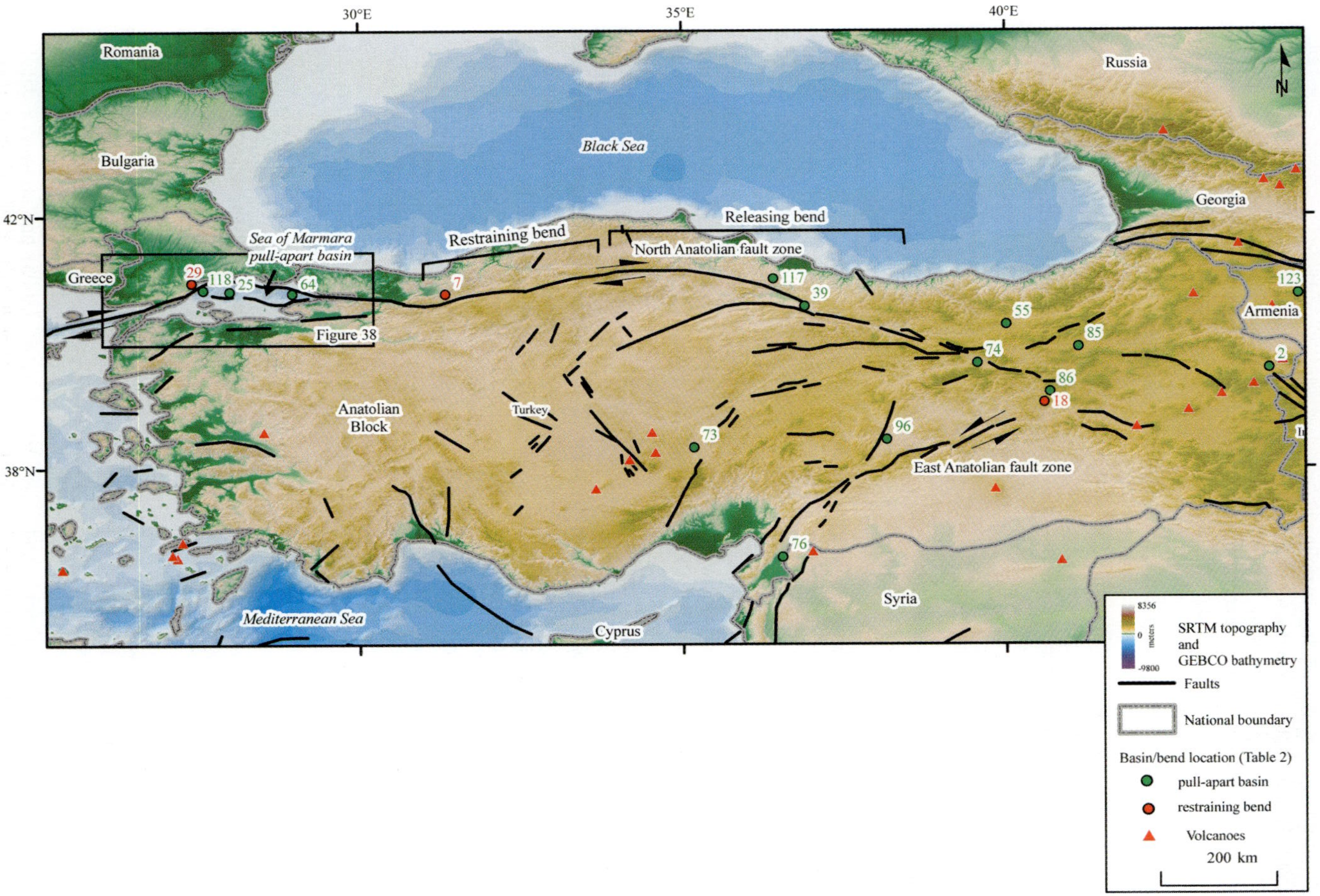

Fig. 37. (**a**) Regional tectonic map of the North and East Anatolian strike-slip fault systems of Turkey, using SRTM topography and GEBCO bathymetry as a base map. Black lines are active faults. GPS studies show that the direction of plate motion of the Anatolian block is roughly east–west and right-lateral for the North Anatolian Zone, and left-lateral for the East Anatolian fault zone. Laterally extensive paired bends are interpreted along both boundaries.

Sea of Marmara region (Fig. 38). In areas to the east, the fault is approximately parallel to the slip vectors constrained by GPS measurements (Reilinger *et al.* 2006).

Bends along the North Anatolian fault zone. Right-lateral offset estimates for the North Anatolian Fault vary widely, but many agree to its right-lateral offset estimate of 25–85 km over the past 5 Ma (cf. Armijo *et al.* 1999). Major bends along the trace of the fault visible at the scale of the map shown in Figure 38 include the Ganos restraining bend; the Sea of Marmara pull-apart basins; the Niksar Z-shaped pull-apart; and the Erzincan rhomboidal pull-apart (Barka & Gulen 1989; Fig. 38; Table 2). Smaller pull-apart and fault wedge basins, and their relationships to earthquake ruptures, are described by Barka & Kadinsky-Cade (1988), Barka (1992), Lettis *et al.* (2002) and Harris *et al.* (2002).

Bends in the Sea of Marmara. This is one of the most intensively surveyed and studied strike-slip bend areas in the world, because of the likely occurrence of a future strike-slip rupture on the North Anatolian fault zone near the large population centre and strategic port of Istanbul, Turkey, on the Bosporus Strait linking the Sea of Marmara with the Black Sea (Fig. 38). The last rupture, in 1999, on a century-long series of westward-propagating ruptures, affected the Izmit area east of Istanbul (Harris *et al.* 2002; Lettis *et al.* 2002; Fig. 38). In this area, the North Anatolian fault zone bifurcates into two strands: a northern strand is marked by an array of three spindle to lazy-Z-shaped pull-apart basins in the Sea of Marmara, and a large restraining bend along the Gallipoli Peninsula and Ganos Mountain marked by en échelon folding (Armijo *et al.* 1999; Fig. 38).

There have been many surveys of the three pull-aparts in the Sea of Marmara, with most controversy focused on whether the basins are active or have been bypassed by through-going strike-slip fault traces (Okay *et al.* 2000; Rangin *et al.* 2004; Seeber *et al.* 2006). A straighter and more southerly strand of the fault called the southern strand of the North Anatolian fault zone passes south of the Sea of Marmara. GPS surveys show that motion is roughly east–west and parallel to the slight arc in the strike-slip faults (Fig. 38).

Paired-bend interpretation. The pattern of adjacent restraining and releasing bends shown on Figure 38 can be interpreted in terms of the same paired-bend model used for the Californian and Lebanese bends. Deviation of the northern strand by the restraining bend, is returned to the original trend of the fault trace by the opening of the pull-apart array in the Sea of Marmara. The southern strand may represent a single trace of the North Anatolian fault prior to this northern deviation, since GPS vectors indicate that most of the present-day motion is occurring on the northern trace of the fault (Fig. 38).

Bends along the East Anatolian fault zone. The East Anatolian fault zone exhibits a very discontinuous fault trace, with one gentle restraining bend near Bingol and one rhomboidal pull-apart at Lake Hazar (Hempton & Dunne 1983; Table 2). Smaller pull-apart and fault wedge basins and their relationships to earthquake ruptures are described by Barka & Kadinsky-Cade (1988) and Barka (1992).

Tectonic setting and geological development of bends along the active and ancient Alpine–Carpathian strike-slip fault systems and inverted faults of the European foreland

Introduction. The Alpine–Carpathian collision consisted of a west to east, diachronous continent–continent collision from France in western Europe to Romania in central Europe (Royden *et al.* 1982; Ratschbacher *et al.* 1991; Fig. 39). This collision produced lateral escape of continental fragments towards uncollided areas in the east – along with indent-related strike-slip faults within the orogenic belt and pull-apart basins, and restraining bends at complex fault bends. In addition to deformation within the orogenic belt, far-field effects of the collision reactivated existing basement structural features of various types as far north as the British Isles and the North Sea. In this section, I review a few of the better-studied restraining and releasing bends formed as a result of this orogenic event.

Vienna Basin. Terminal collisions in western areas and continuing collisions in eastern areas caused the overriding Alps–Pannonian plate to fragment along large strike-slip fault systems (Royden *et al.* 1982; Fodor 1995; Mann *et al.* 1995; Fig. 40a). A major block boundary formed between an inactive and terminally collided area west of the Vienna area and the Pannonian plate that continued to move to the NE east of the Vienna area (Fig. 40b). As offset lengthened along this boundary, a large rhomboidal basin formed beneath the site of the present-day city of Vienna, Austria (Fig. 40c).

Surface mapping of scarps, subsurface mapping and well-log correlation by Decker *et al.* (2005) and Hinsch *et al.* (2005) has allowed a detailed picture of this rhomboidal basin with its distinctive twin depocentres characteristic of rhomboidal

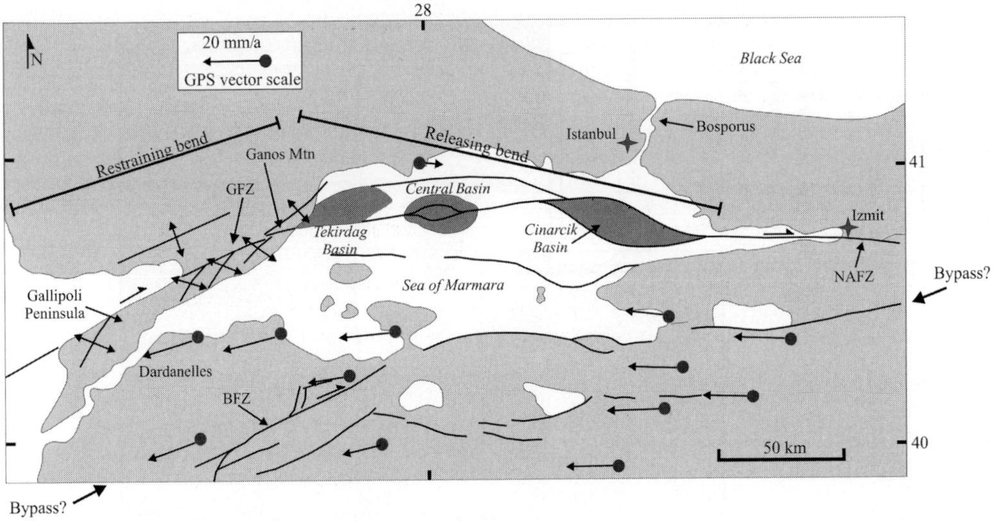

Fig. 38. Schematic map of the active traces of the North Anatolian fault system in western Turkey, modified from Armijo *et al.* (2005), and their relation to GPS plate motions, from McClusky *et al.* (2000). See the regional tectonic map in Figure 37 for location. Inferred paired restraining- and releasing bends are indicated on both subparallel fault systems. Note how the coastlines of the Sea of Marmara mirror the coastal and offshore paired-bend structure.

pull-aparts (Mann *et al.* 1983; Fig. 7d). Other pull-aparts have been described to the south (Sachsenhofer *et al.* 2003) and residual seismic activity and surface faulting continues on all of these structures to the present day.

Bends in the North Sea and British Isles. Many pre-existing structural grains were reactivated during the Early Cenozoic as a result of the Alpine collision. Some of these examples have been particularly well imaged by subsurface seismic and well data collected by the oil industry. For example, the Pijnacker oil field offshore of the Netherlands in the North Sea originated as an Early Cretaceous, right-lateral rhomboidal pull-apart that was subsequently inverted as a lazy-S gentle restraining bend when movement on the strike-slip fault reversed and became left-lateral (Fig. 41).

Dooley *et al.* (1999) show a rhomboidal pull-apart mapped from 3D seismic data in the subsurface of the Netherlands (Fig. 42a). The interior of the basin contains transverse normal faults that accommodate basin extension as predicted in the standard pull-apart model (Fig. 7). The master faults bounding the basin also exhibit a change in major throw as the basin depocentre is crossed (Fig. 42a).

Another example from Dooley *et al.* (1999) shows that a half-graben formed during pre-Jurassic and Jurassic rifting was modified by post-Early Cretaceous right-lateral motion on the Great Glen fault zone (Fig. 42b). Reactivation probably related to the Tertiary Alpine reactivation event produced deepening of the basin along steeper master fault strands.

According to Holloway & Chadwick (1986), the rhomboidal Petrockstow and Bovey basins formed by about 6 km of left-lateral offset during the Early Tertiary, possibly as a result of extension in the offshore area of southern England. In Mid-Tertiary times, slip on the fault reversed to right-lateral and produced reverse faulting at the margins of the Petrockstow and Bovey basins (Fig. 43c). The likely mechanism of the Mid-Tertiary event was a far-field effect of the Alpine shortening event.

Tectonic setting and geological development of active bends of the Iranian Plateau

Active strike-slip faults and fault bends are common in central and eastern Iran (Fig. 44), and include the site of the classic mapping study by Tchalenko & Ambraseys (1970) of shear structures formed along the 25-km-long rupture zone of the 1968 Dasht-e-Bayaz earthquake (Fig. 45). Active deformation in Iran is related to about 25 mm/a of active convergence between the Arabian plate and Eurasia; the Arabia plate acts as a rigid indentor and produces cross-cutting strike-slip faults that separate eastern Iran from Afghanistan (cf. the Neyband, Gowk, Sistan fault zones of Walker & Jackson 2004; Fig. 44). Some shortening is absorbed by folding in the Zagros fold-and-thrust belt, whilst the remainder is absorbed by conjugate

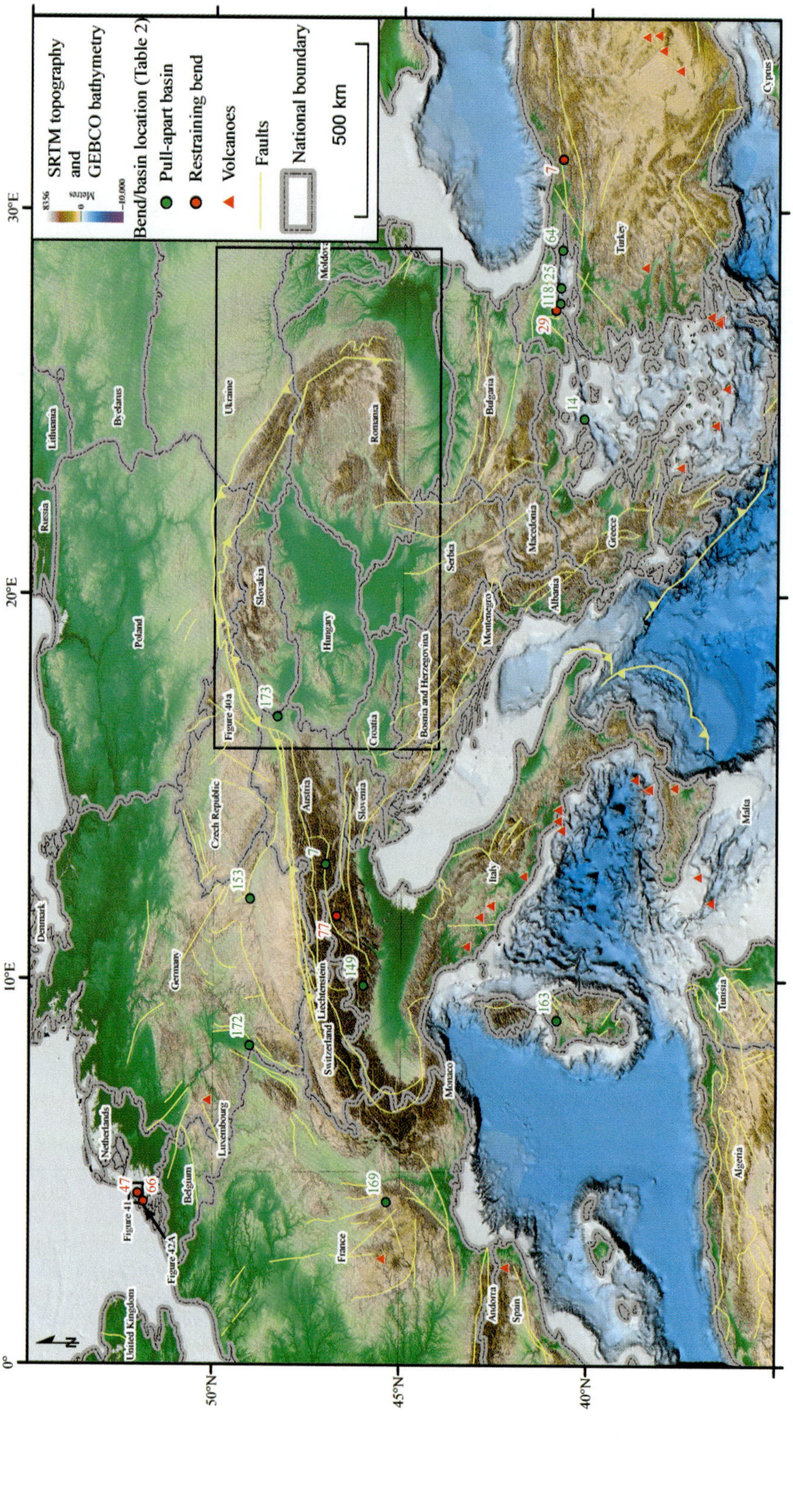

Fig. 39. Regional tectonic map of Cenozoic to recent strike-slip fault systems of Europe, using SRTM topography and GEBCO bathymetry as a base map. Yellow lines are active faults taken from compilations by Levine (1995) and Coffin *et al.* (1997).

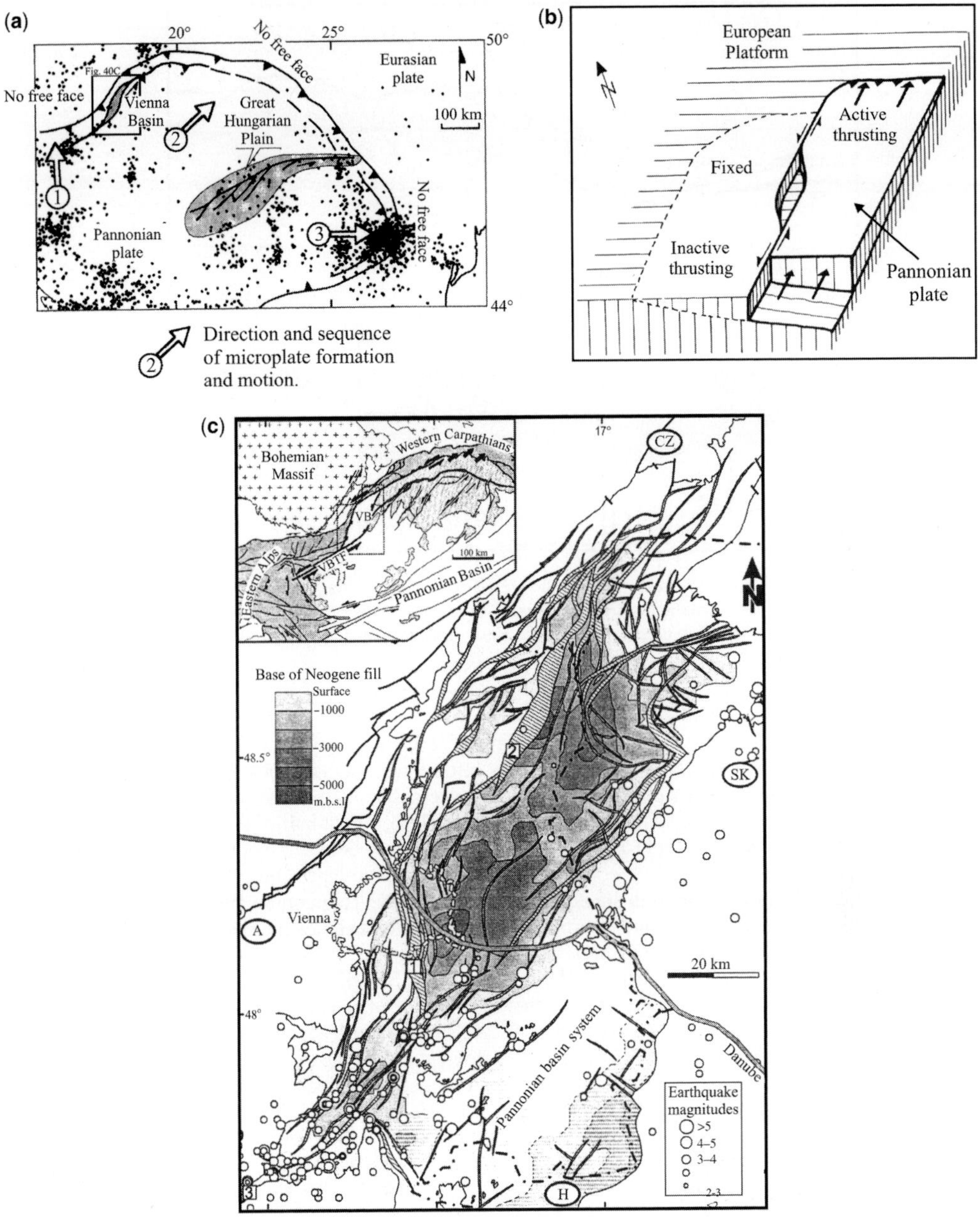

Fig. 40. (**a**) Tectonic setting of the Vienna Basin during west–east oblique collision, from Mann *et al.* (1995). (**b**) Block diagram, showing relation between the Vienna pull-apart basin and underlying nappes and thrust faults (from Royden 1985). (**c**) Structural setting and seismicity of the Vienna pull-apart basin, from Hinsch *et al.* (2005). Inset shows main structural units of the Eastern Alpine–Carpathian region. Key Abbreviations: VB, Vienna Basin; VBTF, Vienna Basin transfer fault. The map shows the faulted pre-Neogene basement surface and earthquake epicentres since 1201 BC. Labelled points: 1, Leopoldsdorf Fault; 2, Steinberg Fault; 3, transition to the Mur–Murz Fault. A, Austria; CZ, Czech Republic; H, Hungary; SK, Slovakia.

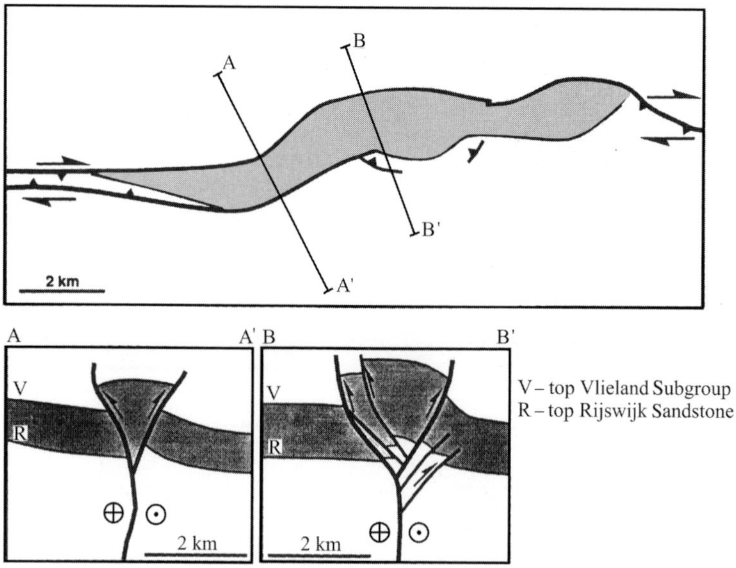

Fig. 41. (**a**) Structure map of the Pijnacker field in the west Netherlands, using subsurface mapping from Racero-Baema & Drake (1996) (see the location of the map on Fig. 39). This structure formed as an Early Cretaceous rhomboidal pull-apart basin, and was inverted during the Early Tertiary. (**b**) Cross-sections based on seismic-reflection profiles.

strike-slip faults and block rotations (Walker & Jackson 2004). The left-lateral Dasht-e-Bayaz Fault is antithetic to the overall right-lateral shear zone and accommodates a component of clockwise block rotation within the overall right-lateral shear zone (Fig. 45).

Whilst a paired bend is present at the scale of the Dasht-e-Bayaz rupture, few pull-aparts are observed at more regional scales in this area. Restraining bends commonly connect different strands within the various shear zones, such as the lazy-S gentle bend along the Purang right-lateral strike-slip fault in eastern Iran (drawn from Fig. 7 of Walker & Jackson 2004) (Fig. 46a). The *M*6.6 2003 Bam earthquake, which caused over 40 000 deaths, originated on a buried right-lateral fault parallel to the Gowk fault zone in eastern Iran. Imagery studies show that the rupture occurred at the edge of a gentle restraining bend (Fu *et al.* 2004).

Tectonic setting and geological development of active bends of the Tibetan Plateau, western China and Mongolian Altai

Continental collision effects are most prominent in central and eastern Asia, where the India indentor is forcing tectonic blocks to the east and SE along indent-linked strike-slip faults (Fig. 47) (Tapponier *et al.* 1982). The northern and eastern boundaries of the escaping continental fragments are mainly left-lateral faults, whereas the southern and western boundaries are mainly right-lateral faults (Fig. 47a). About half of the 50 mm/a of northward convergence between India and Asia is absorbed by the horizontal extrusion of the escaping fragments (Avouac & Tapponnier 1993). The area affected by collision extends into western China, Mongolia, Thailand, the Andaman Sea, and perhaps as far north as Siberia.

Figure 47a shows a kinematic model for the deformation of Asia, based on various types of geological, earthquake, and tectonic data compiled by Avouac & Tapponnier (1993) and redrawn by Yeats *et al.* (1997). The rotation senses of major blocks are given along the major directions of block motions. An obvious characteristic of the indent-linked strike-slip faults is that they are generally curved and discontinuous and therefore contrast with the more regular continental-boundary faults discussed earlier. Another characteristic is the absence of large-scale pull-apart zones or paired bends.

Fault bends of the Tibetan Plateau and western China. The largest fault on the Tibetan Plateau is the 600-km-long left-lateral Altyn Tagh Fault, that is believed to have initiated in the early Eocene (Fig. 47b). Cowgill *et al.* (2004*a*, *b*) studied the isolated Akato Tagh gentle bend occurring at a

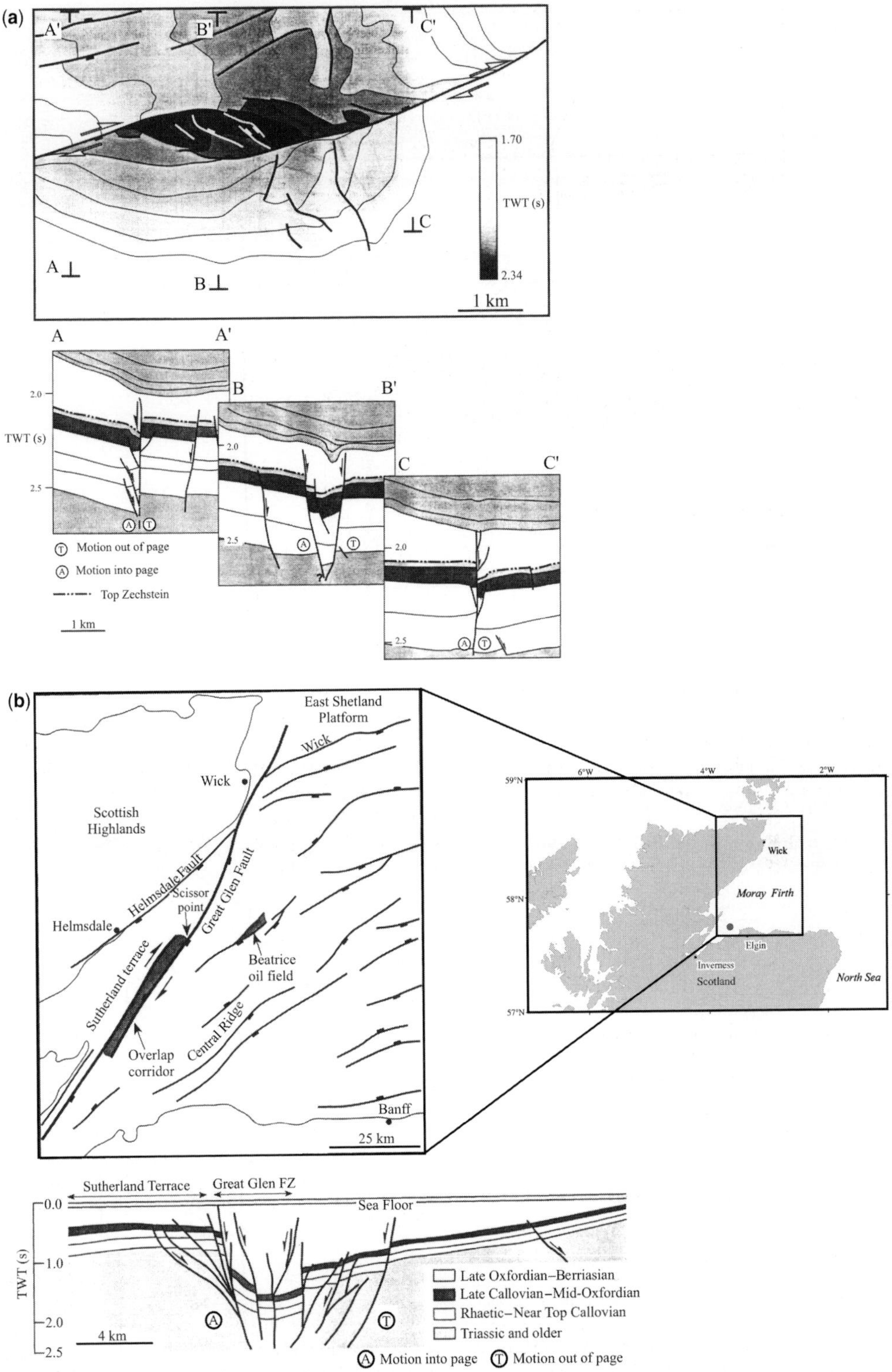

Fig. 42. (**a**) Time-structure map and cross-sections from a rhomboidal basin in the Netherlands, using 3D seismic-reflection data from Dooley *et al.* (1999) and subsurface mapping from Richard *et al.* (1995). (See location of the map on Fig. 39.) (**b**) Map and cross-section of a rhomboidal pull-apart basin of Late Jurassic age in the Inner Moray Firth, from Dooley *et al.* (1999), using subsurface mapping of Underhill (1991).

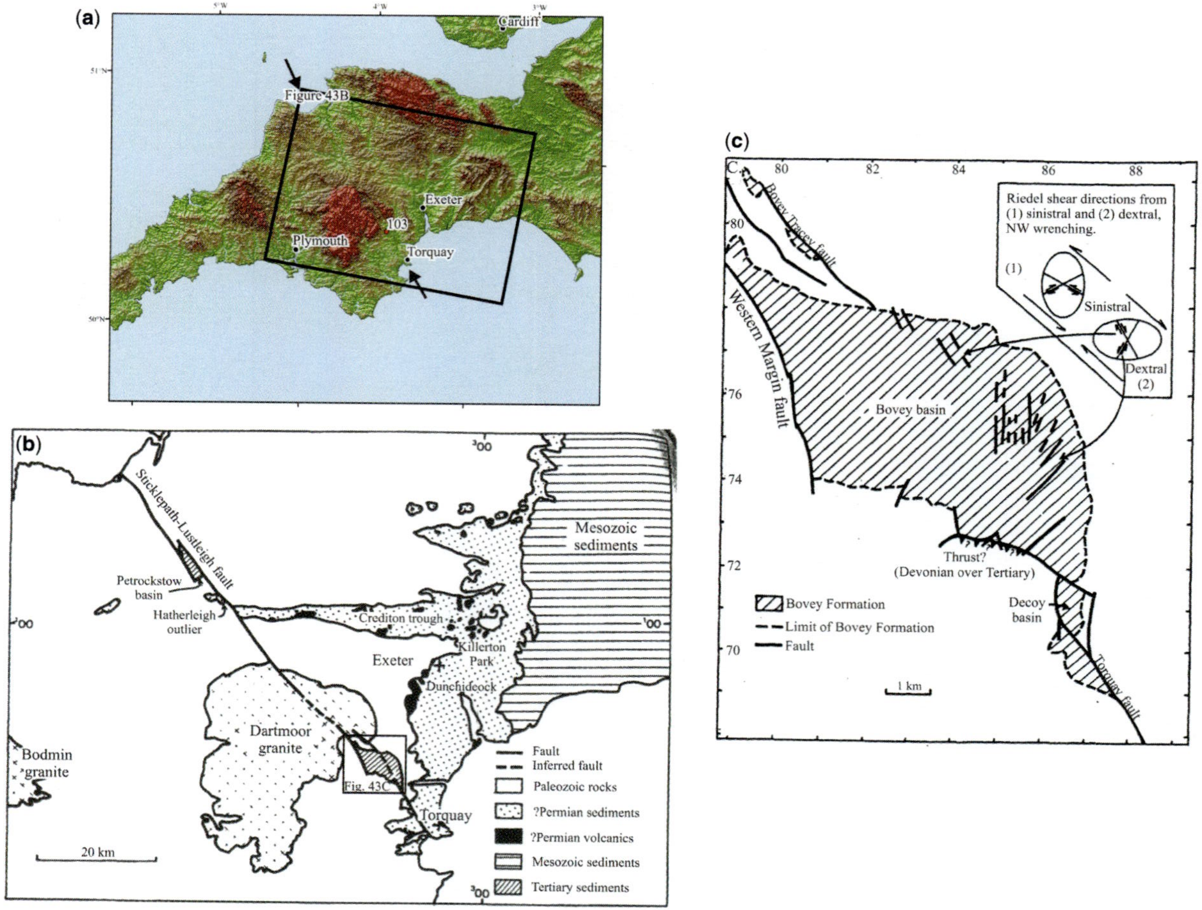

Fig. 43. (**a**) SRTM topography and GEBCO bathymetry used as a base map. (**b**) Simplified geological map of part of SW England, showing the Sticklepath–Lustleigh fault system, modified from Holloway & Chadwick (1986). (**c**) Geological map of the Bovey Basin, the largest basin along the Sticklepath–Lustleigh fault system. The basin formed by left-lateral offset on the left-stepping Bovey Tracey–Torquay faults and the Western margin fault. Minor Late Tertiary reactivation of the Bovey Tracey–Torquay faults is indicated by the presence of minor faults that have been interpreted as Riedel Shears.

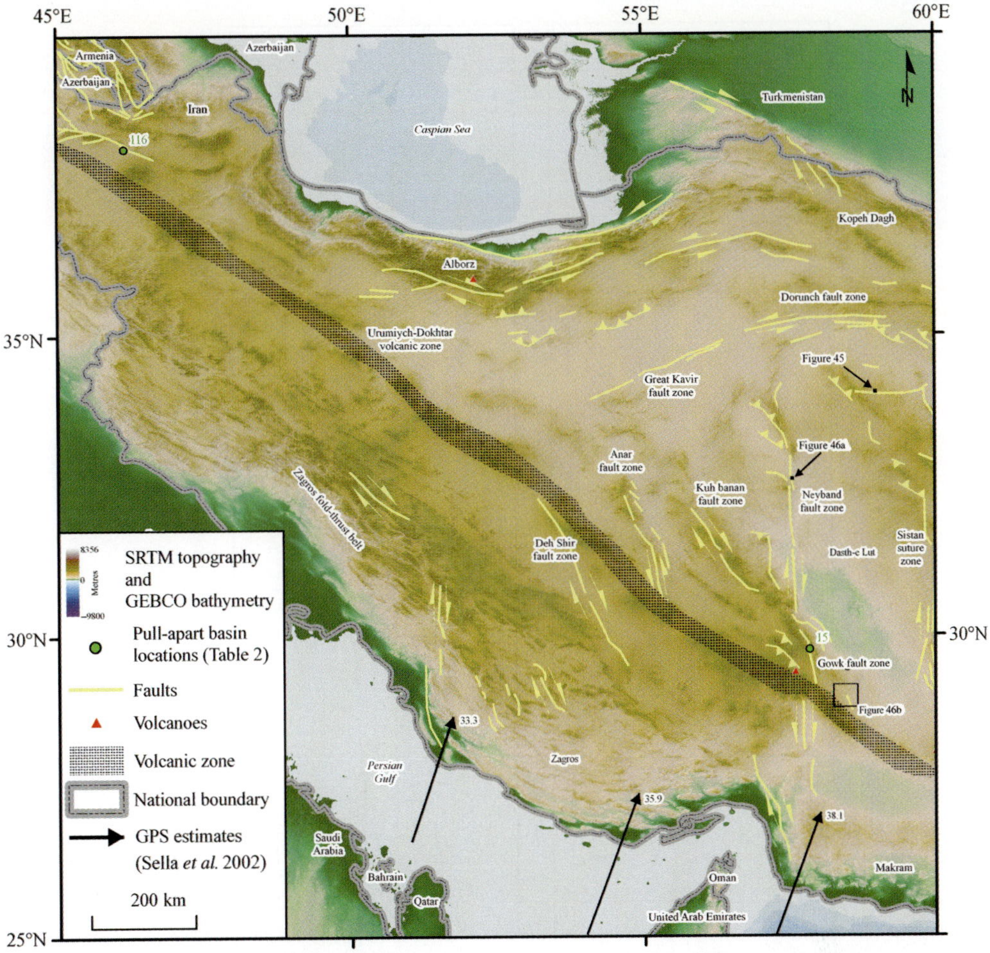

Fig. 44. Regional tectonic map of the active to Cenozoic strike-slip fault systems of Iran, using SRTM topography and GEBCO bathymetry as a base map. Yellow lines are active faults, taken from compilations by Walker & Jackson (2004) and Karakhanian *et al.* (2004). GPS-based directions of Arabia–Eurasia plate motion are from Sella *et al.* (2002).

20-km-wide fault separation (Fig. 48). Their main conclusions were that the bend and its zone of transpressional deformation probably formed recently, as the result of linkage of an en échelon set of shear fractures during the recent development of this fault strand.

Pull-apart basins are more common on right-lateral faults with NE trends, like the Karakoram fault zone, where an array of two large lazy-Z shape basins are present (Murphy & Burgess 2006); the Gulu fault zone, where one rhomboidal pull-apart is present (Edwards & Ratschbacher 2005); and the Haiyuan fault zone where an array of five rhomboidal pull-aparts are present (Zhang *et al.* 1989; Figs 47b, 49).

Mongolian and Gobi Altai. Restraining bends of various structural styles are common in the Mongolian and Gobi Altai and are described in detail by Cunningham (2003*a*,*b*; 2007). Four of these bends are illustrated on Figure 50, and they exhibit some characteristics in common with bends on the indent-linked strike-slip faults of the Tibetan Plateau. These bends are found in mechanically weak Palaeozoic accretionary belts that separate older cratonic areas and are subject to far-field stresses induced by the collision of India with Eurasia.

Restraining bends in this setting have distinctive characteristics. First, they tend to be isolated structures; second, the master faults cannot be traced far

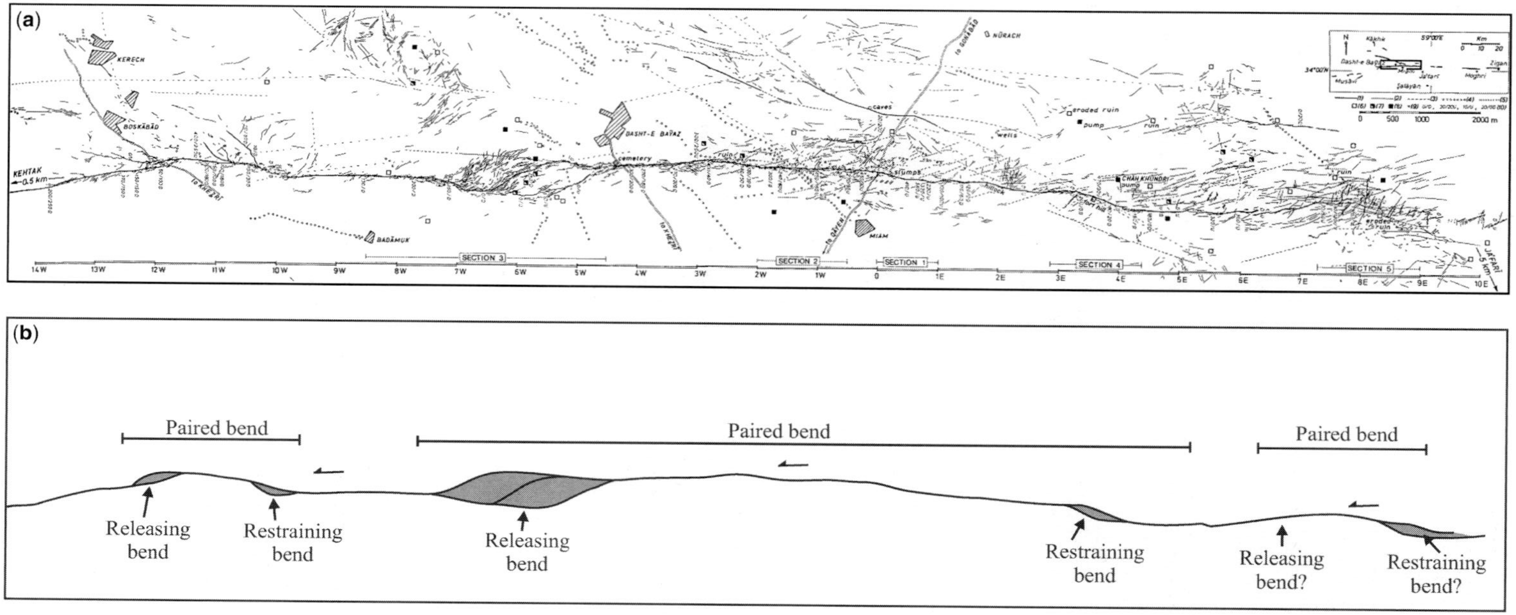

Fig. 45. (**a**) Fracture pattern in Quaternary alluvium that was mapped by Tchalenko & Ambraseys (1970), following the *M*7.2 Dasht-e-Bayaz earthquake of 31 August 1968 in Iran (location shown on Fig. 44). (**b**) Interpretation, showing the paired-bend structure.

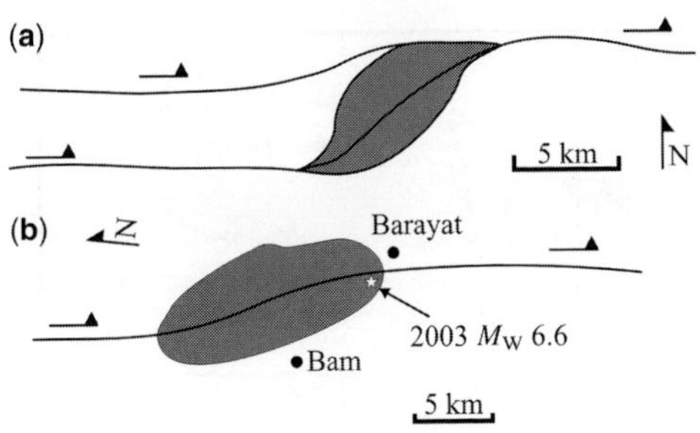

Fig. 46. (**a**) Sketch of the restraining bend connecting parallel strike-slip fault zones, taken from a Landsat image presented by Walker & Jackson (2004) for the Purang right-lateral strike-slip fault of Iran. (**b**) Sketch from the imagery of Fu *et al.* (2004) of a restraining bend near the epicentre of the 2003 M_w 6.6 Bam earthquake (star) that killed 40 000 people and left 30 000 injured.

as continuous strike-slip faults; and, third, pull-apart basins and paired bends are not recognized.

Tectonic setting and geological development of active bends of SE Asia

Several strike-slip fault systems extend SE through Mynamar, Thailand and Viet Nam, and may accommodate the escape of tectonic wedges into this region (cf. Morley 2002). Morley (2007) and Smith *et al.* (2007) have described a 150-km-long, lazy-Z restraining bend (the Chainat Complex) on the 500-km-long left-lateral Mae Ping fault zone.

Tectonic setting and geological development of active bends of the Lake Baikal region, north-central Asia

Oligocene to Recent rifting and basin formation in Lake Baikal (Fig. 51) can be explained by a step-over in a roughly east–west striking left-lateral fault zone that closely follows the southern edge of the Siberian craton. Extension in this region was long considered to be a distant, indentation and escape effect of the India collision, but Calais *et al.* (2003) have recently challenged this idea, using GPS results from the region. Their view is that trench forces from the western Pacific may dominate over the distant indent-linked India event.

Lake Baikal Basin. This seismically active basin is a lazy-S-shaped pull-apart basin occupied by a deep lake. The basin formed in the Oligocene and remains an active feature up to the present day (Mats 1993; Petit 1998; Calais *et al.* 2003).

Tunka Basin. The Oligocene to Recent Tunka Basin forms one of the many satellite basins surrounding Lake Baikal, and is an excellent example of a lazy-Z-shaped pull-apart basin (Larroque *et al.* 2001; Arjannikova 2004; Fig. 52).

Tectonic setting and geological development of ancient bends of the mid-continent of North America

Strike-slip faulting was pervasive in the interior of North America during the Late Palaeozoic collision between northern South America and the southern edge of North America and during the Late Cretaceous–Early Tertiary Laramide event associated with shallow subduction or collision along the western edge of North America (Kluth & Coney 1981; Marshak *et al.* 2003; Fig. 53a–c). A modern analogue of this type of distributed intracontinental deformation would be the modern India–Asia collision (Fig. 53c).

Owl Creek and Quealy Dome restraining bend. Bends in this area of Wyoming appear to be reactivated by the NE-directed Laramide event (Fig. 53d). The Owl Creek uplift is a restraining bend along an east–west strike-slip fault (Paylor & Yin 1993; Fig. 54a & b). The Quealy Dome formed between two left-lateral strike-slip faults related to the Laramide Orogeny (Fig. 54c & d) and resembles bends created in laboratory models (McClay & Bonora 2001).

Arbuckle Mountains and Ardmore Basin. Deformation and restraining-bend formation in this area was related to the inversion of the Cambrian Southern Oklahoma aulacogen during the Late Palaeozoic Ancestral Rockies Orogeny (Marshak *et al.* 2003; Figs 53 & 55).

Tectonic setting and description of natural bends on strike-slip faults in zones of active and ancient oblique subduction

In his pioneer study, Fitch (1972) first pointed out the role of 'trench-linked' (Woodcock 1986) or 'trench-parallel' (Yeats *et al.* 1997) strike-slip faults in accommodating the strike-slip component of oblique subduction zones. The basic principle is that a vertical strike-slip fault can concentrate shear in a much more efficient way than distributing shear across a much larger subhorizontal subduction zone interface. Fitch (1972) also noted that trench-linked faults probably nucleate on weak

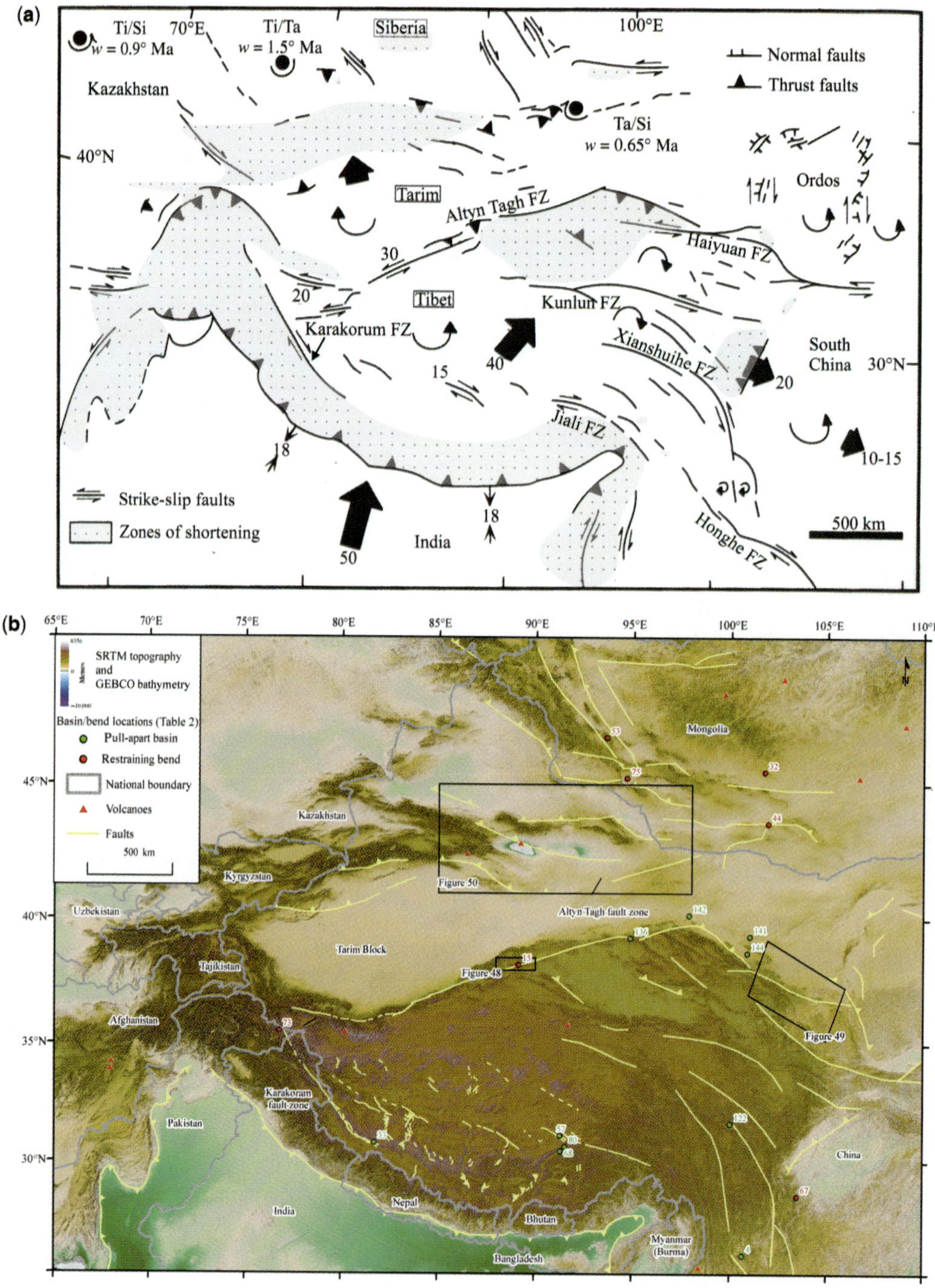

Fig. 47. (**a**) Block rotation model for the deformation of Tibet, from Yeats *et al.* (1997), redrawn from Avouac & Tapponnier (1993). Large arrows show block directions, small arrows show directions of rotation, and grey stippled zones show areas of shortening. (**b**) Regional tectonic map of active to Cenozoic strike-slip fault systems of the Tibetan Plateau and Central Asia, using SRTM topography as a base map. Yellow lines are active faults taken from compilation by Howard *et al.* (2003) and Levine (1995). Boxes show locations of the detailed maps in Figures 48–50.

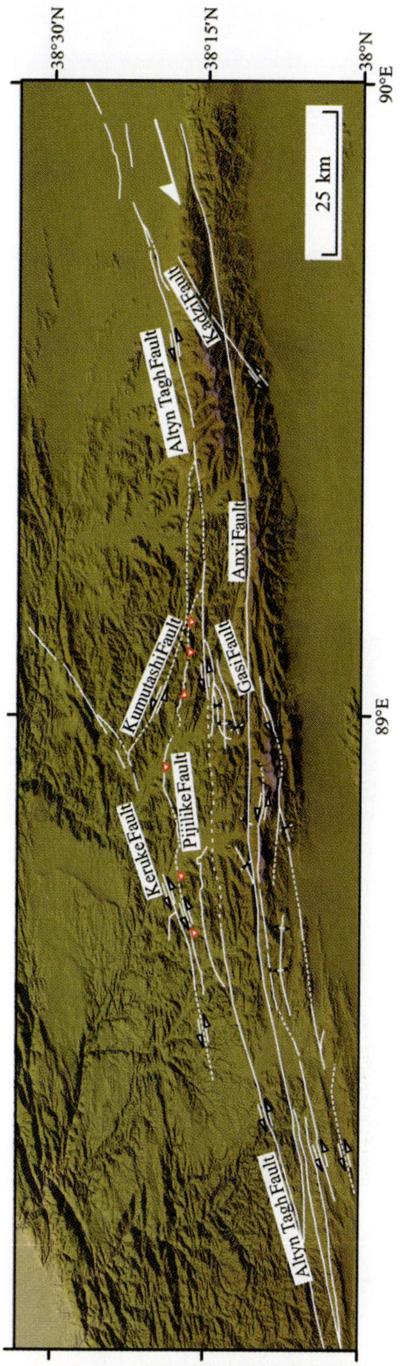

Fig. 48. Major faults of the Akato Tagh restraining bend along the Altyn Tagh fault zone, shown on an STRM base map and modified from Cowgill *et al.* (2004*a*, *b*). See Figure 47 for location.

zones in the crust formed by heated and thinned crust of the volcanic arc itself.

Jarrard (1986) summarized all aspects of trench-parallel strike-slip faults, and introduced the term 'fore-arc sliver plate' to refer to the microplate defined by the trench axis and the trench-parallel strike-slip fault. His review of trench-linked strike-slip faults has been updated with more recent data by Yeats *et al.* (1997). In this section, I will present some examples of trench-linked strike-slip faults with restraining- and releasing-bend segments, in order to compare these bends with those along continental-boundary and indent-linked strike-slip systems. In general, trench-linked faults are highly dependent on the trend of the subduction margin, and will change their character and rate as a function of changes in the strike of the subduction zone. For that reason, trench-linked faults tend to be less continuous and more segmented than continental-boundary faults.

Tectonic setting and geological development of active bends of the Sumatran strike-slip fault system, Indonesia

The right-lateral Sumatran fault system extends 1900 km as a sinusoidal and segmented fault zone (Sieh & Natawidjaja 2000) from its terminus as a Z-shaped pull-apart basin in the Sunda Strait between the islands of Sumatra and Java (Pramumijoyo & Sebrier 1991; Leglemann *et al.* 2000) to an extreme rhomboidal pull-apart basin floored by oceanic crust in the Andaman Sea (Curray 2005; Fig. 56). The rate of right-lateral fault slip passes from zero at its terminus in the Sunda Strait to a rate of 60 mm/a in the Andaman Sea.

A 300-km-wide fore-arc sliver plate occupies the space between the Sumatran trench and deformation front and the Sumatran Fault (Fig. 56). The overall shape of the fault is sinusoidal, with the northern half of the fault concave to the SW and the southern half of the fault concave to the NE. Sieh & Natawidjaja (2000) note that the sinusoidal trace of the Sumatran Fault is mimicked by the sinusoidal trace of the Sumatran deformation front, indicating a close genetic link between the subduction process and the geometry of the trench-linked fault. The question is whether deformation of the subducted slab influences the fault geometry in the overriding plate, or whether the fault geometry of the overriding plate has influenced the subducting slab.

Near its southern end, the fault curves towards the offshore trench and the fore-arc sliver plate thins in map view. Sieh & Natawidjaja (2000) used these relationships to calculate about 100 km of stretching of the fore-arc sliver plate parallel to

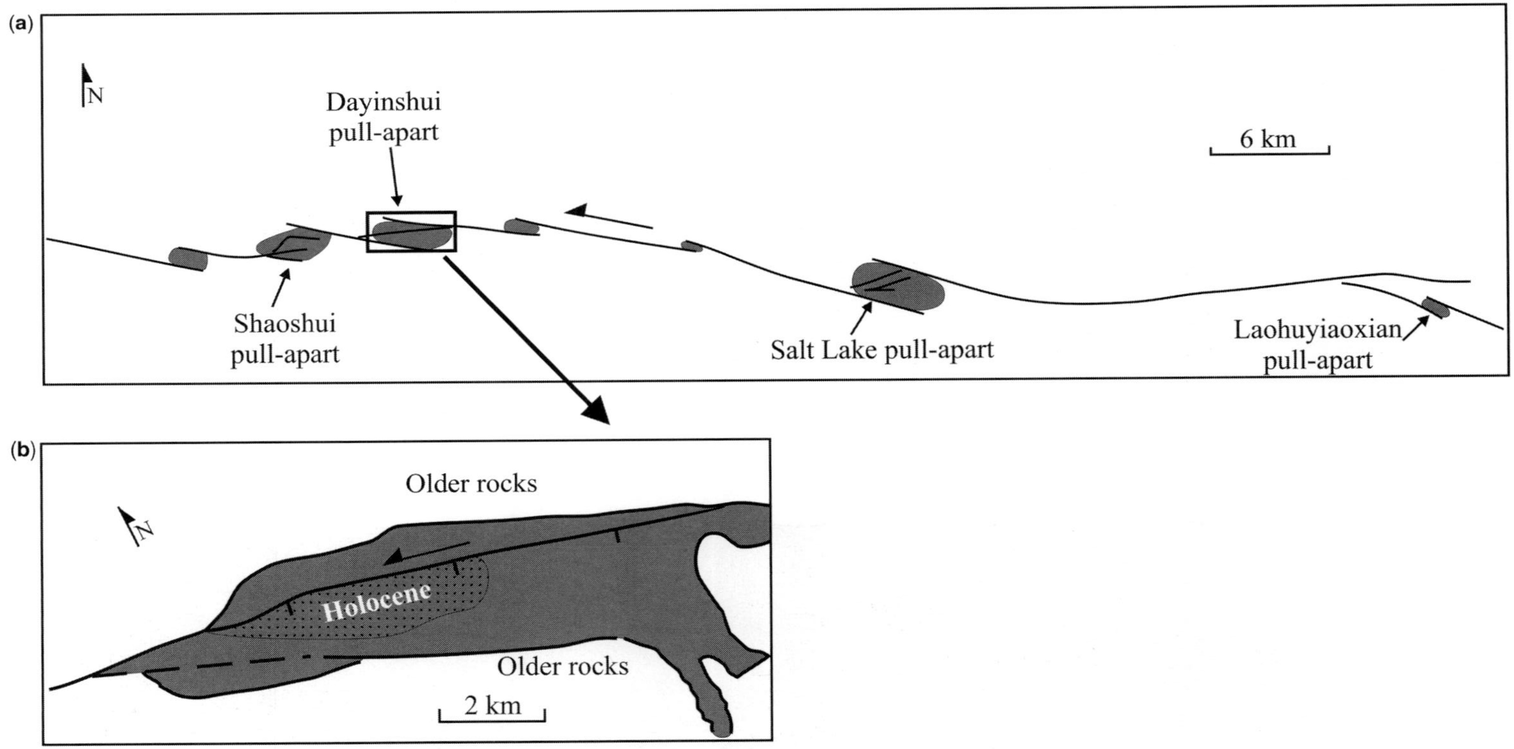

Fig. 49. (**a**) Map of an array of seven pull-apart basins along the Haiyuan Fault, mapped by Zhang *et al.* (1989). See Figure 47 for location. (**b**) The Dayinshui Basin shows 'pull-apart extinction', or an active fault cutting obliquely across the basin as observed in other basins.

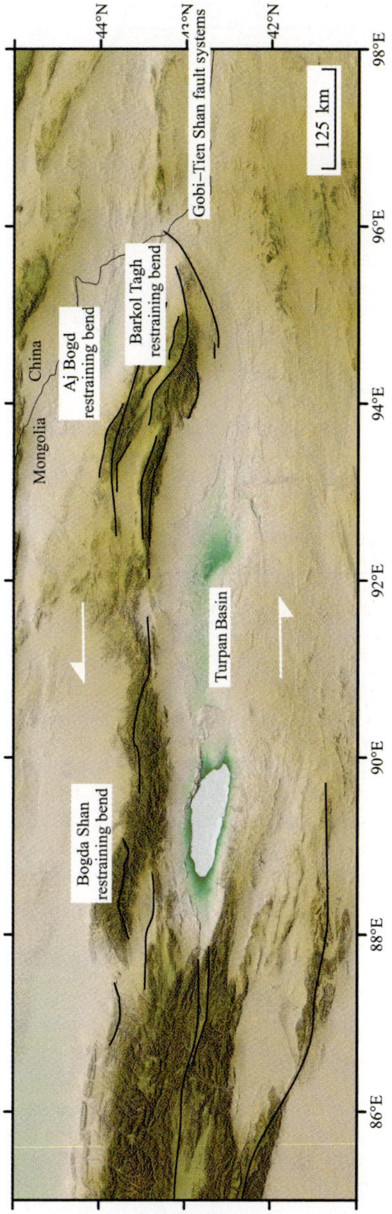

Fig. 50. Restraining bends along the Gobi–Tien Shan fault system in the Gobi Altai, southern Mongolia, modified from Cowgill *et al.* (2004*a*, *b*). See Figure 48 for location.

the trend of the Sumatran Fault. More orthogonal subduction in the area south of the Sunda Strait along the Java Trench does not require the formation of a trench-linked strike-slip fault along that segment of the trench.

Bends. A paired bend is present in the central segment of the Sumatran Fault, and is defined by a large lazy-S restraining bend with an adjacent array of four pull-apart basins showing a range of spindle-shaped to rhomboidal morphologies (Fig. 57a). This bend overlies the approximate down-dip projection of the Investigator fracture zone, a positive bathymetric feature on the down-going plate (Fig. 56). A possible bypass fault is present near the most curved part of the restraining bend (Fig. 57a).

The second-largest and best-studied of the pull-apart basins in the paired bend is the Singkarak pull-apart, which is occupied by a large lake (Fig. 57b). The 23-km length of the basin is shown by Sieh & Natawidjaja (2000) to have formed as a typical rhomboidal pull-apart (Fig. 7). Its length of 23 km is taken by them to be the approximate magnitude for the most recent phase of offset on the Sumatran Fault.

Tectonic setting and geological development of active bends of Atacama strike-slip fault system, Chile

As reviewed by Dewey & Lamb (1992) fore-arc tectonics and trench-linked strike-slip faulting are highly dependent on the orientation of the subduction zone and the obliquity of subduction along the Andean margin. The Atacama fault zone of northern Chile forms a linear zone of Quaternary left-lateral strike-slip faulting that stretches for a distance of 1100 km at an average distance of 30–50 km from the coastline (Fig. 58). A parallel, 700-km-long fault, the Coastal Scarp, exhibits evidence for Quaternary normal faulting. The Atacama Fault, has a long history of left-lateral strike-slip displacement that has been dated back to the Jurassic and Early Cretaceous (Scheuber & Andriessen 1990). Mesozoic extensional duplexes in deeper crustal rocks have been described along the Atacama fault zone by Cembrano *et al.* (2005), along with strike-slip-related block rotations (Arrigada 2003).

Armijo & Thiele (1990) point out that the observed evidence for Quaternary left-lateral shear and normal faulting contradicts the plate-tectonic prediction of right-lateral shear for oblique subduction along this part of the subduction zone (Fig. 58). They propose an alternative block model related to left-lateral shearing of blocks northward into the Arica bend (Atacama fault zone) and normal faulting above the underlying thrust stacks (Coastal

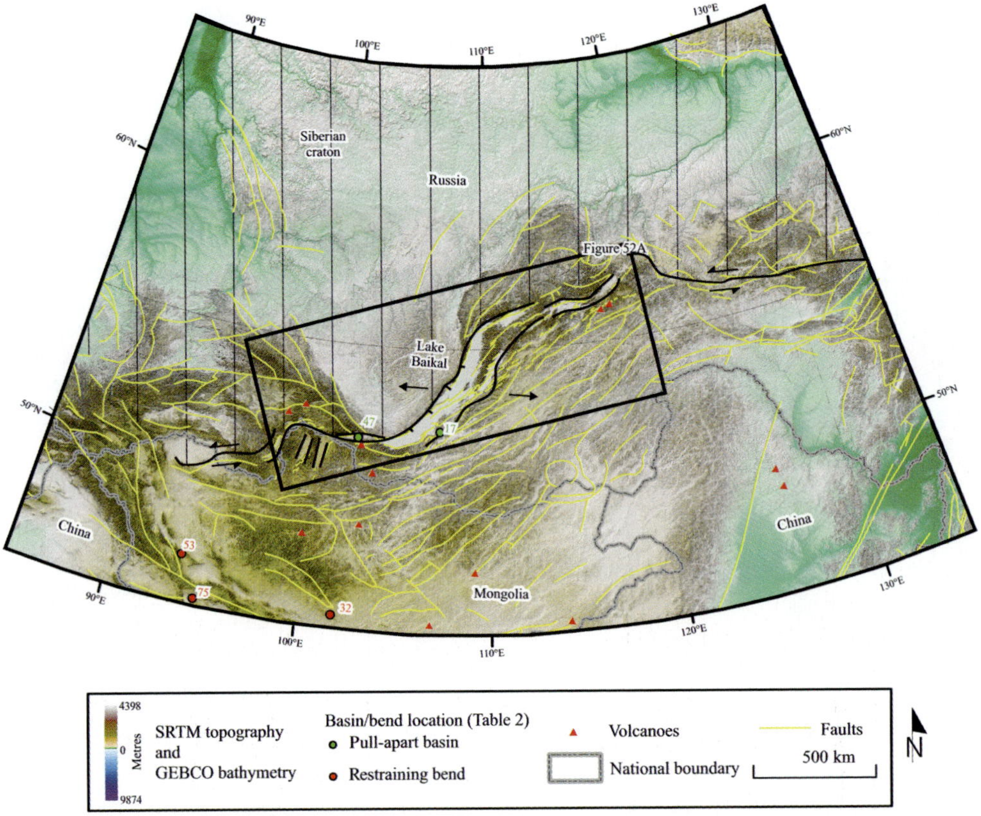

Fig. 51. Tectonic setting of the Lake Baikal pull-apart basin in eastern Russia. Faults are from a compilation by Howard *et al.* (2003), and Levine (1995).

fault zone), to explain their observations. Another active trench-linked fault in southern Chile, the right-lateral Liquine–Ofqui strike-slip fault zone, has the correct shear sense for the direction of oblique subduction (Thomson 2002).

Bends. As in the case of Sumatra, the strike-slip and normal faults of the Chilean margin closely follow the changing trend of the Chile Trench, although the trace of the Atacama fault zone has a more sinusoidal trace than the trench itself (Fig. 58). The Cerro de Mica lazy-Z restraining bend is formed at an 8-km-wide fault separation on two faults adjacent to the present-day trace of the Atacama fault zone (Fig. 58). The internal structure of the Cerro de Mica is comparable with laboratory bends of the same type created by McClay & Bonora (2001). No pull-aparts have been reported from the Atacama fault zone.

The Salina de Fraile lazy-S pull-apart (Reijs & McClay 2003) formed on a left-lateral fault in the Argentinian Andes, presumably as a response to the same tectonic mechanism for left-lateral faulting proposed by Armijo & Thiele (1990; Fig. 59a). The basin exhibits oblique normal faults at its edge, and a cross-basin strike-slip fault zone similar to those described in young pull-aparts by Mann *et al.* (1983) and Zhang *et al.* (1989), and to those produced by Dooley & McClay (1997) in laboratory experiments.

Tectonic setting and geological development of active bends of the Kurile Islands strike-slip fault system, Kurile Islands and northern Japan

Kimura (1986) first proposed the existence of a fore-arc sliver adjacent to the southern Kurile Arc that is driven by the oblique subduction of Pacific crust (Fig. 60). Using data from both onland and offshore, he described a submarine rift at the NE end of the sliver, and a small fold–thrust belt that extended on land into northern Japan (Hokkaido Island; Fig. 61). DeMets (1992) used earthquake focal

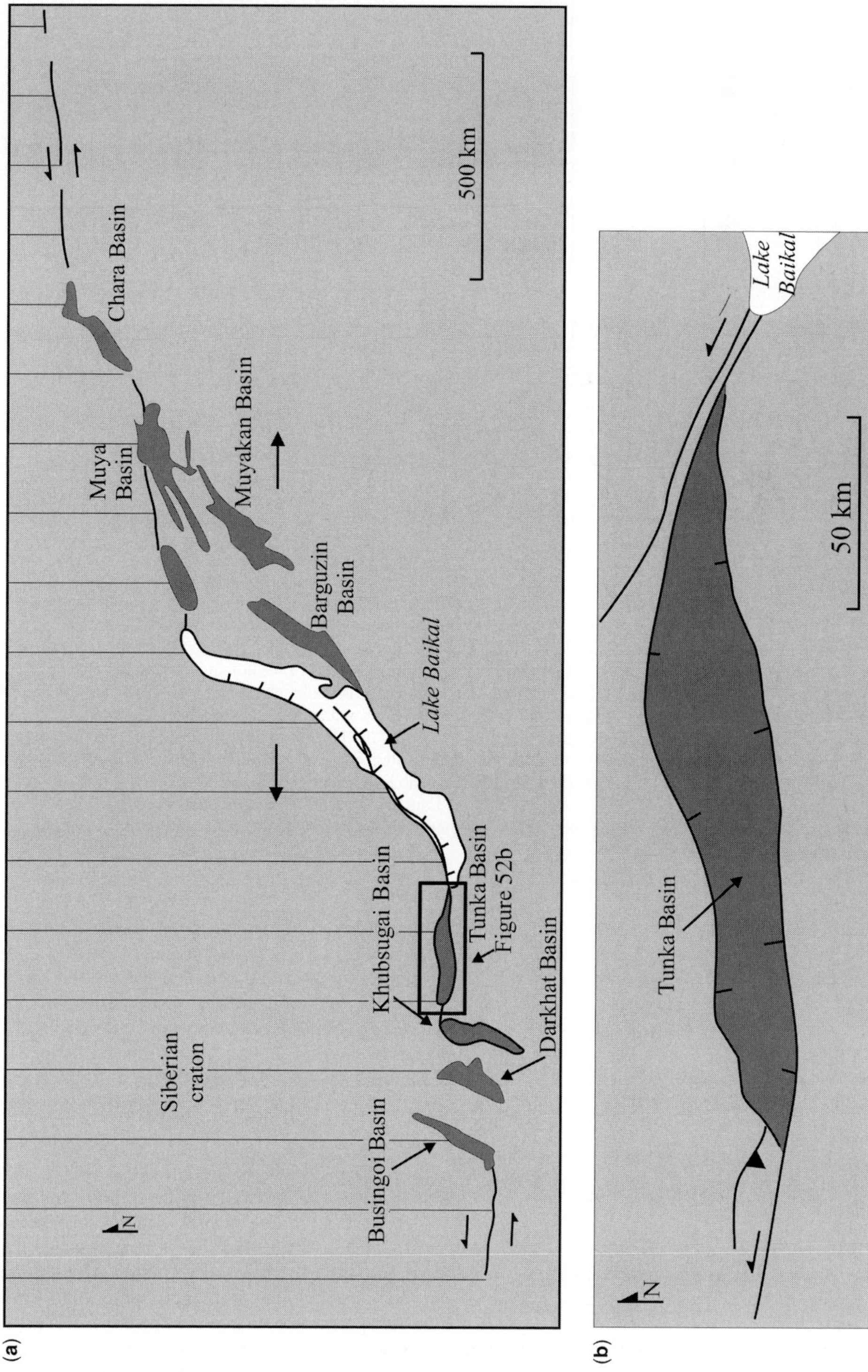

Fig. 52. (**a**) Schematic map of the Lake Baikal pull-apart basin, modified from Larroque *et al.* (1999). (**b**) Schematic map of the Tunka pull-apart basin, modified from Larroque *et al.* (1999).

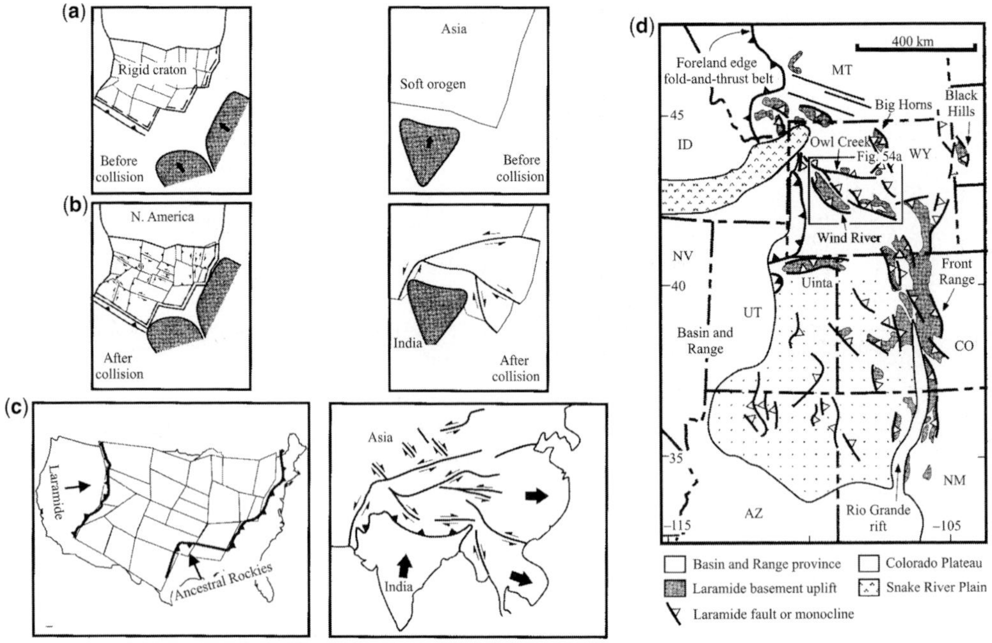

Fig. 53. (**a**) Comparison maps from Marshak *et al.* (2003) of the interior platform of the United States at the time of the Alleghenian–Ouachita Orogeny (left), and the eastern Eurasia Orogeny produced by indentation by the Indian continent (right). The interior platform of the US is a rigid craton whose upper crust has been broken into a mosaic of blocks by faults. In eastern Eurasia, the southern margin of the continent is a soft Phanerozoic orogen. (**b**) During the Alleghenian–Ouachita collision, the craton was strained only slightly, so that the internal crustal blocks moved only slightly. In eastern Eurasia, crustal blocks underwent eastward lateral escape as the lithosphere was strained significantly. (**c**) The mosaic of crustal blocks in the interior platform of the US contrasts with the major regional faults in eastern Eurasia. (**d**) Map of the US, showing the locations of the Alleghenian–Ouachita Orogeny responsible for the Palaeozoic Ancestral Rockies, and the Laramide Orogeny responsible for reactivation of the previously formed Ancestral Rockies along with formation of new structures in the Late Cretaceous and Early Cenozoic. (**d**) Regional map from Marshak *et al.* (2003), showing the distribution of Laramide basement uplifts and structures of the Colorado Plateau and the Rocky Mountains.

mechanisms to support the existence of the Kurile fore-arc sliver and a 6–11 mm/a rate for its SW translation relative to the North American plate.

Bends. As in the case of the Sumatran Fault, the unnamed left-lateral strike-slip fault bounding the sliver curves strongly into the rifted area and may indicate enhanced crustal thinning within the accretionary wedge (Fig. 61). One rhomboidal restraining bend is shown by Kimura (1986) along this fault adjacent to the Kurile Islands of Etoroph and Urup.

Tectonic setting and geological development of active bends on strike-slip faults in Japan

Oblique subduction beneath the central part of Japan induces left-lateral faulting along the Median Tectonic Line, the longest strike-slip fault in Japan (Fig. 62). The Median Tectonic Line formed along a terrane boundary within the crust, and gentle restraining bends in the fault trace are marked by elongate outcrops of blueschist rocks (Mann & Gordon 1996). Part of this shortening is accommodated by an extensive zone of right-lateral faults, which are much shorter in length but distributed over a larger area (Fig. 62).

Bends. The 1995 Kobe earthquake – which, at a cost in excess of 200 billion US dollars, was one of the most expensive natural disasters in history – ruptured a zone that is subparallel to the Median Tectonic Line (Fig. 62). Aftershocks outline a rhomboidal pull-apart along this fault near the epicentral area (Sato *et al.* 1998). The adjacent Rokko Range uplift may be a restraining bend paired to this releasing bend (Sangawa 1986).

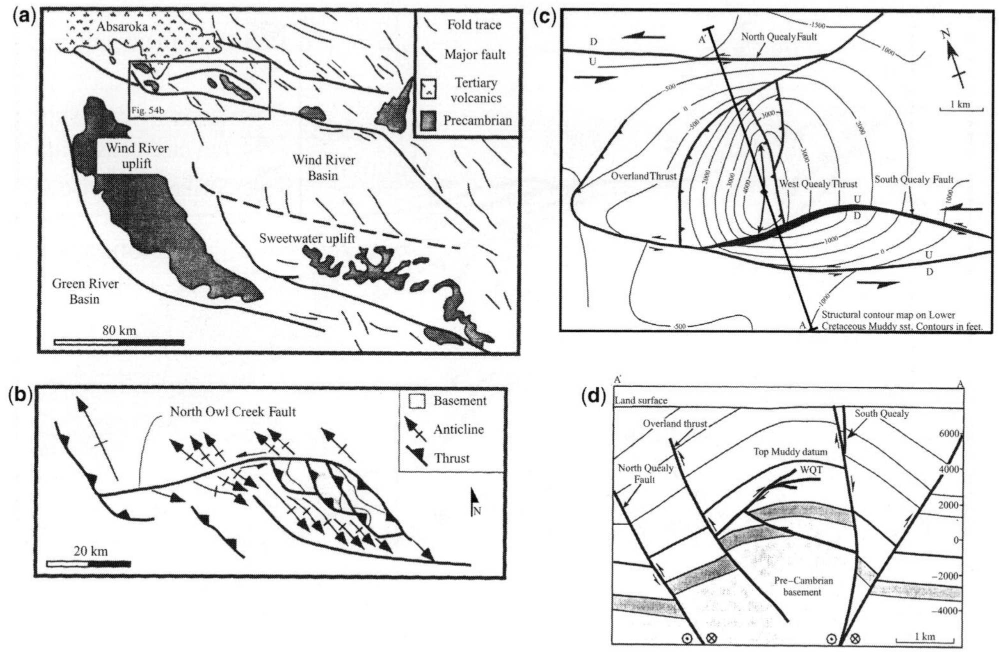

Fig. 54. (**a**) Map from Marshak *et al.* (2003), redrawn from Stone (1969), showing en échelon folds suggestive of strike-slip faulting through Wyoming formed during the Laramide Orogeny. (**b**) Map from Marshak *et al.* (2003), redrawn from Paylor & Yin (1993), showing en échelon folds in a restraining bend along the North Owl Creek Fault in Wyoming. (**c**) Map from Stone (1995) in McClay & Bonora (2001), showing the Quealy restraining bend in the Laramie Basin, Wyoming. (**d**) Cross-section through the Quealy restraining bend, from Stone (1995) in McClay & Bonora (2001).

Tectonic setting and geological development of active bends along the Central American fore-arc sliver

The presence of strike-slip faults along the volcanic line of the Central America Arc, as a possible indicator of the edge of a fore-arc sliver, was first proposed on the basis of earthquakes, by Harlow & White (1985; Fig. 63a) and White & Harlow (1993). DeMets (2001) placed kinematic constraints on the fore-arc sliver using earthquake focal mechanisms, and estimated a rate of 9 mm/a for the northward displacement of the sliver. Unlike Sumatra, the Kuriles and Japan, the fore-arc sliver is driven by less than 10° of oblique plate convergence.

Bends. For many years, there was no geological evidence for active strike-slip faults parallel the volcanic arc. Carr (1976) and Corti *et al.* (2005) identified active right-lateral faults in the El Salvador strike-slip fault system that roughly parallels the volcanic arc. The fault zone contains one prominent rhomboidal pull-apart Basin, the Rio Lempa Basin (Fig. 63b). To the south, in Nicaragua, Girard & van Wyk de Vries (2005) have proposed similar pull-apart basins, although these features are largely covered and have not been mapped in the subsurface.

Tectonic setting and geological development of active bends of the Philippine fault system

Introduction. The 1200-km-long, left-lateral Philippine strike-slip fault system is a trench-linked strike-slip fault related to oblique subduction of the Philippine Sea plate beneath Eurasia (Fitch 1972; Barrier *et al.* 1991; Fig. 64). Motion at a rate of 1–5 mm/a on the Philippine fault zone and Philippine Trench occurred in the Late Neogene as a back-thrusting response to underthrusting of the Palawan Block along the western edge of the block.

Bends. The fault has a pronounced arcuate shape that is convex to the NE and may reflect the

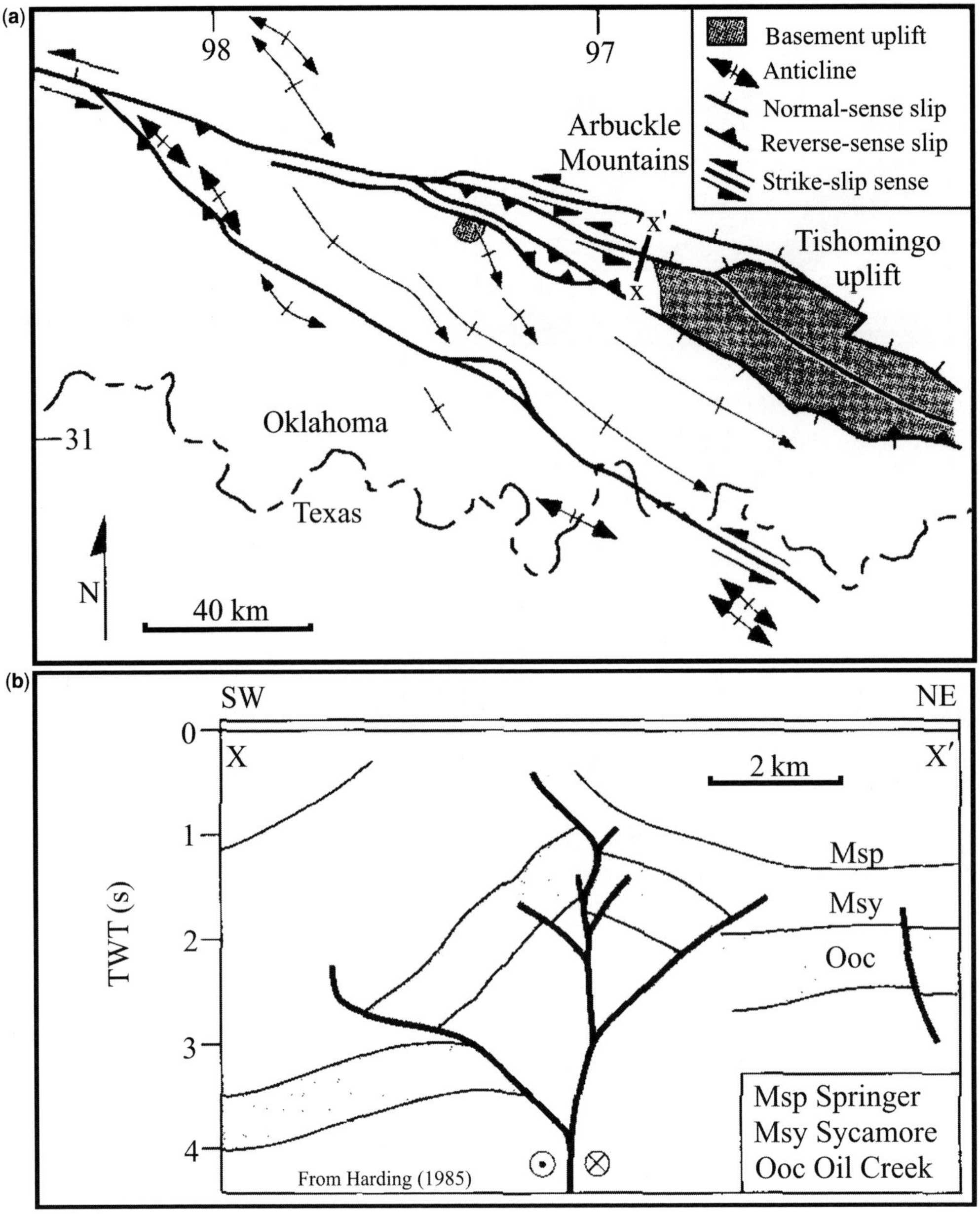

Fig. 55. (**a**) Map of part of the southern Oklahoma aulacogen, including the Arbuckle Mountains and the Ardmore Basin. (**b**) Flower-structure interpretation of a seismic line across the Ardmore Basin (at X–X′). Modified from Harding (1985).

collision of the Palawan Block (Fig. 64). Geological studies (Barrier *et al.* 1991) and GPS studies (Aurelio 2000) show that motion along this curved part of the fault is almost pure strike-slip. The northern terminus of the fault is characterized by an apparent large restraining-bend structure (Fig. 64). An inverted releasing bend containing Late Miocene–Early Pliocene sedimentary rocks has been described from the northern end of the fault system (Ringenbach *et al.* 1990).

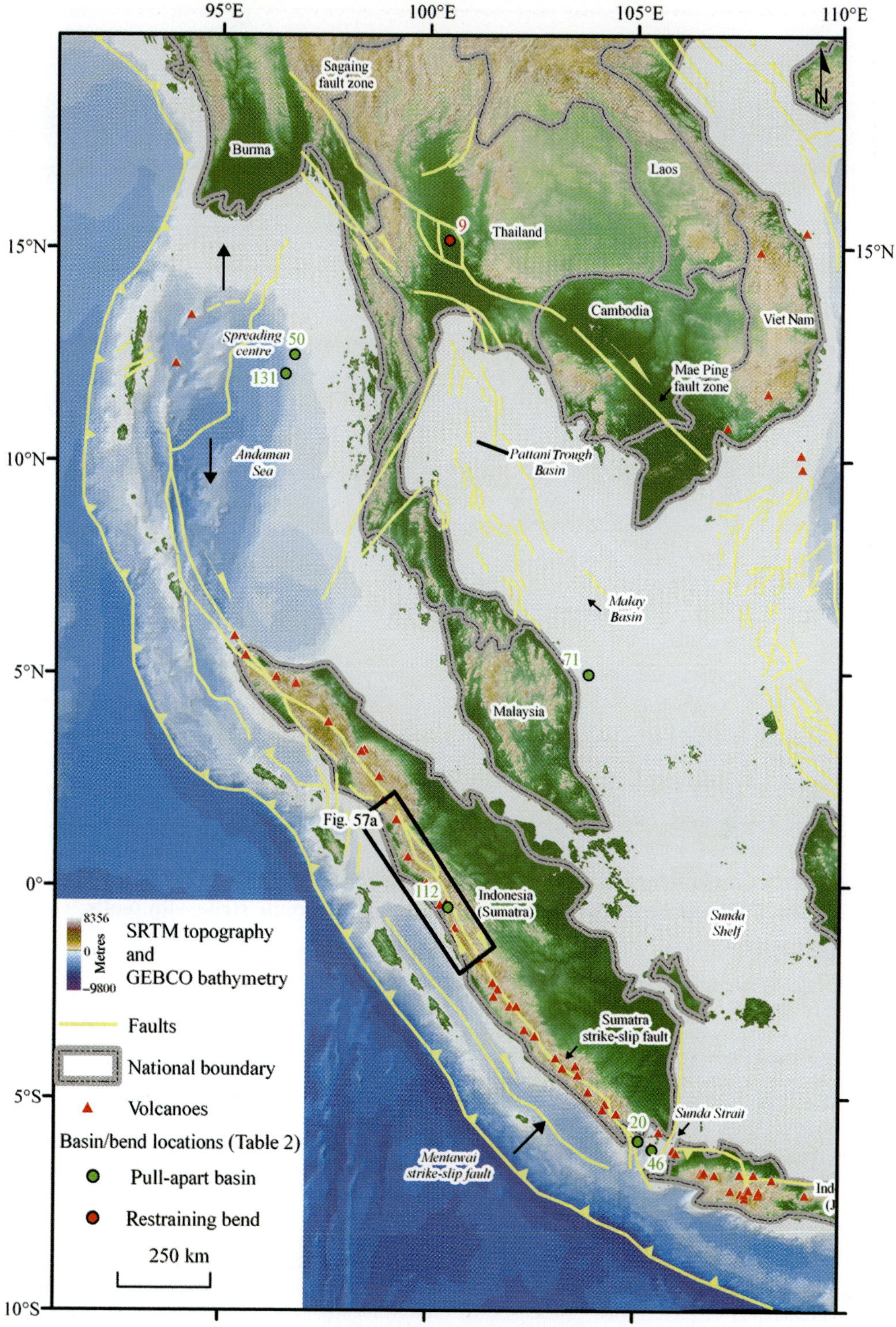

Fig. 56. Regional tectonic map of active to Cenozoic strike-slip fault systems of Sumatra and SE Asia, using SRTM topography and GEBCO bathymetry as a base map. Yellow lines are active faults taken from compilation by Levine (1995), and Coffin *et al.* (1997). Box shows the location of the detailed map in Figure 57.

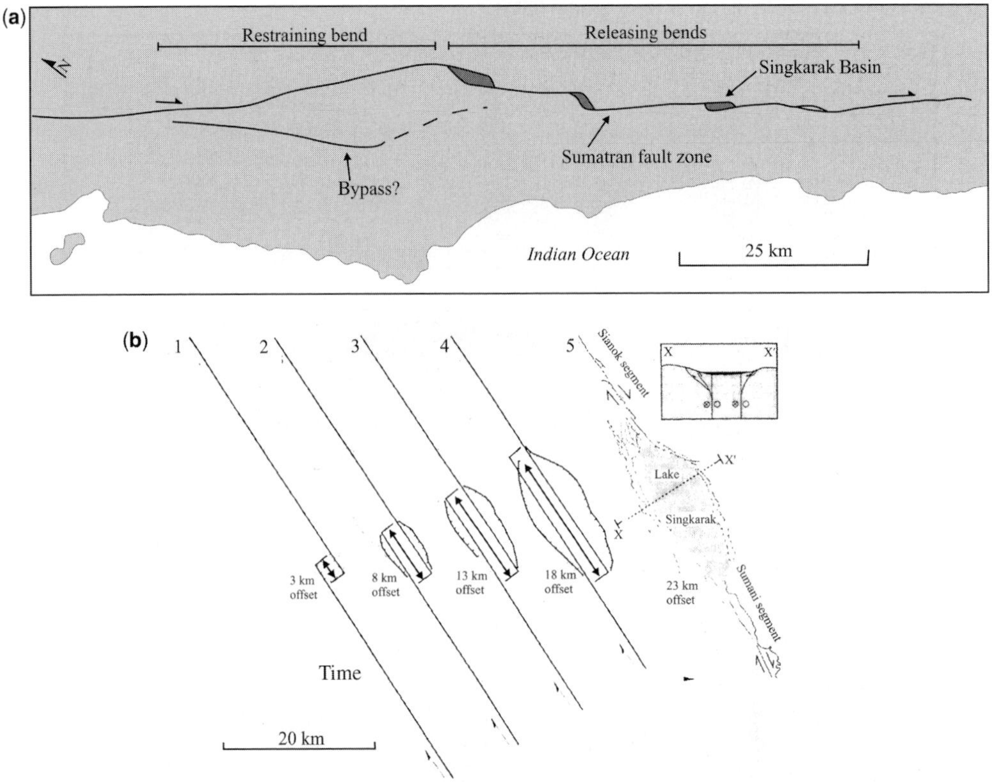

Fig. 57. (**a**) Sketch map of a large paired-bend structure along the Sumatran fault zone, from mapping by Sieh & Natawidjaja (2000). (**b**) Evolutionary model by Sieh & Natawidjaja (2000), showing the progressive opening of the Lake Singkarak pull-apart basin along the Sumatran fault zone.

Tectonic setting and geological development of ancient bends of the Berengian margin

The Berengian margin, or the eastern margin of the Bering Sea adjacent to Alaska, was mapped by the oil industry during routine hydrocarbon exploration in the 1970s and 1980s (Worrall 1991). This wealth of subsurface seismic-reflection and drill data revealed a 1000-km-long, right-lateral strike-slip system active from during the Palaeocene and Eocene (Fig. 65). As in Sumatra and other modern arcs, this margin was characterized by a strike-slip fault parallel to a subduction zone and defined a fore-arc sliver.

Bends. Subsurface mapping has shown several excellent examples of paired-bend geometries (St George pull-apart basin and Black Hills restraining bend, Fig. 66a) and isolated rhomboidal pull-aparts (Fig. 66b) and sharp restraining bends (Fig. 66c).

Tectonic setting and descripition of bends on cratonic strike-slip faults

I am adding a fifth category of bends to the other classes proposed by Woodcock (1986; Table 1). I call this fifth class the 'bends on cratonic strike-slip faults'. These are defined as bends in cratonic areas that are far removed from active plate boundaries, yet are still subject to intraplate stresses. These bends, which are remarkably similar to sharp restraining bends, are defined mainly on the basis of their intraplate earthquake activity, and in many cases seismically reactivate known basement structures that generally correspond with failed rifts (Gangopadhyay & Talwani 2003). Cratonic settings of strike-slip faults differ from the areas of Central Asia discussed by Cunningham (2007) because the Central Asian examples that are all found in mechanically weak Palaeozoic accretionary belts that separate areas of old, cold cratonic areas and are subject to distant stressed induced by continental collisions.

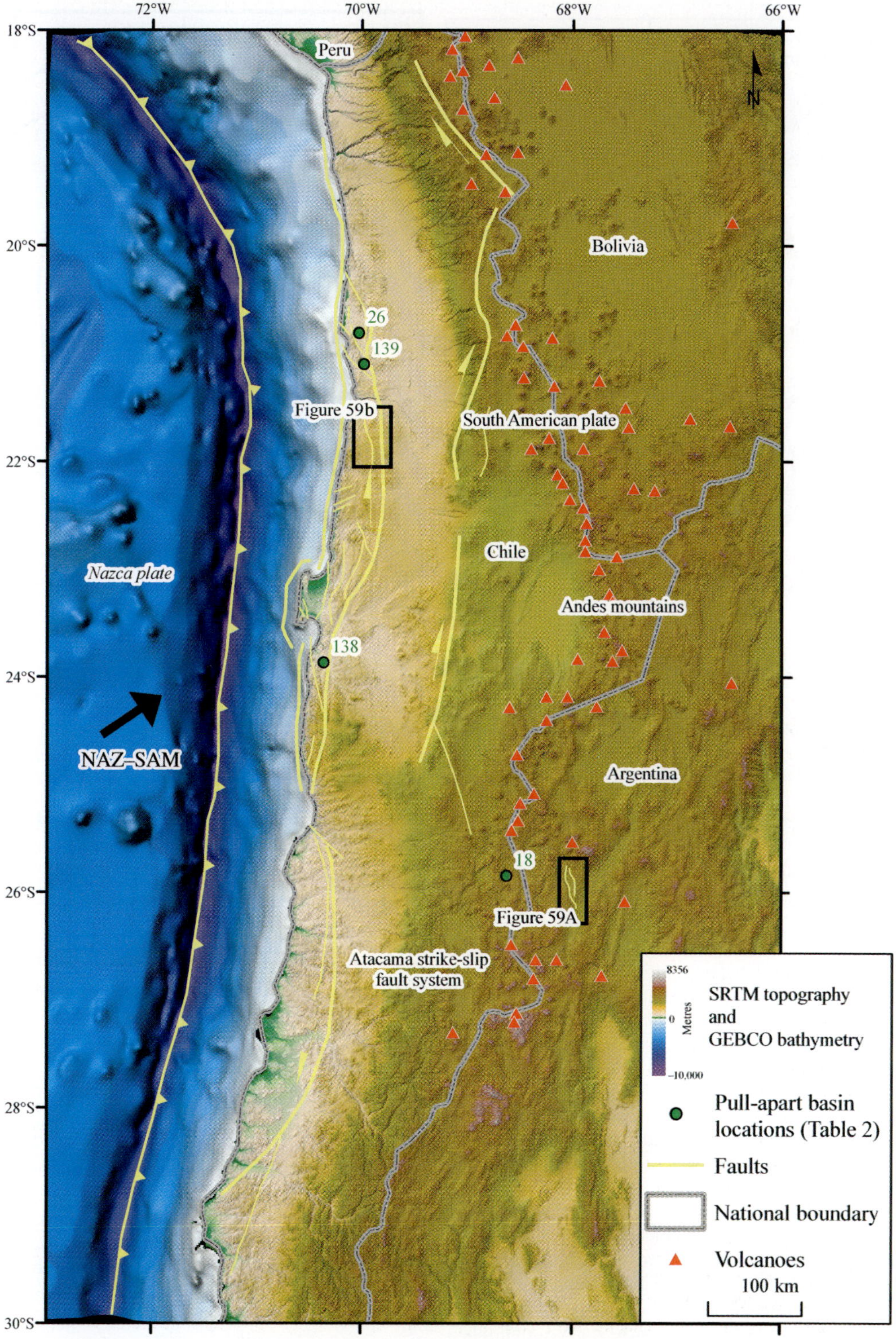

Fig. 58. Regional tectonic map of active to Cenozoic strike-slip fault systems of Chile, using SRTM topography and GEBCO bathymetry as a base map. Yellow lines are active faults taken from compilations by Levine (1995) and Armijo & Thiele (1990). Boxes show locations of detailed maps in Figures 59A & 58B.

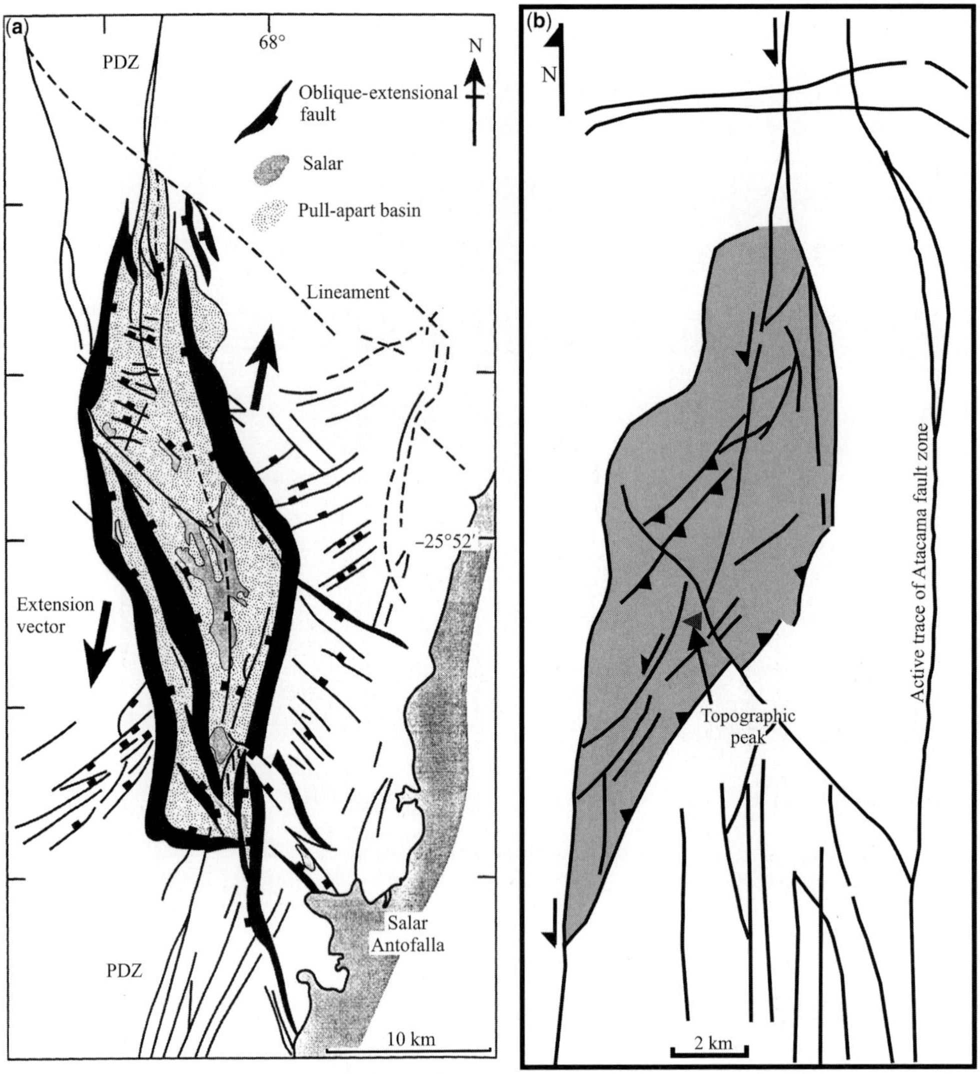

Fig. 59. (**a**) Map of the Salina del Fraille releasing bend in the northern Argentinian Andes, from Dooley & McClay (1997). (**b**) Map of the Cerro de la Mica restraining bend along the Atacama fault system in northern Chile (from McClay & Bonora 2001).

New Madrid area, USA. This is the most seismically active part of the mid-continent of North America and was the site of three or four large earthquakes in the period of 1811–1812, all with a body-magnitude of 7.0 or greater (Marshak *et al.* 2003). Earthquake epicentres define a bend in a fault that has reactivated older faults in the underlying Reelfoot Rift, a NNE-trending failed rift into continental crust of North America. The rift formed in Late Proterozoic to Early Cambrian times (Fig. 67a). Focal mechanisms indicate that earthquakes on the NE and SW fault segments are right-lateral strike-slip, and that earthquakes on the central segment are thrusts with the NE side up (Gangopadhyay & Talwani 2003). The deformation is consistent with reactivation of the ancient faults by the present-day NE–SW-trending maximum compressive stress trajectory of the central part of the North American continent (Marshak *et al.* 2003; Fig. 67a).

Middleton Place–Summerville seismic zone, South Carolina. This cluster of seismicity was the

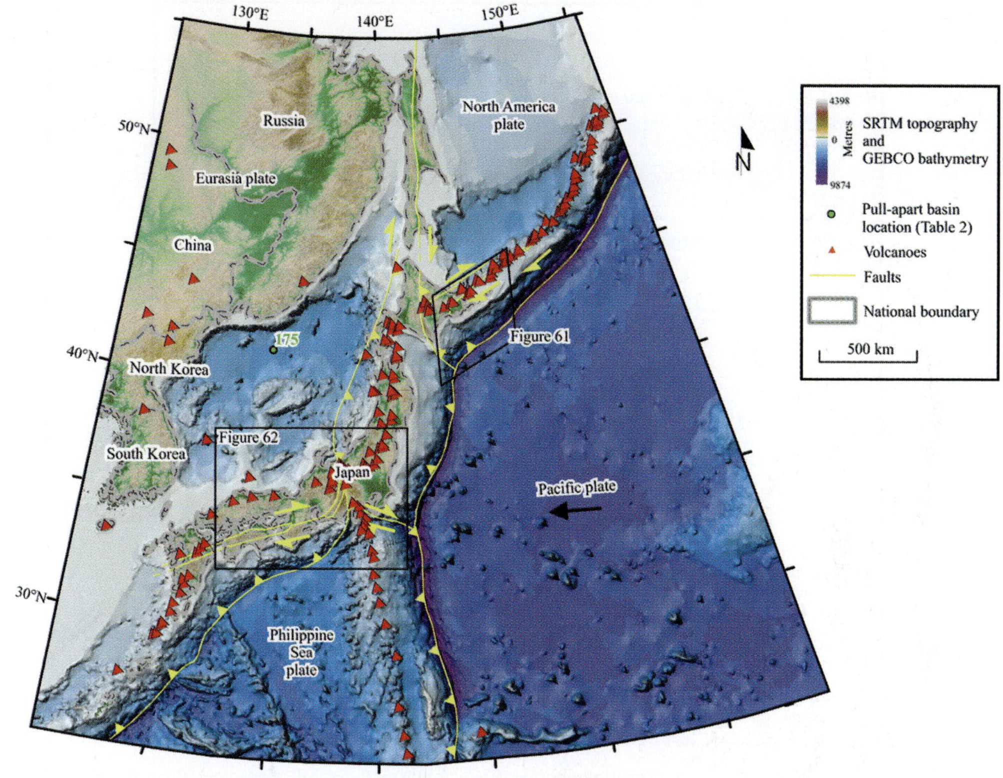

Fig. 60. Regional tectonic map of active to Cenozoic strike-slip fault systems of Japan, using SRTM topography and GEBCO bathymetry as a base map. Yellow lines are active faults, taken from compilations by Mann & Gordon (1996) and Coffin *et al.* (1997). Boxes show the locations of detailed maps in Figures 61 & 62.

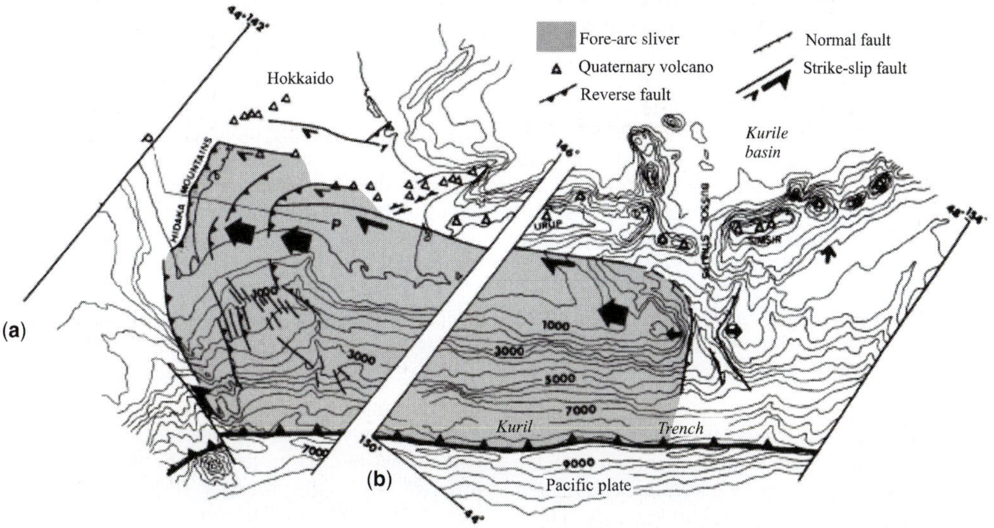

Fig. 61. Active fore-arc sliver (shaded in grey) along the SW Kurile Arc north of Japan, from Kimura (1986). Oblique subduction of the Pacific plate beneath the Kurile arc drives the sliver to the SW into a mini-collision zone with the northern island of Japan (Hokkaido). An unnamed active strike-slip fault connects the extensional zone in the Bussol Strait with a collisional zone on and near the island of Hokkaido in northern Japan.

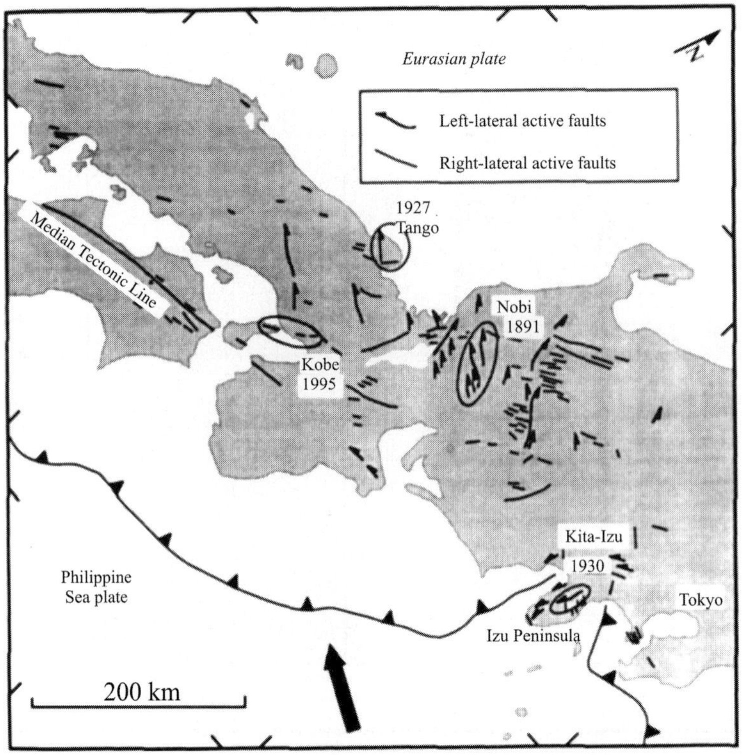

Fig. 62. Many active faults in Japan are driven by oblique subduction right-lateral displacement along the Median Tectonic Line (modified from Yeats *et al.* (1997). Ellipses show the locations of major damaging shallow strike-slip earthquakes related to deformation of the overriding plate in Japan, including the fault responsible for the 1995 Kobe earthquake, one of the most economically devastating earthquakes in history.

source of the 1886 Charleston (South Carolina) earthquake, and is currently the most active zone of seismicity in the South Carolina coastal plain (Gangopadhyay & Talwani 2005; Fig. 67B). As in the case of the New Madrid seismic zone, this cluster overlies a junction of faults related to the Mesozoic South Georgia rift basin. Earthquake focal mechanisms and modelling show that the NE and SW fault segments are right-lateral strike-slip, and that earthquakes on the central segment are thrusts with the NE side up (Gangopadhyay & Talwani 2005). The present-day maximum compressive stress trajectory of the eastern part of the North American continent is NE–SW and consistent with this kinematic interpretation.

Significance and relevance to strike-slip bends. Both bends have been explained in terms of the pre-existing faults acting as 'stress concentrators,' which cause anomalous buildups in their vicinity within a localized volume of weak crust (Gangopadhyay & Talwani 2005). Gangopadhyay & Talwani (2003) have compiled 39 examples of similar tectonic settings of seismic areas at fault intersections mostly within failed rift structures on continental margins far from the effects of active plate boundaries. One fault is typically longer than the other, and is favourably oriented for failure – given the direction of the maximum compressive stress. The recurrence interval of major earthquakes on these zones is much longer than similar faults in active plate boundaries.

Discussion

A major point of this lengthy introduction was to show that the structural styles of about 200 well-documented releasing and restraining bends are remarkably similar, despite the diverse nomenclature that has arisen to describe them and their different strike-slip settings (Fig. 2, Table 1). Releasing bends in all settings far outnumber restraining bends. In this discussion, I will point out a few of the major differences seen in bends formed in the four tectonic settings; speculate on

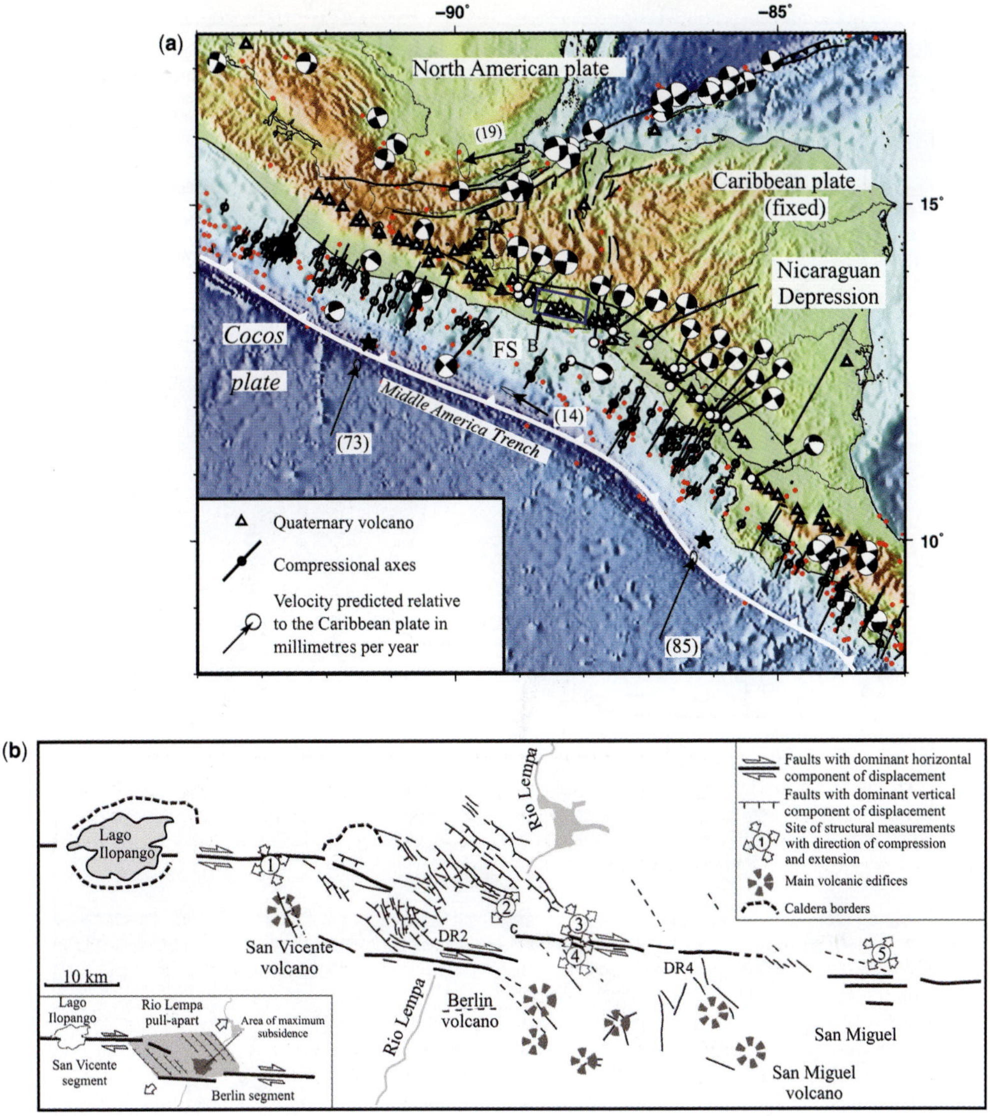

Fig. 63. (**a**) Regional tectonic map of active to Cenozoic strike-slip fault systems of Central America, using SRTM topography and ETOPO bathymetry as a base map. Black lines are active faults, taken from a compilation by DeMets (2001). (**b**) Schematic structural map of interpreted pull-apart by Corti *et al.* (2005).

their tectonic origins; and suggest some possible areas for future work.

Pull-apart development

The geological model proposed by Crowell (1974*a*, *b*) and modified by Mann *et al.* (1983) has been applied with success to many releasing bends worldwide (Fig. 7). The largest problem area is why do the strike-slip faults become curved in the first place for the releasing bend to form? In continental-boundary settings, many pull-aparts occur as part of a paired bend. Paired bends or fault zone ripout structures may form by a process of fault abandonment and reoccupation, as outlined by Swanson (2005; Fig. 12b). One open question is whether the restraining bend or the releasing bend appears first to form a paired bend. My proposal in this paper and in Mann *et al.* (2007*a*) is that the restraining bend forms first as a result of the

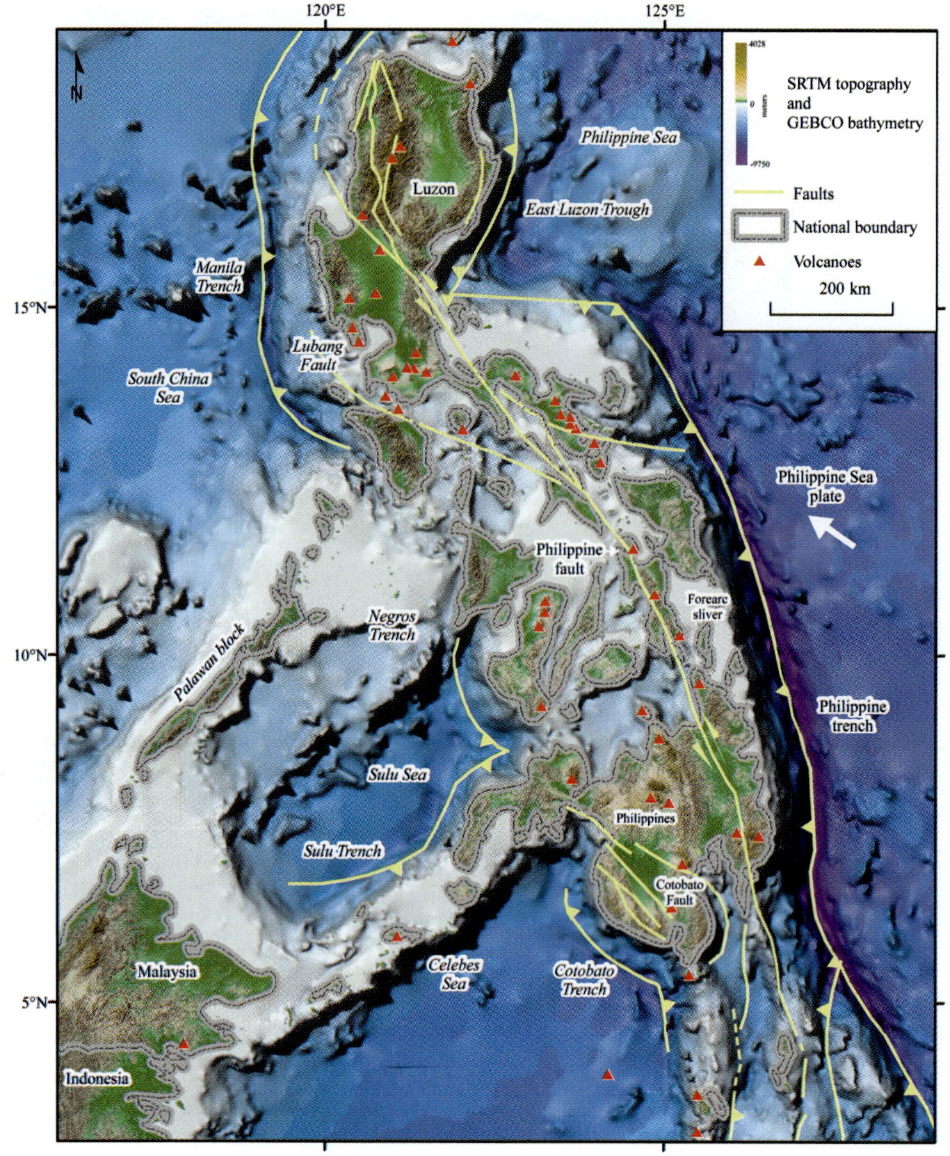

Fig. 64. Regional tectonic map of active to Cenozoic strike-slip fault systems of the Philippines archipelago, using SRTM topography and GEBCO bathymetry as a base map. Yellow lines are active faults, taken from a compilation by Coffin *et al.* (1997). GPS vector taken from Barrier *et al.* (1991).

interaction of strike-slip faults with a pre-existing basement feature. This proposal could be tested by careful dating of the initiation of adjacent pull-aparts and restraining bends that make up a paired bend.

Another pull-apart problem is why do basins become extinct by developing a cross-basin fault (Zhang *et al.* 1989)? One idea is that as faults adjust to changing plate kinematics, new faults may form to allow the fault to remain in a slip-parallel orientation (Fig. 8a). Another idea is that bends on faults can form as a natural consequence of shear-zone evolution, but this model would only apply to new faults with little offset (Fig. 12a).

Tectonic settings appear to account for few differences in the style of opening of pull-apart

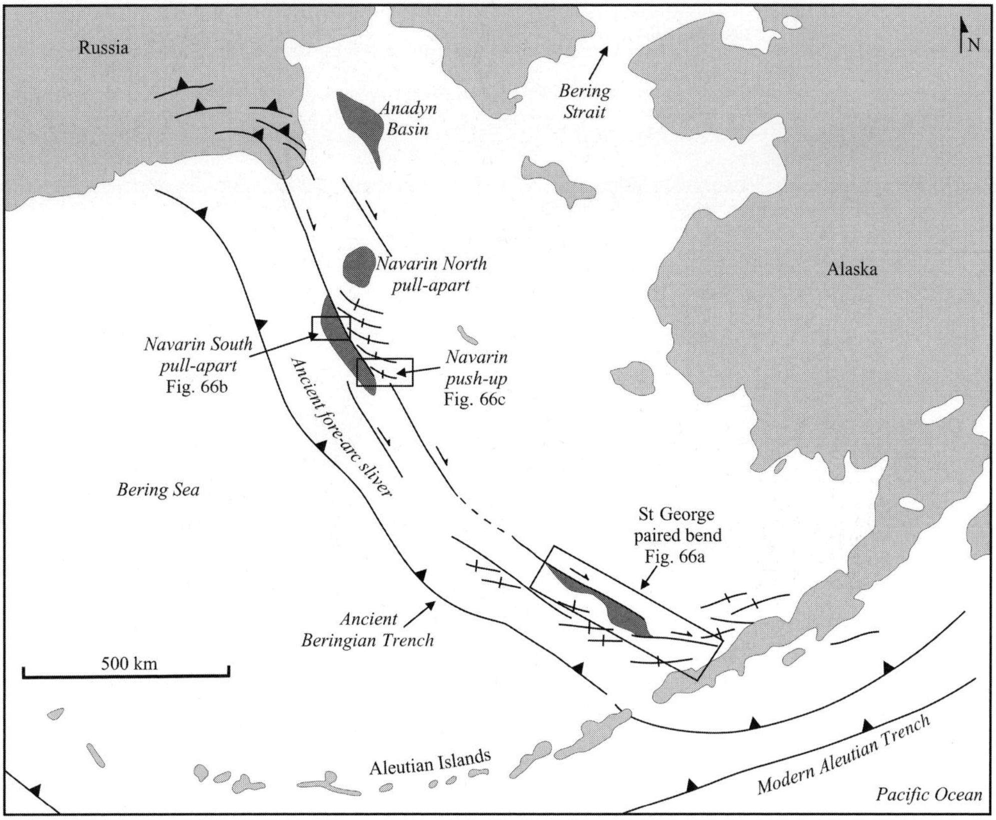

Fig. 65. Sketch map from Worrall (1991), showing Palaeogene strike-slip faults and bends along the Bering Sea margin.

basins. Larger basins appear to reflect larger angular differences between the original curved trend of the fault and the plate vector. Pull-aparts appear much less common in indent-linked systems like the Himalayas and the Tibetan Plateau (Fig. 47a), and are more common on long continental boundaries like the San Andreas (Fig. 15) and Caribbean faults (Fig. 23).

Restraining-bend development

Restraining-bend models are far less evolved and tested than those for releasing bends. The choices include the shear-zone model that would be relevant only for a newly formed fault; the sidewall ripout model which Swanson (2005) has shown to be relevant to many natural fault settings independent of their scale; and the paired-bend model proposed here.

In the paired-bend model, the formation of a restraining bend is followed by the formation of an adjacent releasing bend; the releasing bend forms as the strike-slip fault readjusts to its bent configuration (Fig. 8c). I speculate that the final stage in this process is the formation of a 'shortcut' or 'bypass' fault, which effectively bypasses the area of active bend deformation and returns the fault to an original, or at least straighter orientation. The reason why the strike-slip fault returns to its original, straight trace after being diverted at the bend is not clear. One explanation is that it is easier for the fault to reoccupy a zone of weakness parallel to the interplate slip vector (that is, its former fault trace) than breaking a new fault. A second explanation is that increased convergence at the restraining bend causes the paired-bend area to lock and be 'lopped off' by the more efficient bypass fault. A third explanation is based on the conservation of crustal volume in the bends area: crustal thickening in the restraining-bend area is compensated by thinning in the adjacent pull-apart area in order to maintain a crustal balance.

The paired-bend model is consistent with one of the fundamental differences between continental

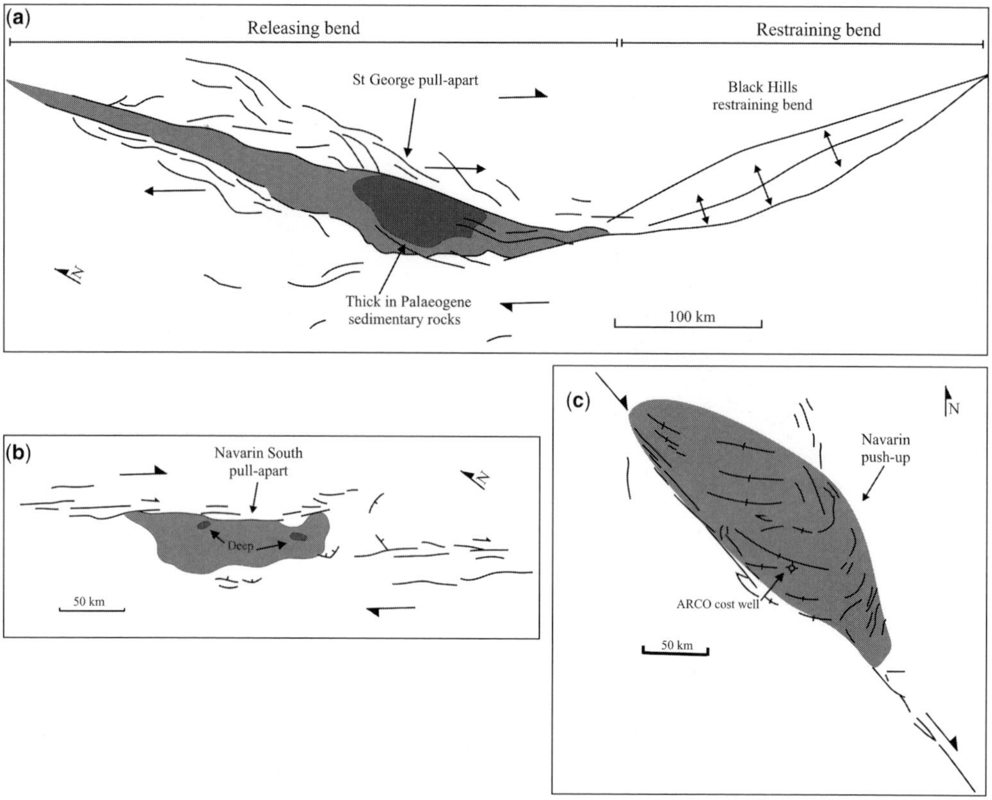

Fig. 66. Sketch maps modified from maps by Worrall (1991), showing details of the Bering Sea strike-slip zone. Location shown on Figure 65. (**a**) Paired-bend geometry of the St George pull-apart basin and the Black Hills restraining bend. (**b**) Navarin South pull-apart basin. (**c**) Navarin restraining bend.

strike-slip faults and oceanic transform faults: the rarity of restraining bends along the main fault trace, or 'principal displacement zone' of oceanic transform faults (Garfunkel 1986). I attribute the rarity of restraining bends along oceanic transform faults to the lack of pre-existing crustal heterogeneities in the oceanic crust on which restraining fault bends nucleate. Oceanic transform faults are instead characterized by divergent fault structures (for example, pull-apart basins, wide fault valleys) which probably form as a direct consequence of changes in relative plate motion (Garfunkel 1986; Fig. 8a).

This process of restraining-bend formation, uplift, rapid erosion, fault readjustment, and releasing-bend formation appears to be the dominant tectonic mechanism for active deformation and fault-controlled sedimentation along many of the fault systems reviewed here and may, therefore, provide a predictive model for interpreting other active and ancient strike-slip plate boundaries. The key to proving this hypothesis lies in precise age determinations of the initial age of uplift of the restraining bend and the age of initial subsidence of the releasing bend. Because active restraining bends generally form land areas, detailed geological mapping and stratigraphic studies are needed to determine their uplift history. Marine seismic surveys and deep-sea drilling are needed to determine the history of adjacent releasing bends, which commonly occur in deep-water areas.

Conclusions

The main conclusions and questions raised by this compilation include the following:

1. There are systematic structural relationships between releasing and restraining-bend segments along active strike-slip faults. Previous workers have not emphasized synchronous tectonic activity at paired restraining and releasing bends, because paired bends commonly

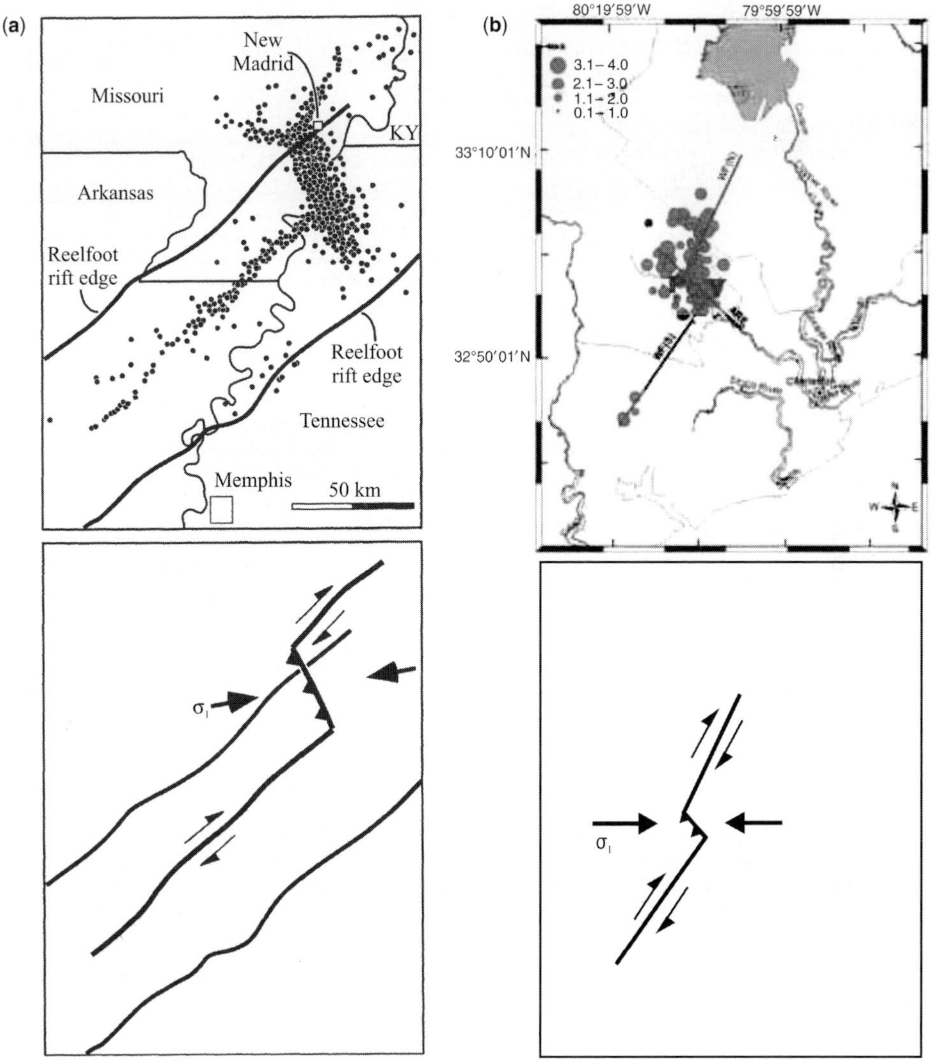

Fig. 67. (**a**) Map showing the location of earthquake epicentres in the New Madrid seismic zone in southern Missouri. Sketch showing restraining-bend structure based on fault-plane solutions (from Marshak *et al.* 2003). (**b**) Map showing the location of earthquake epicentres in the Middleton Place Summerville seismic zone (from Gangopadhyay & Talwani 2005). Sketch showing restraining-bend structure based on fault-plane solutions (modified from Gangopadhyay & Talwani 2005).

extend over land and sea areas with along-strike lengths of tens to hundreds of kilometres.
2. In paired-bend systems, some restraining bends appear to originate first as a result of interference between the strike-slip faults and pre-existing structures in the basement. In the case of Jamaica and Lebanon, pre-existing structures correspond with extinct rifts that strike at an angle of about 45° to the strike-slip fault.
3. The rarity of restraining bends along oceanic transform faults probably reflects the lack of pre-existing crustal heterogeneities in oceanic crust.
4. Which comes first: the restraining or releasing bend? The formation of two major restraining bends (Jamaica, Lebanon) at the sites of extinct rifts suggests that bends may originate first, as a result of strike-slip reactivation.

5. Why does the strike-slip fault crossing a paired bend return to its original trace after being diverted at the bend? One explanation is that it is easier for the fault to reoccupy a zone of weakness parallel to the interplate slip vector (that is, the old fault zone) than breaking a new fault through an unfaulted area. A second explanation is that increased convergence at the restraining bend causes the paired-bend area to lock and be 'lopped off' by the more efficient bypass fault.
6. Is the uplift history of the restraining-bend area consistent with the subsidence history of the releasing bend? In many areas, there are not enough data from either the releasing- or restraining-fault area to show clearly that the two areas deformed simultaneously. The paired-bend model is testable in active strike-slip zones by combined marine and land studies capable of precise age determinations of the time of uplift of the restraining bend and the time of subsidence of its paired releasing bend.

I would like to thank K. Burke and J. Dewey for introducing me to the world of strike-slip faults and pull-apart basins in their graduate tectonics courses at the State University of New York at Albany. I would also like to thank L. Bingham for her assistance with map-making and manuscript preparation. Special thanks go to D. Cunningham, M. Gordon, T. Dooley, M. Swanson and L. Lavier for their helpful reviews. The authors acknowledge the financial support for this publication provided by the University of Texas at Austin's Geology Foundation and Jackson School of Geosciences. This paper is University of Texas Institute for Geophysics Contribution No. 1866.

References

ACOSTA, J. & UCHUPI, E. 1996. Transtensional tectonics along the South Scotia ridge, Antarctica. *Tectonophysics*, **267**, 31–56.

ADIYAMAN, O. & CHOROWICZ, J. 2002. Late Cenozoic tectonics and volcanism in the northwestern corner of the Arabian plate: a consequence of the strike-slip Dead Sea fault zone and the lateral escape of Anatolia. *Journal of Volcanology and Geothermal Research*, **117**, 327–345.

ALLEN, 1998. Transtensional deformation in the evolution of the Bohai Basin, northern China. *In:* HOLDSWORTH, R. E., STRACHAN, R. A. & DEWEY, J. F. (eds) *Continental Transpression and Transtensional Tectonics*. Geological Society, London, Special Publications, **135**, 215–229.

ALLEN, P. A. & ALLEN, J. R. 2005. *Basin Analysis: Principles and Applications*. Vol. 2. Blackwell Publishing, Malden, MA.

ALLEN, C. R., ZHUOLI, L., HONG, Q., HUEZE, W., HUAWEI, Z. & WEISHI, H. 1991. Field study of a highly active fault zone: the Xianshuihe fault of southwestern China. *Geological Society of America Bulletin*, **103**, 1178–1199.

ALLEN, M. B., ALSOP, G. I. & ZHEMCHUZHNIKOV, V. G. 2001. Dome and basin refolding and transpressive inversion along the Karatau Fault System, southern Kazakstan. *Journal of the Geological Society, London*, **158**, 83–95.

ALLEN, M. B., MACDONALD, D. I. M., VINCENT, S. J., BROUET–MENZIES, C. & XUN, Z. 1997. Early Cenozoic two–phase extension and Late Cenozoic thermal subsidence and inversion of the Bohai Basin, northern China. *Marine and Petroleum Geology*, **14**, 951–972.

ANADON, P., CABRERA, L., GUIMERA, J. & SANTANACH, P. 1985. Paleogene strike-slip deformation and sedimentation along the southeastern margin of the Ebro Basin. *In*: BIDDLE, K. T. & CHRISTIE-BLICK, N. (eds) *Strike–Slip Deformation, Basin Formation, and Sedimentation*. SEPM Special Publications, **37**, 303–318.

ANDERSON, R. S. 1990. Evolution of the Santa Cruz mountains by advection of crust past a San Andreas fault bend. *Science*, **249**, 397–401.

ANDERSON, R. S. 1994. Evolution of the Santa Cruz Mountains, California, through tectonic growth and geomorphic decay. *Journal of Geophysical Research*, **99**, 20 161–20 179.

ANDREWS, D. J., OPPENHEIMER, D. H. & LEIENKAEMPER, J. J. 1993. The Mission link between the Hayward and Calaveras faults. *Journal of Geophysical Research*, **98**, 12 083–12 095.

ANGELIER, J., BERGERAT, F., BELLOU, M. & HOMBERG, C. 2004. Co-seismic strike-slip fault displacement determined from push-up structures: the Selsund Fault case, South Iceland. *Journal of Structural Geology*, **26**, 709–724.

ARBENZ, J. 1984. *Oil Potential of the Dead Sea Area: Seismic Oil Exploration, Tel Aviv, Reports*, **84-111**, 1–14.

ARGUS, D. F. & GORDON, R. G. 2001. Present tectonic motion across the Coast Ranges and San Andreas fault system in central California. *Geological Society of America Bulletin*, **113**, 1580–1592.

ARJANNIKOVA, A., LARROQUE, C., RITZ, J.-F., DEVERCHERE, J., STEPHAN, J. F., ARJANNIKOV, S. & SAN'KOV, V. 2004. Geometry and kinematics of recent deformation in the Mondy–Tunka area (southwesternmost Baikal rift zone, Mongolia–Siberia). *Terra Nova*, **16**, 265–272.

ARMIJO, R. & THIELE, R. 1990. Active faulting in northern Chile: ramp stacking and lateral decoupling along a subduction plate boundary? *Earth and Planetary Science Letters*, **98**, 40–61.

ARMIJO, R., MEYER, B., HUBERT, A. & BARKA, A. 1999. Westward propagation of the North Anatolian fault into the northern Aegean: timing and kinematics. *Geology*, **27**, 267–270.

ARMIJO, R., MEYER, B., NAVARRO, S., KING, G. & BARKA, A. 2002. Asymmetric slip partitioning in the Sea of Marmara pull-apart: a clue to propagation processes of the North Anatolian Fault? *Terra Nova*, **14**, 80–86.

ARMIJO, R., PONDARD, N. *ET AL.* 2005. Submarine fault scarps in the Sea of Marmara pull-apart (North

Anatolian fault): implications for seismic hazard in Istanbul. *Geochemistry Geophysics Geosystems*, **6**, 1–29.

ARMIJO, R., TAPPONNIER, P. & HAN, T. 1989. Late Cenozoic right-lateral strike-slip faulting in southern Tibet. *Journal of Geophysical Research*, **94**, 2787–2838.

ARMIJO, R., TAPPONNIER, P., MERCIER, J. & HAN, T. 1986. Quaternary extension in Southern Tibet: field observations and tectonic implications. *Journal of Geophysical Research*, **B91**, 13 803–13 872.

ARRIAGADA, C. 2003. Paleogene clockwise tectonic rotations in the forearc of central Andes, Antofagasta Region, Northern Chile. *Journal of Geophysical Research*, **108**, 2032, doi: 10.1029/2001JB001598, 2003.

ASPLER, L. B. & DONALDSON, J. A. 1985. The Nonacho Basin (Early Proterozoic), Northwest Territories, Canada; sedimentation and deformation in a strike-slip setting. *In*: BIDDLE, K. T. & CHRISTIE-BLICK, N. (eds) *Strike-Slip Deformation, Basin Formation, and Sedimentation*. SEPM Special Publications, **37**, 193–210.

AUDEMARD, F. A. 2006. Surface rupture of the Cariaco July 09, 1997 earthquake on the El Pilar fault, northeastern Venezuela. *Tectonophysics*, **424**, 19–39.

AUDEMARD, F. E. & SERRANO, I. 2001. Future petroliferous provinces of Venezuela. *In*: DOWNEY, M., THREET, J. & MORGAN, W. (eds) *Petroleum Provinces of the Twenty-First Century*. AAPG Memoir, **74**, 353–372.

AURELIO, M. A. 2000. Shear partitioning in the Philippines: constraints from Philippine fault and global positioning system data. *The Island Arc*, **9**, 584–597.

AVOUAC, J.-P. & TAPPONNIER, P. 1993. Kinematic model of active deformation in Central Asia. *Geophysical Research Letters*, **20**, 895–898.

AYDIN, A. & NUR, A. 1982. Evolution of pull-apart basins and their scale independence. *Tectonics*, **1**, 91–105.

AYDIN, A. & NUR, A. 1985. The types and role of stepovers in strike-slip tectonics. *In*: BIDDLE, K. T. & CHRISTIE-BLICK, N. (eds) *Strike-Slip Deformation, Basin Formation, and Sedimentation*. SEPM Special Publications, **37**, 35–44.

AYDIN, A. & PAGE, B. M. 1984. Diverse Pliocene–Quaternary tectonics in a transform environment, San Francisco Bay region, California. *Geological Society of America Bulletin*, **95**, 1303–1317.

AYDIN, A., SCHULTZ, R. A. & CAMPAGNA, D. 1990. Fault-normal dilation in pull-apart basins: implications for the relationship between strike-slip faults and volcanic activity. *Annales Tectonicae*, **IV**, 45–52.

BABB, S. & MANN, P. 1999. Structural and sedimentary development of a Neogene transpressional plate boundary between the Caribbean and South America plates in Trinidad and the Gulf of Paria. *In*: MANN, P. (eds) *Caribbean Basins*. Sedimentary Basins of the World, Elsevier Science, Amsterdam, **4**, 495–557.

BAHAT, D. 1983. New aspects of rhomb structures. *Journal of Structural Geology*, **5**, 591–601.

BAKUN, W., AAGAARD, B. *ET AL*. 2005. Implications for prediction and hazard assessment from the 2004 Parkfield earthquake. *Nature*, **437**, 969–974.

BALLANCE, P. F. & READING, H. G. 1980. *Sedimentation in Oblique-Slip Mobile Zones*. Blackwell Scientific, Oxford, 265 pp.

BARKA, A. 1992. The North Anatolian fault zone. *Annalae Tectonicae*, **6**, 164–195.

BARKA, A. A. & GULEN, L. 1989. Complex evolution of the Erzincan Basin (Eastern Turkey). *Journal of Structural Geology*, **11**, 275–283.

BARKA, A. & KADINSKY-CADE, 1988. Strike-slip fault geometry in Turkey and its influence on earthquake activity. *Tectonics*, **7**, 663–684.

BARKA, A., SERDAR AKYUZ, H., COHEN, H. A. & WATCHORN, F. 2000. Tectonic evolution of the Niksar and Tasova–Erbaa pull-apart basins, North Anatolian fault zone: their significance for the motion of the Anatolian block. *Tectonophysics*, **322**, 243–264.

BARNES, P. M. 2005. Releasing and restraining bends on the strike-slip Alpine and Boo Boo faults, New Zealand: new structural insights and slip rates from high-resolution multibeam bathymetric data. *In*: CUNNINGHAM, D. & MANN, P. (eds) *Tectonics of Strike-Slip Restraining and Releasing Bends in Continental and Oceanic Settings*, 28–30 September 2005, Geological Society, London.

BARNES, P. M., SUTHERLAND, R., DAVY, B. & DELTEIL, J. 2001. Rapid creation and destruction of sedimentary basins on mature strike-slip faults: an example from the offshore Alpine Fault, New Zealand. *Journal of Structural Geology*, **23**, 1727–1739.

BARNES, P. M., SUTHERLAND, R. & DELTEIL, J. 2005. Strike-slip structure and sedimentary basins of the southern Alpine Fault, Fiordland, New Zealand. *Geological Society of America Bulletin*, **117**, 411–435.

BARRIER, E., HUCHON, P. & AURELIO, M. 1991. Philippine fault: a key for Philippine kinematics. *Geology*, **19**, 32–35.

BARTOV, Y. & SAGY, A. 2004. Late Pleistocene extension and strike-slip in the Dead Sea basin. *Geological Magazine*, **141**, 565–572.

BASILE, C. & BRUN, J. P. 1999. Transtensional faulting patterns ranging from pull-apart basins to transform continental margins: an experimental investigation. *Journal of Structural Geology*, **21**, 23–37.

BEANLAND, S. 1995. Role of the North Island shear belt in the Hikurangi margin. *In*: (eds) *New Zealand Geophysical Society Symposium Abstracts*. New Zealand Geophysical Society, **1995**.

BEAUDOIN, B. 1994. Lower–crustal deformation during terrane dispersion along strike-slip faults. *Tectonophysics*, **232**, 257–266.

BELT, E. 1968. Post-Acadian rifts and related facies, eastern Canada. *In*: ZEN, E., WHITE, W., HADLEY, J. & THOMPSON, J. (eds) *Studies in Appalachian Geology, Northern and Maritime*. Wiley, Interscience, New York, 95–113.

BEN-AVRAHAM, Z. 1985. Structural framework of the Gulf of Elat (Aqaba), northern Red Sea. *Journal of Geophysical Research*, **90**, 703–727.

BEN-AVRAHAM, Z. 1992. Development of Asymmetric basins along continental transform faults. *Tectonophysics*, **215**, 209–220.

BEN-AVRAHAM, Z. & TIBOR, G. 1993. The northern edge of the Gulf of Elat. *Tectonophysics*, **226**, 319–331.

BEN-AVRAHAM, Z. & ZOBACK, M. D. 1992. Transform-normal extension and Asymmetric basins: an alternative to pull-apart models. *Geology*, **20**, 423–426.

BEN-AVRAHAM, Z., ALMAGOR, G. & GARFUNKEL, Z. 1979. Sediments and structure of the Gulf of Elat (Aqaba)–Northern Red Sea. *Sedimentary Geology*, **23**, 239–267.

BEN-AVRAHAM, Z., TEN BRINK, U. S. & CHARRACH, J. 1990. Transverse faults at the northern end of the southern basin of the Dead Sea graben. *Tectonophysics*, **180**, 37–47.

BENKHELIL, J., MASCLE, J. & TRICART, P. 1995. The Guinea continental margin: an example of a structurally complex transform margin. *Tectonophysics*, **248**, 117–137.

BERGER, B. R. 2007. The 3-D fault and vein architecture of strike-slip releasing and restraining bends: evidence from volcanic-center-related mineral deposits. *In*: CUNNINGHAM, W. D. & MANN, P. (eds) *Tectonics of Strike-Slip Restraining and Releasing Bends*. Geological Society, London, Special Publications, **290**, 447–471.

BERRYMAN, K. R., BEANLAND, S., COOPER, S., CUTTEN, H., NORRIS, R. & WOOD, P. 1992. The Alpine fault, New Zealand: variation in Quaternary structural style and geomorphic expression. *Annalae Tectonicae*, **6**, 126–163.

BERTOLUZZA, L. & PEROTTI, C. R. 1997. A finite-element model of the stress field in strike-slip basins: implications for the Permian tectonics of the Southern Alps (Italy). *Tectonophysics*, **280**, 185–197.

BIDDLE, K. T. & CHRISTIE-BLICK, N. 1985. *Strike-Slip Deformation, Basin Formation, and Sedimentation*. SEPM Special Publications, **37**, 386 pp.

BILHAM, R. & WILLIAMS, P. 1985. Sawtooth segmentation and deformation processes on the southern San Andreas Fault, California. *Geophysical Research Letters*, **12**, 557–560.

BILICH, A., FROHLICH, C. & MANN, P. 2001. Global seismicity characteristics of subduction-to-strike-slip transitions. *Journal of Geophysical Research*, **106**, 19 443–19 452.

BISCHOFF, J. L. & HENYEY, T. L. 1974. Tectonic elements of the central part of the Gulf of California. *Geological Society of America Bulletin*, **85**, 1893–1904.

BLYTHE, A. E., BURBANK, D. W., FARLEY, K. A. & FIELDING, E. J. 2000. Structural and topographic evolution of the Central Transverse Ranges, California, from apatite fission-track, (U–Th)/He and digital elevation model analyses. *Basin Research*, **12**, 97–114.

BOEHM, M. C. 1984. An overview of the lithostratigraphy, biostratigraphy, and paleoenvironments of the late Neogene San Felipe marine sequence, Baja California, Mexico. *In*: FRIZZELL, V. A. JR. (eds) *Geology of the Baja California Peninsula, Pacific section*. SEPM Special Publications, **39**, 253–265.

BOHANNON, R. G. & HOWELL, D. G. 1982. Kinematic evolution of the junction of the San Andreas, Garlock, and Big Pine faults, California. *Geology*, **10**, 358–363.

BONATTI, E., CHARMAK, A. & HONNOREZ, J. 1979. Tectonic and igneous emplacement of crust in oceanic transform zones. *In*: TALWANI, M. P., HARRISON, C. & HAYES, D. (eds) *Maurice Ewing Series*. American Geophysical Union, **2**, 239–248.

BRADLEY, D. C. 1983. Tectonics of the Acadian Orogeny in New England and adjacent Canada. *Journal of Geology*, **91**, 381–400.

BRANKMAN, C. & AYDIN, A. 2004. Uplift and contractional deformation along a segmented strike-slip fault system: the Gargano Promontory, southern Italy. *Journal of Structural Geology*, **26**, 807–824.

BREW, G., LUPA, J., BARAZANGI, M., SAWAF, T., AL-IMAM, A. & ZAZA, T. 2001. Structure and tectonic development of the Ghab basin and the Dead Sea fault system, Syria. *Journal of the Geological Society, London*, **158**, 665–674.

BRIAIS, A., PATRIAT, P. & TAPPONNIER, P. 1993. Updated interpretation of magnetic anomalies and seafloor spreading stages in the South China Sea: implications for the Tertiary tectonics of SE Asia. *Journal of Geophysical Research*, **98**, 6299–6328.

BRIDWELL, R. J. 1975. Sinuosity of strike-slip fault traces. *Geology*, **3**, 630–632.

BROWN, N. H. & SIBSON, R. H. 1989. Structural geology of the Octillo Badlands antidilational fault jog, southern California. *In*: SCHWARTZ, D. P. & SIBSON, R. H. (eds) *Fault Segmentation and Controls of Rupture Initiation and Termination*. US Geological Survey Open File Reports, **89-315**, 94–110.

BROWN, N. N., FULLER, M. D. & SIBSON, R. H. 1991. Paleomagnetism of the Ocotillo Badlands, southern California, and implications for slip transfer through an antidilational fault jog. *Earth and Planetary Science Letters*, **102**, 277–288.

BROWN, R. & VEDDER, J. 1967. Surface tectonic fractures along the San Andreas fault. *In*: BROWN, R., VEDDER, J. *ET AL*. (eds) *The Parkfield-Cholame, California, Earthquakes of June–August, 1966 — Surface Geologic Effects, Water-Resource Aspects, and Preliminary Seismic Data*. US Geological Survey Professional Papers, **579**, US Government Printing Office, Washington, D.C., 2–23.

BROWN, R. D. 1991. *Quaternary Deformation*. World Wide Web Address: http://education.usgs.gov/california/pp1515/chapter4.html.

BROWN, R. D., VEDDER, J. G. *ET AL*. (eds). 1967. *The Parkfield–Cholame California, Earthquakes of June–August 1966 – Surface Geologic Effects, Water-Resources Aspects, and Preliminary Seismic Data*. USGS, United States Government Printing Office. Geological Survey Professional Papers, **579**.

BUCKNAM, R. C. & HANCOCK, P. L. 1992. *Major Active Faults of the World*. Results of IGCP Project, **206**, 284.

BURCHFIEL, D. C. & STEWART, J. H. 1966. 'Pull-apart' origin of the central segment of Death Valley, California. *Geological Society of America Bulletin*, **77**, 431–442.

BURCHFIEL, B. C., HODGES, K. V. & ROYDEN, L. H. 1987. Geology of Panamint Valley–Saline Valley pull-apart system, California: palinspastic evidence for low-angle geometry of a Neogene range-bounding fault. *Journal of Geophysical Research*, **92**, 10 422–10 426.

BURGMANN, R., ARROWSMITH, J. R., DUMITRU, T. & MCLAUGHLIN, R. 1994. Rise and fall of the southern Santa Cruz Mountains, California, from fission tracks, geomorphology, and geodesy. *Journal of Geophysical Research*, **99**, 20 181–20 202.

BURGMANN, R., HILLEY, G., FERRETTI, A. & NOVALI, F. 2006. Resolving vertical tectonics in the San Francisco Bay area from permanent scatterer InSAR and GPS analysis. *Geology*, **34**, 221–224.

CABRERA, L., ROCA, E. & SANTANACH, P. 1988. Basin formation at the end of a strike-slip fault: the Cerdanya Basin (eastern Pyrenees). *Journal of the Geological Society, London*, **145**, 261–268.

CALAIS, E. & MERCIER DE LEPINAY, B. 1991. From transtension to transpression along the northern Caribbean plate boundary off Cuba: implications for the recent motion of the Caribbean plate. *Tectonophysics*, **186**, 329–350.

CALAIS, E., MAZABRAUD, Y., MERCIER DE LEPINAY, B., MANN, P., MATTIOLI, G. & JANSMA, P. 2002. Strain partitioning and fault slip rates in the northeastern Caribbean from GPS measurements. *Geophysical Research Letters*, **29**, 1–4.

CALAIS, E., VERGNOLLE, M., SAN'KOV, V., LUKHNEV, A., MIROSHNITCHENKO, A., AMARJARGAL, S. & DEVERCHERE, J. 2003. GPS measurements of crustal deformation in the Baikal–Mongolia area (1994–2002): implications for current kinematics of Asia. *Journal of Geophysical Research*, **108**, B10, 2501, doi: 1029/2002JB002373, 2003.

CAMPAGNA, D. J. & AYDIN, A. 1991. Tertiary uplift and shortening in the Basin and Range: the Echo hills, Southeastern Nevada. *Geology*, **19**, 485–488.

CAMPBELL, J. K. 1992. The Wairau and Alpine faults – structure and tectonics of the S-bend junction. *In*: CAMPBELL, J (eds) *Geological Society of New Zealand Annual Conference, Christchurch, New Zealand.* Geological Society New Zealand Miscellaneous Publications, **63B**, 10–31.

CAREY, W. 1958. A tectonic approach to continental drift. *In*: CAREY, S. W. (eds) *Continental Drift: a Symposium.* University of Tasmania, Hobart, 177–355.

CARR, M. J. 1976. Underthrusting and Quaternary faulting in northern Central America. *Geological Society of America Bulletin*, **87**, 825–829.

CEMBRANO, J., GONZALEZ, G., ARANCIBIA, G., AHUMADA, I., OLIVARES, V. & HERRERA, V. 2005. Fault zone development and strain partitioning in an extensional strike-slip duplex: a case study from the Mesozoic Atacama fault system, Northern Chile. *Tectonophysics*, **400**, 105–125.

CHEN, W. S., RIDGWAY, K. D., HORNG, C. S., CHEN, Y. G., SHEA, K. S. & YEH, M. G. 2001. Stratigraphic architecture, magnetostratigraphy, and incised-valley systems of the Pliocene–Pleistocene collisional marine foreland basin of Taiwan. *Geological Society of America Bulletin*, **113**, 1249–1271.

CHOROWICZ, J., CHOTIN, P. & GUILLANDA, R. 1996. The Garzon fault: active southwestern boundary of the Caribbean plate in Colombia. *International Journal of Earth Sciences*, **85**, 1437–3254.

CHOROWICZ, J., DHONT, D. & GUNDOGDU, N. 1999. Neotectonics in the eastern north anatolian fault region (Turkey) advocates crustal extension: mapping from SAR ERS imagery and digital elevation model. *Journal of Structural Geology*, **21**, 511–532.

CHRISTIE-BLICK, N. & BIDDLE, K. T. 1985. Deformation and basin formation along strike-slip faults. *In*: BIDDLE, K. T. & CHRISTIE-BLICK, N. (eds) *Strike-Slip Deformation, Basin Formation, and Sedimentation.* SEPM Special Publications, **37**, 1–34.

CLAYTON, L. 1966. Tectonic depressions along the Hope fault, a transcurrent fault in North Canterbury, New Zealand. *New Zealand Journal of Geology and Geophysics*, **9**, 95–104.

COBBOLD, P. R., DAVY, P. *ET AL.* 1993. Sedimentary basins and crustal thickening. *Sedimentary Geology*, **86**, 77–89.

COFFIN, M. F., GAHAGAN, L. M. & LAWVER, L. A. 1997. *Present-day Plate Boundary Digital Data Compilation.* University of Texas Institute for Geophysics Technical Reports, **174**, 5.

CORSINI, M., VAUCHEZ, A. & CABY, R. 1996. Ductile duplexing at a bend of a continental-scale strike-slip shear zone: example from NE Brazil. *Journal of Structural Geology*, **18**, 385–394.

CORTI, G., CARMINATI, E., MAZZARINI, F. & GARCIA, M. O. 2005. Active strike-slip faulting in El Salvador, Central America. *Geology*, **33**, 989–992.

COWAN, D. S., CLADOUHOS, T. T. & MORGAN, J. K. 2003. Structural geology and kinematic history of rocks formed along low-angle normal faults, Death Valley, California. *Geological Society of America Bulletin*, **115**, 1230–1248.

COWAN, H. A. 1990. Late Quaternary displacements on the Hope Fault at Glynn Wye, North Canterbury. *New Zealand Journal of Geology and Geophysics*, **33**, 285–293.

COWAN, H. A. 1992. *Structure, seismicity and tectonics of the Porters Pass–Amberley fault zone, North Canterbury, New Zealand.* PhD thesis, University of Canterbury, New Zealand.

COWAN, H. A. 1994. *Field Guide to New Zealand Active Tectonics – IASPEI 94.* 27th General Assembly of the International Association of Seismology and Physics of the Earth's Interior, Royal Society of New Zealand, Wellington, New Zealand, 88 pp.

COWAN, H. A., NICOL, A. & TONKIN, P. 1996. A comparison of historical and paleoseismicity in a newly formed fault zone and a mature fault zone, North Canterbury, New Zealand. *Journal of Geophysical Research*, **101**, 6021–6036.

COWAN, H. A., WALTER, P., DART, R. & MACHETTE, M. 2000. Maps and databases of Quaternary faults and folds in Panama and Costa Rica and adjoining offshore regions. *EOS, Transactions of the American Geophysical Union*, **81**, 1183–1184.

COWGILL, E., ARROWSMITH, J. R., YIN, A., XIAOFENG, W. & ZHENGLE, C. 2004*b*. The Akato Tagh bend along the Altyn Tagh Fault, northwest Tibet, 2: active deformation and the importance of transpression and strain hardening within the Altyn Tagh system. *Geological Society of America Bulletin*, **116**, 1443–1464.

COWGILL, E., YIN, A., ARROWSMITH, J. R., XIAOFENG, W. & SHUANHONG, Z. 2004*a*. The Akato Tagh bend along the Altyn Tagh fault, northwest Tibet, 1: Smoothing by vertical-axis rotation and the effect of topographic stresses on bend-flanking faults. *Geological Society of America Bulletin*, **116**, 1423–1442.

CROWELL, J. C. 1974*a*. Origin of Late Cenozoic basins in southern California. *In*: DICKINSON, W. (ed.) *Tectonics and Sedimentation*. SEPM Special Publications, **22**, 190–204.

CROWELL, J. C. 1974*b*. Sedimentation along the San Andreas fault, California. *In*: DOTT, R. H. JR. & SHAVER, R. H. (eds) *Modern and Ancient Geosynclinal Sedimentation*. SEPM Special Publications, **19**, 292–303.

CROWELL, J. C. 1976. Implications of crustal stretching and shortening of coastal Ventura Basin, California. *In*: HOWEL, D. G. (ed.) *Aspects of the History of the California Continental Borderland: Pacific Section*. AAPG Miscellaneous Publications, **24**, 365–382.

CROWELL, J. C. & LINK, M. H. 1982. *Geologic History of Ridge Basin, Southern California*. SEPM, Los Angeles, 304 pp.

CRUIKSHANK, K. M., ZHAO, G. & JOHNSON, A. M. 1991. Duplex structures connecting fault segments in Entrada sandstone. *Journal of Structural Geology*, **13**, 1185–1196.

CUNNINGHAM, A. P., BARKER, P. F. & TOMLINSON, J. S. 1998. Tectonics and sedimentary environment of the North Scotia Ridge region revealed by side-scan sonar. *Journal of the Geological Society, London*, **155**, 941–956.

CUNNINGHAM, D. 1995. Orogenesis at the southern tip of the Americas: the structural evolution of the Cordillera Darwin metamorphic complex, southernmost Chile. *Tectonophysics*, **244**, 197–229.

CUNNINGHAM, D., DAVIES, S. & BADARCH, G. 2003*a*. Crustal architecture and active growth of the Sutai Range, western Mongolia: a major intracontinental, intraplate restraining bend. *Journal of Geodynamics*, **36**, 169–191.

CUNNINGHAM, D., DIJKSTRA, A. H., HOWARD, J., QUARLES, A. & BADARCH, G. 2003*b*. Active intraplate strike-slip faulting and transpressional uplift in the Mongolian Altai. *In*: STORTI, F., HOLDSWORTH, R. E. & SALVINI, F. (eds) *Intraplate Strike-Slip Deformation Belts*. Geological Society, London, Special Publications, **210**, 65–87.

CUNNINGHAM, D., KLEIPEIS, K. A., GOSE, W. A. & DALZIEL, I. W. D. 1991. The Patagonian Orocline: new paleomagnetic data from the Andean magmatic arc in Tierra del Fuego, Chile. *Journal of Geophysical Research*, **96**, 16 061–16 067.

CUNNINGHAM, D., OWEN, L. A., SNEE, L. W. & JILIANG, L. 2003*c*. Structural framework of a major intracontinental orogenic termination zone: the easternmost Tien Shan, China. *Journal of the Geological Society, London*, **160**, 575–590.

CUNNINGHAM, D., WINDLEY, B. F., DORJNAMJAA, D., BADAMGAROV, G. & SAANDAR, M. 1996*b*. A structural transect across the Mongolian western Altai: active transpressional mountain building in Central Asia. *Tectonics*, **15**, 142–156.

CUNNINGHAM, W. D. 1993. Strike-slip faults in the southernmost Andes and the development of the Patagonian orocline. *Tectonics*, **12**, 169–186.

CUNNINGHAM, W. D. 2007. Structural and topographic characteristics of restraining bend mountain ranges of the Altai, Gobai Altai and Easternmost Tien Shan. *In*: CUNNINGHAM, W. D. & MANN, P. (eds) *Tectonics of Strike-Slip Restraining and Releasing Bends*. Geological Society, London, Special Publications, **290**, 219–237.

CUNNINGHAM, W. D. & MANN, P. 2007. Tectonics of strike-slip restraining and releasing bends. *In*: CUNNINGHAM, W. D. & MANN, P. (eds) *Tectonics of Strike-Slip Restraining and Releasing Bends*. Geological Society, London, Special Publications, **290**, 1–12.

CUNNINGHAM, W. D., WINDLEY, B. F., DORJNAMJAA, D., BADAMGAROV, G. & SAANDAR, M. 1996*a*. Late Cenozoic transpression in southwestern Mongolia and the Gobi Altai–Tien Shan connection. *Earth and Planetary Science Letters*, **140**, 67–81.

CURRAY, J. R. 2005. Tectonics and history of the Andaman Sea region. *Journal of Asian Earth Sciences*, **25**, 187–232.

CURTIS, M. 1999. Structural and kinematic evolution of a Miocene to Recent sinistral restraining bend: the Montejunto massif, Portugal. *Journal of Structural Geology*, **21**, 39–54.

DACZKO, N. R., WERTZ, K. L., MOSHER, S., COFFIN, M. F. & MECKEL, T. A. 2003. Extension along the Australian–Pacific transpressional transform plate boundary near Macquarie Island. *Geochemistry Geophysics Geosystems, G3*, **4**, 22.

DAERON, M. 2005. *Role, cinematique et comportement sismique a long terme de la faille de Yammouneh*. PhD, Institute de Physique du Globe de Paris, Paris.

DAERON, M., KLINGER, Y., TAPPONNIER, P., ELIAS, A., JACQUES, E. & SURSOCK, A. 2005. Sources of the large A.D. 1202 and 1759 Near East earthquakes. *Geology*, **33**, 529–532.

DAVIS, G. & REYNOLDS, S. 1996. *Structural Geology of Rocks and Regions*, 2nd edn. John Wiley, New York, 776 pp.

DAVIS, T. & DUEBENDORFER, E. 1987. *Strip Map of the Western Big Bend Segment of the San Andreas Fault*, Geological Society of America Map and Chart Series **MC-60**.

DECKER, K., PERESSON, H. & HINSCH, R. 2005. Active tectonics and Quaternary basin formation along the Vienna Basin transform fault. *Quaternary Science Reviews*, **24**, 307–322.

DE PAOLA, N., HOLDSWORTH, R. E., COLLETTINI, C., MCCAFFREY, K. J. W. & BARCHI, M. R. 2007. The structural evolution of dilational stepovers in regional transtensional zones. *In*: CUNNINGHAM, W. D. & MANN, P. (eds) *Tectonics of Strike-Slip Restraining and Releasing Bends*. Geological Society, London, Special Publications, **290**, 433–445.

DELONG, S. E., DEWEY, J. & FOX, P. L. 1979. Topographic and geologic evolution of fracture zones. *Journal of the Geological Society, London*, **136**, 303–310.

DEMETS, C. 1992. Oblique convergence and deformation along the Kuril and Japan trenches. *Journal of Geophysical Research*, **97**, 17 615–17 625.

DeMets, C. 2001. A new estimate for present-day Cocos–Caribbean plate motion: implications for slip on the Central American volcanic arc. *Geophysical Research Letters*, **28**, 4043–4046.

DeMets, C. & Wiggins-Grandison, M. 2007. Deformation of Jamaica and motion of the Gonave microplate from GPS and seismic data. *Geophysical Journal International*, **168**, 362–378.

DeMets, C., Jansma, P. *et al.* 2000. GPS geodetic constraints on Caribbean–North America plate motion. *Geophysical Research Letters*, **27**, 437–440.

Deng, Q., Wu, D., Zhang, P.-Z. & Chen, S. 1986. Structure and deformational character of strike-slip fault zones. *Pure and Applied Geophysics*, **124**, 203–223.

Dewey, J. F. & Lamb, S. H. 1992. Active tectonics of the Andes. *Tectonophysics*, **205**, 79–95.

Dibblee, T., Jr. 1968. Displacements on the San Andreas fault system in the San Gabriel, San Bernardino, and San Jacinto Mountains, southern California. *In*: Dickinson, W. & Grantz, A. (eds) *Proceedings of a Conference on Geologic Problems of the San Andreas Fault System, California*. Stanford University Publications in Geological Sciences, **11**, 260–276.

Dickinson, W. R. 1996. *Kinematics of Transrotational Tectonism in the California Tranverse Ranges and Its Contribution to Cumulative Slip Along the San Andreas Transform Fault System*. Geological Society of America Special Papers, **305**, 1–14.

Dixon, T., Decaix, J. *et al.* 2002. Seismic cycle and rheological effects on estimation of present-day slip rates for the Agua Blanca and San Miguel–Vallecitos faults, northern Baja California, Mexico. *Journal of Geophysical Research*, **107**, B10, 2226, doi: 10.1029/2000JB000099, 2002.

Dolan, J. F., Mullins, H. T. & Wald, D. J. 1998. *Active Tectonics of the North-Central Caribbean: Oblique Collision, Strain Partitioning, and Opposing Subducted Slabs*. Geological Society of America Special Papers, **326**, 1–61.

Dooley, T. & McClay, K. R. 1997. Analog modeling of pull-apart basins. *AAPG Bulletin*, **81**, 1804–1826.

Dooley, T., McClay, K. R. & Bonora, M. 1999. 4D evolution of segmented strike-slip fault systems: applications to NW Europe. *In*: Fleet, A. J. & Boldy, S. A. R. (eds) *Petroleum Geology of Northwest Europe: Proceedings of the 5th Conference*. Geological Society, London, 215–225.

Du, Y. & Aydin, A. 1995. Shear fracture patterns and connectivity at geometric complexities along strike-slip faults. *Journal of Geophysical Research*, **100**, 18 093–18 102.

Dziak, R., Fox, C. G. & Embley, R. W. 1991. Relationship between the seismicity and geologic structure of the Blanco transform fault zone. *Marine Geophysical Researches*, **13**, 203–208.

Edwards, M. A. & Ratschbacher, L. 2005. Seismic and aseismic weakening effects in transtension: field and microstructural observations on the mechanics and architecture of a large fault zone in south-eastern Tibet. *In*: Bruhn, D. & Burlini, L. (eds) *High Strain Zones: Structure and Physical Properties*. Geological Society, London, Special Publications, **245**, 109–141.

Ego, F., Sebrier, M., Lavenu, A., Yepes, H. & Egues, A. 1996. Quaternary state of stress in the northern Andes and the restraining bend model for the Ecuadorian Andes. *Tectonophysics*, **259**, 101–116.

Ehrhardt, A., Hubscher, C., Ben-Avraham, Z. & Gajewski, D. 2005. Seismic study of pull-apart-induced sedimentation and deformation in the northern Gulf of Aqaba (Elat). *Tectonophysics*, **396**, 59–79.

Einsele, G. 1986. Interaction between sediments and basalt injections in young Gulf of California-type spreading centers. *International Journal of Earth Sciences*, **75**, 197–208.

Embley, R. W. & Wilson, D. S. 1992. Morphology of the Blanco transform fault zone, northeast Pacific: implications for its tectonic evolution. *Marine Geophysical Researches*, **14**, 25–45.

Embley, R., Kulm, V., Massoth, G., Abbott, D. & Holmes, M. 1987. Morphology, structure, and resource potential of the Blanco transform fault zone. *In*: Scholl, D., Grantz, A. & Vedder, J. (eds) *Geology and Resource Potential of the Continental Margin of Western North America and Adjacent Ocean Basins – Beaufort Sea to Baja, California*. AAPG, 549–562.

Erlstrom, M., Thomas, S. A., Deeks, N. & Sivhed, U. 1997. Structure and tectonic evolution of the Tornquist Zone and adjacent sedimentary basins in Scania and the southern Baltic Sea area. *Tectonophysics*, **271**, 191–215.

Escalona, A. & Mann, P. 2003. Three-dimensional structural architecture and evolution of the Eocene pull-apart basin, central Maracaibo basin, Venezuela. *Marine and Petroleum Geology*, **20**, 141–161.

Escalona, A. & Mann, P. 2006. Tectonic controls of the right-lateral Burro Negro tear fault on Paleogene structure and stratigraphy, northeastern Maracaibo Basin. *AAPG Bulletin*, **90**, 479–504.

Eusden, J. D., Pettinga, J. R. & Campbell, J. K. 2000. Structural evolution and landscape development of a collapsed transpressive duplex on the Hope Fault, North Canterbury, New Zealand. *New Zealand Journal of Geology and Geophysics*, **43**, 391–404.

Eusden, J. D., Pettinga, J. R. & Campbell, J. K. 2005. Structural collapse of a transpressive hanging-wall fault wedge, Charwell region of the Hope Fault, South Island, New Zealand. *New Zealand Journal of Geology and Geophysics*, **48**, 295–309.

Eyal, Y., Eyal, M., Bartov, Y., Steinitz, G. & Folkman, Y. 1986. The origin of the Bir Zreir rhomb-shaped graben, eastern Sinai. *Tectonics*, **5**, 267–277.

Feigl, K. L., Agnew, D. C. *et al.* 1993. Space geodectic measurement of crustal deformation in central and southern California 1984–1992. *Journal of Geophysical Research*, **98**, 21 677–21 712.

Feng, Q. & Lees, J. M. 1998. Microseismicity, stress and fracture in the Coso geothermal field, California. *Tectonophysics*, **289**, 221–238.

Fisher, M. A. 1982. Petroleum geology of Norton basin, Alaska. *AAPG Bulletin*, **66**, 286–301.

Fisher, M. A., Normak, W. R., Langenheim, V. E., Calvert, A. J. & Sliter, R. 2004. The offshore Palos Verdes fault zone near San Pedro, Southern California. *Bulletin of the Seismological Society of America*, **94**, 506–530.

FITCH, T. J. 1972. Plate convergence, transcurrent faults, and internal deformation adjacent to Southeast Asia and the Western Pacific. *Journal of Geophysical Research*, **77**, 4432–4460.

FITZGERALD, P. G., STUMP, E. & REDFIELD, T. F. 1993. Late Cenozoic uplift of Denali and its relation to relative plate motion and fault morphology. *Science*, **259**, 497–499.

FLINCH, J. F., RAMBARAN, V., ALI, W., DE LISA, V., HERNANDEZ, G., RODRIGUES, K. & SAMS, R. 1999. Structure of the Gulf of Paria pull-apart basin (Eastern Venezuela–Trinidad). *In*: MANN, P. (ed.) *Sedimentary Basins of the World: Caribbean Basins*. Elsevier, Amsterdam, 477–494.

FODOR, L. 1995. From transpression to transtension: Oligocene–Miocene structural evolution of the Vienna basin and the East Alpine–Western Carpathian junction. *Tectonophysics*, **242**, 151–182.

FORNARI, D., GALLO, D., EDWARDS, M., MADSEN, J., PERFIT, M. & SHOR, A. 1989. Structure and topography of the Siqueiros transform fault system: evidence for the development of intra-transform spreading centers. *Marine Geophysical Researches*, **11**, 263–299.

FOSSEN, H. & TIKOFF, B. 1998. Extended models of transpression and transtension, and application to tectonic settings. *In*: HOLDSWORTH, R., STRACHAN, R. & DEWEY, J. (eds) *Continental Transpression and Transtensional Tectonics*. Geological Society, London, Special Publications, **135**, 15–33.

FOX, P. J. & GALLO, D. G. 1984. A tectonic model for ridge–transform–ridge plate boundaries: implications for the structure of oceanic lithosphere. *Tectonophysics*, **104**, 205–242.

FREUND, R. 1971. The Hope fault: a strike-slip fault in New Zealand. *New Zealand Geological Survey Bulletin*, **86**, 1–49.

FREUND, R. 1974. Kinematics of transform and transcurrent faults. *Tectonophysics*, **21**, 93–134.

FREUND, R., GARFUNKEL, Z., ZAK, I., GOLDBERG, M., WEISSBROD, T. & DERIN, B. 1970. The shear along the Dead Sea rift. *Philosophical Transactions of the Royal Society of London, Series A*, **267**, 107–130.

FU, B., NINOMIYA, Y., LEI, X., TODA, S. & AWATA, Y. 2004. Mapping active fault associated with the 2004 Mw 6.6 Bam (SE Iran) earthquake with ASTER 3D images. *Remote Sensing of Environment*, **92**, 153–157.

FUIS, G., MOONEY, W., HEALY, J., MCMECHAN, G. & LUTTER, W. 1982. *Crustal Structure of the Imperial Valley Region, in the Imperial Valley, California, Earthquake of October 15, 1979. US Geological Survey Professional Papers*, **1254**.

FUNEDDA, A. & OGGIANO, G. 2005. Tertiary strike-slip basins related to releasing geometry inherited by the Variscan basement: an example from northern Sardinia (Italy). *In*: CUNNINGHAM, D. & MANN, P. (eds) *Tectonics of Strike-Slip Restraining and Releasing Bends in Continental and Oceanic Settings*, 28–30 September 2005, Geological Society, London, England.

GALINDO-ZALDIVAR, J., BALANYA, J. C. *ET AL.* 2002. Active crustal fragmentation along the Scotia–Antarctic plate boundary east of the South Orkney Microcontinent (Antarctica). *Earth and Planetary Science Letters*, **204**, 33–46.

GALINDO-ZALDIVAR, J., JABALOY, A., MALDONADO, A. & GALDEANO, C. S. D. 1996. Continental fragmentation along the South Scotia Ridge transcurrent plate boundary (NE Antarctic Peninsula). *Tectonophysics*, **258**, 275–301.

GAMOND, J. F. 1987. Bridge structures as sense of displacement criteria in brittle fault zones. *Journal of Structural Geology*, **9**, 609–620.

GANGOPADHYAY, A. & TALWANI, M. 2003. Symptomatic features of intraplate earthquakes. *Seismological Research Letters*, **74**, 863–883.

GANGOPADHYAY, A. & TALWANI, P. 2005. Fault intersections and intraplate seismicity in Charleston, South Carolina: insights from a 2-D numerical model. *Current Science*, **88**, 1609–1616.

GARFUNKEL, Z. 1981. Internal structure of the Dead Sea leaky transform (rift) in relation to plate kinematics. *Tectonophysics*, **80**, 81–108.

GARFUNKEL, Z. 1986. Review of oceanic transform activity and development. *Journal of the Geological Society, London*, **143**, 775–784.

GARFUNKEL, Z. 2005. The structure and history of the Dead Sea pull-apart (rhomb graben). *In*: CUNNINGHAM, D. & MANN, P. (eds) *Tectonics of Strike-Slip Restraining and Releasing Bends in Continental and Oceanic Settings*, 28–30 September 2005, Geological Society, London, England.

GARFUNKEL, Z. & BEN-AVRAHAM, Z. 1996. The structure of the Dead Sea basin. *Tectonophysics*, **266**, 155–176.

GARFUNKEL, Z., ZAK, I. & FREUND, R. 1981. Active faulting in the Dead Sea Rift. *Tectonophysics*, **80**, 1–26.

GIBBS, A. D. 1989. Structural styles in basin formation. *In*: TANKARD, A. J. & BALKWILL, H. R. (eds) *Extensional Tectonics and Stratigraphy of the North Atlantic Margins*. AAPG Memoir, **46**, 81–93.

GINER-ROBLES, J. L., GONZALEZ-CASADO, J. M., GUMIEL, P., MARTIN-VELAZQUEZ, S. & GARCIA-CUEVAS, C. 2003. A kinematic model of the Scotia Plate (SW Atlantic Ocean). *Journal of South American Earth Sciences*, **16**, 179–191.

GIRARD, G. & VAN WYK DE VRIES, B. 2005. The Managua Graben and Las Sierras–Masaya volcanic complex (Nicaragua); pull-apart localization by an intrusive complex: results from analogue modeling. *Journal of Volcanology and Geothermal Research*, **144**, 37–57.

GIRDLER, R. W. 1989. A. M. Quennell: father of transform faults and poles of rotation? *EOS, Transactions, American Geophysical Union*, **70**, 193, 199, 205.

GOLDFINGER, C., KULM, L. V. D., YEATS, R. S., HUMMON, C., HUFTILE, G. J., NIEM, A. R. & MCNEILL, L. C. 1996. Oblique strike-slip faulting of the Cascadia submarine forearc: the Daisy Bank fault zone off Central Oregon. *In*: BEBOUT, G. E., SCHOLL, D. W., KIRBY, S. H. & PLATT, J. P. (eds) *Subduction Top to Bottom*. American Geophysical Union, Geophysical Monograph Series, **96**, 65–74.

GOLKE, M., CLOETINGH, S. & FUCHS, K. 1994. Finite element modeling of pull-apart basin opening. *Tectonophysics*, **240**, 45–57.

GOMEZ, F., KHAWLIE, M., TABET, C., DARKAL, A. N., KHAIR, K. & BARAZANGI, M. 2006. Late Cenozoic uplift along the northern Dead Sea transform in Lebanon and Syria. *Earth and Planetary Science Letters*, **241**, 913–931.

GOMEZ, F., NEMER, T., MEGHRAOUI, M. & BARAZANGI, M. 2005. Active tectonics of the Lebanese restraining bend along the Dead Sea fault (Lebanon and Syria): strain partitioning, earthquakes, and internal deformation of the Arabian plate. *In*: CUNNINGHAM, D. & MANN, P. (eds) *Tectonics of Strike-Slip Restraining and Releasing Bends in Continental and Oceanic Settings*, 28–30 September 2005, Geological Society, London, England.

GOMEZ, F., NEMER, T., TABET, C., KHAWLIE, M., MEGHRAOUI, M. & BARAZANGI, M. 2007. Strain partitioning of active transpression within the Lebanese restraining bend of the Dead Sea fault (Lebanon and SW Syria). *In*: CUNNINGHAM, W. D. & MANN, P. (eds) *Tectonics of Strike-Slip Restraining and Releasing Bends*. Geological Society, London, Special Publications, **290**, 285–303.

GORDON, M. B. & MUEHLBERGER, W. R. 1994. Rotation of the Chortís block causes dextral slip on the Guayape fault. *Tectonics*, **13**, 858–872.

GOVERS, R. & WORTEL, M. J. R. 2005. Lithosphere tearing at STEP faults: response to edges of subductions. *Earth and Planetary Science Letters*, **236**, 505–523.

GRAYMER, R. W., LANGENHEIM, V. E., SIMPSON, R. W., JACHENS, R. C. & PONCE, D. A. 2007. Relatively simple through-going fault planes at large-earthquake depth may be concealed by surface complexity of strike-slip faults. *In*: CUNNINGHAM, W. D. & MANN, P. (eds) *Tectonics of Strike-Slip Restraining and Releasing Bends*. Geological Society, London, Special Publications, **290**, 189–201.

GUIRAUD, M. & SEGURET, M. 1985. A releasing solitary overstep model for the Late Jurassic–Early Cretaceous (Wealdian) Soria strike-slip basin (northern Spain). *In*: BIDDLE, K. T. & CHRISTIE-BLICK, N. (eds) *Strike-Slip Deformation, Basin Formation, and Sedimentation*. SEPM Special Publications, **37**, 159–176.

GUZMÁN SPEZIALE, M. 2001. Active seismic deformation in the grabens of northern Central America and its relationship to the relative motion of the North America–Caribbean plate boundary. *Tectonophysics*, **337**, 39–51.

GUZMÁN-SPEZIALE, M. & MENESES-ROCHA, J. J. 2000. The North America–Caribbean plate boundary west of the Motagua–Polochic fault system: a fault jog in southeastern Mexico. *Journal of South American Earth Sciences*, **13**, 459–468.

HARDING, T. 1985. Seismic characteristics and identification of negative flower structures, positive flower structures, and positive structural inversion. *AAPG Bulletin*, **69**, 582–600.

HARDING, T. 1990. Identification of wrench faults using subsurface structural data: Criteria and pitfalls. *AAPG Bulletin*, **74**, 1590–1609.

HARDING, T., VIERBUCHEN, R. & CHRISTIE-BLICK, N. 1985. Structural styles, plate-tectonic settings, and hydrocarbon traps of divergent (transtensional) wrench faults. *In*: BIDDLE, K. T. & CHRISTIE-BLICK, N. (eds) *Strike-Slip Deformation, Basin Formation and Sedimentation*. SEPM Special Publications, **37**, 51–77.

HARLOW, D. H. & WHITE, R. A. 1985. Shallow earthquakes along the volcanic chain in Central America: evidence for oblique subduction. *Earthquake Notes*, **55**, 28.

HARLOW, G. E., HEMMING, S. R., AVE LALLEMANT, H. G., SISSON, V. B. & SORENSEN, S. S. 2004. Two high-pressure–low-temperature serpentinite-matrix melange belts, Motagua fault zone, Guatemale: a record of Aptian and Maastrichtian collisions. *Geology*, **32**, 17–20.

HARRIS, R., DOLAN, J., HARTLEB, R. & DAY, S. 2002. The 1999 Izmit, Turkey, earthquake: a 3D dynamic stress transfer model of intraearthquake triggering. *Bulletin of the Seismological Society of America*, **92**, 245–255.

HATCHER, R. 1995. *Structural Geology: Principles, Concepts, and Problems*, 2nd edn. Prentice Hall, New Jersey, 525 pp.

HATCHER, R., ZIETZ, I., REGAN, R. & ABU-AJAMIEH, A. 1981. Sinistral strike-slip motion on the Dead Sea rift: confirmation from new magnetic data. *Geology*, **9**, 458–462.

HEARNE, B. C. JR., MCLAUGHLIN, R. J. & DONNELLY-NOLAN, J. M. 1988. Tectonic framework of the Clear Lake Basin, California. *Geological Society of America Special Papers*, **214**, 9–19.

HEIMANN, A., & RON, H. 1987. Young faults in the Hula pull-apart basin, central Dead Sea transform. *Tectonophysics*, **141**, 117–124.

HEIMANN, A., EYAL, M. & EYAL, Y. 1990. The evolution of the Barahta rhomb-shaped graben, Mount Hermon, Dead Sea transform. *Tectonophysics*, **180**, 101–110.

HEMPTON, M. R. 1983. The evolution of thought concerning sedimentation in pull-apart basins. *In*: BOARDMAN, S. J. (eds) *Revolution in the Earth Sciences*, Kendall/Hunt Publishers, Dubuque, Iowa, 167–180.

HEMPTON, M. R. 1987. Constraints on Arabian plate motion and extensional history of the Red Sea. *Tectonics*, **148**, 687–705.

HEMPTON, M. & DUNNE, L. 1983. Sedimentation in pull-apart basins: active examples in eastern Turkey. *Journal of Geology*, **92**, 513–530.

HEMPTON, M. R. & NEHER, K. 1986. Experimental fracture, strain, and subsidence patterns over en echelon strike-slip faults: implications for the structural evolution of pull-apart basins. *Journal of Structural Geology*, **8**, 597–605.

HENSTOCK, P., LEVANDER, A. & HOLE, J. 1997. Deformation in the lower crust of the San Andreas fault system in northern California. *Science*, **278**, 650–653.

HENYEY & BISCHOFF, J. L. 1973. Tectonic elements of the northern part of the Gulf of California. *Geological Society of America Bulletin*, **84**, 315–330.

HINSCH, R., DECKER, K. & WAGREICH, M. 2005. 3-D mapping of segmented active faults in the southern Vienna Basin. *Quaternary Science Reviews*, **24**, 321–336.

HOLDSWORTH, R., STRACHAN, R. & DEWEY, J. 1998. *Continental Transpressional and Transtensional Tectonics*. Geological Society of London, London, 360.

HOLE, J. A., THYBO, H. & KLEMPERER, S. L. 1996. Seismic reflections from the near-vertical San Andreas Fault. *Geophysical Research Letters*, **23**, 237–240.

HOLLOWAY, S. & CHADWICK, R. A. 1986. The Sticklepath–Lustleigh Fault zone: Tertiary sinistral reactivation of a Variscan dextral strike-slip fault. *Journal of the Geological Society, London*, **143**, 447–452.

HONG, J. & MIYATA, T. 1999. Strike-slip origin of Cretaceous Mazhan Basin, Tan–Lu fault zone, Shandong, east China. *The Island Arc*, **8**, 80–91.

HOSSACK, J. R. 1984. The geometry of listric growth faults in the Devonian basins of Sunnfjord, W. Norway. *Journal of the Geological Society, London*, **141**, 629–637.

HOWARD, J. P., CUNNINGHAM, W. D., DAVIES, S. J., DIJKSTRA, A. H. & BADARCH, G. 2003. The stratigraphic and structural evolution of the Dzereg Basin, western Mongolia: clastic sedimentation, transpressional faulting and basin destruction in an intraplate, intracontinental setting. *Basin Research*, **15**, 45–72.

HURWITZ, S., GARFUNKEL, Z., BEN-GAI, Y., REZNIKOV, M., ROTSTEIN, Y. & GVIRTZMAN, H. 2002. The tectonic framework of a complex pull-apart basin: seismic reflection observations in the Sea of Galilee, Dead Sea transform. *Tectonophysics*, **359**, 289–306.

INGERSOLL, R. V. & BUSBY, C. J. 1995. Tectonics of sedimentary basins. *In*: BUSBY, C. J. & INGERSOLL, R. V. (eds) *Tectonics of Sedimentary Basins*. Blackwell Science, Oxford, 1–51.

INGERSOLL, R. V. & RUMELHART, P. E. 1999. Three-stage evolution of the Los Angeles Basin, southern California. *Geology*, **27**, 593–596.

ITOH, Y., TAKEMURA, K. & KAMATA, H. 1998. History of basin formation and tectonic evolution at the termination of a large transcurrent Sault system: Deformation mode of central Kyushu Japan. *Tectonophysics*, **284**, 135–150.

JAIMES, M. 2003. *Paleogene to Recent tectonic and paleogeographic evolution of the Cariaco basin, Venezuela*. MSc thesis, the University of Texas at Austin.

JAMISON, W. 1991. Kinematics of compressional fold development in convergent wrench terranes. *Tectonophysics*, **190**, 209–232.

JANSMA, P. E., LOPEZ, A., MATTIOLI, G. S., DEMETS, C., DIXON, T. H., MANN, P. & CALAIS, E. 2000. Neotectonics of Puerto Rico and the Virgin Islands, northeastern Caribbean, from GPS geodesy. *Tectonics*, **9**, 1021–1037.

JANY, I., SCANLON, K. M. & MAUFFRET, A. 1990. Geological interpretation of combined Seabeam, Gloria and seismic data from Anegada Passage (Virgin Islands, North Caribbean). *Marine Geophysical Researches*, **12**, 173–196.

JARRARD, R. D. 1986. Relations among subduction parameters. *Reviews of Geophysics*, **24**, 217–284.

JOHNSON, C. & HADLEY, D. 1976. Tectonic implications of the Brawley earthquake swarm, Imperial Valley, California, January, 1975. *Bulletin of the Seismological Society of America*, **66**, 1133–1144.

JOHNSON, S. Y. 1985. Eocene strike-slip faulting and nonmarine basin formation in Washington. *In*: BIDDLE, K. T. & CHRISTIE-BLICK, N. (eds) *Strike-Slip Deformation, Basin Formation, and Sedimentation*. SEPM Special Publications, **37**, 283–302.

KARAKHANIAN, A. S., TRIFONOV, V. G. *ET AL.* 2004. Active faulting and natural hazads in Armenia, eastern Turkey and northwestern Iran. *Tectonophysics*, **380**, 189–219.

KARIG, D. E. 1979. Material transport within accretionary prisms and the 'Knocker' Problem. *Journal of Geology*, **88**, 27–37.

KELLER, V. A., HALL, S. H., DART, C. J. & MCCLAY, K. R. 1995. The geometry and evolution of a transpressional strike-slip system: the Carboneras fault, SE Spain. *Journal of the Geological Society*, **152**, 395–351.

KIM, Y., PEACOCK, D. C. P. & SANDERSON, D. 2004. Fault damage zones. *Journal of Structural Geology*, **26**, 503–517.

KIMURA, G. 1986. Oblique subduction and collision: forearc tectonics of the Kuril Arc. *Geology*, **14**, 404–407.

KINGMA, J. T. 1958. Possible origin of piercement structures, local unconformities, and secondary basins in the Eastern Geosyncline, New Zealand. *New Zealand Journal of Geology and Geophysics*, **1**, 269–274.

KLUTH, C. & CONEY, P. J. 1981. Plate tectonics of the ancestral Rocky Mountains. *Geology*, **9**, 10–15.

KOCYIGIT, A. & EROL, O. 2001. A tectonic escape structure: Erciyes pull-apart basin, Kayseri, central Anatolia, Turkey. *Geodinamica Acta*, **14**, 133–145.

KRAPEZ, B. & EISENLOHR, B. 1998. Tectonic settings of Archaean (3325–2775 Ma) crustal–supracrustal belts in the West Pilbara Block. *Precambrian Research*, **88**, 173–205.

LACROIX, S., SAWYER, E. W. & CHOWN, E. H. 1998. Pluton emplacement within an extensional transfer zone during dextral strike-slip faulting: an example from the Late Archaean Abitibi greenstone belt. *Journal of Structural Geology*, **20**, 43–59.

LALLEMAND, S. & JOLIVET, L. 1986. Japan Sea: a pull-apart basin? *Earth and Planetary Science Letters*, **76**, 375–389.

LANEY, S. E. & GATES, A. E. 1996. Three-dimensional shuffling of horses in a Strike-slip duplex: an example from the Lambertville Sill, New Jersey. *Tectonophysics*, **258**, 53–70.

LANGENHEIM, V. E., JACHENS, R. C., MORTON, D. M., KISTLER, R. W. & MATTI, J. C. 2004. Geophysical and isotopic mapping of preexisting crustal structures that influenced the location and development of the San Jacinto fault zone, southern California. *Geological Society of America Bulletin*, **116**, 1143–1157.

LARROQUE, C., RITZ, J.-F. *ET AL.* 1999. Present-day kinematics of the Tunka Basin (West Baikal, Siberia): transtension or transpression? Some new insights from morphotectonics and neotectonics. *EOS, Transactions of the American Geophysical Union*, **80**, 1023–1024.

LARROQUE, C., RITZ, J. F. *ET AL.* 2001. Interaction compression et extension à la limite Mongolie–Siberie: analyse preliminaire des deformations recentes et actualles dans la Bassin de Tunka. *Comptes Rendus de l' Académic des Sciences, Paris* **332**, 177–184.

LAZAR, M., BEN-AVRAHAM, Z. & SCHATTNER, U. 2006. Formation of sequential basins along a strike-slip fault – geophysical observations from the Dead Sea basin. *Tectonophysics*, **421**, 53–69.

LEBRUN, J.-F., LAMARCHE, G., COLLOT, J.-Y. & DELTIEL, J. 2000. Abrupt strike-slip fault to subduction transition: the Alpine Fault–Puysegur trench connection, New Zealand. *Tectonics*, **17**, 688–706.

LEES, J. M. 2002. Three-dimensional anatomy of a geothermal field, Coso, southeast-central California. *In*: GLAZNER, A. F., WALKER, J. D. & BARTLEY, J. M. (eds) *Geologic Evolution of the Mojave Desert and Southwestern Basin and Range*. GSA Memoir, **195**, 259–276.

LEGG, M., BORRERO, J. & SYNOLAKIS, C. 2004. Tsunami hazards associated with the Catalina fault in southern California. *Earthquake Spectra*, **20**, 917–950.

LEGG, M. R., GOLDFINGER, C., KAMERLING, M. J., CHAYTOR, J. D. & EINSTEIN, D. E. 2007. Morphology, structure and evolution of California Continental Borderland restraining bends. *In*: CUNNINGHAM, W. D. & MANN, P. (eds) *Techtonics of Strike-Slip Restraining and Releasing Bends*. Geological Society, London, Special Publications, **290**, 143–168.

LELGEMANN, H., GUTSCHER, M.-A., BIALAS, J., FLUEH, E. R. & WEINREBE, W. 2000. Transtensional basins in the Western Sunda Strait. *Geophysical Research Letters*, **27**, 3545–3548.

LEMISZKI, P. J. & BROWN, L. D. 1988. Variable crustal structure of strike-slip fault zone as observed on deep seismic reflection profiles. *Geological Society of America Bulletin*, **100**, 665–676.

LENSEN, G. 1958. A method of graben and horst formation. *Journal of Geology*, **66**, 579–587.

LE PICHON, X. & FRANCHETEAU, J. 1978. A plate-tectonic analysis of the Red Sea–Gulf of Aden area. *Tectonophysics*, **46**, 369–406.

LE PICHON, X., SENGÖR, A. M. C. *ET AL*. 2001. The active Main Marmara Fault. *Earth and Planetary Science Letters*, **192**, 595–616.

LEROY, S., DELEPINAY, B. M., MAUFFRET, A. & PUBELLIER, M. 1996. Structural and tectonic evolution of the eastern Cayman Trough (Caribbean sea) from seismic reflection data. *AAPG Bulletin*, **80**, 222–247.

LEROY, S., MAUFFRET, A., PATRIAT, P. & DE LEPINAY, B. M. 2000. An alternative interpretation of the Cayman trough evolution from a reidentification of magnetic anomalies. *Geophysical Journal International*, **141**, 539–557.

LESNE, O., CALAIS, E. & DEVERCHERE, J. 1998. Finite element modelling of crustal deformation in the Baikal rift zone: new insights into the active–passive rifting debate. *Tectonophysics*, **289**, 327–340.

LETTIS, W., BACHHUBER, J. *ET AL*. 2002. Influence of releasing step-overs on surface fault rupture and fault segmentation: examples from the 17 August 1999 Izmit earthquake on the Northern Anatolian Fault, Turkey. *Bulletin of the Seismological Society of America*, **92**, 19–42.

LEVINE, M. J. 1995. *Geo Data Explorer (GEODE)*. US Geological Survey. http://geode.usgs.gov Accessed December 2006.

LINK, M. H., ROBERTS, M. T. & NEWTON, M. S. 1985. Walker Lake Basin, Nevada: an example of Late Tertiary to Recent sedimentation in a basin adjacent to an active strike-slip fault. *In*: BIDDLE, K. T. & CHRISTIE-BLICK, N. (eds) *Strike-Slip Deformation, Basin Formation, and Sedimentation*. SEPM Special Publications, **37**, 105–126.

LITTLE, T. A. 1990. Kinematics of wrench and divergent-wrench deformation along a central part of the Border Ranges fault system, northern Chugach Mountains, Alaska. *Tectonics*, **9**, 585–611.

LITTLE, T. A., COX, S., VRY, J. K. & BATT, G. 2005. Variations in exhumation level and uplift rate along the oblique-slip Alpine Fault, central Southern Alps, New Zealand. *Geological Society of America Bulletin*, **117**, 707–723.

LODOLO, E., MENICHETTI, M., BARTOLE, R., BEN-AVRAHAM, Z., TASSONE, A. & LIPPAI, H. 2003. Magallanes–Fagnano continental transform fault (Tierra del Fuego, southernmost South America). *Tectonics*, **22**, 1076.

LOWELL, J. 1985. *Structural Styles in Petroleum Exploration*. Oil and Gas Consultants International, Tulsa, OK, 460 pp.

LUTZ, A., DORSEY, R., HOUSEN, B. & JANECKE, S. 2006. Stratigraphic record of Pleistocene faulting and basin evolution in the Borrego badlands, San Jacinto fault zone, southern California. *Geological Society of America Bulletin*, **118**, 1377–1397.

LUYENDYK, B. P. 1991. A model for Neogene crustal rotations, transtension, and transpression in Southern California. *Geological Society of America Bulletin*, **103**, 1528–1536.

MCBRIDE, J. H., BARAZANGI, M., BEST, J., AL-SAAD, D., SAWAF, T., AL-OTRI, M. & GEBRAN, A. 1990. Seismic reflection structure of intracratonic Palmyride fold–thrust belt and surrounding Arabian Platform, Syria. *AAPG Bulletin*, **74**, 238–259.

MCCLAY, K. & BONORA, M. 2001. Analog models of restraining stepovers in strike-slip fault systems. *AAPG Bulletin*, **85**, 233–260.

MCCLUSKY, S., BALASSANIAN, S. *ET AL*. 2000. Global positioning system constraints on plate kinematics and dynamics in the Eastern Mediterranean and Caucasus. *Journal of Geophysical Research*, 5695–5719.

MCCLUSKY, S., REILINGER, R., MAHMOUD, S., SARI, D. & TEALEB, A. 2003. GPS constraints on Africa (Nubia) and Arabia plate motions. *Geophysical Journal International*, **155**, 5–22.

MCCULLOCH, D. S. 1987. Regional geology and hydrocarbon potential of offshore Central California. *In*: SCHOLL, D. W., GRANTZ, A. & VEDDER, J. G. (eds) *Geology and Resource Potential of the Continental Margin of Western North America and Adjacent Basins – Beaufort Sea to Baja California*. Circum-Pacific Council of Energy and Mineral Resources, Earth Science Series, **6**, 353–401.

MCKENZIE, D. 1972. Active tectonics of the Mediterranean Region. *Geophysical Journal of the Royal Astronomical Society*, **30**, 109–185.

MCLAUGHLIN, R. & NILSEN, T. H. 1982. Neogene nonmarine sedimentation and tectonics in small pull-apart basins of the San Andreas fault system, Sonoma County, California. *Sedimentology*, **29**, 865–876.

MANAKER, D., MICHAEL, A. & BÜRGMANN, R. 2005. Subsurface structure and kinematics of the

Calaveras–Hayward Fault stepover from three-dimensional V_p and seismicity, San Francisco Bay region, California. *Bulletin of the Seismological Society of America*, **95**, 446–470.

MANN, P. 1997. Model for the formation of large, transtensional basins in zones of tectonic escape. *Geology*, **25**, 211–214.

MANN, P. & BURKE, K. 1984. Neotectonics of the Caribbean. *Reviews of Geophysics and Space Physics*, **22**, 309–362.

MANN, P. & BURKE, K. 1990. Transverse intra-arc rifting: Palaeogene Wagwater Belt, Jamaica. *Marine and Petroleum Geology*, **17**, 410–427.

MANN, P. & GORDON, M. B. 1996. Tectonic uplift and exhumation of blueschist belts along transpressional strike-slip fault zones. *In*: BEBOUT, G. E., SCHOLL, D. W., KIRBY, S. H. & PLATT, J. P. (eds) *Subduction Top to Bottom*. American Geophysical Union, Geophysical Monograph Series, **96**, 143–154.

MANN, P., HEMPTON, M. R., BRADLEY, D. C. & BURKE, K. 1983. Development of pull-apart basins. *Journal of Geology*, **91**, 529–554.

MANN, P., TAYLOR, F. W., BURKE, K. & KULSTAD, R. 1984. Subaerially exposed Holocene coral reef, Enriquillo Valley, Dominican Republic. *Geological Society of America Bulletin*, **95**, 1084–1092.

MANN, P., DRAPER, G. & BURKE, K. 1985. Neotectonics of a strike-slip restraining bend system, Jamaica. *In*: BIDDLE, K. T. & CHRISTIE-BLICK, N. (eds) *Strike-Slip Deformation, Basin Formation, and Sedimentation*. SEPM Special Publications, **37**, 211–226.

MANN, P., TYBURSKI, S. A. & ROSENCRANTZ, E. 1991. Neogene development of the Swan Islands restraining bend complex, Caribbean Sea. *Geology*, **19**, 823–826.

MANN, P., TAYLOR, F., EDWARDS, R. & KU, T. 1995. Actively evolving microplate formation by oblique collision and sideways motion along strike-slip faults: an example from the northeastern Caribbean plate boundary. *Tectonophysics*, **246**, 1–69.

MANN, P., PRENTICE, C. S., BURR, G., PENA, L. R. & TAYLOR, F. W. 1998. Tectonic geomorphology and paleoseismology of the Septentrional fault system, Dominican Republic. *Geological Society of America Special Papers*, **326**, 63–123.

MANN, P., MCLAUGHLIN, P. P., JR., VAN DEN BOLD, W. A., LAWRENCE, S. R. & LAMAR, M. E. 1999. Tectonic and eustatic controls on Neogene evaporitic and siliciclastic deposition in the Enriquillo Basin, Dominican Republic. *In*: MANN, P. (ed.) *Caribbean Basins*, Sedimentary Basins of the World, **4**, Elsevier Science, Amsterdam, 287–342.

MANN, P., CALAIS, E., RUEGG, J.-C., DEMETS, C., JANSMA, P. & MATTIOLI, G. 2002. Oblique collision in the northeastern Caribbean from GPS measurements and geological observations. *Tectonics*, **21**, 1057–1080.

MANN, P., GAHAGAN, L. & GORDON, M. 2003. Tectonic setting of the world's giant oil and gas fields. *In*: HALBOUTY, M. (ed.) *Giant Oil Fields of the Decade 1990–1999*. AAPG Memoir, **78**, 15–105.

MANN, P., ESCALONA, A. & CASTILLO, M. 2006. Regional geologic and tectonic setting of the Maracaibo supergiant basin, western Venezuela. *AAPG Bulletin*, **90**, 445–478.

MANN, P., DEMETS, C. & WIGGINS-GRANDISON, M. 2007*a*. Toward a better understanding of the Late Neogene strike-slip restraining bend in Jamaica: Geodetic, geologic, and seismic constraints. *In*: CUNNINGHAM, W. D. & MANN, P. (eds) *Tectonics of Strike-Slip Restraining and Releasing Bends*. Geological Society, London, Special Publications, **290**, 239–253.

MANN, P., ROGERS, R. & GAHAGAN, L. 2007*b*. Overview of plate tectonic history and its unresolved tectonic problems. *In*: BUNDSCHUH, J. & ALVARADO, G. E. (eds) *Central America: Geology, Resources and Hazards*. Taylor and Francis/Balkema, Leiden, The Netherlands, **1**, 201–237.

MARSHAK, S., NELSON, W. J. & MCBRIDE, J. H. 2003. Phanerozoic strike-slip faulting in the continental interior platform of the United States: examples from the Laramide Orogen, Midcontinent, and Ancestral Rocky Mountains. *In*: STORTI, F., HOLDSWORTH, R. E. & SALVINI, F. (eds) *Intraplate Strike-Slip Deformation Belts*. Geological Society, London, Special Publications, **210**, 159–184.

MASSELL, C., COFFIN, M. F. *ET AL*. 2000. Neotectonics of the Macquarie Ridge Complex, Australia–pacific plate boundary. *Journal of Geophysical Research*, **105**, 13 457–13 480.

MATS, V. D. 1993. The structure and development of the Baikal rift depression. *Earth Science Reviews*, **34**, 81–118.

MATTAUER, M. & MATTE, P. 1998. Le bassin Stephanian de St-Etienne ne resulte pas d'une extension tardi-hercynienne generalisee: c'est un bassin pull-apart en relation avec un decrochement dextre. *Geodinamica Acta*, **11**, 23–31.

MATTERN, F. 2001. Permo-Silesian movements between Baltica and Western Europe: tectonics and 'basin families'. *Terra Nova*, **13**, 368–375.

MAY, S. R., EHMAN, K. D., GRAY, G. G. & CROWELL, J. C. 1993. A new angle on the tectonic evolution of the Ridge basin, a 'strike-slip' basin in southern California. *Geological Society of America Bulletin*, **105**, 1357–1372.

MECKEL, T. A., MANN, P., MOSHER, S. & COFFIN, M. F. 2005. Influence of cumulative convergence on lithospheric thrust fault development and topography along the Australian–Pacific plate boundary south of New Zealand. *Geochemistry Geophysics Geosystems*, **6**, 20.

MENARD, H. W. & ATWATER, T. 1968. Changes in the direction of sea floor spreading. *Nature*, **219**, 463–467.

MENESES-ROCHA, J. J. 2001. Tectonic evolution of the Ixtapa graben, an evolution of a strike-slip basin of southeastern Mexico: implications for regional petroleum systems. *In*: BARTOLINI, C., BUFFLER, R. T. & CANTÚ-CHAPA, A. (eds) *The Western Gulf of Mexico Basin*. AAPG Memoir, **75**, 183–216.

MERRITTS, D. & BULL, W. 1989. Interpreting Quaternary uplift rates at the Mendocino triple junction, from uplifted marine terraces. *Geology*, **17**, 1020–1024.

MILLER, B. V., NANCE, R. D. & MURPHY, J. B. 1995. Kinematics of the Rockland Brook Fault, Nova Scotia: implications for the interaction of the

Meguma and Avalon terranes. *Journal of Geodynamics*, **19**, 253–270.

Miller, R. 1994. A mid-crustal contractional stepover zone in a major strike-slip system, North Cascades, Washington. *Journal of Structural Geology*, **16**, 47–60.

Milnes, A. 1994. Aspects of 'strike-slip' or wrench tectonics – an introductory discussion. *Norsk Geologisk Tidsskrift*, **74**, 129–133.

Miskelly, T. E. Jr. 2002. *Structural framework and depositional systems of a complex rift and strike-slip margin: blocks CI-104 and 105, Ivory Coast, West Africa.* MS thesis, University of Texas at Austin, 128 pp.

Moody, J. & Hill, M. 1956. Wrench-fault tectonics. *Geolegial Society of America, Bulletin*, **67**, 1207–1246.

Moores, E. & Twiss, R. 1995. *Tectonics*. W. H. Freeman, New York, 415 pp.

Morley, C. K. 2002. A tectonic model for the Tertiary evolution of strike-slip faults and rift basins in SE Asia. *Tectonophysics*, **347**, 189–215.

Morley, C. K., Smith, M., Carter, A., Charusiri, P. & Chantraprasert, S. 2007. Evolution of deformation styles at a major restraining bend, constraining from cooling histories, Mae Ping Fault zone, Western Thailand. *In*: Cunningham, W. D. & Mann, P. (eds) *Tectonics of Strike-Slip Restraining and Releasing Bends*. Geological Society, London, Special Publications, **290**, 325–349.

Mount, V. & Suppe, J. 1987. State of stress near the San Andreas fault: implications for wrench tectonics. *Geology*, **15**, 1357–1372.

Mouslopoulou, V., Nicol, A., Little, T. A. & Walsh, J. J. 2007. Terminations of large strike-slip faults: an alternative model from New Zealand. *In*: Cunningham, W. D. & Mann, P. (eds) *Tectonics of Strike-Slip Restraining and Releasing Bends*. Geological Society, London, Special Publications, **290**, 387–415.

Muehlberger, W. R. & Gordon, M. B. 1987. Observations on the complexity of the East Anatolian Fault, Turkey. *Journal of Structural Geology*, **9**, 899–903.

Murphy, M. A. & Burgess, W. P. 2006. Geometry, kinematics, and landscape characteristics of an active transtension zone, Karakoram fault system, southwest Tibet. *Journal of Structural Geology*, **28**, 268–283.

Namson, J. & Davis, T. 1988. Structural transect of the western Transverse Ranges, California: implications for lithospheric kinematics and seismic risk evaluation. *Geology*, **16**, 675–679.

Neugebauer, J. 1994. Closing-up structures, alternatives to pull-apart basins: the effect of bends in the North Anatolian fault, Turkey. *Terra Nova*, **6**, 359–365.

Nicholson, C., Seeber, L. & Williams, P. 1986. Seismicity and fault kinematics through the eastern Transverse Ranges, California: block rotation, strike-slip faulting, and low-angle thrusts. *Journal of Geophysical Research*, **91**, 4891–4908.

Nicholson, C., Seeber, L., Williams, P. & Sykes, L. 1985. Seismic deformation along the southern San Andreas fault, California: implications for conjugate slip, rotational block tectonics. *Tectonics*, **91**, 4891–4908.

Norris, R. J. & Cooper, A. F. 1995. Origin of small-scale segmentation and transpressional thrusting along the Alpine Fault, New Zealand. *Geological Society of America Bulletin*, **107**, 231–240.

Oglesby, D. 2005. The dynamics of strike-slip stepovers with linking dip-slip faults. *Bulletin of the Seismological Society of America*, **95**, 1604–1622.

Okay, A. I., Demirbağ, E., Kurt, H., Okay, N. & Kuşçu, İ. 1999. An active, deep marine strike-slip basin along the North Anatolian fault in Turkey. *Tectonics*, **18**, 129–147.

Okay, A. I., Kaşlilar-Özcan, A., İmren, C., Boztepe-Güney, A., Demirbağ, E. & Kuşçu, İ. 2000. Active faults and evolving strike slip basins in the Marmara Sea, northwest Turkey: a multi-channel seismic reflection study. *Tectonophysics*, **321**, 189–218.

Oldow, J. S. 2003. Active transtensional boundary zone between the western Great Basin and Sierra Nevada block, western U.S. Cordillera. *Geology*, **31**, 1033–1036.

Owen, L. A., Cunningham, D., Windley, B. F., Badamgarov, G. & Dorjnamjaa, D. 1999. The landscape evolution of Nemegt Uul: a late Cenozoic transpressional uplift in the Bogi Altai, southern Magnolia. *In*: Smith, B. J., Whalley, W. B. & Warke, P. A. (eds) *Uplift, Erosion and Stability: Perspectives on Long-term Landscape Development*. Geological Society, London, Special Publications, **162**, 201–218.

Page, B. M. 1990. Evolution and complexities of the transform system in California, USA. *Annales Tectonicae*, **4**, 53–69.

Parkinson, C. & Dooley, T. 1996. Basin formation and strain partitioning along strike-slip fault zones. *Bulletin of the Geological Survey of Japan*, **47**, 427–436.

Parsons, T., Bruns, T. R. & Sliter, R. 2005. Structure and mechanics of the San Andreas–San Gregorio fault Junction, San Francisco, California. *Geochemistry Geophysics Geosystems*, **6**, Q01009, doi: 10.1029/2004-GC000838.

Parsons, T., Sliter, R. *et al.* 2003. Structure and mechanics of the Hayward–Rodgers Creek fault stepover, San Francisco Bay, California. *Bulletin of the Seismological Society of America*, **93**, 2187–2200.

Paterson, B. & Campbell, J. K. 1998. Engineering projects at Arthurs Pass. *In*: Laird, M. (eds) *Geological Society of New Zealand Annual Conference, Christchurch, New Zealand*. Geological Society New Zealand Miscellaneous Publications, **101B**, 2.1–2.28.

Paylor, E. D. I. & Yin, A. 1993. Left-slip evolution of the North Owl Creek fault system, Wyoming, during Laramide shortening. *Geological Society of America, Special Papers*, **280**, 229–242.

Pelayo, A. M. & Wiens, D. A. 1989. Seismotectonics and relative plate motions in the Scotia Sea region. *Journal of Geophysical Research*, **94**, 7293–7320.

Pennington, W. D., Mooney, W. D., Van Hissenhoven, R., Meyer, R. P. & Ramirez, J. 1979. Results of a reconnaissance microearthquake survey of Bucaramanga, Colombia. *Geophysical Research Letters*, **6**, 65–68.

PEREZ, O. J., BILHAM, R. *ET AL.* 2001. Velocity field across the southern Caribbean plate boundary and estimates of Caribbean/South-American plate motion using GPS geodesy 1994–2000. *Geophysical Research Letters*, **28**, 2987–2990.

PERSAUD, P., STOCK, J., STECKLER, M., MARTIN-BARAJAS, A., DIEBOLD, J., GONZALEZ-FERNANDEZ, A. & MOUNTAIN, G. 2003. Active deformation and shallow structure of the Wagner, Consag, and Delfin basins, northern Gulf of California, Mexico. *Journal of Geophysical Research*, **108**, **B7**, 2355, doi: 10.1029/2002JB001937, 2003.

PETIT, C. 1998. Style of active intraplate deformation from gravity and Seismicity data: the Baikal Rift, Asia. *Terra Nova*, **10**, 160–169.

PETIT, C., DEVERCHERE, J., HOUDRY, F., SANKOV, V. A., MELNIKOVA, V. I. & DELVAUX, D. 1996. Present-day stress field changes along the Baikal rift and tectonic implications. *Tectonics*, **15**, 1171.

PFLAKER, G. 1976. Tectonic aspects of the Guatemala earthquake of 4 February 1976. *Science*, **193**, 1201–1208.

PINDELL, J. L. & BARRETT, S. F. 1990. Geologic evolution of the Caribbean: a plate-tectonic perspective. *In*: DENGO, G. & CASE, J. E. (eds) *The Geology of North America. Vol. H, The Caribbean Region*. Geological Society of America, Boulder, Colorado, 405–432.

PITMAN, W. C. & ANDREWS, J. A. 1985. Subsidence and thermal history of small pull-apart basins. *In*: BIDDLE, K. T. & CHRISTIE-BLICK, N. (eds) *Strike-Slip Deformation, Basin Formation and Sedimentation.* SEPM Special Publications, **37**, 45–49.

POCKALNY, R. 1997. Evidence of transpression along the Clipperton Transform; implications for processes of plate boundary reorganization. *Earth and Planetary Science Letters*, **146**, 449–464.

POWELL, R. E. & WELDON, R. J. 1992. Evolution of the San Andreas Fault. *Annual Review of Earth and Planetary Science*, **20**, 431–468.

PRAMUMIJOYO, S. & SEBRIER, M. 1991. Neogene and Quaternary fault kinematics around the Sunda Strait area, Indonesia. *Journal of Southeast Asian Earth Sciences*, **6**, 137–145.

PRENTICE, C., MANN, P., PENA, L. & BURR, G. 2003. Slip rate and earthquake recurrence along the central Septentrional fault, North American–Caribbean plate boundary, Dominican Republic. *Journal of Geophysical Research*, **108**, **B3**, 2149, doi: 10.1029/2001JB000442, 2003.

QUENNELL, A. M. 1956. Tectonics of the Dead Sea rift. *Proceedings, Congreso Geologico Internacional, 20th, Asociacion de Servicios Geologicos Africanos, Mexico City*, 385–405.

QUENNELL, A. M. 1958. The structural and geomorphic evolution of the Dead Sea Rift. *Quarterly Journal of the Geological Society, London*, **114**, 1–24.

RACERO-BAEMA, A. & DRAKE, S. J. 1996. Structural style and reservoir development in the West Netherlands oil province. *In*: RONDEEL, H. E., BATJES, D. A. J. & NIEWENHUIS, W. H. (eds) *Geology of Gas and Oil Under the Netherlands*, Kluwer, Amsterdam, 211–227.

RAHE, B., FERRILL, D. & MORRIS, A. 1998. Physical analog modeling of pull-apart basin evolution. *Tectonophysics*, **285**, 21–40.

RANGIN, C., LE PICHON, X., DEMIRBAG, E. & IMREN, C. 2004. Strain localization in the Sea of Marmara: propagation of the North Anatolian fault in a now inactive pull-apart. *Tectonics*, **23**, TC2014, doi: 10.1029/2002TC001437, 2004.

RATSCHBACHER, L., RILLER, U., MESCHEDE, M., HERRMAN, U. & FRISCH, W. 1991. Second look at suspect terranes in southern Mexico. *Geology*, **19**, 1233–1236.

RECHES, Z. 1987. Mechanical aspects of pull-apart basins and push-up swells with application to the Dead Sea transform. *Tectonophysics*, **141**, 75–88.

REIJS, J. & MCCLAY, K. 1998. Salar Grande pull-apart basin, Atacama fault system, northern Chile. *In*: HOLDSWORTH, R. E., STRACHAN, R. A. & DEWEY, J. F. (eds) *Continental Transpression and Transtensional Tectonics.* Geological Society, London, Special Publications, **135**, 127–141.

REIJS, J. & MCCLAY, K. 2003. The Salina del Fraile pull-apart basin, NW Argentina. *In*: STORTI, F. & SALVINI, F. (eds) *Intraplate Strike-Slip Deformation Belts*. Geological Society, London, Special Publications, **210**, 197–209.

REILINGER, R., MCCLUSKEY, S. *ET AL.* 2006. GPS constraints on continental deformation in the Africa–Arabia–Eurasia continental collision zone and implications for the dynamics of plate interactions. *Journal of Geophysical Research*, **111**, 26.

RICHARD, P. D., NAYLOR, M. A. & KLOOPMAN, A. 1995. Experimental models of strike slip tectonics. *Petroleum Geoscience*, **1**, 71–80.

RINGENBACH, J.-C., STEPHAN, J. F., MALETERRE, P. & BELLON, H. 1990. Structure and geological history of the Lepanto–Cervantes releasing bend on the Abra River Fault, Luzon Central Cordillera, Philippines. *Tectonophysics*, **183**, 225–241.

RODGERS, D. A. 1979. Vertical deformation, stress accumulation, and secondary faulting in the vicinity of the Transverse Ranges of Southern California. *California Division of Mines and Geology*, **203**, 54.

RODGERS, D. A. 1980. Analysis of pull-apart basin development produced by en echelon strike-slip faults. *In*: BALLANCE, P. & READING, H. (eds) *Sedimentation in Oblique-Slip Mobile Zones*. Special Publications of the International Association of Sedimentologists, **4**, 27–41.

ROESKE, S. M., SNEE, L. W. & PAVLIS, T. L. 2003. Dextral-slip reactivation of an arc–forearc boundary during Late Cretaceous–early Eocene oblique convergence in the northern Cordillera. *In*: SISSON, V. B., ROESKE, S. M. & PAVLIS, T. L. (eds) *Geology of a Transpressional Orogen Developed During Ridge–Trench Interaction Along the North Pacific Margin.* Geological Society of America, Special Papers, **371**, 141–169.

ROGERS, R. & MANN, P. 2007. Transtensional deformation of the western Caribbean–North America plate boundary zone. *In*: MANN, P. (ed.) *Geologic and Tectonic Development of the Caribbean Plate*

Boundary in Northern Central America. Geological Society of America Special Papers., **428**, 37–64.

RON, H., NUR, A. & EYAL, Y. 1990. Multiple strike-slip fault sets: a case study from the Dead Sea Transform. *Tectonics*, **9**, 1421–1431.

ROSENCRANTZ, E. 1995. Opening of the Cayman Trough and the evolution of the northern Caribbean Plate boundary. *Geological Society of America Annual Meeting. New Orleans*, **27**, 153.

ROSENCRANTZ, E., ROSS, M. & SCLATER, J. 1988. Age and spreading history of the Cayman trough as determined from depth, heat flow, and magnetic anomalies. *Journal of Geophysical Research*, **93**, 2141–2157.

ROTSTEIN, Y. & BARTOV, Y. 1989. Seismic reflection across a continental transform: an example from a convergent segment of the Dead Sea Rift. *Journal of Geophysical Research*, **94**, 2902–2912.

ROYDEN, L. H., HORVATH, F. & BURCHFIEL, B. C. 1982. Transform faulting, extension, and subduction in the Carpathian Pannonian region. *Geological Society of America Bulletin*, **93**, 717–725.

ROYDEN, L. H. 1985. The Vienna Basin: a thin-skinned pull-apart basin. *In*: BIDDLE, K. T. & CHRISTIE-BLICK, N. (eds) *Strike-Slip Deformation, Basin Formation, and Sedimentation*. SEPM Special Publications, **37**, 283–302.

ROYDEN, L. H. 1993. Evolution of retreating subduction boundaries formed during continental collision. *Tectonics*, **12**, 629–638.

RYANG, W. H. & CHOUGH, S. K. 1999. Alluvial-to-lacustrine systems in a pull-apart margin: southwestern Eumsung Basin (Cretaceous), Korea. *Sedimentary Geology*, **127**, 31–46.

SACHSENHOFER, R., BECHTEL, A., REISCHENBACHER, D. & WEISS, A. 2003. Evolution of lacustrine systems along the Miocene Mur–Murz fault system (Eastern Alps, Austria) and implications on source rocks in pull-apart basins. *Marine and Petroleum Geology*, **20**, 83–110.

SACKS, P. E., MALO, M., TRZCIENSKI, W. E. J., PINCIVY, A. & GOSSELIN, P. 2004. Taconian and Acadian transpression between the internal Humber Zone and the Gaspé Belt in the Gaspé Peninsula: tectonic history of the Shickshock Sud fault zone. *Canadian Journal of Earth Science*, **41**, 635–653.

SANDERSON, D. & MARCHINI, R. D. 1984. Transpression. *Journal of Structural Geology*, **6**, 449–458.

SANGAWA, A. 1986. The history of fault movement since late Pliocene in the central part of southwest Japan. *In*: REILLY, W. I. & HARFORD, B. E. (eds) *Recent Crustal Movements of the Pacific Region*. Royal Society of New Zealand Bulletins, **24**, 75–85.

SANTANACH, P., FERRUS, B., CABRERA, L. & SAEZ, A. 2005. Origin of a restraining bend in an evolving strike-slip system: The Cenozoic As Pontes basin (NW Spain). *Geologica Acta*, **3**, 225–239.

SARIBUDAK, M., SANVER, M., SENGOR, A. & GORUR, N. 1990. Paleomagnetic evidence for substantial rotation of the Almacik flake within the North Anatolian fault zone, NW Turkey. *Geophysical Journal International*, **102**, 563–568.

SAROLGU, F., EMRE, O. & KUSCU, I. 1992. The East Anatolian fault zone of Turkey. *Annalae Tectonicae*, **6**, 99–125.

SATO, H., HIRATA, H., ITO, T., TSUMURA, N. & IKAWA, T. 1998. Seismic reflection profiling across the seismogenic fault of the 1995 Kobe earthquake, southwestern Japan. *Tectonophysics*, **286**, 19–30.

SCHATTNER, U., BEN-AVRAHAM, Z., LAZAR, M. & HUEBSCHER, C. 2006. Tectonic isolation of the Levant basin offshore Galilee–Lebanon – effects of the Dead Sea fault plate boundary on the Levant continental margin, eastern Mediterranean. *Journal of Structural Geology*, **28**, 2049–2066.

SCHEUBER, E. & ANDRIESSEN, P. A. M. 1990. The kinematic and geodynamic significance of the Atacama fault zone, Northern Chile. *Journal of Structural Geology*, **12**, 243–257.

SCHOLZ, C. H. 1977. Transform fault systems of California and New Zealand: similarities in their tectonic and seismic types. *Journal of the Geological Society, London*, **133**, 215–229.

SCHUBERT, C. 1980. Late Cenozoic pull-apart basins, Boconó fault zone, Venezuelan Andes. *Journal of Structural Geology*, **2**, 463–468.

SCHUBERT, C. 1982. Origin of Cariaco basin, southern Caribbean Sea. *Marine Geology*, **47**, 3345–3360.

SCHULTE, S. M. & MOONEY, W. D. 2005. An updated global earthquake catalogue for stable continental regions: reassessing the correlation with ancient rifts. *Geophysical Journal International*, **161**, 707–721.

SCHUMACHER, M. E. 2002. Upper Rhine Graben: role of preexisting structures during rift evolution. *Tectonics*, **21**, 1006, doi: 10.1029/2001TC9000022, 2002.

SCHWARTZ, S., ORANGE, D. & ANDERSON, R. 1990. Complex fault interactions in a restraining bend, southern Santa Cruz mountains, California. *Journal of Geophysical Research*, **17**, 1207–1210.

SEARLE, M. 1986. GLORIA investigations of oceanic fracture zones: comparative study of the transform fault zone. *Journal of the Geological Society, London*, **143**, 743–456.

SEARLE, M. & PHILLIPS, R. 2005. Transpressional uplift of middle and deep crustal rocks by constraining bends along the Karakoram Fault in the K2–Gasherbrum range. *In*: CUNNINGHAM, D. & MANN, P. (eds) *Tectonics of Strike-Slip Restraining and Releasing Bends in Continental and Oceanic Settings*, 28–30 September 2005, Geological Society, London, England.

SEEBER, L., CORMIER, M.-H., MCHUGH, C., EMRE, O., POLONIA, A. & SORLIEN, C. 2006. Rapid subsidence and sedimentation from oblique slip near a bend on the North Anatolian transform fault in the Marmara Sea, Turkey. *Geology*, **34**, 933–936.

SEEBER, L., EMRE, O., CORMIER, M., SORLIEN, C., MCHUGH, C. & POLONIA, A. 2004. Uplift and subsidence from oblique slip: the Ganos–Marmara bend of the North Anatolian transform, western Turkey. *Tectonophysics*, **391**, 239–258.

SEGALL, P. & POLLARD, D. D. 1980. Mechanics of discontinuous faults. *Journal of Geophysical Research*, **85**, 4337–4350.

SELLA, G. F., DIXON, T. H. & MAO, A. 2002. REVEL: a model for recent plate velocities from space geodesy.

Journal of Geophysical Research, **107**, B4, 2081, doi: 10.1029/2000JB000033, 2002.

SENGÖR, A., GORUR, N. & SAROGLU, F. 1985. Strike-slip faulting and related basin formation in zones of tectonic escape: Turkey as a case study. *In*: BIDDLE, K. T. & CHRISTIE-BLICK, N. (eds) *Strike-Slip Deformation, Basin Formation, and Sedimentation*. SEPM Special Publications, **37**, 227–264.

SERPA, L., DE VOOGD, B., WRIGHT, L., WILLEMIN, J., OLIVER, J., HAUSER, E. & TROXEL, B. W. 1988. Structure of the Central Death Valley pull-apart basin and vicinity from COCORP profiles in the Southern Great Basin. *Geological Society of America Bulletin*, **100**, 1437–1450.

SEYMEN, I. 1975. *Tectonic characteristics of the North Anatolian fault zone in Kelkit valley [Kelkit Valdisi kesiminde Kuzey Anadolu Fay zonunum tektonik ozelligi]*. PhD, Istanbul Technical University, Istanbul, Turkey, 192 pp.

SHARP, R. V. 1975. En echelon fault patterns of the San Jacinto Fault zone. *In*: CROWELL, J. C. (ed.) San Andreas fault in southern California; a guide to San Andreas fault from Mexico to Carrizo Plain. *Special Reports, California Division of Mines and Geology*, **118**, 147–152.

SHARP, R. V. & CLARK, M. M. 1972. Geologic evidence of previous faulting near the 1968 rupture on the Coyote Creek Fault. *In*: *The Borrego Mountain Earthquake of April 9, 1968*. US Geological Survey Professional Papers, **787**, US Government Printing Office, Washington, D. C., 131–140.

SHEDLOCK, K. M., BROCHER, T. M. & HARDING, S. T. 1990. Shallow structure and deformation along the San Andreas Fault in Cholame Valley, California, based on high-resolution reflection profiling. *Journal of Geophysical Research*, **95**, 5003–5020.

SHELLEY, D. & BOSSIERE, G. 2001. The Ancenis Terrane: an exotic duplex in the Hercynian belt of Armorica, western France. *Journal of Structural Geology*, **23**, 1597–1614.

SIBSON, R. H. 1985. Stopping of earthquake ruptures at dilational fault jogs. *Nature*, **316**, 248–251.

SIBSON, R. H. 1986*a*. Brecciation processes in fault zones. *Pure and Applied Geophysics*, **124**, 159–175.

SIBSON, R. H. 1986*b*. Rupture interaction with fault jogs. *Geophysical Monograph*, **37**, 157–167.

SIBSON, R. H. 1987. Earthquake rupturing as a mineralizing agent in hydrothermal systems. *Geology*, **15**, 701–704.

SIEH, K. & NATAWIDJAJA, D. 2000. Neotectonics of the Sumatran Fault, Indonesia. *Journal of Geophysical Research*, **105**, 28 295–228 326.

SIMS, D., FERRILL, D. A. & STAMATAKOS, J. A. 1999. Role of a ductile decollement in the development of pull-apart basins: experimental results and natural examples. *Journal of Structural Geology*, **21**, 533–554.

SMALLEY, R., Jr. & KENDRICK, E. 2003. Geodetic determination of relative plate motion and crustal deformation across the Scotia–South America Plate boundary in eastern Tierra del Fuego. *G3*, **4**, p.

SMITH, K. 2004. The North Sea Silverpit crater: impact structure or pull-apart basin? *Journal of the Geological Society, London*, **161**, 593–602.

SMITH, M., CHANTRAPRASERT, S., MORLEY, C. K. & CARTWRIGHT, I. 2007. Structural geometry and timing of deformation in the Chainat Duplex, Thailand. *In*: CUNNINGHAM, W. D. & MANN, P. (eds) *Tectonics of Strike-Slip Restraining and Releasing Bends*. Geological Society, London, Special Publications, **290**, 305–323.

SORLIEN, C. C., KAMERLING, M. J., SEEBER, L. & BRODERICK, K. G. 2006. Restraining segments and reactivation of the Santa Monica–Dume–Malibu Coast fault system, offshore Los Angeles, California. *Journal of Geophysical Research*, **111**, B11402 doi: 10.1029/2005JB003632, 2006.

SOWERS, J., UNRUH, J., LETTIS, W. & RUBIN, T. 1994. Relationship of the Kickapoo fault to the Johnson Valley and Homestead Valley faults, San Bernardino County, California. *Bulletin of the Seismological Society of America*, **84**, 528–536.

SPOTILA, J. A. & ANDERSON, K. B. 2004. Fault interaction at the junction of the Transverse Ranges and Eastern California shear zone: a case study of intersecting faults. *Tectonophysics*, **379**, 43–60.

SPOTILA, J. A., FARLEY, K. A., YULE, J. D. & REINERS, P. W. 2001. Near-field transpressive deformation along the San Andreas fault zone in southern California, based on exhumation constrained by (U–Th)/He dating. *Journal of Geophysical Research*, **106**, 30 909–30 922.

STEEL, R. J. & GLOPPEN, T. G. 1980. Late Caledonian (Devonian) basin formation, western Norway: signs of strike-slip tectonics during infilling. *In*: BALLANCE, P. F. & READING, H. G. (eds) *Sedimentation in Oblique-Slip Mobile Zones*, International Association of Geologists Special Publication. International Association of Sedimentologists, **4**, 79–103.

STEWART, S. A. & ALLEN, P. J. 2005. 3D seismic reflection mapping of the Silverpit multi-ringed crater, North Sea. *Geological Society of America Bulletin*, **117**, 354–368.

STONE, D. S. 1969. Wrench faulting and Rocky Mountain tectonics. *The Mountain Geologist*, **6**, 67–79.

STONE, D. S. 1995. Structure and kinematic genesis of the Quealy wrench duplex: transpressional reactivation of the Precambrian Cheyenne Belt in the Laramie Basin, Wyoming. *AAPG Bulletin*, **79**, 1349–1376.

STORTI, F., HOLDSWORTH, R. & SALVINI, F. 2003. Intraplate strike-slip deformation belts. *In*: STORTI, F., HOLDSWORTH, R. & SALVINI, F. (eds) *Intraplate Strike-slip Deformation Belts*. Geological Society, London, Special Publications, **210**, 1–14.

SWANSON, M. T. 1989*a*. Sidewall ripouts in strike-slip faults. *Journal of Structural Geology*, **11**, 933–948.

SWANSON, M. T. 1989*b*. Extensional duplexing in the York Cliffs strike-slip fault system, southern coastal Maine. *Journal of Structural Geology*, **12**, 499–512.

SWANSON, M. T. 2005. Geometry and kinematics of adhesive wear in brittle strike-slip fault zones. *Journal of Structural Geology*, **27**, 871–887.

SYLVESTER, A. G. (compiler) 1984. *Wrench Fault Tectonics*. AAPG Reprint Series, **28**, 374.

SYLVESTER, A. G. 1988. Strike-slip faults. *Geological Society of America Bulletin*, **100**, 1666–1703.

SYLVESTER, A. G. & SMITH, R. 1976. Tectonic transpression and basement controlled deformation in the

San Andreas fault zone, Salton trough, California. *AAPG Bulletin*, **60**, 2081–2102.

TALBOT, C. J. 2005. Pull-apart emplacement of the Margeride granitic complex (French Massif Central). Implications for the late evolution of the Variscan Orogen. *Journal of Structural Geology*, **27**, 1610–1629.

TALEBIAN, M., FIELDING, E. J. *ET AL*. 2004. The 2003 Bam (Iran) earthquake: rupture of a blind strike-slip fault. *Geophysical Research Letters*, **31**, L11611, doi: 10.1029/2004GL020058.

TALWANI, P. 1999. Fault geometry and earthquakes in continental interiors. *Tectonophysics*, **305**, 371–379.

TAPPONNIER, P., PELTZER, G., LEDAIN, A., ARMIJO, R. & COBBOLD, P. R. 1982. Propagating extrusion tectonics in Asia – new insights from simple experiments with plasticine. *Geology*, **10**, 611–616.

TATAR, O., PIPER, J. D. A., PARK, R. G. & GURSOY, H. 1995. Palaeomagnetic study of block rotations in the Niksat overlap region of the North Anatolian fault zone, Central Turkey. *Tectonophysics*, **244**, 251–266.

TCHALENKO, J. S. & AMBRASEYS, N. N. 1970. Structural analysis of the Dashat-e Bayaz (Iran) earthquake fractures. *Geological Society of America Bulletin*, **81**, 41–60.

TEN BRINK, U. S. & BEN-AVRAHAM, Z. 1989. The anatomy of a pull-apart basin: seismic reflection observations of the Dead Sea. *Tectonics*, **8**, 333–350.

TEN BRINK, U. S., KATZMAN, R. & LIN, J. 1996. Three-dimensional models of deformation near strike-slip faults. *Journal of Geophysical Research*, **101**, 16 205–16 220.

TEN BRINK, U. S., BEN-AVRAHAM, Z. *ET AL*. 1993. Structure of the Dead Sea pull-apart basin from gravity analysis. *Journal of Geophysical Research*, **98**, 21 877–21 894.

TEN BRINK, U. S. & RYBAKOV, M. 1999. Anatomy of the Dead Sea transform; does it reflect continuous changes in plate motion? *Geology*, **27**, 887–890.

THOMAS, C., LIVERMORE, R. & POLLITZ, F. 2003. Motion of the Scotia Sea plates. *Geophysical Journal International*, **155**, 789–804.

THOMSON, S. 2002. Late Cenozoic geomorphic and tectonic evolution of the Patagonian Andes between latitudes 42S and 46S: an appraisal based on fission-track results from the transpressional intra-arc Liquine–Ofaqui fault zone. *Geological Society of America Bulletin*, **114**, 1159–1173.

TITUS, S. J., FOSSEN, H., PEDERSEN, R. B., VIGNERESSE, J. L. & TIKOFF, B. 2002. Pull-apart formation and strike-slip partitioning in an obliquely divergent setting, Leka ophiolite, Norway. *Tectonophysics*, **354**, 101–119.

TRENKAMP, R., KELLOGG, J., FREYMUELLER, J. & MORA, H. 2002. Wide plate margin deformation, southern Central America and northwestern South America, CASA GPS observations. *Journal of South American Earth Sciences*, **15**, 157–171.

TWISS, R. & MOORES, E. 1992. *Structural Geology*. W. H. Freeman, New York, 532 pp.

UMHOEFER, P. J. & STONE, K. A. 1996. Description and kinematics of the SE Loreta basin fault array, Baja California Sur, Mexico: a positive field test of oblique models. *Journal of Structural Geology*, **18**, 595–614.

UNDERHILL, J. R. 1991. Implications of Mesozoic–Recent basin development in the western Inner Moray Firth, UK. *Marine and Petroleum Geology*, **8**, 359–369.

UNRUH, J., HUMPHREY, J. & BARRON, A. 2003. Transtensional model for the Sierra Nevada frontal fault system, eastern California. *Geology*, **31**, 327–330.

VAN DER STRAATEN, H. C. 1993. Neogene strike-slip faulting in southeastern Spain: the deformation of the pull-apart basin of Abaran. *Geologie en Mijnbouw*, **71**, 205–225.

VAN DISSEN, R. & YEATS, R. S. 1991. Hope Fault, Jordan Thrust, and uplift of the Seward Kaikoura Range, New Zealand. *Geology*, **19**, 393–396.

VELANDIA, F., ACOSTA, J., TERRAZA, R. & VILLEGAS, H. 2005. The current tectonic motion of the Northern Andes along the Algeciras Fault System in SW Colombia. *Tectonophysics*, **399**, 313–329.

VINCENT, S. J. & ALLEN, M. B. 1999. Evolution of the Minle and Chaoshui Basins, China: implications for Mesozoic strike-slip basin formation in Central Asia. *Geological Society of America, Bulletin*, **111**, 725–742.

VOGT, P. R., LOWRIE, A., BRACEY, D. R. & HEY, R. N. 1976. Subduction of aseismic oceanic ridges: effects on shape, seismicity, and other characteristics of consuming plate boundaries. *Geological Society of America, Special Papers*, **172**, 1–59.

WAKABAYASHI, J. 2007. Stepovers that migrate with respect to affected deposits: field characteristics and speculation on some details of their evolution. *In*: CUNNINGHAM, W. D. & MANN, P. (eds) *Tectonics of Strike-Slip Restraining and Releasing Bends*. Geological Society, London, Special Publications, **290**, 169–188.

WAKABAYASHI, J., HENGESH, J. V. & SAWYER, T. L. 2004. Four-dimensional transform fault processes: progressive evolution of step-overs and bends. *Tectonophysics*, **392**, 279–301.

WALDRON, J. W. F. 2004. Anatomy and evolution of a pull-apart basin, Stellarton, Nova Scotia. *Geological Society of America Bulletin*, **116**, 109–127.

WALKER, R. & JACKSON, J. 2004. Active tectonics and late Cenozoic strain distribution in central and eastern Iran. *Tectonics*, **23**, TC5010, doi:10.1029/2003TC001529.

WALLEY 1998. Some outstanding issues in the geology of Lebanon and their importance in the tectonic evolution of the Levantine Region. *Tectonophysics*, **298**, 37–62.

WALLS, C., ROCKWELL, T., MUELLER, K., BOCK, Y., WILLIAMS, S., PFANNER, J., DOLAN, J. & FANG, P. 1998. Escape tectonics in the Los Angeles metropolitan region and implications for seismic risk. *Nature*, **394**, 356–360.

WARD, S. N. 1994. A multidisciplinary approach to seismic hazard in southern California. *Seismological Society of America Bulletin*, **84**, 1293–1309.

WEAVER, C. S. & HILL, D. P. 1978. Earthquake swarms and local crustal spreading along major strike slip faults in California. *Pure and Applied Geophysics*, **117**, 51–64.

WEBER, J. C., DIXON, T. H. *ET AL*. 2001. GPS estimate of relative motion between the Caribbean and South American plates, and geologic implications for Trinidad and Venezuela. *Geology*, **29**, 75–78.

WELLMAN, H. W. 1955. New Zealand Quaternary tectonics. *Geologische Rundschau*, **43**, 248–257.

WESNOUSKY, S. G. 1988. Seismological and structural evolution of strike-slip faults. *Letters to Nature*, **335**, 340–343.

WESNOUSKY, S. G. 2005. The San Andreas and Walker Lane fault systems, western North America: transpression, transtension, cumulative slip and the structural evolution of a major transform plate boundary. *Journal of Structural Geology*, **27**, 1505–1512.

WESTAWAY, R. 1995. Deformation around stepovers in strike-slip fault zones. *Journal of Structural Geology*, **17**, 831–846.

WESTAWAY, R. 2004. Kinematic consistency between the Dead Sea Fault Zone and the Neogene and Quaternary left-lateral faulting in SE Turkey. *Tectonophysics*, **391**, 203–237.

WHITE, R. A. & HARLOW, D. H. 1993. Destructive upper crustal earthquakes of Central America since 1900. *Seismological Society of America Bulletin*, **83**, 1115–1142.

WILCOX, R. E., HARDING, T. P. & SEELY, D. R. 1973. Basin wrench tectonics. *AAPG Bulletin*, **57**, 74–96.

WILSON, J. T. 1965. A new class of faults and their bearing on continental drift. *Nature*, **207**, 343–347.

WOOD, R. A., TETTINGA, J. R., BANNISTER, S., LAMARCHE, G. & MCMORRAN, T. J. 1994. Structure of the Hanmer strike-slip basin, Hope fault, New Zealand. *Geological Society of America Bulletin*, **106**, 1459–1473.

WOODCOCK, N. H. 1986. The role of strike-slip fault systems at plate boundaries. *Philosophical Transactions of the Royal Society of London, Series A*, **317**, 13–29.

WOODCOCK, N. H. 2004. Life span and fate of basins. *Geology*, **32**, 685–688.

WOODCOCK, N. H. & FISCHER, M. 1986. Strike-slip duplexes. *Journal of Structural Geology*, **8**, 725–735.

WOODCOCK, N. H. & SCHUBERT, C. 1989. Continental strike-slip tectonics. *In*: HANCOCK, P. L. (ed.) *Continental Tectonics*. Pergamon, Oxford, 251–263.

WORRALL, D. M. 1991. *Tectonic History of the Bering Sea and the Evolution of Tertiary Strike-Slip Basins of the Bering Shelf*. Geological Society of America, Boulder, Colorado, 120 pp.

YEATS, R. S. 1981. Quaternary flake tectonics of the California Transverse Ranges. *Geology*, **9**, 16–20.

YEATS, R. S. & BERRYMAN, K. R. 1987. South Island, New Zealand, and Transverse Ranges, California: a seismotectonic comparison. *Tectonics*, **6**, 363–376.

YEATS, R. S., HUFTILE, G. J. & STITT, L. T. 1994. Late Cenozoic tectonics of the east Ventura basin, Transverse Ranges, California. *AAPG Bulletin*, **78**, 1040–1074.

YEATS, R. S., SIEH, K. & ALLEN, C. R. 1997. *The Geology of Earthquakes*. Oxford University Press, New York, 568 pp.

ZAK, I. & FREUND, R. 1981. Asymmetry and basin migration in the Dead Sea rift. *Tectonophysics*, **80**, 27–38.

ZAMPIERI, D. & MASSIRONI, M. 2005. The Mt. Tormeno restraining stepover: an example of multiple deformation in the eastern Southern Alps, Italy. *In*: CUNNINGHAM, D. & MANN, P. (eds) *Tectonics of Strike-Slip Restraining and Releasing Bends in Continental and Oceanic Settings*. London, England, 28–30 September 2005, Geological Society.

ZHANG, P.-Z., BURCHFIEL, B. C., CHEN, S. & DENG, Q. 1989. Extinction of pull-apart basins. *Geology*, **17**, 814–817.

ZHANG, Y., DONG, S. & SHI, W. 2003. Cretaceous deformation history of the middle Tan-Lu fault zone in Shandong Province, eastern China. *Tectonophysics*, **363**, 243–258.

ZILBERMAN, E., AMIT, R., HEIMANN, A. & PORAT, N. 2000. Changes in Holocene paleoseismic activity in the Hula pull-apart basin, Dead Sea rift, northern Israel. *Tectonophysics*, **321**, 237–252.

ZOBACK, M. L., JACHENS, R. C. & OLSON, J. A. 1999. Abrupt along-strike change in tectonic style: San Andreas fault zone, San Francisco Peninsula. *Journal of Geophysical Research*, **104**, 10 719–10 742.

ZOLNAI, G. 1991. *Continental Wrench-Tectonics and Hydrocarbon Habitat*: *Tectonique Continentale en Cisellement*. AAPG Continuing Education Course Note Series, **30**, Unpaginated.

Morphology, structure and evolution of California Continental Borderland restraining bends

M. R. LEGG[1], C. GOLDFINGER[2], M. J. KAMERLING[3], J. D. CHAYTOR[2] & D. E. EINSTEIN[4]

[1]*Legg Geophysical, Huntington Beach, CA 92647, USA (e-mail: mrlegg@attglobal.net)*

[2]*College of Oceanic and Atmospheric Sciences, Oregon State University, Corvallis, OR 97331, USA*

[3]*Venoco, Inc., Carpinteria, CA 93013, USA*

[4]*Southern California Earthquake Center, Los Angeles, CA 90089, USA*

Abstract: Exceptional examples of restraining and releasing bend structures along major strike-slip fault zones are found in the California continental Borderland. Erosion in the deep sea is diminished, thereby preserving the morphology of active oblique fault deformation. Long-lived deposition of turbidites and other marine sediments preserve a high-resolution geological record of fault zone deformation and regional tectonic evolution. Two large restraining bends with varied structural styles are compared to derive a typical morphology of Borderland restraining bends. A 60-km-long, 15° left bend in the dextral San Clemente Fault creates two primary deformation zones. The southeastern uplift involves 'soft' turbidite sediments and is expressed as a broad asymmetrical ridge with right-stepping en echelon anticlines and local pull-apart basins at minor releasing stepovers along the fault. The northwest uplift involves more rigid sedimentary and possibly igneous or metamorphic basement rocks creating a steep-sided, narrow and more symmetrical pop-up. The restraining bend terminates in a releasing stepover basin at the NW end, but curves gently into a transtensional releasing bend to the SE. Seismic stratigraphy indicates that the uplift and transpression along this bend occurred within Quaternary times. The 80-km-long, 30–40° left bend in the San Diego Trough–Catalina fault zone creates a large pop-up structure that emerges to form Santa Catalina Island. This ridge of igneous and metamorphic basement rocks has steep flanks and a classic 'rhomboid' shape. For both major restraining bends, and most others in the Borderland, the uplift is asymmetrical, with the principal displacement zone lying along one flank of the pop-up. Faults within the pop-up structure are very steep dipping and sub-vertical for the principal displacement zone. In most cases, a Miocene basin has been structurally inverted by the transpression. Development of major restraining bends offshore of southern California appears to result from reactivation of major transform faults associated with Mid-Miocene oblique rifting during the evolution of the Pacific–North America plate boundary. Seismicity offshore of southern California demonstrates that deformation along these major strike-slip fault systems continues today.

Restraining bends are present along strike-slip faults where fault curvature or offset en échelon fault segments tend to impede smooth lateral motion of opposing crustal blocks (Crowell 1974). On right-lateral faults, as along the Pacific–North America transform plate boundary, a restraining bend exists where the fault curves or steps to the left when following the fault trace. Crowding of crustal material by lateral movement into the fault bend produces uplift and crustal thickening by folding and thrust or reverse faulting adjacent to the principal displacement zone (PDZ) of the active strike-slip fault. Such zones are called transpressional (Harland 1971) or convergent strike-slip fault zones (Biddle & Christie-Blick 1985; Sylvester 1988). In contrast, releasing bends or transtensional zones exist where the fault bends or steps to the right for dextral systems. Large-scale transtension results in crustal thinning and basin formation by normal faulting and subsidence adjacent to the PDZ. In the simple fault bend model, deformation is expected to concentrate adjacent to the maximum fault curvature (Fig. 1). This paper examines the morphology and shallow-crustal structure of restraining bends along active strike-slip faults in the southern California region, with a focus on the offshore area, i.e. the California Continental Borderland. From comparison of analogue models of restraining bend geometry and progressive evolution to well-defined Borderland examples,

From: CUNNINGHAM, W. D. & MANN, P. (eds) *Tectonics of Strike-Slip Restraining and Releasing Bends.* Geological Society, London, Special Publications, **290**, 143–168.
DOI: 10.1144/SP290.3 0305-8719/07/$15.00

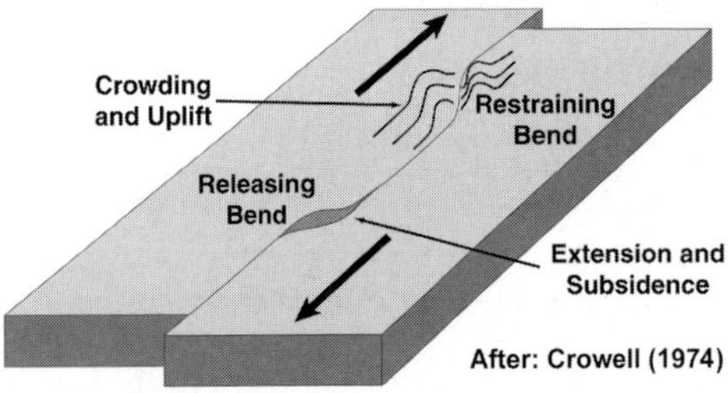

Fig. 1. Material crowded into a restraining bend along a strike-slip fault results in convergence, folding and reverse faulting that creates a local uplift. In contrast, extension and subsidence occurs at a releasing bend.

a better understanding of the structural development and tectonic evolution of restraining bends is derived.

Restraining bend geometry is mechanically unfavourable for strike-slip faulting (Segall & Pollard 1980). In homogeneous media, fault linkages between en échelon and discontinuous fault segments are more likely to form within a releasing geometry, where local extension favours crack growth and propagation. Nevertheless, irregular fault geometry produces abundant restraining bends along strike-slip faults. Bends range in scale from localized jogs in earthquake surface ruptures to crustal-scale uplifts with surface deformation exceeding lengths of 100 km along the fault (Crowell 1974; Sylvester & Smith 1976; Mann *et al.* 1985; Anderson 1990; Butler *et al.* 1998). Consequently, special crustal conditions must be involved to form restraining bends. Possible conditions include pre-existing structural fabric and other crustal heterogeneity, and changing strain fields and related stress fields that alter deformation styles on existing fault systems or create new faults to accommodate the evolving strain field (cf. Dewey *et al.* 1998). Careful studies of well-defined fault bends are needed to deduce the processes involved in restraining-bend formation and evolution.

Finite deformation within long-lived restraining bends results in pronounced topographic expression, called push-up or 'pop-up' structures (cf. Stone 1995; Dewey *et al.* 1998; McClay & Bonora 2001). In contrast, releasing bends create basins, which become filled with sediments that tend to smooth and obscure their morphology. Deformation along oblique strike-slip fault segments tends to occur over broad zones (Wilcox *et al.* 1973; Schreurs & Colletta 1998; Withjack & Jamison 1986; McClay & Bonora 2001), commonly many kilometres wide. The pop-up morphology, even though modified by erosion or other destructive processes, can provide a direct measure of the accumulated deformation along the restraining bend (cf. Wakabayashi 2007). Basin-filling sedimentary sequences record the history of deformation along both restraining and releasing fault bends. Quantification of the bend evolution and inference of the larger-scale processes along the more regional strike-slip fault system are possible using geophysical techniques, like seismic reflection profiling, to measure and map the deformation.

Large restraining bends in active strike-slip faults impede crustal block motion, locally enhancing the accumulation of tectonic stresses that may produce major earthquakes, e.g. 1857 Fort Tejon, California (Sieh 1978); 1989 Loma Prieta, California (Plafker & Galloway 1989; Schwartz *et al.* 1994); and 1999 Izmit and Duzce, Turkey (Aydin & Kalafat 2002; Harris *et al.* 2002). Detailed investigations of mainshock and aftershock sequences for restraining-bend earthquakes provide important data regarding the deeper crustal structure (Seeber & Armbruster 1995). However, if restraining bends are locked between large earthquakes, seismicity in the bend area may be low, and other geophysical methods must be used to evaluate deep bend structure (cf. Langenheim *et al.* 2005).

Many bend structures are buried under thick sedimentary blankets, as in the Los Angeles basin (Fig. 2), and changing tectonic conditions associated with an evolving plate boundary tend to obscure the processes directly related to the restraining bend evolution. The uplift, folding and faulting associated with large restraining bends often form excellent traps for hydrocarbons. Numerous productive oil fields, especially in southern California (Harding 1973, 1974; Wright 1991), exist along active strike-slip faults and are associated with restraining bends. Subsurface samples from wells and boreholes provide stratigraphic control that supplements geophysical data for interpretation of the structural geometry and deformation history of restraining bends.

California Continental Borderland

The California Continental Borderland (Fig. 2) is a mostly submerged part of the Pacific–North America dextral transform plate boundary that exhibits a basin-and-ridge physiography (Shepard & Emery 1941; Moore 1969). Right-slip on irregular fault traces has produced numerous restraining bend pop-ups that exhibit distinctive seafloor morphology. The submarine basins of the Borderland range in depth from a few hundred metres to more than 2000 m and are variably filled with clastic sediments from the adjacent mainland and offshore islands. Erosion is greatly diminished in these deep basins compared with subaerial regions, so that pop-up morphology is well preserved on the

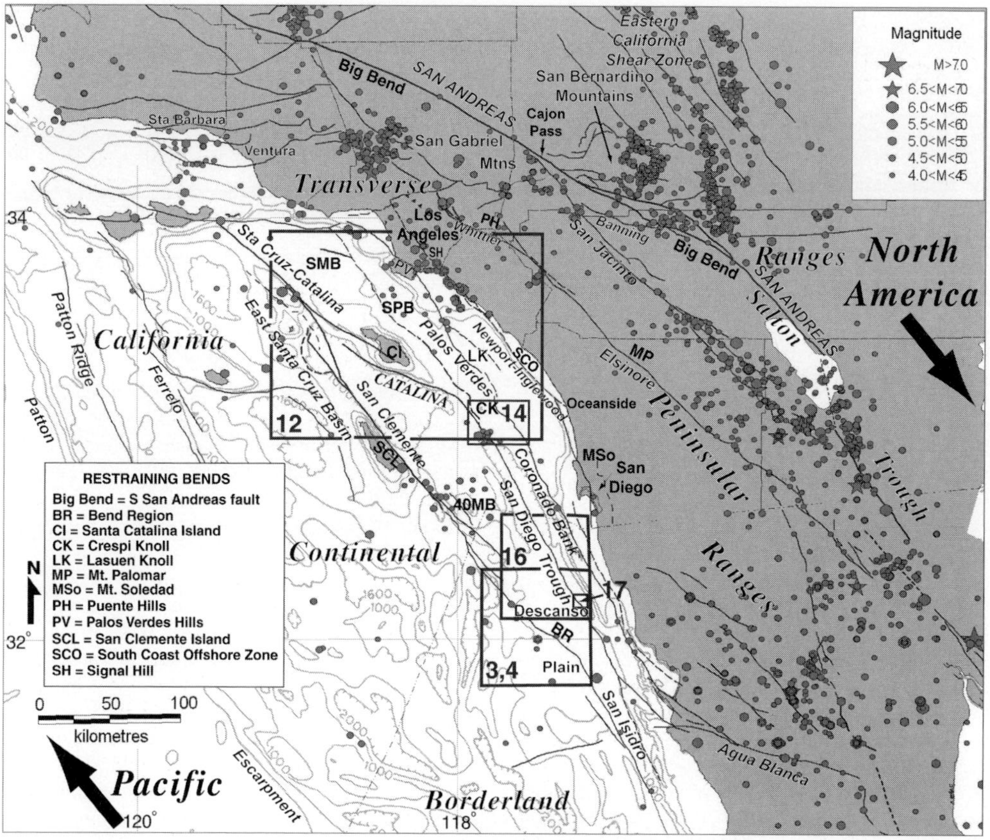

Fig. 2. Location map of seismicity in southern California region and major physiographic provinces. Major fault zones are labelled at some of the more significant restraining bends in the region. Large arrows indicate relative plate-motion vectors for transform faulting between the Pacific and North American plates. Contours in the offshore area are in metres. Rectangles labelled with figure numbers outline locations of detailed maps. SMB, Santa Monica Basin, SPB, San Pedro Basin; 40MB, Fortymile Bank.

seafloor. Deformation history is recorded in well-bedded turbidite and other sedimentary sequences that fill these submarine basins. Using multi-beam bathymetry and seismic reflection profiling, several prominent restraining bend pop-ups within the Borderland are examined, with a focus on two large structural culminations: the San Clemente Fault bend region and the Santa Catalina Island uplift. From our observations, the typical morphology of Borderland restraining bends is described, followed by presentation of a model for the initiation and geological evolution of Borderland restraining bends.

Specific examples of restraining bend pop-ups located within the California Continental Borderland are described and compared with the general model. Where data allow, the age of the bend uplift is estimated and the history of deformation is inferred. These interpretations lead to several conclusions regarding the formation and development of restraining bends along the California continental margin and the Pacific–North America transform plate boundary.

Data

High-resolution contour and shaded relief maps created from processed multi-beam swath bathymetry (Fig. 3) and, where necessary, older single-beam echo-sounder records provide excellent images of the seafloor morphology of strike-slip fault bend structures (NOS 1999; Goldfinger *et al.* 2000; K. Macdonald 2003, pers. comm.). Lateral resolution of mapped features is about 50 to 100 m for areas of multibeam coverage in this study where water depths exceed 1000 m. Areas mapped using single-beam soundings have somewhat lower spatial resolution, mostly within about

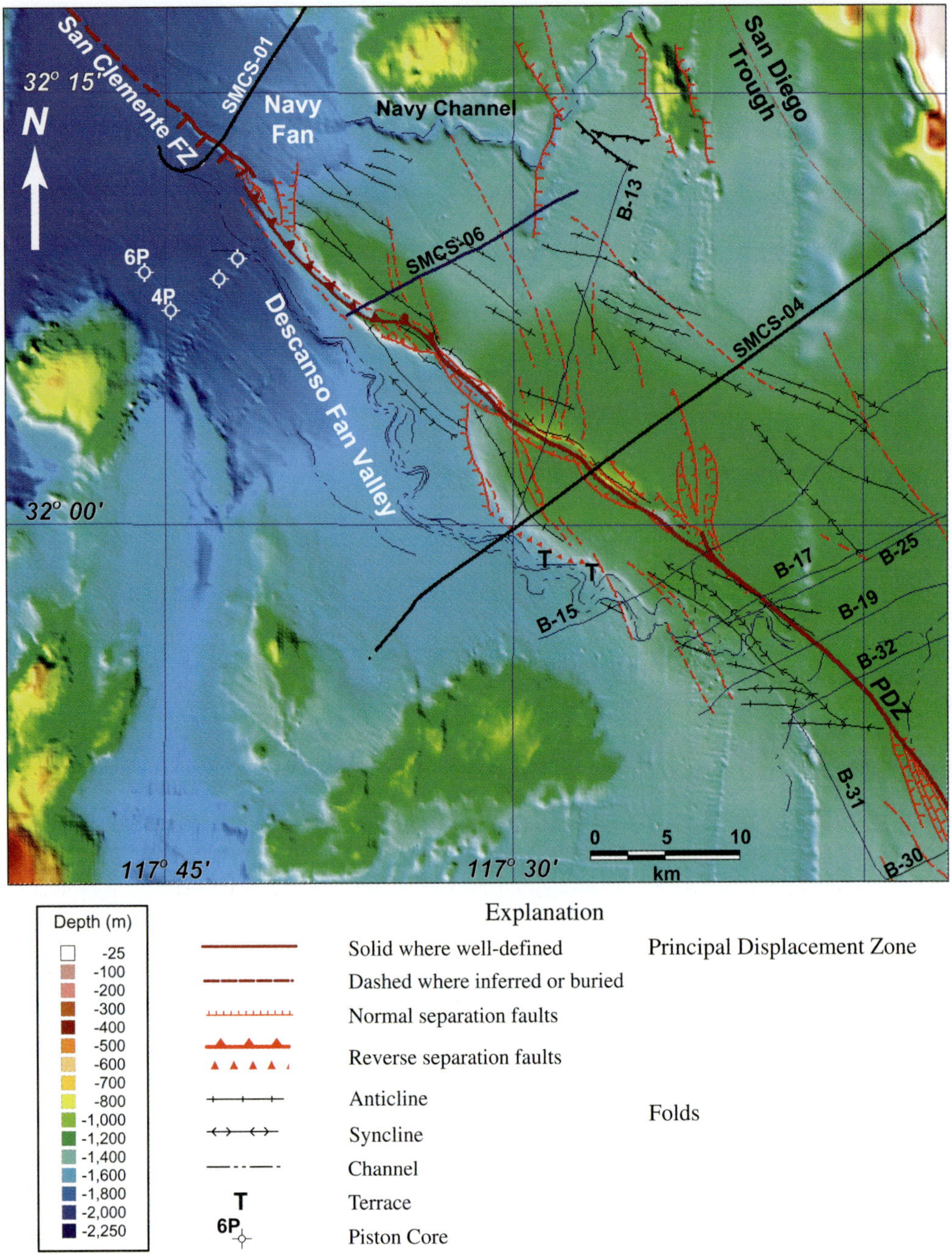

Fig. 3. Map of geological structures along the bend region of the San Clemente Fault (see Fig. 2 for map location). Bathymetry is a compilation of multi-beam and older echo-soundings. Two distinct uplifts, one cored by bedrock, the other in turbidite sediments, comprise the pop-up structure. The principal displacement zone (PDZ) cuts a relatively straight path through the restraining bend, with a gentle curve at the SE end and a right-stepover at the NW end. Branch and secondary faults within the pop-up are generally reverse separation, probably oblique-slip, and trend north to NW, subparallel to the PDZ. Folds are also subparallel to the PDZ within the bend, but trend east–west beyond the transpressional section at the SE end.

500 m. Vertical resolution decreases with increasing depth, but is usually within 2% of the water depth for absolute depth, about 20 to 40 m for the Borderland basins, and within about 5 to 10 m for relative depths within a basin. Oblique three-dimensional shaded relief views (such as Fig. 4) emphasize the seafloor uplift and tectonic geomorphology associated with strike-slip faults and restraining and releasing bends.

High-resolution analogue seismic-reflection profiles acquired using airgun, sparker and 3.5 kHz transducer sources (Vedder *et al.* 1974; Legg 1985) are used to interpret the sub-seafloor character of the faults and prominent stratigraphic sequences associated with the major submarine fans of the Borderland (Figs 5–6). Although the analogue profiles are unmigrated, the dip of shallow reflecting horizons is shallow ($<5°$), and steep faults are recognized by reflector terminations or diffractions (cf. Tucker & Yorston 1973). Structural contours and isopachs of shallow sedimentary sequences were mapped based on the seismic interpretations. Sequence ages were estimated based on sedimentation rates derived from shallow piston cores in the Borderland basins (Emery 1960; Heath *et al.* 1976; Dunbar 1981; Legg 1985), and used to infer the deformation history along the fault bends. Additional stratigraphic control was provided from outcrop samples, dart cores, and borehole data where older sedimentary, volcanic and metamorphic bedrock is exposed in the pop-up (Vedder *et al.* 1974; Vedder 1990).

Digitally recorded, high-resolution, multichannel seismic profiles (MCS) across major segments of the fault bends are used to image the deeper fault structure and sediment deformation, and to measure fault dips and offsets (Figs 7–8; Bohannon *et al.* 1990, K. Macdonald, 2003, pers. comm.). Navigation for these recent seismic surveys was provided by GPS, and geographical positional accuracy within about 10 m is estimated. Conventional two-dimensional seismic data-processing schemes were used for the MCS profiles to prepare common mid-point stack and frequency-wavenumber migrated images. Seismic

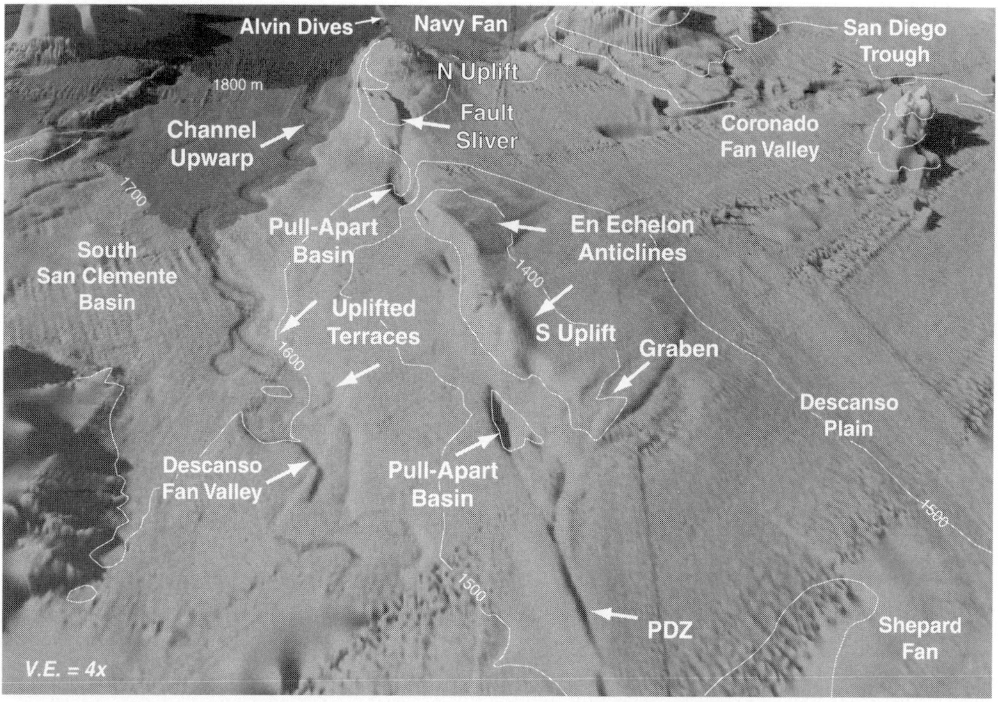

Fig. 4. Perspective shaded-relief view looking NW along the bend region of the San Clemente Fault. The principal displacement zone cuts across the image from bottom right centre to upper left centre. Prominent tectonic geomorphic features are identified that indicate right-slip fault character and seafloor uplift due to oblique convergence along the restraining bend. The bedrock uplift in the NW is steep-sided and more symmetrical than the broad asymmetrical SE uplift in the turbidites. Bathymetric contour values are in metres; the distance along fault is about 60 km from bottom to top.

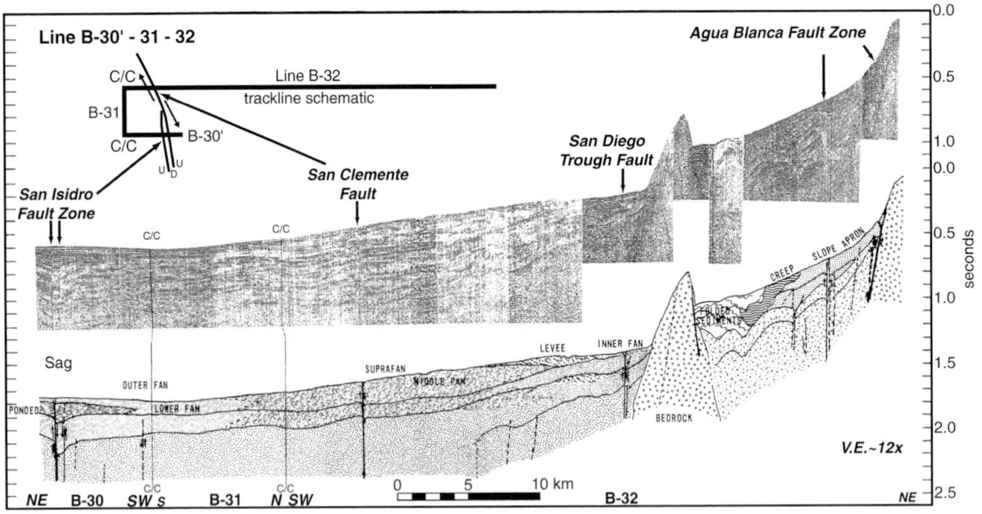

Fig. 5. Sparker seismic profile (line B-30[1]–31-32; see Fig. 3 for profile location) across the Descanso Plain, showing the character of well-defined strike-slip faults away from restraining bends. Two course changes between lines are indicated (C/C). San Clemente Fault is straight and narrow with a strike parallel to the relative plate-motion vector (320°). San Isidro Fault zone is a releasing fault bend, striking about 330°, creating a transtensional sag or stepover basin evident between en échelon fault traces. The San Isidro fault zone represents the southern continuation of the San Clemente fault zone west of Baja California. San Diego Trough Fault is also a straight and narrow right-slip fault across this profile; minor transpression is evident at depth for the Agua Blanca fault zone.

velocities used for migration are based on stacking velocities, seismic refraction and wide-angle reflection profiling in the area (Moore 1969; Shor *et al.* 1976).

Borderland restraining bend morphology

A generalized model of the restraining bend pop-up morphology (Fig. 9) is derived from interpretations of data along the San Clemente Fault bend region in the Descanso Plain offshore of NW Baja California (Figs 2–4). This model represents a double bend, where the fault bends first to the left and then bends back to the original strike. A transpressional zone forms within the restraining double bend and is manifest as a prominent seafloor uplift that may be a complex structure consisting of right-stepping en echelon anticlines with intervening subsidiary extensional strike-slip basins and diverging oblique-slip fault zones (Figs 3–4). The uplift is broadly asymmetrical, with the principal displacement zone (PDZ) of the active strike-slip fault located to one side of the axis of uplift. Typically, the PDZ is vertical for well-defined strike-slip faults in the bend. Reverse faults that probably accommodate oblique-slip exist along the flanks of the uplift and trend subparallel to the PDZ. Beyond the ends of the restraining double bend, oblique extension forms strike-slip basins that may occur as pull-apart (stepover) or sag (releasing bend) basins.

North-northwest trending dip-slip (oblique-slip?) faults that diverge from the PDZ could be interpreted as antithetic Riedel shears (Fig. 9), but their observed trend is about midway between that predicted for synthetic and antithetic faults (e.g. Wilcox *et al.* 1973). Instead, they trend subparallel to the predicted trend for extension fractures. Although superficially these appear to form grabens bounded by normal faults (Fig. 4), based on seismic profiles they are upthrusts with reverse separation (Fig. 7, SMCS04, shot 4250). According to Billings (1972, p. 198), an upthrust is a high-angle fault 'along which the relatively uplifted block has been the active element.' The upthrusts elevate the pop-up and create grabens in places where local uplift lags that of adjacent blocks. Some reverse faults in Borderland restraining bends trend east–west as predicted from the wrench fault model, but most trend subparallel to the PDZ. This indicates strain partitioning, with reverse faults accommodating shortening and vertical faults accommodating strike-slip.

North- to NW-trending normal-separation (oblique-normal?) faults also exist in the sediments above an acoustic basement block to the west of the PDZ (Figs 3 & 7); these result from stretching of the

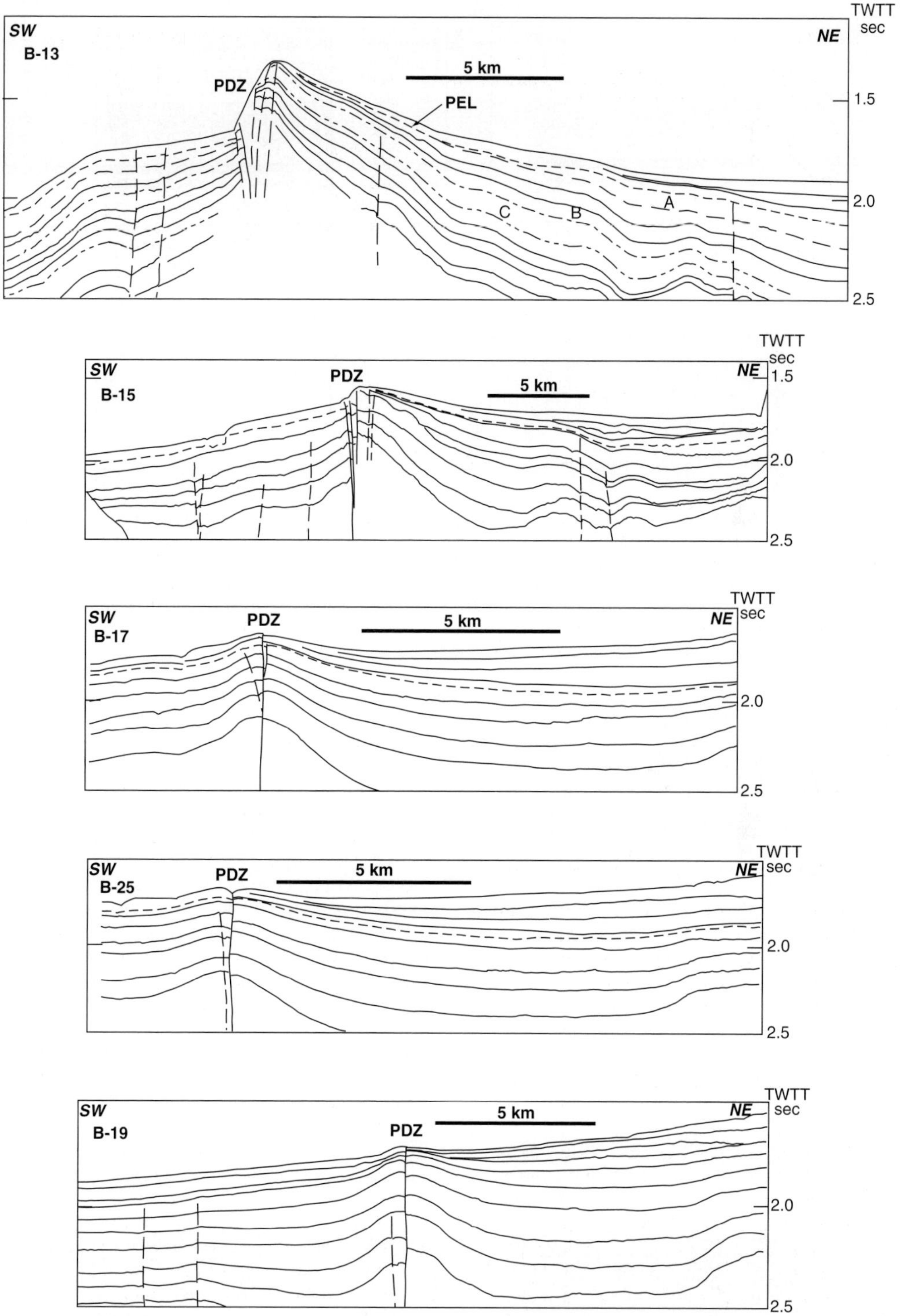

Fig. 6. Line drawings of sparker seismic profiles (see Fig. 3 for profile locations) across the restraining bend region of the San Clemente Fault show a broad zone of faulting and uplift due to transpression, compared with simple parallel right-slip (cf. Fig. 5). Significant stratigraphic horizons are shown to indicate the increasing deformation with depth and time. Uplift rates for the pop-up structure are derived from horizons PEL, A, B and C. The sequence stratigraphic unit PEL is a Late Quaternary hemipelagic layer based on acoustic transparency (Figs 7 & 11).

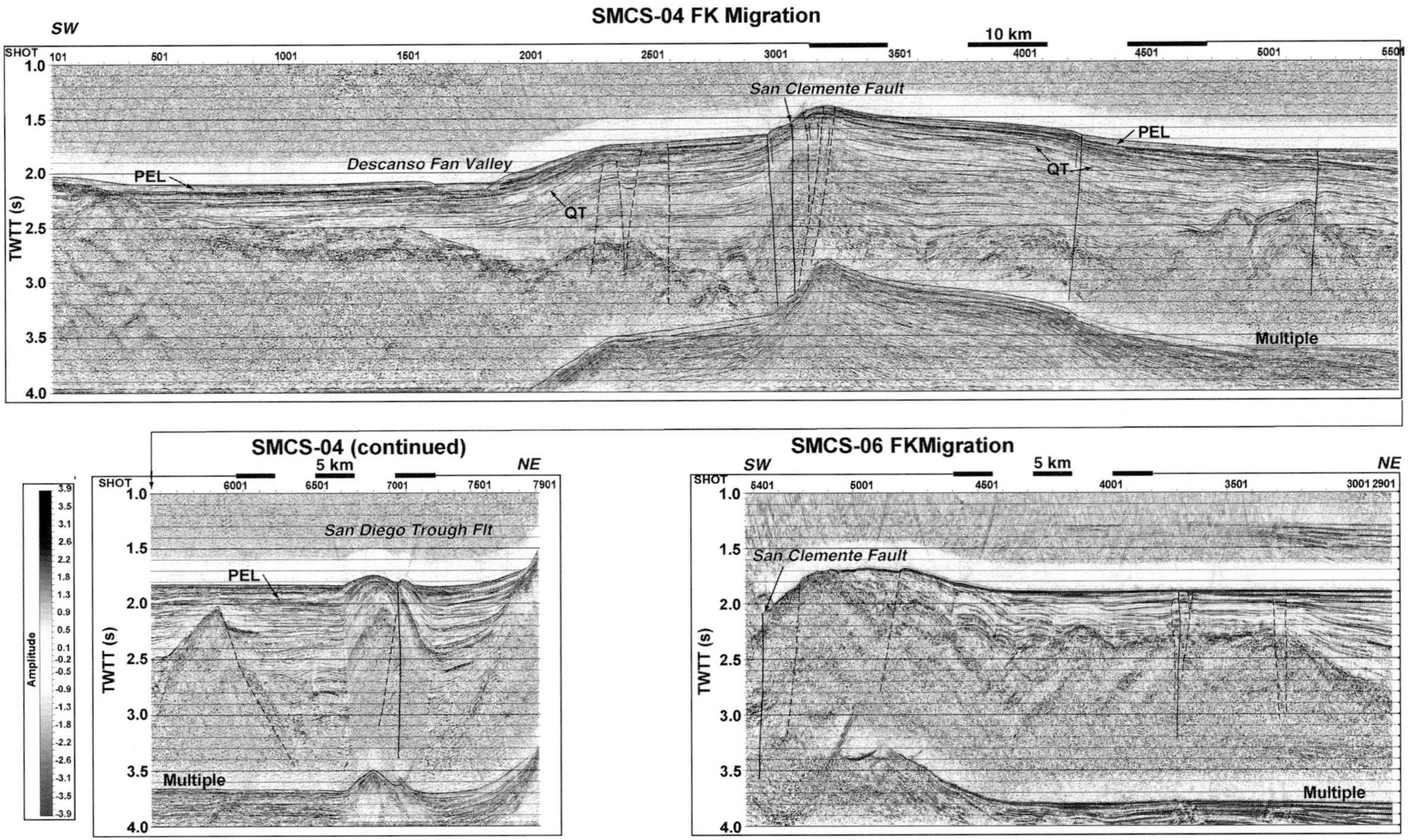

Fig. 7. High-resolution MCS profiles across the bend region of the San Clemente Fault (see Fig. 3 for profile locations) showing the broad zone of deformation in the pop-up structure. Data are post-stack migrated, 12-fold with 1.56 m CMP trace spacing. The principal displacement zones for the San Clemente and San Diego Trough faults are bold solid lines; other faults are solid where well-defined, and dashed where buried or less well-defined. Late Quaternary sequence PEL is hemipelagic based on its acoustic transparency. Some thin turbidite sand layers are found in this unit, based on piston-core samples from the nearby San Clemente basin and Navy fan (Dunbar 1981). Horizon QT represents the top of the pre-uplift turbidite sediments, and records the beginning of transpression.

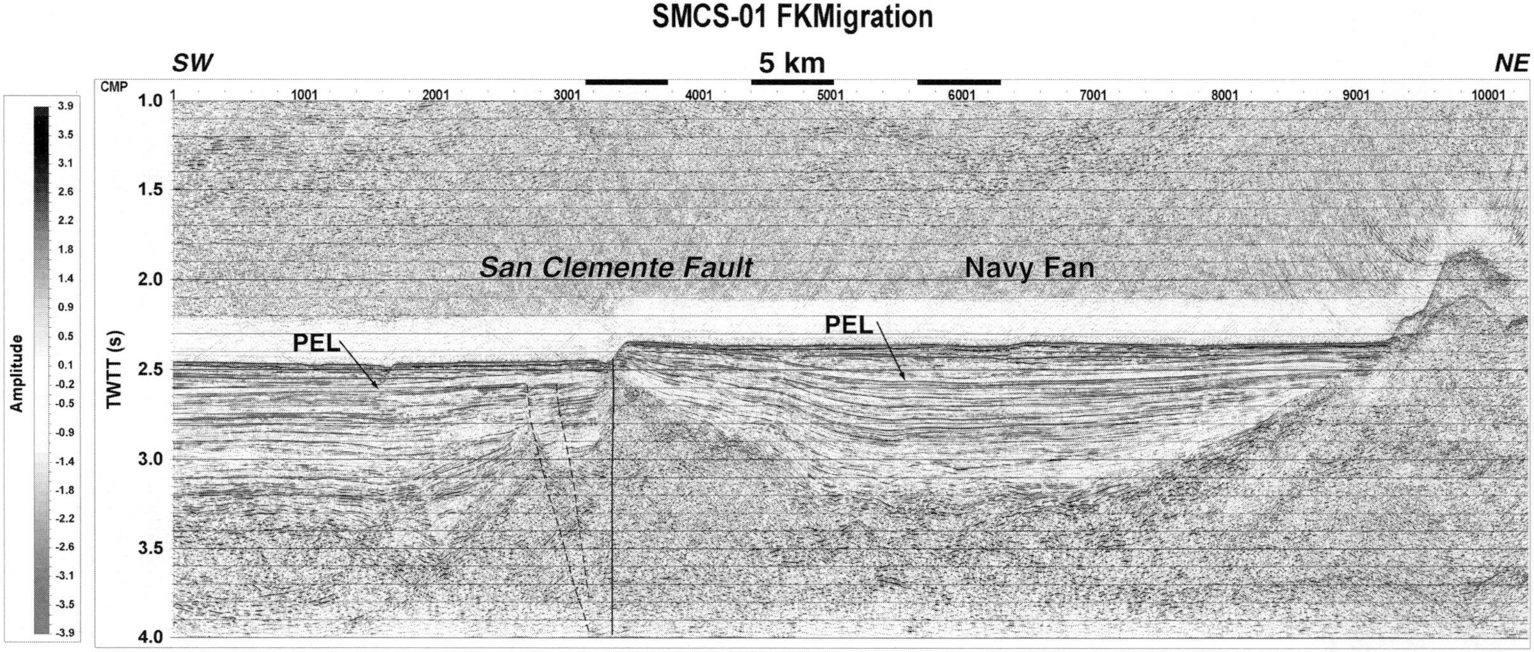

Fig. 8. High-resolution MCS profile across the north end of the San Clemente Fault bend region (see Fig. 3 for profile location). Data processing is the same as for Figure 7. DSV Alvin submersible dives observed vertical seafloor scarps, 1 to 3 m high in mud on the 100-m-high San Clemente Fault escarpment near this profile (Fig. 10). Hemipelagic unit PEL is buried beneath young turbidites of the Navy Fan, but may crop out in the escarpment.

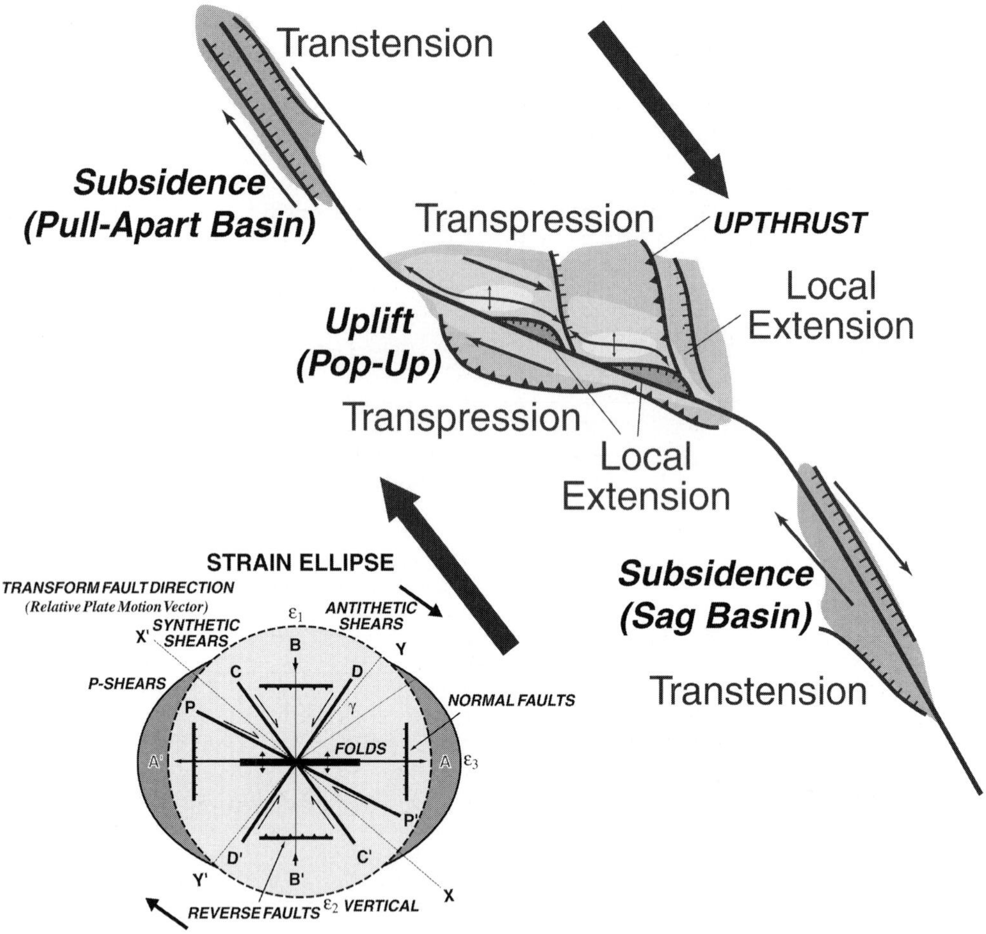

Fig. 9. Typical morphology and structure of restraining double bends in southern California, based on the San Clemente Fault bend region. Some features may be missing on other restraining bends, but some features may be more pronounced than in the San Clemente Fault example. Most Borderland restraining double bends have transtensional zones at the ends. A strain ellipse is shown to highlight the expected structural character for different trends in a zone of NW-directed dextral shear (after Wilcox *et al.* 1973). Contrary to the expected strain patterns, the north-trending faults in the San Clemente Fault bend region are steeply dipping upthrusts, not normal faults.

sedimentary cover as the basement pushes upward due to overall transpression. In other areas, such as the Palos Verdes Hills (Fig. 2), shallow 'keystone' grabens form by extension along the crest of the transpressional uplift (Woodring *et al.* 1946; Francis *et al.* 1999).

In contrast, northwest-trending faults that bound the extensional basins beyond the ends of the restraining double bend are subparallel to the PDZ and commonly exist along monoclinal sags that dip inward to the PDZ (Fig. 5, San Isidro fault zone). This also indicates strain partitioning between normal (oblique?) faults on the flanks of the PDZ that accommodate extension, and a vertical PDZ that accommodates strike-slip. Most faults observed in seismic profiles across Borderland restraining bends have steep dips, measured at 70 to 80° for secondary oblique-slip(?) faults and vertical for the PDZ.

San Clemente Fault bend region

The San Clemente fault zone, with an overall length exceeding 600 km, is an active member of the larger San Andreas fault system that defines the dextral Pacific–North America (PAC–NOAM) transform plate boundary (Fig. 2; Legg 1985; Legg *et al.* 1989). The average strike of the San Clemente

Fault is about 320° (Figs 2–3), parallel to Recent PAC–NOAM relative motion (Minster & Jordan 1978; Demets *et al.* 1990). Segments of the San Clemente Fault that strike 320° are simple, narrow and well-defined in character (Fig. 5, San Isidro Fault), consistent with 'parallel strike-slip' (Legg 1985). Fault segments that bend to the right or are more north-trending are extensional (transtensional), and those that bend to the left or more west-trending are contractional (tranpressional). The San Clemente fault zone includes a 60-km-long restraining bend that exhibits prominent seafloor uplift in the 1300-m-deep Descanso Plain offshore of northwest Baja California, Mexico (Figs 3–4; Legg 1985; Legg & Kennedy 1991; Goldfinger *et al.* 2000). The average strike in the bend region is 308°, about 12° oblique to the left, resulting in a zone of convergent right-slip. The fault (PDZ) appears to curve gently to the left into the restraining bend at the southeast end, but terminates and steps to the right at the NW end of the bend.

The San Andreas Fault accommodates only about 50% of the total right-lateral plate boundary slip expected in the Salton Trough region (about 25 m/ka; Keller *et al.* 1982; Weldon & Sieh 1985; Harden & Matti 1989). The remaining slip must be accommodated on other faults within the 200-km-wide plate boundary in the California/Mexico border region. Peninsular Ranges faults, such as the San Jacinto and Elsinore, accommodate about 12 and 5 m/ka of right-slip (WGCEP 1995), leaving as much as 10 to 20% of the predicted relative plate motion available for faults west of the coastline. At present, the rate of right-slip on the San Clemente Fault is uncertain. Offset Quaternary seafloor morphology including submarine fan and channel features (Legg 1985; Legg *et al.* 1989; this paper), and GPS geodesy (Larson 1993; Bennett *et al.* 1997) suggest that between 1 and 10 m/ka of right-slip occurs on offshore faults.

The total Neogene and younger displacement along the San Clemente fault zone is also unknown. Movement on the San Clemente fault zone is inferred to accommodate major right-lateral components of Neogene transtension (Atwater 1970; Legg 1991; Lonsdale 1991; Crouch & Suppe 1993; Bohannon & Geist 1998). Values exceeding the 40 km originally suggested by Shepard & Emery (1941), from juxtaposing San Clemente Island alongside Fortymile Bank, have been proposed (Goldfinger *et al.* 2000). The rim of an inferred Middle Miocene crater (caldera?) appears to be offset at least 60 km in a dextral sense (Legg *et al.* 2004*b*). Deformation associated with the restraining bend may provide quantitative limits on Late Cenozoic right-slip across the San Clemente fault zone, as well as providing better understanding of restraining bend structure and evolution.

Bend morphology and fault geometry

Transpression in the bend region creates a broad asymmetrical anticlinorium, with the greatest uplift located on the northeast flank of the principal displacement zone (PDZ, Figs 3–4). Even though uplift is asymmetrical in cross-section, basin sediments are raised on both sides of the San Clemente Fault PDZ. This contrasts with fold-and-thrust belts, where crust in the footwall is downwarped and thrust beneath the hanging-wall block. Rather than being concentrated at the point of greatest fault-trace curvature (e.g. Fig. 1; Crowell 1974), the area of greatest uplift is located near the centre of the transpressional fault segment.

The broad seafloor uplift comprises two major segments separated by a saddle (Figs 3–4). Local peaks and narrow troughs exist along the fault in each pop-up. Of the two major uplifts, the northern is more symmetrical, about 6 km wide, with steep slopes, of 10 to 30°, on both the SW and northeast flanks, although steeper slopes exist where the PDZ curves around the SW flank. A narrow (1.5 km), steep-sided ridge on the SW edge of the northern uplift probably represents a fault sliver squeezed upward and offset along the PDZ. A series of youthful anticlines, with axes subparallel to the uplift and fault trend, folds the sediments and seafloor at a 1 km wavelength between the NW uplift and the Navy Channel. The southern uplift is broad and asymmetrical about 17 km wide, with the SW flank bordered by steep slopes inferred to be fault scarps, and the NE flank appearing as a more gentle, $<5°$, bedding-parallel dip-slope (Figs 6–7). Two elongate peaks centred on the southern uplift are separated by a saddle and are offset en échelon to the right. The southeastern end of the uplift consists of a low, elongate ridge, locally cut by an axial trough aligned along the PDZ (Fig. 6, line B-25).

The major active traces (PDZ) of the San Clemente Fault appear as NW-trending seafloor scarps and narrow linear valleys or troughs, generally located along the southwest flank of the major uplifts. These scarps change vergence, from NE to SW along strike, including uphill-facing scarps in some locations (Figs 3–4), typical of strike-slip faults. Two closed depressions, about 4 km long by 1 km wide, and a third smaller depression, lie at stepovers along the major fault traces. Bordered by right-stepping en echelon faults, these are interpreted as stepover (pull-apart) basins formed along the right-slip fault zone. At the NW end of the bend region, the main fault curves and steps to the right around the NW uplift, and then cuts straight

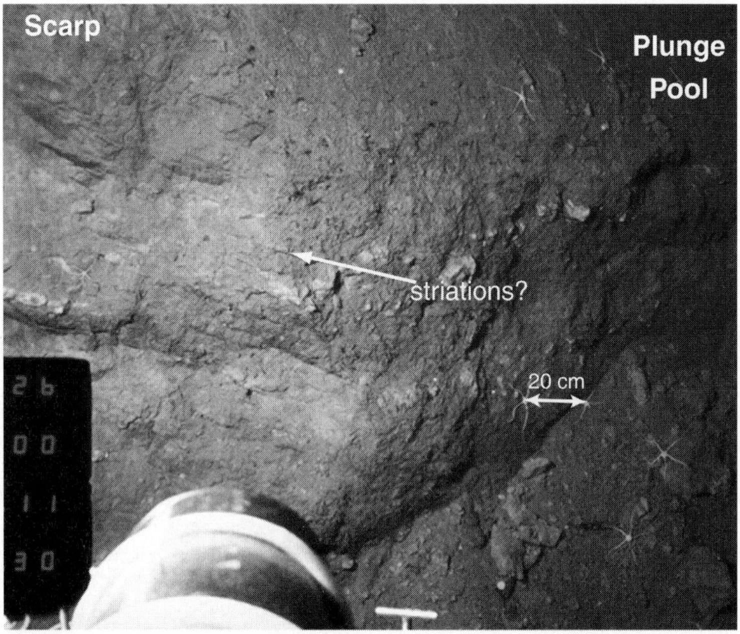

Fig. 10. Photograph from the DSV Alvin of fault scarp along the San Clemente Fault crossing the Navy fan (location in Figs 4 & 16). Subhorizontal lineations may be slickensides consistent with strike-slip. The plunge pool is cut into a scarp by deep-water (*c.* 1800 m) turbidity currents overflowing the scarp from the Navy fan. The scarp is composed of mud and layers of shells associated with ancient benthic communities at former cold seeps. Photo by C. Goldfinger (2000).

across the Navy Fan, forming a prominent scarp, 30 to 100 m high and 7 km long. This scarp is split by a semicircular embayment, about 0.5 km wide, that is inferred to be a plunge pool associated with turbidity currents on the Navy Fan (Fig. 10). Another 4 km to the northwest, this scarp dies out and the gentle slope of the Navy Fan merges into San Clemente Basin. Despite the local trace irregularities, the PDZ cuts a relatively straight 308° path across the bend region, typical of a vertical strike-slip fault. Seismic profiles confirm the vertical dip of the PDZ (Figs 6–8).

Several branch and secondary fault scarps appear to curve away from and trend parallel to the mostly linear fault scarps of the PDZ. A large, 8-km-long, north-trending east-facing set of scarps curves away from the main fault zone for 5 to 10 km at the SE end of the major uplift (Fig. 4, graben). Similar north-trending branch faults exist near the ends of the peak uplift areas (Fig. 3). The broader complex fault pattern is interpreted to represent the surface expression of a flower structure (palm-tree structure), commonly observed in cross-section along strike-slip faults (Wilcox *et al.* 1973; Harding 1985; Sylvester 1988; McClay & Bonora 2000). The systematic location and pattern of the major branch faults at the ends of peak uplift areas suggests that pre-existing structure controls the uplift geometry.

A prominent submarine channel, Descanso Fan Valley, bounds the 5-km-wide shelf along the SW flank of the southern uplift (Figs 3–4 & 7). Descanso Fan Valley, about 0.5 to 1.2 km wide, has broad meanders and uplifted relict terraces or abandoned levées along its northeast flank. Growth of the restraining bend uplift has forced the channel farther to the SW, leaving the older channel terraces and levées behind Fig. 4). The thalweg of the main channel is upwarped about 20 to 30 m where it passes the flank of the NW uplift. Hemipelagic sediments drape across the channel, conforming to the ancient seafloor channel morphology (PEL, Fig. 7, SMCS-04). Thus, tectonic upwarping continued after active channel erosion or aggradation occurred, probably during a Late Pleistocene lowstand or interglacial transgression when turbidite deposition was more frequent in the area (Dunbar 1981).

The broad, gently dipping uplift and shelf between Descanso Fan Valley and the PDZ are cut by NNW-trending fault scarps. Steep fault dips, 70 to 90°, combined with normal separation observed in the MCS profiles (Fig. 7, SMCS-04) suggest that these are oblique-normal faults. The faults prominent near a 'knee' in the uplift SW of the PDZ are controlled by deeper acoustic basement structure that appears to be bent and forced upward due to transpression. Thus, shallow sub-seafloor oblique-extension exists directly above a basement transpressional hinge. The north trend of these scarps is consistent with west-directed extension in a NW-trending zone of dextral shear, but more likely represents the stretching of the strata above the basement pop-up. A similar NNW-trending zone of faulting on the north side of the southern uplift forms a seafloor graben, although multichannel seismic data show that the most prominent of these faults is a high-angle oblique(?) reverse fault (Fig. 7, SMCS-04, shot 4250). A deeper acoustic basement antiform also lies beneath this graben, and may correlate with the acoustic basement hinge on the southwest flank, but offset to the northwest by about 16 km of right-slip. Oblique rifting of the Inner Borderland during the Neogene evolution of the PAC–NOAM transform fault boundary could have created this north-trending structural fabric that controls the younger faulting.

Deformation history

Turbidite sedimentary sequences from nearby submarine fans (Figs 3–7) record the Late Cenozoic history of transpressional uplift along the San Clemente Fault in the bend region. Shallow sequences on the NE flank form a growth wedge with progressive northward tilting and thickening away from the PDZ (Units PEL and QT, Figs 6–7 & 11, Profile B-13). These divergent sequences lap on to or pinch out against deeper sequences and result from continued Late Quaternary uplift. Deeper sequences below Horizons 'A', 'B' and QT (Figs 6–7 & 11, Profile B-13) have relatively uniform thickness and tilt, implying pre-uplift deposition on a relatively flat-lying basin floor. Late Quaternary sedimentation rates derived from piston cores in the area (4P and 6P, Fig. 11; Dunbar 1981; Legg 1985; Lyle *et al.* 1997; Janik 2001) are used to estimate the maximum age of the deepest growth wedge, about 1.0–1.5 Ma. Sequence PEL is estimated to be about 460 $\pm$ 240 ka (Smith & Normark 1976; Legg 1985). Using the maximum observed structural relief, 700 m observed on line B-13 (Figs 4 & 11), the minimum uplift rate is estimated as 0.47 to 0.70 m/ka. The lateral slip rate is likely to be much greater for this predominantly strike-slip fault. Considering that the sedimentation rate for the turbidites below sequence PEL probably exceeds the hemipelagic rate, the initiation of uplift may be significantly younger and the uplift rate faster than these preliminary estimates.

Sedimentary sequences on the SW flank are relatively parallel with uniform thickness, with little or no onlap of deeper sequences (Figs 6–7). A shallow basin, 85 m deep, lies perched beneath the hemipelagic sequence PEL, and is inferred to have formed in the vicinity of a modern pull-apart basin between the peak uplift along the PDZ and the zone of oblique faulting to the SW above the basement hinge (Figs 3 & 7). Tilting of these basin sediments and asymmetry of the pop-up may show that this SW shelf area entered the high-relief section of the bend region at about the beginning of PEL deposition, i.e. *c.* 0.5 Ma. This could represent the time when the basement at the hinge moved into the local restraining bend at the SE side of the largest and highest peak in the bend region, about 7 km to the SE of its current position. Alternatively, tilting began when the hinge entered the local bend 14 km farther SE near the second seafloor peak. In this scenario, the initial uplift began at the farther SE peak, with the basin forming in the releasing bend saddle between the two peaks, followed by tilting and uplift when entering the second local restraining bend. At present, it is uncertain whether this buried basin is an older pull-apart, like the ones evident in the seafloor morphology.

Profiles of uplift measured relative to the flat-lying sequences away from the fault are plotted (Fig. 11, Profile B-13) to show uplift history. Relative uplift increases toward the fault, and with depth and age from the seafloor to horizon C on the northeast flank. Uplift below the shelf on the southwest flank is more uniform across the shelf, with some increase adjacent to the fault and variable increases with depth in the section. Dextral tectonic transport (advection) of the deeper and older turbidites into the bend region from the SE accounts for their uniform thickness and tilt, whereas later sequences like PEL are predominantly hemipelagic and drape over the folded and uplifted seafloor. The structural relief and inferred uplift rate on the SW side of the fault are about one-half that to the NE (Fig. 11), resulting in the asymmetry of the anticlinorium.

Santa Catalina Island

The Catalina Fault forms an 80-km-long restraining double bend (cf. Crowell 1974) between the Santa Cruz–Catalina Ridge and San Diego Trough fault zones (Figs 2 & 12). Uplift due to oblique convergence along this transpressional fault has produced Santa Catalina Island and the wide submerged shelf and slope surrounding the island. Like the San Clemente Fault bend region, the uplift is greatest

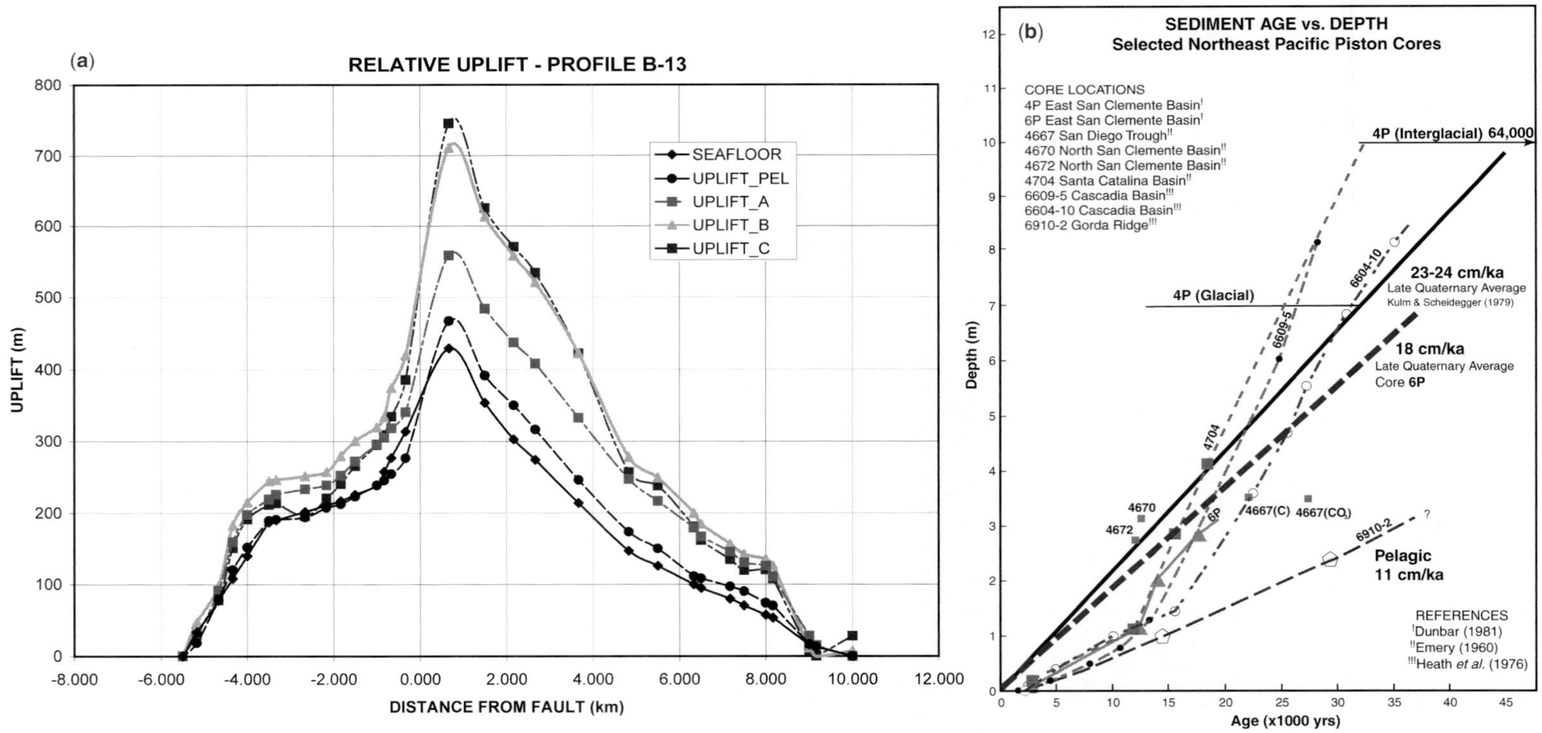

Fig. 11. (**a**) Chart showing relative uplift for sedimentary sequence horizons on profile B-13 (see Fig. 3 for profile location). Relative uplift for horizons B and C is roughly the same, implying that uplift in the bend region initiated sometime after deposition of horizon B. (**b**) Chart showing sediment age v. depth in selected piston cores from southern California and Pacific Northwest deep-water continental margins. Average sedimentation rates for turbidites in the San Clemente Basin provide estimates of age for sequence PEL. Piston cores 4P and 6P were obtained near the Navy Fan, where seismic-reflection profiles across core locations provide sediment thickness estimates. The pelagic sedimentation rate can be used for age estimates of hemipelagic drape in elevated areas where the seafloor is relatively isolated from turbidite deposition. This slow rate (0.1 m/ka) also provides a minimum rate useful for estimating the maximum age of the sedimentary sequences after compaction corrections are applied.

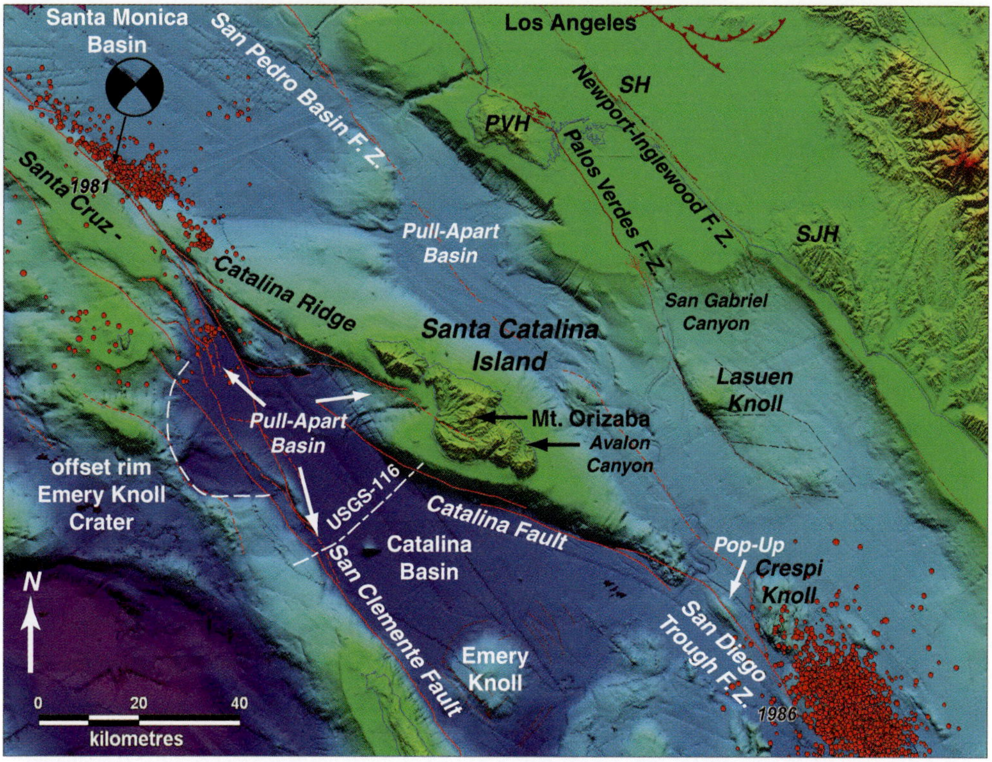

Fig. 12. Shaded relief map of Santa Catalina Island and vicinity, where several restraining-bend pop-ups and releasing-bend basins exist along major fault zones. Epicentres for two moderate earthquakes (1981 Santa Barbara Island, *M* 6.0; 1986 Oceanside, *M* 5.8) and aftershocks bound the Santa Catalina Island restraining bend (locations by Astiz & Shearer 2000; focal mechanism from Corbett 1984). Other restraining-bend pop-ups include the Palos Verdes Hills (PVH) and Lasuen knoll along the Palos Verdes fault zone, and Signal Hill (SH) and possibly the San Joaquin Hills (SJH) along the Newport–Inglewood fault zone. Small pop-ups and pull-apart basins in the vicinity of Crespi knoll are shown in Figure 14. Total relief across the Catalina Fault is almost 2000 m, from Catalina Basin to Mt Orizaba. From 60 to 72 km of right-slip on San Clemente Fault is inferred from offset of Emery Knoll crater rim (Legg *et al.* 2004*b*).

near the middle of the transpressional fault segment. The principal displacement zone (PDZ) lies along the steep escarpment on the SW flank of the uplift. To the NW, the Catalina Fault merges with the San Clemente Fault along the Santa Cruz–Catalina Ridge.

The Catalina restraining bend is comparable in scale and geometry to that of the San Bernardino Mountains segment of the San Andreas Fault in the region of the Big Bend where the San Andreas cuts across the southern California Transverse Ranges (Fig. 2). Indeed, the broad-scale geometry of the San Clemente fault system offshore of southern California is remarkably similar to that of the southern San Andreas Fault, from the Cajon Pass to the Salton Sea. Such geometric similarity may result from important similarities in the processes that create these major fault bend structures along the PAC–NOAM transform-fault plate boundary.

Bend morphology and fault geometry

Santa Catalina Island and its broad submerged platform resemble the classic ‘rhomboid’ pop-up structures observed in analogue models of restraining stepovers (McClay & Bonora 2001). Widespread outcrops of Catalina Schist metamorphic rocks and Miocene volcanic and plutonic rocks on the island show that the Santa Catalina pop-up involves geological basement, which produces a relatively narrow (20 km maximum width) and steep-sided morphology. Like the San Clemente Fault bend region, local irregularities in the PDZ along the Catalina Fault result in distinct secondary structures. A large, 8- to 9-km releasing (right) stepover along the NW one-third of the island creates a pronounced jog in the escarpment and sharp narrowing of the island uplift (Fig. 12). This stepover has almost 25 km of fault overlap, and major SW-trending stream channels on Santa Catalina Island are deflected to the right by the zone of dextral shear that continues to the NW as the Catalina Ridge Fault. Although a physiographic basin is not apparent, the jog in the Catalina Escarpment is considered to represent an elevated pull-apart basin formed in this releasing fault stepover. A triangular-shaped fault-bounded crustal block at the southern end of the Santa Cruz–Catalina Ridge fits nicely into the gap created by this jog, after removing up to 17 km of right-slip on the Catalina Ridge Fault.

The NE flank of the Catalina pop-up is less steep (<10°) than the Catalina Escarpment (almost 20°). The relatively flat floor of Catalina basin to the SW lies at depths of 1000 to 1300 m, whereas the Santa Monica and San Pedro basins are about 800 to 900 m deep. Thus, the pop-up creates a tectonic dam that traps turbidite sediments to the NW, leaving Catalina basin relatively sediment-starved (cf. Gorsline & Emery 1959). Only minor faults are mapped along the NE flank of the Catalina pop-up (Vedder *et al.* 1986).

Deformation history

Based on high-resolution multi-channel seismic profiles along the Catalina Ridge, the principal traces of the San Clemente and Santa Catalina faults have vertical to subvertical dip. Uplift predicted by a seven-segment elastic dislocation fault model of the Santa Catalina Island pop-up, matches the island/platform morphology using fault depths reaching 16 km. In the model, fault dips range from 60 to 70° for the WNW-trending oblique-slip segments beneath the island, and 80 to 90° for the NW-trending right-slip segments along Catalina Ridge and in the San Diego Trough (Legg *et al.* 2004*a*). Tomographic models of wide-angle seismic-reflection data are consistent with high-angle faulting along the SW flank of Santa Catalina Island cutting through the upper crust (ten Brink *et al.* 2000).

Catalina basin is relatively flat-floored, with sediments covering a highly irregular bedrock surface (Vedder *et al.* 1974). Shallow sedimentary sequences adjacent to the base of the Catalina Escarpment appear to thicken near the major fault stepover (Fig. 13). These sequences are elevated and probably represent a slope apron deposit from the north end of the island (Teng 1985). Distal turbidite sediments from the San Gabriel submarine canyon create a low-relief, gently SW-sloping sequence at the SE end of the basin. Sedimentary sequences elsewhere in the basin along the Catalina Escarpment are relatively uniform in thickness and uplifted. Locally thick sequences exist along the SW flank of the Catalina basin, where right stepovers along the San Clemente Fault create pull-apart basins. Deeper basin-fills with triangular cross-sections adjacent to the Catalina Fault appear to be associated with ancient pull-apart basins formed during the Neogene transtensional rifting of the Inner Borderland rather than the post-Miocene transpression at the restraining bend. Like the San Clemente Fault bend region, Late Quaternary uplift is evident on both sides of the Catalina Fault. Convergence results in crustal thickening at a subvertical oblique-slip fault, without

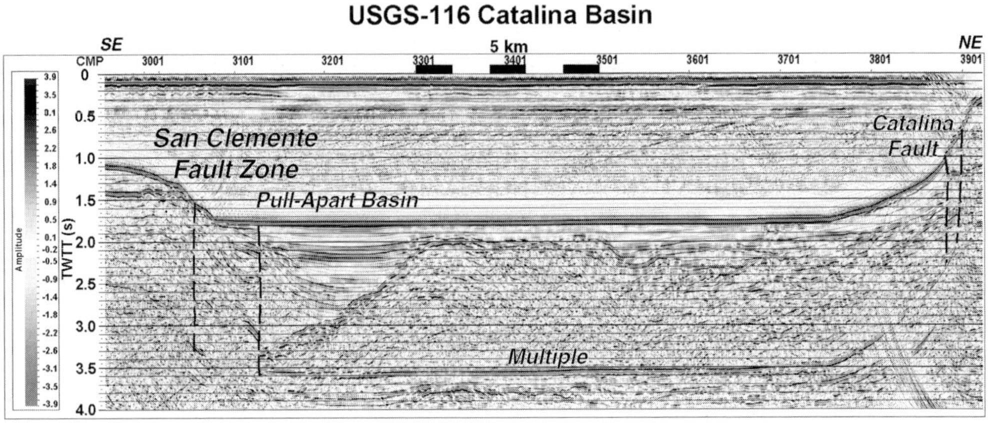

Fig. 13. Seismic-reflection profile USGS-116 across the Catalina basin (see Fig. 12 for profile location). Note the thin sediment cover over an irregular basement surface. A pull-apart basin exists where the San Clemente Fault steps to the NE to eventually merge with the Catalina Fault. The major faults have subvertical dips, typical of strike-slip faults. Convergence across the Catalina Fault has elevated Santa Catalina Island, and uplift occurs on both sides of the PDZ. Seismic data from USGS (J. Childs 2005, pers. comm.) FK migration at 4800 fps velocity was applied to 22-fold USGS stacked data.

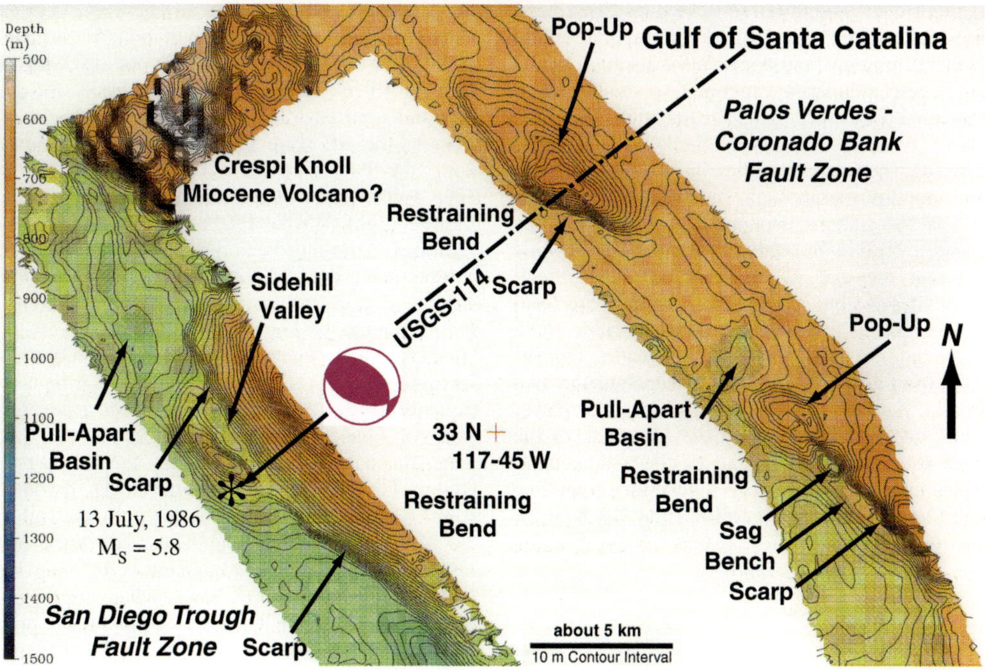

Fig. 14. SeaBeam swath bathymetry over restraining bend pop-ups and releasing stepover pull-apart basins in the northern San Diego Trough area (see Fig. 2 for map location). The southern California offshore area has many paired sets of restraining bends and releasing stepovers along the major NW-trending right-slip fault zones. The 1986 Oceanside earthquake exhibited oblique-reverse movement on a fault plane parallel to the restraining bend in the San Diego Trough fault zone (Hauksson & Jones 1988). Aftershocks spanned the region between the San Diego Trough and Coronado Bank fault zones, and included some events with strike-slip focal mechanisms.

underthrusting or basin subsidence at a reverse or thrust fault.

The lack of prominent uplifted marine terraces is noteworthy for Santa Catalina Island (Davis 2004) creating controversy regarding its uplift history. Well-defined submerged marine terraces are apparent in the bathymetry (Fig. 12; Emery 1958), whereas the uplift of sediments around the base of the escarpment surrounding the island and elevated fluvial terraces along Avalon canyon (Davis 2004) suggest Late Cenozoic uplift. Lower bathyal to abyssal microfossils from latest Miocene to Early Pliocene tuffaceous sandstone and siltstone exposed on the island (Vedder *et al.* 1979), imply that as much as 2 km of post-Miocene uplift has occurred. The topographic relief across the Catalina Fault measures about 1950 m, from Catalina basin to Mt Orizaba; upon adding another 300 m of sedimentary fill in the basin, structural relief exceeds 2200 m. Unfortunately, GPS and other geodetic data are inadequate at present to verify uplift or subsidence in recent history.

Seismicity along the Catalina Fault is minor, with a few small events (magnitude *c*. 3) recorded during the past 73 years of seismograph operation (Figs 2 & 12). More abundant activity is mapped to the NE in the San Pedro basin. Moderate earthquakes (M 5.8 to 6.0) ruptured at each end of the Catalina Fault bend during the 1980s (Corbett 1984; Hauksson & Jones 1988). The mainshocks and their abundant aftershock sequences may signify that the intervening transpressional fault segment is locked and accumulating strain for future large earthquakes ($M > 7$) if rupture involves most of the intervening fault segment (Legg *et al.* 2004*a*).

Other Inner Borderland restraining bends

Regionally, the California Continental Borderland abounds with restraining and releasing bend structures along the several major right-slip faults (Fig. 2). Two prominent restraining bend pop-up structures along the Palos Verdes Fault in the southern California coastal area include Lasuen Knoll and the Palos Verdes Hills (Fig. 12, Borrero *et al.* 2004; Ward & Valensise 1994). A transtensional zone, inferred to be a releasing bend trough along the Palos Verdes Fault, separates these two large pop-ups and controls the location of the San Gabriel submarine canyon. Two smaller transpressional uplifts are found farther south along the Palos Verdes–Coronado Bank fault zone (Fig. 14). The northern feature has the classic rhomboid pop-up morphology observed in analogue models (McClay & Bonora 2001), whereas the southern feature appears as a small 1-km-wide bump at the end of an elongate ridge. Small closed depressions that probably represent local stepover basins (sags) occur along the ridge south of the pop-up, and a larger 2-km-wide stepover basin (pull-apart) is inferred to separate the two restraining bend pop-ups. The principal displacement zone (PDZ) lies along the SW flank of the uplift for the southern three of these restraining bends, and is delineated by a seafloor scarp. The Palos Verdes Hills Fault follows the NE flank of the pop-up at the SW edge of the Los Angeles basin. A steep, subvertical fault bounds the SW flank of the small rhomboidal pop-up with a sediment-filled half-graben-shaped basin to the SW, and uplifted Pliocene and older sedimentary rocks to the NE (Fig. 15). A vertical PDZ exists along the Palos Verdes Fault under San Pedro Bay, according to exploration industry seismic interpretations (Wright 1991).

Some restraining bend pop-ups appear as more symmetrical double humps in cross-section (Fig. 6, line B-25) and as pecan-shaped double bumps on the seafloor, like the feature west of Crespi knoll, between the SE end of the Catalina Fault bend and the 1986 Oceanside earthquake sequence (Fig. 12). A pull-apart basin is located south of Crespi knoll (Fig. 14) where the San Pedro Basin fault zone splits from the San Diego Trough Fault. Miocene volcanic rocks dredged from Crespi knoll, adjacent to this pull-apart, may

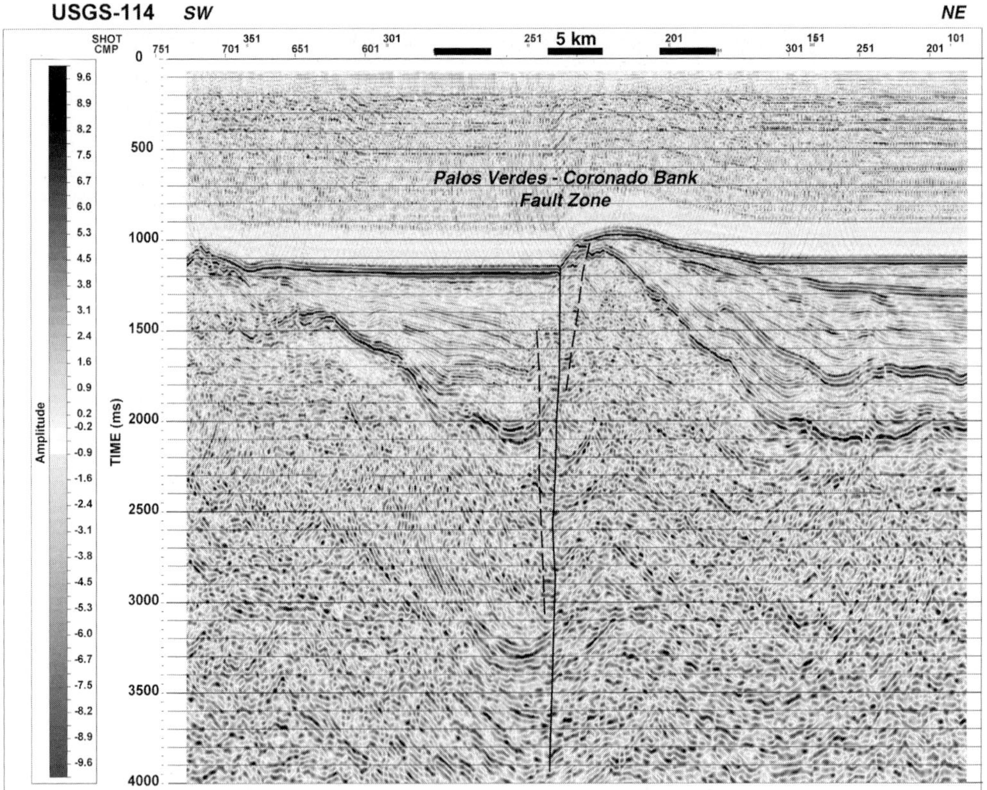

Fig. 15. Migrated 44-channel seismic-reflection profile USGS-114 across the Palos Verdes–Coronado Bank fault zone in the Gulf of Santa Catalina (see Fig. 14 for profile location; processing by C. Sorlien 1992). A small restraining-bend pop-up is juxtaposed against a small half-graben basin. Onlap of sedimentary sequences against tilted strata in both the basin and the NE flank of the uplift records contemporaneous pop-up growth and basin subsidence. Seafloor outcrops sampled on the uplift scarp from DSV Alvin were of Pliocene age (Kennedy *et al.* 1985).

indicate reactivation of an ancient extensional structure formed during the Mid-Miocene oblique rifting of the Inner Borderland. For these pop-up configurations, a near-vertical PDZ bisects the uplift, and strata dip away from the PDZ on both sides; uplift occurs on both sides of the PDZ. It is proposed that strike-slip has juxtaposed two formerly asymmetrical pop-up features to form the local double bump, as these small features are observed at the ends of larger restraining bend uplifts. Alternatively, uplift across a subvertical fault is eroded along the axial PDZ where material is weaker than on the uplift flanks, or these features are narrow crestal grabens where extension occurs above the peak transpressional uplift.

The San Diego Trough Fault has at least two other small restraining bends south of the large Catalina Fault double bend. A NW-trending ridge along the west edge of Coronado Fan Valley, at the base of the Coronado Escarpment, has been interpreted as a natural levée, although seismic reflection profiles show a rock core within this ridge (Shepard & Dill 1966). High-resolution swath bathymetry shows that relatively straight seafloor scarps along the San Diego Trough Fault are offset to the left, so that we interpret this ridge as a restraining stepover pop-up structure (Fig. 16). Uplift along the fault has forced the upper Coronado Fan Valley to turn south for about 20 km along the base of the Coronado Escarpment. The northern 14 km meanders between the inferred pop-up and the escarpment; the southern 6 km is very straight – possibly controlled by a north-trending releasing segment of the San Diego Trough Fault.

Another 22 km south along the San Diego Trough Fault, and directly east of the northern bend region of the San Clemente Fault, is a small 2.3-km-wide by 3.8-km-long seafloor pop-up (Figs 2 & 17). This pop-up exists at a single bend involving a 12° change in fault strike. Locally, there is a small $<2°$ change in fault strike, and a

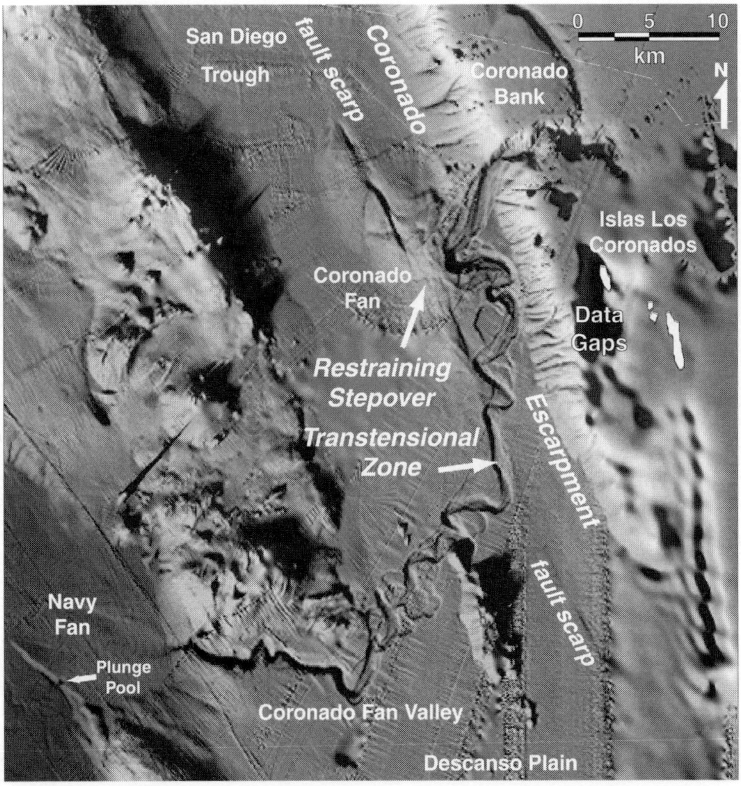

Fig. 16. Shaded relief image of Coronado canyon and the fan valley area (see Fig. 2 for map location). The San Diego Trough Fault is evident as the linear seafloor scarp cutting across the eastern side of the San Diego Trough. A left stepover in the fault zone creates an uplift that confines the Coronado Fan Valley to the east below the Coronado Escarpment. A NNW-trending fault segment forms a transtensional zone that directs the channel to the south of the restraining stepover.

500 m right (releasing) stepover from the SW to NE fault segments. Also, the SE fault appears to branch into a west-trending (281°) scarp that bounds the south edge of the pop-up structure. The high-resolution multi-channel seismic profile shows about 700 m of structural relief on this pop-up (Fig. 7, SMCS-04), although only about 70 m extends above the turbidite fill as seafloor relief. In the broader view, the SE fault segment curves gently into the area of the pop-up, from a strike of 329° to a strike of about 319°, then steps to the right about 500 m to the NW fault segment with a strike of 317°. The west-trending branch that forms the seafloor scarp results from contraction at the termination of the SE fault segment. Thus, the pop-up structure results from the combination of the change in fault trend, i.e. restraining bend, plus the termination of the SE fault segment. A similar combination may be responsible for uplift of Lasuen knoll (Fig. 12), but more detailed subsurface mapping is required to resolve that larger structure and its deformation history.

Review of the Borderland restraining-bend structure

A broad sampling of restraining bend pop-up structures, and associated releasing bend or stepover pull-apart basins, exist in the California Continental Borderland, where more than 20 Ma of evolution of the dextral Pacific–North America transform fault system has been sustained. Some of the restraining bend structures resemble those observed in analogue models, exhibiting the classic rhomboid shape, but others have distinct differences. In particular, there is an overall right-stepping en echelon pattern to the faulting, which is contrary to the typical left-stepping pattern observed in the analogue wrench fault experiments for right-slip faulting. For example, along the San Clemente

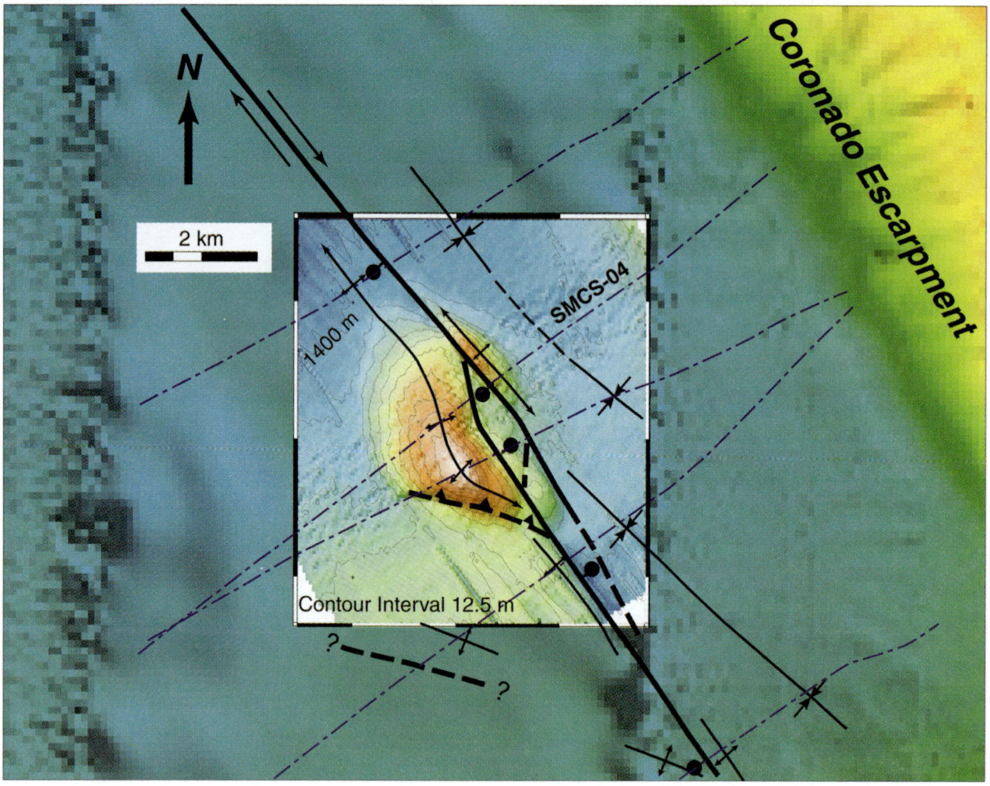

Fig. 17. High-resolution bathymetry and structural interpretation of small restraining-bend pop-up along the San Diego Trough Fault (see Fig. 2 for map location). Well-defined faults are solid with bar and ball placed on the downthrown side where seafloor is offset. Dashed faults are buried or inferred, and are queried where uncertain. Dash–dot lines mark the location of seismic profiles used to map buried faults and folds. Line SMCS-04 is shown in Figure 7. Note that a local releasing fault stepover (right en échelon offset) exists within the restraining bend (left bend).

fault zone, major fault segments step to the right in the vicinity of Navy Fan (Fig. 3), near Fortymile Bank (Fig. 2), and west of Santa Catalina Island (Figs 2 & 12). Similarly, the Catalina Fault exhibits a major right stepover along the SW side of Santa Catalina Island (Fig. 12). This right-stepping pattern resembles that of the Gulf of California, and is considered to be related to the oblique-rifting of the Inner Borderland during the Neogene time (Legg & Kamerling 2004). Two large restraining double bends display major uplift concentrated along the oblique transpressional fault segment – not at the bend where fault curvature is maximal. In contrast, a small restraining bend with releasing stepover on the San Diego Trough Fault exhibits local uplift related to a combination of fault segment termination and to the minor change in fault strike. Many transpressional pop-up structures are manifest as double bumps straddling the principal displacement zone (PDZ) of the major fault. Where soft sediments are involved, uplift exists over a broad area, with gentle seafloor slopes away from the PDZ. Where more rigid bedrock is involved, uplift is more narrow and steep-sided.

In all cases, the dip of the main strike-slip fault is very steep (70° to 80°) to vertical. This observation is the most diagnostic feature of strike-slip faults, even where a flower or palm-tree structure may be absent. The steep dip is manifest at the seafloor or ground surface as a very straight fault trace, even where high topographic relief is crossed. Uplift in the pop-up occurs on both sides of the PDZ, perhaps due to the subvertical fault dip. Typically, this uplift is asymmetrical as shortening becomes greater to one side of the PDZ. Subsidence of the footwall, common to reverse faults, is uncommon in Borderland strike-slip restraining bends.

The surface expression of flower structure appears as branching and secondary fault scarps; small pull-apart (stepover) basins; and larger-scale graben and oblique-fault branch scarps that may extend several kilometres away from the PDZ. Structural relief observed in seismic profiles across the active pop-up structures ranges from several hundred metres to more than two kilometres for Borderland restraining bends. The larger restraining bends approach 100 km lengths and are possibly locked due to enhanced normal stress in transpression between large ($M > 7$) earthquakes.

Tectonic evolution of Borderland restraining bends

Four observations common to Borderland restraining bends lead to a simple model for their tectonic evolution (Fig. 18). First, the strike of the principal displacement zone (PDZ) in the major restraining bends is parallel to the Miocene Pacific–North America (PAC–NOAM) relative motion vector(s). Catalina and Whittier faults trend about 290°, Palos Verdes Hills Fault trends about 300°, and San Clemente Fault bend region trends about 308° (Fig. 2). The clockwise rotation of these trends possibly results from the clockwise rotation of the relative plate motion vector through late Cenozoic times (Atwater & Stock 1998). Second, the major faults within the restraining bend pop-up have very steep to vertical dips. This is more consistent with formation as a strike-slip fault rather than a normal or reverse fault (Anderson 1951). Third, the pop-up structures for the major restraining bends have structurally inverted Miocene basins. Although not a sediment-filled basin, Santa Catalina Island is a volcanic centre that formed where extension thinned the crust, substantially exhuming the Catalina Schist subduction complex and providing access to upper-mantle magmatic sources (Vedder *et al.* 1979; Bohannon & Geist 1998; ten Brink *et al.* 2000). Fourth, there is an overall right-stepping en echelon character to the major right-slip fault pattern of the Borderland. Locally, this pattern persists as the pull-apart (right stepover) basin within the restraining bend pop-up. Regionally, this pattern is apparent as the major fault segments step to the right across larger pull-apart or stepover basins, as best expressed in the San Clemente fault system (Fig. 2).

Phase 1 – oblique rifting and formation of transform faults and spreading centres

The California Continental Borderland formed during the Neogene development of the PAC–NOAM transform plate boundary (Atwater 1970; Legg 1991; Lonsdale 1991; Crouch & Suppe 1993). During Mid-Miocene times, oblique rifting of the western Transverse Ranges away from the continental margin formed the Inner Borderland Rift (Legg 1991; Crouch & Suppe 1993). Pre-existing structural fabric from Mesozoic to Early Cenozoic subduction controlled the orientation of the rift, which has a trend of 330° as expressed by the San Diego Trough, roughly parallel to the modern coastline. For a displacement vector about 30° oblique to the rift trend, i.e. 290° to 308°, a complex pattern of faults forms, including mostly strike-slip faults, both dextral and sinistral, as well as extensional faults (Fig. 18, Phase 1 Withjack & Jamison 1986). New right-slip faults created to accommodate the PAC–NOAM transform motion trend NW, subparallel to the rift margins. Normal faults formed along a north–south trend.

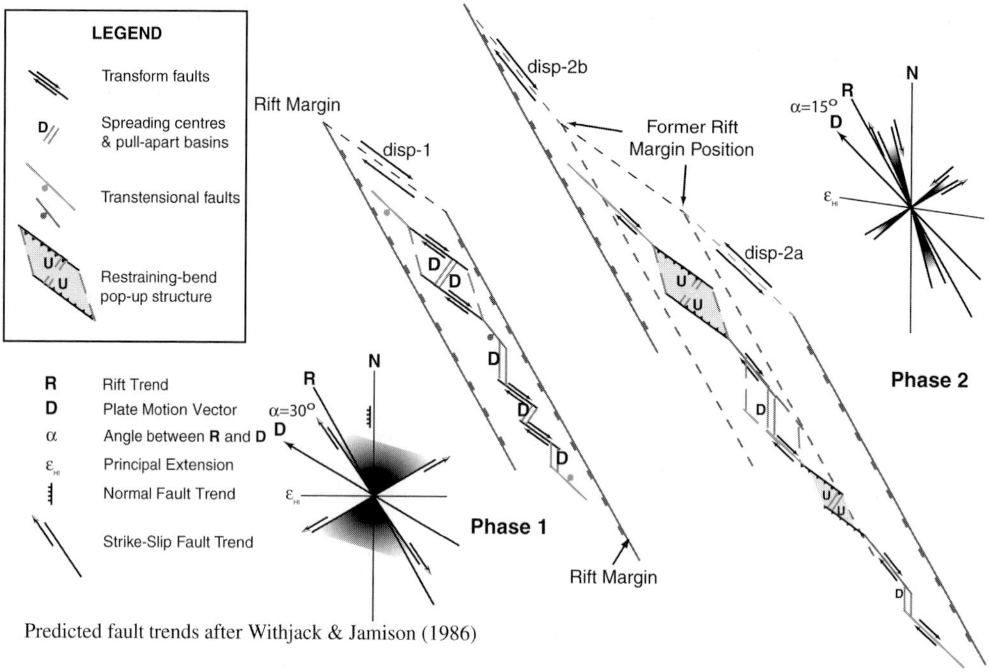

Fig. 18. Simplified model for generation of restraining bends in an evolving dextral transform plate boundary. Initial oblique rifting along transform boundary creates a series of right-stepping en echelon transform faults linked by pull-apart basins. The structural fabric from the prior tectonic boundary style (subduction for California) controls the geometry of the rift formed due to a relative motion vector more westerly than the ancient structural trend. Some new strike-slip faults will have trends oriented clockwise to the transform vector, including synthetic shears. Transtensional basins form along these faults and at the fault stepovers (pull-apart basins). Clockwise rotation of the transform plate-motion vector results in transpression along the original transform faults, and former transtensional faults may become parallel right-slip transform faults. Transtensional basins become structurally inverted and form restraining-bend pop-up structures.

However, when displacement exceeded several kilometres, most of the relative motion became concentrated on a few major right-slip faults that grew parallel to the relative plate-motion vector and linked north-trending continental rift centres. Seafloor spreading, with creation of new oceanic crust along faults orthogonal to the transform faults, did not occur, because the rift initiated in the thickened crust of a former subduction zone accretionary wedge, and extension was insufficient to create zero-thickness lithosphere.

Creation of new transform faults in the Inner Borderland Rift was facilitated by thinned continental crust that was also thermally weakened by active volcanism during Mid-Miocene times (Vedder *et al.* 1974; Weigand 1994). The transform-fault trend parallels the displacement vector, whereas synthetic Riedel shears with a more northern trend would become transtensional. Most significantly, a right-stepping en échelon pattern of right-slip transform faults linking north-trending pull-apart basins and incipient seafloor-spreading centres was created. This fault pattern resembles the modern Gulf of California transform fault system (Lonsdale 1985) and differs from the classical 'wrench fault tectonics' where dextral strike-slip results in a left-stepping en echelon pattern of synthetic Riedel shears (Wilcox *et al.* 1973; Withjack & Jamison 1986; McClay & White 1995).

Phase 2 – clockwise rotation of the relative plate-motion vector, transpression, and basin inversion

In Late Miocene times, the PAC–NOAM relative plate motion vector shifted clockwise (Fig. 18, Phase 2, Atwater & Stock 1998). Existing transform faults with the old relative motion trend, to the west or counterclockwise to the new trend, became transpressional. Other right-slip faults with trends more parallel to the new plate-motion vector became pure strike-slip, ceasing transtensional basin formation. Stepover (pull-apart) basins, previously formed between en échelon transform faults, stopped subsiding and became structurally inverted due to transpression. With a rift trend more closely aligned with the relative plate motion vector, formation of new faults would tend to favour strike-slip on synthetic Riedel shears that could also grow into transform faults parallel to the displacement vector. Following the plate-boundary jump inland to the modern San Andreas fault system, at about 6 Ma, creation of the major southern California restraining bend (Fig. 2, 'big bend') increased the NE-directed shortening across the Inner Borderland. This latter episode is considered responsible for enhanced restraining bend pop-up formation, such as along the Palos Verdes Hills Fault and at Santa Catalina Island, and may represent a further clockwise rotation of the displacement vector (Fig. 18, Disp-2b). The Late Quaternary Pasadenan Orogeny (Wright 1991) further increased contractional strain with north-directed shortening between the western Transverse Ranges and the northern Borderland, enhancing transpressional uplift along major Borderland right-slip faults including the Santa Cruz–Catalina Ridge, Palos Verdes, and Whittier faults.

Conclusions

Borderland restraining bends exist at a wide range of scales, from a few kilometres to more than 100 km in length. Morphology of Borderland restraining-bend pop-ups is well preserved due to the low erosion rates in deep offshore basins. In plan view, Borderland pop-up morphology is generally polygonal and often rhomboidal, as observed in many analogue models. Pop-up structures involving turbidite sediments are smoother and more gently sloping than pop-ups in more rigid bedrock and basement material. The uplift is usually asymmetrical, with the principal displacement zone (PDZ) aligned along one side of the peak uplift. Unlike low-angle to moderately dipping thrust and reverse faults, where underthrusting occurs, shortening normal to the vertical strike-slip fault results in uplift on both sides of the fault. Borderland pop-ups typically involve structural inversion of ancient sedimentary basins, and extreme uplift in some large restraining bends exposes volcanic and metamorphic basement rocks. In contrast to analogue models of pop-up internal structure, the PDZ and major branch and secondary faults within the restraining bend have very steep to vertical dips. This is considered a result of evolution within an active transform fault plate boundary. An overall right-stepping en echelon fault pattern combined with the basin inversion at major restraining bends, where the PDZ trends parallel to the Miocene Pacific–North America relative plate-motion vector, suggests that Borderland restraining bends formed initially as transform faults in an oblique rift similar to the modern Gulf of California. Subsequent clockwise rotation of the relative motion vector led to creation of new transform faults parallel to the evolving plate-motion vector, and initiated transpression and basin inversion along the earlier transform faults at their pull-apart basin or incipient spreading-centre intersections.

We appreciate the careful reviews by A. Sylvester, P. Mann, M. Gordon and D. Cunningham that helped to improve this manuscript, and special thanks go to editors P. Mann and D. Cunningham for their patient support whilst completing the final manuscript. Summer intern H. Wang helped to measure the uplift at the San Clemente Fault bend region. The University of California provided support for a student cruise led by K. Macdonald to acquire new multi-beam bathymetry and high-resolution seismic data offshore of Baja California. M. Barth ensured that the seismic data were of the highest quality, and K. Broderick helped to process some of the multi-beam data. This research was supported by the Southern California Earthquake Center. SCEC is funded by NSF Cooperative Agreement EAR-0106924 and USGS Cooperative Agreement 02HQAG0008. The SCEC contribution number for this paper is 954. Support in part was also provided by USGS award numbers 01HQGR0017 (MRL) and 01HQGR0018 (CG).

References

ANDERSON, E. M. 1951. *The dynamics of Faulting*. Oliver and Boyd, Edinburgh, 206 pp.

ANDERSON, R. S. 1990. Evolution of the northern Santa Cruz Mountains by advection of crust past a San Andreas fault bend. *Science*, **249**, 397–401.

ASTIZ, L. & SHEARER, P. M. 2000. Earthquake locations in the inner Continental Borderland offshore southern California. *Seismological Society of America Bulletin*, **90**, 425–449.

ATWATER, T. 1970. Implications of plate tectonics for the Cenozoic evolution of western North America. *Geological Society of America Bulletin*, **81**, 3515–3536.

ATWATER, T. M. & STOCK, J. 1998. Pacific–North America plate tectonics of the Neogene southwestern United States – an update. *International Geologic Review*, **40**, 375–402.

AYDIN, A. & KALAFAT, D. 2002. Surface ruptures of the 17 August and 12 November 1999 Izmit and Duzce earthquakes in northwestern Anatolia, Turkey: their tectonic and kinematic significance and the associated damage. *Seismological Society of America Bulletin*, **92**, 95–106.

BENNETT, R. A., RODI, W. & REILINGER, R. E. 1996. Global Positioning System constraints on fault slip rates in southern California and northern Baja, Mexico. *Journal of Geophysical Research*, **101**, 21 943–21 960.

BIDDLE, K. T. & CHRISTIE-BLICK, N. (eds) 1985. *Strike-slip Deformation, Basin Formation, and Sedimentation.* SEPM Special Publications, **37**, 386 pp.

BILLINGS, M. P. 1972. *Structural Geology.* 3rd edn, Prentice-Hall. Englewood Cliffs, New Jersey, 606 pp.

BOHANNON, R. G. & GEIST, E. 1998. Upper crustal structure and Neogene tectonic development of the California Continental Borderland. *Geological Society of America Bulletin*, **110**, 779–800.

BOHANNON, R., EITTREIM, S. *ET AL.* 1990. A seismic-reflection study of the California Continental Borderland [abstract]. *Transactions of the American Geophysical Union*, **71**, 1631.

BORRERO, J. C., LEGG, M. R. & SYNOLAKIS, C. E. 2004. Tsunami sources in the southern California bight. *Geophysical Research Letters*, **31**, L13211.

BUTLER, R. W. H., SPENCER, S. & GRIFFITHS, H. M. 1998. The structural response to evolving plate kinematics during transpression: evolution of the Lebanese restraining bend of the Dead Sea Transform. *In*: HOLDSWORTH, R. E., STRACHAN, R. A. & DEWEY, J. F. (eds) *Continental Transpressional and Transtensional Tectonics*, Geological Society, London, Special Publications, **135**, 81–106.

CORBETT, E. J. 1984. *Seismicity and crustal structure studies of southern California: tectonic implications from improved earthquake locations.* PhD thesis California Institute of Technology, Pasadena, California, 231 pp.

CROUCH, J. K. & SUPPE, J. 1993. Late Cenozoic tectonic evolution of the Los Angeles basin and inner California borderland: a model for core complex-like crustal extension. *Geological Society of America Bulletin*, **105**, 1415–1434.

CROWELL, J. C. 1974. Origin of late Cenozoic basins in southern California. *In*: DICKINSON, W. R. (ed.) *Tectonics and Sedimentation.* SEPM Special Publications, **22**, 190–204.

DAVIS, P. 2004. The marine terrace enigma of Catalina Island – an uplifting experience? *In*: LEGG, M., DAVIS, P. & GATH, E. (eds) *Geology and Tectonics of Santa Catalina Island and the California Continental Borderland*, South Coast Geological Society Field Trip Guidebooks, **32**, Santa Ana, California, 115–121.

DEMETS, C., GORDON, R. G., ARGUS, D. F. & STEIN, S. 1990. Current plate motions. *Geophysical Journal International*, **101**, 425–478.

DEWEY, J. F., HOLDSWORTH, R. E. & STRACHAN, R. A. 1998. Transpression and transtension zones. *In*: HOLDSWORTH, R. E., STRACHAN, R. A. & DEWEY, J. F. (eds) *Continental Transpressional and Transtensional Tectonics*. Geological Society, London, Special Publications, **135**, 1–14.

DUNBAR, R. B. 1981. *Sedimentation and the history of upwelling and climate in high fertility areas of the northeastern Pacific Ocean.* PhD thesis San Diego, Scripps Institution of Oceanography.

EMERY, K. O. 1958. Shallow submerged marine terraces of southern California. *Geological Society of America Bulletin*, **69**, 39–60.

EMERY, K. O. 1960. *The Sea off Southern California, a Modern Habitat of Petroleum.* John Wiley, New York, 366 pp.

FRANCIS, R. D., SIGURDSON, D. R., LEGG, M. R., GRANNELL, R. B. & AMBOS, E. L. 1999. Student participation in an offshore seismic reflection study of the Palos Verdes fault, California Continental Borderland. *Journal of Geo-Science Education*, **47**, 22–30.

GOLDFINGER, C., LEGG, M. & TORRES, M. 2000. New mapping and submersible observations of recent activity of the San Clemente fault [abstract]. *EOS, Transactions of the American Geophysical Union*, Fall Meeting, San Francisco, **81**, 1068.

GORSLINE, D. S. & EMERY, K. O. 1959. Turbidity current deposits in San Pedro and Santa Monica Basins off southern California. *Geological Society of America Bulletin*, **70**, 279–290.

HARDEN, J. W. & MATTI, J. C. 1989. Holocene and late Pleistocene slip rates on the San Andreas fault in Yucaipa, California, using displaced alluvial-fan deposits and soil chronology. *Geological Society of America Bulletin*, **101**, 1107–1117.

HARDING, T. P. 1973. Newport–Inglewood trend, California – an example of wrench style deformation. *AAPG Bulletin*, **57**, 97–116.

HARDING, T. P. 1974. Petroleum traps associated with wrench faults. *AAPG Bulletin*, **58**, 1290–1304.

HARDING, T. P. 1985. Seismic characteristics and identification of negative flower structures, positive flower structures, and positive structural inversion. *AAPG Bulletin*, **69**, 585–600.

HARLAND, W. B. 1971. Tectonic transpression in Caledonian Spitzbergen. *Geological Magazine*, **108**, 27–42.

HARRIS, R. A., DOLAN, J. F., HARTLEB, R. & DAY, S. M. 2002. The 1999 Izmit, Turkey, earthquake: a 3D dynamic stress transfer model for intraearthquake triggering. *Seismological Society of America Bulletin*, **92**, 256–266.

HAUKSSON, E. & JONES, L. M. 1988. The July 1986 Oceanside ($M_L = 5.3$) earthquake sequence in the continental borderland, southern California. *Seismological Society of America Bulletin*, **78**, 1885–1906.

HEATH, G. R., MOORE, T. C., Jr. & DAUPHIN, J. P. 1976. Late Quaternary accumulation rates of opal, quartz, organic carbon, and calcium carbonate in the Cascadia Basin area, northeast Pacific. *Geological Society of America Memoir*, **145**, 393–409.

JANIK, A. 2001. Pleistocene sedimentation in the outer basins of the California borderlands from digitally-acquired 3.5 kHz subbottom profiles and ODP Leg 167 drilling [abstract]. *EOS, Transactions of the American Geophysical Union*, **82**, F740.

KELLER, E. A., BONKOWSKI, M. S., KORSCH, R. J. & SHLEMON, R. J. 1982. Tectonic geomorphology of the San Andreas fault zone in the southern Indio Hills, Coachella Valley, California. *Geological Society of America Bulletin*, **93**, 46–56.

KENNEDY, M. P., CLARKE, S. H., GREENE, H. G. & LONSDALE, P. F. 1985. Observations from DSRV ALVIN of Quaternary faulting on the southern California continental margin. *US Geological Survey Open-File Report*, 85–39, 26 pp.

KULM, L. D. & SCHEIDEGGER, K. F. 1979. Quaternary sedimentation on the tectonically active Oregon continental slope. *In*: DOYLE, L. J. & PILKEY, O. H. (eds) *Geology of Continental Slopes.* Society of Economic

Paleontologists & Mineralogists Special Publication, **27**, 247–263.

LANGENHEIM, V. E, JACHENS, R. C., MATTI, J. C., HAUKSSON, E., MORTON, D. M. & CHRISTENSEN, A. 2005. Geophysical evidence for wedging in the San Gorgonio Pass structural knot, southern San Andreas fault zone, southern California. *Geological Society of American Bulletin*, **117**, 1554–1572.

LARSON, K. M. 1993. Application of the Global Positioning System to crustal deformation measurements, 3, Result from the southern California borderlands. *Journal of Geophysical Research*, **98**, 21 713–21 726.

LEGG, M. R. 1985. *Geologic structure and tectonics of the inner continental borderland offshore northern Baja California, Mexico*. PhD thesis, University of California, Santa Barbara, California, 410 pp.

LEGG, M. R. 1991. Developments in understanding the tectonic evolution of the California Continental Borderland. *In*: OSBORNE, R. H. (ed.) *From Shoreline to Abyss*, SEPM Shepard Commemorative Volumes, **46**, 291–312.

LEGG, M. R. & KAMERLING, M. J. 2004. Fault trends and the evolution of the Pacific–North America transform in southern California. *EOS, Transactions of the American Geophysical Union*, Fall Meeting, San Francisco.

LEGG, M. R. & KENNEDY, M. P. 1991. Oblique divergence and convergence in the California Continental Borderland. *In*: ABBOTT, P. L. & ELLIOTT, W. J. (eds) *Environmental Perils of the San Diego Region*. San Diego Association of Geologists Guidebooks, 1–16.

LEGG, M. R., BORRERO, J. C. & SYNOLAKIS, C. E. 2004*a*. Tsunami hazards associated with the Catalina Fault in southern California. *Earthquake Spectra*, **20**, 917–950.

LEGG, M. R., LUYENDYK, B. P., MAMMERICKX, J., DE MOUSTIER, C. & TYCE, R. C. 1989. Sea Beam survey of an active strike-slip fault – the San Clemente fault in the California Continental Borderland. *Journal of Geophysical Research*, **94**, 1727–1744.

LEGG, M. R., NICHOLSON, C., GOLDFINGER, C., MILSTEIN, R. & KAMERLING, M. J. 2004*b*. Large enigmatic crater structures offshore southern California. *Geophysical Journal International*, **159**, 803–815.

LONSDALE, P. 1985. A transform continental margin rich in hydrocarbons, Gulf of California. *AAPG Bulletin*, **69**, 1160–1180.

LONSDALE, P. 1991. Structural patterns of the Pacific floor offshore of Peninsular California. *In:* DAUPHIN, P. & NESS, G. (eds) *The Gulf and Peninsular province of the Californias*. *AAPG* Memoir **47**, 87–125.

LYLE, M., KOIZUMI, I. *ET AL*. 1997. *Proceedings of the Ocean Drilling Program, Initial Reports*, **167**, ch. 5, Site 1011, 86–127.

MCCLAY, K. & BONORA, M. 2001. Analog models of restraining stepovers in strike-slip fault systems. *AAPG Bulletin*, **85**, 233–260.

MCCLAY, K. & WHITE, M. J. 1995. Analogue modelling of orthogonal and oblique rifting. *Marine and Petroleum Geology*, **12**, 137–151.

MANN, P., DRAPER, G. & BURKE, K. 1985. Neotectonics of a strike-slip restraining bend system, Jamaica. *In*: HOLDSWORTH, R. E., STRACHAN, R. A. & DEWEY, J. F. (eds) *Continental Transpressional and Transtensional Tectonics*, Geological Society, London, Special Publications, **135**, 211–226.

MINSTER, J. B. & JORDAN, T. H. 1978. Present-day plate motions. *Journal of Geophysical Research*, **83**, 5331–5334.

MOORE, D. G. 1969. *Reflection Profiling Studies of the California Continental Borderland – Structure and Quaternary Turbidite Basins*. Geological Society of America Special Papers, **107**, 142 pp.

National Ocean Survey (NOS) 1999. *Hydrographic Survey Data*. National Oceanic and Atmospheric Administration, National Geophysical Data Center, Boulder, CO.

PLAFKER, G. & GALLOWAY, J. P. 1989. Lessons Learned from the Loma Prieta, California, Earthquake of October 17, 1989. *US Geological Survey Circular*, **1045**, 48 pp.

SCHREURS, G. & COLLETTA, B. 1998. Analogue modelling of faulting in zones of continental transpression and transtension. *In*: HOLDSWORTH, R. E., STRACHAN, R. A. & DEWEY, J. F. (eds) *Continental Transpressional and Transtensional Tectonics*. Geological Society, London, Special Publications, **135**, 59–79.

SCHWARTZ, S. Y., ORANGE, D. L. & ANDERSON, R. S. 1994. Complex fault interactions in a restraining bend on the San Andreas fault, southern Santa Cruz Mountains, California. *In*: SIMPSON, R.W. (ed.) *The Loma Prieta, California, Earthquake of October 17, 1989 – Tectonic Processes and Models*. US Geological Survey Professional Papers, **1550-F**, F49–F54.

SEEBER, L. & ARMBRUSTER, J. G. 1995. The San Andreas fault system through the Transverse Ranges as illuminated by earthquakes. *Journal of Geophysical Research*, **100**, 8285–8310.

SEGALL, P. & POLLARD, D. 1980. Mechanics of discontinuous faults. *Journal of Geophysical Research*, **84**, 4337–4350.

SHEPARD, F. P. & DILL, R. F. 1966. *Submarine Canyons and Other Sea Valleys*. Rand McNally & Company, Chicago, 381 pp.

SHEPARD, F. P. & EMERY, K. O. 1941. Submarine topography of the California coast – canyons and tectonic interpretation. *Geological Society of America Special Papers* **31**, 171 pp.

SHOR, G. G., Jr, RAITT, R. W. & MCGOWAN, D. D. 1976. Seismic refraction studies in the southern California borderland, 1949–1974. *Scripps Institute of Oceanography References*, **76–13**, 70 pp.

SIEH, K. E. 1978. Slip along the San Andreas fault associated with the Great 1857 earthquake. *Seismological Society of America Bulletin*, **68**, 1421–1448.

SMITH, D. L. & NORMARK, W. R. 1976. Deformation and patterns of sedimentation, south San Clemente Basin, California borderland. *Marine Geology*, **22**, 175–188.

STONE, D. S. 1995. Structure and kinematic genesis of the Quealy wrench duplex: transpressional reactivation of the Precambrian Cheyenne belt in the Laramie basin, Wyoming. *AAPG Bulletin*, **79**, 1349–1376.

SYLVESTER, A. G. 1988. Strike-slip faults. *Geological Society of America Bulletin*, **100**, 1666–1703.

SYLVESTER, A. G. & SMITH, R. R. 1976. Tectonic transpression and basement controlled deformation in the

San Andreas fault zone, Salton Trough, California. *AAPG Bulletin*, **60**, 2081–2102.

TEN BRINK, U. S., ZHANG, J., BROCHER, T. M., OKAYA, D. A., KLITGORD, K. D. & FUIS, G. S. 2000. Geophysical evidence for the evolution of the California Inner Continental Borderland as a metamorphic core complex. *Journal of Geophysical Research*, **105**, 5835–5857.

TENG, L. S. 1985. Seismic stratigraphic study of, the California Continental Borderland basins: structure, stratigraphy, and sedimentation. PhD thesis, University of Southern California, Los Angeles, 197 pp.

TUCKER, P. M. & YORSTON, H. J. 1973. *Pitfalls in Seismic Interpretation.* Society of Exploration Geophysicists Monographs, **2**, Tulsa, OK, 50 pp.

VEDDER, J. G. 1990. Maps of California Continental Borderland showing compositions and ages of bottom samples acquired between 1968 and 1979. *US Geological Survey Miscellaneous Field Studies Map MF-2122*, three sheets, scale 1:250 000.

VEDDER, J. G., BEYER, L. A., JUNGER, A., MOORE, G. W., ROBERTS, A. E., TAYLOR, J. C. & WAGNER, H. C. 1974. Preliminary report on the geology of the Continental Borderland of southern California. *US Geological Survey, Miscellaneous Field Studies Map MF-624*, scale 1:500 000.

VEDDER, J. G., GREENE, H. G., CLARKE, S. H. & KENNEDY, M. P. 1986. Geologic map of the mid-southern California continental margin. California Division of Mines and Geology, *California Continental Margin Geologic Map Series*, Area 2 of 7, sheet 1 of 4, scale 1:250 000.

VEDDER, J. G., HOWELL, D. G. & FORMAN, J. A. 1979. Miocene strata and their relation to other rocks, Santa Catalina Island, California. *In*: ARMENTROUT, J. M., COLE, M. R. & TERBEST, H., JR (eds). *Cenozoic Paleogeography of the Western United States*, SEPM, Pacific Coast Paleogeography Symposium, **3**, 239–256.

WAKABAYASHI, J. 2007. Step-overs that migrate with respect to affected deposits: field characteristics and speculation on some details of their evolution. *In*: CUNNINGHAM, W. D. & MANN, P. (eds) *Tectonics of Strike-Slip Restraining and Releasing Bends*. Geological Society, London, Special Publications, **290**, 169–188.

WARD, S. N. & VALENSISE, G. 1994. The Palos Verdes terraces, California: bathtub rings from a buried reverse fault. *Journal of Geophysical Research*, **99**, 4485–4494.

WEIGAND, P. 1994. Petrology and geochemistry of Miocene volcanic rocks from Santa Catalina and San Clemente Islands, California. *In*: HALVORSON, W. L. & MAENDER, G. J. (eds) *The Fourth California Islands Symposium: Update on the Status of Resources*, Santa Barbara Museum of Natural History, Santa Barbara, CA, 267–280.

WELDON, R. J., II & SIEH, K.E. 1985. Holocene rate of slip and tentative recurrence interval for large earthquakes on the San Andreas fault, Cajon Pass, southern California. *Geological Society of America Bulletin*, **96**, 793–812.

WGCEP, Working Group on California Earthquake Probabilities. 1995. Seismic hazards in southern California: probable earthquakes, 1994 to 2024. *Seismological Society of America Bulletin*, **85**, 379–439.

WILCOX, R. E., HARDING, T. P. & SEELEY, D. R. 1973. Basic wrench tectonics. *AAPG Bulletin*, **57**, 74–96.

WITHJACK, M. O. & JAMISON, W. R. 1986. Deformation produced by oblique rifting. *Tectonophysics*, **126**, 99–124.

WOODRING, W. P., BRAMLETTE, M. N. & KEW, W. S. W. 1946. Geology and paleontology of the Palos Verdes Hills, California. *US Geological Survey Professional Paper*, **207**, 145 pp.

WRIGHT, T. L. 1991. Structural geology and tectonic evolution of the Los Angeles basin. *In*: BIDDLE, K. T. (ed.) *Active Margin Basins. AAPG* Memoirs, **52**, 13–134.

Stepovers that migrate with respect to affected deposits: field characteristics and speculation on some details of their evolution

J. WAKABAYASHI

Department of Earth and Environmental Sciences, California State University, Fresno, 2576 E. San Ramon Ave. MS ST-24, Fresno, CA 93740-8039, USA
(e-mail: jwakabayashi@csufresno.edu)

Abstract: Traditionally, geologists have viewed strike-slip stepover regions as progressively increasing in structural relief with increasing slip along the principal displacement zones (PDZs). In contrast, some stepover regions may migrate along the strike of the PDZs with respect to deposits affected by them, leaving a 'wake' of formerly affected deposits trailing the active stepover region. Such stepovers generate comparatively little structural relief at any given location. For restraining bends of this type, little exhumation and erosion takes place at any given location. Another characteristic of migrating stepovers is local tectonic inversion that may migrate along the strike of the PDZs. This is most easily observed for migrating releasing bends where the wake is composed of former pull-apart basin deposits that have been subject to shortening and uplift. This type of basin inversion occurs along the San Andreas Fault, wherein the wake is affected by regional transpression. Some migrating stepovers may evolve by propagation of the PDZ on one side of the stepover, and shut-off of the PDZ on the other side. Possible examples of migrating stepovers are present along the northern San Andreas fault system at scales from metres (sag ponds and pressure ridges) to tens of kilometres (large basins and transpressional uplifts). Migrating stepovers and 'traditional' stepovers may be end members of stepover evolutionary types, and the ratio of wake length to the amount of slip along the PDZs during stepover development measures the 'migrating stepover component' of a given stepover. For a 'pure' migrating type, the wake length may be equal to or greater than the PDZ cumulative slip during the time of stepover evolution, whereas for a 'pure' traditional type, there would be no wake.

Bends and stepovers occur along all strike-slip systems (e.g. Crowell 1974*a*, *b*; Christie-Blick & Biddle 1985). To aid discussion one can define stepover terms as follows: (1) main strike-slip faults or bounding faults also known as principal displacement zones or PDZs, and (2) transverse or relay structures that accommodate the transfer of slip between the PDZs on either side of the stepover region (Fig. 1). Geological features related to stepovers and bends have received considerable attention from researchers (e.g. Crowell 1974*b*; Mann *et al.* 1983; Christie-Blick & Biddle 1985; Westaway 1995; Dooley & McClay 1997). Studies of the evolution of stepover features have traditionally considered stepovers as features that progressively increase in structural relief with increasing slip on the PDZs connected to them (e.g. Mann *et al.* 1983; Dooley & McClay 1997; Dooley *et al.* 1999; McClay & Bonora 2001); such stepovers will be referred to herein as 'traditional' stepovers. More recently, Wakabayashi *et al.* (2004) presented field evidence for a type of stepover that appears to have migrated with respect to the affected deposits rather than increased in structural relief with greater slip accommodation; such stepovers will be referred to herein as 'migrating' stepovers.

In this paper, I will review some field examples presented by Wakabayashi *et al.* (2004) and present an updated discussion speculating on the evolution of such structures. The new material presented in this paper, compared with Wakabayashi *et al.* (2004) includes the following:

1. A discussion of a full spectrum of hypothetical migrating stepover types with different migration alternatives and evaluation of different reference frames (whereas the earlier paper discussed only one type of migrating stepover).
2. The earlier paper considered migrating stepovers as a counter-example to 'traditional' stepovers, whereas herein a unifying model is proposed with migrating stepovers and 'traditional' stepovers as end members of stepover evolution types.
3. More detailed maps are provided for the field examples, and additional diagrams are provided to aid in visualization of the various stepover models proposed.

From: CUNNINGHAM, W. D. & MANN, P. (eds) *Tectonics of Strike-Slip Restraining and Releasing Bends*. Geological Society, London, Special Publications, **290**, 169–188.
DOI: 10.1144/SP290.4 0305-8719/07/$15.00

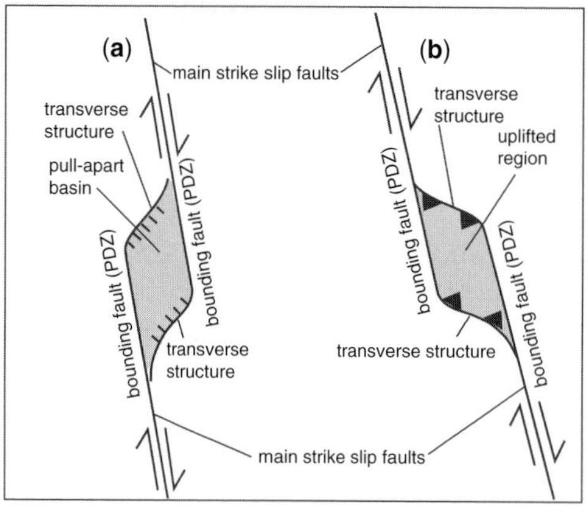

Fig. 1. Simple classification of some structures associated with stepovers or bends along strike-slip faults. An idealized releasing stepover (**a**), and restraining stepover (**b**), are shown.

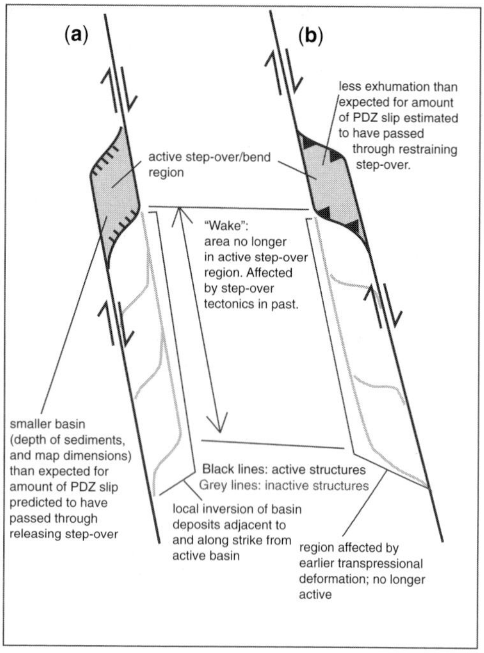

Fig. 2. Diagram illustrating some of the features associated with migrating stepovers described in the text.

Field examples of migrating stepovers from the San Andreas fault system

General field characteristics

Field observations along the San Andreas fault system of coastal California suggest that some types of stepovers or bends migrate with respect to formerly affected deposits (Wakabayashi *et al.* 2004). Some field observations suggesting migrating stepovers are as follows (shown schematically on Fig. 2):

1. Structural relief of a stepover region is much smaller than expected for the estimated amount of slip accommodated by the structure during its lifetime. For releasing stopovers, this means a smaller basin than expected, and for restraining bends this means much less uplift and exhumation than expected. Note that the amount of expected slip through the stepover region is not necessarily the total amount of slip on the PDZ, because a stepover may form much later than the fault (the Olema Creek Formation example presented below may be an example).
2. Tectonic inversion has occurred out-of-phase with known regional tectonic changes.
3. Former basinal deposits are found along the strike of a PDZ adjacent to an active pull-apart basin (or, more generally, an active transtensional basin), forming a 'wake' (by analogy to the wake of a ship) that appears to mark the earlier presence of a pull-apart environment.
4. For some stepovers, propagation of PDZs has occurred, and some also exhibit progressive along-strike dying out of activity on a PDZ.

The San Andreas fault system

The dextral San Andreas fault system (SAFS) of coastal California accommodates 75–80% (38–40 mm/a) of present Pacific–North American plate motion (e.g. Argus & Gordon 1991), and 70% (540–590 km) of the dextral displacement that has accumulated across the plate boundary over the last 18 Ma (Atwater & Stock 1998). Regional fault-normal convergence across the northern San Andreas system (NSAFS) occurs at less than 10% of the dextral slip rate (Argus & Gordon 1991, 2001). Although this regional convergence contributes to some of the shortening seen along the NSAFS, the most prominent transpressional features are largely driven by local restraining bends or stepovers along strike-slip faults (e.g. Aydin & Page 1984; Bürgmann *et al.* 1994; Unruh & Sawyer 1995).

Subduction, associated with the development of the Franciscan Complex, occurred along the western margin of North America prior to the initial interaction between an East Pacific Rise and the subduction zone at about 28 Ma, but the SAFS did not become established until about

18 Ma (Atwater & Stock 1998). Since 18 Ma, SAFS has progressively lengthened and accumulated dextral displacement, as the Mendocino triple junction (fault–fault–trench triple junction), migrated NE and the Rivera triple junction (ridge–fault–trench triple junction) migrated SE (e.g. Atwater & Stock 1998). In addition to the Franciscan and related rocks, the Coast Range basement also includes the Salinian block, a continental fragment composed of granitic and high-grade metamorphic rocks that has been translated into the Coast Ranges from the SE along SAFS faults and faults that predated the transform regime (Page 1981). The Franciscan structural grain is defined by a series of nappe sheets, composed of both coherent and mélange units (Wakabayashi 1992, 1999). The nappe sheets have been folded about subhorizontal fold axes that trend more westerly than the strike of the SAFS faults. Most of the major faults of the NSAFS cut Franciscan basement or form the contacts between Franciscan and Salinian basement, although some cut across Salinian basement (Page 1981).

The SAFS comprises multiple dextral strands, including the San Andreas Fault. The distribution of active structures and dextral slip rates has shifted irregularly throughout the history of the transform-fault system, as new faults have formed, old ones have shut off, and slip rate distribution within the system changed (e.g. Powell 1993; Wakabayashi 1999). Deposition of Late Cenozoic sedimentary and volcanic rocks has occurred during the development of the transform margin, and these deposits are critical for evaluating tectonic processes associated with the SAFS (e.g. Page 1981).

There are many examples of transtensional (pull-apart) basins and local transpressional structures related to stepovers and bends along the various dextral faults of the SAFS (e.g. Aydin & Page 1984; Nilsen & McLaughlin 1985; Crowell 1974*a*, *b*, 1987; Yeats 1987). Local transpressional and transtensional geological features along strike-slip faults occur across a range of scales, from tens of kilometres for large basins and push-up regions, to metres for sag ponds and small pressure ridges (Crowell 1974*a*).

I will review examples of migrating stepovers and bends from the NSAFS (see Fig. 3 for locations). For releasing structures, such as pull-apart basins, the basins have migrated along bounding strike-slip faults (PDZs) with respect to their former basinal deposits. For restraining structures, such as fold and thrust belts or transpressional welts, the features have migrated along bounding strike-slip faults with respect to deposits that were deformed by these structures. The deposits interpreted to have been affected by stepover tectonics

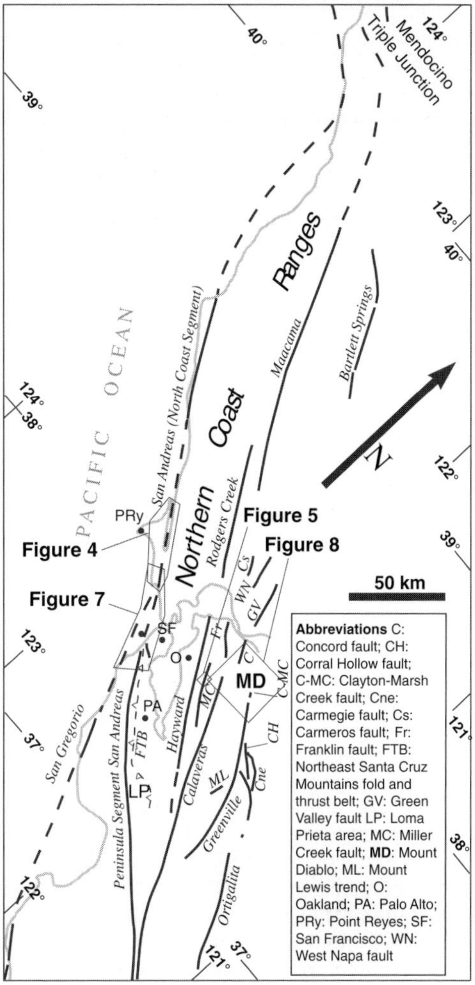

Fig. 3. The northern San Andreas Fault system, showing major dextral strike-slip faults and other features discussed in the text. Note that the Northeast Santa Cruz Mountains fold and thrust belt (FTB) is not the only fold and thrust belt in this area, but this specific belt is shown because it is specifically discussed in the text. Adapted from Wakabayashi (1999).

that now lie outside of an active stepover area are referred to as the 'wake'.

Migrating releasing bends

Because transtensional tectonics within the NSAFS are a local consequence of releasing stepovers and bends in an otherwise transpressional setting (for the last *c.* 8 Ma, Atwater & Stock 1998; Argus & Gordon 2001), the migration of such a stepover away from pull-apart basin deposits results in the subsequent uplift and shortening (locally driven inversion) of such deposits. Below I will describe pull-apart basins that may have migrated with respect to their deposits at three scales: tens of kilometres, kilometres, and meters (but not presented in that order).

Tomales Bay depocentre and Olema Creek Formation

The Olema Creek Formation (OCF), composed of loosely consolidated muds, sands, silts, and gravels, is about 110–185 ka old and crops out in a 3.5-km-long by 0.5-km-wide belt south of Tomales Bay, bounded to the west by the active San Andreas Fault and to the east by the 'eastern boundary fault', a strand of the San Andreas Fault system that has not been active in Holocene times (Grove *et al.* 1995; Grove & Niemi 2005; Fig. 4). The San Andreas Fault in this area is north of the junction of the San Gregorio and San Andreas Faults, and consequently has the combined displacement of the two strands, which is about 210 km (22–36 km for the San Andreas, 180 km for the San Gregorio; Wakabayashi 1999). Although the San Andreas Fault south of the San Gregorio–San Andreas Fault intersection has only been active for 1.5 to 2 Ma, the San Gregorio Fault may have been active since the inception of the San Andreas fault system at about 18 Ma. Thus the San Andreas Fault in the vicinity of the OCF may be about 18 Ma. old. This section of the San Andreas Fault separates Franciscan Complex basement on the east from Salinian basement on the west.

The OCF was deposited in estuarine, deltaic and fluvial environments similar to the head of the modern Tomales Bay and the coastal flat associated with its associated feeder streams, but is now exposed at elevations up to 70 m above sea-level, and deformed with beds tilted to dips of up to 65° (Grove *et al.* 1995; Grove & Niemi 2005). The steeper dips (up to 65°) are associated with the southernmost exposures of the OCF, whereas the northernmost exposures have dips of 5–10° (Grove & Niemi 2005). The OCF regionally dips northward ('shingles') at shallow angles, so that the deposits young northwestward (Grove *et al.* 1995; Grove & Niemi 2005; Fig. 4 inset). Because there is no evidence that a regional change in plate motions took place after 100 ka, the change in tectonic regime that affected the OCF must have been a local one. Moreover, the northward shingling of the OCF, and the progressively greater deformation toward the southern limit of exposure, also reflect a local rather than regional tectonic inversion, as well as a process that has migrated northward along the strike of the San Andreas Fault. The slip rate of the San Andreas Fault since the initial deposition of the Olema Creek Formation has been estimated as about 25 mm/a (Niemi & Hall 1992; Grove & Niemi 2005), so about 4.8 km of dextral slip has accumulated on the San Andreas Fault during that time. The subsiding environment along the San Andreas Fault, an environment in which the OCF was deposited, may have been related to a releasing bend or stepover, possibly between the San Andreas Fault and the eastern boundary fault. It may not have (or may not be) a true pull-apart, but may instead be a transtensional basin between the San Andreas and the eastern boundary fault or equivalent bounding structure. The subsiding area may have migrated northward to Tomales Bay, leaving a wake of deposits outside of the active depocentre and subject to shortening (Grove *et al.* 1995). The San Andreas Fault appears to step right within Tomales Bay, based on the geometry of the shoreline and the position of the San Andreas Fault north of Tomales Bay (Fig. 4). A right step may be present in northern Tomales Bay, corresponding with the deepest part of the bay (where the −36-foot contours are shown on Fig. 4). The topographic valley occupied by southern Tomales Bay may be related to the above-mentioned stepover, or it may also be a result of generally transtensional reach of the fault in the southern part of Tomales Bay. The presence, along the eastern shore of Tomales Bay, of the emergent 125–155-ka Millerton Formation (Fig. 4), which includes marine and estuarine deposits, suggests that there is no direct connection between subsidence in the southern Tomales Bay and the probable right step in the northern part of the bay.

No transverse faults have been identified within the OCF. Fold axes oblique or at high angles to the San Andreas Fault have been mapped within the OCF exposures (Grove *et al.* 1995), but these are probably associated with the uplift and shortening of the deposits rather than being related to transverse structures formed during deposition. Because the present-day sea-level is a high stand (e.g. Vail *et al.* 1977; Keller & Pinter 1996, 194–196), the emergence of the OCF is a product of vertical tectonics, rather than a lowering of sea-level since deposition. As noted above, the vertical tectonics appear to have been local, and the locus of uplift, as well as the deposition that preceded it, appears to have migrated northward along the San Andreas Fault. The southern edge of the modern depocentre analogue of the OCF deposits appears to be 10 km NW of the southern limit of the exposed OCF. Consequently, the length of the wake associated with depositional environment of the OCF is estimated to be about 10 km (Fig. 4).

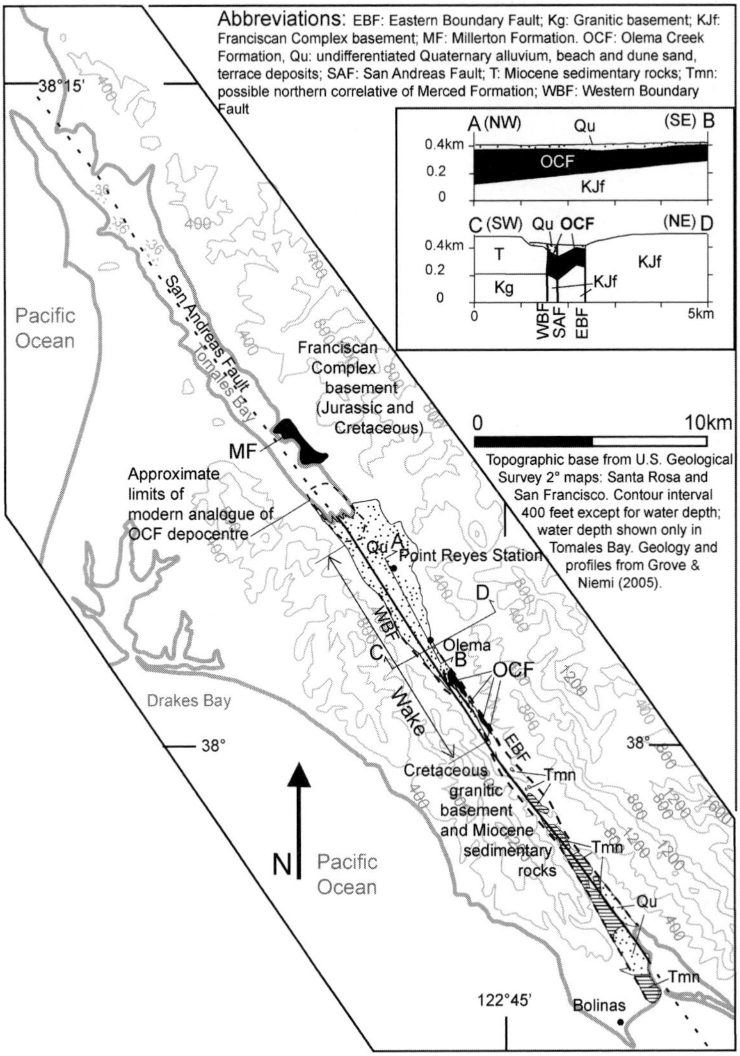

Fig. 4. Geology associated with the Olema Creek Formation and associated structures.

Inverted graben along the Miller Creek Fault

The Miller Creek Fault is a dextral-reverse fault between the Calaveras and Hayward faults that is presently moving at a low slip rate (probably a few tenths of a mm/a), but during earlier times it apparently represented a major strand of the SAFS (Wakabayashi & Sawyer 1998*a*; Wakabayashi 1999; Figs 3 & 5). The Miller Creek Fault may have initiated movement 10–12 Ma ago and has a cumulative dextral displacement of 30–50 km (Wakabayashi 1999). Basement rocks are not exposed along the Miller Creek Fault because they are overlain by Miocene deposits, but the basement probably consists of Franciscan Complex, Great Valley Group, and related rocks (e.g. Page 1981). Although the Miller Creek Fault has been considered a reverse or thrust fault in the past (e.g. Wakabayashi *et al.* 1992), palaeoseismic and field evidence suggests that the fault has been a strike-slip fault during the Quaternary (Wakabayashi & Sawyer 1998*b*). The evidence supporting Quaternary strike-slip movement includes subhorizontal slickenlines on fault surfaces in a palaeoseismic trench, and the steep fault dip ascertained from the trace of the fault over topography (Fig. 5). A palaeoseismic trench across this fault at a ridgetop saddle revealed a graben, filled with

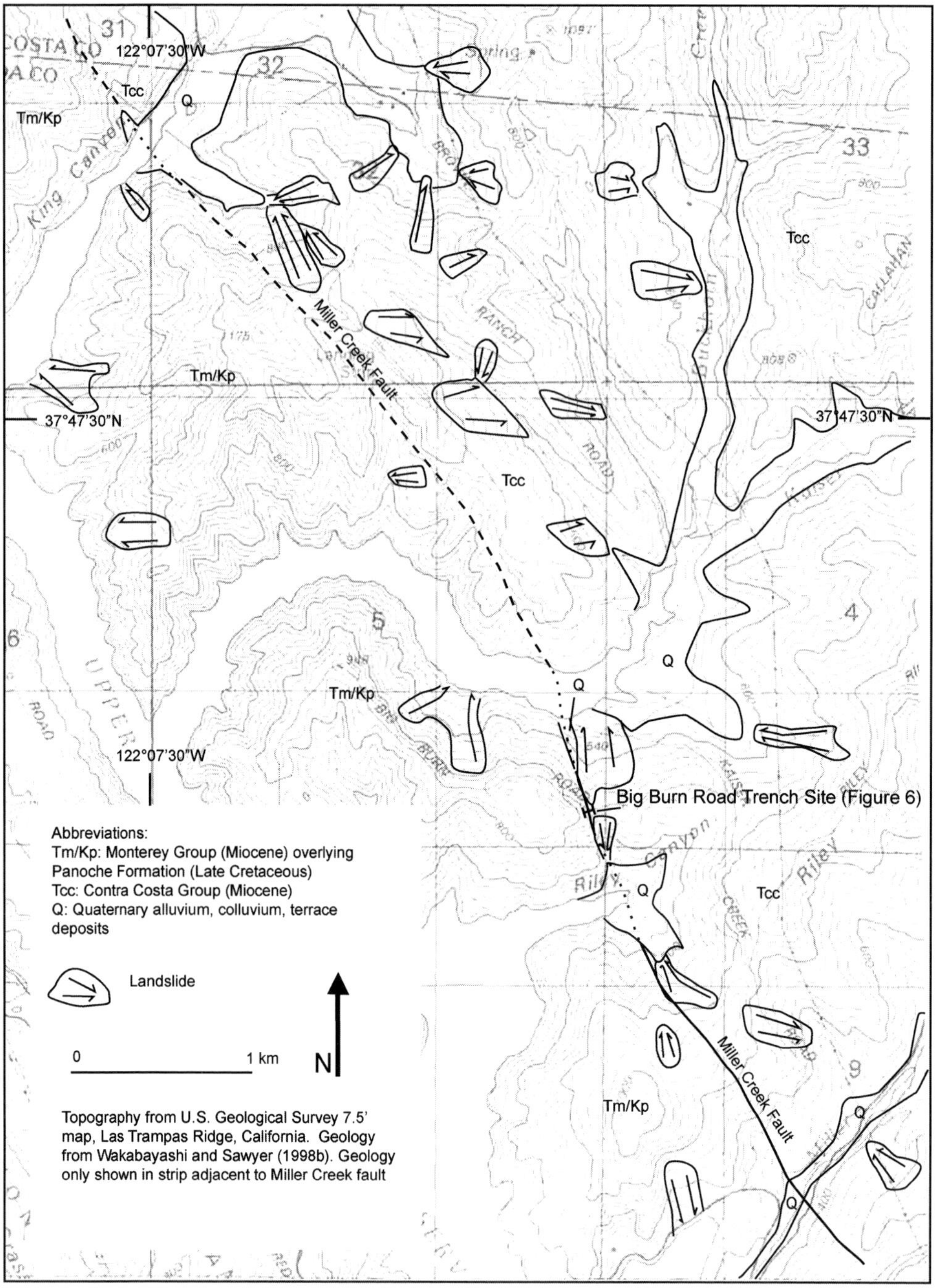

Fig. 5. The Miller Creek Fault in the vicinity of the Big Burn Road trench site shown in Figure 6.

Late Quaternary colluvium, that is bounded by Late Miocene bedrock (graben is bounded by faults F3 and F2 in Fig. 6; Wakabayashi & Sawyer, 1998*a*, *b*). The apparent separation of the eastern graben-bounding faults (fault F3) near the ground surface (i.e. reflecting the most recent movement) is reverse, rather than normal (Fig. 6), indicating a reversal of separation sense (inversion) in the

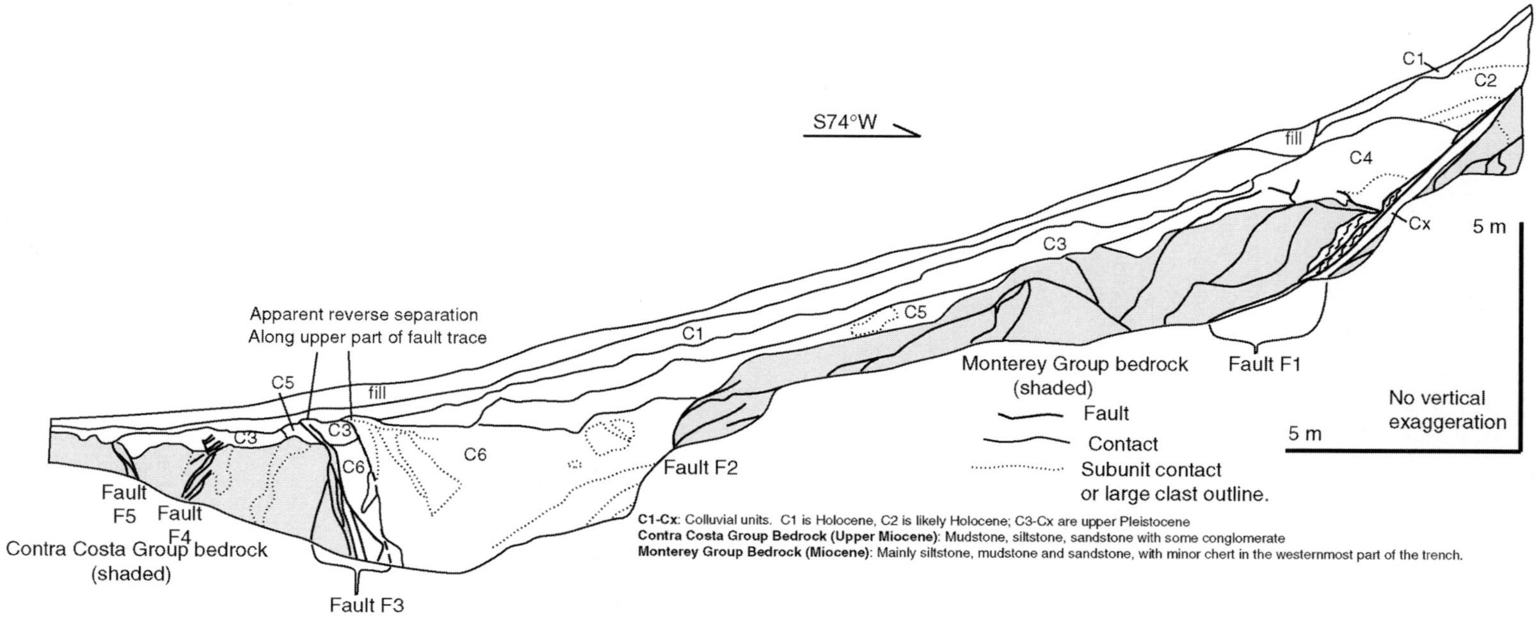

Fig. 6. Trench log of the Miller Creek Fault at the Big Burn Road trench site.

latest Pleistocene. In addition, the colluvium in the graben contains abundant Claremont chert – a lithology not present directly upslope from the saddle. This can reflect either: (1) considerable erosion of the ridge, removing upslope Claremont chert since deposition of the colluvium in the graben, or (more likely) (2) strike-slip movement of the graben material relative to the western side of the graben since deposition, moving the graben away from a position wherein Claremont chert was upslope. A possible modern analogue of the graben has not been found north or south of the trench site, probably because the trace of the fault away from the ridgetop saddle is covered by young landslides or deep colluvium (Fig. 5). Such landslides and colluvium would accumulate at a rate far in excess of the likely tectonic subsidence associated with a several-metre-scale pull-apart along the fault; slip rates on the fault have been estimated at a few tenths of a millitre per year at most (Wakabayashi & Sawyer 1998*b*). The local inversion noted in the trench suggests that the graben deposits may be a small-scale analogue of the Olema Creek Formation, but the lack of an identified 'wake' or modern depocentre, precludes a direct comparison.

Merced/San Gregorio depocentre and the Merced Formation

The Plio-Pleistocene Merced Formation, consisting of lightly cemented marine sandstones and siltstones, crops out in a belt ≤2.5 km wide, extending 19 km along the east side of the San Andreas Fault (Brabb & Pampeyan 1983; Fig. 7). Right separation of basement and Late Cenozoic features along the San Andreas Fault on the San Francisco Peninsula (Peninsula San Andreas Fault) is 22 to 36 km, and if the long-term slip rate is similar to the Holocene rate, then the fault did not begin moving until 2 Ma or later (Wakabayashi 1999). The basement cut by the San Andreas Fault along this reach is composed of Franciscan Complex rocks.

The Merced Formation records basinal deposition, and subsequent uplift and shortening along the San Andreas Fault. The relationship between the Merced Formation and its proposed analogue depocentre west of the Golden Gate is somewhat more complicated than the Olema Creek Formation example, because the intersection of the San Gregorio Fault occurs offshore and may influence tectonic subsidence and associated deposition (Bruns *et al.* 2002).

There is some controversy concerning the age of the basal Merced Formation, but it is probably 2 Ma or younger, and the youngest Merced may be 500 ka or younger (Wakabayashi *et al.* 2004). The Merced Formation and the overlying basal part of the Colma Formation are interpreted as sediments that were deposited in trangressive–regressive cycles related to eustatic sea-level fluctuations within a subsiding basin, with deposition keeping pace with tectonic subsidence (Clifton *et al.* 1988).

The Merced Formation is now exposed at elevations up to 200 m on the San Francisco Peninsula, with dips as steep as vertical (Bonilla 1971; Brabb & Pampeyan 1983). No transverse faults have been identified within the Merced Formation. No regional evidence exists for a <500 ka regional change in plate motions, and neither is there evidence for shortening initiating throughout this region at <500 ka, so the inversion of the Merced Formation appears to be a local phenomenon.

Deposition of the Merced Formation appears to be related to a stepover along the San Andreas Fault. Slip on the San Andreas Fault steps 3 km right to the Golden Gate Fault, forming an active pull-apart basin off the Golden Gate (Bruns *et al.* 2002; Fig. 7). Although there is clearly a net right stepover of the San Andreas Fault from its reach on the San Francisco Peninsula to where it comes on land north of the Golden Gate, the overall pattern of fault-slip transfer is more complex. If the active pull-apart is a product of slip transfer from the San Andreas Fault to the Golden Gate Fault, then slip must then step left again to the north on to the on-land San Andreas Fault north of the Golden Gate (Fig. 7). Indeed, Geist & Zoback (2002) suggest that the tsunami associated with 1906 earthquake along this reach of the San Andreas Fault was a product of vertical seafloor movement along the releasing stepover restraining stepover pair noted above. The Golden Gate Fault appears to be the offshore continuation of the San Bruno fault. Zoback *et al.* (1999) showed a basin in the right-step region, based on seismic reflection and gravity studies, and identified extensional focal mechanisms in the area. Seismic and potential field data suggest that the San Andreas Fault directly south of the active pull-apart has little separation, little evidence for recent activity, and did not move until the Late Quaternary, whereas the Golden Gate Fault has much greater cumulative separation (Jachens & Zoback 1999; Zoback *et al.* 1999; Bruns *et al.* 2002). The onshore, southern continuation of the Golden Gate Fault, the San Bruno Fault, is no longer active (Hengesh & Wakabayashi 1995*a*; McGarr *et al.* 1997). One interpretation of the activity history of the two faults is that the offshore San Andreas Fault south of the pull-apart is very young, having recently propagated northward to that position, whereas the southern part of the Golden Gate Fault is becoming dormant as the San Andreas propagates (Wakabayashi *et al.* 2004); however, an alternative

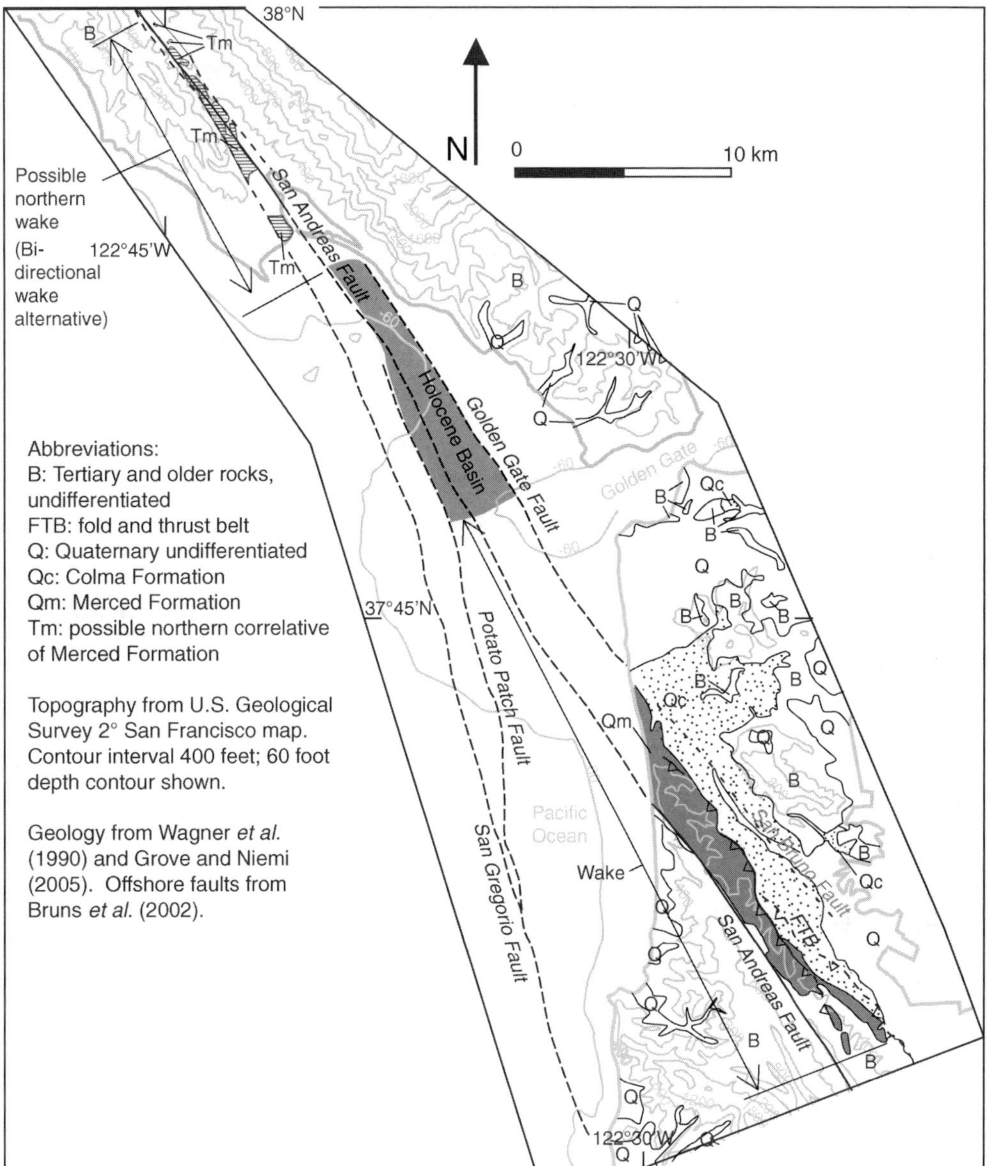

Fig. 7. Geology of the Merced Formation and related structures.

explanation is evaluated in the discussion section of this paper.

No fault scarp breccias, analogous to those found in some strike-slip bains (Crowell 1982), have been identified within the Merced Formation, possibly owing to high rates of sedimentation that kept pace with vertical tectonics in the depocentre (Wakabayashi *et al.* 2004). This is consistent with the bathymetry that appears to show a lobe of deposition west of the Golden Gate, but no expression of the Holocene pull-apart (Fig. 7).

The southern limit of the exposed Merced Formation on the San Francisco Peninsula is presently about 27 km south of the southern margin of the Holocene basin, similar to the total offset on the Peninsula San Andreas Fault. This would be defined as the wake length associated with this stepover (Fig. 7). The along-strike length of the modern depocentre is up to 15 km. These relationships suggest that the active pull-apart migrated with respect to its wake (the Merced Formation) at about the same rate as the slip rate on the Peninsula San Andreas Fault. In other words, the pull-apart migrated with the Pacific plate side of the fault. Wakabayashi *et al.* (2004) suggested that the migration of the stepover was accommodated by the San Andreas Fault propagating northward, whilst the Golden Gate–San Bruno Fault progressively shut off activity. This migration process will be evaluated further in the discussion section of this paper.

Deposits similar to the Merced Formation are present on the Marin Peninsula, north of the Golden Gate (Tm on Fig. 7). However these deposits appear to be older (Clark *et al.* 1984) and they may be related to a basin other than that in which the Merced Formation was formed. Alternatively, they may be related to the same depocentre, having moved north of the depocentre as a consequence of processes such as a change in the position of the San Gregorio Fault to a more easterly location, displacing basinal deposits northward relative to the depocentre at the slip rate of the San Gregorio Fault (Hengesh & Wakabayashi 1995*b*), or they may be part of a northern wake that is a consequence of northward migration of the stepover at less than the slip rate of the San Andreas Fault (the mechanism will be evaluated in the discussion section).

Migrating restraining bends

Lack of evidence for large structural relief associated with most restraining bends and steps

The northern SAFS has several left (restraining) stepovers, slip transfers, or bends which have accommodated tens of kilometres of slip (Wakabayashi 1999). Examples include the Hayward–Calaveras slip transfer zone; the northern termination of the Green Valley Fault; and the northernmost part of the SAFS in the region of the Mendocino triple junction (Fig. 3). If the same family of transverse structures accommodated all of the slip during the evolution of the restraining bend region, then considerable structural relief should result, associated with significant shortening, exhumation, and rock uplift. Using an assumption based on material movement through a stepover region that remains parallel to the PDZ, Wakabayashi *et al.* (2004), suggested that the predicted amount of rock uplift would be 6 km and more if the same transverse structures had remained active during the history of the Mt Diablo restraining stepover, the Calaveras–Hayward restraining slip transfer zone, and the connection between the eastern SAFS faults to the Mendocino triple junction. Late Cenozoic exhumation exceeding 2–3 km should be associated with Late Cenozoic apatite fission-track ages, based on the estimated geothermal gradients along the NSAFS, but restraining stepovers are not associated with such young apatite fission-track ages, with the exception of the Loma Prieta region (Wakabayashi *et al.* 2004).

The lack of structural or thermochronological evidence for large amounts of structural relief associated with most restraining stepovers in the NSAFS suggests that the restraining bends and stepovers may have migrated with respect to rocks originally deformed in these areas – analogous to the migrating releasing stepovers discussed above. Some examples are described below. In contrast to migrating releasing stepovers that leave basinal deposits behind them, the wakes (material formerly affected by the stepover) of migrating restraining stepovers/bends record deformation, uplift, exhumation and erosion.

Mount Diablo restraining stepover

The Mount Diablo restraining left stepover lies between the Greenville Fault to the south and the Concord Fault to the north (Unruh & Sawyer 1995, 1997), i.e. along the easternmost active dextral faults of the SAFS in the San Francisco Bay area (Figs 3 & 8). Mount Diablo is the most prominent topographic landmark in the San Francisco Bay region, with an elevation more than 500 m greater than the highest ridges outside of the stepover area, and it is underlain by an active fold-and-thrust belt called the Mount Diablo fold-and-thrust belt (MFTB), whose strike is oblique to that of the Concord and Greenville faults (Fig. 8). Comparable slip rates on the PDZs and the MFTB are consistent with the evolution of the fold and thrust belt within a restraining stepover as originally proposed by Unruh & Sawyer (1995; data summarized in Wakabayashi *et al.* 2004). The total Late Cenozoic dextral slip that has transferred through the Mount Diablo stepover is at least 18 km, the displacement on this reach of the Greenville fault (Wakabayashi 1999; Fig. 3). The age of initiation of the Concord and Greenville faults is not well constrained. If the average slip rates for the faults are assumed to equal the Holocene slip rate, then the faults would be about 6 Ma old, but there is evidence that the eastern faults of the San Andreas fault system had much higher slip rates prior to about 2 Ma (Wakabayashi 1999). The Greenville and Concord faults cut Franciscan basement.

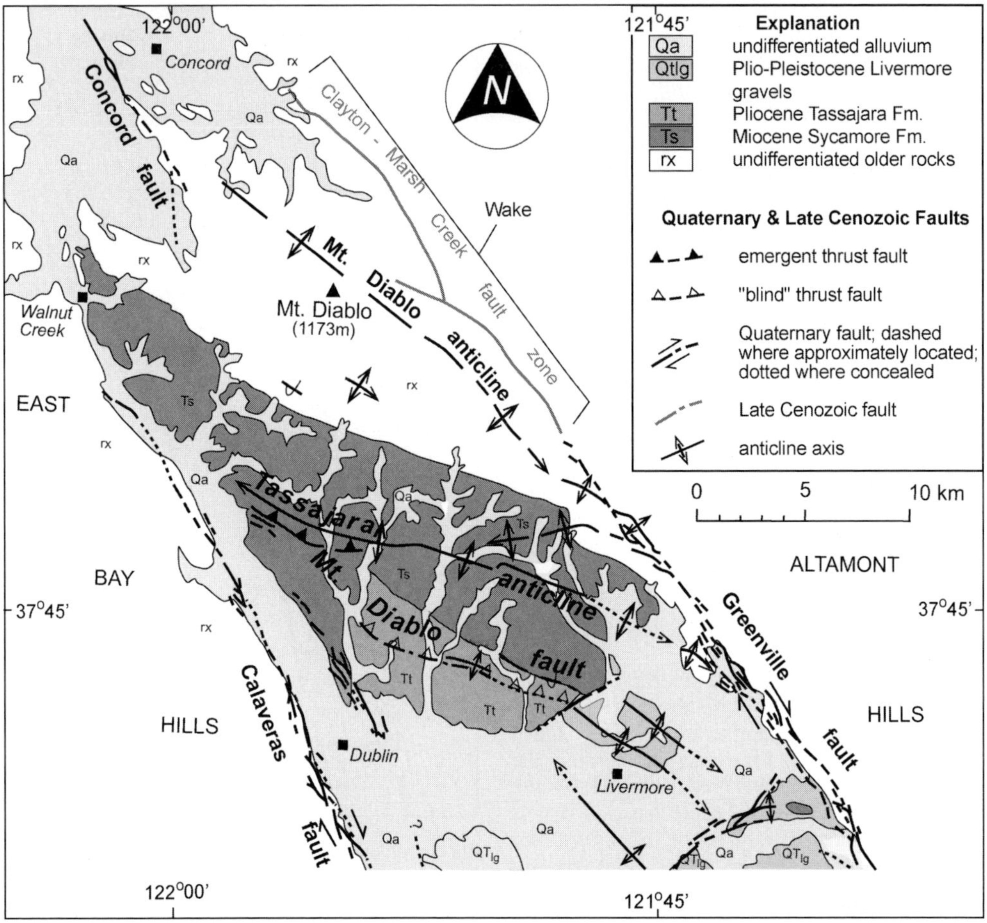

Fig. 8. Geology of the Mt Diablo restraining stepover area. Modified from Wakabayashi *et al.* (2004).

Wakabayashi *et al.* (2004) estimated that the stepover geometry and the amount of slip accommodated on the PDZs should have resulted in at least 6 km of rock uplift at Mt Diablo if the stepover did not migrate with respect to the affected deposits. In contrast, apatite fission-track ages, that predate the transform regime (T. A. Dumitru data in Unruh 2000), indicate exhumation of less than 2–3 km. The northern end of the eastern PDZ, the Greenville–Clayton–Marsh Creek Fault, has not been active in latest Quaternary times, whereas the southernmost part of the Concord Fault exhibits characteristics of an immature (young) fault (Wakabayashi *et al.* 2004). These observations and field data indicating southward (foreland) propagation of the MFTB suggest that the stepover has migrated southward with respect to affected deposits. The 'wake' of this stepover may be 15 km long or longer, based on the inactive length of the Clayton–Marsh Creek fault system.

Transfer of slip from eastern faults of the San Andreas Fault system to the Mendocino triple junction

The largest-scale restraining bend or stepover in the NSAF may occur near the northern terminus of the system at the Mendocino triple junction (Figs 3 & 9). San Andreas-age dextral faults are not present north of the Mendocino triple junction (Kelsey & Carver 1988). In the northernmost SAFS, 230–250 km of dextral slip, the aggregate amount of displacement of faults east of the San Andreas Fault (eastern faults), must transfer westward to the Mendocino triple junction, otherwise there would be enormous slip incompatibilities along the eastern faults, with zero displacement at their northern tips and a large displacements to the south (Wakabayashi 1999). Transfer of slip from the eastern faults to the Mendocino triple junction is a restraining slip transfer or stepover. For any range of transverse structure dips and strikes, this stepover geometry predicts an unrealistic amount of rock uplift for which there is no evidence (tens of kilometres, Wakabayashi *et al.* 2004). The actual geometry of transverse structures in the triple-junction region may be complex. Some strike-slip faults south of the triple junction change into thrust faults north of the triple junction (Kelsey & Carver 1988). The uplift rates are higher in the triple-junction region than to the north along the subducting margin or to the south along the transform boundary (Merritts & Bull 1989), and this uplift does not appear to be limited to that directly associated with reverse movement along the structures noted above. Tens of kilometres south of the triple junction, there are discrete high-angle faults that strike more westerly than the dextral strands of the San Andreas Fault system, and cut Franciscan nappe structures and later out-of-sequence thrust faults that imbricate the Franciscan nappes. Late Cenozoic deposits are not present along or across these structures, so it is difficult to verify whether these are palaeo-transverse structures as suggested by Wakabayashi (1999) and Wakabayashi *et al.* (2004).

The eastern faults, such as the Hayward–Rodgers Creek–Maacama trend, and the Green Valley Fault, die out northward as well-defined faults (Figs 3 & 9). This may be because the eastern faults are young and propagating northward, whilst slip transfers to the triple junction that is migrating at about 25 mm/a NW relative to the westernmost of the eastern faults (Wakabayashi 1999; Wakabayashi *et al.* 2004). In order to transfer slip to the migrating triple junction, new transverse faults must continue to form (Fig. 9). Thus, the stepover region progressively migrates so that large-scale displacement or structural relief has not developed on any given transverse structure. The northernmost SAFS may be another example of a restraining stepover that has migrated with respect to the rocks deformed within the stepover. It is difficult to estimate the length of the wake, because the boundaries of the active stepover area are not well defined, and because the nature of Late Cenozoic deformation in the region SE of the triple junction has not been determined, owing to the lack of Late Cenozoic deposits. The 'wake' of the stepover region may be 200–250 km long, based on the presence of possible old transverse structures cutting basement as far as south as the inboard 10 Ma contour in Figure 9; the lack of such structures in the San Francisco Bay area; and the temporal–spatial distribution of slip on the eastern fault strands (Wakabayashi 1999). This wake corresponds only with the activity history of the eastern faults, not the older Mendocino triple junction itself.

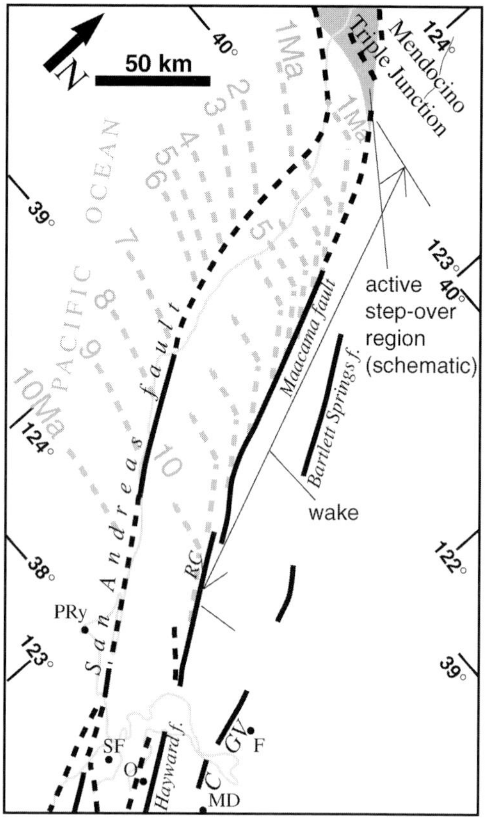

Fig. 9. Migration of the restraining transfer zone to the Mendocino triple junction. Approximate past positions of the triple junction (west of the the San Andreas Fault), and transverse structures (east of the San Andreas Fault; this is the wake) shown in greyed dashed lines with corresponding age designations. Longitude/latitude and other reference points are valid only for present, given that they would differ upon restoration to the different time frames. Abbreviations, in addition to those given in Figure 3: F, Fairfield; RC, Rodgers Creek fault; TV, Tolay volcanics. Adapted from Wakabayashi (1999).

Discussion: models for migrating stepovers

Migrating stepovers: speculation on modes of migration

I have interpreted the above stepover and bend regions as having migrated with respect to deposits that were originally within the stepover or bend regions. Although alternatives such as regional plate-motion changes do not appear to explain the field relations (for examples involving basin deposits), other types of processes require further explanation. Block rotation may account for local inversion, but it should not leave a wake of deformation or inverted deposits parallel to a PDZ. Because the Miller Creek Fault example does not have an identified wake or modern analogue depocentre, block rotation is an alternative mechanism to explain the inversion of Late Pleistocene graben deposits there. Another alternative that may explain some of the types of field relations described may be reorganization of an evolving transform fault. By creating new faults, removing bends and irregularities, and other changes in distribution of slip, it may be possible to locally invert deposits, as material may shift from a locally transtensional mode to a transpressional one, or vice versa. Similarly to the block rotation mechanism, however, the rearrangement of an evolving transform fault should not create wakes on one or both sides of a stepover region along the strike of a fault.

Below, I present a simple model for migrating stepover evolution. It is probably obvious that this model is vastly simpler than real stepovers. For simplicity, I have used PDZs that do not change through time other than propagating in some cases; two transverse structures bounding releasing stepovers (for simple rhombochasm-type pull-apart geometry); and one transverse structure for restraining bends (it is simply the bent PDZ; other off-PDZ structures are not explicitly defined). It is hoped that this cartoon simplification does not overly simplify the problem to the point that the model fails to address geological reality.

For stepovers to have migrated with respect to affected deposits as I have proposed above, new transverse structures associated with the bend or stepover regions must have progressively formed in the direction of the migration (Fig. 10b frame sequence B-1 to B-2b or B-2c; Fig. 10c frame sequence C-1 to C-2b). If the same transverse structures migrated instead of new structures forming, the material bounded by them would have migrated with them and, if the structures were still active, the stepover or bend would have grown in structural relief (Fig. 10b frame sequence B-1 to B-2a; Fig. 10c frame sequence C-1 to C-2a). A migrating stepover probably has an early stage of development in which a stepover region grows to a certain size (Fig. 10a) before migration begins with respect to formerly affected deposits, in contrast to progressively growing stepover regions that maintain the same set of transverse structures and continue to grow (Fig. 10b & 10c). The nature of fault propagation and migration of restraining bends and releasing bends have important geometric differences, and so they are discussed separately below.

In addition to the creation of new transverse structures and the shut-off of old ones, migration of releasing bends suggests that the strike-slip fault on one side of the pull-apart may propagate in the direction of basin migration (PDZ1 in Fig. 10b, frame B-2b), whereas the fault on the other side of the basin will progressively shut off (PDZ2 in Fig. 10b, frame B-2b). The propagation and shut-off of bounding strike-slip faults is consistent with the seismic reflection interpretation of Bruns *et al.* (2002) for the relationship of the Merced Formation to offshore pull-apart structures. In some (or many) cases, transverse structures may not be discrete faults, but may be zones of distributed deformation (perhaps above blind normal or oblique faults) accommodating the differential displacement between reaches of PDZs with differing senses of movement (for example, pure strike-slip v. normal-oblique). On the other hand, it is clear from examples presented here and from global examples of releasing bends that actual basin geometry is usually vastly more complex than the schematic model pull-apart shown in Figure 9b, so other types of kinematic links may be associated with the observed fault separations.

The mechanism illustrated in Figure 10b, frame B-2b, predicts that the tip of the trailing PDZ moves faster than the slip rate of the fault relative to Block b, meaning that the overall migration rate of the pull-apart is also faster than the slip rate, and the tip of the fault propagates into the Block a. The seismic reflection data of Bruns *et al.* (2002) show evidence for recent offset, but little cumulative vertical separation, along the northernmost Peninsula San Andreas Fault that forms the western boundary of the Merced Formation and either the western boundary or one of the western faults bounding the analogue Holocene pull-apart basin (Fig. 7). Wakabayashi *et al.* (2004) interpreted these data to indicate that the northernmost Peninsula San Andreas Fault was young as a result of propagation of the tip of the fault. Alternatively the seismic evidence may suggest that the northernmost Peninsula San Andreas is long-lived but exhibits little cumulative displacement because most of the slip has transferred to the eastern PDZ by that point. For the Olema Creek Formation, the depocentre (associated with a transtensional reach along the San Andreas Fault or pull-apart) may have migrated 10 km relative to its former deposits, whereas if it migrated at the slip rate of the San Andreas Fault, then the migration distance would have been about 4.8 km.

In the case of the Merced Formation, the original releasing stepover structure may have been created at about the time of the creation of the Peninsula San Andreas Fault at about 2 Ma, based on the estimate for the age of the basal Merced Formation and the estimate for the age of inception of slip on the Peninsula San Andreas Fault from division of the slip amount by the Holocene slip rate. In contrast, the structure related to the deposition of the Olema Creek Formation may have formed at about 200 ka, based on the age of the oldest Olema Creek Formation deposits, on a fault that may have formed at 18 Ma. Thus, the two examples appear to represent migrating stepovers that formed at different times relative to the evolution of the strike-slip that they were/are associated with.

Figure 10b, frame B-2b, may illustrate only an end-member case of how releasing stepovers migrate. In the case illustrated, the stepover migrates at a rate slightly faster than the slip rate

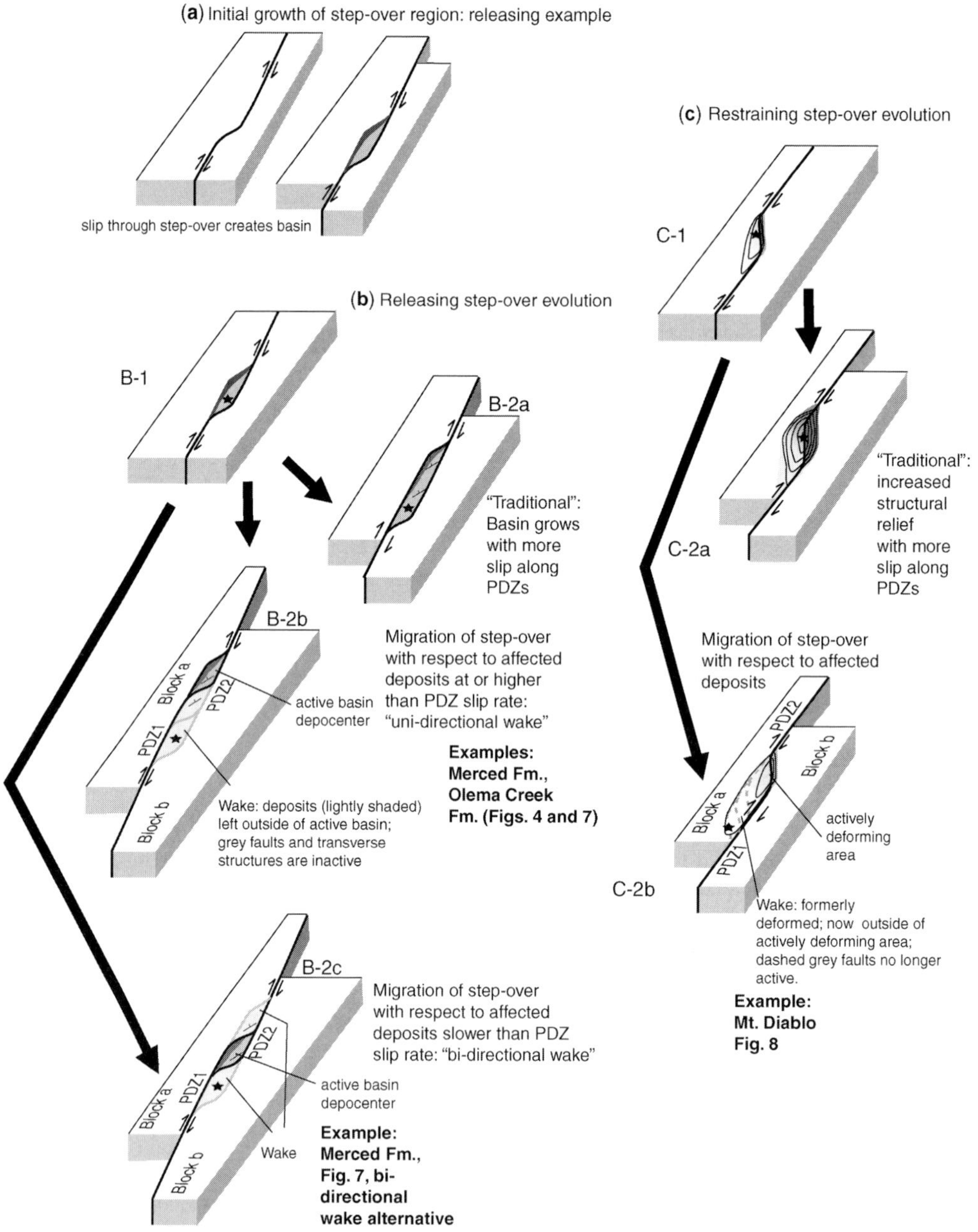

Fig. 10. Progressive evolution of releasing and restraining stepover and bend regions along strike-slip faults. The star in each diagram is a reference point that represents a point within deposits affected by the stepover.

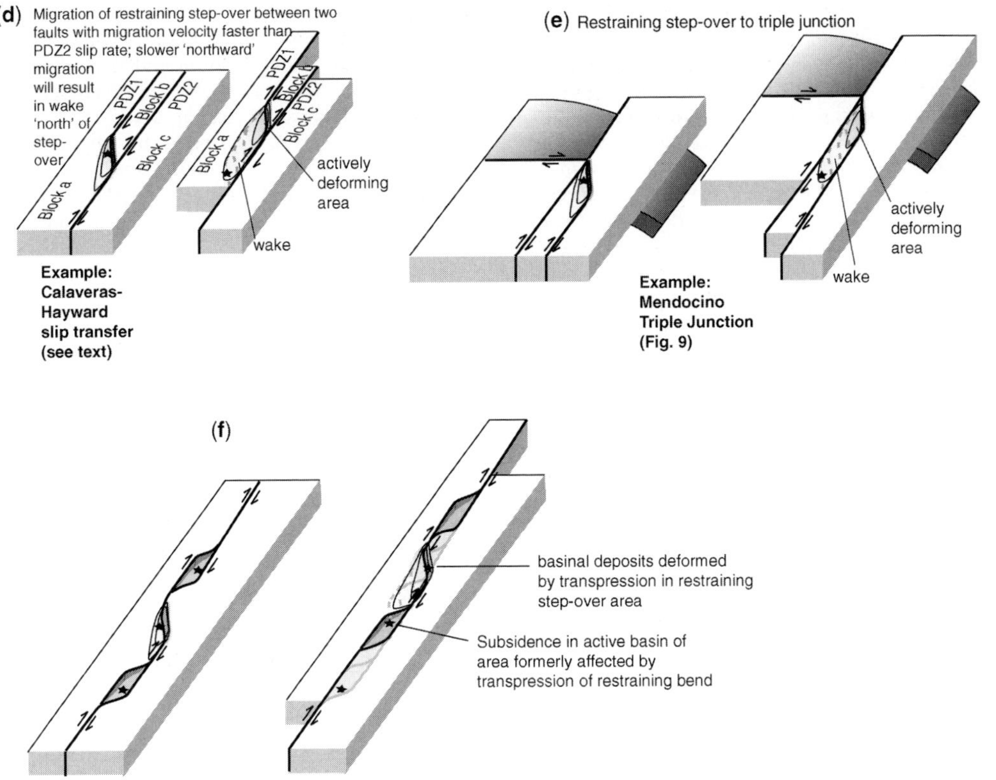
(d) Migration of restraining step-over between two faults with migration velocity faster than PDZ2 slip rate; slower 'northward' migration will result in wake 'north' of step-over
PDZ1
Block b
PDZ2
Block c
Block a
actively deforming area
wake
Example: Calaveras-Hayward slip transfer (see text)
(e) Restraining step-over to triple junction
actively deforming area
wake
Example: Mendocino Triple Junction (Fig. 9)
(f)
basinal deposits deformed by transpression in restraining step-over area
Subsidence in active basin of area formerly affected by transpression of restraining bend

Fig. 10. *Continued.*

of Block a relative to Block b (i.e. faster than the slip rate of the fault). This example was based on interpretations of the Merced and Olema Creek formations, but there is no reason to believe that the opposite case cannot occur in which a pull-apart migrates southeastward along a NW-striking dextral fault at or slightly faster than the rate of the eastern side of the fault relative to the western side. Such an example would leave a wake NW of the pull-apart basin. Intermediate examples would be those in which the pull-apart migration rate lags behind the fault slip rate. Such examples should have 'bi-directional' wakes, or wakes both NW and SE of the pull-apart basins (Fig. 10b, frame B-2c). The Merced-like strata in Marin County (Tm in Fig. 7) have been interpreted to be a part of the Merced Formation faulted northward by eastward stepping of the San Gregorio Fault (Hengesh & Wakabayashi 1995*b*) or older deposits unrelated to the Merced Formation (Wakabayashi *et al.* 2004). An alternative explanation is that the strata represent the northern wake of the migrating Merced Formation pull-apart, and that the pull-apart has a bi-directional wake. The uncertainty in the amount of PDZ slip (22 to 36 km) is large enough to be consistent with either the bi-directional or uni-directional wake alternatives.

In the case of the regionally transpressive San Andreas fault system, the wake of releasing step-overs is subject to shortening and uplift (positive inversion) as a consequence of the regionally transpressive regime. As noted by Wakabayashi *et al.* (2004), such inversion may occur along a neutral transform, or even transtensional regime, if a migrating restraining stepover follows the migrating releasing stepover and interacts with the wake of the former (Fig. 10f). In addition, inversion of releasing stepover wakes may occur in a neutral transform system depending on the geometry of crustal/lithospheric flow into the stepover region. This is because some crustal or lithospheric flow toward the pull-apart area probably occurs, otherwise a pull-apart will create a void that extends as deep as the depth of strike-slip movement.

For releasing stepovers, a model nearly identical to that illustrated in Fig. 10b B-2b (uni-directional wake with a propagating PDZ) was proposed for the Dead Sea pull-apart, arguably the world's most studied pull-apart basin, by ten Brink & Ben-Avraham (1989). More recently, Lazar *et al.* (2006) offered an alternative model for the Dead Sea basin, based on new subsurface geological and geophysical data, in which multiple sub-basins form and grow as strike slip progresses on en échelon segments of strike-slip faults. Lazar *et al.* (2006) propose that new transverse structures can progressively form as a result of slip on the PDZ displacing a bounding and converging strike-slip segment relative to the basins, rather than as a result of PDZ propagation. Although such a model may better explain the evolution of the Dead Sea basin than a migrating stepover model, it may be a bit more difficult to reconcile with the interpreted development of wakes proposed for the examples reviewed here.

Migrating restraining bends (Fig 10c, Frame sequence C-1 to C-2b) differ from the migrating releasing stepovers of Figure 10b (Frame sequence B-1 to B-2b) in that fault tips propagate opposite to the direction of the slip of the block that the fault is propagating into when considered relative to the block on the other side of the old PDZ (this will be simply described as 'propagating opposite to slip') rather than in the same direction (herein referred to as 'propagating ahead of slip'). Because of this, migrating restraining stepovers should always form uni-directional wakes, regardless of the migration velocity and direction. For cases involving a restraining stepover along a single strike-slip fault, fault propagation – coupled with progressive fault dormancy of the paired PDZ – must occur in order for the structure to migrate with respect to the affected deposits. Cases involving multiple faults, such as the Calaveras–Hayward Fault slip transfer zone and the Mendocino triple junction present different situations. For cases similar to that of the Calaveras–Hayward Fault slip transfer (not discussed in detail here, but discussed in Wakabayashi *et al.* 2004), illustrated in Figure 10d, propagation of the Hayward Fault (PDZ1 in Fig. 10d) will only occur if the slip-transfer region is migrating northwestward slower than the rate of the western side of the eastern fault north of the slip transfer (i.e. migrating slower than Block b in Fig. 10d). In the Calaveras–Hayward case, this would mean migrating northwesterly relative to the east side of the Calaveras Fault (Block c of Fig. 10d) slower than the 6 mm/a slip rate of the Calaveras Fault north of the slip transfer. In such a case, the Hayward Fault would propagate southeastward and the wake would form NW of the slip-transfer region. If the northwestward migration rate of the slip transfer zone is faster than Block b (Fig. 10d), then a wake should be formed SE of the migrating slip-transfer zone, and the southern part of the Hayward fault will progressively cease activity, as illustrated in the example of Figure 10d.

For the Mendocino triple-junction example (Fig. 9 & Fig. 10e), similar to the general restraining stepover example, the eastern faults of the San Andreas Fault propagate 'opposite' slip and the more inboard the fault, the faster the fault propagation rate, because the relative migration rate of the triple junction is the sum of the slip rates of all faults to the west.

Migrating stepovers and progressively growing stepovers as end members of general stepover types

As noted by Wakabayashi *et al.* (2004), both migrating and 'traditional' (progressively increasing structural relief) types of stepovers appear to be common worldwide. It is reasonable to believe that many stepovers may exhibit combinations of these characteristics. In other words, many stepovers may exhibit evidence of progressive increases in structural relief, but also have wakes of material formerly – but no longer – affected by the stepover. Migrating stepovers as described herein may represent one end member of stepover evolution, and stepovers that progressively grow in structural relief represent another end member. I suggest that we might attempt to classify the type of stepover evolution on the basis of how much of the slip along the PDZs has been accommodated by structural relief development v. wake development. If the wake length is greater than or equal to the estimated PDZ slip through the stepover region, then the PDZ slip has been accommodated by wake development and stepover migration. Comparatively short wakes for large amounts of PDZ slip would suggest that most of the slip has been accommodated by structural relief development in the stepover region. A problem with such a classification is that there are probably many examples where stepovers are much younger than the PDZs that they form along (the Olema Creek Formation is one likely example). In those cases, it may be difficult to determine the amount of slip along the PDZs that has been associated with stepover evolution. Three examples reviewed above for which I can estimate wake lengths (the Merced Formation, Mount Diablo, and the Mendocino triple junction) appear to have wake lengths that are close to the estimated amount of slip that has passed through the stepover region (Table 1). Thus, those three examples would be close to the 'migrating' end member among the larger family of stepovers. A fourth example, the Olema Creek Formation, with a PDZ displacement of 4.8 km and a wake length of 10 km, may represent a case where a releasing

Table 1. *PDZ displacement through stepover v. wake length*

Stepover	PDZ displacement (km)	Wake length (km)
Merced Formation	22–36	27 (uni-directional) 39 (bi-directional)
Olema Creek Formation	4.8	10
Mount Diablo	18	15
Eastern faults of San Andreas fault system: connection to Mendocino triple junction	230–250	200–250

bend migrated at a velocity greatly in excess of the PDZ slip rate. It too would be classed as a purely migrating type of stepover.

The Loma Prieta restraining bend may exhibit characteristics between the 'traditional' and migrating end-members of stepover/bend type. The exhumation associated with the Loma Prieta bend is large enough that Late Cenozoic apatite fission-track ages have been obtained from the area (Bürgmann *et al.* 1994), indicating much more significant structural relief than that associated with other restraining stepovers or bends along the NSAFS. The restraining bend may be associated with the fold and thrust belt that marks the eastern range-front of the San Francisco Peninsula, east of the San Andreas Fault (FTB in Figs 3 & 7). The northern limit of the fold and thrust belt is the northern limit of the Merced Formation exposures described above. The belt thins northward and the deformation at the northernmost limit of the belt appears to have begun within the last few hundred thousand years or so, whereas the belt shows evidence of activity for at least 2 Ma to the south (Kennedy 2002). Thus, it appears that the fold and thrust deformation is propagating northwestward along the San Andreas Fault. This propagation may be a consequence of progressive increase in structural relief of the Loma Prieta restraining bend area, or it may result from the northward migration of the Loma Prieta bend relative to affected deposits. The detailed evaluation of the spatial distribution and kinematics of aftershocks of the 1989 Loma Prieta earthquake by Twiss & Unruh (2007) may provide evidence for migration of the Loma Prieta stepover. Their analysis shows that the aftershocks of the 1989 earthquake define a blind dextral fault with a reverse component that forms several en échelon segments. This is suggestive of a young structure. Although the creation of a new master fault in the stepover region may be interpreted as the creation of a new transverse structure in a migrating restraining bend, it may simply be a consequence of rearrangement of faults in the stepover region rather than a product of systematic and progressive migration of restraining-bend deformation.

Migrating stepovers: complications and proposed field tests of mechanisms

I have reviewed field evidence for migrating stepovers and proposed models for their evolution. The field examples presented are certainly not perfect. Only the Mount Diablo example appears to have a well-defined active stepover region with active transverse structures, as well as evidence for propagating and inactive PDZs and possible older transverse structures. The Mendocino triple-junction example appears to have evidence for an active stepover region and propagating faults, but older transverse structures have yet to be confirmed. No transverse structures have been identified for any of the releasing bend examples, and neither is there sufficient geochronological data to indicate a younging of the age of the basal strata of the exposed deposits in the direction of migration, as would be predicted by the model.

To further test these models, additional geochronological data will be useful to see whether there is along-strike younging of the base of basinal deposits making up wakes of proposed migrating releasing bends. Field investigation should be able to determine whether progressively younger transverse structures (blind or otherwise) occur in the direction of migration. For restraining stepovers, analysis of uplift rates (by analysis of stream terrace deposits, for example) should show high recent uplift rates in the active stepover region, with slower rates in the past, whereas the wake region should show lower recent uplift rates but a period of fast uplift rates in the past. It may be possible to track a progressively migrating pulse of rapid uplift by examining a series of transverse drainages. This paper has examined only the most surficial aspect of stepover evolution and has neglected the mechanisms for accommodating the stepover evolution at the mid-crustal and lower levels. The deeper evolution of such structures will require evaluation of detailed seismic and potential field data in conjunction with surface geology and geodetic data.

Conclusions

Migrating stepovers, as defined herein, migrate with respect to affected deposits. Wakes of formerly affected deposits allow recognition of migrating stepovers. Many stepovers exhibit characteristics associated with migration with respect to affected deposits, whereas many others show evidence of significant increase in structural relief with time. A migrating stepover that experienced minimal structural-relief increase over its evolution may represent one end member of stepovers, whereas a stepover that has progressively increased in structural relief and time and not developed a wake may represent another end member. It is likely that many stepovers exhibit characteristics of both end members, and comparative importance of migration v. structural relief growth may be gauged by the ratio between the wake length and PDZ slip accumulated during stepover development. High wake length to PDZ slip ratios are indicative of stepovers that have formed largely by migration with respect to their deposits. Speculation as to what processes or physical variables control the type of stepover evolution is beyond the scope of this paper. The models for stepover development presented here are rather simplistic compared with the complex geology seen at many stepover regions. Nonetheless, relatively simple field tests can be used to evaluate the validity of these models.

I have benefited from discussions on this subject with many colleagues, especially the participants at the Geological Society of London conference on tectonics of strike-slip releasing and restraining bends, and T. Bruns, U. S. ten Brink, M. L. Zoback, C. Busby, K. Grove and T. Atwater. I am grateful for detailed, constructive and thought-provoking reviews by P. Mann and M. Legg.

References

ARGUS, D. F. & GORDON, R. G. 1991. Current Sierra Nevada–North America motion from very long baseline interferometry: implications for the kinematics of the western United States. *Geology*, **19**, 1085–1088.

ARGUS, D. F. & GORDON, R. G. 2001. Present tectonic motion across the Coast Ranges and San Andreas fault system in central California. *Geological Society of America Bulletin*, **113**, 1580–1592.

ATWATER, T. & STOCK, J. 1998. Pacific–North America plate tectonics of the Neogene southwestern United States: an update. *International Geology Review*, **40**, 375–402.

AYDIN, A. & PAGE, B. M. 1984. Diverse Pliocene–Quaternary tectonics in a transform environment, San Francisco Bay region, California. *Geological Society of America Bulletin*, **95**, 1303–1317.

BEN-AVRAHAM, Z. & ZOBACK, M. D. 1992. Transform-normal extension and asymmetric basins: an alternative to pull-apart models. *Geology*, **20**, 423–426.

BONILLA, M. G. 1971. Preliminary geologic map of the San Francisco South quadrangle and part of the Hunters Point quadrangle. *US Geological Survey Miscellaneous Field Studies Map*, **MF-574**.

BRABB, E. E. & PAMPEYAN, E. H. 1983. Geologic map of San Mateo County, California. *US Geological Survey Miscellaneous Investigation Series Map*, **I-1257**.

BRUNS, T. R., COOPER, A. K., CARLSON, P. R. & MCCULLOCH, D. S. 2002. Structure of the submerged San Andreas and San Gregorio fault zones in the Gulf of the Farallones off San Francisco, CA, from high-resolution seismic reflection data. *In*: PARSONS, T. (ed.) *Crustal Structure of the Coastal and Marine San Francisco Bay Region*. US Geological Survey Professional Papers, **1658**, 79–119.

BÜRGMANN, R., ARROWSMITH, R., DUMITRU, T. & MCLAUGHLIN, R. 1994. Rise and fall of the southern Santa Cruz Mountains, California, from fission tracks, geomorphology and geodesy. *Journal of Geophysical Research*, **99**, 20 181–20 202.

CHRISTIE-BLICK, N. & BIDDLE, K. T. 1985. Deformation and basin formation along strike-slip faults. *In*: BIDDLE, K. T. & CHRISTIE-BLICK, N. (eds) *Strike-Slip Deformation, Basin Formation, and Sedimentation*. SEPM Special Publications, **37**, 1–34.

CLARK, J. C., BRABB, E. E., GREENE, H. G. & ROSS, D. C. 1984. Geology of Point Reyes peninsula and implications for San Gregorio fault history. *In*: CROUCH, J. K. & BACHMAN, S. B. (eds) *Tectonics and Sedimentation Along the California Margin. Pacific Section*, SEPM Special Publications, **38**, 67–86.

CLIFTON, H. E., HUNTER, R. E. & GARDNER, J. V. 1988. Analysis of eustatic, tectonic, and sedimentologic influences on transgressive and regressive cycles in the late Cenozoic Merced Formation, San Francisco, California. *In*: PAOLA, C. & KLEINSPEHN, K. L. (eds) *New Perspectives of Basin Analysis*. Springer Verlag, New York, 109–128.

CROWELL, J. C. 1974*a*. Sedimentation along the San Andreas Fault, California. *In:* DOTT, R. H. & SHAVER, R. H. (eds) *Modern and Ancient Geosynclinal Sedimentation*. SEPM Special Publications, **19**, 292–303.

CROWELL, J. C. 1974*b*. Origin of late Cenozoic basins of southern California. *In*: DICKINSON, W. R. (ed.) *Tectonics and Sedimentation*. SEPM Special Publications, **22**, 190–204.

CROWELL, J. C. 1982. The Violin Breccia, Ridge Basin, southern California. *In*: CROWELL, J. C. & LINK, M. H. (eds) *Geologic History of Ridge Basin, Southern*

California. Pacific Section, SEPM, Los Angeles, California, 89–98.

CROWELL, J. C. 1987. Late Cenozoic basins of onshore southern California: complexity is the hallmark of their tectonic history. *In*: INGERSOLL, R. V. & ERNST, W. G. (eds) *Cenozoic Basin Development of Coastal California*. Englewood Cliffs, New Jersey, Prentice-Hall, Rubey Volumes, **VI**, 206–241.

DOOLEY, T. & MCCLAY, K. 1997. Analog modeling of pull-apart basins. *AAPG Bulletin*, **81**, 1804–1826.

DOOLEY, T., MCCLAY, K. R. & BONORA, M. 1999. 4D Evolution of segmented strike-slip fault systems: applications to NW Europe, *In*: FLEET, A. J. & BOLDY, S. A. R. (eds) *Geological Society of London, Proceedings of the 5th Conference–Petroleum Geology of NW Europe*, 215–225.

GEIST, E. L. & ZOBACK, M. L. 2002. Examination of the tsunami generated by the 1906 San Francisco Mw = 7.8 earthquake, using new interpretations of the offshore San Andreas fault. *In*: PARSONS, T. (ed.) *Crustal Structure of the Coastal and Marine San Francisco Bay Region*. US Geological Survey Professional Papers, **1658**, 29–42.

GROVE, K. & NIEMI, T. M. 2005. Late Quaternary deformation and slip rates in the northern San Andreas fault zone at Olema Valley, Marin County, California. *Tectonophysics*, **401**, 231–250.

GROVE, K., COLSON, K., BINKIN, M., DULL, R. & GARRISON, C. 1995. Stratigraphy and structure of the late Pleistocene Olema Creek Formation, San Andreas fault zone north of San Francisco, California. *In*: SANGINES, E. M., ANDERSEN, D. W. & BUISING, A. W. (eds) *Recent Geologic Studies in the San Francisco Bay Area*. Pacific Section, SEPM, **76**, 55–76.

HENGESH, J. V. & WAKABAYASHI, J. 1995*a*. Quaternary deformation between Coyote Point and Lake Merced on the San Francisco Peninsula: Implications for evolution of the San Andreas fault. *Final Technical Report, US Geological Survey National Earthquake Hazards Reduction Program, Reston, Virginia*. Award No. **1434-94-G-2426**.

HENGESH, J. V. & WAKABAYASHI, J. 1995*b*. Dextral translation and progressive emergence of the Pleistocene Merced basin and implications for timing of initiation of the San Francisco Peninsula segment of the San Andreas fault. *In*: SANGINES, E. M., ANDERSEN, D. W. & BUISING, A. W. (eds) *Recent Geologic Studies in the San Francisco Bay Area*. Pacific Section, SEPM, **76**, 47–54.

JACHENS, R. C. & ZOBACK, M. L. 1999. The San Andreas Fault in the San Francisco Bay region, California: structure and kinematics of a young plate boundary. *International Geology Review*, **41**, 191–205.

KELLER, E. A. & PINTER, N. 1996. *Active Tectonics, Earthquakes, Uplift, and Landscape*. Prentice Hall, New Jersey, 338 pp.

KELSEY, H. M. & CARVER, G. A. 1988. Late Neogene and Quaternary tectonics associated with northward growth of the San Andreas transform fault, northern California. *Journal of Geophysical Research*, **93**, 4797–4819.

KENNEDY, D. G. 2002. Neotectonic character of the Serra Fault, northern San Francisco Peninsula, California. M.Sc. thesis, San Francisco State University, San Francisco, CA, 100 pp.

LAZAR, M., BEN-AVRAHAM, Z. & SCHATTNER, U. 2006. Formation of sequential basins along a strike-slip fault – geophysical observations from the Dead Sea basin. *Tectonophysics*, **421**, 53–69.

MCCLAY, K. R. & BONORA, M. 2001. Analogue models of restraining stepovers in strike-slip fault systems. *American Association of Petroleum Geologists Bulletin*, **85**, 233–260.

MCGARR, A. F., JACHENS, R. C., JAYKO, A. S. & BONILLA, M. G. 1997. Investigation of the San Bruno fault near the proposed expansion of the Bay Area Rapid Transit line from Colma to San Francisco International Airport, San Mateo County, California, *US Geological Survey Open File Reports*, **97–429**.

MANN, P., HEMPTON, M. R., BRADLEY, D. C. & BURKE, K. 1983. Development of pull-apart basins. *Journal of Geology*, **91**, 529–554.

MERRITTS, D. & BULL, W. B. 1989. Interpreting Quaternary uplift rates at the Mendocino triple junction, northern California, from uplifted marine terraces. *Geology*, **17**, 1020–1024.

NIEMI, T. M. & HALL, N. T. 1992. Late Holocene slip rate and recurrence of great earthquakes on the San Andreas fault in northern California. *Geology*, **20**, 195–198.

NILSEN, T. H. & MCLAUGHLIN, R. J. 1985. Comparison of tectonic framework and depositional patterns of the Hornelen strike-slip basin of Norway and the Ridge and Little Sulphur Creek strike-slip basins of California. *In*: BIDDLE, K. T. & CHRISTIE-BLICK, N. (eds) *Strike-Slip Deformation, Basin Formation, and Sedimentation*. SEPM, Special Publications, **37**, 79–104.

PAGE, B. M. 1981. The southern Coast Ranges. *In*: ERNST, W. G. (ed) *Geotectonic Development of California*. Englewood Cliffs, New Jersey, Prentice-Hall, Rubey Volumes, **I**, 329–417.

POWELL, R. E. 1993. Balanced palinspastic reconstruction of pre-late Cenozoic paleogeography, southern California: geologic and kinematic constraints on evolution of the San Andreas fault system. *In*: POWELL, R. E., WELDON, R. J., II & MATTI, J. C. (eds) *The San Andreas Fault System: Displacement, Palinspastic Reconstruction and Geologic Evolution*. Geological Society of America Memoirs, **178**, 1–106.

TEN BRINK, U. S. & BEN-AVRAHAM, Z. 1989. The anatomy of a pull-apart basin: seismic reflection observations of the Dead Sea basin. *Tectonics*, **8**, 333–350.

TWISS, R. J. & UNRUH, J. R. 2007. Structure, deformation, and strength of the Loma Prieta fault, northern California, USA, as inferred from the 1989–1990 Loma Prieta aftershock sequence. *Geological Society of America Bulletin*, **119**, 1079–1106.

UNRUH, J. R. 2000. Characterization of blind seismic sources in the Mt. Diablo–Livermore region, San Francisco Bay area, California. *Final Technical Report, US Geological Survey National Earthquake Hazards Reduction Program, Reston, Virginia*, Award No. **99-HQ-GR-0069**.

UNRUH, J. R. & SAWYER, T. L. 1995. *Late Cenozoic Growth of the Mt. Diablo Fold and Thrust Belt, Central Contra Costa County, California and Implications for Transpressional Deformation of the Northern*

Diablo Range. Pacific Section Convention, American Association of Petroleum Geologists and Society of Economic Paleontologists and Mineralogists, 47.

UNRUH, J. R. & SAWYER, T. L. 1997. Assessment of blind seismogenic sources, Livermore Valley, eastern San Francisco Bay region. *Final Technical Report, US Geological Survey, National Earthquake Hazards Reduction Program, Reston, Virginia*, Award No. **1434-95-G-2611**.

VAIL, P. R., MITCHUM, R. M. JR. & THOMPSON, S., III 1977. Seismic stratigraphy and global changes in sea level; part 4, global cycles of relative changes in sea level. *In*: PAYTON, C. E. (ed.) *Seismic Stratigraphy: Applications to Hydrocarbon Exploration*. AAPG Memoirs, **26**, 83–97.

WAGNER, D. L., BORTUGNO, E. J. & MCJUNKIN, R. D. (compilers). 1990. *Geologic Map of the San Francisco–San José quadrangle*. California Division of Mines and Geology Regional Map Series, Map **5A** (geology).

WAKABAYASHI, J. 1992. Nappes, tectonics of oblique plate convergence, and metamorphic evolution related to 140 million years of continuous subduction, Franciscan Complex, California. *Journal of Geology*, **100**, 19–40.

WAKABAYASHI, J. 1999. Distribution of displacement on, and evolution of, a young transform fault system: the northern San Andreas fault system, California. *Tectonics*, **18**, 1245–1274.

WAKABAYASHI, J. & SAWYER, T. L. 1998*a*. Holocene (?) oblique slip along the Miller Creek fault, eastern San Francisco Bay Area, Eos, **79**, 45 (Fall Meeting Supplement), F613.

WAKABAYASHI, J. & SAWYER, T. L. 1998*b*. Paleoseismic investigation of the Miller Creek fault, eastern San Francisco Bay area, California. *Final Technical Report, US Geological Survey National Earthquake Hazards Reduction Program, Reston, Virginia*, Award No. **1434-HQ-97-GR-03141**.

WAKABAYASHI, J., HENGESH, J. V. & SAWYER, T. L. 2004. Four-dimensional transform fault processes: progressive evolution of step-overs and bends. *Tectonophysics*, **392**, 279–301.

WESTAWAY, R. 1995. Deformation around stepovers in strike-slip fault zones. *Journal of Structural Geology*, **17**, 831–846.

YEATS, R. S. 1987. Changing tectonic styles in Cenozoic basins of southern California. *In*: INGERSOLL, R. V. & ERNST, W. G. (eds) *Cenozoic Basin Development of Coastal California*. Englewood Cliffs, New Jersey, Prentice-Hall, Rubey, Volumes **VI**, 284–298.

ZOBACK, M. L., JACHENS, R. C. & OLSON, J. A. 1999. Abrupt along-strike change in tectonic style: San Andreas fault zone, San Francisco Peninsula. *Journal of Geophysical Research*, **104**, 10 719–10 742.

Relatively simple through-going fault planes at large-earthquake depth may be concealed by the surface complexity of strike-slip faults

R. W. GRAYMER, V. E. LANGENHEIM, R. W. SIMPSON, R. C. JACHENS & D. A. PONCE

345 Middlefield Rd., MS 973, Menlo Park, CA 94025, USA (e-mail: rgraymer@usgs.gov)

Abstract: At the surface, strike-slip fault stepovers, including abrupt fault bends, are typically regions of complex, often disconnected faults. This complexity has traditionally led geologists studying the hazard of active faults to consider such stepovers as important fault segment boundaries, and to give lower weight to earthquake scenarios that involve rupture through the stepover zone. However, recent geological and geophysical studies of several stepover zones along the San Andreas fault system in California have revealed that the complex nature of the fault zone at the surface masks a much simpler and direct connection at depths associated with large earthquakes (greater than 5 km). In turn, the simplicity of the connection suggests that a stepover zone would provide less of an impediment to through-going rupture in a large earthquake, so that the role of stepovers as segment boundaries has probably been overemphasized. However, counter-examples of fault complexity at depth associated with surface stepovers are known, so the role of stepovers in fault rupture behaviour must be carefully established in each case.

Major active strike-slip faults are often characterized by major (kilometre-scale) discontinuities, or stepovers. The surface fault expression within these stepover zones can be exceptionally complex. Primary fault strands or primary displacement zones (PDZs) may not directly connect, or connect through abrupt bends, and through-going fault strain may be accommodated by multiple oblique faults and (or) folds.

In many analyses of seismic hazards, these stepover zones are considered to be important fault segment boundaries (for example, King & Nabelek 1985; Sibson 1985, 1986). Because of the complex and disrupted nature of the stepover zones, segment-based rupture probability studies usually assign lower weight to scenarios that require rupture through a stepover (for example, the Working Group on California Earthquake Probabilities 1999, 2003). Because of their potential importance for earthquake dynamics, many quantitative studies of stepovers have been made (see Harris & Day 1993 for a review of the literature). Recent numerical modelling studies (for example, Harris *et al.* 1991; Harris & Day 1993) suggest that stepover width plays an important role in the potential for through-going rupture.

However, recent studies of the dominantly right-lateral strike-slip San Andreas fault system (Fig. 1) in central and southern California, USA, have revealed that at depths associated with major earthquakes (greater than 5 km), the connection through some stepovers may be simpler and more direct than that expressed at the surface. Fault rupture through such stepovers may be more likely than previously suggested, perhaps resulting in less-frequent, but larger earthquakes.

Below we outline the geological and geophysical evidence for simplicity at depth associated with several restraining and releasing stepovers in the San Andreas fault system. We also discuss a counter-example revealed by geophysical studies in southern California. Lastly, we look at the palaeoseismic and historic earthquake evidence for faults associated with simple-at-depth stepovers in order to examine the possibility of an earthquake that has ruptured through a stepover.

Throughout this paper we make extensive use of hypocentre locations relocated using the double difference technique to define the active fault surface at depth. In a recent application of this technique, Schaff *et al.* (2002) state that location errors for the relocated hypocentres are typically one to two orders of magnitude smaller than the 1.2 km horizontal and 2.5 km vertical average 95% confidence estimates for the standard catalogue data in northern California. However, the relative location of closely spaced relocated hypocentres is likely to be better than absolute location of relocated hypocentres, so the analysis that follows is based on the shape of the fault at depth as defined by the relative location of the hypocentres, rather than details of the position of the fault surface at depth, relative to surficial features, as defined by the absolute location of the hypocentres.

From: CUNNINGHAM, W. D. & MANN, P. (eds) *Tectonics of Strike-Slip Restraining and Releasing Bends.* Geological Society, London, Special Publications, **290**, 189–201.
DOI: 10.1144/SP290.5 0305-8719/07/$15.00

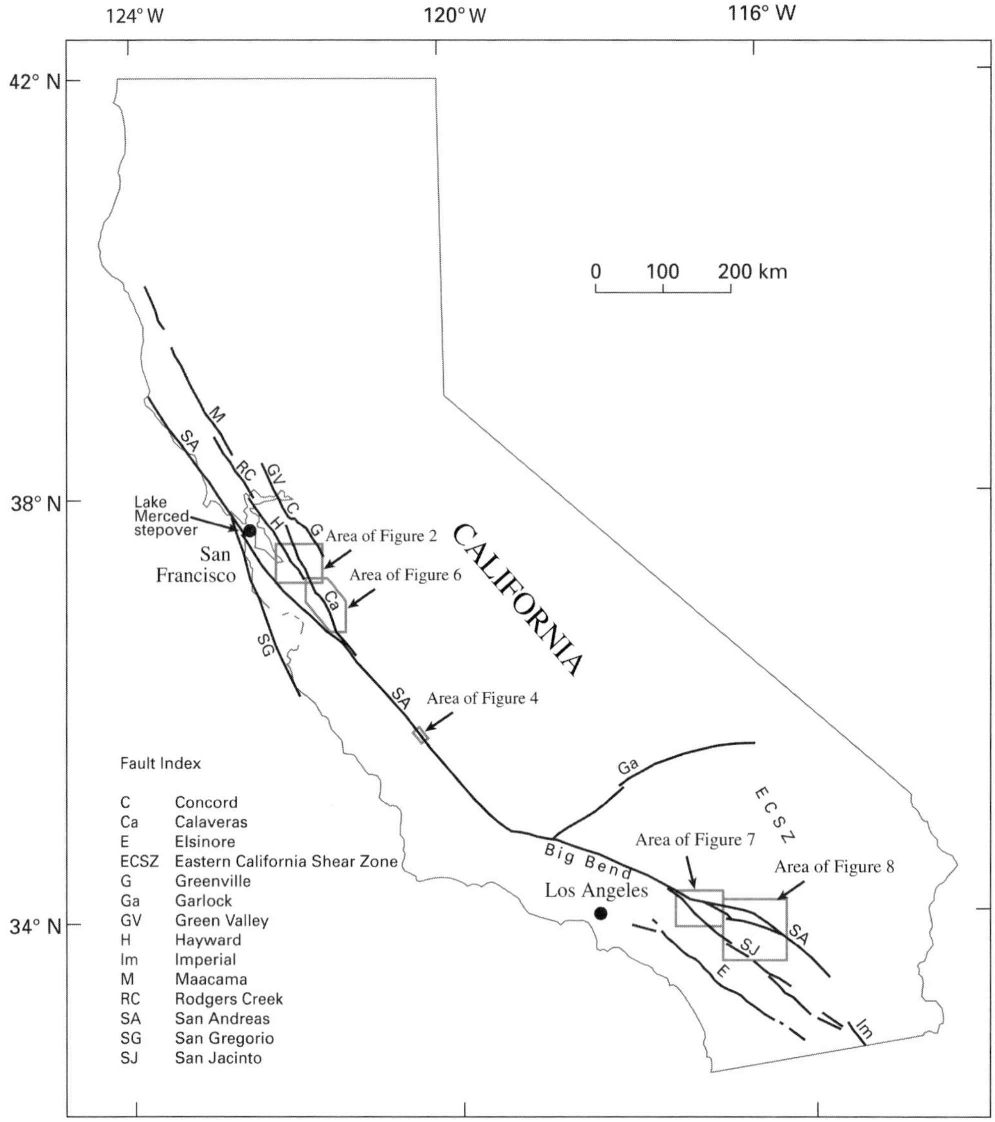

Fig. 1. Index map of the San Andreas Fault system in California, showing the location of the areas discussed in the text.

Restraining stepovers – Hayward–Calaveras Fault stepover

In the San Francisco Bay region, the San Andreas fault system includes several dominantly right-lateral strands, including the San Andreas, Calaveras and Hayward faults (Fig. 1). At the Earth's surface, the Hayward Fault splays from the southern Calaveras Fault, and about two-thirds of the total Calaveras slip is transferred on to the Hayward Fault, through a 25-km-long, 5-km-wide left step in the dominantly right-lateral faults (Fig. 2). This stepover is characterized by multiple oblique reverse faults (Fig. 2), with no single, simple direct surface connection between the main fault strands or PDZs.

However, at fault depths associated with large earthquakes (greater than 5 km), double-difference relocated microseismicity (Figs 2 & 3; Waldhauser & Ellsworth 2002; Hardebeck *et al.* 2004) reveals that the fault connections are much simpler, with a single continuous east-dipping fault at depth (Simpson *et al.* 2003, 2004; Manaker *et al.* 2005). Cross-sectional views of seismicity and geology suggest that the surface complexity is probably limited to the upper 3 to 5 km in the stepover

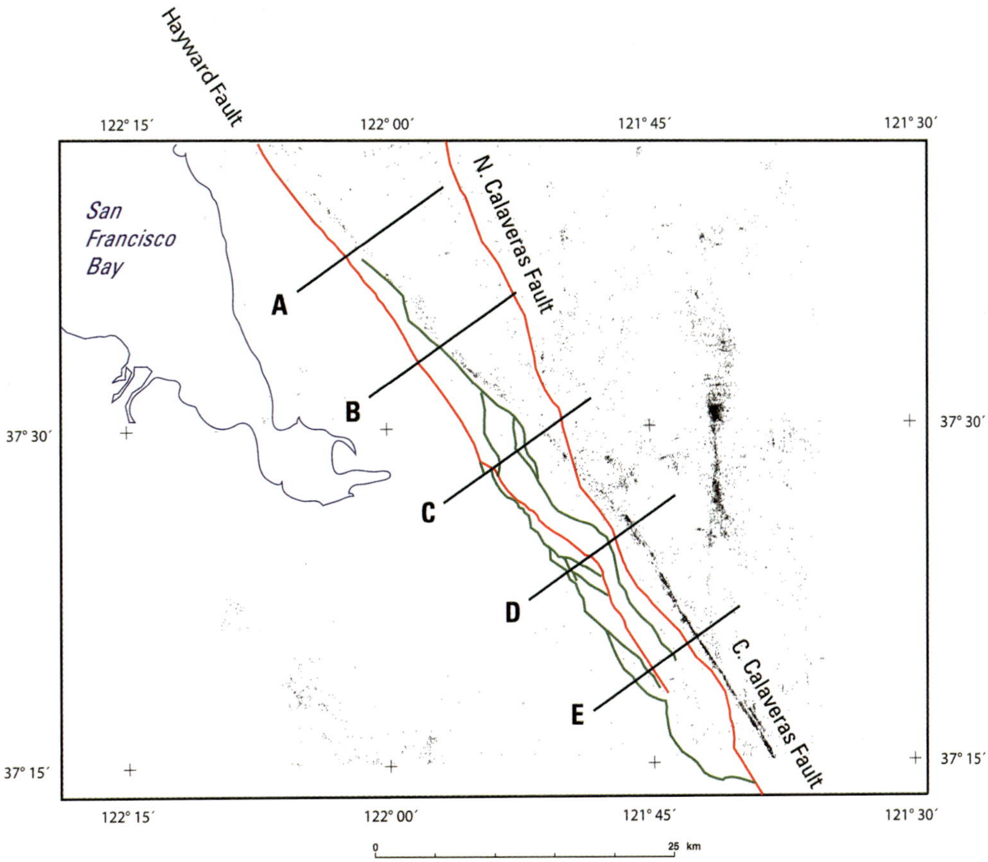

Fig. 2. Map of the southern Hayward and northern and central Calaveras faults (in red) in the southern San Francisco Bay region, California, including the left (restraining) stepover of 3–5 km width. Oblique faults forming the complex surface zone of deformation within the stepover are shown in green, the 1984 through 2000 double-difference relocated seismicity below 5 km is shown as black dots. Note that the deep seismicity defines a simple fault connection with a 12° westward bend. The locations of the cross-sections (A through E) in Figure 3 are also shown. (Modified from Bryant 1982; Graymer *et al.* 1996; Wentworth *et al.* 1998; Simpson *et al.* 2004.)

zone (Fig. 3). Seismicity occurs in diffuse and discontinuous clumps and probably reflects sporadic activity on the multiple oblique faults in this near-surface volume. Below 5 km, cross-sections of seismicity strongly suggest a single through-going structure connecting the central Calaveras Fault to the Hayward Fault. This picture is supported by geophysical studies of the region (Ponce *et al.* 2004; Manaker *et al.* 2005), which show a steeply east-dipping Hayward Fault joining with a steeply east-dipping central Calaveras Fault.

Releasing stepovers

San Andreas Fault, Cholame Valley

In Cholame Valley south of Parkfield in central California, the main creeping trace of the right-lateral San Andreas Fault makes a releasing right stepover bend across an elongate (rhombochasm?) valley (Fig. 4). This 3.5-km-wide stepover lies at the north end of a locked segment of the fault last ruptured in the great *M* 7.9 Fort Tejon earthquake of 1857. Viewed in a larger context, a straight edge laid along the San Andreas Fault in Central California reveals that a 35 km section of the Parkfield reach of the San Andreas trace north of the Cholame Valley stepover appears to be warped to the NE, perhaps by non-elastic behaviour of the crust at the transition from creeping to locked behaviour (Simpson *et al.* 2006).

Aftershocks of the 1966 and 2004 *M c.* 6 Parkfield earthquakes and background earthquakes in the intervening period (Figs 4 & 5) lie several kilometres to the SW of the main creeping trace (Bakun *et al.* 2005), approximately under the straight-edge

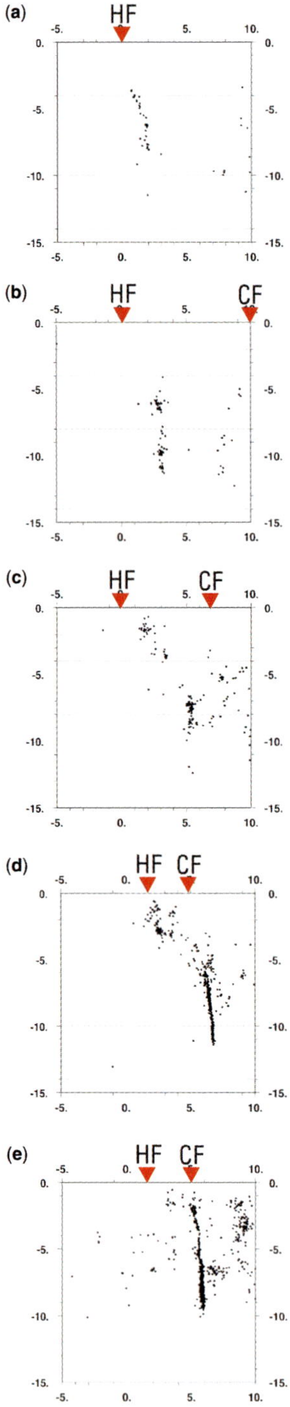

Fig. 3. Cross-sections of 1984–2000 double-difference relocated seismicity through the left (restraining) stepover between the Hayward Fault (HF) and Calaveras Fault (CF). No vertical exaggeration is shown; the horizontal and vertical scale is in kilometres; the horizontal scale zero-point is along the projection of the Hayward Fault north of the stepover; and hypocentres projected into the section are within 2.5 km of the section plane. Note that the broad, complex zone of deformation above 5 km is well illuminated by seismicity in section D, whereas in all sections the seismicity below 5 km defines a through-going single structure (modified from Simpson *et al.* 2004).

line connecting San Andreas segments to north and south of the Parkfield reach with an approximately three-degree bend in trend, and under the creeping SW Fracture Zone, which experienced post-seismic offset in the 2004 event. The complexity shown by the surface traces and the discordance of the location and geometry of the main trace relative to the seismicity is probably limited to the upper few kilometres. The cross-sections of seismicity strongly suggest a bifurcation of active slip in the upper 6 kilometres on to both the SW Fracture Zone and the main San Andreas Fault (Fig. 5d), perhaps in a flower structure, whereas below 6 km the seismicity is more readily fitted with a single through-going fault surface.

San Jacinto Fault, San Bernardino Valley

Analysis of geophysical data in San Bernardino Valley in southern California (Anderson *et al.* 2004) suggests that the simplicity at depth within releasing stepovers may eventually lead to the evolution of a simpler fault trace at the surface. Inversion of gravity data indicates a rhombohedral-shaped basin beneath San Bernardino Valley that is bisected by the neotectonic trace of the right-lateral strike-slip San Jacinto Fault (Anderson *et al.* 2004; Fig. 6). The absence of surface faults associated with the bounding faults of the pull-apart basin indicates that the basin, reflecting a 5-km-wide extensional step, presumably resulted from an earlier history of faulting. The evolution of a simpler surface trace is consistent with sandbox models of releasing stepovers (Dooley & McClay 1997).

San Bernardino Valley is also the location where slip on the San Jacinto Fault may be transferred on to the San Andreas Fault in an apparent extensional stepover, as proposed by Morton & Matti (1993). Focal mechanisms in the valley show predominantly normal motion (Jones 1988; Hauksson 2000), consistent with slip stepping to the right across the valley. Anderson *et al.* (2004), however, argued that most, if not all, of the 25 km of cumulative strike-slip displacement on the San Jacinto Fault (documented to the SE of San Bernardino Valley) exits the valley at its north end on several distinct strands of the modern San

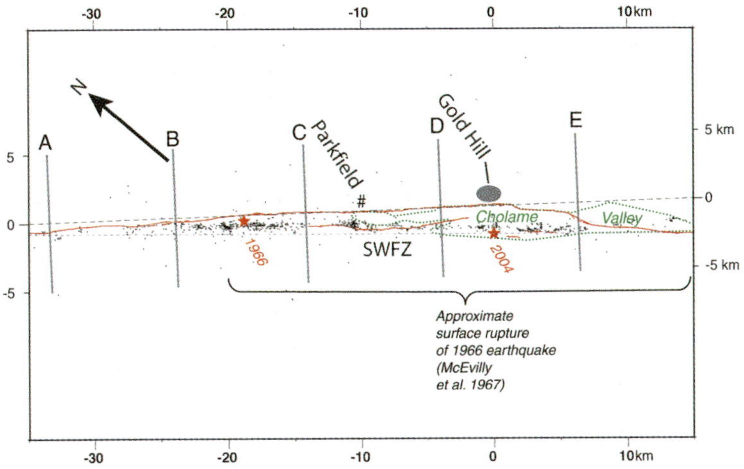

Fig. 4. Map of the San Andreas fault zone near Parkfield, California, including the 3.5-km right (releasing) step in the active fault traces and the SW Fracture Zone (SWFZ) shown with red lines. Also shown are the double-difference relocated seismicity below 6 km from 2004 aftershocks (small black dots), the epicentres of the 1966 and 2004 earthquakes (red stars), the extent of surface rupture from the 1966 earthquake (McEvilly *et al.* 1967), and the outline of Cholame Valley (green dotted line), possibly a pull-apart basin associated with the right step, and the locations of the cross-sections shown in Figure 5. The dashed line follows the average azimuth of the San Andreas Fault surface traced in the Parkfield area (N42°W), and marks the centre of the cross-sections shown in Figure 5. Note that the seismicity defines a relatively narrow, very gently bent (3°), fault connection. The scale is in kilometres, the azimuth of the upper and lower figure boundary is N40°W (modified from Simpson *et al.* 2006).

Jacinto Fault. This would seem to preclude a large amount of slip on the San Jacinto Fault stepping to the NE across the valley to directly link on to the San Andreas Fault, at least during most of the history of movement on the fault during the past 1.5 Ma (Matti & Morton 1993). Double-difference relocated seismicity in the stepover region (Hauksson & Shearer 2005), although less focused on narrow, discrete fault strands than the examples from central and northern California, shows one and possibly two SW-dipping zones of seismicity beneath the strike-slip basin below 10 to 12 km (Fig. 6); selected focal mechanisms indicate normal right-lateral oblique faulting. The deeper, more concentrated zone of seismicity projects up to the trace of the San Andreas Fault and may reflect a very young, simpler connection at depth between the San Jacinto and San Andreas faults.

Complexly bent faults – Central Calaveras Fault

Along the southern reach of the central Calaveras Fault (Fig. 7), the surface expression of the PDZ displays (from south to north) a left bend of 7° (L7 on Fig. 7), a right bend of 30° (R30), a left bend of 30° (L30), a left bend of 18° (L18), a right bend of 12° (R12), a right bend of 26° (R26) and a left bend of 30° (L30), over a distance of about 40 km. The northern two bends form a releasing bend or right step about 3 km wide, with a small associated pull-apart basin at San Felipe Valley (Chuang *et al.* 2002). Faulting along the 40 km reach is complex, including numerous short, disconnected fault strands, but no identifiable primary through-going surface trace. Shallow seismicity along this reach is generally diffuse and locally complex (Schaff *et al.* 2002).

Deeper seismicity, however, defines a relatively simple narrow planar fault surface through most of this area. The fault surface dips very steeply east to NE and forms a notably straight line in map view over most of its length (Fig. 7), in contrast to the multiple bends in the mapped surface traces. Previous workers (Reasenberg & Ellsworth 1982; Schaff *et al.* 2002) have pointed out two potential stepovers on the fault at depth (Fig. 6), but those authors correctly point out that neither of the proposed deep stepovers is related to any of the surface bends. The northernmost of the proposed deep stepovers is located approximately at the point where the Silver Creek Fault impinges on to the Calaveras Fault, so this discordance may reflect minor young movement on the Silver Creek Fault.

Other possibly simple-at-depth stepovers

Recent work has suggested that surface complexity may mask deep simplicity for several additional

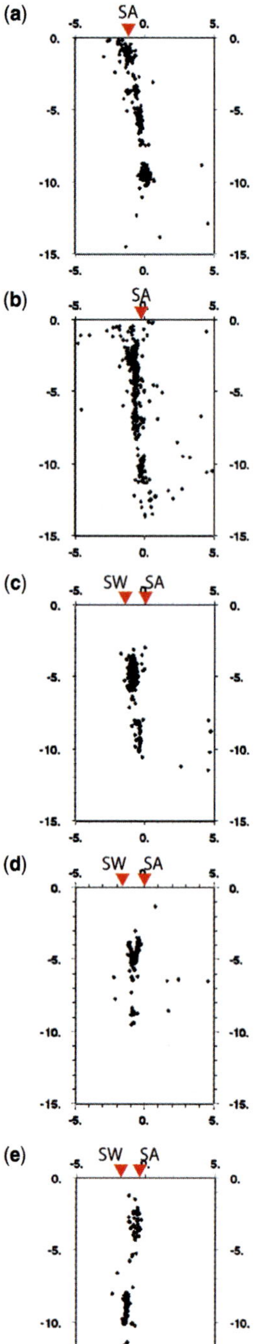

Fig. 5. Cross-sections of 1985 through 2004 double-difference relocated seismicity through the right (releasing) stepover in the right-lateral San Andreas fault zone near Parkfield, California. The surface position of the San Andreas Fault (SA) and SW Fracture Zone (SW) are marked with red triangles. The multi-strand nature of the stepover zone in the upper 5 km is well defined in section D, whereas in all sections the seismicity below 5 km suggests a single, through-going structure. Relocated seismicity from Bakun *et al.* (2005).

faults worldwide. Preliminary analysis of shallow fault dip based on a gravity inversion suggests that another simple-at-depth extensional stepover on the San Andreas Fault system may characterize the Lake Merced right step in the San Andreas Fault in San Francisco (Fig. 1; Zoback 2003). Simple-at-depth extensional stepovers have also been proposed for the North Anatolian Fault associated with the 1999 Izmit, Turkey, earthquake (Aochi & Madariaga 2003 although Harris *et al.* (2002) suggest that the narrowness of the stepovers (less than 2 km) is responsible for the through-going rupture) the Nojima–Rokko Fault system associated with the 1995 Kobe, Japan, earthquake (Wald 1996), and the suite of five faults associated with the 1992 Landers, California, earthquake (Felzer & Beroza 1999). Examination of each of these cases is beyond the scope of this paper, but it seems likely that the simple at-depth phenomenon described herein is not restricted to the San Andreas fault system.

A counter-example

A counter-example to the idea that faults within stepover regions become less complex at depth is the restraining bend in San Gorgonio Pass (Fig. 8), east of the San Bernardino releasing stepover. The San Gorgonio Pass restraining bend is located near the eastern end of the Big Bend (Fig. 1), where the San Andreas Fault steps 15 km to the left over a distance of approximately 20 km. The active strand of the San Andreas Fault has migrated in time and space during the past 5 Ma (Matti & Morton 1993) and geological mapping (Lawson *et al.* 1908; Noble 1932; Allen 1957; Matti & Morton 1993; Yule & Sieh 2003) does not indicate a continuous fault strand between the currently active strands of the San Andreas Fault, the San Bernardino segment of the San Andreas Fault (SBSSAF on Fig. 8) NW of the stepover, and the Coachella Valley segment of the Banning Fault (CVSBF on Fig. 8) to the SE.

Seismicity within the San Gorgonio restraining bend is diffuse, and is distributed on both strike-slip and thrust faults (Jones *et al.* 1986; Seeber & Armbruster 1995; Magistrale & Sanders 1996; Carena *et al.* 2004). The complexity of faulting at the surface apparently continues to the base of

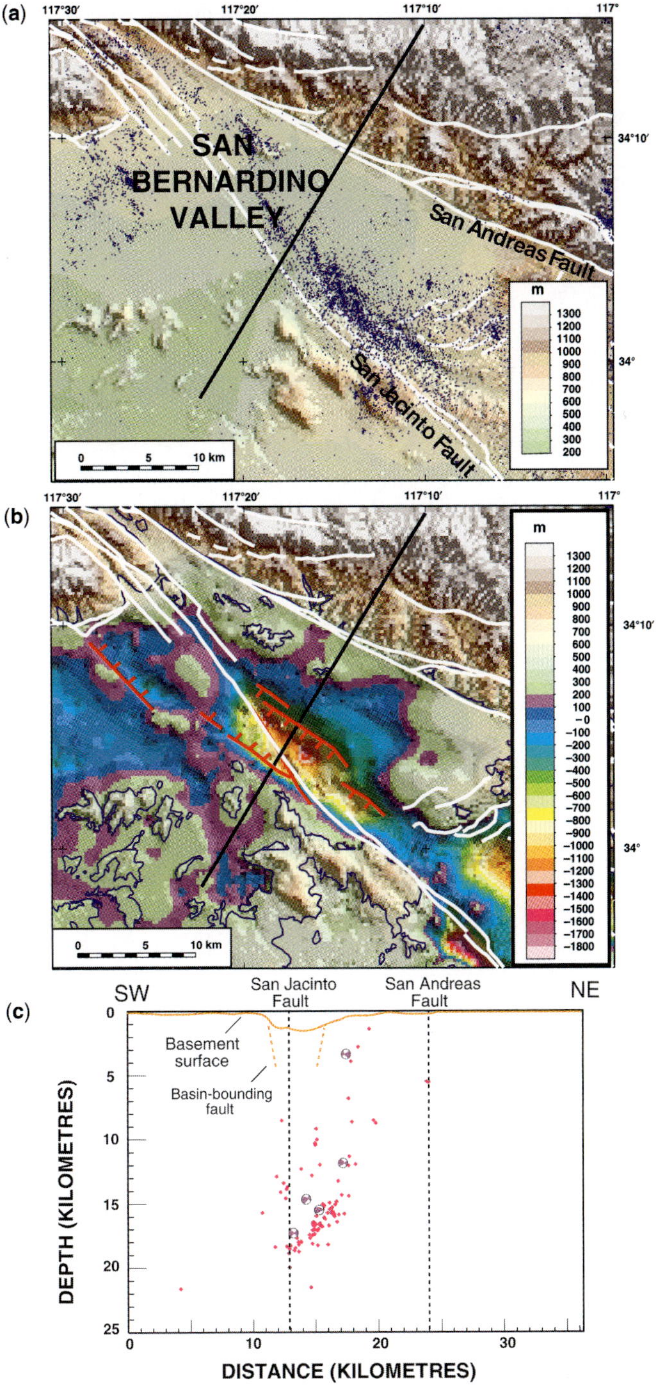

Fig. 6. Shaded-relief topography (**a**), basement topography defined by inversion of gravity data (**b**), and seismicity along a NE–SW profile (**c**) in the San Bernardino Valley region. Faults from Matti & Morton (1993) are shown in white on (a) and (b). The black line on (a) and (b) is the location of profile shown in (c). Dark-blue dots on (a) show double-difference relocated seismicity from Hauksson & Shearer (2005). Red faults on (b) are interpreted from analysis of gravity and magnetic data (Anderson *et al.* 2004). Double-difference relocated seismicity from Hauksson & Shearer (2005) within 1 km of the profile is shown on (c) as magenta dots. Beachballs are focal mechanisms consistent with normal faulting within 1 km of the profile from Hauksson (2000).

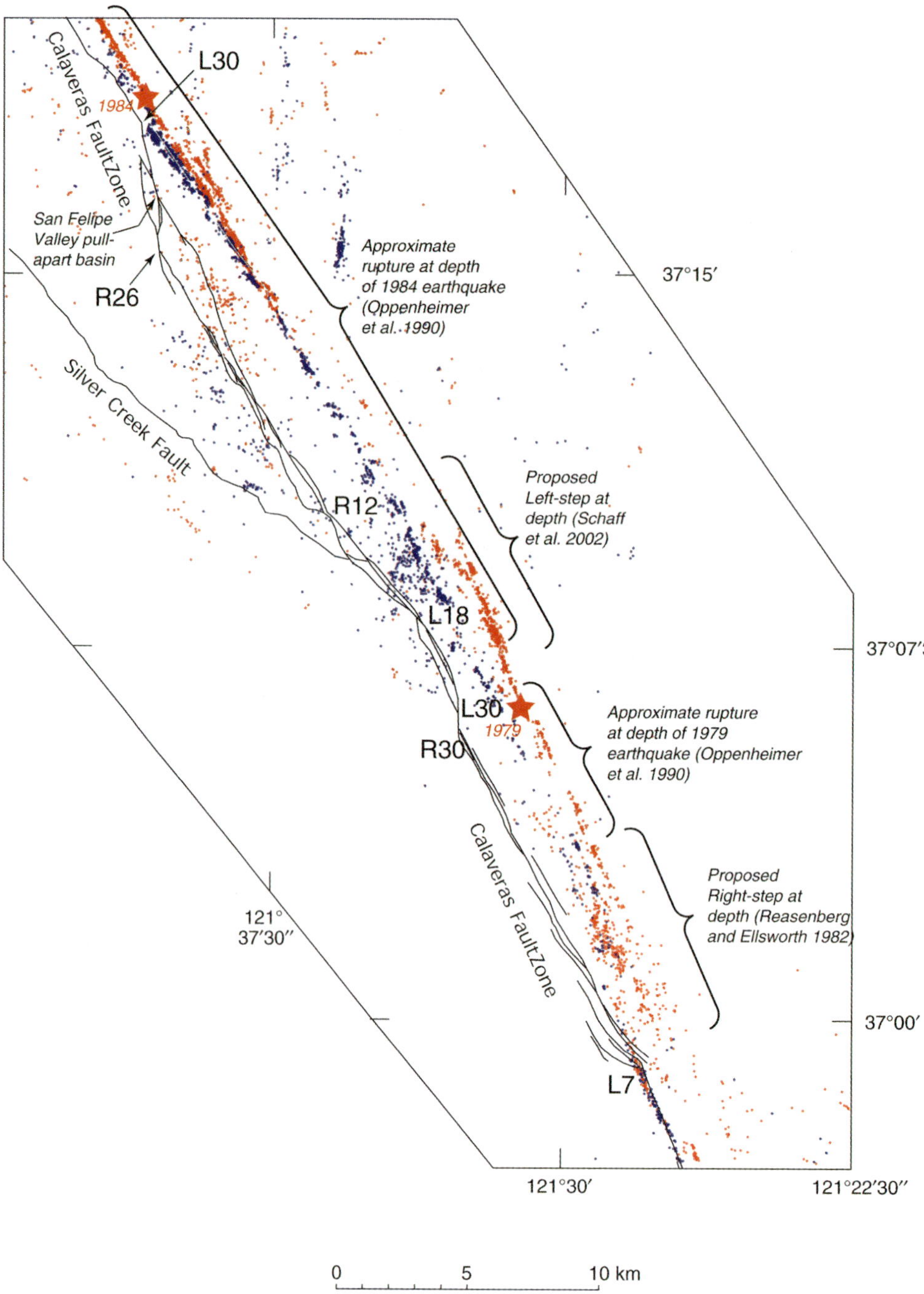

Fig. 7. Simplified map of the right-lateral southern Calaveras Fault showing right (releasing) and left (restraining) bends. The microseismicity below 5 km is shown in red, above 5 km in blue. The epicentres of the 1979 and 1984 earthquakes are shown as red stars. The rupture length of those earthquakes (Oppenheimer *et al.* 1990) is also shown, along with the deep discontinuities in the fault surface proposed by Reasenberg & Ellsworth (1982) and Schaff *et al.* (2002). Note that although the surface fault zone is complex, and the shallow microseismicity is in places diffuse, the deep microseismicity defines a single, east-dipping, gently bent, through-going fault over most of its length, proposed discontinuities at depth are unrelated to surface discontinuities, and the 1984 earthquake rupture extends through multiple bends in the surface fault (modified from Wentworth *et al.* 1998; Simpson *et al.* 2004).

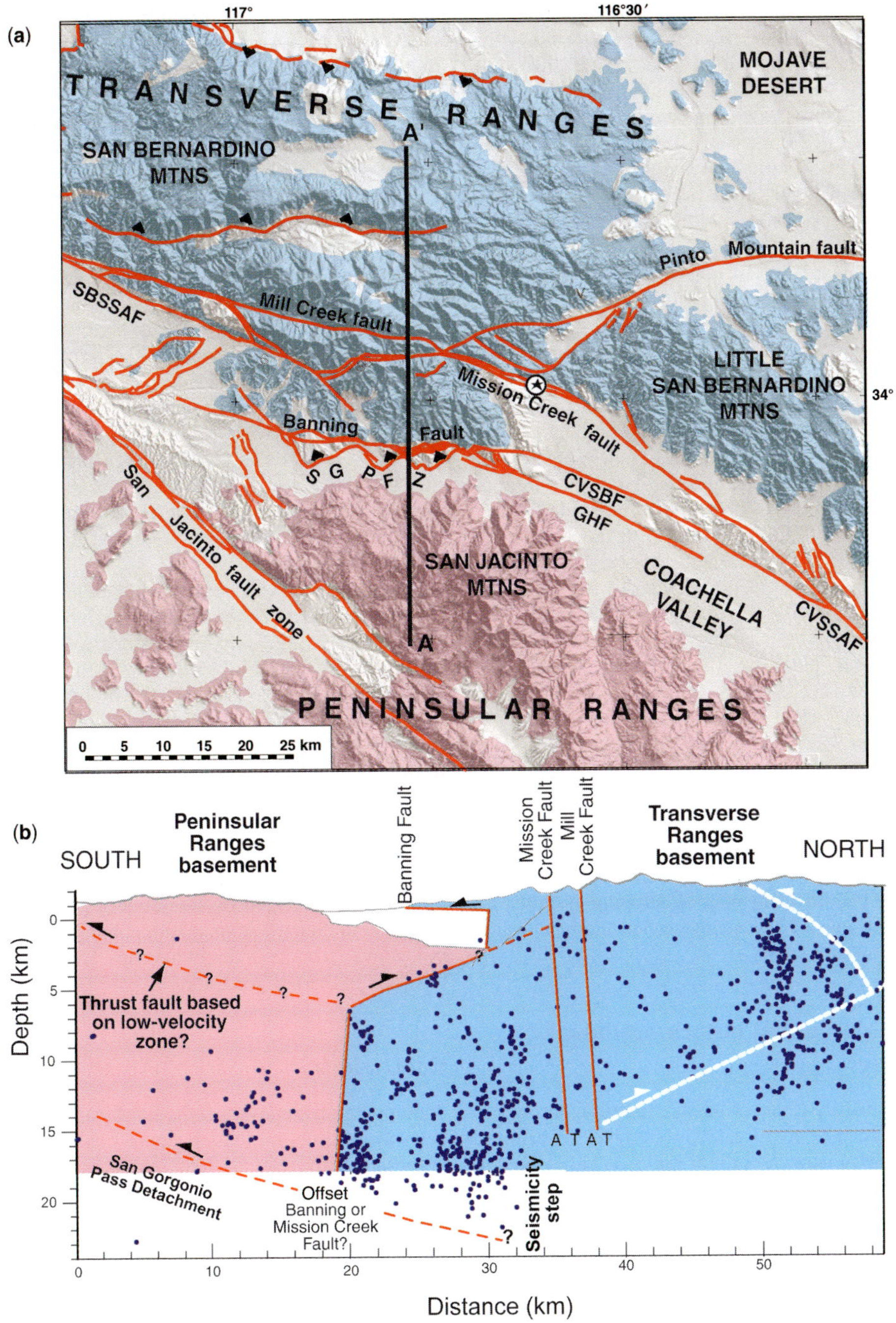

Fig. 8. Simplified geological map (**a**) of the San Gorgonio Pass region, modified from Matti & Morton (1993). Pink shows the extent of Peninsular Ranges basement terrane outcrops; blue shows the extent of Transverse Ranges basement terrane outcrops. Strands of the San Andreas Fault Zone include the Banning, Mission Creek, Mill Creek, San Bernardino segment (SBSSAF); the Coachella Valley segment of the Banning Fault (CVSBF), the Garnet Hill Fault (GHF), and the Coachella Valley segment of the San Andreas Fault (CVSSAF). SGPFZ, San Gorgonio Pass Fault zone. The star marks the location of the 1986 M_L 5.9 North Palm Springs earthquake. The location of cross-section (**b**) is shown by a thick black line. The geometry of the cross-section is constrained by geological and geophysical analysis in Langenheim *et al.* (2005). Dark-blue dots are locations of double-difference relocated seismicity within 1 km of the profile.

seismicity at depths of 15 to 20 km (Fig. 8). Double-difference relocated seismicity does not indicate a simpler fault configuration at depth (Hauksson & Shearer 2005). The diffuse nature of seismicity in San Gorgonio Pass appears to reflect very complex three-dimensional brittle deformation along interlaced strike-slip and thrust faults. Analysis of aeromagnetic and gravity data defines intertonguing thrust wedges within San Gorgonio Pass, with wedging of Peninsular Ranges batholithic crust northward above Transverse Ranges basement at seismogenic depths (Fig. 8(b); Langenheim *et al.* 2005). Given the multilayered crustal deformation occurring in this region, it is not surprising that the San Andreas Fault exhibits a complex rupture pattern throughout the crust of the restraining stepover. The inability to identify a single, through-going strand of the San Andreas Fault using seismicity does not preclude a through-going rupture, but the seismicity does not support a simpler fault at depth within this stepover. Aftershocks of the 1986 M_L 5.9 North Palm Springs earthquake (Fig. 8) indicated not only the main rupture plane (right-lateral strike-slip on an approximately 45° dipping plane) but also a gently north-dipping décollement plane at the base of the rupture (Nicholson *et al.* 1986), informally named the San Gorgonio detachment. Deformation may eventually bypass the San Andreas Fault in the San Gorgonio Pass area by distributing slip on to the Eastern California Shear Zone and the San Jacinto Fault. The Eastern California Shear Zone accommodates 20 to 25% of Pacific–North America plate motion (Savage *et al.* 1990; Sauber *et al.* 1994), whereas the San Jacinto Fault may be slipping as much as twice the rate of the southern San Andreas Fault (Johnson *et al.* 1994).

Palaeoseismic and historic earthquake evidence

The possibility of an earthquake rupturing through a stepover can be investigated by studying the historic and palaeoseismic record for faults on either side of a stepover. Although palaeoseismic data are insufficient to prove through-going rupture, or even to strongly suggest it in most cases, it can show whether such a rupture is possible.

On the Hayward–Calaveras system, the most recent earthquake (1868) is thought to be limited to the Hayward Fault (Lawson 1908; Yu & Segall 1996), but there is no palaeoseismic evidence for the timing of the most recent surface-breaking earthquake on the central Calaveras (Working Group on California Earthquake Probabilities 2003), and direct observations of surface rupture and triangulation data from 1868 are not detailed or complete enough to preclude any central Calaveras rupture, so it is not possible to interpret a rupture through the stepover related to this or earlier quakes.

On the San Andreas Fault near Parkfield, historic earthquakes (1966 and 2004) appear to have occurred entirely within the stepover region, although they both appear to have ruptured through the sharp releasing step in Cholame Valley (Fig. 4; McEvilly *et al.* 1967; Thurber *et al.* 2006). Likewise, the 1857 earthquake ruptured through the releasing step in Cholame Valley at least as far north as Parkfield, and palaeoseismic data north and south of the right stepover, although poorly constrained, are compatible with the model of a through-going prehistoric rupture. South of the stepover, the most recent prehistoric earthquake post-dates a radiocarbon date of 1058 to 1291 AD (Stone *et al.* 2002), whereas north of the stepover the most recent prehistoric earthquake post-dates a radiocarbon date of 1440 to 1640 AD (Toké *et al.* 2004).

On the central Calaveras Fault, King & Nabelek (1985) suggested that the rupture length for both the 1979 (Coyote Lake) and 1984 (Morgan Hill) earthquakes were limited by bends in the surface fault. However, as shown on Fig. 6, the rupture at depth for both earthquakes clearly passed significant bends in the surface trace, including the pull-apart basin at San Felipe Valley in 1984, and the southern tip of the 1979 rupture formed where the surface trace was relatively straight. The southward rupture propagation at depth in both earthquakes was much more likely to be limited by deep discontinuities as suggested by Reasenberg & Ellsworth (1982) and Schaff *et al.* (2002), but, as pointed out above, the deep discontinuities are unrelated to bends in the surface trace.

Conclusions

Although fault stepovers at the Earth's surface have been used to define fault-segment boundaries, the observation of relative fault simplicity at depth (greater than 5 km) under some stepovers suggests that large quakes may be nominally affected by the complexity seen at the surface. However, not all stepovers are accompanied by simplicity at depth; some appear to be associated with complex faulting at depth. Other former stepovers seem to have been superseded by simple faults at the surface. One appealing structural model, supported by sandbox models of pull-apart basins, is one in which faults are initially prone to near-surface complexity (stepovers), while being simple at depth. If the deep fault remains stable through geological time, then the surface complexity is abandoned in favour of surface simplicity. If, on the other hand, the deep fault locus changes with time, the result can be fault complexity both at depth and surface. The geological setting of the fault may also influence its simplicity or complexity.

Whatever the underlying mechanism, the role of the stepover as a fault-segment boundary that will affect fault rupture does not have an obvious correlation with surface complexity. Deep discontinuities unrelated to surface complexity may be far more important in determining rupture length. Before stepovers are used in earthquake assessment, they must be evaluated using relocated microseismicity, potential field geophysics, and(or) palaeoseismic data. Present usage probably assigns undue importance to stepovers as impediments to through-going fault rupture, and therefore systematically underestimates the probablility of less-frequent, larger-magnitude earthquakes.

References

ALLEN, C. R. 1957. San Andreas fault zone in San Gorgonio Pass, southern California. *Geological Society of America Bulletin*, **68**, 319–350.

ANDERSON, M., MATTI, J. & JACHENS, R. 2004. Structural model of the San Bernardino basin, California, from analysis of gravity, aeromagnetic, and seismicity data. *Journal of Geophysical Research*, **109**, B04404 10.1029/2003JB002544, 1–20.

AOCHI, H. & MADARIAGA, R. 2003. The 1999 Izmit, Turkey, earthquake: nonplanar fault structure, dynamic rupture process, and strong ground motion. *Bulletin of the Seismological Society of America*, **93**, 1249–1266.

BAKUN, W. H., AAGAARD, B. *ET AL*. 2005. Implications for prediction and hazard assessment from the 2004 Parkfield earthquake. *Nature*, **437**, 969–974.

BRYANT, W. A. 1982. Southern Hayward Fault zone, Alameda and Santa Clara Counties, California. *In*: HART, E. W., HIRSCHFELD, S. E. & SCHULZ, S. S. (eds) *Proceedings, Conference on Earthquake Hazards in the Eastern San Francisco Bay Area*. California Division of Mines and Geology Special Publications, **62**, 35–44.

CARENA, S., SUPPE, J. & KAO, H. 2004. Lack of continuity of the San Andreas Fault in southern California: three-dimensional fault models and earthquake scenarios. *Journal of Geophysical Research*, **109**, B04313 10.1029/2003/JB002643.

CHUANG, F. C., JACHENS, R. C., WENTWORTH, C. M. & SANGER, E. A. 2002. Young strike-slip basin on the Calaveras Fault in San Felipe Valley, California. *Geological Society of America Abstracts with Programs*, **35/5**, p. 99.

DOOLEY, T. & MCCLAY, K. 1997. Analog modeling of pull-apart basins. *AAPG Bulletin*, **81**, 1804–1826.

FELZER, K. R. & BEROZA, G. C. 1999. Deep structure of a fault discontinuity. *Geophysical Research Letters*, **26**, 2121–2124.

GRAYMER, R. W., JONES, D. L. & BRABB, E. E. 1996. *Preliminary Geologic Map Emphasizing Bedrock Formations in Alameda County, California: A Digital Database*. United States Geological Survey Open-File Reports, **96–252**, World Wide Web Address: http://pubs.usgs.gov/of/1996/of96-252.

HARDEBECK, J. L., MICHAEL, A. J. & BROCHER, T. M. 2004. Seismic velocity structure and seismotectonics of the Hayward Fault System, East San Francisco Bay, California. *Eos, Transactions, American Geophysical Union*, **85/47**, Fall Meeting Supplement, Abstract S34A-05.

HARRIS, R. A. & DAY, S. M. 1993. Dynamics of fault interaction: parallel strike-slip faults. *Journal of Geophysical Research*, **98**, 4461–4472.

HARRIS, R. A., ARCHULETA, R. J. & DAY, S. M. 1991. Fault steps and the dynamic rupture process: 2-d numerical simulations of a spontaneously propagating shear fracture. *Geophysical Research Letters*, **18**, 893–896.

HARRIS, R. A., DOLAN, J. F., HARTLEB, R. & DAY, S. M. 2002. The 1999 Izmit, Turkey, earthquake: a 3D dynamic stress transfer model of intraearthquake triggering. *Bulletin of the Seismological Society of America*, **92**, 245–255.

HAUKSSON, E. 2000. Crustal structure and seismicity distribution adjacent to the Pacific and North America plate boundary in southern California. *Journal of Geophysical Research*, **105**, 13 875–13 903.

HAUKSSON, E. & SHEARER, P. 2005. Southern California hypocenter relocation with waveform cross-correlation, part 1: results using the double-difference method. *Bulletin of the Seismological Society of America*, **95**, 896–903.

JOHNSON, H. O., DUNCAN, A. C. & WYATT, F. K. 1994. Present-day deformation in southern California. *Journal of Geophysical Research*, **99**, 23 951–23 974.

JONES, L. 1988. Focal mechanisms and the state of stress on the San Andreas fault in southern California. *Journal of Geophysical Research*, **93**, 8869–8891.

JONES, L., HUTTON, L., GIVEN, D. & ALLEN, C. 1986. The North Palm Springs, California, earthquake sequence of July 1986. *Bulletin of the Seismological Society of America*, **76**, 1830–1837.

KING, G. C. P. & NABELEK, J. 1985. Role of fault bends in the initiation and termination of earthquake rupture. *Science*, **228**, 984–987.

LANGENHEIM, V. E., JACHENS, R. C., MATTI, J. C., HAUKSSON, E., MORTON, D. M. & CHRISTENSEN, A. 2005. Geophysical evidence for wedging in the San Gorgonio structural knot, southern San Andreas fault zone, southern California. *Geological Society of America Bulletin*, **117**, 1554–1572.

LAWSON, A. C. 1908. *The California Earthquake of April 18, 1906*. Carnegie Institute Washington Publications, **87/1**.

MCEVILLY, T. V., BAKUN, W. H. & CASADAY, K. B. 1967. The Parkfield, California earthquake of 1966. *Bulletin of the Seismological Society of America*, **57**, 1221–1244.

MAGISTRALE, H. & SANDERS, C. 1996. Evidence from precise earthquake hypocenters for segmentation of the San Andreas fault in San Gorgonio Pass. *Journal of Geophysical Research*, **101**, 3031–3041.

MANAKER, D. M., MICHAEL, A. J. & BURGMANN, R. 2005. Subsurface structure and kinematics of the Calaveras–Hayward fault stepover from three-dimensional V_P and seismicity, San Francisco Bay region, California. *Bulletin of Seismological Society of America*, **95**, 446–470.

MATTI, J. C. & MORTON, D. M. 1993. Paleogeographic evolution of the San Andreas fault in southern California: a reconstruction based on a new cross-fault correlation. *In*: POWELL, R. E., WELDON, R. E., II & MATTI, J. C. (eds) *The San Andreas Fault System: Displacement, Palinspastic Reconstruction, and Geologic Evolution*. Geological Society of America Memoirs, **178**, 107–159.

MORTON, D. M. & MATTI, J. C. 1993. Extension and contraction within an evolving divergent strike-slip fault complex: the San Andreas and San Jacinto fault zones at their convergence in southern California. *In*: POWELL, R. E., WELDON, R. E., II & MATTI, J. C. (eds) *The San Andreas Fault System: Displacement, Palinspastic Reconstruction, and Geologic Evolution*. Geological Society of America Memoirs, **178**, 217–230.

NICHOLSON, C., SEEBER, L., WILLIAMS, P. & SYKES, L. 1986. Seismicity and fault kinematics through the eastern Transverse Ranges, California: block rotation, strike-slip faulting, and low-angle thrusts. *Journal of Geophysical Research*, **91**, 4891–4908.

NOBLE, L. F. 1932. The San Andreas rift in the desert region of southeastern California. *Carnegie Institute of Washington Yearbook*, **31**, 355–363.

OPPENHEIMER, D. H., BAKUN, W. H. & LINDH, A. G. 1990. Slip partitioning of the Calaveras Fault, California, and prospects for future earthquakes. *Journal of Geophysical Research*, **95**, 8483–8498.

PONCE, D. A., SIMPSON, R. W., GRAYMER, R. W. & JACHENS, R. C. 2004. Gravity, magnetic, and high-precision relocated seismicity profiles suggest a connection between the Hayward and Calaveras Faults, Northern California. *Geochemistry, Geophysics, Geosystems – G^3*, **5**, 39 pp.

REASENBERG, P. & ELLSWORTH, W. L. 1982. Aftershocks of the Coyote Lake, California, earthquake of August 6, 1979: a detailed study. *Journal of Geophysical Research*, **87**, 10 637–10 655.

SAUBER, J. W., THATCHER, W., SOLOMON, S. C. & LISOWSKI, M. 1994. Geodetic slip rate for the eastern California shear zone and recurrence of Mojave desert earthquakes. *Nature*, **367**, 264–266.

SAVAGE, J. C., LISOWSKI, M. & PRESCOTT, W. H. 1990. An apparent shear zone trending north-northwest across the Mojave Desert into Owens Valley, eastern California. *Geophysical Research Letters*, **17**, 2113–2116.

SCHAFF, D. P., BOKELMANN, G. H. R., BEROZA, G. C., WALDHAUSER, F. & ELLSWORTH, W. L. 2002. High-resolution image of Calaveras Fault seismicity. *Journal of Geophysical Research*, **107/B9**, 2186, doi:10.1029/2001JB000633.

SEEBER, L. & ARMBRUSTER, J. G. 1995. The San Andreas fault system through the Transverse Ranges as illuminated by earthquakes. *Journal of Geophysical Research*, **100**, 8285–8310.

SIBSON, R. H. 1985. Stopping of earthquake ruptures at dilational fault jogs. *Nature*, **316**, 248–251.

SIBSON, R. H. 1986. Rupture interactions with fault jogs. *In*: DAS, S., BOATWRIGHT, J. & SCHOLZ, C. H. (eds) *Earthquake Source Mechanics*. American Geophysical Union Geophysical Monographs, **37**, 157–167.

SIMPSON, R. W., BARALL, M., LANGBEIN, J., MURRAY, J. R. & RYMER, M. J. 2006. San Andreas Fault geometry in the Parkfield, California region. *Bulletin of the Seismological Society of America*, **96/4b**, 528–537.

SIMPSON, R. W., GRAYMER, R. W., JACHENS, R. C., PONCE, D. A. & WENTWORTH, C. M. 2003. Geometry of the Hayward Fault and its connection to the Calaveras Fault inferred from relocated seismicity. *In*: PONCE, D. A., BURGMANN, R., GRAYMER, R. W., LIENKAEMPER, J. J., MOORE, D. E. & SCHWARTZ, D. P. (eds) *Proceedings of the Hayward Fault Workshop, Eastern San Francisco Bay area, California*. US Geological Survey Open-File Reports, **OF 03-0485**, World Wide Web Address: http://pubs.usgs.gov/of/2003/of03-485/.

SIMPSON, R. W., GRAYMER, R. W., JACHENS, R. C., PONCE, D. A. & WENTWORTH, C. M. 2004. *Cross-sections and Maps showing Double-difference Relocated Earthquakes from 1984–2000 Along the Hayward and Calaveras Faults, California*. US Geological Survey Open-File Reports, **OF 2004-1083**, World Wide Web Address: http://pubs.usgs.gov/of/2004/1083.

STONE, E. M., GRANT, L. B. & ARROWSMITH, J. R. 2002. Recent rupture history of the San Andreas Fault southeast of Cholame in the northern Carrizo Plain, California. *Bulletin of the Seismological Society of America*, **92**, 983–997.

THURBER, C., ZHANG, H., WALDHAUSER, F., HARDEBECK, J., MICHAEL, A. & EBERHART-PHILLIPS, D. 2006. Three-dimensional compressional wavespeed model, earthquake relocations, and focal mechanisms for the Parkfield, California, region. *Bulletin of the Seismological Society of America*, **96/4b**, 538–549.

TOKÉ, N. A., ARROWSMITH, J. R., CROSBY, C. J. & YOUNG, J. J. 2004. *Preliminary Paleoseismology Results from the Parkfield, CA, Segment of the San Andreas Fault*. Southern California Earthquake Center Annual Meeting, Proceedings and Abstracts, **14**. World Wide Web Address: http://activetectonics.la.asu.edu/Parkfield/paleoseis.html.

WALD, D. J. 1996. Slip history of the 1995 Kobe, Japan, earthquake determined from strong motion, teleseismic, and geodetic data. *Journal of Physics of the Earth*, **44**, 489–503.

WALDHAUSER, F. & ELLSWORTH, W. L. 2002. Fault structure and mechanics of the Hayward fault, California, from double-difference earthquake locations. *Journal of Geophysical Research*, **107**, doi:10.1029/2000JB000084.

WENTWORTH, C. M., BLAKE, M. C., JR, MCLAUGHLIN, R. J. & GRAYMER, R. W. 1998. *Preliminary Geologic Map of the San Jose 30 × 60 Minute Quadrangle, California: A Digital Map Image*. United States Geological Survey Open-File Report, **98-795**, World Wide Web Address: http://pubs.usgs.gov/of/1998/of98-795/.

Working Group on California Earthquake Probabilities 1999. *Earthquake Probabilities in the San Francisco Bay Region; 2000 to 2030, a Summary of Findings*. United States Geological Survey Open-File Reports, **OF 99-517**, World Wide Web Address: http://geopubs.wr.usgs.gov/open-file/of99-517.

Working Group on California Earthquake Probabilities 2003. *Earthquake Probabilities in the San Francisco Bay Region; 2002–2031*. United States Geological Survey Open-File Reports, **OF 03-214**, World Wide Web Address: http://geopubs.wr.usgs.gov/open-file/of03-214/.

YU, E. & SEGALL, P. 1996. Slip in the 1868 Hayward earthquake from the analysis of historical triangulation data. *Journal of Geophysical Research*, **101**, 16 101–16 118.

YULE, D. & SIEH, K. 2003. Complexities of the San Andreas fault near San Gorgonio Pass: implications for large earthquakes. *Journal of Geophysical Research*, **108**, 9–1 to 9–23.

ZOBACK, M. L. 2003. Relationship between mapped fault stepovers and earthquake fault planes at depth. *Eos, Transactions, American Geophysical Union*, **84/46**, Fall Meeting Supplement, Abstract T12E-03.

Extensional deformation and development of deep basins associated with the sinistral transcurrent fault zone of the Scotia–Antarctic plate boundary

F. BOHOYO[1,6], J. GALINDO-ZALDÍVAR[2], A. JABALOY[2], A. MALDONADO[3], J. RODRÍGUEZ-FERNÁNDEZ[3], A. SCHREIDER[4] & E. SURIÑACH[5]

[1]*Instituto Geológico y Minero de España. C/ La Calera, 1, 28760, Tres Cantos, Madrid, Spain (e-mail: f.bohoyo@igme.es)*

[2]*Departamento de Geodinámica, Universidad de Granada. 18071 Granada, Spain*

[3]*Instituto Andaluz Ciencias de la Tierra, CSIC/Universidad Granada. Facultad de Ciencias, 18002 Granada, Spain*

[4]*P. P. Shirshov Institute of Oceanology, Russian Academy of Sciences, 23, Krasikova 117218 Moscow, Russia*

[5]*Departament de Geologia Dinàmica i Geofísica, Universitat de Barcelona, 08028 Barcelona, Spain*

[6]*Geological Science Division. British Antarctic Survey, High Cross, Mandigley Road, Cambridge, CB3 0ET, UK (e-mail: fbo@pcmail.nerc-bas.ac.uk)*

Abstract: The Scotia–Antarctic plate boundary extends along the southern branch of the Scotia Arc, between triple junctions with the former Phoenix plate to the west (57°W) and with the Sandwich plate to the east (30°W). The main mechanism responsible for the present arc configuration is the development of the Scotia and Sandwich plates from 30–35 Ma, related to breakup of the continental connection between South America and the Antarctic Peninsula. The Scotia–Antarctic plate boundary is a very complex tectonic zone, because both oceanic and continental elements are involved. Present-day sinistral transcurrent motion probably began 8 Ma ago. The main active structures that we observed in the area include releasing and restraining bends, with related deep extensional and compressional basins, and probable pull-apart basins. The western sector of the plate boundary crosses fragmented continental crust: the Western South Scotia Ridge, with widespread development of pull-apart basins and releasing bends deeper than 5000 m, filled by asymmetrical sedimentary wedges. The northern border of the South Orkney microcontinent, in the central sector, has oceanic and continental crust in contact along a large thrust zone. Finally, the eastern sector of the South Scotia Ridge is located within Discovery Bank, a piece of continental crust from a former arc. On its southern border, strike-slip and normal faults produce a 5500-m-deep trough that may be interpreted as a pull-apart basin. In the eastern and western South Scotia Ridge, despite extreme continental-crustal thinning, the basins show no development of oceanic crust. This geometry is conditioned by the distinctive rheological behaviour of the crust involved, with the bulk concentration of deformation within the rheologically weaker continental blocks.

The development of the Scotia Arc between the major South America and Antarctic plates constitutes the main tectonic event in the SW Atlantic since the Oligocene. The former continental connection between South America and the Antarctic Peninsula was broken during the development of the Scotia Arc, leading to dispersal of continental blocks and the joining of the Atlantic and Pacific oceans (Barker & Burrell 1977; King & Barker 1988; Barker *et al.* 1991; Livermore *et al.* 1994; Aldaya & Maldonado 1996; Galindo-Zaldívar *et al.* 1996; Maldonado *et al.* 1998; Barker 2001). The internal part of this tectonic arc is formed by two minor plates – the Scotia and the Sandwich plates – mainly composed of oceanic crust and separated by the East Scotia Ridge, an active spreading axis (Fig. 1). The northern and southern branches of the Scotia Arc constitute the east–west-oriented boundaries of these minor plates, accommodating the present-day relative sinistral motion between the South America and Antarctic plates. However, the eastern and western boundaries of the tectonic

From: CUNNINGHAM, W. D. & MANN, P. (eds) *Tectonics of Strike-Slip Restraining and Releasing Bends.* Geological Society, London, Special Publications, **290**, 203–217.
DOI: 10.1144/SP290.6 0305-8719/07/$15.00

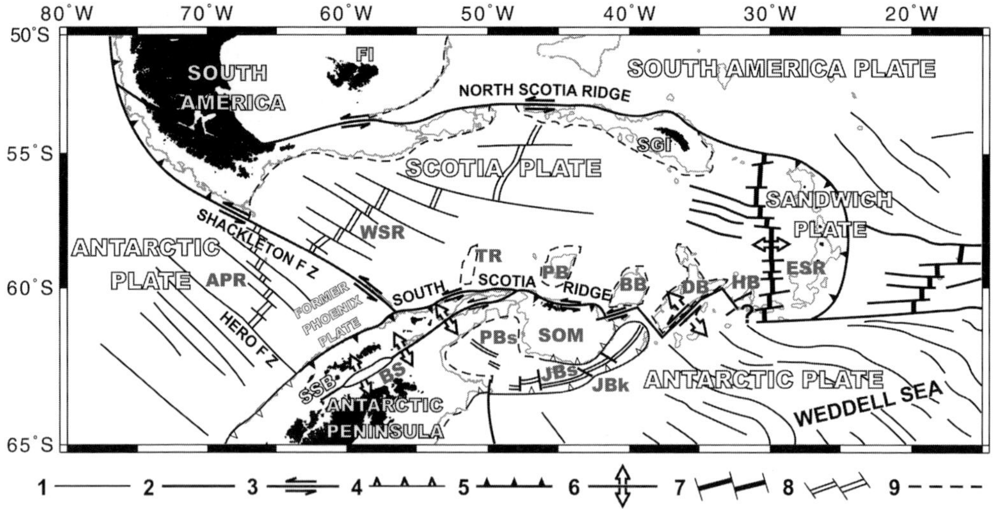

Fig. 1. Geological setting of the study area within the Scotia Arc (modified from Galindo-Zaldívar *et al.* 2002). 1, inactive fracture zone; 2, active fracture zone; 3, transform or transcurrent fault; 4, inactive subduction zone or reverse fault; 5, active subduction zone; 6, rift; 7, active spreading axis; 8, inactive spreading axis; 9, continental–oceanic crustal boundary; APR, Antarctic–Phoenix Ridge; BB, Bruce Bank; DB, Discovery Bank; ESR, East Scotia Ridge; FI, Falkland Islands; FZ, fracture zone; HB, Herdman Bank; JBk, Jane Bank; JBs, Jane Basin; PB, Pirie Bank; PBs, Powell Basin; SGI, South Georgia Island; SOM, South Orkney microcontinent; SSB, South Shetland Block; TR, Terror Rise; WSR, West Scotia Ridge.

arc have different characteristics. While the eastern boundary is determined by the present subduction of the South America plate below the Sandwich plate (Livermore *et al.* 1997; Larter *et al.* 2003; Livermore 2003), the western boundary is located at the Shackleton fracture zone, which is an intra-oceanic active sinistral transpressive NW–SE-oriented fault zone with sharp positive relief and small basins (Maldonado *et al.* 1998; Livermore *et al.* 2000, 2004; Fig. 1).

The tectonic evolution related to the plate boundary has determined the individualization and assemblage of a large amount of oceanic and continental elements. The southern part of the Scotia plate is a large oceanic basin formed by the oceanic spreading of the West Scotia Ridge and several small basins (Protector, Dove and Scan basins) bounded by thinned continental blocks (Terror Rise, and Pirie, Bruce and Discovery banks). The South Shetland Block constitutes an independent tectonic continental element separated from the northern Antarctic Peninsula by the Bransfield Strait (Aldaya & Maldonado 1996; Fig. 1). The Antarctic plate is also formed by continental and oceanic crusts. Northeastward of Antarctic Peninsula continental crust, lies the small oceanic Powell Basin, formed by the eastward drifting of the South Orkney microcontinent (SOM; King & Barker 1988; Barker *et al.* 1991; Rodríguez-Fernández *et al.* 1994, 1997; Eagles & Livermore 2002). The Jane Basin and Jane Bank, constituting an arc–back-arc system related to the subduction of the oceanic crust of the Weddell Sea, are located at the SE border of the SOM. Towards the east, a complex array of continental blocks south of the Discovery Bank shows evidence of tectonic activity associated with the present-day plate boundary (Maldonado *et al.* 1998; Galindo-Zaldívar *et al.* 2002; Fig. 1).

The plate boundaries around the Scotia Arc are highlighted by the distribution of earthquake epicentres (Fig. 2). Most seismic activity takes place at the eastern part of the arc, where the South America plate bends below the Sandwich plate in an active subduction zone. All deep (>150 km) and most intermediate (50–150 km) events are concentrated here, while the other boundaries are mainly described by shallow events (0–50 km). In addition, a second area of significant seismic activity is located at the Western South Scotia Ridge (WSSR), between the South Shetland Islands and South Orkney Islands (Fig. 2). Analysis of earthquake focal mechanisms indicates a regional stress regime characterized by NE–SW compression, with local stress perturbations (Pelayo & Wiens 1989; Galindo-Zaldívar *et al.* 1996; Giner-Robles *et al.* 2003; Thomas *et al.* 2003).

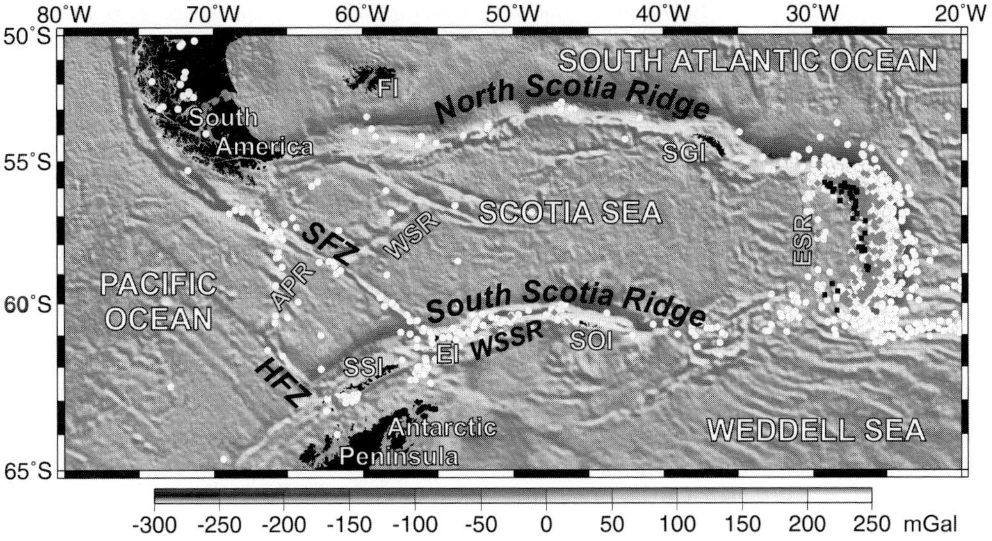

Fig. 2. GEOSAT free-air gravity anomaly map (Livermore *et al.* 1994; Sandwell & Smith 1997) including location of earthquake epicentres (1973–2004) (NEIC – National Earthquake Information Center; USGS http://www.neic.cr.usgs.gov). Shallow events (<50 km), white circles; intermediate (50 to 150 km), grey diamonds, and deep (>150 km), black squares. APR, Antarctic–Phoenix Ridge; EI, Elephant Island; ESR, East Scotia Ridge; FI, Falkland Islands; SFZ, Shackleton fracture zone; SGI, South Georgia Island; SOI, South Orkney Islands; SSI, South Shetland Islands; WSSR, Western South Scotia Ridge.

The aim of this contribution is to analyse in detail and summarize the deformation related to the Scotia–Antarctic plate boundary, focusing mainly on the development of deep extensional releasing-bend- and pull-apart basins within fragments of continental crust. We describe the main structures of the South Scotia Ridge on the basis of bathymetric, gravimetric and multi-channel seismic profiles, together with global gravity and bathymetry derived from satellite images (Sandwell & Smith 1997; Smith & Sandwell 1997). Seismicity data are also taken into account in determining present-day tectonic activity.

Data and methodology

Over the last 15 years, a number of cruises aboard the Spanish Vessel B/O Hespérides have made it possible to gather an important geophysical data-set along the southern branch of the Scotia Arc. This body of knowledge includes multi-channel seismic (MCS), gravity, magnetic and swath bathymetry data recorded along profiles orthogonal and parallel to the Scotia–Antarctic plate boundary. Most of the data used in this work stem from the HESANT92-93 and SCAN97 cruises.

The MCS profiles were obtained with a tuned array of 5 BOLT air guns with a total volume of 22.4 l, using a streamer with a total length of 2.4 km and 96 channels. The shot interval was 50 m. Data were recorded with a DFS V digital system and a sampling record interval of 2 ms with 10-s record lengths. The data were processed in sequence, including migration using a DISCO/FOCUS system; the processed profiles have a fold of 24. Swath bathymetry data were obtained using a SIMRAD EM12 multi-beam sounder.

Gravity data were acquired only during the SCAN97 cruise, with a Bell Aerospace TEXTRON BGM-3 marine gravimeter. The free-air anomaly was determined using Lanzada software (A. Carbó pers. comm. 2006). In addition, free-air gravity and bathymetric data from GEOSAT (Sandwell & Smith 1997; Smith & Sandwell 1997) were considered. Gravity models were developed by means of the GRAVMAG program (Pedley *et al.* 1993) using the free-air anomaly and taking into account the geometry obtained from MCS profiles, which locally also allow determination of the position of the Moho Densities for gravity modelling were attributed according to the mean values of each lithology (Telford *et al.* 1990) and the probable nature of crustal elements, providing for minor variation during modelling to reach the best fit (mantle, 3.35 g cm^{-3}; continental-crust basement, 2.67 g cm^{-3}; intermediate crust, 2.80 g cm^{-3}; oceanic crust basement 2.88 to 3.00 g cm^{-3}; sediments, 2.30 to 2.50 g cm^{-3} and sea-water 1.03 g cm^{-3}). These density values are in agreement with gravity models developed in the southern

Weddell Sea (Ferris et al. 2000), in a similar tectonic setting.

The Scotia–Antarctic plate boundary along the South Scotia Ridge

The southern boundary of the Scotia Arc is formed by a complex chain of high blocks and narrow, small basins developed in stretched continental crust, which together constitute the South Scotia Ridge (SSR), which extends from the NE end of Bransfield Strait as far as Discovery Bank. This continental barrier separates the oceanic crusts of the Scotia and Antarctic plates and represents part of the former continental connection between South America and the Antarctic Peninsula (Fig. 1).

Present-day tectonic deformation in this ridge, as evidenced by seismicity and MCS profiles, is mostly the result of sinistral faults of variable character. From east to west, they deform the central and southern part of Discovery Bank, located at the eastern part of the SSR, then continue along the northern boundary of the South Orkney microcontinent at the central SSR (Fig. 3). At the continental blocks of the western South Scotia Ridge, they show a sinistral transtensive character (Acosta & Uchupi 1996; Galindo-Zaldívar *et al.* 1996); and finally the displacement is transferred to the normal faults in the Bransfield Strait, located between the South Shetland Block and the Antarctic Peninsula (Acosta & Uchupi 1996; Aldaya & Maldonado 1996; Galindo-Zaldívar *et al.* 1996, 2004; González-Casado *et al.* 2000; Figs 1 & 3). The northern boundary of the central and western South Scotia Ridge is defined by a subduction zone or a reverse fault that superposes the continental block of the ridge over the oceanic crust of the Scotia plate. The westward extremity of the South Scotia Ridge is a complex region where the Shackleton fracture zone intersects the ridge. The thickened oceanic crust of the Shackleton fracture zone, with related positive relief that constitutes the boundary between the Antarctic and Scotia plates, is subducted along the northern boundary of the South Shetland continental block. This subduction produces crustal thickening and uplift of Elephant Island, where blueschists crop out (Trouw *et al.* 2000; Figs 1 & 3). Westward, this subduction zone and related South Shetland Trench are more active, separating the Antarctic plate (former Phoenix plate) from the South Shetland Block (Fig. 1).

Discovery Bank

Discovery Bank constitutes the largest continental fragment of the eastern part of the South Scotia Ridge (Figs 3–7). The available data suggest that this high is of a continental nature, as demonstrated by the shallow bathymetry of up to 700 m depth, the presence of large magnetic anomalies, due to massive basic igneous rocks intruded in the continental crust, typical of the Pacific margin of the Antarctic Peninsula (Garrett *et al.* 1987), the seismic tomography velocities (Vuan *et al.* 2005), and seismic facies observed in multi-channel

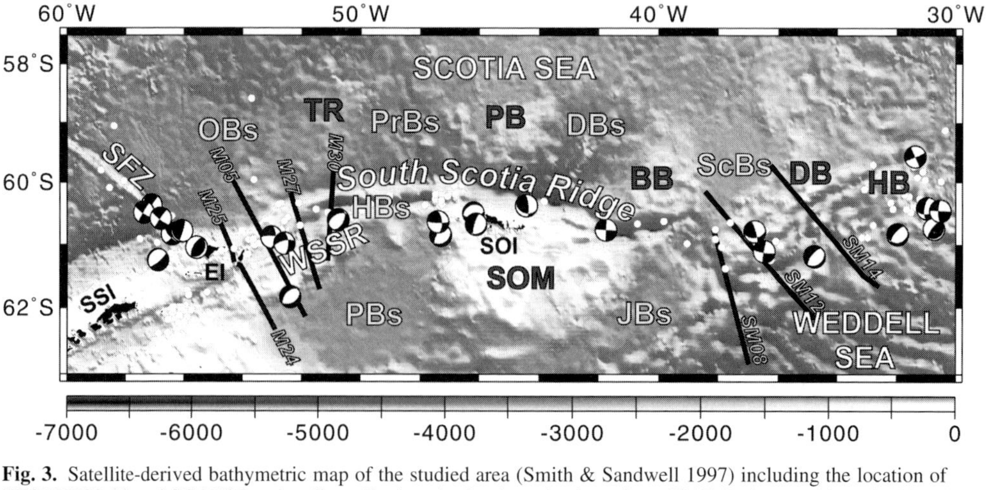

Fig. 3. Satellite-derived bathymetric map of the studied area (Smith & Sandwell 1997) including the location of HESANT92-93 and SCAN97 cruises lines. Earthquake focal mechanisms from CMT (Dzienwenski *et al.* 1981). BB, Bruce Bank; DB, Discovery Bank; DBs, Discovery Basin; EI, Elephant Island; HB, Herdman Bank; JBs, Jane Basin; OBs, Ona Basin; PB, Pirie Bank; PBs, Powell Basin; PrBs, Protector Basin; ScBs, Scan Basin; SFZ, Shackleton fracture zone; SOI, South Orkney Islands; SOM, South Orkney microcontinent; SSI, South Shetland Islands; TR, Terror Rise; WSSR, Western South Scotia Ridge.

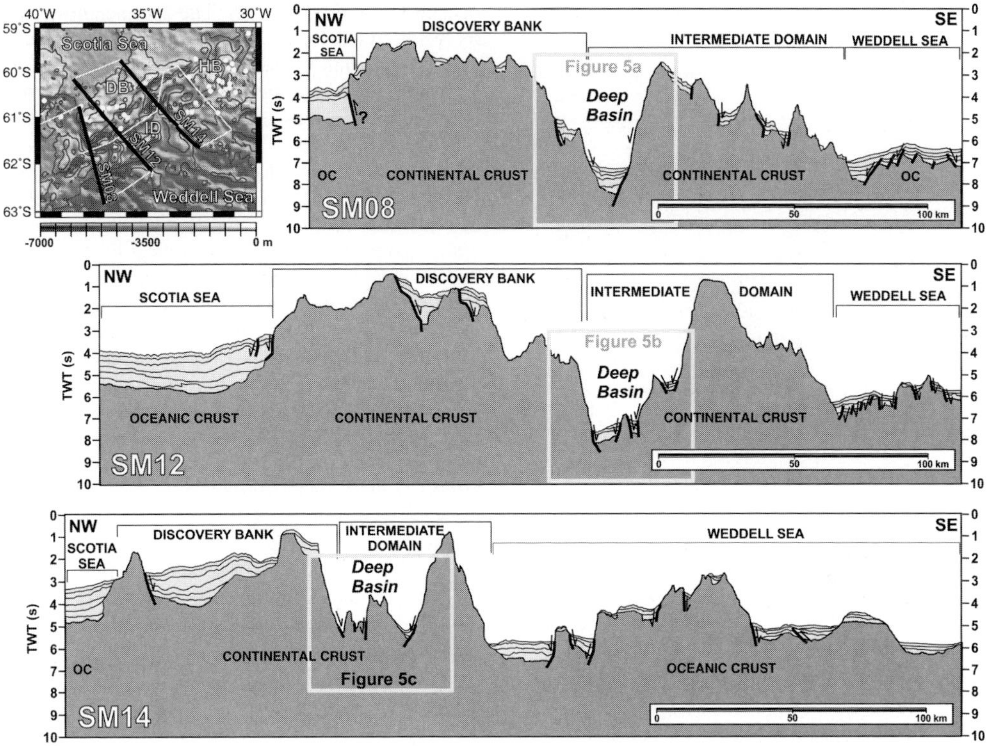

Fig. 4. SM08, SM12 and SM14 multi-channel seismic profiles and interpretation. Location given in Figure 3 and inset map. OC, oceanic crust. The positions of SMC details in Figure 5 are indicated. The bathymetry map includes the location of SMC profiles.

seismic profiles (Galindo-Zaldívar *et al.* 2002; Vuan *et al.* 2005). Previous methods do not, however, directly indicate the geometry of the Moho discontinuity. Gravimetric models developed along several profiles, together with the SMC profiles (Figs 4–6), led us to determine a crustal thickness of at least 14 km in the central part of the bank. North of the Discovery Bank is the oceanic crust of the Scotia Sea; this boundary is now covered by a thick practically undeformed sedimentary sequence (Fig. 4) that might represent an older passive margin. Some reverse or subvertical NE–SW-trending normal faults are identified locally.

The southern margin, meanwhile, is more complex: Discovery Bank is deformed by large normal faults with related scarps reaching up to 4500 m, isolating several perched basins, and developing a narrow, deep basin that may reach −5500 m in depth. Seismic profiles, gravity anomalies and available bathymetric data suggest that faults have ENE–WSW strikes and dip generally SE along the margin. The basin is bathymetrically irregular along strike; whereas to the west (profile SM08) the southern boundary is the sharpest boundary. In the central part (profile SM12) the basin is slightly tilted northward, with irregular borders. To the east (profile SM14), the basin becomes roughly symmetrical with a central high (Figs 4 & 5).

The thickness of the sedimentary record of this basin decreases progressively northeastward (Figs 4 & 5). Although the basement reaches great depths, the seismic profiles (Figs 4 & 5) show the basin to be floored by continental crust. Gravity models confirm that the basement is of a continental or intermediate nature, with a very shallow mantle position (Fig. 6).

Another secondary elongated high, parallel to the basin and to the southern border of the Discovery Bank, is identified, with gravity and MCS data suggesting its correspondence to an intermediate crustal domain (Figs 4 & 6). This high is largely deformed by normal faults. Towards the NE it is very narrow, while southwestward it becomes wider and asymmetrical, with progressive southern slopes where small perched basins are developed.

This intermediate crustal domain is bordered to the south by the oceanic crust of the Weddell Sea, which is slightly deeper than the Scotia Sea

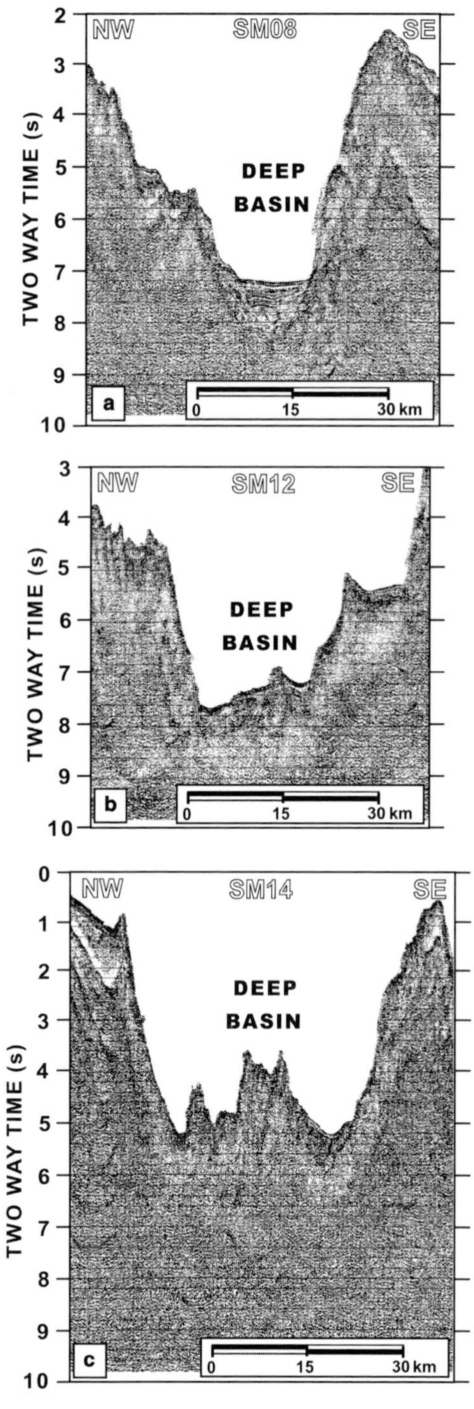

Fig. 5. Enlarged view of SM08 (**a**), SM12 (**b**) and SM14 (**c**) multi-channel seismic profiles showing the deep basin inside Discovery Bank. Locations are given in Figure 4.

oceanic crust (Figs 4 & 6). The sedimentary cover in the Weddell Sea is scarce and probably determined by oceanic currents (Maldonado *et al.* 2005).

South of Discovery Bank, the seismicity and exposed fault scarps indicate Recent and present-day activity on faults that border the Deep Basin (Figs 4–5 & 7). However, the continental–oceanic boundaries (northward with the Scotia Sea and southward with the Weddell Sea) are generally covered by sediments, indicating that these major contacts are tectonically inactive.

Although the geometry of the faults is determined by seismic profiles, gravity anomalies and available bathymetric data, the kinematics of the present-day active faults may be established on the basis of the vertical displacements observed from MCS profiles (Figs 4–5), and the earthquake focal mechanisms established by Pelayo & Wiens (1989) and CMT (Dziewonski *et al.* 1981; Harvard Seismology Centroid Moment Tensor Catalog (Fig. 3). In the eastern and central part of the southern border of the Discovery Bank, Intermediate Domain and Deep Basin, faults have a NE–SW orientation and accommodated normal-sense slip responsible for crustal thinning and basin development. Yet towards the SW part and inside the Discovery Bank, strike-slip earthquake focal mechanisms are dominant. The active faults are probably sinistral ENE–WSW to east–west faults, consistent with one of the nodal planes of the focal mechanisms (Figs 3–5 & 7). If the fault plane were north–south oriented, the fault would be dextral, producing displacements of the basin borders and closure of the eastern part of the Deep Basin. Because this is not supported by the observed data, we have to discard the latter interpretation.

South Orkney microcontinent

The South Orkney microcontinent (SOM) represents the largest (70 000 km^2) continental fragment of the southern part of the Scotia Arc. It is formed by low-grade metamorphic rocks of Triassic age, part of an accretionary wedge, and unconformable Upper Jurassic conglomerates that crop out in the South Orkney Islands (Trouw *et al.* 1997; Fig. 3). In the southern and eastern sector of the SOM and on the eastern margin of the Powell Basin, magnetic anomaly data show a high-amplitude anomaly (500–1000 nT) related to igneous basic rocks that may belong to the Pacific Margin Anomaly (Garrett *et al.* 1987; King & Barker 1988; Suriñach *et al.* 1997) or to the subduction of the Weddell Sea below the SOM (Bohoyo *et al.* 2002). In addition, alkaline basalts, Pliocene

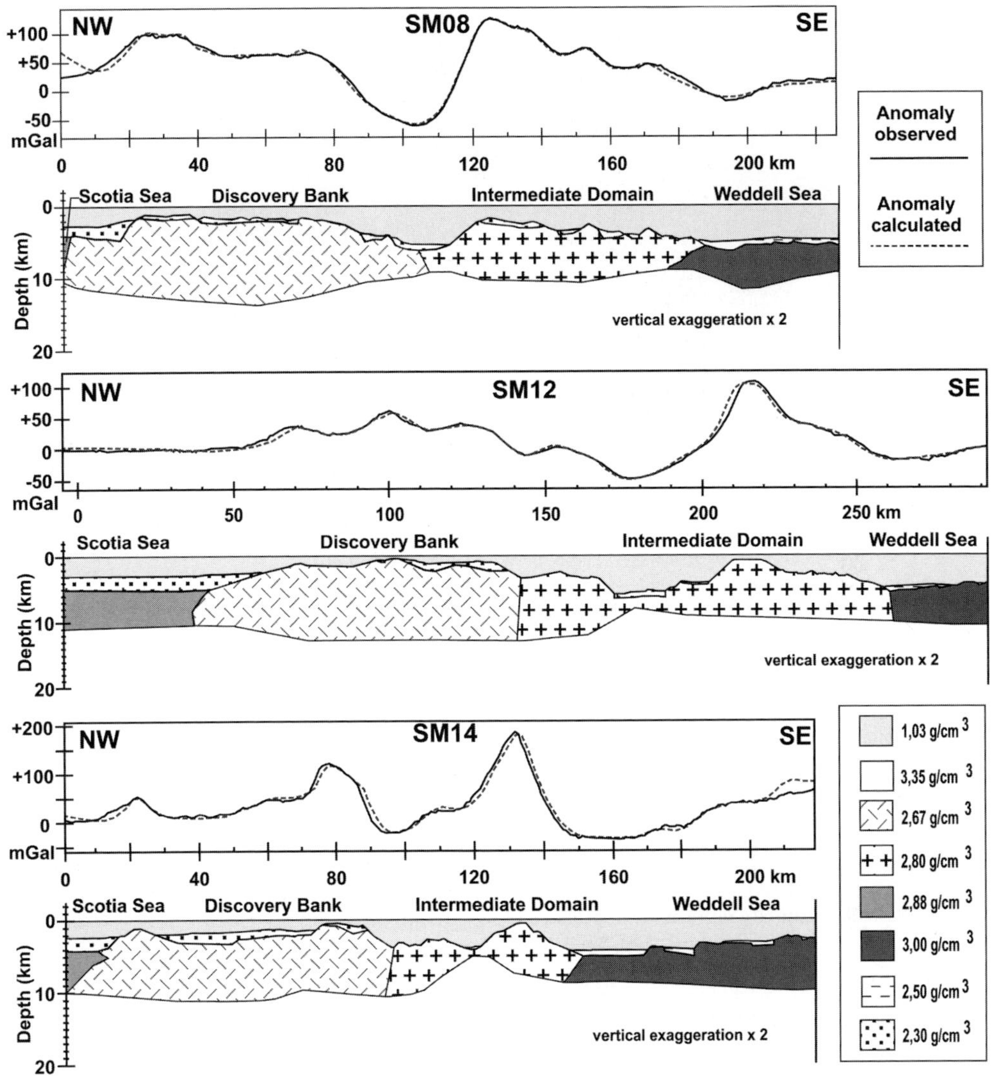

Fig. 6. Free-air gravity profiles and models of the Discovery Bank. See text for discussion.

to Recent in age, have been dredged in the northern and southern margins of the Powell Basin (Barber *et al.* 1991). The SOM is affected by normal faults that produce elongate horst and graben structures that determine the variability of the sedimentary cover from the Eocene to the present day. The more recent basins in the SOM are oriented north–south and are related to Oligocene rifting associated with the opening of the Powell Basin (King & Barker 1988).

The northern margin of the SOM is characterized by an active sinistral transcurrent fault zone affecting a curved plate boundary. Oblique relative motions produce differential strain with compressional, pure strike-slip and extensional sectors, depending on the boundary orientation with respect to the relative plate motion. Generally, the northern border of the SOM is characterized by a linear depression associated with a gravity minimum (Figs 2 & 3), that may be related to thrusting or sinistral transcurrent displacement of the SOM with respect to the Scotia Sea. Between the SOM and Bruce Bank, an east–west oriented depression with seismic activity and a strike-slip earthquake focal mechanism points to a purely sinistral regime on east–west oriented faults (Fig. 3). The NNE margin of the SOM, with WNW–ESE orientation, is characterized by a

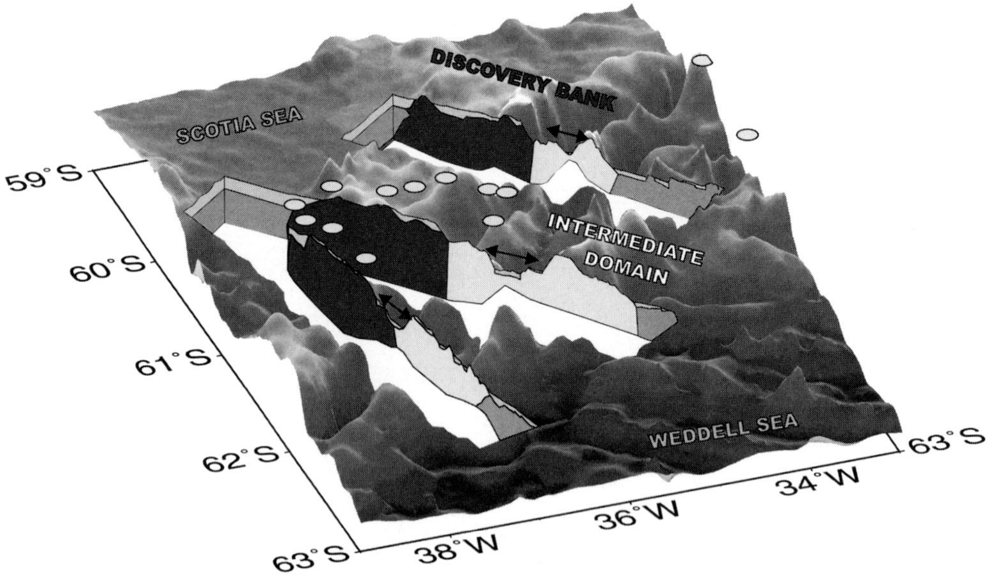

Fig. 7. Three-dimensional block of the Discovery Bank. In cross-sections the oceanic domains and sedimentary cover are represented in medium grey; the Discovery Bank continental domain are shown in dark grey; and the Intermediate Domain is shown in light grey. Earthquake epicentres are also indicated (light-grey circles).

convergent component that produces the subduction of the Scotia plate below the Antarctic plate, as proposed by Kavoun & Vinnikovskaya (1994); Klepeis & Lawver (1996); Lodolo *et al.* (1997); Maldonado *et al.* (1998) and Busetti *et al.* (2000). Towards the NW – even though the seismic activity is located near the continental–oceanic boundary – active faults with seismicity deform the continental margin, and also are located within the SOM, determining a cluster of epicentres (Fig. 3).

The structure of the internal regions of the SOM and the southern margin result from a heterogeneous continental fragmentation process, now inactive, with seismically inactive faults overlapped by uncut sediments. The process of thinning and stretching of continental crust through the eastward drifting of the SOM relative to the Antarctic Peninsula, between the Late Eocene and the Early Miocene (35–18 Ma), opened Powell Basin and defined the western boundary of the SOM. The oceanic spreading occurred between 26 and 17.6 Ma (Coren *et al.* 1997), although older ages have been proposed (29.7 to 21.1 Ma, Eagles & Livermore 2002). Discontinuous NW–SE extinct oceanic ridges are present in the middle part of the basin (King & Barker 1988; Coren *et al.* 1997; Rodríguez-Fernández *et al.* 1997); whereas fragmentation of the southern and eastern boundaries is tied to the development of the Jane Basin and the subduction of the oceanic crust of the Weddell Sea below the southern margin of the SOM (Bohoyo *et al.* 2002; Bohoyo 2004). The inactivity of the eastern, southern and western margins of the SOM, along with the oceanic spreading in Powell and Jane basins, determine that all these tectonic elements are included in the Antarctic plate from 14 Ma onwards (Rodríguez-Fernández *et al.* 1997; Bohoyo *et al.* 2002).

Western South Scotia Ridge

The Western South Scotia Ridge (WSSR; Fig. 3) is formed by continental crust, probably of a nature similar to that of the Antarctic Peninsula, located to the west, and to that of the SOM situated to the east (Fig. 3). Magnetic anomaly data (Suriñach *et al.* 1997) show the southern part of the ridge to be intruded by basic igneous rocks corresponding to the NE prolongation of the PMA (Pacific Margin Anomaly belt, Garrett *et al.* 1987). The basement is generally exposed in the highs (Fig. 8). The ages of the sediments that fill the internal basins are unknown.

The continental ridge is bound to the north by oceanic crust of the Ona and Protector basins, as well as the thinned continental crust of the Terror Rise that gave rise to the SW Scotia Sea (Fig. 3). The oceanic crust of the Ona Basin abyssal plain was formed by the extinct West Scotia Ridge, located in the centre of the Drake Passage (BAS, 1985). The magnetic anomaly chron C8r (27.027 Ma) is well identified (BAS, 1985; Eagles

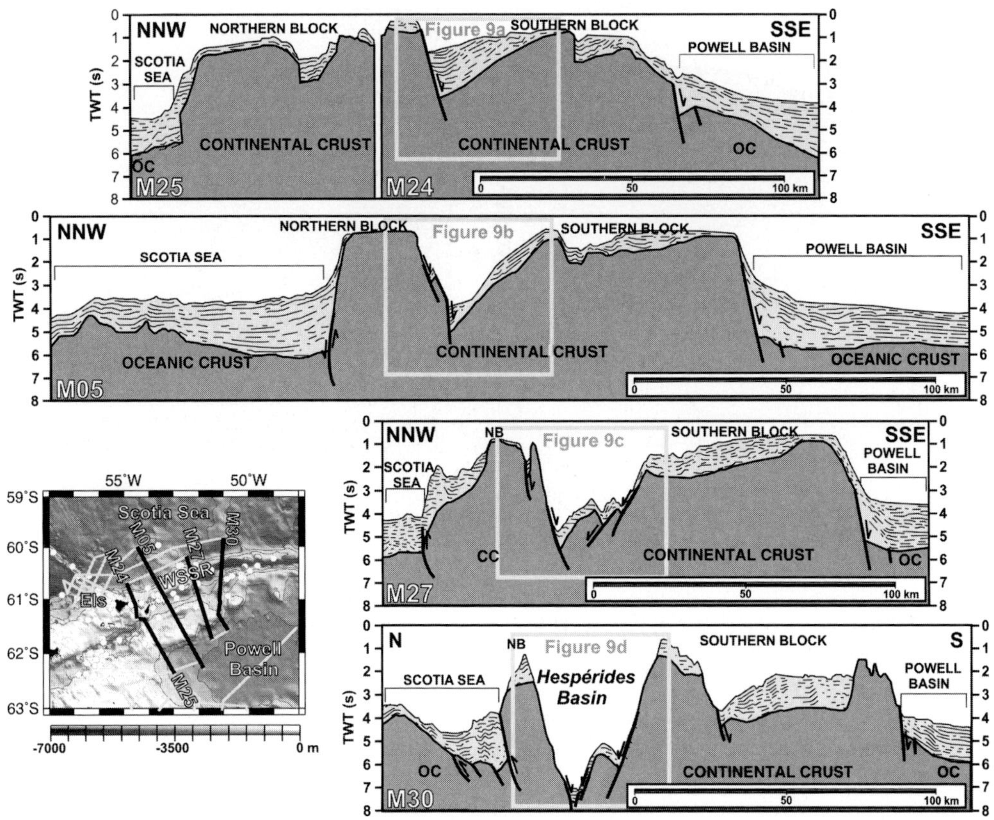

Fig. 8. M25, M24, M05, M27 and M30 multi-channel seismic profiles and interpretation. Location in Figure 3. OC, oceanic crust; CC, continental crust; and NB, Northern Block. The position of SMC details in Figure 9 are indicated. The inset bathymetric map includes the locations of SMC profiles.

et al. 2005), although chron C10 (28.7 Ma) has been recognized closest to WSSR (Lodolo *et al.* 1997; Geletti *et al.* 2005). The Terror Rise (Fig. 3), formed by thinned continental crust, represents a NNE–SSW-elongated elevation between 2000 m and 3500 m in depth separating the Ona and Protector basins. The Protector Basin (Fig. 3) is approximately 250 km wide and corresponds with an abyssal plain with water depths in the range of 3000 to 4000 m. Magnetic anomalies point to oceanic spreading between anomaly chron C5Dn (17.6 Ma) and C5ACn (14 Ma) (Galindo-Zaldívar *et al.* 2006).

The northern boundary of the WSSR continental crust is produced by a southward-dipping reverse fault, related to ancient subduction (Aldaya & Maldonado 1996; Lodolo *et al.* 1997); and locally, by an accretionary prism, covered by the most recent sediments (Fig. 8). Moreover, east of 52° W, there is evidence of recent compressive deformation and development of an east–west (Fig. 3). The reverse faults associated with this boundary have approximately east–west to WNW–ESE strikes (Galindo-Zaldívar *et al.* 1996).

The seismicity and well-exposed fault scarps observed in MCS profiles (Figs 3 & 8) indicate that most of the active deformation occurs in the axial depression formed by several connected basins, crossing the ridge in a slightly oblique sense. This axial depression is characterized by deep basins (e.g. the Hesperides Basin, among others) locally over 5500 m in depth and associated with high seismicity (Figs 3, 8 & 9). The depression is related to fault surfaces dipping southward, with great scarps that in the bathymetry show WSW–ENE and SW–NE strikes. The SW internal basins are characterized by northward tilted blocks and a wedge-shaped infilling typical of half-grabens. The sediments onlap the southern margins, where progressive unconformities develop, whereas they are affected by normal faults on the northern margins (profiles M24–25 and M05; Figs 8–10). Fault geometry and earthquake focal mechanisms indicate an active sinistral transtensive regime for the fault

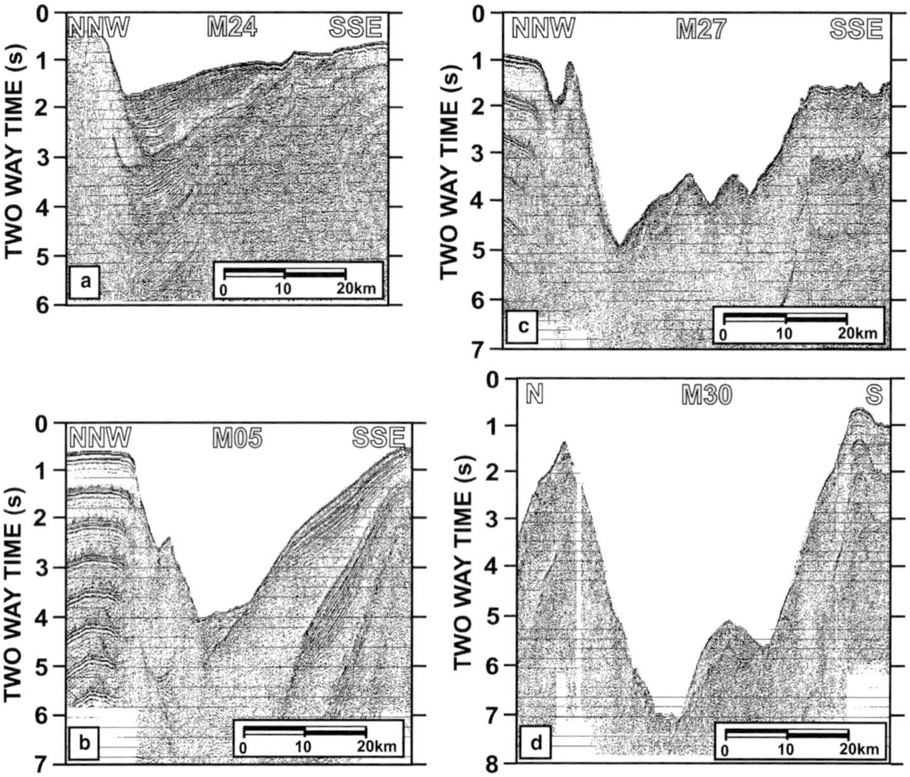

Fig. 9. Enlarged view of M24 (**a**), M05 (**b**), M27 (**c**) and M30 (**d**), multi-channel seismic profiles showing the internal basins and Hespérides Basin inside the Western South Scotia Ridge. Location in Figure 8.

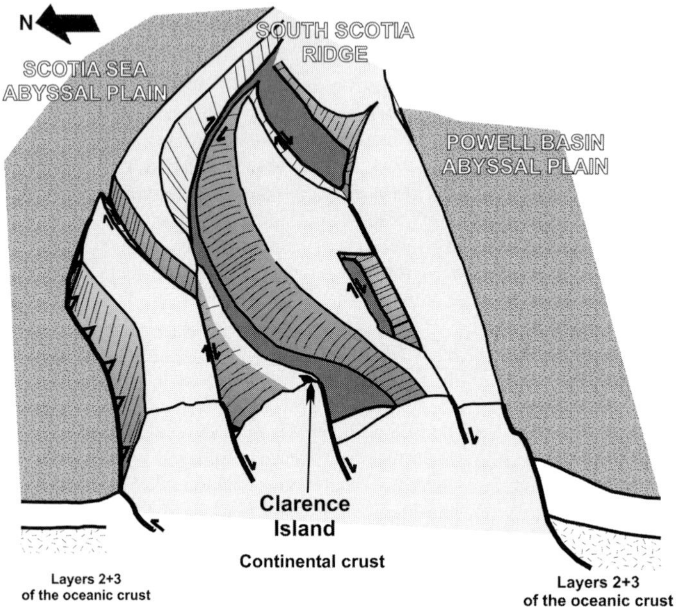

Fig. 10. Three-dimensional block model of the Western South Scotia Ridge.

system, although locally transpressive areas may develop in the eastern part of the ridge (Galindo-Zaldívar *et al.* 1996).

The southern crustal block of the Western South Scotia Ridge is connected with the oceanic crust of the Powell Basin, bounded by an inactive transform fault that extends eastward towards the South Orkney microcontinent. This transform fault was active during the opening of the Powell Basin (King *et al.* 1997; Figs 8 & 10).

The stress field, deduced from the P- and T-axes of earthquake focal mechanisms in the South Scotia Ridge, reveals a steeply inclined maximum compression and subhorizontal extension with a NW–SE trend, in agreement with the transtensive sinistral regime of the deformed zone, i.e. roughly east–west oriented. The axial depression of the WSSR extends to the Bransfield Strait rift and allows the definition of the South Shetland Block as a continental block that lies at the boundary between the Antarctic plate, the Scotia plate and the former Phoenix plate (Aldaya & Maldonado 1996).

Discussion and conclusions

The South Scotia Ridge traces the location of the Scotia–Antarctic plate boundary. A regional NE–SW compression and NW–SE extension in the Scotia Arc is evidenced by the analysis of earthquake focal mechanisms (Pelayo & Wiens 1989; Galindo-Zaldivar *et al.* 1996; Giner-Robles *et al.* 2003; Thomas *et al.* 2003). Furthermore, the entire South Scotia Ridge has been described as a sinistral transcurrent zone in accordance with the east–west orientation and the Scotia Arc stress regime. The tectonic models derived from such research predict the development of variable geological structures, including NW–SE extension in the WSSR and the easternmost SSR (Discovery Bank), and some segments of purely transcurrent sinistral faults at the NNE border of the SOM. Giner-Robles *et al.* (2003) suggest a transtensive regime located NNW of these islands. The fault sets that are in agreement with this general regime are N45°E-oriented normal faults, N135°E reverse faults, and strike-slip faults of intermediate orientations, according to their theoretical slip models.

The tectonic elements located near the plate boundary are heterogeneous in nature and constitute a very complex structure (Table 1). The rocks cropping out on the Antarctic Peninsula, South Shetland and South Orkney Islands (BAS 1985); the scarce direct data from ODP dredges and cores (Barker *et al.* 1984; Barber *et al.* 1991; Mao & Mohr 1995) and the geophysical data (gravity, magnetic, MCS) underline the continental nature of the submerged banks. While the Scotia plate is mainly formed by oceanic crust with submerged continental banks, some continental blocks are highly stretched, like the Terror Rise. In the Antarctic plate there are dominantly continental crustal elements corresponding to the NE promontory of the Antarctic Peninsula and the SOM. Other areas of oceanic crust (Powell Basin and Jane Basin) constitute small oceanic basins older than 14 Ma (Rodríguez-Fernández *et al.* 1997; Bohoyo *et al.* 2002; Bohoyo 2004).

This heterogeneous distribution of tectonic elements gives way to different rheological behaviour on the part of the continental, oceanic and intermediate blocks when subjected to an external stress field. The strength profiles proposed by Ranalli & Murphy (1987) for continental extensional and oceanic regions, indicate that the stress necessary to produce brittle deformation of continental lithosphere is lower than for oceanic lithosphere. The comparatively low resistance to deformation of continental crust favours the regional concentration of plate-boundary-related tectonic deformation within continental fragments

Table 1. *Main submerged banks around the Southern Scotia Arc and published sources that document their continental nature*

Submerged bank	Main references
Western South Scotia Ridge (WSSR)	Garrett *et al.* (1987); Acosta & Uchupi (1996); Galindo-Zaldívar *et al.* (1996); Suriñach *et al.* (1997); Maldonado *et al.* (1998)
South Orkney microcontinent (SOM)	Garrett *et al.* (1987); Barker & King (1988); Bohoyo *et al.* (2002); Bohoyo (2004)
Terror Rise and Pirie Bank (TR & PB)	Bohoyo (2004); Galindo-Zaldívar *et al.* (2006)
Bruce Bank (BB)	Mao & Mohr (1995); SCAN2004 cruise preliminary report
Discovery Bank (DB)	Garrett *et al.* (1987); Galindo-Zaldívar *et al.* (2002); Vuan *et al.* (2005)

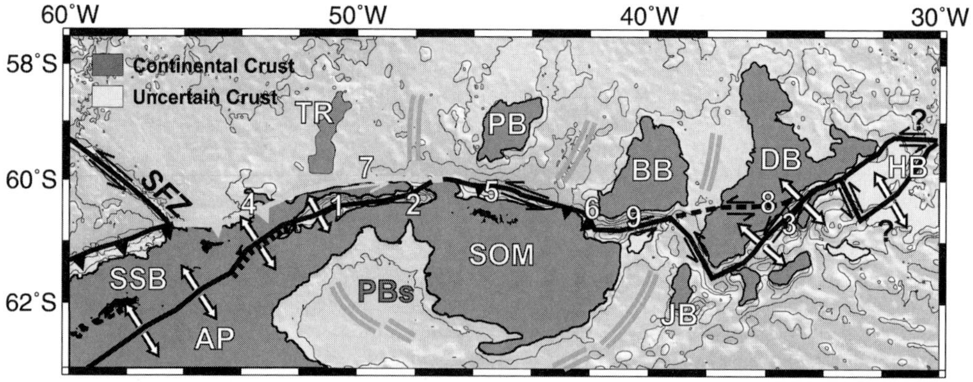

Fig. 11. Geological sketch of the main tectonic structures and the crustal nature of the submerged banks. Numbers referring to main tectonic features are shown in Table 2. Active structures shown as black lines, and inactive structures as grey lines. AP, Antarctic Peninsula; BB, Bruce Bank; DB, Discovery Bank; HB, Herdman Bank; JB, Jane Bank; PB, Pirie Bank; PBs, Powell Basin; SFZ, Shackleton fracture zone; SOM, South Orkney microcontinent; SSB, South Shetland Block; TR, Terror Rise.

(Galindo-Zaldívar *et al.* 2002). The current sinistral transcurrent boundary became active after the opening of all the small basins in the southern Scotia Sea (Galindo-Zaldívar *et al.* 2006).

In the eastern part of the SSR, the seismicity and the active faults are clearly concentrated in Discovery Bank and the Intermediate Domain. Although most of the active faults are associated with the development of the Deep Basin, there is also tectonic activity inside the Discovery Bank, producing incipient E–W scarps and epicentre alignments (Figs 4–5, 7 & 11).

Table 2. *Main tectonic structures associated with the present Scotia–Antarctic plate boundary*

Extensional regime: pull-apart basin and normal faults (location numbers shown in Fig. 11)	Location	Main references
1	Western South Scotia Ridge basins	Galindo-Zaldívar *et al.* (1996); this study
2	Hespérides basin in WSSR	Acosta & Uchupi (1996); Galindo-Zaldívar *et al.* (1996); this study
3	Deep basin on Discovery Bank	Galindo-Zaldívar *et al.* (2002); this study
Compressive regime: subduction, thrusting and folding (ID in map)		
4	WSSR	Galindo-Zaldívar *et al.* (1996); Lodolo *et al.* (1997); this study
5	Northern SOM	Kavoun & Vinnikovskaya (1994); Maldonado *et al.* (1998); Bussetti *et al.* (2000)
6	NE SOM	Leitchenkov pers. comm.
Pure strike-slip regime: transcurrent faults		
7	WSSR	Galindo-Zaldívar *et al.* (1996); this study
8	Discovery Bank	Galindo-Zaldívar *et al.* (2002); this study
9	NE SOM	This study

The active structures continue westward along the northern border of the SOM, where a present-day active oceanic–continental boundary and related trench and compressive structures are recognized (Maldonado *et al.* 1998; Busetti *et al.* 2000) (Fig. 11). Westward, in the WSSR, the contact between the oceanic crust of the Scotia Sea and the northern continental block of WSSR is a reverse fault. However, this part of the oceanic–continental boundary appears inactive, and no well-developed trench is present (Fig. 11). The inactivity along the boundary, and the presence of oceanic crust in the Powell Basin to the south would have favoured the localization of deformation within the WSSR continental blocks.

The distribution of continental crustal blocks between regions of oceanic crust gives rise to a different trend for each sector of the Scotia–Antarctic plate boundary, which is in turn responsible for the development of transtensive, pure strike-slip, and transpressive structures (Table 2). Whereas E–W oriented faults are generally subvertical and accommodate a purely sinistral regime, where the faults are oriented in a more northeasterly direction, they become progressively more transtensional, and conjugate fault sets dipping NW and SE accommodate development of deep basins within the continental blocks. This structural situation is observed both in the eastern part of the SSR, where the Deep Basin develops south of the Discovery Bank, and in the western sector, where an elongated deep basin (Hespérides Basin and other connected basins) divides the WSSR into its northern and southern ridges (Figs 8–11).

These basins are asymmetrical in the WSSR, featuring sedimentary wedge infill above hanging walls that tilt to the north. The Deep Basin located south of the Discovery Bank is more symmetrical. This contrasting behaviour may be due to the geological nature of the deformed blocks, which are continental in the WSSR and include previous structural heterogeneities, but one of an intermediate crustal nature in the Deep Basin of southern Discovery Bank.

The available present-day data do not allow for a more detailed study of these sectors. Interpretations of free-air gravity anomalies for the region (Livermore *et al.* 1994; Sandwell & Smith 1997; Fig. 2) suggest that the deep basins of the WSSR develop mainly as a consequence of releasing bends of fault zones; while in the Deep Basin of Discovery Bank we may invoke a pull-apart basin model confined, at least to the SW, by a transform fault (Fig. 11). We conclude that the behaviour of deep basins developing inside continental blocks is similar: they attain depths of over 5000 m, surpassing the depths of the surrounding oceanic crust, and show no evidence of oceanic spreading. The evolution of these basins, formed in regional transcurrent tectonic regimes, can be clearly differentiated from the classic development of passive margins during continental stretching.

The Scotia–Antarctic plate boundary, along the southern boundary of the Scotia Arc, therefore constitutes an exceptional example of complex interplate deformation driven by regional tectonic stresses and influenced in a complex way by the different rheological behaviour of oceanic and continental crust.

We thank the Commander, officers and crew of the BIO Hesperides for their support in obtaining these data, sometimes under severe conditions. The diligence and expertise of engineer E. Litcheva, who processed the MCS data, is appreciated. The comments of Dr D. Cunningham and Dr M. Menichetti have largely improved the contents of the paper. Spain's CICYT supported this research through project CGL2004-05646/ANT and MEC postdoctoral research programme.

References

ACOSTA, J. & UCHUPI, E. 1996. Transtensional tectonics along the South Scotia Ridge, Antarctica. *Tectonophysics*, **267**, 31–56.

ALDAYA, F. & MALDONADO, A. 1996. Tectonics of the triple junction at the southern end of the Shackleton Fracture Zone (Antarctic Peninsula). *Geo-Marine Letters*, **16**, 279–286.

BARBER, P. L., BARKER, P. F. & PANKHURST, R. J. 1991. Dredged rocks from Powell basin and the South Orkney microcontinent. *In*: THOMSON, M. R. A., CRAME, J. A. & THOMSON, J. W. (eds) Geological evolution of the Antarctica. Cambridge University Press, Cambridge, 361–367.

BARKER, P. F. 2001. Scotia Sea regional tectonics evolution: implications for mantle flow and palaeocirculation. *Earth Science Reviews*, **55**, 1–39.

BARKER, P. F. & BURRELL, J. 1977. The opening of Drake Passage. *Marine Geology*, **25**, 15–34.

BARKER, P. F., BARBER, P. G. & KING, E. C. 1984. An Early Miocene ridge crest–trench collision on the South Scotia Ridge near 36° W. *Tectonophysics*, **102**, 315–332.

BARKER, P. F., DALZIEL, I. W. D. & STOREY, B. C. 1991. Tectonic development of the Scotia Arc region. *In*: TINGEY, R. J. (ed.) *Antarctic Geology*. Oxford University Press, Oxford, 215–248.

BAS 1985. *Tectonic Map of the Scotia Arc*. 1:3 000 000. British Antarctic Survey, Cambridge.

BOHOYO, F. 2004. *Fragmentación continental y desarrollo de cuencas oceánicas en el sector meridional del Arco de Scotia, Antártida*. PhD thesis, Granada University, Spain, 252 pp.

BOHOYO, F., GALINDO-ZALDÍVAR, J., MALDONADO, A., SCHREIDER, A. A. & SURIÑACH, E. 2002. Basin development subsequent to ridge–trench collison: the Jane Basin, Antarctica. *Marine Geophysical Researches*, **23**, 413–421.

BUSETTI, M., ZANOLLA, M. & MARCHETTI, A. 2000. Geological structure of the South Orkney Microcontinent. *Terra Antactica*, **8/2**, 1–8.

COREN, F., CECCONE, G., LODOLO, E., ZANOLLA, C., ZITELLINI, N., BONAZZI, C. & CENTONZE, J. 1997. Morphology, seismic structure and tectonic development of the Powell Basin, Antarctica. *Journal of the Geological Society, London*, **154**, 849–862.

DZIEWONSKI, A. M., CHOU, T. A. & WOODHOUSE, J. H. 1981. Determination of earthquake source parameters from waveform data for studies of global and regional seismicity. *Journal of Geophysical Research*, **86**, 2825–2852.

EAGLES, G. & LIVERMORE, R. A. 2002. Opening history of Powell Basin, Antarctic Peninsula. *Marine Geology*, **185**, 195–202.

EAGLES, G., LIVERMORE, R. A., FAIRHEAD, J. D. & MORRIS, P. 2005. Tectonic evolution of the west Scotia Sea. *Journal of Geophysical Research*, **110**, doi:10.1029/2004JB003154.

FERRIS, J. K., VAUGHAN, A. P. M. & STOREY, B. C. 2000. Relics of a complex triple junction in the Weddell Sea embayment, Antarctica. *Earth and Planetary Science Letters*, **178**, 215–230.

GALINDO-ZALDÍVAR, J., BALANYÁ, J. C. *ET AL.* 2002. Active crustal fragmentation along the Scotia–Antarctic plate boundary east of the South Orkney Microcontinent (Antarctica). *Earth and Planetary Science Letters*, **204**, 33–46.

GALINDO-ZALDÍVAR, J., BOHOYO, F., MALDONADO, A., SCHREIDER, A. A., SURIÑACH, E. & VAZQUEZ, J. T. 2006. Propagating rift during the opening of a small oceanic basin: the Protector Basin (Scotia Arc, Antarctica). *Earth and Planetary Science Letters*, **241**, 398–412.

GALINDO-ZALDÍVAR, J., GAMBÔA, L. A. P., MALDONADO, A., NAKAO, S. & BOCHU, Y. 2004. Tectonic development of the Bransfield Basin and its prolongation to the South Scotia Ridge, northern Antarctic Peninsula. *Marine Geology*, **206**, 267–282.

GALINDO-ZALDÍVAR, J., JABALOY, A., MALDONADO, A. & SANZ DE GALDEANO, C. 1996. Continental fragmentation along the South Scotia Ridge transcurrent plate boundary (NE Antarctic Peninsula). *Tectonophysics*, **242**, 275–301.

GARRETT, S. W., RENNER, R. G. B., JONES, J. A. & MCGIBBON, K. J. 1987. Continental magnetic anomalies and the evolution of the Scotia arc. *Earth and Planetary Science Letters*, **81**, 273–281.

GELETTI, R., LODOLO, E., SCHREIDER, A. A. & POLONIA, A. 2005. Seismic structure and tectonics of the Shackleton Fracture Zone (Drake Passage, Scotia Sea). *Marine Geophysical Researches*, **26**, 17–28.

GINER-ROBLES, J. L., GONZALEZ-CASADO, J. M., GUMIEL, P., MARTIN-VELAZQUEZ, S. & GARCIA-CUEVAS, C. 2003. A kinematic model of the Scotia plate (SW Atlantic Ocean). *Journal of South American Earth Sciences*, **16**, 179–191.

GONZÁLEZ-CASADO, J. M., GINER-ROBLES, J. L. & LÓPEZ-MARTÍNEZ, J. 2000. Bransfield Basin, Antarctic Peninsula: not a normal back-arc basin. *Geology*, **28**, 1043–1046.

KAVOUN, M. & VINNIKOVSKAYA, O. 1994. Seismic stratigraphy and tectonics of the northwestern Weddell Sea (Antarctica) inferred from marine geophysical surveys. *Tectonophysics*, **240**, 299–341.

KING, E. C. & BARKER, P. F. 1988. The margins of the South Orkney microcontinent. *Journal of the Geological Society, London*, **145**, 317–331.

KING, E. C., LEITCHENKOV, G., GALINDO-ZALDÍVAR, J., MALDONADO, A. & LODOLO, E. 1997. Crustal structure and sedimentation in Powell Basin. *In*: BARKER, P. F. & COOPER, A. (eds), *Geology and Seismic Stratigraphy of the Antarctic Margin*. Part 2. American Geophysical Union, Washington, DC, **71**, 75–93.

KLEPEIS, K. A. & LAWVER, L. A. 1996. Tectonics of the Antarctic–Scotia plate boundary near Elephant and Clarence Islands, West Antarctica. *Journal of Geophysical Research*, **101-B9**, 20 211–20 231.

LARTER, R. D., VANNESTE, L. E., MORRIS, P. & SMYTHE, D. K. 2003. Structure and tectonic evolution of the South Sandwich arc. *In*: LARTER, R. D. & LEAT, P. T. (eds) *Intra-Oceanic Subduction Systems: Tectonic and Magmatic Processes*. Geological Society, London, Special Publications, **219**, 255–284.

LIVERMORE, R. A. 2003. Back-arc spreading and mantle flow in the East Scotia Sea. *In*: LARTER, R. D. & LEAT, P. T. (eds) *Intra-Oceanic Subduction Systems: Tectonic and Magmatic Processes*. Geological Society, London, Special Publications, **219**, 255–284.

LIVERMORE, R. A., BALANYÁ, J. C. *ET AL.* 2000. Autopsy on a dead spreading centre: the Phoenix Ridge, Drake Passage, Antarctica. *Geology*, **28**, 607–610.

LIVERMORE, R. A., CUNNINGHAM, A. P., LARTER, R. D. & VANNESTE, L. E. 1997. Subduction influence on magma supply at the East Scotia Ridge. *Earth and Planetary Science Letters*, **150**, 261–275.

LIVERMORE, R. A., EAGLES, G., MORRIS, P. & MALDONADO, A. 2004. Shackleton Fracture Zone: no barrier to early circumpolar ocean circulation. *Geology*, **32**, 797–800.

LIVERMORE, R. A., MCADDO, D. C. & MARKS, K. M. 1994. Scotia Sea tectonics from high-resolution satellite gravity. *Earth and Planetary Science Letters*, **123**, 255–268.

LODOLO, E., COREN, R., SCHREIDER, A. A. & CECCONE, G. 1997. Geophysical evidence of a relict oceanic crust in the South-western Scotia Sea. *Marine Geophysical Researches*, **19**, 439–450.

MALDONADO, A., BARNOLAS, A. *ET AL.* 2005. Miocene to recent contourite drift development in the northern Weddell Sea (Antarctica). *Global and Planetary Change*, **45**, 99–129.

MALDONADO, A., ZITELLINI, N. *ET AL.* 1998. Small ocean basin development along the Scotia–Antarctica plate boundary and in the northern Weddell Sea. *Tectonophysics*, **296**, 371–402.

MAO, S. & MOHR, B. A. R. 1995. Middle Eocene dinocysts from Bruce Bank (Scotia Sea, Antarctica) and their paleoenvironmental and paleogeographic implications. *Review of Palaeobotany and Palynology*, **86**, 235–263.

PEDLEY, R. C., BUBSBY, J. P. & DABEK, Z. K. 1993. *GRAVMAG 1.7 (2.5 D)*, British Geological Survey, London.

PELAYO, A. M. & WIENS, D. A. 1989. Seismotectonics and relative plate motions in the Scotia Sea Region. *Journal of Geophysical Research*, **94**, 7293–7320.

RANALLI, G. & MURPHY, D. C. 1987. Rheological stratification of the lithosphere. *Tectonophysics*, **132**, 281–295.

RODRÍGUEZ-FERNÁNDEZ, J., BALANYA, J. C., GALINDO-ZALDÍVAR, J. & MALDONADO, A. 1994. Margin styles of Powell Basin and their tectonic implications (NE Antarctic Peninsula). *Terra Antartica*, **1**, 303–306.

RODRÍGUEZ-FERNÁNDEZ, J., BALANYA, J. C., GALINDO-ZALDÍVAR, J. & MALDONADO, A. 1997. Tectonic evolution and growth patterns of a restricted ocean basin: the Powell Basin (northeastern Antarctic Peninsula). *Geodinamica Acta*, **10**, 159–174.

SANDWELL, D. T. & SMITH, W. H. F. 1997. Marine gravity anomaly from Geosat and ERS-1 satellite altimetry. *Journal of Geophysical Research*, **102**, 10 039–10 054.

SMITH, W. H. F. & SANDWELL, D. T. 1997. Global seafloor topography from satellite altimetry and ship depth soundings. *Science*, **277**, 1957–1962.

SURIÑACH, E., GALINDO-ZALDÍVAR, J., MALDONADO, A. & LIVERMORE, R. A. 1997. Large amplitude magnetic anomalies in the northern sector of the Powell Basin, NE Antarctic Peninsula. *Marine Geophysical Researches*, **19**, 65–80.

TELFORD, W. M., GELDART, L. P. & SHERIFF, R. E. 1990. *Applied Geophysics* (2nd ed). Cambridge University Press, Cambridge, 770 pp.

THOMAS, C., LIVERMORE, R. A. & POLLITZ, F. 2003. Motion of the Scotia Sea plates. *Geophysical Journal International*, **155**, 789–804.

TROUW, R. A. J., PASSCHIER, C. W., SIMOES, L. S. A., ANDREIS, R. R. & VALERIANO, C. M. 1997. Mesozoic tectonic evolution of the South Orkney Microcontinent, Scotia arc, Antarctica. *Geological Magazine*, **134**, 383–401.

TROUW, R. A. J., PASSCHIER, C. W., VALERIANO, C. M., SIMOES, L. S., PACIULLO, F. V. P. & RIBEIRO, A. 2000. Deformational evolution of a Cretaceous subduction complex: Elephant Island, South Shetland islands, Antarctica. *Tectonophysics*, **319**, 93–110.

VUAN, A., LODOLO, E., PANZA, G. F. & SAULI, C. 2005. Crustal structure beneath Discovery Bank in the Scotia Sea from group velocity tomography and seismic reflection data. *Antarctic Science*, **17**, 97–106.

Structural and topographic characteristics of restraining bend mountain ranges of the Altai, Gobi Altai and easternmost Tien Shan

D. CUNNINGHAM

Department of Geology, University of Leicester, Leicester Road, Leicester, LE1 7RH, UK (e-mail: wdc2@le.ac.uk)

Abstract: Restraining bend mountain ranges are fundamental orogenic elements in the Altai, Gobi Altai and eastern Tien Shan. In this paper, 12 separate restraining bends are reviewed to identify common structural and topographic characteristics. The 12 restraining bends occur in one of three different tectonic settings: (1) strike-slip fault termination zones; (2) at a major strike-slip fault bend where the individual strike-slip fault can be traced continuously from one end of the range to the other; and (3) where two separate strike-slip fault segments converge and overlap. Fault maps of the 12 separate bends reveal that they are all flower or half-flower structures in cross-section, but there is considerable architectural diversity and all have unique individual topographic, structural and dimensional characteristics. Many factors account for the architectural diversity of the restraining bend mountains, especially stepover width, total amounts of strike-slip displacement, reactivation of older structures, tectonic setting, and the angular relation between fault trace and maximum horizontal stress. The stepover sense for regionally important strike-slip faults is controlled by pre-existing basement heterogeneities and is dominantly contractional. Therefore, releasing bends and transtensional basins are largely absent. Throughout the region there is a continuum of mountain range types, from purely contractional ridges to isolated restraining bends along strike-slip-dominated zones. Nucleation, topographic uplift, along- and across-strike growth of the bend, and restraining bend coalescence with adjacent ranges appears to be an important mountain-building process in the Altai, Gobi Altai and eastern Tien Shan; similar processes are likely in other intracontinental transpressional orogens.

Restraining bend mountain ranges are important tectonic landforms that commonly develop where crustal shortening is focused at strike-slip fault terminations, at contractional stepovers between parallel strike-slip segments, or where continuous strike-slip faults bend along strike (Crowell 1974; Biddle & Christie-Blick 1985). Examples are reported from all continents (Mann 2007), marine settings (Pockalny 1997; Dolan & Mann 1998; Legg *et al.* 2004), and even on other planets (McBee *et al.* 2003). They may form small hills, ridges, or major mountain ranges and they may occur along transform boundaries, along intracontinental strike-slip faults and within interplate and intraplate transpressional orogens. They are complex structural zones where oblique deformation often predominates, and where large earthquakes may nucleate (Barka & Kadinsky-Cade 1988; Schwartz *et al.* 1990). In addition, because restraining bend mountain ranges commonly expose crystalline basement rocks due to thrusting and erosion, they may provide the only windows into a region's older crustal history, including its formation and subsequent deformation (e.g. Blue Mountains, Jamaica, Mann *et al.* 1985). Restraining bends may also host important mineral and hydrocarbon resources in their interiors and flanking basins (e.g. Muir 2002; Wright 1991; Song & Cawood 1999).

The structural architecture and topographic evolution of restraining bends depends on many factors. The most important are the following:

1. the total amount of strike-slip displacement accommodated at the bend;
2. whether the bend formed at a strike-slip fault termination, at a gentle bend along a single strike-slip fault, or where parallel strike-slip fault segments overlap (and if there is overlap, the degree of overlap);
3. the original width of the stepover or bend;
4. the existence of pre-existing faults and bulk rheological heterogeneity ('structural grain') in the crust and the degree to which it is reactivated by faults that are responsible for constructing the range;
5. the long-term average slip rate on the master strike-slip fault system;
6. long-term climate patterns and mountain erosion rates that compete with mountain uplift;
7. the amount of strain partitioning of oblique deformation within the bend into separate thrust and strike-slip displacements;

From: CUNNINGHAM, W. D. & MANN, P. (eds) *Tectonics of Strike-Slip Restraining and Releasing Bends.* Geological Society, London, Special Publications, **290**, 219–237.
DOI: 10.1144/SP290.7 0305-8719/07/$15.00

8. the extent of local vertical axis rotations within the bend and the amount of vertical axis rotations in the larger region that the bend occurs within;
9. the extent to which strain hardening processes operate within the system and whether generation of significant topography favours fault abandonment and initiation of new faults;
10. the orientation of maximum horizontal stress ($S_{H_{max}}$) and whether the angle between $S_{H_{max}}$ and the evolving bend is constant through time or variable.

Because all of these factors will be different for every restraining bend, it follows that restraining-bend mountain ranges should be diverse in topography, geomorphology, architecture and evolutionary development. Therefore, sandbox models that attempt to recreate restraining-bend nucleation and growth, and to account for their diversity, should consider all these factors (e.g. McClay & Bonora 2001). In this paper, 12 separate, actively forming restraining bends from the Altai, Gobi Altai and easternmost Tien Shan in Central Asia are analysed and compared in order to document important common features and developmental processes, and to establish their overall importance as major building blocks within larger orogenic belts. Results from previous fieldwork published by the author and others for the Altai (Cunningham *et al.* 1996*a*, 2003*a*; Cunningham 2005), Gobi Altai (Cunningham *et al.* 1996*b*, 1997; Cunningham 1998, 2005; Bayasgalan *et al.* 1999*a*, *b*) and easternmost Tien Shan (Cunningham *et al.* 2003*c*) are combined with new analyses of Landsat TM and SRTM 90-minute topographic data.

Geology and active tectonics of the Altai, Gobi Altai and easternmost Tien Shan

All 12 restraining bends discussed in this paper occur in an intracontinental and intraplate setting in Mongolia and western China and therefore have evolved in a setting very different from that of transform-boundary restraining bends such as the California Transverse Ranges or Lebanon Ranges (Fig. 1). The Central Asian examples discussed here are found within diffuse regions of active transpressional deformation and mountain building which lack a dominant San Andreas or Dead Sea transform-type fault system. Fault displacements in the Altai, Gobi Altai and easternmost Tien Shan are ultimately driven by NNE $S_{H_{max}}$ due to the Indo-Eurasia collision 2500 km to the south (Tapponnier & Molnar 1979; Cobbold & Davy 1988). Late Cenozoic reactivation of the entire region began in the Oligocene, but has been most active since the Late Miocene (approximately 10 Ma–present, Baljinnyam *et al.* 1993; Vassallo *et al.* 2006). Major reviews of the active fault systems and regional seismicity of the Altai, Gobi Altai and easternmost Tien Shan can be found in Baljinnyam *et al.* 1993; Cunningham *et al.* 2003*c*; Bayasgalan *et al.* 2005; Cunningham 2005). GPS data indicate generally NNE-directed crustal displacements in the Altai region (*c.* 5 mm/a) and eastward-directed displacements in the Gobi Altai region (*c.* 3–4 mm/a) relative to a fixed Siberia (Calais *et al.* 2003).

Prior to Late Cenozoic reactivation, the entire region experienced a complex Phanerozoic history, involving multiple Palaeozoic terrane accretion events which amalgamated the crust and caused several major mountain-building events (Badarch *et al.* 2002). These events led to an overall NW–SE basement grain in the Altai, and western Gobi Altai, and a more east–west grain in the easternmost Tien Shan and southern Gobi Altai. During the Jurassic–Cretaceous, most of the eastern Altai and Gobi Altai region experienced crustal extension and rift basin development (Traynor & Sladen 1995). By the end of the Cretaceous and Early Tertiary, the region was stable and quiescent and a widespread erosion surface developed, which is today preserved throughout the region at different topographic levels, including many mountain summits (Berkey & Morris 1924; Devyatkin 1974).

The Gobi Altai has a basin-and-range physiography and is characterized by diffuse active sinistral transpression with several major approximately east–west strike-slip faults cutting the region and linking with NW-striking thrust faults (Fig. 1). Restraining bends along the Bogd fault system comprise the highest mountains in the northern Gobi Altai, whereas in the southern and western Gobi Altai, several other individual restraining bends occur along other important sinistral strike-slip faults (Fig. 2). Active deformation in the Gobi Altai is best described as a regional transpressional fault array with thrust ridges terminating in strike-slip faults and regional discretely spaced strike-slip faults focusing deformation along their stepover zones and termination zones (Fig. 1). The Karlik Tagh Range of the easternmost Tien Shan is a major restraining-bend termination for the Gobi–Tien Shan fault system which links ranges of the southern Gobi Altai with the eastern Tien Shan (Fig. 1).

The Altai is characterized by active dextral transpressional deformation within a 250–300-km-wide NW-trending uplifted belt in western Mongolia, and adjacent areas of China, Russia and Kazakhstan. It contains discrete regional dextral strike-slip faults and linked thrusts which connect in complex patterns in Mongolia to

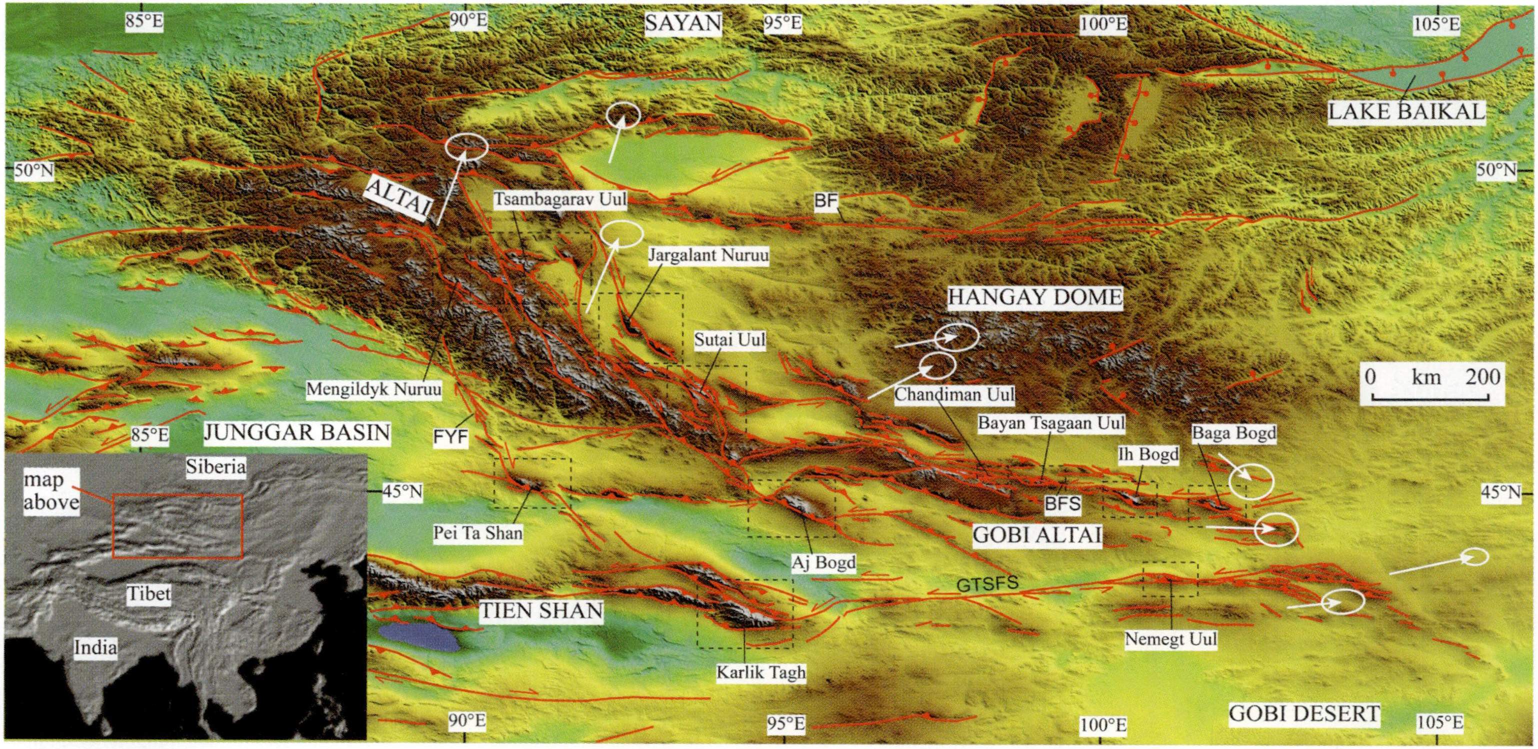

Fig. 1. Digital topographic map showing active fault systems in Altai, Gobi Altai, easternmost Tien Shan, and Hangay Dome regions, Central Asia. Faults taken from Cunningham *et al.* (1996*a*, *b*, 2003*a*, *b*, *c*); Cunningham (1998, 2005) and satellite image analysis. Twelve separate actively forming restraining bend mountain ranges are identified. Locations of DEMs for all 12 ranges shown in Figures 2 & 3 are indicated (boxed areas). GPS vectors with error ellipses taken from Calais *et al.* (2003). GTSFS: Gobi–Tien Shan fault system; BFS: Bogd fault system; FYF: Fu-Yun Fault.

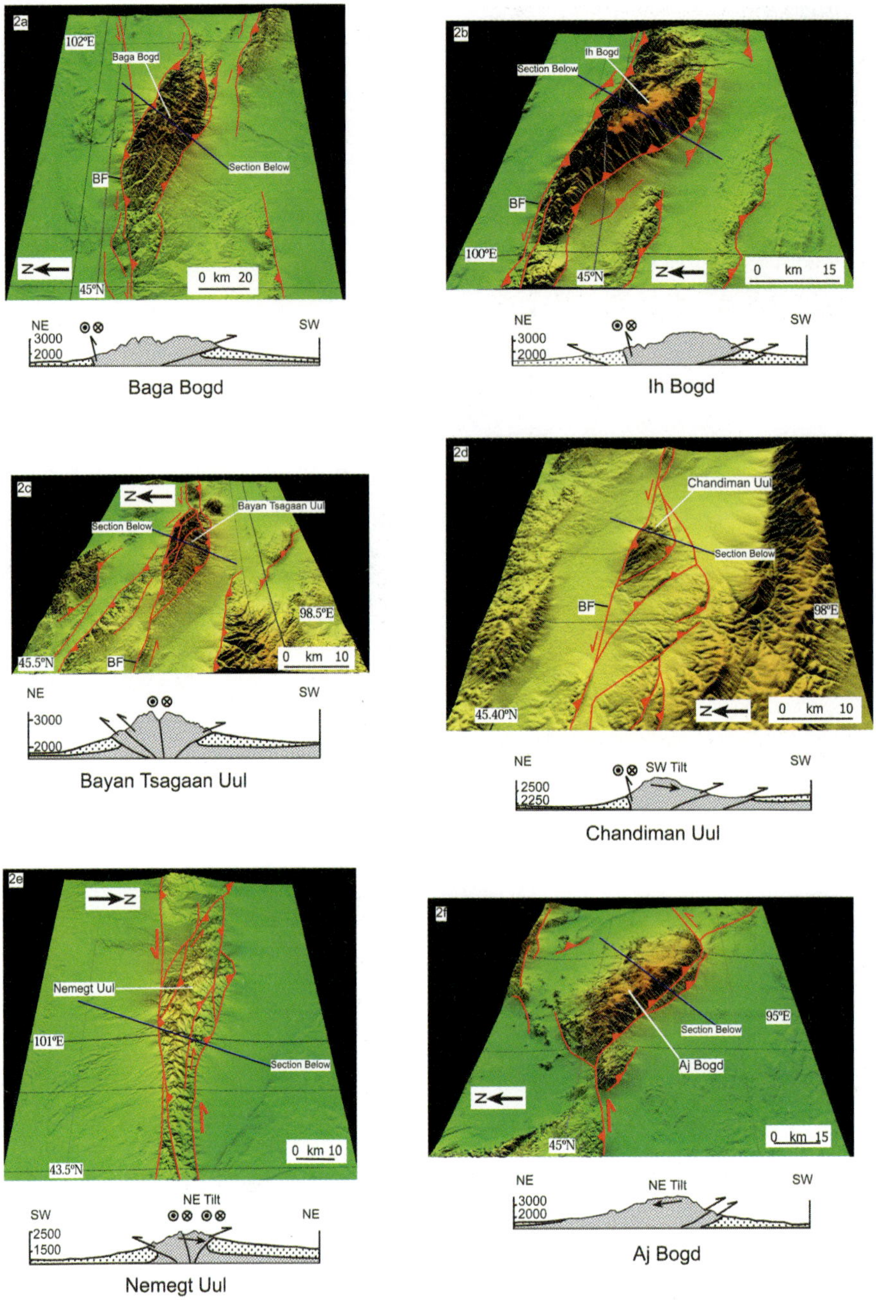

Fig. 2. Six different restraining bend mountain ranges from the Gobi Altai region, Mongolia. For each range, an oblique DEM perspective and topographic cross-section using SRTM90 data are presented. The location of each cross-section is indicated on the map. On cross-sections, a shaded pattern represents Palaeozoic basement rocks, and a stippled pattern equals Mesozoic–Cenozoic sedimentary–basin deposits. Active faults are identified and shown on DEM and cross-section. Some ranges have an obvious SW or NE tilt (labelled in cross-section) as indicated by the orientation of uplifted and preserved peneplain surfaces and internal asymmetrical drainage lengths. Most marked faults have been ground-truthed (Florensov & Solonenko 1963; Baljinnyam *et al.* 1993; Cunningham *et al.* 1996*b*, 1997; Owen *et al.* 1999; Cunningham, unpublished data 1994–2004; Bayasgalan *et al.* 1999*a*, *b*), except those at Aj Bogd which are interpreted from Landsat imagery only. BF: Bogd Fault.

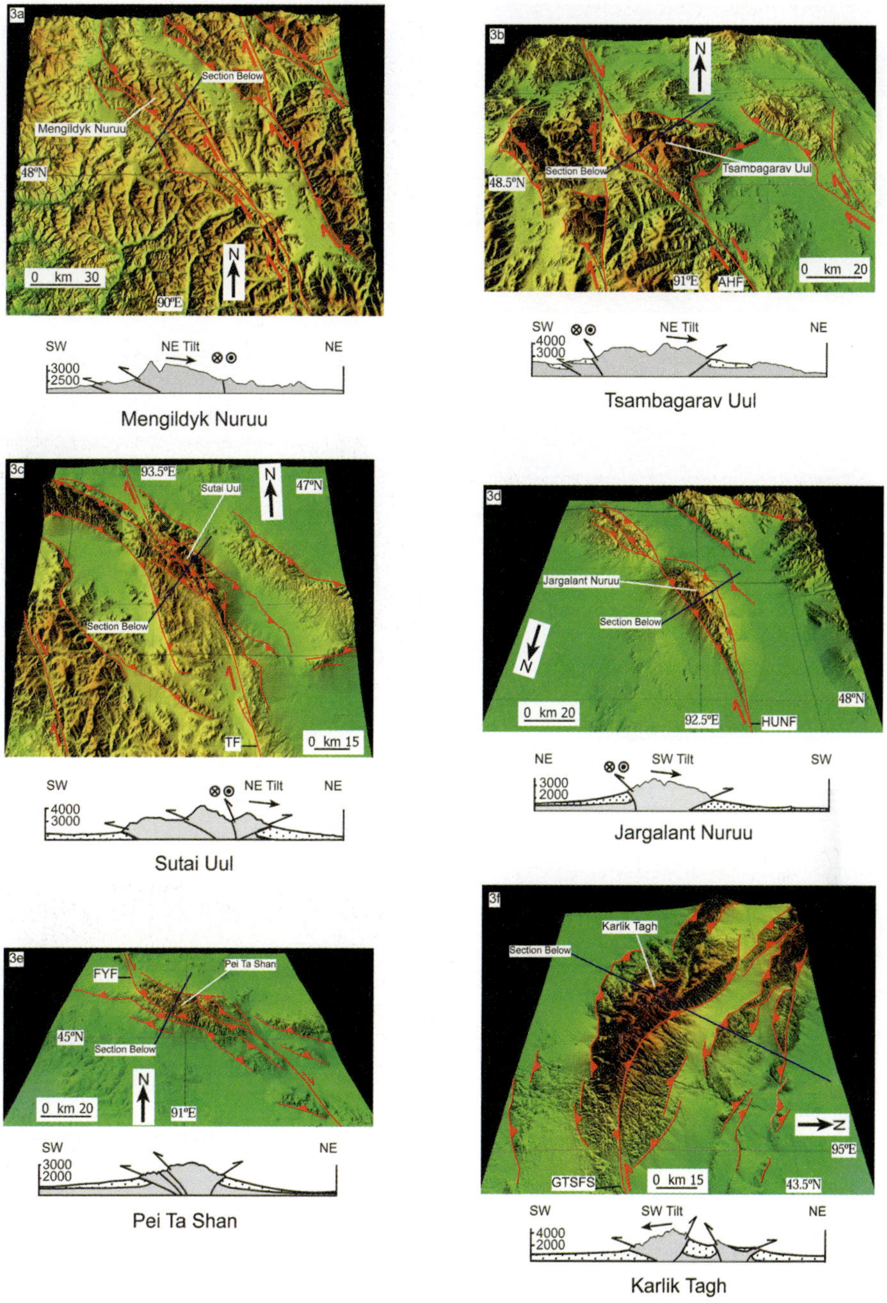

Fig. 3. Six different-restraining bend mountain ranges from the Altai and easternmost Tien Shan regions, Mongolia, and western China. For each range, an oblique DEM perspective and topographic cross-section using SRTM90 data are presented. The locations of each cross-section are indicated on the map. On cross-sections, a shaded pattern represents Palaeozoic basement rocks, and a stippled pattern indicates Mesozoic–Cenozoic sedimentary basin deposits. Active faults are identified and shown on DEM and cross-section. Some ranges have an obvious SW or NE tilt (labelled in cross-section), as indicated by the orientation of uplifted and preserved peneplain surfaces and asymmetrical drainage lengths. Most marked faults have been ground-truthed (Cunningham *et al.* 1996*a*; 2003*a*, *b*, *c*; Cunningham 2005, unpublished data), except those at Pei Ta Shan, which are interpreted from Landsat imagery only. AHF, Ar Hötöl Fault; TF, Tonhil Fault; HUNF, Har Us Nuur Fault; FYF, Fu–Yun Fault; GTSFS, Gobi–Tien Shan fault system.

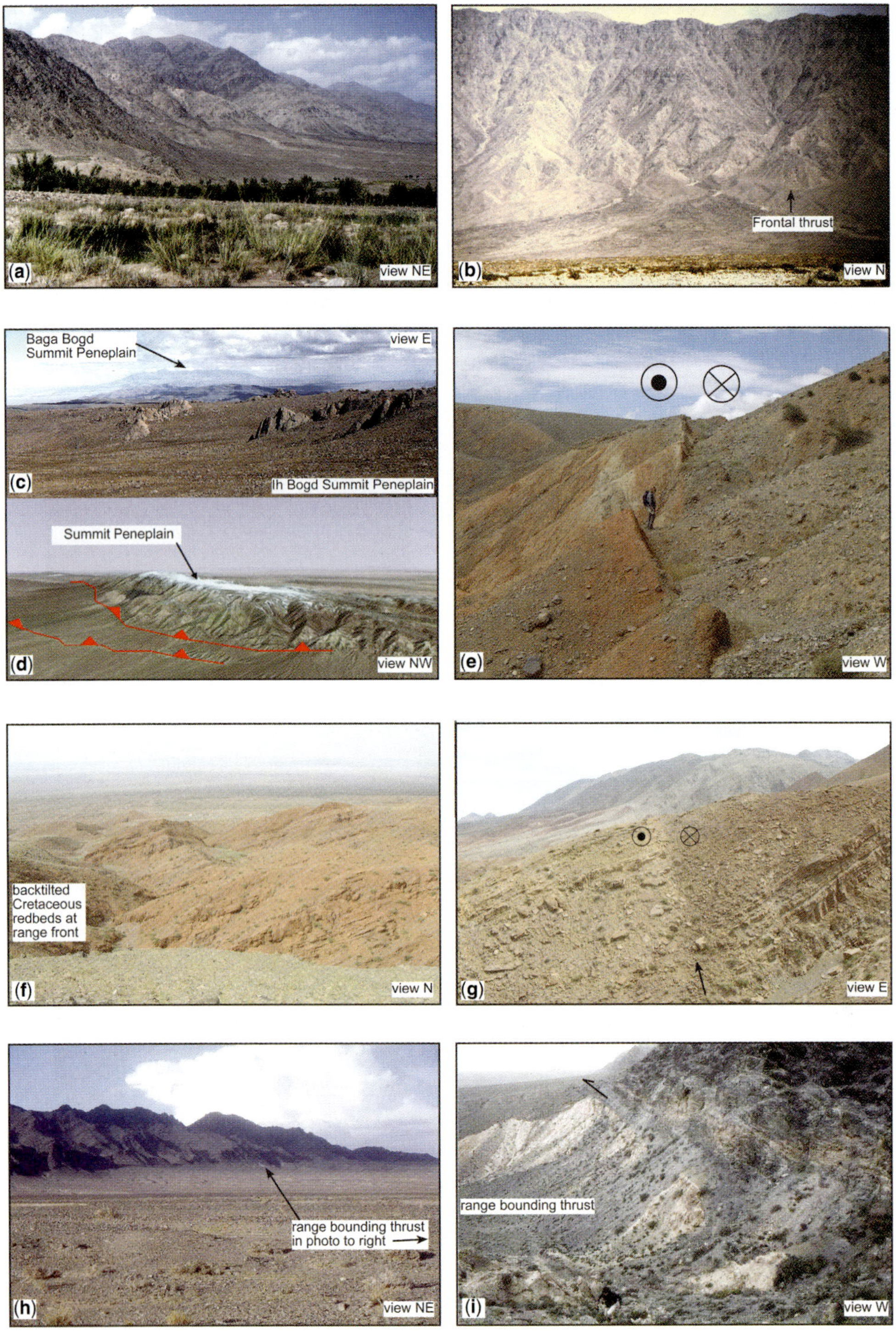

Fig. 4. Typical structural and geomorphological characteristics of restraining bend mountain ranges in the Gobi Altai and Easternmost Tien Shan: (**a**) steep mountain front, deeply incised canyons, and frontal alluvial-fan complex, SE side of Karlik Tagh, easternmost Tien Shan (42°52′32.16″N, 94°33′58.70″E); (**b**) Steep thrust-faulted southern front with

regionally define a transpressional fault array (Cunningham 2005). Oblique-slip faulting is common, as suggested by earthquake focal mechanism solutions and fault data (Bayasgalan *et al.* 2005). Previous work by Cunningham *et al.* (2003*a*) and Cunningham (2005) identified a variety of transpressional fault systems in the Mongolian Altai responsible for the modern range topography. Several major restraining bends occur along the eastern margin and within the central Altai and comprise some of the highest mountains in the region (Fig. 3). In the Chinese and Russian Altai, individual restraining bend mountain ranges have not been documented, perhaps because thrust displacements are more dominant and restraining bends never formed, or because early formed restraining bends have now laterally coalesced with other thrusted ranges and do not stand out as singular topographic and structural culminations.

Structural and geomorphological characteristics of Altai, Gobi Altai and easternmost Tien Shan restraining bends

Typical structural and geomorphological features of some or all of the 12 restraining bend mountain ranges described in this paper (Figs 4 & 5) include the following:

1. a sigmoidal shape for the range when seen in map view (satellite and digital elevation model (DEM) perspective, Figs 2–3).
2. Active deformation zones characterized by strike-slip, thrust and oblique-slip displacements along at least one of the major range fronts (Figs 4b, e, h & i, 5a–f).
3. Low mountain front sinuosity on one or both sides of the range, with major range-bounding faults, steep relief, and frontal alluvial fan complexes (Figs 2–4a, b, h & i, 5a, c, e & f).
4. uplifted and preserved Cretaceous–Palaeocene summit peneplains that provide a useful topographic datum for determining amounts of topographic uplift, the degree of range asymmetry, and which thrust faults have had the most displacement (Figs 2–3, 4c & d & 5g).
5. One or more thrust ridges ('forebergs') in the adjacent alluvial foreland, which trend at an acute angle to the range front (Figs 2–3 & 5b).
6. Overthrusted and back-tilted to locally overturned, Mesozoic–Cenozoic basin-fill at the range front (Figs 4f, & i & 5d).
7. asymmetrical drainage lengths within the range and deep canyons near the range front, so that the lowest valley-floor width–valley-height ratios are near the mountain front (Fig. 4a).
8. A master strike-slip fault that curves into and enters the range from a low-relief flat basin area (Figs 2–3 & 5h).

For each restraining bend, dimensional and geometric characteristics are calculated and compared in Table 1. Some subjectivity is unavoidable in measuring the exact area of uplifted terrain and choosing lines to calculate length/width ratios for each range. Therefore these values should be regarded as approximate, with errors of 10–20%.

Restraining bends of the Gobi Altai

The left-lateral Bogd Fault cuts the northern Gobi Altai region (Fig. 1) and is responsible for nucleating four individual restraining bends at right-stepping bends along its length; these ranges comprise the highest peaks in the NE Gobi Altai. The 1957 *M* 8.3 Gobi Altai earthquake caused surface strike-slip and thrust faulting within a zone approximately 260-km-long by 40-km-wide that includes the Ih Bogd, Baga Bogd and Bayan Tsagaan Uul restraining bends (Florensov & Solonenko 1965; Kurushin *et al.* 1997).

Baga Bogd

Baga Bogd is a sigmoidal-shaped restraining bend near the eastern termination of the sinistral Bogd Fault (Fig. 2a). The geology of the range consists of Precambrian metamorphic and igneous crystalline basement cut by Carboniferous and Permian intrusives and overlapped by Jurassic–Cretaceous clastic sedimentary rocks and basaltic lavas

Fig. 4. (*Continued*) fresh alluvial-fan complex, Karlik Tagh (42°52′42.05″N, 94°02′51.36″E); (**c**) uplifted and preserved Cretaceous–Palaeocene peneplain on the summit of Ih Bogd, Gobi Altai (44°56′32.00″N, 100°18′04.13″E), with Baga Bogd summit peneplain visible in distance to east; (**d**) oblique NW-directed satellite perspective of Ih Bogd summit peneplain and bounding thrust faults; (**e**) 1957 sinistral surface rupture along the Bogd strike-slip fault, NW end of the Ih Bogd restraining bend (45°08′50.70″N, 99°37′21.50″E); (**f**) back-tilted and overthrust Mesozoic clastic sediments along the northern range front of the Ih Bogd restraining bend, northern Gobi Altai; (**g**) wrench zone NW mountain front, Ih Bogd (45°05′14.10″N, 100°03′15.7″E), (**h**) steep linear mountain front, south side of Nemegt Uul (43°37′12.80″N, 100°55′12.92″E), (**i**) range-bounding thrust on the south side of Nemegt Uul restraining bend, southern Gobi Altai, Mongolia (43°37′10.96″N, 100°55′42.56″E). The direction of view from the photographer's perspective is indicated for each photograph.

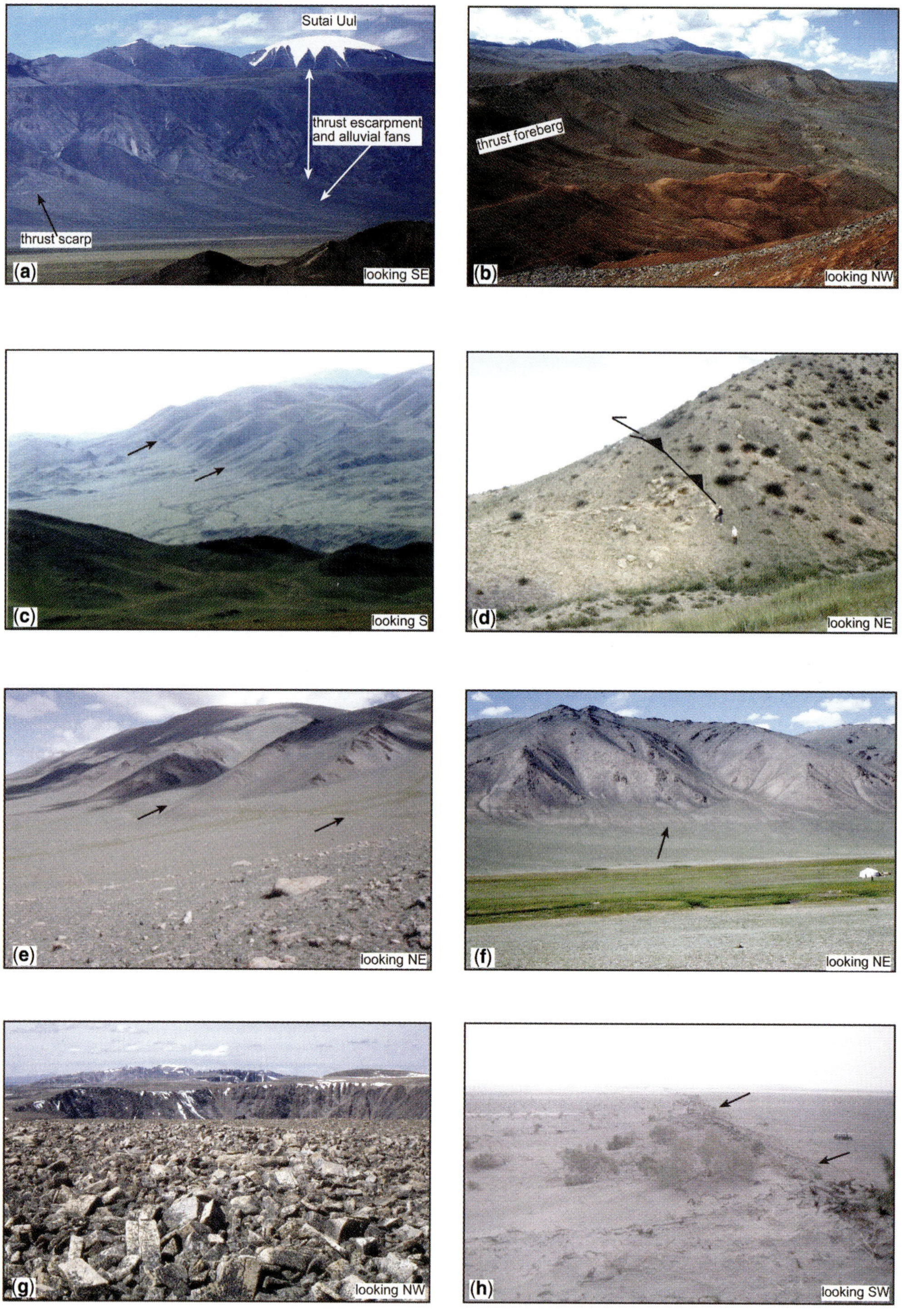

Fig. 5. Typical structural and geomorphological characteristics of restraining bend mountain ranges in the Altai: (**a**) active fault scarp and alluvial-fan complex at the foot of the major thrust escarpment on the SW side of the Sutai Uul restraining bend, Mongolian Altai (46°37′02.04″N, 93°26′30.17″E); (**b**) thrust ridge ('foreberg') on the NE side of the

(Tomurtogoo 1999). The range has bilateral vergence with major thrust faults on the north and south sides of the range that cut Holocene alluvial-fan deposits (Bayasgalan *et al.* 1999*b*). Strike-slip displacements are also accommodated along the north side of the range where two sinistral faults enter the range from the NE and link with frontal thrusts. Two forebergs occur along the north side and a prominent landslide occurs in the northern foreland of the range (Philip & Ritz 1999). The interior of the range is poorly studied, and few Cenozoic faults are recognized on satellite imagery. The highest summit areas are part of a horizontal remnant peneplain (Fig. 4c, in the distance). The range is interpreted to be an asymmetrical flower structure with the strike-slip component of deformation focused along the north side of the range (Fig. 2a).

Ih Bogd

Ih Bogd is the highest range in the Gobi Altai (3957 m) and is a sigmoidal shaped restraining bend along the Bogd Fault west of Baga Bogd (Figs 1 & 2b). The geology of the range consists of Precambrian metamorphic and igneous crystalline basement cut by Carboniferous and Permian intrusives and overlapped by Jurassic–Cretaceous clastic sedimentary rocks and basaltic lavas (Tomurtogoo 1999). The summit comprises a large uplifted Cretaceous–Palaeocene peneplain remnant that slopes gently towards the ESE (Figs 2b, 4c & d). The range is bound by major thrust faults on its southern and NE sides that cut Holocene alluvial fans and overthrust and tilt Mesozoic basin deposits (Figs 2b, 4d, f, & g). Strike-slip displacements are accommodated along the north side of the range and forebergs and zones of wrench deformation also occur along the north front (Figs 2b & 4d–g; Kurushin *et al.* 1997). Much of the interior of the range appears to lack neotectonic deformation, although very little detailed structural mapping has been carried out within the northern half of the range. The range is interpreted to be an asymmetrical flower structure with the strike-slip component of deformation focused along the north side of the range (Fig. 2b).

Bayan Tsagaan Uul

Bayan Tsagaan Uul is an elongate slightly sigmoidal restraining bend west of Ih Bogd along the Bogd fault system (Figs 1 & 2c). The geology of the range consists of Precambrian crystalline basement overlapped by Cambrian–Ordovician–Silurian sedimentary and metasedimentary units which are overlapped by Jurassic–Cretaceous clastic sedimentary rocks (Tomurtogoo 1999). The range is bounded by outward-directed thrusts on its northern and southern sides, but is also internally cut by strike-slip faults and oblique-slip thrusts which cut through the stepover and link the two Bogd Fault segments that enter the range at each end (Fig. 2c). Thus the range has a strike-slip duplex fault geometry. There are no obvious peneplain remnants preserved in the range, because faults that cut the range interior are eroded into major canyons which dissect the topography. Alluvial fans are cut by thrust faults on the north and south sides of the range, and a thrust ridge deforms the foreland NE of the range. The range is interpreted to be a near-symmetrical flower structure with the strike-slip component of deformation focused along faults which cut through the centre of the range (Fig. 2c).

Chandiman Uul

Chandiman Uul is a small restraining bend west of Bayan Tsagaan Uul, that has formed at a minor right step along the Bogd Fault (Figs 1 & 2d). The geology of the range consists of Lower Palaeozoic felsic–intermediate intrusives, orthogneiss and metasedimentary and metavolcanic units (Cunningham *et al.* 1996*b*; Tomurtogoo 1999). The range is topographically and structurally asymmetrical (Fig. 2d). The strike-slip component of deformation is focused along the northern range front where thrusting has also caused uplift and an overall SW tilt to the range. Several other thrust faults have propagated towards the SW along the southern side of the range to accommodate contraction at the bend (Fig. 2d). The range nearly links with Bayan Tsagaan Uul to the east. The range is interpreted to be an asymmetrical flower structure with the strike-slip component of deformation focused along the north side of the range (Fig. 2d).

Fig. 5. (*Continued*) Sutai Uul restraining bend, Mongolian Altai (46°34′55.42″N, 93°56′12.93″E); (**c**) termination of the Har Us Nuur strike-slip fault as a dextral reverse fault along the NE mountain front, Jargalant Range (47°26′01.28″N, 92°54′15.88″E); (**d**) frontal thrust placing Palaeozoic metasedimentary rocks above Mesozoic clastic sediments, SW mountain front, Jargalant Range (47°11′23.83″N, 92°54′15.88″E); (**e**) dextral reverse fault, SW mountain front, Tsambagarav Range (48°32′46.02″N, 90°49′55.92″E); (**f**) active thrust scarp, SW side, Tsambagarav Range (48°32′46.02″N, 90°49′55.92″E); (**g**) summit peneplain (3550 m), Mengildyk Nuruu (48°13′25.30″N, 90°05′04.40″E); (**h**) main trace of the Gobi–Tien Shan fault system, SW Mongolia (43°11′43.26″N, 95°45′13.49″E). The direction of view from the photographer's perspective is indicated for each photo.

Table 1. *Geometric and dimensional characteristics of 12 restraining bend mountain ranges discussed in this study*

Restraining bend	Area (km^2)	L/W (average = 3.13)	Symmetrical or Asymmetrical and tilt direction	Width of step (km)	Minimum vertical uplift (m) (peneplain-constrained)	Right- or left-stepping
Baga Bogd	1240	64/17 = 3.76	A (NE)	25	2000	Right
Ih Bogd	1770	84/25 = 3.36	A (SE)	20	2700	Right
Bayan Tsagaan	710	71/16 = 4.44	S	9	1500	Right
Chandiman	**180**	26/10 = 2.6	A (SW)	**3**	**500**	Right
Nemegt	1010	92/16 = **5.75**	A (NE)	19	1700	Right
Aj Bogd	3900	75/39 = 1.92	A (NE)	48	2600	Right
Sutai	2820	97/31 = 3.13	A (NE)	12	2400	Left
Tsambagarav	1710	57/35 = **1.63**	A (NE)	16	2800	Left
Mengildyk	3100	90/42 = 2.14	A (NE)	37	1500	Left
Jargalant	3820	125/39 = 3.21	A (SW)	44	2200	Left
Pei Ta Shan	3360	96/38 = 2.53	A (NE)	50	2200	Left
Karlik Tagh	**19 000**	280/90 = 3.11	A (SW)	**74**	**4900**	Right

Numbers in bold are maximum and minimum values for each parameter.

Nemegt Uul

Nemegt Uul is an elongate sigmoidal-shaped range at a major right step along the Gobi–Tien Shan left-lateral strike-slip fault system in the southern Gobi Altai (Figs 1 & 2e). The geology of the range consists of Devonian–Carboniferous metasedimentary and metavolcanic units, unmetamorphosed and undated volcanic rocks, an ophiolite complex of probable Devonian age, scattered acid–intermediate intrusive rocks and unconformable Cretaceous clastic sedimentary rocks (Cunningham *et al.* 1996*b*; Owen *et al.* 1999; Tomurtogoo 1999; Rippington *et al.* 2006). The range has active thrust faults along the northern and southern fronts that cut and deform Quaternary alluvial deposits and older Cretaceous clastic sediments (Figs 2e & 4h & i; Owen *et al.* 1999). The range is topographically asymmetrical, with the highest elevations in the southern half of the range and long drainages running northward out of the range. The north side of the range is bordered by a major alluvial-fan complex which stores the majority of sediment removed from the range (Fig. 2e). The range contains Late Cenozoic thrusts and oblique-slip faults, and is a NE-tilted asymmetrical flower structure in cross-section (Fig. 2e; Owen *et al.* 1999; Rippington *et al.* 2006). A major fault that cuts diagonally across the western part of the range transfers the strike-slip displacements from faults that enter the range from the NE to faults that exit the range to the NW (Fig. 2e).

Aj Bogd

Aj Bogd is an isolated massif that has formed at a major right step along an unnamed left-lateral strike-slip fault in the far western Gobi Altai (Figs 1 & 2f). The geology of the range consists dominantly of Devonian and Carboniferous intrusive, volcanic and sedimentary rocks (Tomurtogoo 1999). The range is topographically asymmetrical with major thrust displacements along the southern front and the entire range is tilted to the NE (Fig. 2f). A major alluvial-fan complex borders the southern front and thrust forebergs have propagated to the SW into the range foreland. The summit is a NE-tilted Cretaceous–Palaeocene peneplain remnant. The range is fault bounded on the NW side and cut internally by NE-striking faults; however the range is best described as a SW-directed thrust block at a major strike-slip stepover (Fig. 1 & 2f).

Restraining bends of the Altai and easternmost Tien Shan

Mengildyk Uul

Mengildyk Nuruu is a restraining bend at the termination of an unnamed dextral strike-slip fault system in the interior of the Mongolian Altai (Figs 1 & 3a). The geology of the range consists of Cambro-Ordovician greenschist-grade metasedimentary rocks and umetamorphosed sedimentary and volcanic rocks cut by a large Devonian granitic complex (Zaitsev 1978; Cunningham *et al.* 2003*a*). The range is bounded on its SW side by a major thrust fault that has elevated the range, whereas the NE side of the range is not fault-bounded and the range is regionally tilted to the NE. The strike-slip fault that enters the range from the SE links with the range-bounding thrust on the SW side of the range, but also cuts into the range and continues for at least 40 km (Fig. 3). The range has a plateau summit, which is interpreted to be a Late Cretaceous–Palaeogene peneplain remnant (Fig. 5g).

Tsambagarav Uul

Tsambagarav Uul is the third-highest mountain range in Mongolia (4193 m), and is a restraining bend at a gentle left step along the Ar Hötöl dextral strike-slip fault system (Figs 1 & 3b). The geology of the range consists of Cambro-Ordovician basement phyllites and schists cut by extensive Silurian and Devonian intrusive complexes (Zaitsev 1978; Cunningham *et al.* 2003*a*). The range has an unusual triangular shape in plan view, with thrust faults bounding all sides of the range (Fig. 3b). The summit is an ice-capped plateau that is interpreted as a Late Cretaceous–Palaeocene peneplain remnant. Drainage length asymmetries and the regional tilt of the summit peneplain suggest that the entire range is tilted towards the NE due to major thrust displacements along the SW front (Figs 3b & 5e & f). Along the SW range front, dextrally displaced Quaternary glacial deposits, a prominent scarp, and en echelon tension gashes indicate that the SW front is active and has accommodated strike-slip and thrust displacements (Cunningham *et al.* 2003*a*).

Sutai Uul

Sutai Uul is a major restraining bend and topographic culmination in the SE Altai that developed at a left step along the Tonhil dextral strike-slip fault (Figs 1 & 3c). The geology of the range consists of Vendian–Cambrian metavolcanics; Ordovician–Silurian metasediments; Devonian–Carboniferous volcanic and sedimentary rocks; undifferentiated intrusive units, and Jurassic–Cretaceous clastic sedimentary rocks (Zaitsev 1978; Cunningham *et al.* 2003*b*). Structural studies of the range reveal that it is an actively forming asymmetrical flower structure with an overall NE tilt due to greater thrust displacements along the SW side of the range (Fig. 3c; Cunningham *et al.* 2003*b*). Strike-slip displacements are accommodated within the range core and not along the mountain fronts. The NE and SW range fronts are bounded by active thrust faults as indicated by fresh thrust scarps that cut bordering alluvial fan complexes (Fig. 5a). The NE side of the range contains two major thrust forebergs that have propagated into the adjacent basin and which are linked by an oblique-slip thrust in the basin centre, suggesting that the range is developing by strike-slip duplexing within its bounding foreland (Figs 3c & 5b; Howard *et al.* 2006). The range has an overall sigmoidal shape in plan view, and the summit is a broad ice-capped plateau that is a Late Cretaceous–Palaeogene peneplain remnant (Fig. 3c).

Jargalant Nuruu

Jargalant Nuruu is a prominent restraining bend mountain range on the easternmost edge of the Mongolian Altai at the SE termination of the Har Us Nuur dextral strike-slip fault (Figs 1 & 3d). The geology of the range consists of Vendian–Cambrian metsedimentry and metavolcanic rocks and Carboniferous intrusives overlapped by Jurassic–Cretaceous clastic sedimentary rocks (Zaitsev 1978; Cunningham *et al.* 1996*a*). The NE range front has accommodated both thrusting and strike-slip displacements, and drainage and topographic asymmetries indicate that the range is tilted SW, presumably due to greater thrust displacements on the NE side (Figs 3d & 5c). The SW range front is marked by overthrusted Mesozoic and Cenozoic basin fill, and deformation has progressed well into the range foreland (Fig. 5d; Cunningham *et al.* 1996*a*; Howard *et al.* 2003). Both range fronts have very low mountain-front sinuosity, suggesting that they are bounded by active faults (Fig. 3d).

Pei Ta Shan

Pei Ta Shan is an isolated massif on the Mongolian Chinese border. It formed at a restraining bend along the Fu-Yun dextral strike-slip fault system (Figs 1 & 3e). It also is the westernmost range along an east–west line of transpressional uplifts between the southern Altai and the easternmost Tien Shan (Fig. 1). The range thus occupies a complex structural position between opposing north–south right-lateral and east–west left-lateral fault systems, and no modern structural studies of its tectonic evolution were located during a literature search for the range. The geology of the range consists of Carboniferous and Devonian sedimentary and volcanic rocks (Tomurtogoo 1999). Based on examination of satellite data, the range appears to have several major thrust faults along its SW side, and the alluvial fan complex that borders the range is deformed and cut by a major thrust fault, suggesting the range front is still active (Fig. 3e). The NE side is also sharply defined on imagery, suggesting that it too is fault-bound. The range appears to be an asymmetrical flower structure, with its greatest thrust displacements along the SW side of the range (Fig. 3e).

Karlik Tagh

Karlik Tagh is a major restraining bend at the SW termination of the sinistral Gobi–Tien Shan fault system (Figs 1 & 3f). The range rises up to 4900 m above the surrounding region, and marks the eastern limit of the Tien Shan. It thus constitutes a major intracontinental orogenic termination zone (Cunningham *et al.* 2003*c*). The geology of the range consists of Ordovician basement schists and phyllites; Devonian sedimentary and volcanic rocks; and extensive Carboniferous volcanic and sedimentary rocks cut by numerous intrusives ranging from acid to basic compositions (Cunningham *et al.* 2003*c*). The range is thrust-bounded on both its southern and northern sides, and it is flanked by major alluvial-fan complexes (Figs 3f & 4a & b). The range contains a prominent ice-cap and summit peneplain that are regionally tilted towards the SW, suggesting that thrust displacements on the NNE sides of the range are greatest (Fig. 3f). The major thrust fault on the northern side of the range curves 90° from a NW strike in the NW to a NE strike in the NE, and it links continuously to the Gobi–Tien Shan strike-slip fault (Figs 3f & 5h). Several other thrust ridges linked with strike-slip faults occur north of Karlik Tagh, suggesting that Karlik Tagh and areas north of it comprise a regional horsetail splay termination for the Gobi–Tien Shan left-lateral strike-slip fault system (Fig. 3f; Cunningham *et al.* 2003*c*). From a central Tien Shan perspective, Karlik Tagh and the areas directly north of it comprise the transitional zone from an orogen dominated by thrusting to an oblique-transpressional system (Fig. 1).

Discussion

The 12 restraining bends reviewed in this paper show significant diversity in their fault architecture, topographic expression, map view form, cross-sectional structure and dimensional values (Figs 2–3 & 6; Table 1). They vary significantly in uplifted area from the smallest bend: Chandiman Uul (*c.* 180 km^2) to the largest bend: Karlik Tagh (*c.* 19,000 km^2). They are longer than they are wide, varying from an approximately 5.75:1 ratio (Nemegt Uul) to a 1.5–2.0:1 ratio (Tsambagarav Uul and Aj Bogd), with length/width ratios

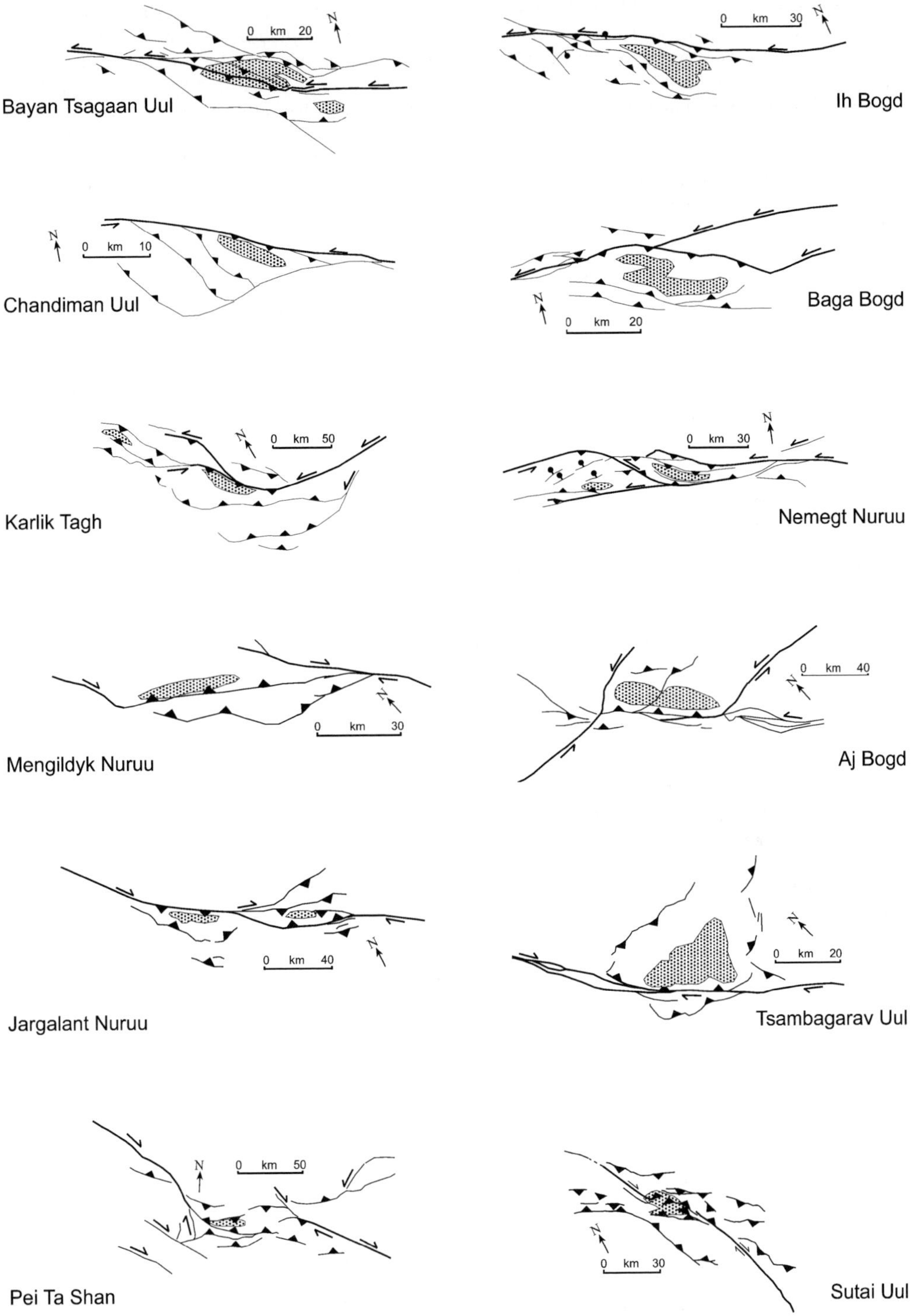

Fig. 6. Twelve fault maps for individual restraining bends reviewed in this study. Areas of highest summit peaks and summit peneplains are shaded. Scales for each map differ, as indicated.

averaging around 3:1 for all 12 bends. Most are topographically asymmetrical ranges, as evidenced by tilted summit peneplains, asymmetrical drainage lengths, and asymmetrically positioned drainage divides. All Gobi Altai bends and Karlik Tagh are right stepping, whereas all Altai bends and Pei Ta Shan are left stepping (Table 1; Figs 1–3). The width of the bend or termination zone varies from only 3 km (Chandiman Uul), to 74 km for Karlik Tagh, with an average width of 30 km for the 12 bends. Vertical topographic uplift as constrained by relict peneplain surfaces varies from 500 m to 4900 m. The uplifted peneplain remnants correlate with regional base level outside of the uplifted areas, so the topographic uplift values for each restraining bend that contains a summit peneplain are regarded as maximum values for Late Cenozoic exhumation during construction of the mountain range.

The 12 restraining bends described in this paper occur in one of three different tectonic settings with respect to the master strike-slip fault that runs into the range (Fig. 7). Karlik Tagh, Mengildyk Nuruu and Jargalant Nuruu occur at strike-slip fault termination zones where strike-slip motion is terminally accommodated by thrusting and uplift. Bayan Tsagaan Uul, Ih Bogd, Baga Bogd, Chandiman Uul and Aj Bogd each occur at a major strike-slip fault bend, where the individual strike-slip fault can be traced continuously from one end of the range to the other, but bends along its length. Nemegt Uul, Sutai Uul and Pei Ta Shan occur where two separate strike-slip fault segments converge and overlap. In the majority of cases, the major strike-slip fault motion is accommodated along one of the range fronts. However, in Bayan Tsagaan Uul, Nemegt Uul, Mengildyk Nuruu and Sutai Uul, major strike-slip motion is

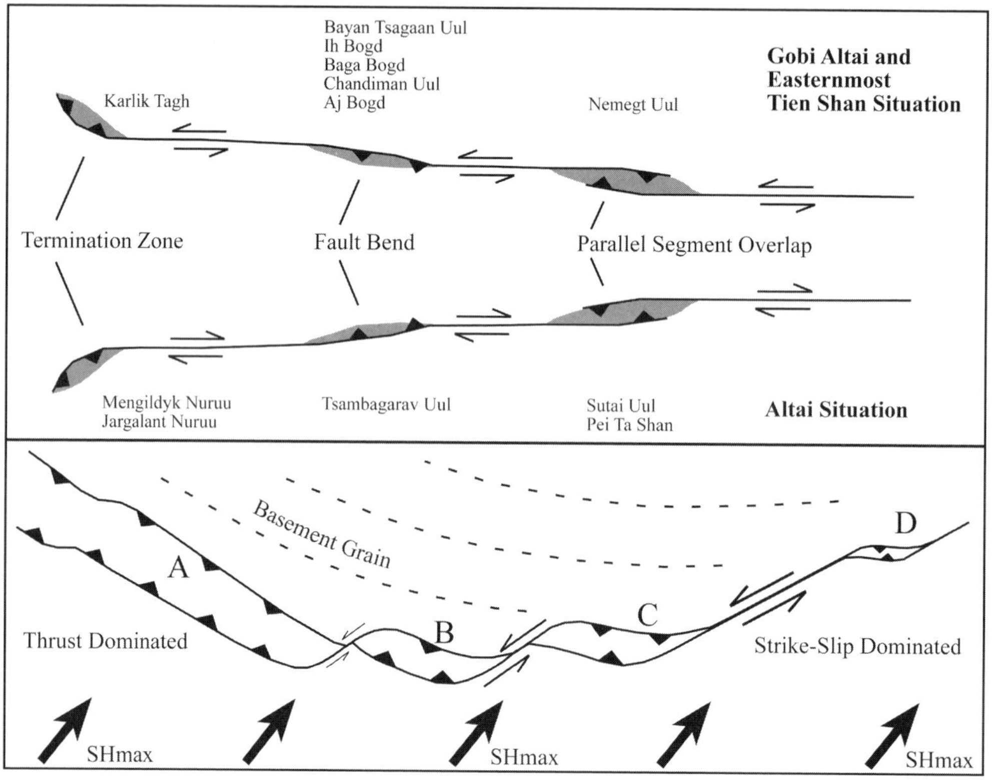

Fig. 7. Top: three separate tectonic settings where the Altai, Gobi Altai and easternmost Tien Shan restraining bends have formed. Bottom: possible continuum of transpressional mountain ranges from thrust-dominated ranges with subordinate strike-slip transfer faults (A and B) to true restraining bends where strike-slip faults between ranges have lengths equal or greater to the thrusted widths of the ranges (C and D). It is recommended that only C and D should be considered as true restraining bends, because strike-slip displacements are likely to be equal or greater than thrust displacements in these ranges. In the Altai and Gobi Altai, the entire continuum of transpressional ranges exists, from thrust-dominated ridges with little if any strike-slip faulting to isolated individual strike-slip-dominated restraining bends (Fig. 1).

accommodated along faults that cut through the core of the range (Figs 2–3 & 6). In cases where the majority of the strike-slip motion is accommodated along a range front, some component of thrusting is also likely to be present to account for the mountain front relief. However, in 10 of the 12 bends studied, some of the contractional deformation at the restraining bend is partitioned and accommodated by other thrust faults, especially along the front on the opposite side of the range (e.g. the southern side of Ih Bogd or Baga Bogd; Fig. 2a & b). The exceptions are Aj Bogd and Mengildyk Nuruu, where thrust faults are found on one side of the range only (Figs 2f & 3a). Based on conservative extrapolation of surface fault dips to depth, all 12 bends in cross-section have either a half-flower or full-flower structural geometry. Their structural asymmetry is matched by their overall topographic asymmetry and internal drainage length asymmetries.

It is difficult to calculate the total amount of strike-slip displacement along each master strike-slip fault system associated with each restraining bend, because unambiguous piercing points are not yet documented in the region. Based on simple plots that correlate fault length with total offset, approximations of total strike-slip offsets can be made (using a power-law where $D = 0.03L^{1.06}$; Cowie & Scholz 1992). The Gobi–Tien Shan Fault can be traced for over 600 km (Fig. 1) and is therefore likely to have accommodated between 25–30 km total left-lateral offset. The Bogd, Fu-Yun, Har Us Nuur and Ar Hotol Faults can all be traced for at least 400 km, and are likely to have accommodated total left-lateral offsets in the order of 15–20 km. Strike-slip faults that enter the Aj Bogd, Sutai and Mengildyk restraining bends are all approximately 200 km in length and probably accommodated between 5 and 10 km total offset (Fig. 1). However, the total amount of strike-slip offset does not have to be equal to total thrust displacements within each restraining bend, because:

1. some bends may form at different times within the overall strike-slip history of the system;
2. some older, earlier-formed bends that accommodated some of the displacement may even be short-lived and abandoned (Cowgill *et al.* 2004); and
3. the vertical component of displacement at the bend will vary depending on whether the bend is gentle or sharp.

Thus, these values of strike-slip offsets should be regarded as maximum values for thrust displacements accommodated within each restraining bend. Cunningham *et al.* (2003*c*) estimated 10–15 km of total thrust displacements within the Karlik Tagh restraining bend, based on tilted and offset remnant peneplain surfaces. This value contrasts with 25–30 km of expected sinistral strike-slip displacement along the Gobi–Tien Shan fault system, which terminates at the bend. Block rotations and distributed oblique deformation within the Karlik Tagh region could account for the discrepancy, or displacement estimates in Cunningham *et al.* (2003*c*) are conservatively low.

The role of fault reactivation in Late Cenozoic construction of the 12 restraining bends reported here is suspected, but so far has only been found to be locally important. Most range-bounding faults lack evidence for an older displacement history, and within Ih Bogd and Nemegt Uul, Mesozoic normal faults are passively uplifted within the range without showing evidence for any younger reactivation (author's unpublished field data, 2003–2005). Exceptions to this rule occur along the eastern edge of the Altai, where Howard *et al.* (2003, 2006) argue on geometric and stratigraphic grounds that Quaternary thrust faults along the NE margin of the Sutai restraining bend and Quaternary thrust faults along the SW margin of the Jargalant restraining bend are reactivated Mesozoic normal faults (Figs 3c & d). Similarly, the dextral thrust fault bounding the SW side of the Tsambagarav restraining bend contains structural evidence for brittle normal-fault reactivation and a two-stage Mesozoic–Cenozoic history (Cunningham *et al.* 2003*a*). In Karlik Tagh, the major frontal thrusts along the north and south sides are both reactivated Triassic ductile thrust zones as dated by $^{40}Ar/^{39}Ar$ geochronology (Cunningham *et al.* 2003*c*). However, all major ranges in the central and southern Mongolian Altai, Gobi Altai and easternmost Tien Shan are cored by Palaeozoic basement rocks (often exposed as preserved peneplaned erosion surfaces; Figs 4c & d) and significant reactivation of older rift-related normal faults is therefore unlikely because rift hanging-wall sedimentary strata are nowhere found at significantly elevated positions (i.e. >2700 m).

Reactivation of a bulk rheological heterogeneity in the upper crust (structural grain) consisting of faults, shear zones and metamorphic fabrics is suggested by the stepover sense for all 12 restraining bends. East–west sinistral strike-slip faults in the Gobi Altai are linked at stepover zones by basement grain-parallel ESE–WNW faults. In the Altai, NNW–SSE striking dextral strike-slip faults are linked at stepover zones by basement grain-parallel NW–SE faults. The lack of transtensional basins along the east–west sinistral strike-slip systems of the Gobi Altai (which would form at a left step instead of right step) and along the NNW–SSE dextral strike-slip faults of the Altai (requiring a right step instead of a left step) is presumably

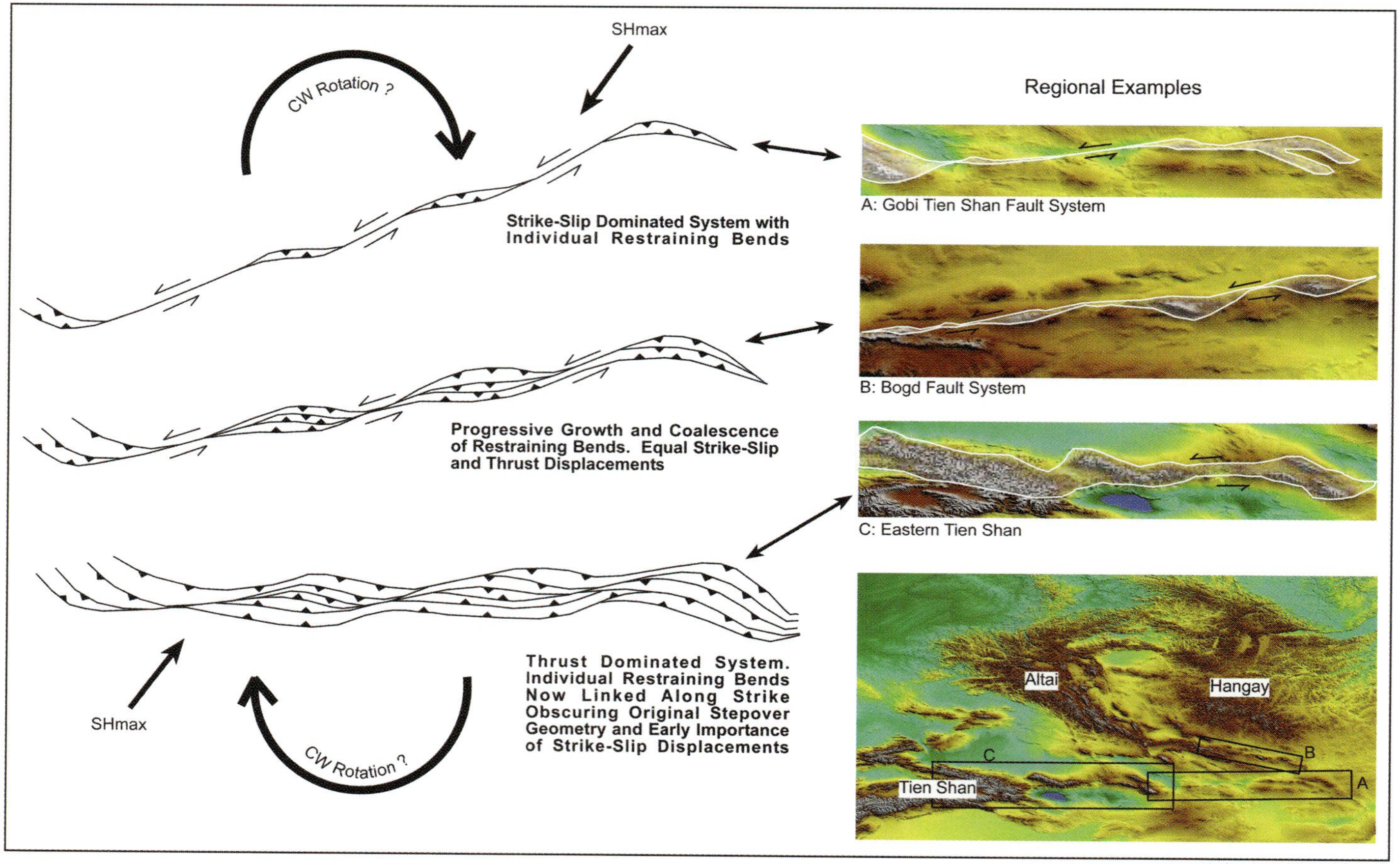

Fig. 8. Evolution of restraining bend mountain ranges with progressive fault displacements and possible vertical axis rotations of the entire system relative to a fixed $S_{H_{max}}$. With progressive displacements, ranges will widen and lengthen through time and may coalesce along strike. Progressive clockwise vertical-axis rotations of the entire system will increasingly promote thrust displacements and increasingly discourage strike-slip displacements. Thus, the kinematics of the entire system will evolve into a more contraction-dominated orogen, possibly obscuring the early importance of strike-slip displacements. Early stages in this process (top) have structural and topographic similarities with the Gobi–Tien Shan fault system, and may be comparable. Middle stages have structural and topographic similarities with the northern Gobi Altai restraining bends along the Bogd fault system. Advanced stages may correspond with the eastern Tien Shan. Note, vertical axis rotations are not required for the three-stage evolutionary process shown here – the system will evolve as shown, as long as convergence occurs oblique to the deformation belt. However, vertical-axis rotations are likely to accompany crustal deformation in diffuse regions of continental strike-slip faulting (Bayasgalan *et al.* 1999*a*; Walker & Jackson 2004).

because strike-slip fault nucleation and propagation across the older basement grain at a high angle are mechanically inhibited.

Transpressional deformation in the Altai and Gobi Altai regions varies from thrust-dominated domains to strike-slip-dominated domains. In strike-slip-dominated regions, each restraining bend stands out as an individual structural and topographic culmination. However in thrust-dominated regions, the mountain ranges are longer and more topographically continuous, and the strike-slip faults are subordinate transfer zones between offset thrust ridges. Thus, there is a continuum of possible transpressional mountain ranges (Fig. 7), and it seems sensible to only use the term 'restraining bend' when the length of the strike-slip fault that enters the range is equal or greater than the thrusted width of the range. In high mountainous regions of the central and northern Altai and western Gobi Altai, individual restraining bends are not obvious (Fig. 1) and the mountains are thrust dominated; strike-slip faulting is subordinate to absent (Cunningham 2005). However, it is speculated here that the earliest stages of uplift may have involved important strike-slip displacements and individual restraining bends may have nucleated. Any restraining bends that initially formed may now be overprinted and enlarged by younger thrust displacements that have widened and lengthened the range, leading to linkage and coalescence with other ranges (this is seen today with the Sutai, Mengildyk and Tsambagarav ranges, which although distinguishuable as separate massifs, are topographically linked to neighbouring ranges). In addition, if vertical-axis rotations have occurred since restraining bend nucleation, the faults and blocks that they bound may have rotated to orientations less favourable for strike-slip movements and more favourable for thrusting, and the early-formed strike-slip faults will be overprinted and obscured by younger thrust sheets as the ranges grow along strike and perpendicular to strike. In any case, with or without vertical axis rotations, different stages in the restraining-bend growth process can be observed in the Gobi Altai and easternmost Tien Shan (Fig. 8; Cunningham *et al.* 1996*b*; Cunningham *et al.* 2003*c*).

The fault maps for all 12 restraining bends indicate that they all have individual architectural characteristics (Fig. 6). Differences in fault patterns are likely due to differences in total strike-slip displacement approaching the bend and passing through the bend, differences in stepover width, differences in pre-existing basement structural heterogeneity and orientation of basement structures relative to $S_{H_{max}}$, the possible role of vertical axis rotations and changing fault kinematics through time, possible strain hardening processes within the fault system and significant topography generation which will lead to abandonment of some faults (Cowgill *et al.* 2004), and whether the bend formed at a strike-slip fault termination, gentle bend, or parallel segment overlap zone. For these reasons, the restraining bends in the Altai, Gobi Altai and easternmost Tien Shan differ markedly from simple sandbox model examples (McClay & Bonora 2001) although they share the overall common characteristics of having elongate sigmoidal shapes, flower structure cross-sections, linked thrust, oblique-slip and strike-slip faults, and they are localised nodes of uplift, exhumation and erosion flanked by clastic basins.

The low erosion rates in the Altai, Gobi Altai and easternmost Tien Shan (precipitation rates vary from 50 mm/a in the southern Gobi Altai to 350 mm/a in high Altai with 100–200 mm/a average for entire region; Mongolian National Atlas, 1990) provide a unique opportunity to quantitatively determine the maximum topographic uplift within each bend. Preserved summit peneplains provide limits on maximum uplift and exhumation. If the regional climate was more humid, then greater erosion of the restraining bend core would lower the range topography and therefore inhibit development of new thrust faults, because uplifted topography produced by existing thrust faults would be balanced by erosion, and the system would achieve a steady state and new faults would not form (cf. Willett *et al.* 2001). The simple conclusion is that for a given horizontal strain rate, restraining bends in dry regions may tend to grow larger and form more significant mountain ranges than in humid regions. Arid zone restraining bends may also show more advanced stages of restraining bend evolution. Therefore, regions like the Altai and Gobi Altai offer an excellent opportunity to study restraining bends at different stages of evolution.

The 12 restraining bends described here comprise fundamental orogenic elements in the Altai, Gobi Altai and easternmost Tien Shan. They are important sites of mountain-range nucleation and oblique deformation. Other mountain ranges along the perimeter of the Indo-Eurasia deformation field: such as the Hindu Kush, and the western spurs of the Tien Shan, Tarbagatay, Sayan and Stanovoy ranges where active oblique deformation predominates, are likely regions in which to study similar processes of restraining-bend development within an overall transpressional deformation regime (Tapponnier & Molnar 1979; Cobbold & Davy 1988; Cunningham 2005). Restraining-bend nucleation, topographic uplift, along-strike growth, and potential coalescence with neighbouring ranges may also have been an important mountain-building process in other intracontinental transpressional orogens, and workers investigating modern and ancient examples should consider this possibility.

This work draws on research supported by NERC Grant NER/D/S2003/00671 and Royal Society Grant 574006.G503/22077/SM awarded to D. Cunningham. Reviews by L. Webb and R. Walker are much appreciated.

References

ANON. 1990. *Mongolian National Atlas*, Ulan Bator and Moscow.

BADARCH, G., CUNNINGHAM, D. & WINDLEY, B. 2002. A new terrane subdivision for Mongolia: Implications for the Phanerozoic crustal growth of Central Asia. *Journal of Asian Earth Sciences*, **21**, 87–110.

BALJINNYAM, I., BAYASGALAN, A. *ET AL*. 1993. Ruptures of major earthquakes and active deformation in Mongolia and its surroundings: Boulder, Colorado. *Geological Society of America Memoirs*, **181**, 1–62.

BARKA, A. A. & KADINSKY-CADE, K. 1988. Strike-slip fault geometry in Turkey and its influence on earthquake activity. *Tectonics*, **7**, 663–684.

BAYASGALAN, A., JACKSON, J. & MCKENZIE, D. 2005. Lithosphere rheology and active tectonics in Mongolia: relations between earthquake source parameters, gravity and GPS measurements, 2005. *Geophysical Journal International*, **163**, 1151–1179.

BAYASGALAN, A., JACKSON, J., RITZ, J. -F. & CARRETIER, S. 1999*a*. Field examples of strike-slip fault terminations in Mongolia and their tectonic significance. *Tectonics*, **18**, 394–411.

BAYASGALAN, A., JACKSON, J., RITZ, J. -F. & CARRETIER, S. 1999*b*. 'Forebergs' flower structures and the development of large intracontinental strike-slip faults: the Gurvan Bogd fault system in Mongolia, *Journal of Structural Geology*, **21**, 1285–1302.

BERKEY, C. P. & MORRIS, F. K. 1924. The peneplanes of Mongolia. *American Museum Novitates*, **136**, 1–11.

BIDDLE, K. T. & CHRISTIE-BLICK, N. 1985. Glossary – strike-slip deformation, basin formation and sedimentation. *In*: BIDDLE, K. T. & CHRISTIE-BLICK, N. (eds), *Strike-slip Deformation, Basin Formation and Sedimentation*, SEPM Special Publications, **37**, 375–384.

CALAIS, E., VERGNOLLE, M., SAN'KOV, V., LUKHNEV, A., MIROTSHNITCHENKO, A., AMARJARGAL, S. & DÉVERCHERE, J. 2003. GPS measurements of crustal deformation in the Baikal–Mongolia area (1994–2002): implications for current kinematics of Asia. *Journal of Geophysical Research*, **108**, 2501, doi:10.1029/2002JB002373, 2003.

COBBOLD, P. R. & DAVY, P. R. 1988. Indentation tectonics in nature and experiment. 2. Central Asia. *Bulletin of the Geological Institute University of Uppsala*, N.S., 14, 143–162.

COWGILL, E., ARROWSMITH, J. R., YIN, A., XIAOFENG, W. & ZHENGLE, C. 2004. The Akato Tagh bend along the Altyn Tagh fault, northwest Tibet 2: active deformation and the importance of transpression and strain hardening within the Altyn Tagh system. *GSA Bulletin*, **116**, 1443–1464.

COWIE, P. A. & SCHOLZ, C. H. Displacement-length scaling relationships for faults: data synthesis and conclusion. *Journal of Structural Geology*, **14**, 1149–1156.

CROWELL, J. C. 1974. Sedimentation along the San Andreas fault, California. *In*: DOTT, R. H. JR & SHAVER, R. H. (eds), *Modern and Ancient Geosynclinal Sedimentation*. SEPM Special Publications, **19**, 292–303.

CUNNINGHAM, D., DAVIES, S. & BADARCH, G. 2003*b*. Crustal architecture and active growth of the Sutai Range, western Mongolia: a major intracontinental, intraplate restraining bend. *Journal of Geodynamics*, **36**, 169–191.

CUNNINGHAM, D., DIJKSTRA, A., HOWARD, J., QUARLES, A. & BADARCH, G. 2003*a*. Active intraplate strike-slip faulting and transpressional uplift in the Mongolian Altai. *In*: STORTI, F., HOLDSWORTH, R. E. & SALVINI, F. (eds), *Intraplate Strike-slip Deformation Belts*. Geological Society, London, Special Publications, **210**, 65–87.

CUNNINGHAM, D., OWEN, L. A., SNEE, L. & LI JILIANG, 2003*c*. Structural framework of a major intracontinental orogenic termination zone: the easternmost Tien Shan, China, *Journal of the Geological Society, London*, **160**, 575–590.

CUNNINGHAM, D. 2005. Active intracontinental transpressional mountain building in the Mongolian Altai: defining a new class of orogen. *Earth and Planetary Science Letters*, **240**, 436–444.

CUNNINGHAM, W. D. 1998. Lithospheric controls on late Cenozoic construction of the Mongolian Altai. *Tectonics*, **17**, 891–902.

CUNNINGHAM, W. D., WINDLEY, B. F., DORJNAMJAA, D., BADAMGAROV, G. & SAANDAR, M. 1996*a*. Structural transect across the Mongolian Altai: active transpressional mountain building in central Asia. *Tectonics*, **15**, 142–156.

CUNNINGHAM, W. D., WINDLEY, B. F., DORJNAMJAA, D., BADAMGAROV, G. & SAANDAR, M. 1996*b*. Late Cenozoic transpression in southwestern Mongolia and the Gobi Altai–Tien Shan connection. *Earth and Planetary Sciences*, **140**, 67–82.

CUNNINGHAM, W. D., WINDLEY, B. F., OWEN, L. A., BARRY, T., DORJNAMJAA, D., BADAMGARAV, G. & SAANDAR, M. 1997. Geometry and style of partitioned deformation within a late Cenozoic transpressional zone in the eastern Gobi Altai Mountains, Mongolia. *Tectonophysics*, **277**, 285–306.

DEVYATKIN, E. V. 1974. Structures and formational complexes of the Cenozoic activated stage (in Russian). *In*: *Tectonics of the Mongolian People's Republic*, Moscow, Nauka, 182–195.

DOLAN, J. F. & MANN, P. 1998. *Active strike-slip and collisional tectonics of the northern caribbean plate boundary zone*. Geological Society America Special Papers, **326**.

FLORENSOV, N. A. & SOLONENKO, V. P. (eds) 1963. *The Gobi–Altai Earthquake*. Moscow, Akademiya Nauk USSR (in Russian; English translation, 1965, US Department of Commerce, Washington, DC), 424 pp.

HOWARD, J., CUNNINGHAM, D. & DAVIES, S. 2006. Competing processes of clastic deposition and compartmentalised inversion in an actively evolving transpressional basin, western Mongolia. *Journal of the Geological Society, London*, **163**, 657–670.

HOWARD, J., CUNNINGHAM, D., DAVIES, S., DIJKSTRA, A. & BADARCH, G. 2003. Stratigraphic and structural evolution of the Dzereg Basin, Mongolia. *Basin Research*, **15**, 45–72.

KURUSHIN, R. A., BAYASGALAN, A. *ET AL*. 1997. *The Surface Rupture of the 1957, Gobi–Altay, Mongolia, Earthquake*. Geological Society of America Special Papers, **320**, 1–144.

LEGG, M. R., KAMERLING, M. J. & FRANCIS, R. D. 2004. Termination of strike-slip faults at convergence zones within continental transform boundaries: examples from the California Continental Borderland. *In*: GROCOTT, J., MCCAFFREY, K. J. W., TAYLOR, G. & TIKOFF, B. (eds), *Vertical Coupling and Decoupling in the Lithosphere*, Geological Society, London, Special Publications, **227**, 65–82.

MCBEE, J. H., HARTMANN, D. & COLLINS, G. C. 2003. Strain across ridges on Europa, *Lunar and Planetary Sciences Conference*, **XXXIV**, abstract 1783.

MCCLAY, K. & BONORA, M. 2001. Analog models of restraining stepovers in strike-slip fault systems. *AAPG Bulletin*, **85**, 233–260.

MANN, P., DRAPER, G. & BURKE, K. 1985. Neotectonics of a strike-slip restraining bend system, Jamaica. *In*: BIDDLE, K. T. & CHRISTIE-BLICK, N. (eds), *Strike-Slip Deformation, Basin Formation and Sedimentation*, SEPM Special Publications, **37**, 211–226.

MANN, P. 2007. Global catalogue, classification and tectonic origin of active restraining-and releasing bends, on strike-slip fault systems. *In*: CUNNINGHAM, W. D. & MANN, P. (eds) *Tectonics of Strike-Slip Restraining and Releasing Bends*. Geological Society, London, Special Publications, **290**, 13–142.

MUIR, T. L. 2002. The Hemlo gold deposit, Ontario, Canada: principal deposit characteristics and constraints on mineralization. *Ore Geology Reviews*, **21**, 1–66.

OWEN, L. A., CUNNINGHAM, W. D., WINDLEY, B. F., BADAMGAROV, J. & DORJNAMJAA, D. 1999. The landscape evolution of Nemegt Uul: a late Cenozoic transpressional uplift in the Gobi Altai, southern Mongolia. *In*: SMITH, B. J., WHALLEY, W. B. & WARKE, P. A. (eds) *Uplift, Erosion and Stability: Perspectives on Long-term Landscape Development*. Geological Society, London, Special Publications, **162**, 229–238.

PHILIP, H. & RITZ, J.-F. 1999. Gigantic paleolandslide associated with active faulting along the Bogd fault (Gobi Altay, Mongolia). *Geology*, **27**, 211–214.

POCKALNY, R. 1997. Evidence of transpression along the Clipperton Transform: implications for processes of plate boundary organization. *Earth and Planetary Science Letters*, **146**, 449–464.

RIPPINGTON, S., CUNNINGHAM, D. & ENGLAND, R. 2006. Polyphase deformation and intraplate mountain building at the Nemegt Uul restraining bend in the Gobi Altai Mountains, Mongolia. *Geophysical Research Abstracts*, **8**, 06486.

SCHWARTZ, S. Y., ORANGE, D. L. & ANDERSON, R. S. 1990. Complex fault interactions in a restraining bend on the San Andreas Fault, southern Santa Cruz Mountains, California. *Geophysical Research Letters*, **17**, 1207–1210.

SONG, T. & CAWOOD, P. A. 1999. Multistage deformation of linked fault systems in extensional regions: an example from the northern Perth basin, western Australia. *Australian Journal of Earth Sciences*, **46**, 897.

TAPPONNIER, P. & MOLNAR, P. 1979. Active faulting and Cenozoic tectonics of the Tien Shan, Mongolia, and Baykal regions. *Journal of Geophysical Research*, **84**, 3425–3459.

TOMURTOGOO, O. 1999. *Geological Map of Mongolia*, 1:1 000 000, Mongolian Academy of Science, Institute of Geology and Mineral Resources.

TRAYNOR, J. J. & SLADEN, C. 1995. Tectonic and stratigraphic evolution of the Mongolian People's Republic and its influence on hydrocarbon geology and potential. *Marine and Petroleum Geology*, **12**, 35–52.

VASSALLO, R., JOLIVET, M. *ET AL*. 2006. Chronology and uplift rates of the relief in the Altay and the Gobi–Altay mountain ranges (Mongolia). *Geophysical Research Abstracts*, **8**, 00573.

WALKER, R. & JACKSON, J. 2004. Active tectonics and late Cenozoic strain distribution in central and eastern Iran. *Tectonics*, **23**, TC5010, doi:10.1029/2003TC001529, 2004.

WILLETT, S. D., SLINGERLAND, R. & HOVIUS, N. 2001. Uplift, shortening, and steady state topography in active mountain belts. *American Journal of Science*, **301**, 455–485.

WRIGHT, T. L. 1991. Structural geology and tectonic evolution of the Los Angeles basin, California. *In*: BIDDLE, K. T. (ed.) *Active Margin Basins*. AAPG Memoirs, **52**, 35–134.

ZAITSEV, N. S. 1978. *Geological Map of the Mongolian Altai* 1:500 000 scale, Academy of Sciences of the USSR, Academy of Sciences, Mongolian People's Republic, Combined Soviet-Mongolian Scientific Research Geological Expedition. Nauka, Moscow (in Russian).

Toward a better understanding of the Late Neogene strike-slip restraining bend in Jamaica: geodetic, geological, and seismic constraints

P. MANN[1], C. DEMETS[2] & M. WIGGINS-GRANDISON[3]

[1]*Institute for Geophysics, Jackson School of Geosciences, 4412 Spicewood Springs Road, Bldg 600, University of Texas at Austin, Austin, Texas, 78759, USA (e-mail: paulm@ig.utexas.edu)*

[2]*Department of Geology and Geophysics, University of Wisconsin, 1215 West Dayton Street, Madison, WI 53706, USA*

[3]*Earthquake/Seismic Unit, Department of Geography and Geology, University of West Indies, Mona, Kingston 7, Jamaica, West Indies*

Abstract: We describe the regional fault pattern, geological setting and active fault kinematics of Jamaica, from published geological maps, earthquakes and GPS-based geodesy, to support a simple tectonic model for both the initial stage of restraining-bend formation and the subsequent stage of bend bypassing. Restraining-bend formation and widespread uplift in Jamaica began in the Late Miocene, and were probably controlled by the interaction of roughly east–west-trending strike-slip faults with two NNW-trending rifts oriented obliquely to the direction of ENE-trending, Late Neogene interplate shear. The interaction of the interplate strike-slip fault system (Enriquillo-Plantain Garden fault zone) and the oblique rifts has shifted the strike-slip fault trace *c.* 50 km to the north and created the 150-km-long by 80-km-wide restraining bend that is now morphologically expressed as the island of Jamaica. Recorded earthquakes and recent GPS results from Jamaica illustrate continued bend evolution during the most recent phase of strike-slip displacement, at a minimum GPS-measured rate of 8 ± 1 mm/a. GPS results show a gradient in left-lateral interplate strain from north to south, probably extending south of the island, and a likely gradient along a ENE–WSW cross-island profile. The observed GPS velocity field suggests that left-lateral shear continues to be transmitted across the Jamaican restraining bend by a series of intervening bend structures, including the Blue Mountain uplift of eastern Jamaica.

Regional significance of restraining bends

Restraining bends are a relatively common structural and morphological feature along both active and ancient intercontinental strike-slip faults, and occur at a variety of scales ranging from tens of metres to tens of kilometres (Crowell 1974; Gomez *et al.* 2007; Mann 2007) (Fig. 1). Regardless of their scale, interplate 'gentle' or curved restraining bends (Crowell 1974) consist of a convergent fault segment of variable width, geological complexity and topographic elevation, that is misaligned with the direction of the regional interplate slip vector (Mann & Gordon 1996). For example, along the San Andreas fault system, the Transverse Ranges restraining bend occurs at a significant misalignment of about 20° with the regional interplate slip vector between the Pacific plate and North America plate (Fig. 1A). Accommodating about 35 mm/a of right-lateral slip, the San Andreas system is a relatively simple linear boundary along most of its length, but becomes much wider (*c.* 150 km) at the Transverse Ranges, where there are multiple, seismogenic, subparallel thrusts and strike-slip fault strands (Matti & Morton 1993). Restraining bends in intraplate areas of deformation like central Asia are more complex in that that they do not show paired geometries, and they cannot be related in a convincing way to the plate motions (Mann 2007; Cunningham 2007). Instead, reactivation of underlying basement structures seems to play a large role in their evolution.

South of the Transverse Ranges in the Salton Sea and Gulf of California, the trace of the main plate boundary strike-slip fault is again oblique to the direction of interplate slip, but curving in the opposite direction. In this orientation, the misaligned fault trace in this topographically depressed or submarine region is characterized by a broad, 100–150-km-wide zone of oblique opening or 'transtension', which in some areas has progressed to the point of forming localized zones of volcanism or short, oceanic spreading centres (Fig. 1a).

From: CUNNINGHAM, W. D. & MANN, P. (eds) *Tectonics of Strike-Slip Restraining and Releasing Bends.* Geological Society, London, Special Publications, **290**, 239–253.
DOI: 10.1144/SP290.8 0305-8719/07/$15.00

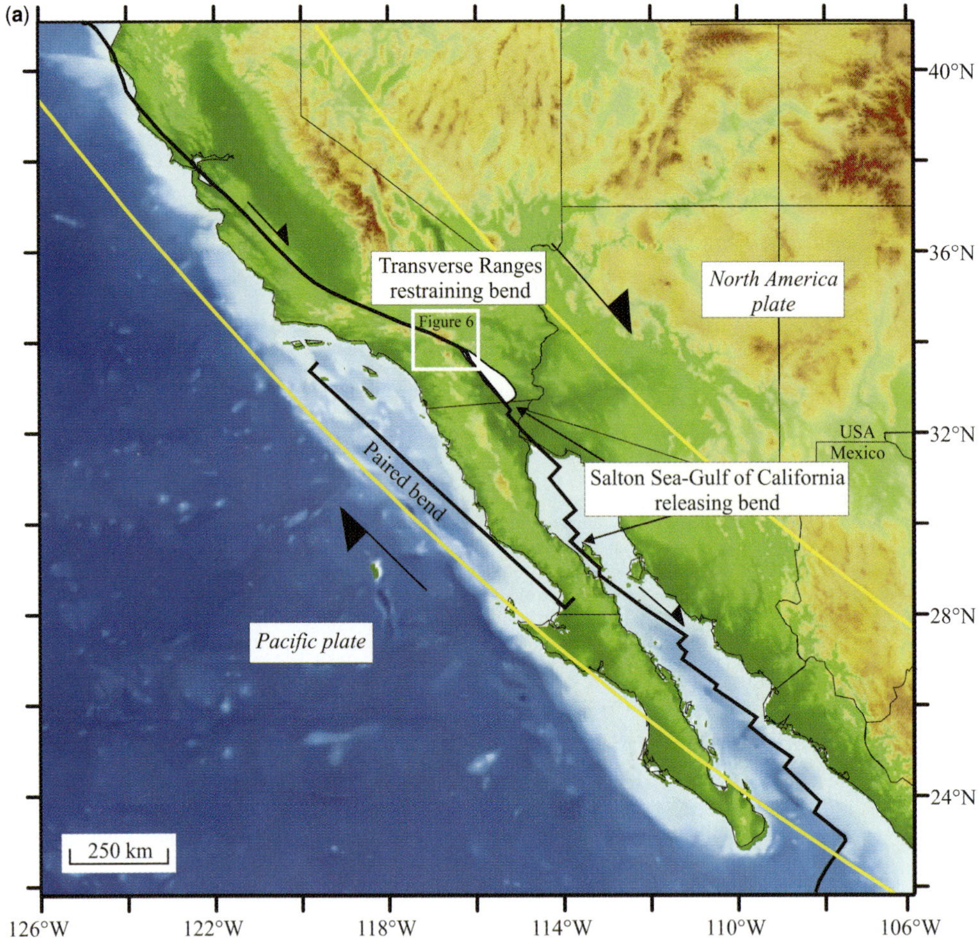

Fig. 1. (**a**) Major restraining and releasing fault bends along the active right-lateral strike-slip San Andreas fault system separating the Pacific and North America plates (direction of plate motion is shown by yellow circles about the pole of rotation from DeMets *et al.* 1994; topographic base is from NASA SRTM data). The Transverse Ranges represent a major restraining-bend segment that is flanked to the south by the Salton Sea–Gulf of California releasing bend (area in white surrounding the Salton Sea is highly extended and below sea-level). The adjacent Transverse Ranges restraining bend and the Salton Sea–Gulf of California releasing bends are termed a 'paired bend'. (**b**) Major restraining and releasing fault bends along the active left-lateral strike-slip northern Caribbean fault system separating the North America and Caribbean plates (direction of plate motion is shown by yellow circles about the pole of rotation from DeMets *et al.* 2000). The island of Jamaica represents a major restraining bend at a right-step in the Enriquillo–Plantain Garden fault zone (EPGFZ) and the Walton fault zone (WFZ). The Jamaica restraining bend is flanked to the west by the West Jamaica releasing bend. The adjacent Jamaican restraining bend and the West Jamaica releasing bend are termed a 'paired bend'. A parallel strike-slip fault zone consists of the Oriente fault zone (OFZ) and the Swan Islands fault zone (SIFZ). The active Mid-Cayman spreading centre (MCSC) is a 100-km-long left-step between the two faults.

The overall effect of the obliquely convergent fault segment in the Transverse Ranges of California and the adjacent obliquely divergent fault segments in the Imperial Valley and Gulf of California is a 'paired bend', or adjacent restraining and releasing-bend segments (Mann 2007). The paired bend gives the southern San Andreas–Gulf of California fault system a gently undulating appearance when viewed on a regional scale (Fig. 1a).

A similar, undulating strike-slip fault pattern composed of adjacent or paired restraining and releasing bends is present along the northern Caribbean plate boundary zone (Fig. 1b). This interplate strike-slip boundary consists of a

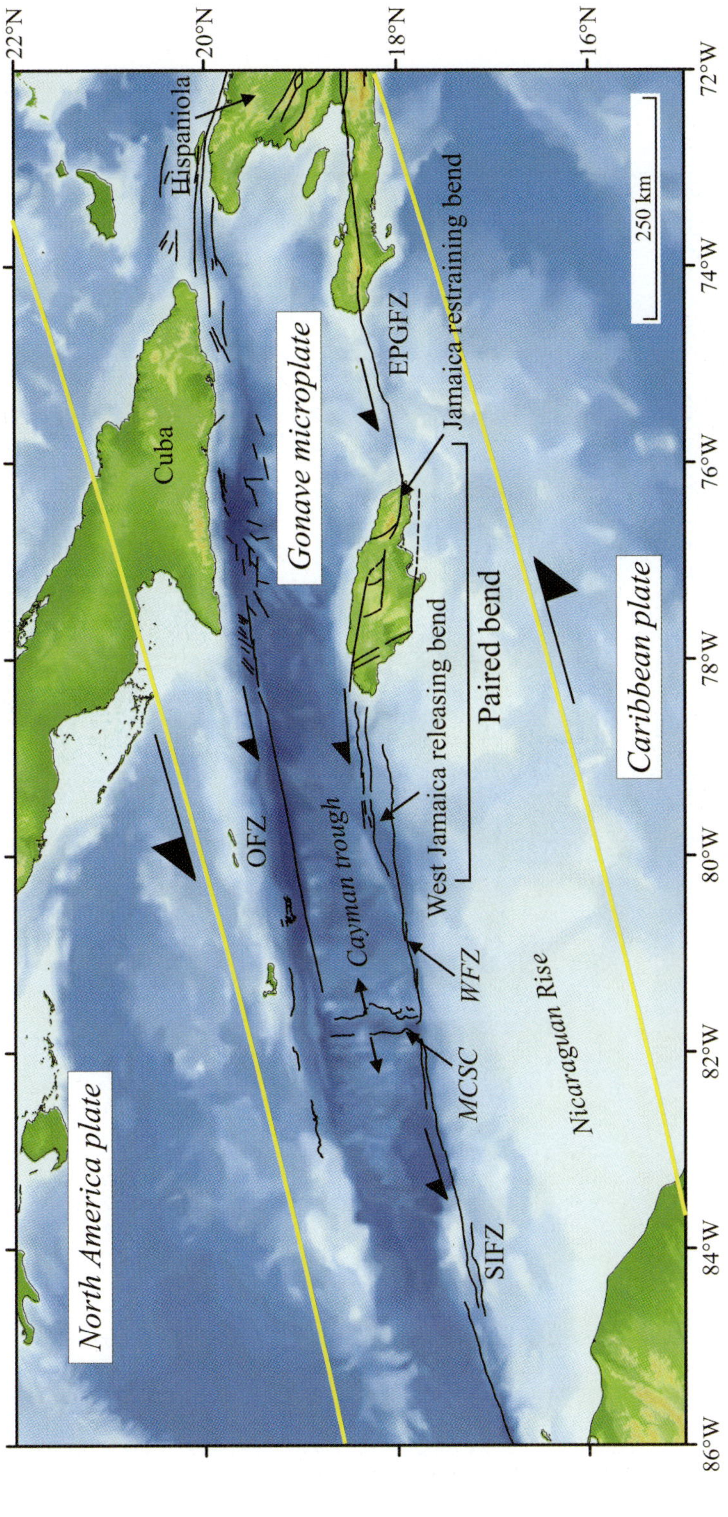

Fig. 1. *(Continued)*

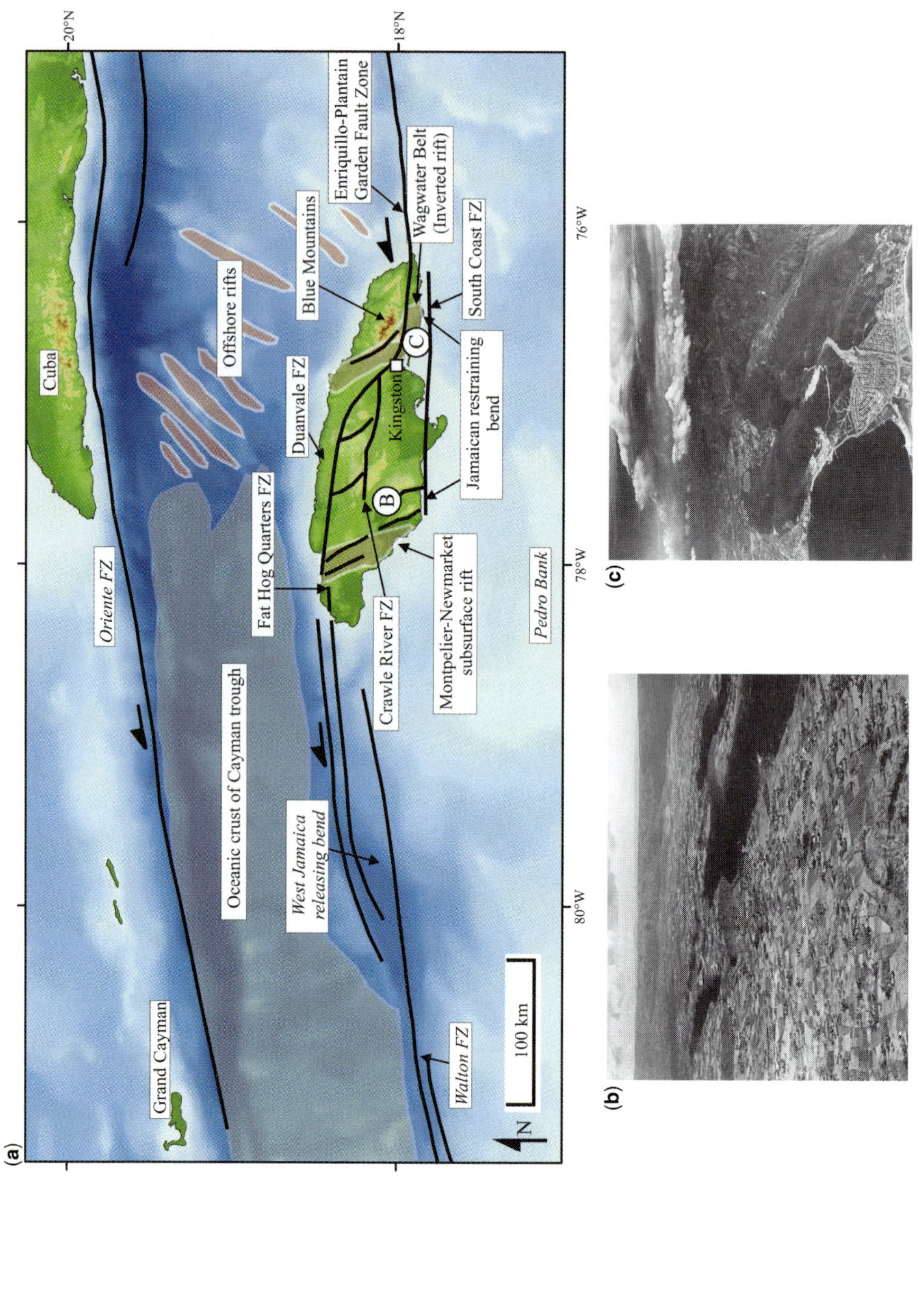

Fig. 2. For caption see p.243.

100–250-km-wide, seismogenic zone of mainly left-lateral strike-slip deformation extending over 3000 km along the northern edge of the Caribbean plate (Burke *et al.* 1980; Calais & Mercier de Lepinay 1990). Prominent restraining bends occur on the islands of Hispaniola (Mann *et al.* 2002) and Jamaica (Mann *et al.* 1985; Rosencrantz & Mann, 1991) (Fig. 2), near the Swan Islands of Honduras along the southern edge of the Cayman Trough (Mann *et al.* 1991), and in northern Central America (Mann & Gordon 1996). Several of these areas constitute paired bends with the familiar undulating fault shape that juxtaposes restraining and releasing bends (Mann 2007) (Fig. 1b).

Swanson (2005) has compiled geological and structural information on gently undulating strike-slip fault traces worldwide. He attributes the adjacent curving traces of strike-slip faults ('paired bends' in this paper) at outcrop to regional scales to adhesive wear, or enhanced friction, along a straight fault plane followed by the nucleation and abrupt lateral shift of the active strike-slip fault to a new parallel fault trace. The deviated fault trace forms a characteristic 5- to 500-km-long, map-view feature of strike-slip faults, which he calls a 'sidewall ripout'. One faulted edge of the deviated 'sidewall ripout' constitutes a releasing bend, whilst the other side constitutes a restraining bend.

Tectonic origin of restraining bends

Two main questions that remain on the origin and evolution of restraining bends are:

(1) what tectonic process is responsible for producing the curvilinear traces in the strike of the fault that ultimately produce alternating restraining and releasing bends; and
(2) how do areas of alternating restraining- and releasing-bend segments evolve with progressive strike-slip displacement?

For the first question, two possible answers are that pre-existing crustal structures influence the strike of the propagating strike-slip fault. For example, a rift or former suture zone might divert the fault from a straight orientation. A second possibility would be that the mechanical properties of the fault itself might cause the fault to lock, accumulate strain, and either form a more curving fault trace (Bridwell 1975) and/or abandon the locked trace and develop an adjacent, parallel fault (Swanson 2005). For the second question, bypassing of bends would appear to be a fundamental property of strike-slip faults as the faults attempt to maintain a straight trace.

In this paper we describe the regional fault pattern, geological setting, and active fault kinematics of Jamaica, from published geological maps, earthquakes and GPS-based geodesy, in order to gain a better understanding of the Jamaican fault restraining bend. Evolution of the restraining-bend faults in Jamaica has important implications for the assessment of seismic risk on this rugged, 10 991 km^2 island with a population of 2.6 million and a long historical record of destructive earthquakes and tsunamis (Wiggins-Grandison & Atakan 2005). An understanding of the present-day stage of restraining-bend development in Jamaica would improve our conceptual basis for understanding those Jamaican faults that might be the source of future large earthquakes.

Tectonic and geological setting of the Jamaican restraining bend

Tectonic setting

The island of Jamaica is one of only two places where strike-slip faults that carry Caribbean–North America–Gonave microplate motion come onshore in the Greater Antilles islands of the northern Caribbean (Cuba, Jamaica, Hispaniola and Puerto Rico) (Fig. 1b). The second place is the neighbouring island of Hispaniola (Haiti and Dominican Republic) where large strike-slip faults are well exposed and can be traced as continuous features for most of the length of the island (Mann *et al.* 1995; Mann *et al.* 1998) (Figs 1b & 2).

Fig. 2. (*Continued*) (**a**) Topography of Jamaica from the NASA SRTM data-set combined with the GEBCO bathymetry of its offshore area. Labelling identifies the major mapped faults onland (Mann *et al.* 1985) and offshore (Rosencrantz & Mann 1991) in the Jamaican paired-bend system. On- and offshore Palaeocene–Early Eocene rift features in Jamaica associated with the east–west opening of the Early Eocene oceanic Cayman Trough are shown in a pale-brown colour. Onland rifts include the now inverted and highly deformed Wagwater Belt of eastern Jamaica, and the subsurface and undeformed Montpelier–Newmarket Rift of western Jamaica. Oceanic crust of the Cayman Trough is shown by the grey area. (**b**) Oblique aerial view (photograph taken by S. Tyndale-Biscoe) of the Spur Tree fault zone of western Jamaica, showing the prominent scarp and plateau-like character of the elevated shallow-water limestone of Oligocene–Miocene age. The location of the photograph is shown by the letter 'B' on the map in (a). (**c**) Oblique aerial view (by S. Tyndale-Biscoe) of the inverted normal fault-bounding the western edge of the deformed Wagwater Belt. Inversion of the Wagwater Belt has induced folding of the adjacent Mio-Pliocene limestones. The location of the photograph is shown by the letter 'C' on the map in (a).

Located in the northern Caribbean Sea at the NW end of the Nicaragua Rise, Jamaica consists of an emergent Cretaceous-age oceanic volcanic arc and volcanogenic sedimentary rocks, overlain by 5–7 km of Tertiary carbonate rocks (Lewis & Draper 1990). Seismic velocities yield an island-arc crustal thickness of 25–30 km, with most locally recorded earthquakes concentrated from depths of 15 to 30 km (Wiggins-Grandison 2003, 2004) (Fig. 3).

At the longitude of Jamaica, Caribbean–North America plate motion of 19 ± 2 mm/a (DeMets *et al.* 2000; Weber *et al.* 2001) is carried by two parallel strike-slip faults, the Oriente transform fault immediately south of Cuba and the Walton/Plantain Garden/Enriquillo faults, which lie 100–150 km farther south (Figs 1b & 2). Mann *et al.* (1995) and Mann *et al.* (2002) postulate that lithosphere between these faults constitutes an independent Gonave microplate that has formed in response to the ongoing collision between the leading edge of the Caribbean plate in Hispaniola and the Bahama carbonate platform. GPS measurements in Hispaniola (Dixon *et al.* 1998; Calais *et al.* 2002; Mann *et al.* 2002) and measurements in Jamaica described by DeMets & Wiggins-Grandison (2007) indicate that roughly half of Caribbean–North America motion (8–14 mm/a) is carried by the Enriquillo–Plantain Garden–Walton faults, consistent with the microplate hypothesis. A key question of major relevance to seismic risk assessment in Jamaica is how strike-slip motion on the Plantain Garden Fault of SE Jamaica is transferred to the island to link with the Walton and other faults making up a large releasing bend west of the island (Figs 1b & 2).

Topographic and geological setting

The Jamaican restraining bend consists of a topographically uplifted area in eastern Jamaica that is bounded at its southern edge by the Enriquillo–Plantain Garden Fault, a transitional area of lower

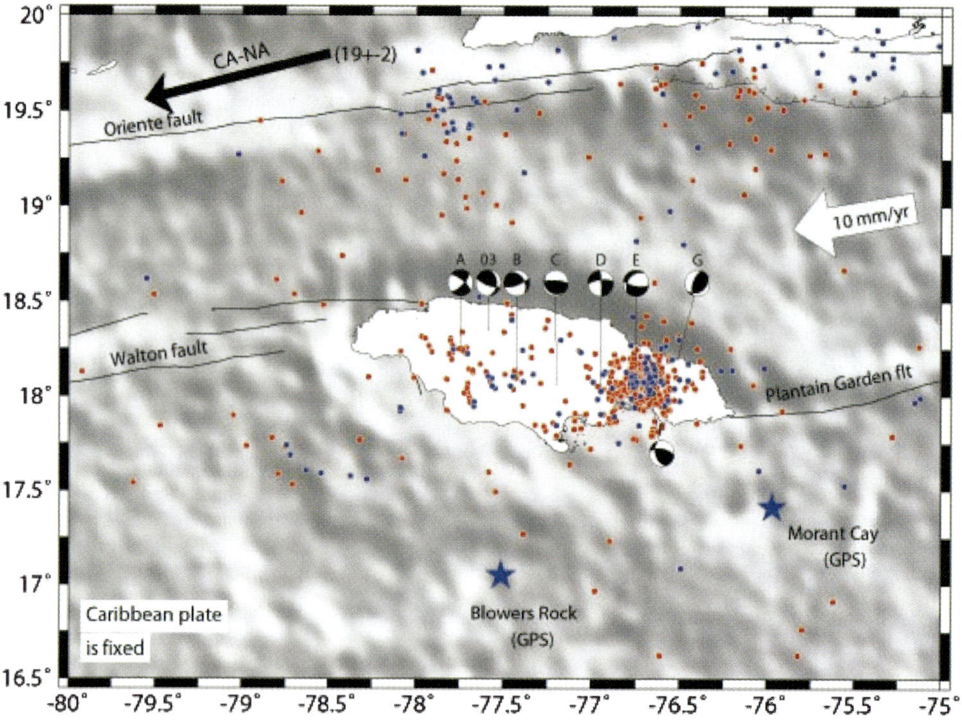

Fig. 3. 1963 to 2004 teleseismicity (blue) and microseismicity (red) of the Jamaica restraining bend on GEBCO bathymetric base map. Micro-earthquakes with magnitudes >1 were recorded by the local Jamaican network (Wiggins-Grandison 2001). Velocities represent motions of the North America plate (black arrow) and the Gonave microplate (white arrow), whose motion is derived from the GPS velocities shown in Figure 4. Faults on and near Jamaica must accommodate a total of 10 (± 2) mm/a of slip. Note that the seismic zone is largely restricted to the island of Jamaica, so the southern edge of the Gonave microplate is believed to lie near the south coast of the island. In-progress GPS sites on Blowers Rock and Morant Cay are therefore assumed to be part of the stable Caribbean plate.

topography in western Jamaica, and an offshore releasing bend, or pull-apart basin, along the Walton fault zone (Rosencrantz & Mann 1991). The West Jamaica releasing bend forms where the plate boundary curves towards a more east–west trend (Figs 1b & 2). Mann *et al.* (1985) and Mann *et al.* (1990) propose that two Palaeogene rifts – that are highly oblique to the EW direction of active plate motion – may be the crustal features responsible for diverting the intersecting plate boundary strike-slip faults from their expected east–west strike directions parallel to the small circles of rotation about the Caribbean pole of rotation (DeMets *et al.* 2000).

In western and central Jamaica, the landscape forms a relatively flat, elevated plateau that exposes karsted Oligocene–Miocene carbonate rocks (Fig. 2b). Faults form prominent scarps in these carbonate lithologies and exhibit topographic relief up to 600 m (Horsfield 1974; Wadge & Dixon 1984; Mann *et al.* 1985) (Fig. 2b). The Palaeogene rifts are subsurface features known from oil exploration both on and offshore of Jamaica (Arden 1975), but in eastern Jamaica, the Wagwater Rift is completely inverted by reverse faulting along its former normal-faulted margins, and elevated into a mountain range (Mann *et al.* 1985; Mann & Burke 1990) (Fig. 2c). Reaching 2.5 km above sea-level, the steep-sided Blue Mountain restraining bend is directly adjacent to the deformed Wagwater Belt and dominates the island's topography and seismicity (Fig. 3d). Its anomalously high elevation and enhanced seismic activity indicate that the Blue Mountains may continue to play an important role in transferring a significant part of the strike-slip displacement northward across the Jamaican restraining bend.

Historic and modern seismicity of Jamaica

Modern records of Jamaican earthquakes date from British settlement of the island in the 1600s. Jamaica has experienced 13 earthquakes of Mercalli intensity VII–X since 1667 (Wiggins-Grandison 2001). The island's most devastating earthquake, a MMI X that occurred on 7 June 1692, caused extensive liquefaction of the island's southern alluvial plains where most of its population was (and still is) concentrated. The 1692 event killed roughly one-quarter of the inhabitants of Port Royal, Jamaica's principal city at the time (Fig. 3). A century ago, the 14 January 1907 Kingston earthquake (MMI IX) killed 1000 and left 90 000 homeless in the capital city. Today, more than one-third of Jamaica's 2.6 million inhabitants and the country's economic base are concentrated in the southern area of the country near the capital city of Kingston (Fig. 2a). Kingston and the surrounding densely populated plains are underlain by thick, unconsolidated alluvium deposited by Jamaica's southward-flowing rivers, and are subject to liquefaction effects and seismic-wave focusing above basinal bedrock topography (Wiggins-Grandison *et al.* 2003). Despite the obvious earthquake hazards, none of Jamaica's Late Holocene faults have been systematically mapped or trenched to reveal which of the main faults may have ruptured during these destructive historical earthquakes.

Teleseisms and relocated microseisms are concentrated primarily along the geomorphically prominent Blue Mountain restraining bend of eastern Jamaica (Fig. 2b), but earthquakes also occur along other topographically prominent faults in the central and western areas of the island (Burke *et al.* 1980; Wadge & Dixon 1984; Mann *et al.* 1985) (Fig. 3). Nearly all relocated earthquakes are found between depths of 12 and 27 km, remarkably deep in comparison with continental settings such as northern and central California, where almost all seismicity is confined above crustal depths of 11–12 km (e.g. Castillo & Ellsworth 1993). Focal mechanisms reveal an island dominated by east–west-directed, left-lateral shear with a lesser north–south convergent component (Wiggins-Grandison 2003; Wiggins-Grandison & Atakan 2005). These focal mechanisms are in excellent agreement with our new GPS velocity field and roughly east–west-trending fold axes that affect the Cretaceous and Tertiary stratigraphic section on the island (e.g. Lewis & Draper 1990).

Initial GPS results for the period 1999–2005

Installation and occupation of an island-wide GPS network was initiated in 1999, and by mid-2001 consisted of the present 20-station network (Fig. 4). By early 2004, all 20 GPS sites had been occupied a sufficient number of times (three to six) over a sufficiently long time-span to define the first-order deformation pattern of the island. The GPS data and velocities presented herein are fully described by DeMets & Wiggins-Grandison (2007), and represent the first description of the island's present velocity field. We summarize these data briefly below and refer interested readers to DeMets & Wiggins-Grandison (2007) for a more complete description and interpretation of these data.

The GPS data were analysed using GIPSY analysis software from the Jet Propulsion Laboratory (JPL); free-network satellite orbits and satellite clock offsets obtained from JPL; a precise point

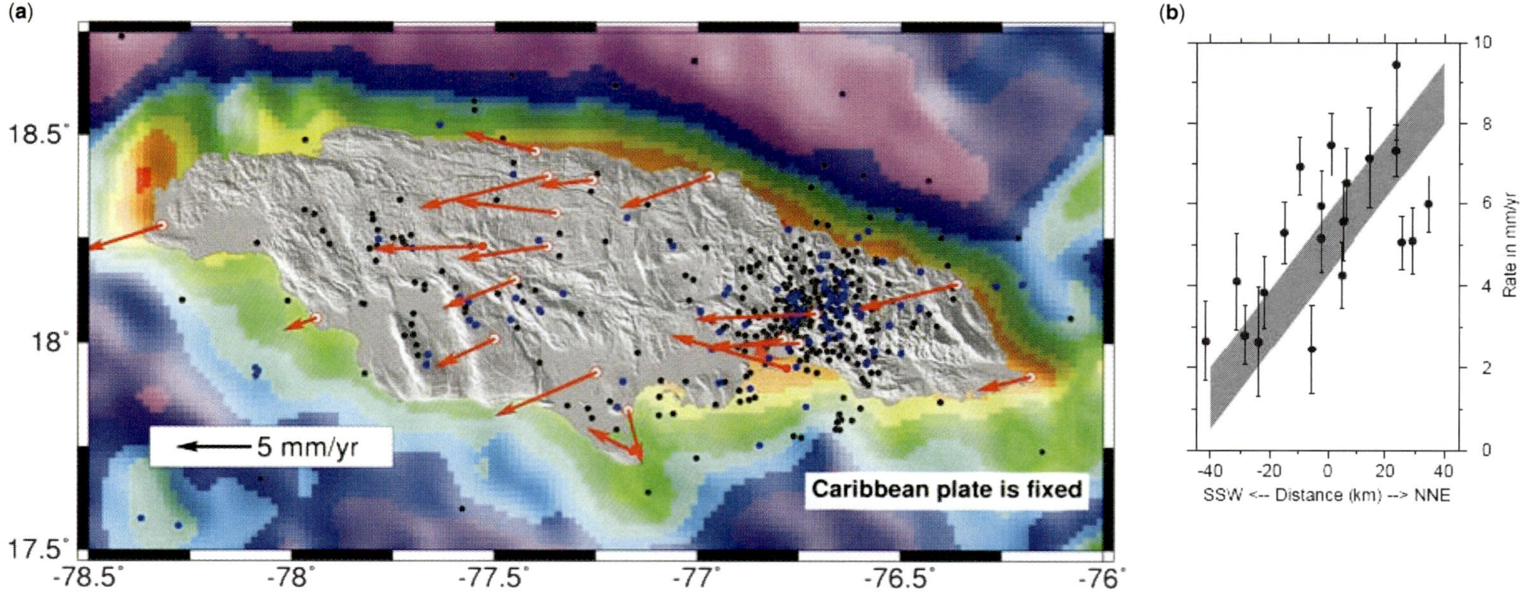

Fig. 4. (**a**) GPS velocity field relative to Caribbean plate (geodetic reference frame is ITRF2000) superimposed on 90-metre Space Shuttle Topographic Radar Mission (SRTM) topography. Velocity uncertainties are omitted for clarity, but are typically $\pm$ 2–3 mm/a at the 1D, 1-sigma level. (**b**) GPS rate components parallel to S75°W projected on to N15°E transect of island. A transect from ENE to WSW (not shown) exhibits a similar gradient.

positioning analysis strategy (Zumberge *et al.* 1997); and resolution of phase ambiguities when possible. Continuous and semi continuous data at three stations were used to estimate and remove common-mode, non-tectonic noise from all station time series. All of the GPS coordinate time series are well-behaved and have the usual levels of daily scatter (2–4 mm in the northern component, and 3–5 mm in the eastern component). Linear regression of the coordinate time-series yields well-constrained velocities at all 20 GPS sites. Individual site velocity uncertainties are estimated using the Mao *et al.* (1999) algorithm.

Figure 4 shows our most recent GPS velocity field after removing the motion of the Caribbean plate predicted by an angular velocity vector that best fits the velocities of 15 sites in the Caribbean plate interior (DeMets *et al.* 2007). Several useful conclusions can be drawn from the GPS velocity field independent of any modelling. One first-order conclusion is that none of Jamaica moves as part of the Caribbean plate interior (Fig. 4). GPS sites instead move westward relative to the Caribbean plate at a maximum rate of 8 ± 1 mm/a, representing a minimum estimate for Gonave microplate motion relative to the Caribbean plate (Fig. 1b). GPS site directions are uniformly parallel to the southern boundary of the Gonave microplate, indicating that this boundary is dominated by active left-lateral shear as predicted from geological and geomorphological studies (Burke *et al.* 1980; Wadge & Dixon 1984; Mann *et al.* 1985) (Fig. 2b & c).

The velocity field exhibits significant gradients from NNE–SSW and ENE–WSW (Fig. 4), indicating that deformation is two-dimensional. These gradients are a likely consequence of distributed elastic strain from one or more locked faults that transfer slip across the restraining bend. Finally, our data provisionally suggest the existence of a GPS velocity gradient of 2 ± 1 mm/a across the topographically high and seismically active Blue Mountain restraining bend (Fig. 4).

Implications of GPS results for specific faults in Jamaica

If faults in Jamaica are frictionally locked to depths of 20–30 km, as relocated earthquakes suggest (Wiggins-Grandison 2004), then the observed GPS velocity gradients (e.g. Fig. 4) are attributable to elastic strain accumulation and will require detailed modelling to interpret. However, useful inferences can already be drawn without resorting to modelling. With respect to the Caribbean plate interior, the midpoint of the 8 mm/a north–south velocity gradient that we observe occurs just south of the Crawle River left-lateral strike-slip fault zone, close to the latitudinal mid-point of the island (Fig. 2a). Constraints on seafloor-spreading rates across the Cayman spreading centre (Rosencrantz & Mann 1991) (Fig. 1b), and plate circuit closures suggest that Gonave–Caribbean motion cannot be significantly faster than 8–10 mm/a – approximately what we are measuring across Jamaica. Assuming that elastic strain accumulates symmetrically in a laterally homogeneous crust, as seems likely in the thick volcanic arc typical of Jamaican crust, then the fact that the midpoint of our observed velocity gradient occurs in central Jamaica argues against models in which most or all long-term fault slip in Jamaica is focused predominantly along the southern or northern coasts of the island (Fig. 2a).

Modelling will be required to determine whether the data can be used to distinguish between alternative deformation models or possibly a simple single-fault model in which most fault slip is concentrated along faults in central Jamaica. Our data do not appear to be consistent with a model in which faults in the Blue Mountain restraining bend of eastern Jamaica transfer most or all slip from the Plantain Garden fault northward to the Duanvale and Fat Hog Quarters faults along the northern coast of Jamaica, as suggested by Mann *et al.* (1985) (Fig. 2a). Given the topographic and seismic prominence of the Blue Mountains, as well as the prevalence of fault scarps affecting carbonate rocks of Oligocene–Miocene age in the west-central area of Jamaica (Wadge & Dixon 1984) (Fig. 2b & c), this conclusion is an unexpected result (Fig. 2b).

Within the uncertainties, the GPS velocity field (Fig. 4a) and its associated gradients (Fig. 4b) permit partitioning of slip between the east–west-trending Duanvale fault of northern Jamaica; the Crawle River fault zone of central Jamaica; and the South Coast Fault of southern Jamaica (Fig. 2a). Smaller velocity uncertainties are needed to determine whether sudden changes in GPS velocities coincide with any of these faults, as might be expected if any of them are creeping. If, as seems more likely, the faults are locked, then careful modelling will be required to define the range of slip rates and models that are capable of describing the observed site velocities.

From syntheses of satellite imagery and field mapping, various tectonic models have been proposed for Jamaica, including:

(1) a broad, east–west-striking left-lateral shear zone in which several parallel strike-slip faults are active (Burke *et al.* 1980; Wadge and Dixon 1984); and
(2) a right-stepping restraining bend connecting two parallel, left-lateral strike-slip faults (Mann *et al.* 1985).

The widespread microseismicity; fault scarps in Neogene carbonate rocks of the central and western parts of the island (Fig. 2a & b); and two-dimensional gradients in the GPS velocity field, all support a model in which deformation of the island is accommodated by multiple fault step-overs that transfer strike-slip motion across the restraining-bend via a series of intervening active thrust faults. In such a restraining-bend model, slip along the Plantain Garden and South Coastal faults would transfer gradually northward to the Crawle River and Duanvale fault zones, with the summed slip rates for the three faults equalling a minimum of 8 ± 1 mm/a at any given longitude. This model may explain the prevalence of arcuate, north–south-trending, scissor-like scarps that presumably link the east–west-striking strike-slip faults (Wadge & Dixon 1984; Mann *et al.* 1985) (Fig. 4).

Discussion

Combining the GPS and earthquake results with Late Neogene geological data allows a longer term (*c.* 10 Ma) view of how the Jamaican bend continues to evolve and influence the geomorphology, fault kinematics and present-day seismicity of the island (Fig. 3). We propose several stages in the development of the Jamaican paired-bend system (Fig. 5) and compare these stages with better-studied restraining bends along the San Andreas fault system in southern California (Fig. 1a & 6).

Early stages of Jamaican paired-bend development

In our proposed model, the initial stage of paired-bend development occurs when the east–west-striking Enriquillo–Plantain Garden fault zone propagates westward into the Jamaican region and encounters the NE-trending Wagwater and Newport–Montpelier rifts of Palaeogene age (Fig. 5a). The formation of the Enriquillo–Plantain Garden fault zone is attributed to the Miocene–Recent collision between the leading edge of the Caribbean plate in Hispaniola and the Bahama carbonate platform (Mann *et al.* 1995) (Fig. 2a). Formation of the Enriquillo–Plaintain Garden Fault led to the detachment of the Gonave microplate from the NE corner of the Caribbean plate (Fig. 2a). Intersection of the east–west-striking Enriquillo Plantain Garden strike-slip fault with the Wagwater Rift of eastern Jamaica is proposed to have led to its Early to Middle Miocene–Recent inversion and the deviation of the fault trace to a more NW curvature of the fault in eastern Jamaica (Fig. 5b).

A widespread unconformity of Early Miocene age in the carbonate section of Jamaica may date the onset of convergent deformation and uplift in this part of the island (Eva & McFarlane 1979). Green (1977) suggested that continued uplift and strike-slip faulting in eastern Jamaica is marked by a Late–Middle Miocene unconformity. Along the NE coast of Jamaica near Buff Bay, an abrupt facies change occurs between white chalky limestone and grey or brown marl of Late Miocene age (Blow 1969). This facies change may reflect the early inversion of the Wagwater Rift and uplift of the Blue Mountain part of the restraining bend. From the Late Miocene to the present day, uplift has been progressively propagating westward, as shown by the east to west gradient in topographic and erosional level (Fig. 2a). Miocene to Pliocene carbonate rocks around the periphery of the Blue Mountains become progressively conglomeratic in character, and reflect the continued and perhaps accelerated Late Neogene uplift of the Blue Mountains segment of the restraining bend (Horsfield 1974; Mann *et al.* 1985) (Fig. 2a & c).

As the Blue Mountain restraining bend developed, we envision activity along faults in the north-central part of Jamaica, such as the Duanvale fault zone (Figs 2a & 7c). The Duanvale Fault would link the Jamaican restraining bend in eastern Jamaica to the West Jamaica releasing bend (Fig. 2a). In the offshore area of western Jamaica, Rosencrantz & Mann (1991) mapped recent sea-floor fault-breaks roughly parallel to the Duanvale fault-zone. Due to a lack of core data and high-quality seismic-reflection profiles, we have no direct constraints on the age of initiation of the West Jamaica releasing bend but, based on the paired-bend concept, we would predict its age of initiation to be roughly that of the Miocene Blue Mountain uplift, or Early to Middle Miocene.

Later stages of paired-bend development

Earthquakes (Fig. 3) and GPS results (Fig. 4) provide us with insights into how the Jamaican paired bend has continued to evolve to the present day. Recorded seismicity is focused on the area where the Enriquillo-Plantain Garden fault zone intersects the Wagwater inverted rift (Fig. 4). However, a band of earthquakes along the southern coast of the island, along with GPS results, indicates that some left-lateral shearing is accommodated along the southern – rather than northern – parts of the island (Fig. 4). This southern band of east–west-trending seismicity suggests that the previously formed bend structures to the north may be in the initial stages of bypass by the South Coast fault zone, as shown schematically in Figure 5d. Bypass is likely to be a gradual process.

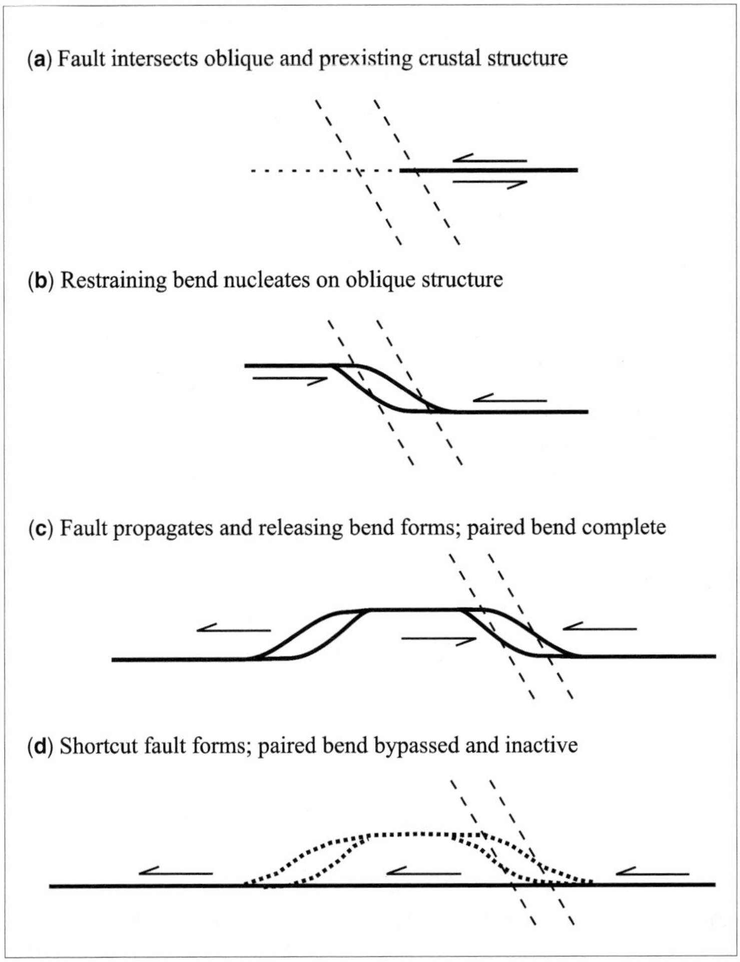

(a) Fault intersects oblique and prexisting crustal structure
(b) Restraining bend nucleates on oblique structure
(c) Fault propagates and releasing bend forms; paired bend complete
(d) Shortcut fault forms; paired bend bypassed and inactive

Fig. 5. Proposed conceptual model for the origin and abandonment of a paired strike-slip bend (adjacent restraining and releasing fault segment), based on compilation of information from the Jamaica restraining bend. (**a**) Strike-slip fault intersects an oblique structure. In the case of the Jamaica bend, the oblique structure is a Palaeogene rift basin (cf. Fig. 2a); to our knowledge, no similar type of pre-existing, oblique structure has been recognized beneath the Transverse Ranges of California (cf. Fig. 1a). (**b**) A restraining bend forms and a strike-slip fault continues to propagate. Formation of a restraining bend at the site of the pre-existing rift leads to basin inversion. (**c**) A releasing bend forms as the fault continues to propagate; the commonly observed map view pattern of a 'paired bend' is now complete. (**d**) A bypass fault forms that leads to eventual abandonment of both the restraining and releasing bends in the paired bend. Previous workers have shown that transfer of slip from the bend-impaired trace of the San Andreas to the San Jacinto faults has reached an advanced stage (i.e. modern slip on the San Andreas Fault is 27 ± 4 mm/a, whilst slip on the San Jacinto fault zone is 8 ± 4 mm/a).

Tectonic comparison with the southern California restraining bend

The San Bernardino restraining bend of the southern San Andreas Fault (Fig. 1a) makes an interesting tectonic comparison with the restraining-bend tectonics that we have described in Jamaica (Fig. 6a & b). In southern California, the San Bernardino restraining-bend fault is associated with 3.5-km-high topography. As in the Blue Mountains of eastern Jamaica, the restraining-bend uplift is domal with a very steep, fault-bounded southern edge and a more inclined and gently dipping northern flank (Fig. 6a & b). Palaeoseismological (Matti & Morton 1993) and shorter-term GPS-based studies (Bennett *et al.* 2004) have shown that the curved and topographically elevated San Bernardino and Indio segments of the San

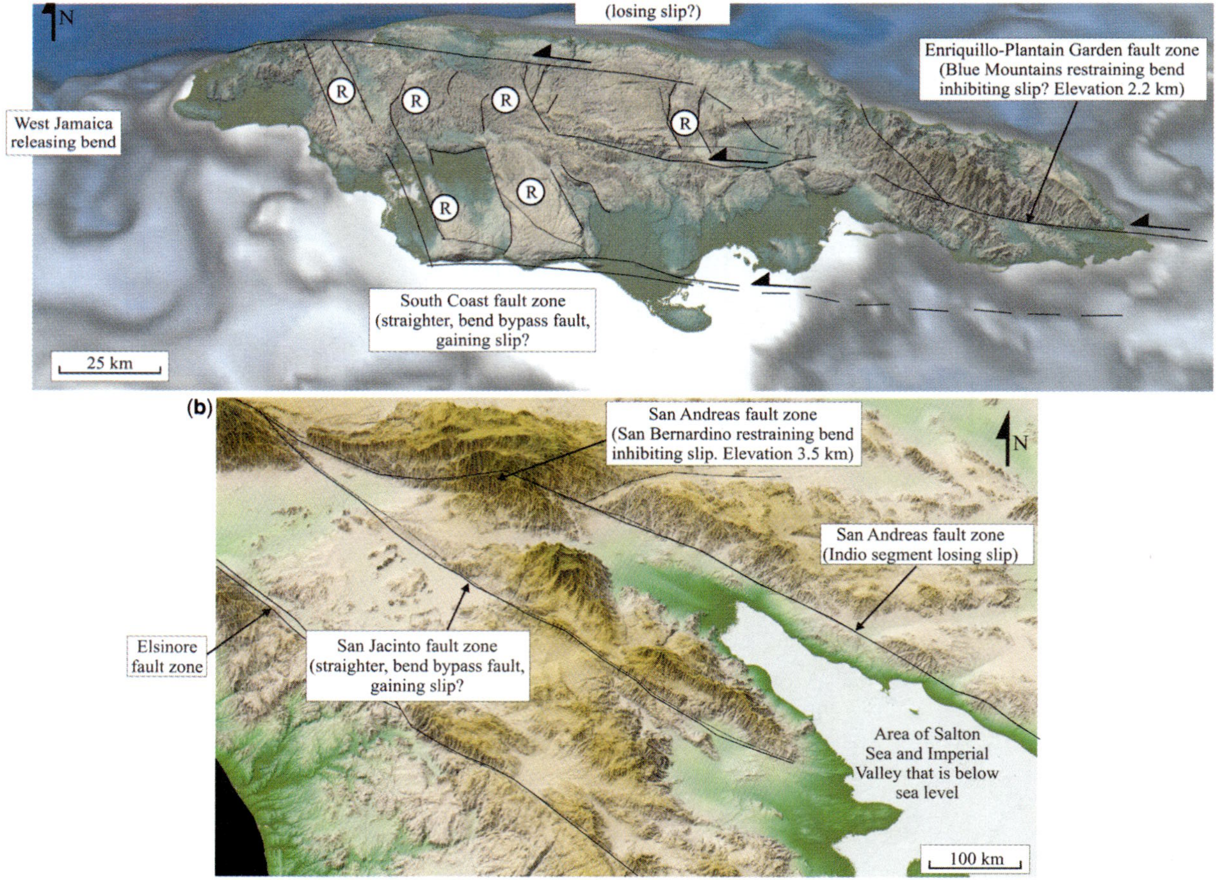

Fig. 6. Comparison of the Jamaica bend with the San Bernardino restraining-bend segment of the San Andreas fault zone and the San Jacinto 'bypass fault' (Bennett *et al.* 2004). (**a**) Earthquakes in southern Jamaica suggest that some of the 8 mm/a of left-lateral shear through Jamaica has shifted to the South Coast fault zone. (**b**) Using geological and GPS data, previous workers have proposed that the inception of the San Jacinto fault zone accompanied the formation of a major restraining bend on the San Bernardino segment of the San Andreas Fault at about 1.5 Ma. This coincidence suggests that the San Jacinto Fault may be acting as a bypass fault that is reducing the rate of strain along the restraining-bend segment (cf. Fig. 7d). This shift may explain the marked difference in seismic strain between the two faults: the San Jacinto Fault has ruptured in several $M > 6$ earthquakes in the last century, whereas the southernmost San Andreas Fault has been remarkably quiescent.

Andreas fault zone are progressively ceding slip to the straighter and topographically lower San Gabriel fault zone. These displacement-rate studies have shown that a change in the slip rate on one fault is matched by an equal and opposite change in the rate on the other (Bennett *et al.* 2004). Hence, there is strong evidence for some level of trade-off, or co-dependence, between the two fault zones. Increased slip transfer will lead to faster slip on the San Jacinto Fault, and eventual abandonment of the San Bernardino restraining bend of the southern San Andreas fault zone. This shift may explain the marked difference in seismic strain between the two faults: the San Jacinto Fault has ruptured in several $M > 6$ earthquakes in the last century, whereas the southernmost San Andreas fault has been remarkably quiescent.

How did the San Bernardino restraining bend nucleate to create the eventual bypass of motion on to the San Jacinto fault zone? To our knowledge, there is no comparable crustal feature like the Wagwater Rift that might have originally led to the curving trace of the San Gabriel fault zone. Swanson (2005) proposed that fault-zone adhesion or increased friction (in the absence of any pre-existing crustal structure) has led to the formation of a 150-km-long and 30-km-wide 'sidewall ripout' structure bounded by the straight San Andreas Fault to the NE and the curved San Gabriel Fault to the SW.

In Jamaica, the geomorphology of the bend area indicates a higher degree of cross-fault relays than in the case of the San Andreas and San Jacinto fault zones (Fig. 6b). On Figure 6b we have marked six, prominent fault-bounded elevations as 'R', or possible relays in which active motion is transferred at stepovers between parallel fault strands. If these relay zones are indeed active, as suggested by their morphology, this observation would indicate that the Jamaican restraining bend is still functioning and has not reached the advanced stage of bypass that has currently been attained by the San Bernardino bend of southern California.

Implications for seismic hazard studies

The proposed model has direct implications for the seismic hazards posed to Jamaica's 2.6 million inhabitants. Figure 7 shows the geographical distribution of the number of times per century that intensities of modified Mercalli VI or greater have been reported in Jamaica from 1880 to 1960 (Shepherd & Aspinall 1980). Like the pattern of recorded seismicity seen on Figure 3, there are higher concentrations of historical earthquakes in the area of the inverted Wagwater Rift of eastern Jamaica and in the subsurface Montpelier Rift of western Jamaica. The close spatial association of the pattern of historical seismicity indicates that these crustal features are an important control on the

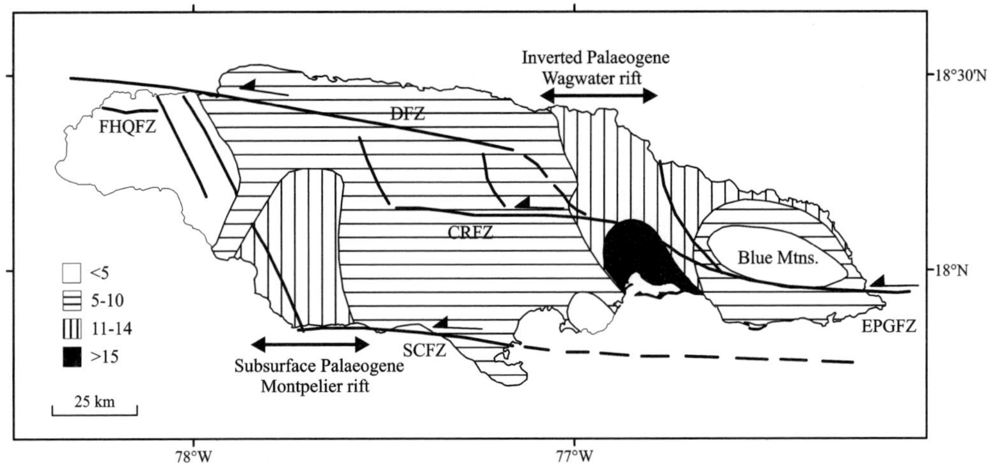

Fig. 7. Geographical distribution of the number of times per century that intensities of modified Mercalli VI or greater have been reported in Jamaica from 1880 to 1960 (modified from Shepherd & Aspinall 1980). Higher concentrations of earthquakes correlate with the area of the inverted Wagwater Rift of eastern Jamaica and the subsurface Montpelier Rift in western Jamaica, and indicate the continued importance of pre-existing crustal structures for the development of paired-bend systems. The areas most affected by the great earthquakes of 1692 and 1907 were along the approximate eastern projection of the South Coast fault zone (SCFZ). A critical question for seismic hazard evaluation is whether motion has shifted to the South Coast Fault or continues to be transferred to more northerly faults like the Crawle River (CRFZ) and Duanvale fault zones (DFZ). EPGFZ: Enriquillo-Plantain Garden fault zone; FHQFZ: Fat Hog Quarters fault zone.

evolution of the Jamaican restraining bend and on the control of present-day earthquakes (Wiggins-Grandison & Atakan 2005).

A key question is whether areas most affected by the destructive earthquakes of 1692 and 1907 were along the approximate eastern projection of the South Coast fault zone, or, alternatively, whether these large earthquakes were related to faulting either in the Wagwater Belt or the adjacent Blue Mountains (Fig. 7). The former scenario would argue for a more advanced bypass stage in restraining-bend evolution (cf. Fig. 5d), while the latter scenario would argue for the continued transfer of slip from the Enriquillo–Plantain Garden fault zone to more northerly faults like the Crawle River and Duanvale fault zones (Fig. 2A). A key avenue of future research will be to identify the presence of Late Holocene faults in both zones, in order to gauge their present level of activity along with their Holocene palaeoseismology.

C. DeMets and M. Wiggins-Grandison were supported by NSF-EAR-0003550. We thank L. Lavier and F. Taylor for helpful reviews. The authors acknowledge the financial support for this publication provided by The University of Texas at Austin's Geology Foundation and Jackson School of Geosciences. UTIG contribution no. 1865.

References

ARDEN, D. D., JR. 1975. Geology and the Nicaragua Rise. *In*: NAIRN, E. E. M. & STEHLI, F. H. (eds) *The Ocean Basins and Margins*, Vol. 3, Plenum, New York, 617–661.

BENNETT, R. A., FRIEDRICH, A. M. & FURLONG, K. P. 2004. Codependent histories of the San Andreas and San Jacinto fault zones from inversion of fault displacement rates, *Geology*, **32**, 961–964.

BLOW, W. H. 1969. Late middle Eocene to Recent planktonic foraminiferal biostratigraphy. *In*: BRONNIMANN, P. & RENZ, H. H. (eds) *Proceedings, International Conference on Planktonic Microfossils, Geneva 1967*, Volume 1, 199–421.

BRIDWELL, R. 1975. Sinuosity of strike-slip fault traces. *Geology*, **3**, 630–632.

BURKE, K., GRIPPI, J. & SENGOR, A. M. C. 1980. Neogene structures in Jamaica and the tectonic style of the northern Caribbean plate boundary zone. *Journal of Geolology*, **88**, 375–386.

CALAIS, E. & MERCIER DE LEPINAY, B. 1990. A natural model of active transpressional tectonics: the en echelon structures of the Oriente Deep along the northern Caribbean transcurrent plate boundary (southern Cuba margin). *Revue de L'Institut Francais du Petrole*, **45**, 147–160.

CALAIS, E., MAZABRAUD, Y., MERCIER DE LEPINAY, B., MANN, P., MATTIOLI, G. & JANSMA, P. 2002. Strain partitioning and fault slip rates in the northeastern Caribbean from GPS measurements, *Geophysical Research Letters*, **106**, 1–9.

CASTILLO, D. A. & ELLSWORTH, W. L. 1993. Seismotectonics of the San Andreas fault system between Point Arena and Cape Mendocino in northern California: implications for the development and evolution of a young transform. *Journal of Geophysical Research*, **98**, 6543–6560.

CROWELL, J. C. 1974. Sedimentation along the San Andreas Fault, California. *In*: DOTT, R. H., JR. & SHAVER, R. H. (eds) *Modern and Ancient Geosynclinal Sedimentation*. Society of Economic Paleontologists and Mineralogists Special Publications, **19**, 292–303.

CUNNINGHAM, W. D. 2007. Structural and topographic characteristics of restraining bend mountain ranges of the Altai, Gobi Altai and easternmost Tien Shan. *In*: CUNNINGHAM, W. D. & MANN, P. (eds) *Tectonics of Strike-Slip Restraining and Releasing Bends*. Geological Society, London, Special Publications, **290**, 219–237.

DEMETS, C. & WIGGINS-GRANDISON, M. 2007. Deformation of Jamaica and motion of the Gonave microplate from GPS and seismic data. *Geophysical Journal International*, **168**, 362–378, doi: 10.1111/j.1365-246X.2006.03236X.

DEMETS, C., MATTIOLI, G., JANSMA, P., ROGERS, R., TENORIO, C. & TURNER, H. L. 2007. Present motion and deformation of the Caribbean plate: constraints from new GPS geodetic measurements from Honduras and Nicaragua, In: MANN, P. (ed.) *Geologic and Tectonic Development of the Caribbean Plate in Northern Central America*. Geological Society of America Special Paper 428, Geological Society of America, Boulder, CO, doi: 10.1130/2007.2428(02).

DEMETS, C., GORDON, R., ARGUS, D. & STEIN, S. 1994. Effect of recent revisions of the geomagnetic reversal timescale on estimates of current plate motions. *Geophysical Research Letters*, **21**, 2191–2194.

DEMETS, C., JANSMA, P. E. *ET AL*. 2000. GPS geodetic constraints on Caribbean–North America plate motion. *Geophysical Research Letters*, **27**, 437–440.

DIXON, T. H., FARINA, F., DEMETS, C., JANSMA, P., MANN, P. & CALAIS, E. 1998. Relative motion between the Caribbean and North American plates and related boundary zone deformation from a decade of GPS observations. *Journal of Geophysical Research*, **103**, 15 157–15 182.

EVA, A. N. & MCFARLANE, N. 1979. Tertiary to early Quaternary facies relationships in Jamaica [abstract] *4th Latin American Geological Congress, Port of Spain, Trinidad and Tobago*.

GREEN, G. W. 1977. Structure and stratigraphy of the Wagwater Belt, Jamaica. *Overseas Geology and Mineral Resources*, **48**, p. 21.

HORSFIELD, W. 1974. Major faults in Jamaica. *Journal of the Geological Society of Jamaica*, **14**, 31–38.

LEWIS, J. & DRAPER, G. 1990. Geology and tectonic evolution of the northern Caribbean margin. *In*: DENGO, G. & CASE, J. E. (eds) *The Geology of North America, Vol. H. The Caribbean Region*. Geological Society of America, Boulder, CO, 77–140.

MANN, P. 2007. Global catalogue, classification, and tectonic origin of restraining- and releasing bends on active and ancient strike-slip fault systems. *In*: CUNNINGHAM, W. D. & MANN, P. (eds) *Tectonics of Strike-Slip Restraining and Releasing Bends*. Geological Society, London, Special Publications, **290**, 13–142.

MANN, P. & BURKE, K. 1990. Transverse intra-arc rifting – Paleogene Wagwater belt, Jamaica. *Marine and Petroleum Geology*, **7**, 410–427.

MANN, P. & GORDON, M. 1996. Tectonic uplift and exhumation of blueschist belts along transpressional strike-slip fault zones. *In*: BEBOUT, G., SCHOLL, D., KIRBY, S. & PLATT, J. (eds) *Dynamics of Subduction Zones*, Geophysical Monographs, **96**, American Geophysical Union, Washington, DC, 143–154.

MANN, P., CALAIS, E., RUEGG, J.-C., DEMETS, C., JANSMA, P. & MATTIOLI, G. S. 2002. Oblique collision in the northeastern Caribbean from GPS measurements and geological observations. *Tectonics*, **37**, 0.1029/2001TC001304.

MANN, P., DRAPER, G. & BURKE, K. 1985. Neotectonics of a strike-slip restraining bend system, Jamaica. *In*: BIDDLE, K. & CHRISTIE-BLICK, N. (eds) *Strike-Slip Deformation, Basin Formation, and Sedimentation*, SEPM Special Publications, **37**, 211–226.

MANN, P., PRENTICE, C. S., BURR, G., PENA, L. R. & TAYLOR, F. W. 1998. Tectonic geomorphology and paleoseismology of the Septentrional fault system, Dominican Republic. *In*: DOLAN, J. F. & MANN, P. (eds) *Active Strike-slip and Collisional Tectonics of the Northern Caribbean Plate Boundary Zone*, Geological Society of America Special Paper, **326**, 63–123.

MANN, P., ROSENCRANTZ, E. & TYBURSKI, S. 1991. Neogene development of the Swan Islands restraining-bend complex, Caribbean Sea. *Geology*, **19**, 823–826.

MANN, P., SCHUBERT, C. & BURKE, K. 1990. Review of Caribbean neotectonics. *In*: DENGO, G. & CASE, J. E. (eds) *The Geology of North America, Volume H, The Caribbean Region*, Geological Society of America, Boulder, CO, 307–338.

MANN, P., TAYLOR, F. W., EDWARDS, R. L. & KU, T. 1995. Actively evolving microplate formation by oblique collision and sideways motion along strike-slip faults: an example from the northeastern Caribbean plate margin. *Tectonophysics*, **246**, 1–69.

MAO, A., HARRISON, C. G. A. & DIXON, T. H. 1990. Noise in GPS coordinate time series. *Journal of Geophysical Research*, **104**, 2797–2816.

MATTI, J. C. & MORTON, D. M. 1993. Paleogeographic evolution of the San Andreas fault in southern California: a reconstruction based on a new cross fault correlation. *In*: CROUCH, J. K. & BACHMAN, S. B. (eds) *Tectonics and Sedimentation Along the California Margin: Pacific Section*, SEPM Special Publications, **38**, 1–16.

ROSENCRANTZ, E. & MANN, P. 1991. SeaMARC II mapping of transform faults in the Cayman Trough, Caribbean Sea. *Geology*, **19**, 690–693.

SHEPHERD, J. & ASPINALL, W. 1980. Seismicity and seismic intensities in Jamaica, West Indies: a problem in risk assessment. *Earthquake Engineering and Structural Dynamics*, **8**, 315–355.

SWANSON, M. 2005. Geometry and kinematics of adhesive wear in brittle strike-slip fault zones. *Journal of Structural Geology*, **27**, 871–887.

WADGE, G. & DIXON, T. H. 1984. A geological interpretation of SEASAT–SAR imagery of Jamaica. *Journal of Geology*, **92**, 561–581.

WEBER, J. C., DIXON, T. H., DEMETS, C., AMBEH, W. B., JANSMA, P., MATTIOLI, G., SALEH, J., SELLA, G., BILHAM, R. & PEREZ, O. 2001. GPS estimate of relative motion between the Caribbean and South American plates, and geologic implications for Trinidad and Venezuela. *Geology*, **29**, 75–78.

WIGGINS-GRANDISON, M. D. 2001. Preliminary results from the new Jamaica seismograph network. *Seismological Research Letters*, **72**, 525–537.

WIGGINS-GRANDISON, M. D. 2003. Advances in Jamaican seismology, Dr. Scient. Thesis, University of Bergen, Norway, 150 pp.

WIGGINS-GRANDISON, M. D. 2004. Simultaneous inversion for local earthquake hypocenters, station corrections, and 1-D velocity model of the Jamaican crust. *Earth and Planetary Science Letters*, **224**, 229–240.

WIGGINS-GRANDISON, M. D. & ATAKAN, K. 2005. Seismotectonics of Jamaica. *Geophysical Journal International*, **160**, 573–580.

WIGGINS-GRANDISON, M. D., KEBEASY, T. R. M. & HUSEBYE, E. S. 2003. Enhanced earthquake risk of Kingston due to wave field excitation in the Liguanea Basin, Jamaica, Caribbean, *Journal of Earth Science*, **37**, 21–32.

ZUMBERGE, J. F., HEFLIN, M. B., JEFFERSON, D. C., WATKINS, M. M. & WEBB, F. H. 1997. Precise point positioning for the efficient and robust analysis of GPS data from large networks. *Journal of Geophysical Research*, **102**, 5005–5017.

Kinematics of the Amanos Fault, southern Turkey, from Ar/Ar dating of offset Pleistocene basalt flows: transpression between the African and Arabian plates

A. SEYREK[1], T. DEMİR[2], M. S. PRINGLE[3], S. YURTMEN[4], R. W. C. WESTAWAY[5], A. BECK[6] & G. ROWBOTHAM[7]

[1]*Department of Soil Science, Harran University, 63300 Şanlıurfa, Turkey*

[2]*Department of Geography, Harran University, 63300 Şanlıurfa, Turkey*

[3]*Scottish Universities' Environmental Research Centre, Rankine Avenue, East Kilbride, Glasgow G75 0QF, UK; Present address: Laboratory for Noble Gas Geochronology, Department of Earth, Atmospheric, and Planetary Sciences, Massachusetts Institute of Technology, 77 Massachusetts Avenue, Building 54-1013, Cambridge, MA 02139-4307, USA*

[4]*Department of Geology, Çukurova University, 01330 Adana, Turkey; Present address: 41 Kingsway East, Westlands, Newcastle-under-Lyme, Staffordshire, ST5 5PY, UK*

[5]*Faculty of Mathematics and Computing, The Open University, Eldon House, Gosforth, Newcastle-upon-Tyne NE3 3PW, UK; Also at: School of Civil Engineering and Geosciences, University of Newcastle-upon-Tyne, Newcastle-upon-Tyne NE1 7RU, UK (e-mail: robwestaway@tiscali.co.uk)*

[6]*Department of Geography, Durham University, South Road, Durham DH1 3LE, UK; Present address: School of Computing, University of Leeds, Leeds LS2 9JT, UK*

[7]*School of Earth Sciences and Geography, Keele University, Keele, Staffordshire ST5 5BG, UK*

Abstract: We report four new Ar/Ar dates and 18 new geochemical analyses of Pleistocene basalts from the Karasu Valley of southern Turkey. These rocks have become offset left-laterally by slip on the N20°E-striking Amanos Fault. The geochemical analyses help to correlate some of the less-obvious offset fragments of basalt flows, and thus to measure amounts of slip; the dates enable slip rates to be calculated. On the basis of four individual slip-rate determinations, obtained in this manner, we estimate a weighted mean slip rate for this fault of 2.89 ± 0.05 mm/a ($\pm 2\sigma$). We have also obtained a slip rate of 2.68 ± 0.54 mm/a ($\pm 2\sigma$) for the subparallel East Hatay Fault farther east. Summing these values gives 5.57 ± 0.54 mm/a ($\pm 2\sigma$) as the overall left-lateral slip rate across the Dead Sea fault zone (DSFZ) in the Karasu Valley. These slip-rate estimates and other evidence from farther south on the DSFZ are consistent with a preferred Euler vector for the relative rotation of the Arabian and African plates of $0.434 \pm 0.012°$ Ma^{-1} about 31.1°N, 26.7°E. The Amanos Fault is misaligned to the tangential direction to this pole by 52° in the transpressive sense. Its geometry thus requires significant fault-normal distributed crustal shortening, taken up by crustal thickening and folding, in the adjacent Amanos Mountains. The vertical component of slip on the Amanos Fault is estimated as *c.* 0.15 mm/a. This minor component contributes to the uplift of the Amanos Mountains, which reaches rates of *c.* 0.2–0.4 mm/a. These slip rate estimates are considered representative of time since. 3.73 ± 0.05 Ma, when the modern geometry of strike-slip faulting developed in this region; an estimated 11 km of slip on the Amanos Fault and *c.* 10 km of slip on the East Hatay Fault have occurred since then. It is inferred that both these faults came into being, and the associated deformation in the Amanos Mountains began, at that time. Prior to that, the northern part of the Africa–Arabia plate boundary was located further east.

Horizontal crustal motions in the eastern Mediterranean region relate to interactions between the African, Arabian, Turkish and Eurasian plates (AF, AR, TR and EU, respectively). The Dead Sea fault zone (DSFZ) forms the AF–AR boundary (Fig. 1); it is well established that its southern part, between Israel and Jordan, is a left-lateral transform-fault zone (e.g. Freund *et al.* 1970;

From: CUNNINGHAM, W. D. & MANN, P. (eds) *Tectonics of Strike-Slip Restraining and Releasing Bends.* Geological Society, London, Special Publications, **290**, 255–284.
DOI: 10.1144/SP290.9 0305-8719/07/$15.00

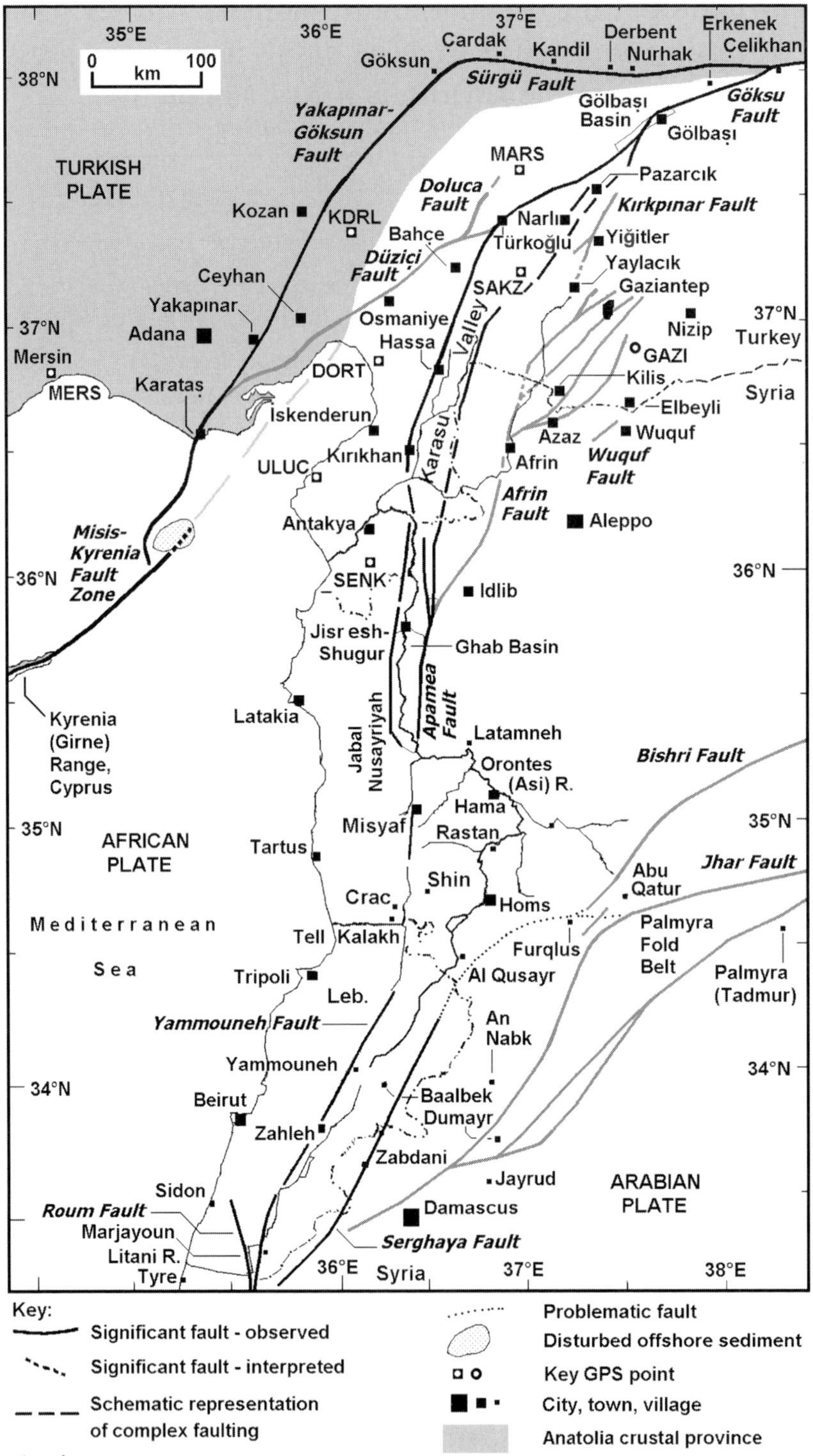

Fig. 1. *Continued.*

Garfunkel 1981). It has recently become clear (Westaway 2003, 2004; Gomez *et al.* 2006) that the northern DSFZ, in western Syria and southern Turkey, is transpressive; moving northward, it steps progressively to the right, away from its Euler pole. As currently defined (Westaway 2003, 2004), the northernmost DSFZ segment, consisting of multiple en échelon fault strands, runs along the Karasu Valley in the Hatay region of southern Turkey. The mainly left-lateral Amanos Fault, bounding the Amanos Mountains at the western margin of the Karasu Valley, between Kırıkhan, Hassa and Türkoğlu (Figs 1 & 2), is the subject of this study.

A geological map and a DEM-based topographic map of the area of Figure 1 were recently published by Gomez *et al.* (2006), and so are not repeated here. More detailed geological and topographic maps of the Karasu Valley and its immediate surroundings were recently published by Westaway *et al.* (2006*a*) and so, likewise, are not repeated here. Such illustrations enable the reader to visualize the topography, and its relationship to geological structure, along the DSFZ, supporting the much more detailed descriptions of key localities (Fig. 3) in the present study. Westaway *et al.* (2006*a*) also showed a map indicating the relationship between the DSFZ and other plate-boundary fault zones in the Middle East and eastern Mediterranean regions; such an illustration is thus not repeated here either.

This study region forms the NW margin of the Arabian Platform, which collided with the Anatolian continental fragment to the north (Fig. 1) during the Cenozoic, following the closure of the Southern Neotethys Ocean. During the Late Cretaceous (Maastrichtian; *c.* 70 Ma; e.g. Dilek & Delaloye 1992), the Hatay ophiolite was obducted on to the Arabian Platform margin. Strike-slip faulting along the AF–AR plate boundary initiated in the early Middle Miocene (e.g. Garfunkel 1981), but the locations of, and slip senses on, active faults within this boundary have subsequently varied over time (e.g. Westaway 2003, 2004). The modern plate-boundary geometry is thought to have existed since *c.* 4 Ma (e.g. Westaway 2003, 2004), with the latest estimate of its initiation being 3.73 ± 0.05 Ma (Westaway *et al.* 2006*a*).

An understanding of the active kinematics of this region is important for several reasons. First, knowledge of slip rates on active faults is necessary to assess the recurrence intervals of large earthquakes, in order to quantify the regional seismic hazard (cf. Meghraoui *et al.* 2003). The most recent large earthquake in this region occurred on 13 August 1822. It had a macroseismic epicentre at 36.7°N, 36.5°E, near Aktepe (Fig. 3); a macroseismic magnitude of 7.5; and involved *c.* 200 km of fault rupture (Ambraseys 1989; Ambraseys & Jackson 1998). However, despite historical records of damage, there is no indication of which fault(s) ruptured during this event. Second, such knowledge is also important for testing regional kinematic models: in this case for the linkage between the AF–AR, AF–TR, and TR–AR plate boundaries (e.g. Westaway 2004). This includes comparison between short-timescale kinematic models from GPS satellite geodesy, such as that by McClusky *et al.* (2000), and longer-timescale geological evidence. Third, the recognition of the transpressive geometry of the northern DSFZ means that it provides scope for testing quantitative models for the distributed deformation associated with transpression, such as that by Westaway (1995). This aspect also relates to the development of the topography and structure of the Amanos Mountains adjoining the northern DSFZ (Figs 2 & 4). It has long been accepted that this mountain range developed as a consequence of the Cretaceous ophiolite obduction (e.g. Schwan 1971, 1972; Tolun

Fig. 1. Map (adapted from Fig. 2 of Westaway 2004, which lists original sources of information) of fault segments of the boundaries between the African, Arabian, and Turkish plates in southern Turkey, western Syria, and Lebanon, in relation to GPS points, from McClusky *et al.* (2000). Faults that are inferred to be significant during the present phase of deformation (post-*c.* 3.7 Ma) are black; those that appear to have been important earlier are grey. The Amanos Fault is shown running southward along the eastern margin of the Amanos Mountains for a distance of *c.* 110 km, from Türkoğlu, past İslahiye and Hassa, to Kırıkhan. K.V. denotes the Karasu Valley. The present study region comprises its *c.* 30-km-long segment between Ceylanlı, *c.* 5 km north of Kırıkhan, and Hassa. The subparallel East Hatay Fault passes through Tahtaköprü, *c.* 12 km farther east, bounding the western margin of the Kurd Dagh Range. The crust of the Anatolian continental fragment is shaded (except where offshore); its southern limit, at the northern margin of the Arabian Platform, marks the suture of the Southern Neotethys Ocean. As currently defined (e.g. Westaway *et al.* 2006*a*), the Göksu Fault, Sürgü Fault and Gölbası-Türkoğlu Fault (between Gölbası and Türkoğlu; illustrated) form the SW end of the East Anatolian fault zone, the boundary between the Turkish and Arabian plates. The 'problematical fault' is the inferred NE continuation of the Serghaya Fault across the Homs area of Syria (Walley 1998), which – if it exists – appears to have not been active since at least the Early Pliocene (Westaway 2004). The two symbols for GPS points represent points located within towns (which are named in Fig. 2), and point GAZI, which is named after Gaziantep but located some distance away from this city.

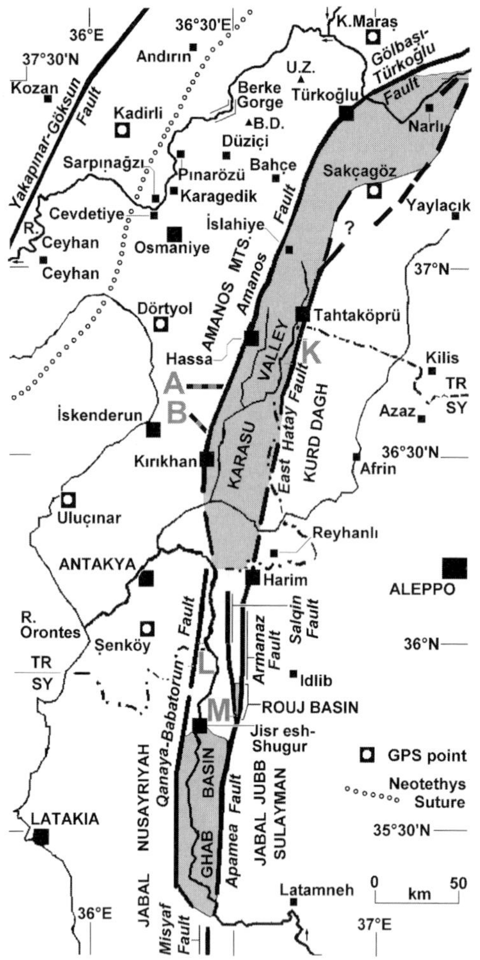

Fig. 2. More detailed map of the study region and its immediate surroundings. Grey shading schematically indicates lowlands forming linear valleys along the DSFZ. The southward continuation of the Karasu Valley, south of Kırıkhan, is called the Amik Basin, whereas its northward continuation is called the Narlı Plain or Aksu Plain, named after the River Aksu, which joins the Ceyhan west of Kahramanmaraş (abbreviated to K. Maraş). Only left-lateral faults considered active during the present phase of slip on the DSFZ are shown (see Fig. 1 for other faults). 'A' and 'B' denote the locations of Fig. 4a & b. 'K', 'L' and 'M' denote piercing points used to measure the total slip on DSFZ strands, as explained in the text. 'B.D.' and 'U.Z.' denote Beşikdüldülü and Uluziyaret, two of the highest summits in the Amanos Mountains (2246 and 2268 m a.s.l., respectively).

& Pamir 1975; Brinkmann 1976, p. 89), its folding being thought to long predate the neotectonic phase. However, this mountain range is aligned along the northern DSFZ (Fig. 2) and contains anticline axes that are also subparallel to this structural trend (Fig. 4). Such observations raise the question of whether some, at least, of this structure might instead relate to the active transpression.

Geological maps of the Karasu Valley have been published many times before (e.g. Çapan *et al.* 1987; Arger *et al.* 2000), and so will not be reproduced here. There are many localities where basalt flows have become offset left-laterally by slip on the Amanos Fault. Dating these basalts can thus indicate the vertical and horizontal slip rates on this fault, thereby addressing each of the above issues. Yurtmen *et al.* (2002) undertook this task using high-precision unspiked K–Ar dating, but found it more difficult than it might at first appear, due to the complexity of this volcanism (and thus the number of possible across-fault flow correlations); inconsistencies and site ambiguities affecting earlier low-precision spiked K–Ar dating; and the non-availability of large-scale topographic maps for much of the region, due to its proximity to the Syrian border (which limited location accuracy and thus the precision to which offset distances could be measured). Yurtmen *et al.* (2002) thus estimated the slip rate on the Amanos Fault as no more than *c.* 1.6 mm/a.

Yurtmen *et al.* (2002) also identified additional localities where future dating could add to constraints on the Amanos Fault slip rate. In the meantime, Tatar *et al.* (2004) published three new spiked K–Ar dates and determined the magnetic polarity of the basalt at many sites in the Karasu Valley. The Yurtmen *et al.* (2002) study predated the realization that the northern DSFZ is transpressive, and so made no attempt to test any kinematic model incorporating transpression. Furthermore, as Westaway (2004) noted, the slip rates, after Yurtmen *et al.* (2002), summed across all documented active left-lateral faults in and around the Karasu Valley, have hitherto significantly underestimated the predicted rate of AF–AR relative motion. Evidently, either additional active faults have remained undocumented, or the existing evidence has led to underestimation of slip rates. The lack of large-scale maps has in the meantime been resolved, due to topographic imagery derived from the Shuttle Radar Topographic Mission (SRTM) and the use of hand-held GPS receivers for location in the field. Along with the importance of the subject matter, these factors justify revisiting the area studied by Yurtmen *et al.* (2002).

Prior to the Yurtmen *et al.* (2002) study, the kinematics of the left-lateral faulting along the Karasu Valley were poorly determined, due to problematic dating (using the spiked K–Ar method, which typically results in large margins of uncertainty for Middle or Late Pleistocene dates) and imprecise documentation of field localities. A great diversity

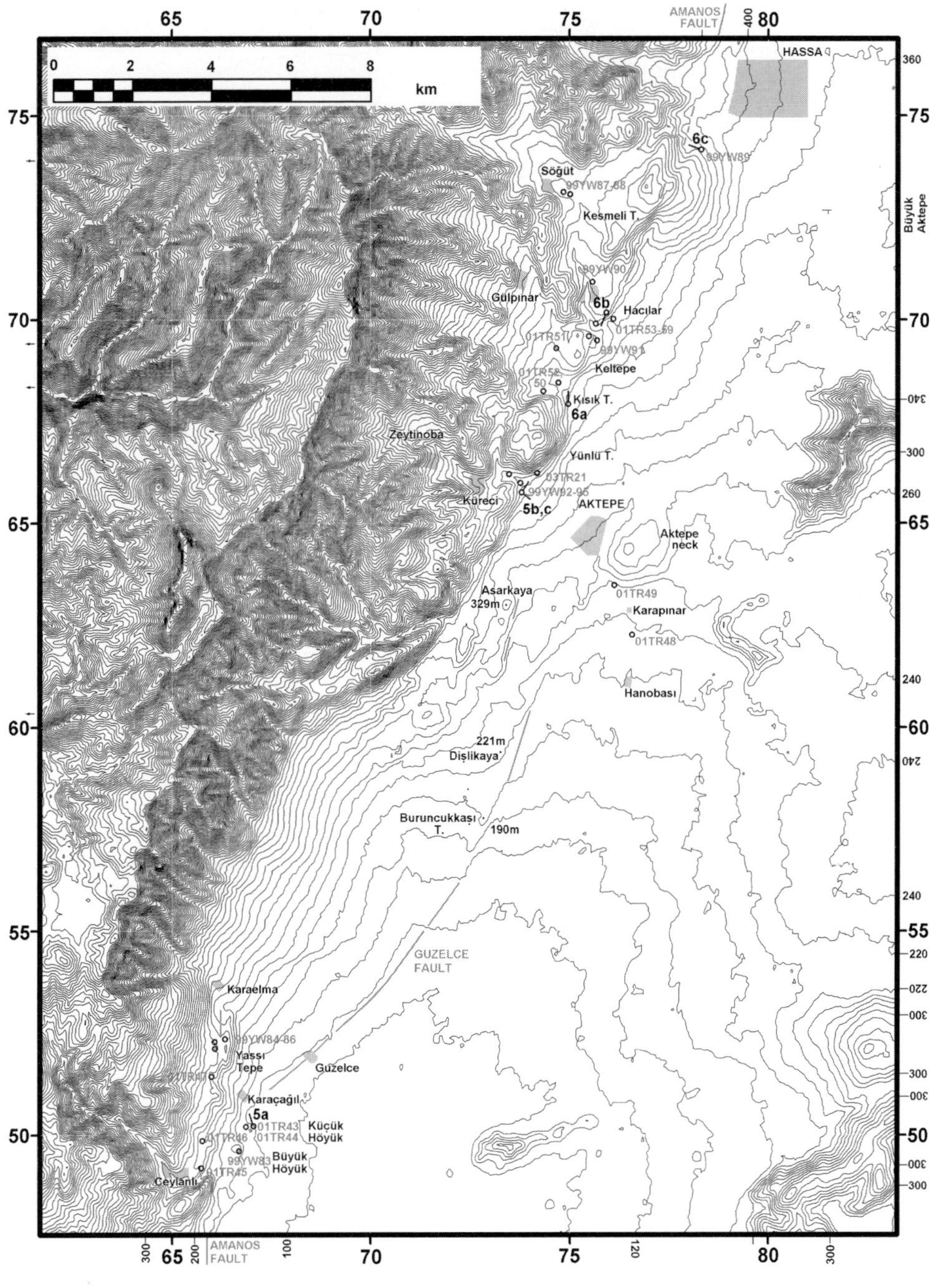

Fig. 3. Topographic map of the present study region, with a 20-m contour interval, based on SRTM data, showing the locations of basalt sampling sites from this study and Yurtmen *et al.* (2002), basaltic necks, and the positions of the Amanos and Guzelce faults. The position of the Amanos Fault is marked with ticks at the edges of the figure, in the Yassı Tepe and Hacılar areas, after Yurtmen *et al.* (2002). Another strand of this fault may run along the foot of the basalt bluff overlooking Karaçağıl, as illustrated (A–B) in Figure 5a. In many other localities the position of the fault is obvious from the range-bounding topography. Elsewhere, notably north of Karaelma, the line of the fault is unclear. Shaded areas next to place-names indicate the approximate extent of the named locality. Arrows indicate rivers draining westward, not eastward into the Karasu Valley. The positions of their headwaters indicate the close proximity of the drainage divide to the eastern margin of the Amanos Mountains. See Westaway *et al.* (2006*a*) for details of data sources and preparation technique.

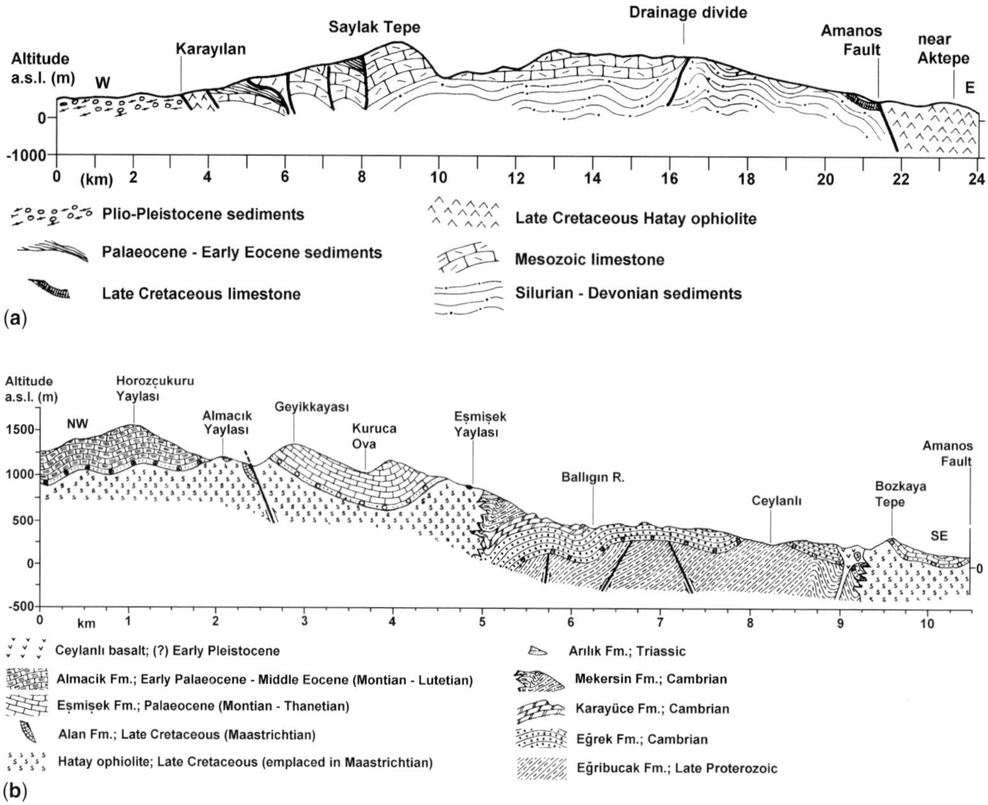

Fig. 4. Cross-sections illustrating the structure of the Amanos Mountains; see Figure 2 for locations. (**a**) West–east profile along *c.* 36.7°N, adapted from cross-section II in plate II of Tolun & Pamir (1975). (**b**) NW–SE profile through Ceylanlı, adapted from cross-section VII in plate I of Atan (1969). The Maastrichtian–Eocene formations depicted are carbonate-dominated; typical lithologies of the older formations are summarized in Figure 12.

of views indeed existed, a popular view (recently restated by Mart *et al.* 2005) being that the Karasu Valley is a graben, bounded to the WNW and ESE by active normal faults (e.g. Schwan 1971). However, although there are obvious topographic escarpments along the margins of this valley (Figs 3 & 5), these are now thought to reflect minor components of vertical slip on predominantly left-lateral faults. Nonetheless, along most of the length of the valley there is little or no actual exposure of any fault surface; this is evidently because for thousands of years the landscape has been degraded by agriculture, and because no large earthquake has occurred for almost two centuries. Yurtmen *et al.* (2002) reviewed the diverse range of hypotheses that became superseded by their study, so such discussion is not repeated here. Tatar *et al.* (2004) have in the meantime proposed that the Karasu Valley is affected by pervasive distributed deformation between the bounding active left-lateral faults. We discount this possibility, as our re-analysis (discussed below) reveals no requirement for any such complexity; the palaeomagnetic declinations that they have measured presumably reflect secular variation of the geomagnetic field rather than vertical-axis rotations.

The layout of this paper is as follows. First we discuss our fieldwork and laboratory analysis. Second, we use the information gained to estimate the slip rates on the Amanos Fault at the localities that have been studied. Finally, we compare this data-set with other evidence, in order to establish the extent of kinematic consistency for the DSFZ as a whole.

Field and laboratory procedures

Fieldwork

Four basalt samples were collected for Ar/Ar dating from critical localities previously identified by Yurtmen *et al.* (2002), illustrated in Figures 5a & b and 6a & b and located in Figure 3. We also collected 14 other basalt samples for geochemical analysis, with the aim of using their geochemistry to correlate fragments of offset basalt flows (cf. Yurtmen *et al.* 2002). To facilitate accurate location, Universal Transverse Mercator (UTM) coordinates of all sample sites and other key localities (referenced to the WGS84 reference frame) were measured using a hand-held GPS receiver. UTM coordinates of some of the sample sites of Yurtmen *et al.* (2002) were likewise measured using GPS, in order to supplement the position fixes on the local maps used in that earlier study (fieldwork for which, in 1999, lacked the benefit of a working hand-held GPS receiver). This check was possible because one person (R.W.) participated in all fieldwork and, in 2001, he could precisely identify sites where he had worked in 1999. The UTM coordinates of each site determined in this manner were found to be typically *c.* 150 m south of the coordinates measured from the 1:25 000-scale maps, due to the different reference frame used in the production of these maps (cf. Westaway *et al.* 2004).

Ar/Ar dating

Initial screening involved inspection of hand specimens and petrographic thin sections. Samples were then crushed, washed in de-ionized water and dilute hydrochloric acid, sieved to a 60–80 μm size fraction, and phenocrysts and xenocrysts were removed by magnetic separation and hand picking. Following irradiation, argon isotopes were measured in the microcrystalline groundmass at the Laboratory for Noble Gas Geochronology, Massachusetts Institute of Technology, using procedures essentially the same as those described by Harford *et al.* (2002). Results are summarized in Table 1 and Figure 7.

All four Ar/Ar dates are tightly constrained, with $\pm 1\sigma$ errors ranging from $<1\%$ for the oldest sample (01TR47) to $<5\%$ for the youngest (01TR53). The MSWD, a measure of the internal consistency between the step-heating splits that contribute to each date, is <1 for three of the dates, and between 1 and 2 for the remaining date (Fig. 7). The extent of concordance between these new dating results and existing age control evidence, and its implications for estimation of the slip rate on the Amanos Fault, will be discussed below.

Geochemical analysis and interpretation

Geochemical analysis of whole-rock basalt samples, following the procedure used by Yurtmen *et al.* (2002), utilized the automated ARL 8420 X-ray fluorescence spectrometer at Keele University (Table 2). Figure 8 shows the classification of these basalts according to their alkali and silica content. As previously noted (e.g. Çapan *et al.* 1987; Yurtmen *et al.* 2002), these basalts are compositionally diverse as a group, but most individual flow units show only limited variation. Most samples are basalt *sensu stricto*, but a few contain >5 wt% of alkali metal oxides and so classify as hawaiites. Sample 99YW83 of Yurtmen *et al.* (2002), from Büyük Höyük neck, in this scheme is classified as a mugearite.

No sample contains normative quartz; all samples contain normative olivine and some contain normative nepheline (Table 2). These basalts are thus classified, according to their normative composition, as olivine tholeiites and alkali olivine basalts (cf. Yurtmen *et al.* 2002). The most mafic flow unit (with the least SiO_2 and the most normative nepheline) is at Hacılar (cf. Yurtmen *et al.* 2002), represented in Figure 8 by samples 99YW90, 01TR53, 01TR54 and 01TR59. As Yurtmen *et al.* (2002) noted, as a result of the left-lateral slip on the Amanos Fault, this flow unit is juxtaposed against much less mafic basalts, which they thought originated from necks at Keltepe or Kısık Tepe farther south (Fig. 3). Combined with dating evidence, this contrast in geochemistry is useful for constraining the kinematics of this fault (Yurtmen *et al.* 2002; see below). The next most mafic basalt seems to be that from the large flow unit in the central Karasu Valley, which originated from the Büyük Aktepe neck in the north (Fig. 3), flowing southward for >20 km past the town of Aktepe (e.g. Çapan *et al.* 1987), and was studied in detail by Polat *et al.* (1997). We sampled this basalt SSE of Aktepe (sample 01TR48; Fig. 8), confirming its mafic characteristics. It can indeed be readily distinguished from other basalts in the vicinity, such as that from the nearby Aktepe neck, represented by sample 01TR49 (e.g. using the much lower SiO_2 and much higher MgO contents; Table 2).

In terms of trace-element geochemistry (Table 2), the Karasu Valley basalts have many characteristics of ocean island basalt (cf. Polat *et al.* 1997; Yurtmen *et al.* 2002), such as high abundances of the light rare-earth elements La (up to 36 ppm) and Ce (up to 80 ppm). They also contain high abundances of incompatible trace elements (e.g. up to 853 ppm Sr) and other characteristics (e.g. low Zr/Nb ratios of *c.* 4) considered indicative of small-degree partial melting of the asthenosphere (such as may occur in ocean islands due to the heating effect of a mantle plume). However, these basalts have clearly not formed in an ocean-island setting. Yurtmen *et al.* (2002) proposed instead that this volcanism has resulted from heating of the mantle lithosphere (due to the increase in Moho temperature that accompanies crustal thickening), causing the remelting of products of

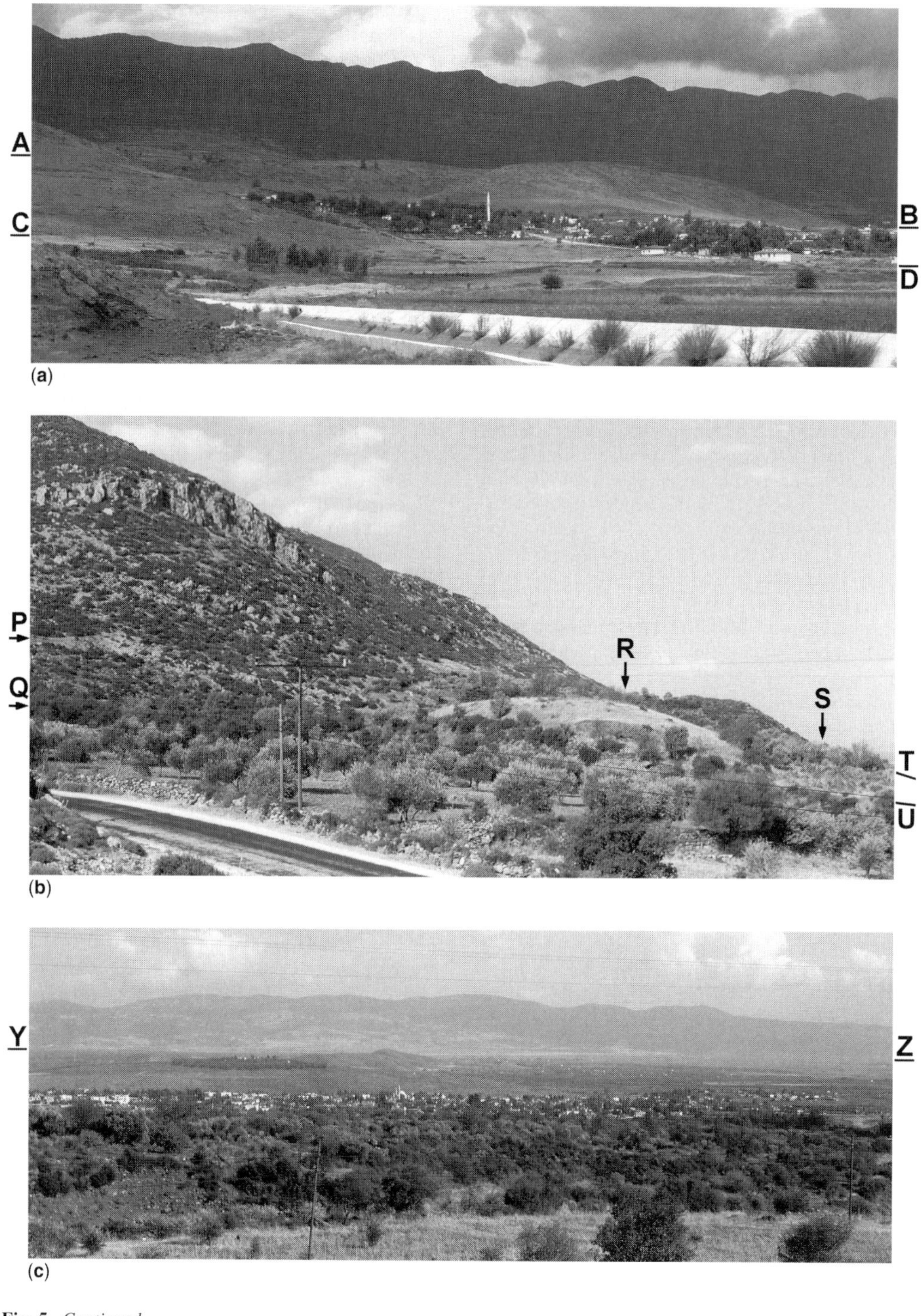

Fig. 5. *Continued.*

earlier small-degree partial melting of the asthenosphere, which had frozen within the mantle lithosphere. Our new data are consistent with this explanation. The transpressive strike-slip faulting now evident in this region provides a mechanism for the required crustal thickening, and can thus be regarded as a potential cause for the volcanism.

In detail, as Yurtmen *et al.* (2002) noted, the Karasu Valley basalts have resulted from the combination of variable (but small) degrees of partial melting of the asthenosphere, followed by variable fractional crystallization and variable (but small) degrees of crustal contamination. Crustal contamination is evident in some samples, for instance from the Ba/La ratios of $\gg 10$ (Table 2). The interplay between effects of partial melting and subsequent fractional crystallization can be illustrated (after Pearce *et al.* 1990) by the variation of Cr against Y (Fig. 9). The compatible element Cr is taken up during early crystallization of basaltic melt (such as may occur at depth in a magma chamber) and so becomes depleted in the residual melt. Based on this criterion, the most 'primitive' magmas (i.e. with the least evidence of fractional crystallization) are those from the large Büyük Aktepe flow unit, represented by our sample 01TR48 and other samples from Polat *et al.* (1997). Conversely, the most evolved magmas are represented by the low-volume flow units at Büyük Höyük (sample 99YW83), Yassı Tepe (samples 99YW84-86), and Küreci (samples 99YW92-95 and 03TR21). This tendency for high-volume flows to have experienced much less fractional crystallization than low-volume flows is the opposite of what has been reported elsewhere (cf. Thompson *et al.* 1990).

The greatest local complexity in the basalt stratigraphy is evident around the left-laterally offset Hacılar river gorge (Figs 5b, 10 & 11). Our seven samples from this area (01TR53-9) reveal three chemically distinct flow units. Sample 01TR53 came from the uppermost basalt flow unit forming the top of the bluff (Fig. 10b) immediately west of the fault at the southern end of this offset reach. Sample 01TR54 came from what appears to be the top of a lower basalt flow within the same flow unit, directly underneath it. Sample 01TR59 was collected east of the fault, from the uppermost basalt flow there, which is observed to cascade over this bluff at an angle of *c.* 25° (Fig. 10b). These three samples are geochemically very similar to each other (Table 2 and Figs 8 & 9) and to sample 99YW90 of Yurtmen *et al.* (2002); they represent the Hacılar basalt.

Sample 01TR58 came from a flat-lying basalt flow, stratigraphically below the one which yielded sample 01TR59, just above river level east of the Amanos Fault (Fig. 10b). We have not observed any corresponding flat-lying basalt in the bluff west of the fault, suggesting that this flow is not of local provenance. This sample is most geochemically similar to sample 01TR50 (Fig. 8),

Fig. 5. (**a**) View NNW from the NE margin of the Küçük Höyük neck at [BA 67099 50157], looking past Karaçağıl to the line of the Amanos Fault at Yassı Tepe, *c.* 2 km away. Yurtmen *et al.* (2002) interpreted this fault, running northward from A, before following a gully on the far side of the Yassı Tepe basalt ridge (which was thus interpreted as a shutter ridge) behind the minaret of Karaçağıl mosque. North of this locality, this interpreted fault is not visible in this view. According to Yurtmen *et al.* (2002), the *c.* 428 ka Yassı Tepe basalt is offset left-laterally by an estimated *c.* 425 m by this fault. This Yassı Tepe basalt overlies more extensive older basalt, offset left-laterally by an estimated *c.* 1000 m, that Tatar *et al.* (2004) showed to be reverse-magnetized and we have dated to *c.* 1195 ka. The line A–B, projecting between Karaçağıl and the near face of Yassı Tepe, may mark a subparallel fault, with a component of downthrow to the east. The contact C–D, between basalt and Karasu Valley alluvium, typically marked in this area by a subdued bluff, is interpreted as the surface trace of Guzelce Fault. Note the concordant summits, locally at *c.* 1500 m a.s.l., of the Amanos mountain range in the background. (**b**) View northward along the Amanos Fault from [BA 73728 65812], on the southern side of the outlet of the Olucak river valley. The limestone mountain on the far side of the valley is Yünlü Tepe. Directly in front of it, near the left edge of the photo, is the downstream end of the basalt flow unit within this valley (P marks its top surface), which is truncated by the Amanos Fault at [BA 73872 66180] (Q). Basalt samples 99YW94 and 99YW95 were collected near river level below this basalt bluff. The line of this fault projects from this bluff into the distance along the escarpment bounding Yünlü Tepe (Q–R). To the right of this is a flat area formed of fluvial gravel, behind and to the right of which (R–S) is the outcrop of basalt, starting around [BA 74151 66226], from which sample 03TR21 was collected. On the right of the photo, at an azimuth of N30°E from the viewpoint, this basalt is observed to be truncated at a second bluff (S), suggesting that another strand of the Amanos Fault may be present. S–T and Q–U indicate the interpreted lines of these two faults as they approach the viewpoint. (**c**) View ESE across the Karasu Valley from the same place as (b). In the middle distance is the town of Aktepe. Behind it (with trees on the summit) is the Aktepe volcanic neck. Behind that is another hill formed by an inlier of the Hatay ophiolite protruding through the basalt and alluvium in the Karasu Valley floor. In the background, rising to *c.* 1000 m a.s.l. and *c.* 18 km away, is the Kurd Dagh mountain range in NW Syria, also formed largely of the Hatay ophiolite and associated Mesozoic sediments. The international border runs along the foot of this mountain range, being demarcated by the Karasu River, which also marks the line of the East Hatay Fault (Y–Z).

(**a**)

(**b**)

(**c**)

Fig. 6. *Continued.*

from Kısık Tepe, west of the Amanos Fault, *c.* 1.8 km farther south. We thus suggest that left-lateral slip has juxtaposed this (sample 01TR50) basalt to its present position.

The remaining three samples were collected from basalt flows just above river level, west of the Amanos Fault. Sample 01TR55 came from a point *c.* 25 m south of the stone bridge over the Kurtlan tributary of the Hacılar River, just upstream of the offset reach of the latter; the other two came from the places indicated in Figure 10, where they stratigraphically underlie the Hacılar basalt. The greatest overall geochemical similarity of these samples is with sample 99YW89 of Yurtmen *et al.* (2002), which was collected *c.* 4.2 km farther north, east of the Amanos Fault. Yurtmen *et al.* (2002) inferred that sample 99YW89 erupted from the Kesmeli Tepe neck north of Hacılar on the west side of the fault (Fig. 11). We presume that another part of the same flow unit flowed southward from its source to the fault line in the Hacılar area, and yielded our three samples (01TR55, 01TR56 and 01TR57), being later buried beneath the younger Hacılar basalt. At the points where samples 01TR56 and 01TR57 were collected, the basalt is observed to dip ESE at *c.* 40°, suggesting that it cascaded across a contemporaneous fault-line bluff. However, corresponding basalt is not found locally east of the fault; it has presumably been displaced farther NNE by the left-lateral slip (see below).

Constraints on horizontal crustal motions, from dating and basalt geochemistry

Hacılar

As noted above (Figs 6b, 10a & b), at Hacılar, the Hacılar river has incised into the young Hacılar basalt (which flowed down this river valley, and is itself offset left-laterally; see above) on both sides of the Amanos Fault; thus, dating the basalt indicates the timescale on which this left-lateral offset has developed. According to Yurtmen *et al.* (2002) this basalt erupted from Camlı Tepe (Fig. 11), SE of Söğüt. This massive basalt shows minimal weathering, indicating a young age. As Yurtmen *et al.* (2002) noted, the *c.* 1.1 Ma whole-rock K–Ar date by Çapan *et al.* (1987) at a site somewhere around Hacılar indicates either inherited argon in their sample or that an older basalt flow was dated (or both), and thus has no bearing on estimation of the slip rate from the river offset.

Yurtmen *et al.* (2002) obtained an unspiked K–Ar date of 196 $\pm$ 12 ka ($\pm 2\sigma$) for the youngest basalt exposed in the river gorge section (their sample 99YW90), *c.* 1 km from the fault offset (Fig. 11). Rojay *et al.* (2001) obtained a spiked K–Ar date of 80 $\pm$ 120 ka ($\pm 2\sigma$) for the youngest basalt near this fault offset (their sample 08; Fig. 11), as the weighted mean of three splits, whose individual ages (all $\pm 2\sigma$) were reported as 130 $\pm$ 130 ka, 70 $\pm$ 72 ka, and 30 $\pm$ 60 ka. The Yurtmen *et al.* (2002) date is concordant with this overall date and with the oldest of these splits, suggesting that the same basalt flow has been dated at both sites. However, the high error margins of this spiked K–Ar dating limit its use in slip-rate determination.

Our new sample 01TR53, collected from the youngest basalt directly adjacent to the fault offset on its African side (Fig. 11), has yielded an Ar–Ar date of 159 $\pm$ 15 ka ($\pm 2\sigma$). This is concordant at the $\pm 3\sigma$ level with the Yurtmen *et al.* (2002) date, suggesting that the same basalt flow is present at both sites, consistent with the observation that no flow front can be seen between the sites. Yurtmen *et al.* (2002) estimated the offset of the

Fig. 6. (**a**) View north from [BA 74919 67956], looking at the escarpment that has formed along the eastern side of Kısık Tepe (A) by the component of downthrow to the east on the Amanos Fault. Behind Kısık Tepe, the southern part of the Keltepe basalt is also visible (B). The surface trace of the Amanos Fault is locally interpreted along the base of the escarpment (C–D). A less-pronounced break of slope within the Keltepe basalt (E) may mark a subsidiary structure. (**b**) View SSW from *c.* [BA 759 751], at the point where the road from Aktepe to Hacılar reaches the top of the Amanos Fault escarpment. The reach of the Hacılar River that has become offset left-laterally by slip on this fault (P–Q) is visible in the foreground. The basalt in the right foreground (in shadow) is part of the Hacılar flow unit that we have dated to 159 $\pm$ 15 ka ($\pm 2\sigma$). The basalt in the middle distance, beyond the offset river reach to the east of the fault (R), yielded sample 99YW91 dated to 476 $\pm$ 16 ka ($\pm 2\sigma$). Beyond it on the western side of the fault, the near skyline (S) is the outline of Keltepe neck, and the far skyline (T) is the north face of Yünlü Tepe (cf. Fig. 5b). South of the Hacılar River, the Amanos Fault can be identified from the alignment of scarps along the faces of Yünlü Tepe (U) and Keltepe (V), which project to the river at a similar break of slope along the line U–V–X, as Yurtmen *et al.* (2002) noted. (**c**) View WNW from [BA 78338 74184], near the point where basalt sample 99YW89 was collected, looking at the escarpment formed by the Amanos Fault. The basalt in the foreground, east of this fault, is inferred to have erupted from Kesmeli Tepe, farther south on the opposite side of the fault. This fault escarpment is formed in Upper Cretaceous limestone, with no basalt locally present on the far side, consistent with the development of significant left-lateral displacement since the basalt erupted. Some blocks of basalt have been recently dumped in the foreground, as a result of land clearance elsewhere, but *in situ* basalt is also present.

Table 1. *Summary of Ar–Ar dating results for Karasu Valley basalts*

Sample	Locality	UTM coordinates	Laboratory no.	^{40}Ar(*r*)/^{39}Ar(*k*) ($\pm 2\sigma$)	Age (ka) ($\pm 2\sigma$)	MSWD	^{39}Ar(*k*) (%)	K/Ca	*n*/*N*
01TR53	Hacılar	BA 75632 69900	B20 RW5F041P	0.2906 ± 0.0266 (±9.16%)	158.7 ± 14.6 (±9.18%)	0.71	79.37	0.357 ± 0.005	7/10
03TR21	Küreci	BA 74151 66226	B17 RW5F041M	0.6796 ± 0.0224 (±3.29%)	385.2 ± 12.9 (±3.35%)	0.24	81.02	0.285 ± 0.004	7/10
01TR51	Kısık Tepe	BA 74511 69310	B19 RW5F041O	0.7446 ± 0.0163 (±2.20%)	411.8 ± 9.4 (±2.28%)	0.24	92.80	0.530 ± 0.007	7/10
01TR47	Karaçağil	BA 66092 51365	B18 RW5F041N	2.1355 ± 0.0252 (±1.18%)	1195.1 ± 15.8 (±1.32%)	1.64	90.74	0.453 ± 0.006	6/10

All samples were of basaltic groundmass, separated as described in the text. ^{40}Ar(*r*)/^{39}Ar(*k*) is the measured ratio of the number of ^{40}Ar atoms produced by radioactive decay of the potassium in the sample to the number of ^{39}Ar atoms produced by irradiation of the potassium in the sample. MSWD is the mean squared weighted deviation, a measure of the scatter between individual age estimates for different heating steps. ^{39}Ar(*k*) (%) is the percentage of the ^{39}Ar in the irradiated sample that is determined to have been produced by irradiation of potassium. K/Ca is the measured ratio of potassium to calcium atoms in the sample, which is used to correct the ^{39}Ar measurements for ^{39}Ar produced by irradiation of calcium. *N* is the total of heating steps measured; *n* is the number of these steps that were used for each age determination. All ages have been calibrated relative to the US Geological Survey Taylor Creek Rhyolite tcr-2a sanidine standard, with an age of 28.34 Ma (after Renne *et al.* 1998). Age calculations assume the decay and isotopic abundance constants from Steiger & Jäger (1977).

Hacılar River by the Amanos Fault (Figs 6b & 10a) as 325 ± 25 m. Dividing this offset by our new age for this basalt gives an estimate for the local slip rate of 2.05 ± 0.37 mm/a ($\pm 2\sigma$). However, rather than being aligned with its present course on the upstream side of the Amanos Fault, it is possible that immediately after the basalt eruption the Hacılar River flowed along what is now the extreme southern margin of its valley, as such a position would have coincided with the southern margin of the basalt flow. If so, the present offset of this river underestimates its true offset, which we estimate as 450 ± 25 m (Fig. 11). With such a slip measurement, the estimated slip rate on the Amanos Fault adjusts to 2.84 ± 0.41 mm/a ($\pm 2\sigma$).

Hassa

As Yurtmen *et al.* (2002) noted, study of the basalt south of Hassa has been difficult, due to the unavailability of the local 1:25000-scale topographic map sheet. The SRTM coverage now available (Fig. 11) makes much more precise analysis possible.

Yurtmen *et al.* (2002) obtained an unspiked K–Ar date of 393 ± 22 ka ($\pm 2\sigma$) for a sample (99YW89) collected near the northern margin of the basalt outcrop south of Hassa, just east of the Amanos Fault (Figs 6c & 11). We established its coordinates as [BA 78338 74184]; the margin of the basalt being an estimated *c.* 100 m farther north.

The SRTM imagery (Fig. 11) now reveals that this basalt forms a substantial edifice, resembling half a cone, with a radius of *c.* 1.5–2 km. In its centre, the basalt surface (reaching *c.* 605 m a.s.l.) is now juxtaposed against a hill, Kalecik Tepe (summit *c.* 665 m a.s.l., *c.* 300 m west of the fault), formed in Late Cretaceous limestone, such that by chance there is locally no east-facing scarp across the Amanos Fault. West of the Amanos Fault, the highest point with a basalt outcrop is Kesmeli Tepe (*c.* 615 m a.s.l. from both the local map and the SRTM data in Fig. 11), *c.* 2 km NNE of Hacılar. Yurtmen *et al.* (2002) tentatively interpreted this as the neck from which the basalt erupted. However, allowing for subsequent downthrow to the east on the Amanos Fault, it is possible that the neck was east of the fault (at the point now east of Kalecik Tepe) from which basalt flowed westward on to what is now Kesmeli Tepe. Basalt seems to have also flowed WNW and northward from the Kesmeli Tepe area, ponding the Hayıtlı River valley to create the alluvial plain evident around the village of Söğüt. This Kesmeli Tepe basalt also forms the land surface westward for *c.* 300 m and southward for *c.* 700 m from Kesmeli Tepe on the African side of the Amanos Fault; beyond these points it has been truncated by

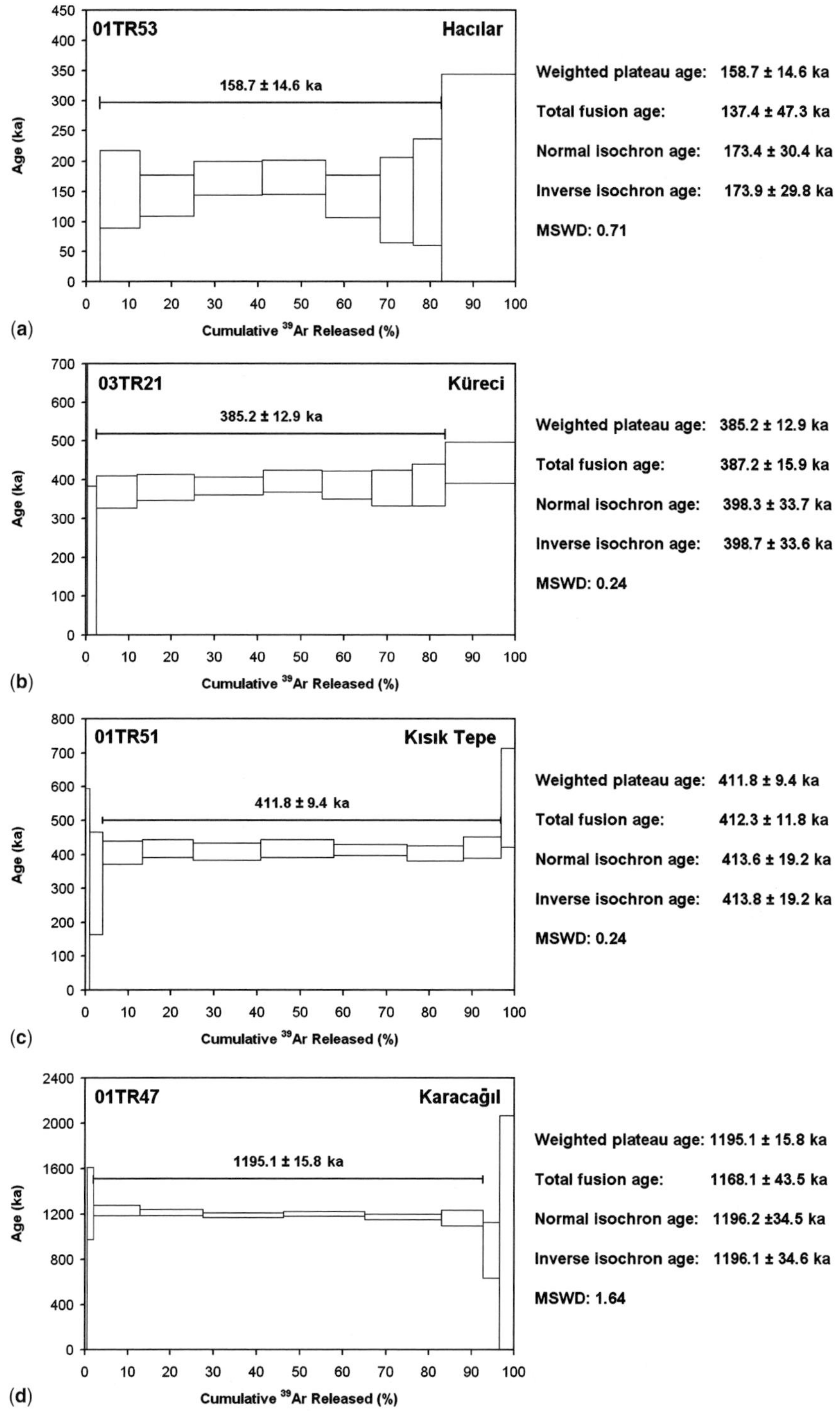

Fig. 7. Ar/Ar age spectra derived from step heating of our four samples. The vertical bars indicate the uncertainty in age derived from each heating step for each sample. The horizontal bars indicate the range of heating steps that contribute to each date. To the right of each diagram the age determinations for the sample are listed, using four different methods, along with the MSWD values (Table 1). The extent of concordance between these different age determinations provides a test of the reliability of each date. For full details see the online supplement at http://www.geolsoc.org.uk/sup18274. A hard copy can be obtained from the Geological Society Library.

Table 2. *Geochemical analyses*

		Major elements (wt%)												Trace elements (ppm)																			Ratios		Norms (wt%)		
Sample	UTM coordinates	SiO_2	TiO_2	Al_2O_3	Fe_2O_3(t)	MnO	MgO	CaO	Na_2O	K_2O	P_2O_5	LOI	Total	Cr	Cu	Ga	Nb	Ni	Pb	Rb	Sr	Th	V	Y	Zn	Zr	Ba	Ce	La	Nd	Cl	S	Ba/La	Zr/Nb	Qz	Ol	Ne
Küçük Höyük																																					
01TR43	BA 67099 50157	48.47	2.21	14.49	11.85	0.15	8.37	8.41	2.75	1.29	0.58	0.83	99.39	240	64	22	33	145	4	21	601	3	199	25	102	164	300	40	28	36	154	123	10.714	4.970	0	11.185	0
01TR44	BA 66948 50342	48.26	2.22	15.66	12.04	0.15	7.71	8.67	3.77	1.35	0.48	0.13	100.45	179	46	23	29	120	4	24	540	0	185	26	90	155	231	25	13	19	45	219	17.769	5.345	0	17.001	2.883
Ceylanlı																																					
01TR45	BA 65834 49424	49.87	2.18	15.37	11.92	0.14	6.76	8.52	3.62	1.18	0.35	0.07	99.97	217	53	21	23	78	4	24	462	1	214	25	98	136	339	14	15	14	51	69	22.600	5.913	0	14.356	0
01TR46	BA 65859 50016	50.54	2.25	15.71	11.75	0.15	5.79	8.78	3.75	1.29	0.55	−0.13	100.44	156	46	21	31	54	6	24	617	0	176	27	97	162	398	61	12	39	118	103	33.167	5.226	0	8.862	0
Karacağil																																					
01TR47	BA 66092 51365	48.99	2.05	15.56	11.90	0.15	6.93	8.71	3.90	1.29	0.82	0.27	100.57	207	61	19	39	135	5	23	848	0	175	27	96	163	285	80	35	36	228	74	8.143	4.179	0	16.059	1.121
Aktepe (Büyük Aktepe basalt (?))																																					
01TR48	BA 76430 62546	46.47	2.25	16.08	13.01	0.16	7.93	9.64	2.90	0.78	0.31	0.07	99.61	258	70	20	25	102	2	8	436	0	205	24	94	121	411	29	14	38	23	118	29.357	4.840	0	20.124	0.668
Aktepe (Aktepe basalt (?))																																					
01TR49	BA 75890 63754	48.30	2.13	17.60	12.22	0.16	5.66	8.48	3.50	1.06	0.27	0.37	99.74	78	35	24	17	26	8	9	352	0	219	29	95	159	514	24	7	22	14	91	73.429	9.353	0	16.302	0
Yünlü Tepe (Küreci basalt)																																					
03TR21	BA 74151 66226	48.56	2.23	16.94	12.39	0.16	5.83	8.67	3.82	1.18	0.31	−0.50	99.60	85	40	24	20	17	6	17	386	4	205	30	96	161	268	16	8	19	6	68	33.500	8.050	0	16.270	2.042
Gülpinar (Keltepe basalt (?))																																					
01TR51	BA 74511 69310	49.51	2.14	16.9	11.77	0.15	5.76	7.90	3.38	1.35	0.32	0.43	99.61	165	37	22	23	50	9	18	397	2	203	30	96	170	219	44	12	28	20	83	18.250	7.391	0	10.332	0
Kisik Tepe																																					
01TR50	BA 74514 68264	48.14	2.05	16.35	11.39	0.15	5.86	9.34	3.39	1.38	0.32	1.18	99.55	132	48	23	22	50	3	17	415	2	191	30	93	166	432	8	21	7	16	370	20.571	7.545	0	14.434	1.422
01TR52	BA 74777 68313	49.68	1.85	15.61	11.38	0.15	7.65	8.43	3.17	1.29	0.25	0.09	99.54	243	52	21	18	84	9	26	365	1	179	25	93	150	382	21	16	17	60	91	23.875	8.333	0	13.605	0
Hacilar (Kisik Tepe basalt (?))																																					
01TR58	BA 75887 69997	48.10	2.23	17.53	11.96	0.15	6.15	8.20	3.27	1.18	0.34	0.63	99.74	154	38	24	24	51	5	10	401	2	222	29	99	176	351	32	25	22	18	105	14.040	7.333	0	14.791	0
Hacilar (Hacilar basalt)																																					
01TR53	BA 75632 69900	45.41	2.82	16.33	13.05	0.16	7.10	10.36	3.08	1.19	0.62	0.17	100.30	108	59	22	38	42	3	12	797	0	199	27	92	174	343	56	34	30	106	107	10.088	4.579	0	17.131	4.470
01TR54	BA 75632 69900	45.59	2.76	16.29	12.98	0.16	6.84	10.42	3.51	1.18	0.61	0.06	100.40	114	57	22	40	43	4	14	794	1	199	27	89	174	436	54	36	31	121	99	12.111	4.350	0	15.953	6.821
01TR59	BA 75989 69965	45.80	2.79	16.36	13.08	0.16	6.89	10.54	3.18	1.18	0.59	0.18	100.74	117	58	21	39	40	9	12	853	0	203	28	93	172	325	62	29	38	112	135	11.207	4.410	0	16.383	4.861
Hacılar (Kesmeli Tepe basalt (?))																																					
01TR55	BA 75430 69837	49.72	2.23	17.40	11.67	0.15	5.46	8.28	3.61	1.44	0.34	0.17	100.48	105	39	25	26	48	6	18	430	2	204	31	94	175	322	30	11	20	32	92	29.273	6.731	0	11.250	0
01TR56	BA 75791 69983	50.00	2.08	16.54	11.40	0.15	6.32	8.60	3.64	1.43	0.32	−0.47	100.01	159	42	24	24	49	15	25	417	2	207	29	92	165	385	27	17	17	23	90	22.647	6.875	0	13.635	0
01TR57	BA 75815 69997	50.44	2.12	16.44	11.47	0.15	6.08	8.63	3.59	1.42	0.32	−0.29	100.37	147	38	21	24	46	9	22	419	2	211	27	94	167	848	20	22	17	17	92	38.545	6.958	0	10.716	0

Analyses of samples by X-ray fluorescence spectrometry was undertaken at Keele University in 2004. LOI means loss on ignition. Qz, Ol and Ne indicate normative quartz, olivine and nepheline. Site coordinates for these samples were measured in the field to the nearest metre, using a hand-held GPS receiver. Sample 01TR54 came from directly below sample 01TR53, in what appears to be a lower basalt flow within the same flow unit. Norm calculations assume a Fe_2O_3/FeO wt% ratio of 0.2 for most samples, or 0.3 for hawaiites (01TR44, 01TR46, 01TR47, 01TR55, 01TR56 and 01TR57), after Middlemost (1989).

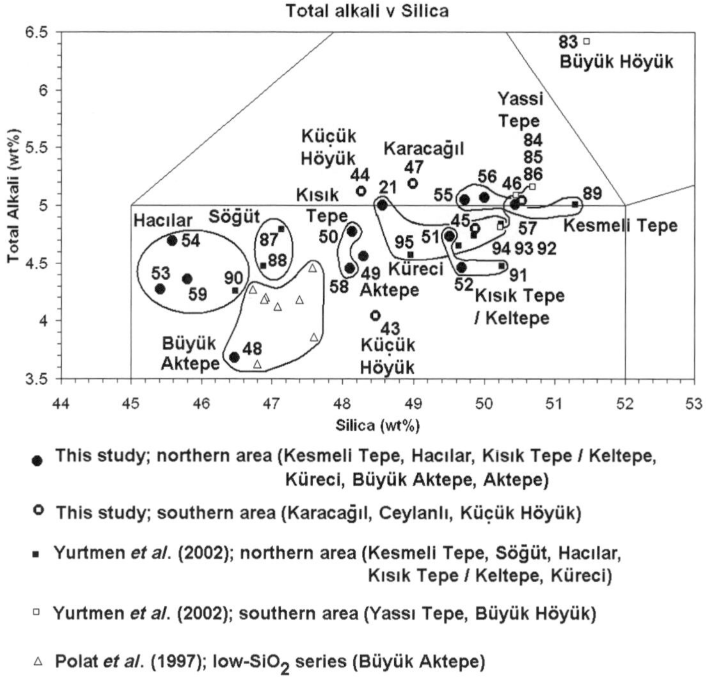

Fig. 8. Total alkali ($Na_2O + K_2O$) v. SiO_2 diagram for Karasu Valley basalts (after Le Bas *et al.* 1986). Samples from Ceylanlı are unlabelled. Samples are labelled in this figure and Fig. 9 using the last two digits of their sample numbers.

fluvial incision and then inset by the younger Hacılar basalt. However, our interpretation of samples 01TR55-7 (see above) suggests that this Kesmeli Tepe basalt persists beneath the Hacılar basalt, west of the Amanos Fault, to a distance of *c.* 2.5 km south of Kesmeli Tepe.

Restoring the highest point of the basalt cone east of the Amanos Fault (*c.* [BA 77455 78080]; L in Fig. 11) against Kesmeli Tepe (*c.* [BA 76355 76910]; 'M' in Fig. 11) indicates 1600 $\pm$ 25 m of subsequent left-lateral slip on this fault. The large volume of basalt produced suggests that this neck was probably active for many tens of thousands of years. Sample 99YW89 came from one of the most distal parts of this flow unit, and in the field it can be seen to almost directly overlie bedrock. However, it cannot be presumed to indicate the start of eruption from this neck; it may simply mark the greatest eruption rate, when basalt flowed farthest from its source. The data-set presented by Yurtmen *et al.* (2002) indicated a concentration of dates around 400–500 ka. The oldest of this group of dates, from Küreci (*c.* 5 km SSW of Kesmeli Tepe), was 553 $\pm$ 20 ka ($\pm 2\sigma$; see below). If this is taken as also marking the start of the Kesmeli Tepe eruption, then the subsequent time-averaged slip rate on the Amanos Fault can be estimated as 2.89 $\pm$ 0.14 mm/a ($\pm 2\sigma$).

Küreci

Our new Ar/Ar date for sample 03TR21, of 385 $\pm$ 13 ka ($\pm 2\sigma$), supplements the data already available from the Küreci area (Fig. 3). As Yurtmen *et al.* (2002) discussed, basalt erupted from a neck near Zeytinoba and flowed eastward down the Küreci river valley, past Küreci village and into the Karasu Valley, crossing the Amanos Fault (Fig. 12). Çapan *et al.* (1987) obtained a spiked K–Ar date of 600 $\pm$ 200 ka ($\pm 2\sigma$) for this basalt just east of Küreci (their sample 18). Rojay *et al.* (2001) added another such date of 190 $\pm$ 100 ka ($\pm 2\sigma$) (their sample 07) for the top of the basalt just west of the Amanos Fault. Yurtmen *et al.* (2002) obtained three unspiked K–Ar dates: 553 $\pm$ 20 ka ($\pm 2\sigma$) for the base of the *c.* 30-m-thick flow unit east of Küreci (sample 99YW92), 325 $\pm$ 14 ka ($\pm 2\sigma$) for the top of this flow unit, also east of Küreci (sample 99YW93), and 253 $\pm$ 14 ka ($\pm 2\sigma$) for basalt in the modern valley floor adjacent to the Amanos Fault (sample 99YW94). Tatar *et al.* (2004) added another spiked K–Ar date, of 620 $\pm$ 260 ka ($\pm 2\sigma$) for the proximal part of this basalt near Zeytinoba (their sample 33); they also established that six samples were normally magnetized, evidently all dating from the Brunhes chron.

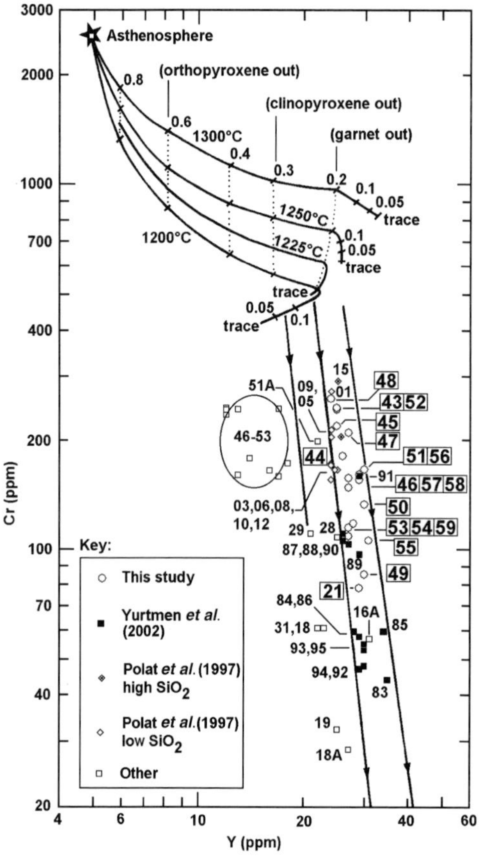

Fig. 9. Chromium v. yttrium diagram highlighting the roles of partial melting and fractional crystallization in the formation of the basalts in the study region. Solid lines are partial melting trends, calculated by Pearce *et al.* (1990) for garnet lherzolite (55% olivine, 20% orthopyroxene, 12.5% clinopyroxene, and 12.5% garnet) at 1200 °C, 1250 °C, and 1300 °C, with each solid phase disappearing at the indicated degree of partial melting. See Pearce *et al.* (1990) for the values of the partition coefficients used and for sources of data. The line for 1225 °C was estimated by Yurtmen *et al.* (2002) by interpolation. Barbed lines indicate mafic crystallization trends, also from Pearce *et al.* (1990).

Sample 03TR21 came from [BA 74151 66226], at the southern limit of basalt outcrop, just east of the Amanos Fault (point D in Fig. 12). Basalt is not present locally across the fault, where outcrop is instead of Late Cretaceous limestone (the Karadağ Limestone of Atan 1969). We measured the northern edge of the basalt in the Küreci valley, just west of the fault, at [BA 73872 66180] (point C in Fig. 12), near where Rojay *et al.* (2001) collected their dating sample (reported by them at [BA 73835 66305]).

Atan (1969) mapped a small outlier of basalt adjoining the Amanos Fault on its eastern side, NE of Yünlü Tepe (Fig. 12). From this map, the northern margin of this outcrop is offset by 1725 ± 50 m from its counterpart west of the Amanos Fault in the Küreci Valley (A–B in Fig. 12). Assuming that this outlier marks the oldest part of the Küreci flow unit, with the same age as our sample 99YW92, the subsequent time-averaged slip rate on the Amanos Fault can be estimated as 3.12 ± 0.21 mm/a ($\pm 2\sigma$). As Figure 12 indicates, this basalt outlier abuts a ridge of ophiolite on its southern side. Once the basalt outlier was displaced away from the Küreci Valley by left-lateral slip on the Amanos Fault, this ophiolite ridge would have been juxtaposed across the mouth of this river valley. It would have thus acted as a shutter ridge, inhibiting later-erupting basalt from flowing out into the Karasu Valley. Ponding of the Küreci basalt by this shutter ridge is presumably the reason why this basalt has attained such a substantial thickness west of the Amanos Fault and yet is virtually absent east of this fault. As noted above, our own fieldwork suggests that the piercing point where the Amanos Fault intersects the northern margin of the Küreci Valley is 'C', not 'B', in Figure. 12, *c.* 100m farther west. If so, then the fault offset adjusts to A–C or *c.* 1800 ± 50m and the associated slip rate (making the same age assumption as before) adjusts to 3.25 ± 0.22mm/a ($\pm 2\sigma$).

Dating the end of this basalt eruption is more problematic. At a relatively late stage, left-lateral slip displaced this ophiolite ridge north of the Küreci Valley, so basalt was again able to flow eastward into the Karasu Valley. However, the limited volume of basalt east of the Amanos Fault suggests that volcanism had more or less ended by this time. Currently, four dates (all $\pm 2\sigma$) potentially constrain the end of this volcanism: 190 ± 100 ka from Rojay *et al.* (2001); 325 ± 14 ka and 253 ± 14 ka from Yurtmen *et al.* (2002); and our new date of 385 ± 13 ka. These dates are not concordant; Yurtmen *et al.* (2002) discussed this issue but could offer no satisfactory explanation. Since all four dates would appear to reflect the youngest volcanism in this locality, it nonetheless seems appropriate to determine their weighted mean: 323 ± 8 ka ($\pm 2\sigma$).

We estimate that the site of our sample 03TR21 is 700 ± 50 m from the point where the Amanos Fault intersects the southern margin of the Küreci Valley ('G' in Fig. 12). This figure thus indicates the left-lateral slip on the fault strand between points D and G since this basalt erupted. Taking the age of this basalt as our weighted mean value of 323 ± 8 ka ($\pm 2\sigma$), we estimate a slip rate on the Amanos Fault of 2.17 ± 0.31 mm/a ($\pm 2\sigma$). As illustrated in Figure 12, a second active fault strand appears to pass east of point D, forming the bluff on the skyline in the extreme right of Figure 5b. The slip rate on this fault can be tentatively estimated as *c.* 1 mm/a, i.e. the difference between the *c.* 3 mm/a slip rate derived from offset A–C (which represents the overall slip rate on the Amanos Fault north of the Küreci Valley) and the *c.* 2 mm/a slip rate derived from offset D–G (which represents the local slip rate on the western strand of the Amanos Fault).

Kısık Tepe

Yurtmen *et al.* (2002) drew attention to two basalt-capped hills, thought to represent necks, west of the Amanos Fault between Hacılar and Küreci, called Keltepe and Kısık Tepe (Figs 3 & 11). They also noted basalt east of the Amanos Fault south of Hacılar which was much more weathered than the nearby Hacılar basalt, and which yielded a much older unspiked K–Ar date of 476 $\pm$ 16 ka ($\pm 2\sigma$) (sample 99YW91). This basalt, which Yurtmen *et al.* (2002) also showed to be chemically distinct from the Hacılar basalt, reaches its maximum thickness east of the Amanos Fault *c.* 250 m south of the 99YW91 site, at Kocaören (*c.* [BA 75565 69485]; 'H' in Fig. 11). Yurtmen *et al.* (2002) suggested that the basalt in this area erupted from Keltepe (*c.* [BA 75255 69125]; 'J' in Fig. 11), from which it is now offset by *c.* 475 m. From the age of sample 99YW91 and this measured offset, they tentatively deduced that this segment of the Amanos Fault has a slip rate of 1.00 $\pm$ 0.14 mm/a ($\pm 2\sigma$).

To check this possibility, we collected three samples of basalt from west of the fault around Kısık Tepe and west of Keltepe. All three are geochemically similar to Yurtmen *et al.*'s (2002) sample 99YW91 (Table 2). Sample 01TR51, collected west of Keltepe, was also Ar/Ar dated to 412 $\pm$ 9 ka ($\pm 2\sigma$).

Local geological mapping (e.g. Atan 1969) indicates that the eruption of the Kısık Tepe and Keltepe basalts ponded the valleys of the Kızılyar and Kurtlan rivers, which flow eastward towards the Karasu Valley. It follows that both eruptions must have been roughly synchronous, otherwise no such ponding would have been possible. The Kurtlan River has subsequently incised through Cretaceous limestone bedrock, around the northern margin of the Keltepe basalt, to form a new course, joining the Hacılar River just west of the Amanos Fault (see Fig. 11, inset). The Kızılyar has instead cut a narrow gorge through the Kısık Tepe basalt adjacent to sample site 01TR52.

We consider it probable that the basalt forming our sample 01TR51 originated from Keltepe, rather than Kısık Tepe. However, the geomorphological argument above suggests that both necks erupted around the same time, so this date can also estimate the timing of eruption of the Kısık Tepe basalt. In view of our earlier estimates for the slip rate on the Amanos Fault, we suggest that the basalt east of the Amanos Fault at Kocaören and at sample sites 99YW91 and 01TR58 (discussed above) is equivalent to that at Kısık Tepe, not Keltepe. Given the greater thickness of basalt and surface altitude (*c.* 465 m at Kocaören; *c.* 435 m at Kısık Tepe; *c.* 440 m west of Kısık Tepe around sample site 01TR50) it is indeed possible that this basalt erupted from a neck at Kocaören, from which it flowed westward across the Amanos Fault into the Kısık Tepe area. Since Kısık Tepe (*c.* [BA 74930 68450]; 'K' in Fig. 11) is *c.* 750 m SSW of Keltepe (Fig. 3), the amount of subsequent left-lateral offset can be estimated as 1225 $\pm$ 25 m.

Although the ages of samples 99YW91 and 01TR51 are not concordant, as for Küreci we form their weighted mean as the 'best estimate' of the timing of local volcanism: 427 $\pm$ 8 ka ($\pm 2\sigma$). Dividing this by the estimated amount of subsequent left-lateral slip (H–K; Fig. 11) gives a slip rate of 2.87 $\pm$ 0.05 mm/a ($\pm 2\sigma$).

Karacağıl

Located *c.* 15 km SSW of Küreci, the Karacağl area is the only other part of the Karasu Valley where we (and Yurtmen *et al.* 2002) have dated any basalts. In a similar way to farther north, the Amanos Fault runs near the front of the Amanos Mountains. North of Karacağıl and south of Karaelma, at Yassı Tepe, this fault (which locally trends north–south) transects basalt that Yurtmen *et al.* (2002) dated to 428 $\pm$ 14 ka ($\pm 2\sigma$) (their sample 99YW84). Yurtmen *et al.* (2002) estimated from the local geomorphology that this Yassı Tepe basalt has become offset left-laterally by 425 $\pm$ 25 m, giving a slip rate of 0.99 $\pm$ 0.14 mm/a ($\pm 2\sigma$).

Near Karacağıl, just east of the Amanos Fault, Parlak *et al.* (1998) obtained a spiked K–Ar date (their sample 51) of 1050 $\pm$ 600 ka ($\pm 2\sigma$). Tatar *et al.* (2004) showed that this basalt is reverse-magnetized, supporting eruption during the Matuyama chron. From the geomorphology, Yurtmen *et al.* (2002) tentatively estimated that this older Karacağıl basalt is offset across the Amanos Fault by 1000 $\pm$ 50m, suggesting a slip rate of 0.95 $\pm$ 0.55mm/a ($\pm 2\sigma$). To improve upon this slip-rate estimate, by redating this basalt, sample 01TR47 was thus collected west of the Amanos Fault in the dry valley of the Çınarlı River, and yielded an Ar/Ar date of 1195 $\pm$ 16 ka ($\pm 2\sigma$). The slip rate thus adjusts to 0.84 $\pm$ 0.08mm/a ($\pm 2\sigma$), consistent with the estimate from the Yassı Tepe basalt.

(a)

(b)

Fig. 10. (**a**) View SSW looking upstream along the offset reach of the Hacılar River from [BA 75979 69965], where sample 01TR59 was collected. The Amanos Fault (which projects along the line A–B) follows the *c.* 80 m high bluff in the left bank of this river, leaving the river at the point where this bank is supported by a retaining wall near the centre of the view (in line with C). To the left of this wall, at the base of the bluff, two basalt flows are visible: a lower one, which is vesicular, and an upper massive one with a rubbly base. This upper flow, which yielded samples 01TR56 and 01TR57 (the latter being collected *c.* 20 m north of the former), dips at *c.* 40° towards S60°E, indicating that the basalt cascaded over a bluff at the fault line. The point where the Hacılar river joins the Amanos Fault

Yurtmen *et al.* (2002) reported a second significant left-lateral fault, which they named the Guzelce Fault (Fig. 3). This fault, recognizable as a bluff where the land surface is downthrown to the SE by up to several tens of metres, trends NE, obliquely away from the Amanos Mountains; its intersection with the Amanos Fault can be projected roughly halfway between Karacağıl and Ceylanlı (Fig. 3).

South of this fault intersection, basalt is present on both sides of the Amanos Fault. However, as there is no continuity of outcrop, no direct estimate of the local left-lateral slip rate is possible. West of the Amanos Fault at Ceylanlı, Çapan *et al.* (1987) and Rojay *et al.* (2001) obtained spiked K–Ar dates of 1730 $\pm$ 200 ka and 1570 $\pm$ 160 ka (both $\pm 2\sigma$), respectively (samples 31 and 02). Parlak *et al.* (1998) reported dates from this area of 2200 $\pm$ 1400 ka, 790 $\pm$ 600 ka, and 400 $\pm$ 400 ka (all $\pm 2\sigma$), Tatar *et al.* (2004) established that the local basalt is reverse-magnetized, consistent with eruption early in the Matuyama chron and with the weighted mean of these five dates, which is 1500 $\pm$ 15 ka ($\pm 2\sigma$). Rojay *et al.* (2001) pointed out that this Ceylanlı basalt outcrop is truncated by the Amanos Fault, which has evidently taken up significant downthrow to the east as well as left-lateral slip. However, there is no outcrop in the Karasu Valley interior that can be matched to this Ceylanlı basalt; its offset counterpart is presumably beneath the alluvium in the valley interior and displaced northward relative to Ceylanlı by an unknown distance.

The only outcrop of basalt east of the Amanos Fault in this area forms the upper parts of the Büyük Höyük and Küçük Höyük necks, which protrude above the surrounding alluvium. Büyük Höyük was dated to 600 $\pm$ 600 ka (spiked K–Ar; $\pm 2\sigma$) by Parlak *et al.* (1998) and to 828 $\pm$ 24 ka (unspiked K–Ar; $\pm 2\sigma$) by Yurtmen *et al.* (2002). The weighted mean of these dates is 828 $\pm$ 24 ka ($\pm 2\sigma$), consistent with the reversed geomagnetic polarity measured by Tatar *et al.* (2004). Küçük Höyük was dated to 660 $\pm$ 80 ka (spiked K–Ar; $\pm 2\sigma$) by Rojay *et al.* (2001), but has not been investigated magnetostratigraphically. As noted above, the geochemistry of the basalts differs significantly both between these necks and relative to the basalts west of the Amanos Fault in the Ceylanlı area. No basis thus exists for correlating the basalts across the Amanos Fault in this area.

The Guzelce Fault

The Guzelce Fault can be traced northeast then NNE of its intersection with the Amanos Fault for *c.* 20 km to near Aktepe, as illustrated in Figure 3. For most of its length, basalt crops out on its western side, with alluvium cropping out below the scarp on its eastern side. This basalt appears to have erupted from the Aktepe neck (Fig. 3), and to have flowed SSW down the Karasu Valley. Rojay *et al.* (2001) dated the basalt at Aktepe neck to 260 $\pm$ 80 ka ($\pm 2\sigma$) (their sample 06). Water-supply boreholes (DSİ 1975; Rojay *et al.* 2001) indicate that basalt (presumably from the same flow unit) is also present in the subsurface east of the Guzelce Fault. In the past quarter of a million years, since the basalt eruption, this fault has thus evidently taken up tens of metres of downthrow to the east, plus a presumably much greater amount of left-lateral slip.

We tentatively reconcile our estimates of the left-lateral slip rate on the Amanos Fault, of *c.* 3 mm/a at the sites north of Aktepe (Küreci, Kısık Tepe, Hacılar, Hassa) and *c.* 1 mm/a in the vicinity of Karacağıl, by presuming that the 'missing' *c.* 2 mm/a is taken up south of Aktepe on the Guzelce Fault. We thus presume that south of the point where the Guzelce Fault splays from

Fig. 10. (*Continued*) (Q in Fig. 6b) is obscured behind D, the northern end of the Kocaören basalt (R in Fig. 6b). The basalt visible in the right foreground, between B and C, on the east side of the Amanos Fault, slopes towards the viewpoint and yielded sample 01TR59, indicating that it is part of the Hacılar basalt. The near skyline to the left of F is formed by the main outcrop of the Hacılar basalt, on the west side of the Amanos Fault. (**b**) View NNE from a point *c.* 50 m SSW of the retaining wall and adjacent basalt flows mentioned above (which are also shown here and link the two photos), looking downstream along the offset reach of the Hacılar River. The viewpoint in (a) is on the skyline at P. The estimated line of the Amanos Fault is marked both on the skyline (Q, adjacent to F in (a)) and in the dry river bed (R); as noted above, the fault leaves the river where its bank is supported by the retaining wall, near the middle of this field of view. Between the skyline and the point where it intersects the river the Amanos Fault can be seen to dip eastward at *c.* 50°. On its eastern (Arabian) side, a flat-lying flow unit of ropy basalt with a vesicular top is just visible in the left bank of the river after it has left the fault line, behind the low river terrace in the foreground. This yielded sample 01TR58. This basalt flow is overlain by very coarse fluvial gravel, mostly comprising limestone clasts, forming a higher river terrace reaching *c.* 15 m above river level. This fluvial gravel is overlain by a wedge of colluvium, then by thin basalt that plunges down the bluff at *c.* 25°, from which sample 01TR59 was collected. The upper end of this *in situ* dipping basalt is separated from its counterpart west of the fault by a *c.* 20-m-wide gap (to the left of Q), where only basalt rubble is present, providing an indication of the heave on the Amanos Fault since the eruption of this basalt occurred.

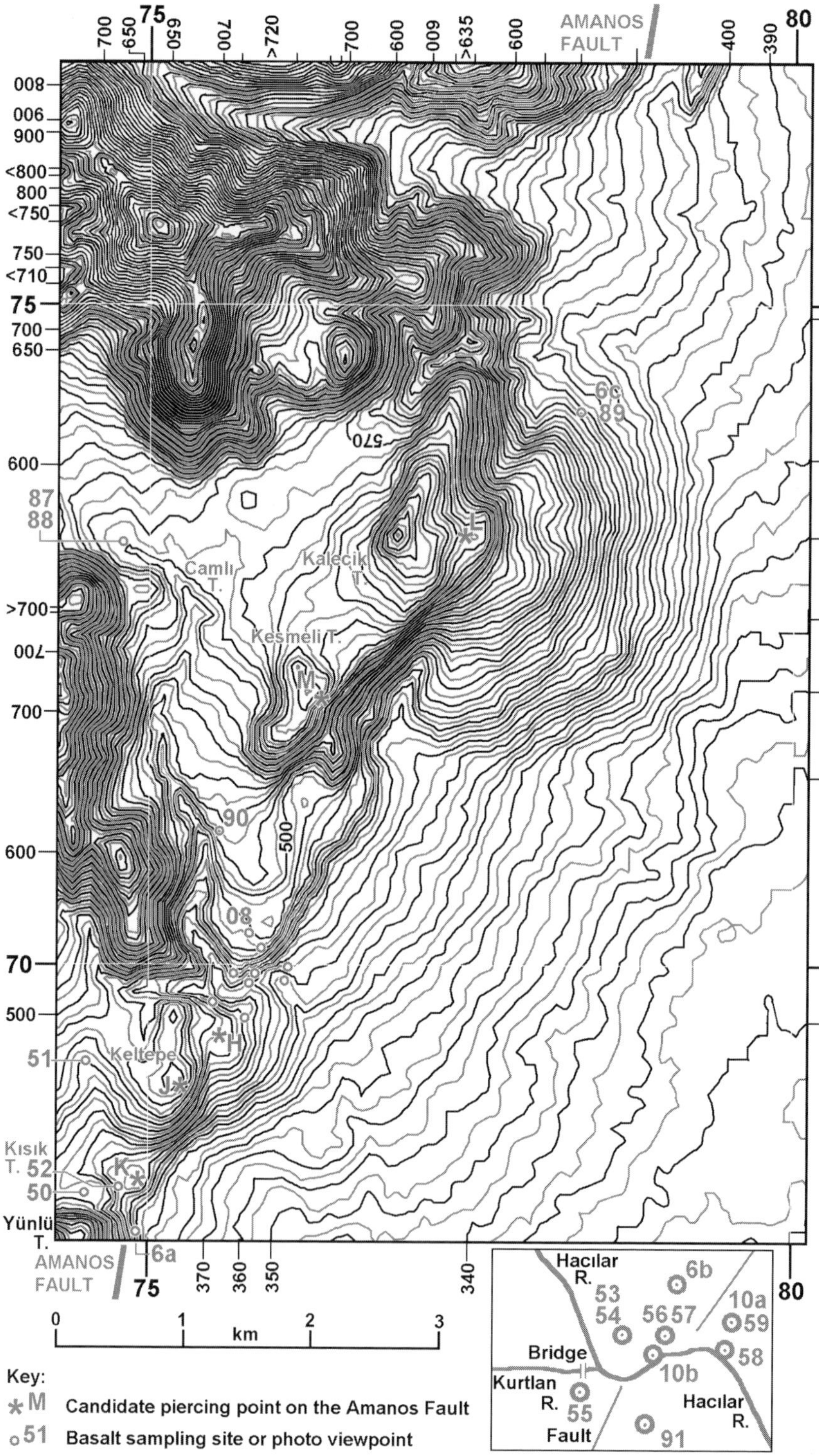

Fig. 11. More detailed SRTM topographic map of the Amanos Fault and its surroundings in the Hassa–Kesmeli Tepe–Hacılar–Keltepe–Kısık Tepe area, with 5-m contours, showing sample locations, photo viewpoints, and candidate piercing points used for slip and slip rate calculation. See Westaway *et al.* (2006*a*) for details of data sources and preparation technique. The inset shows the area of the offset Hacılar river gorge (*c.* [BA 76 70]) at double the scale.

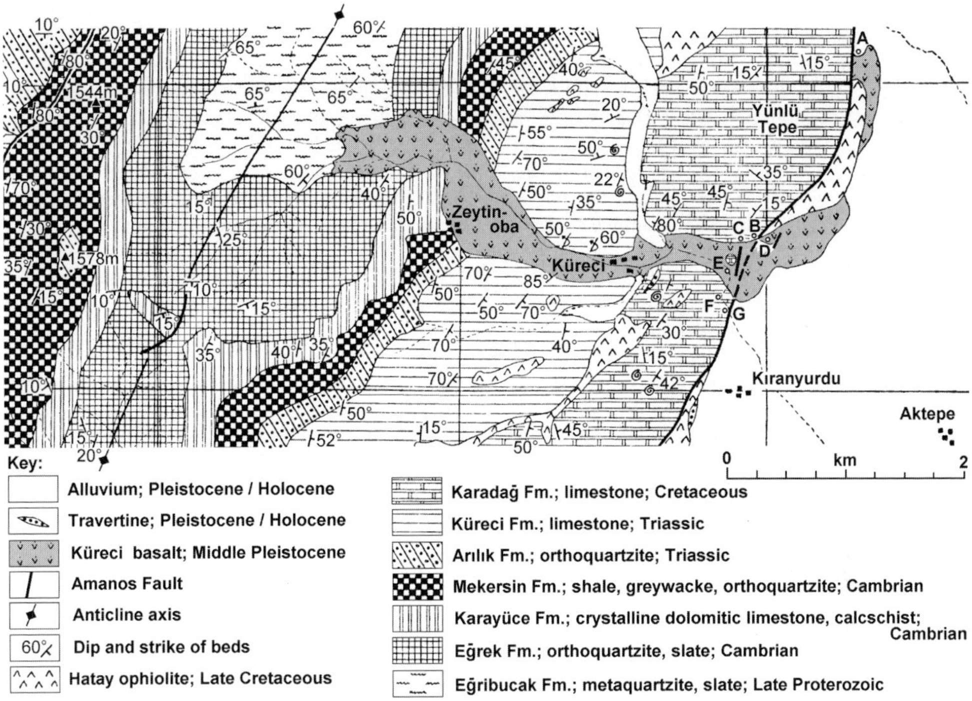

Fig. 12. Geological map of the Küreci basalt flow unit, adapted from part of plate II of Atan (1969), showing his stratigraphic terminology for the area. The meaning of points A, B, C, D and G, illustrated, is discussed in the text. The apparent absence of basalt directly south of point D, not recognized by Atan (1969) but evident from our own fieldwork, is illustrated schematically. Line segments linking C–G and east of D indicate positions of strands of the Amanos Fault deduced from our own fieldwork. Dashed line B–G illustrates Atan's (1969) tentative alternative interpretation. Point E is the location where basalt sample 99YW95 was collected, determined by GPS as [BA 73857 65985]. This site can be reached along a track that leaves the Kıranyurdu–Küreci road at [BA 73703 66045]. Point F is the viewpoint where the photographs in Figure 5b & c were taken. Note the anticline axis suparallel to the Amanos Fault, *c.* 4 km farther west. The Eğribucak inlier of Late Proterozoic basement and the overlying Palaeozoic succession crop out locally, forming the core of the Amanos mountain range. Note that Atan's (1969) formation names are unofficial and that other terminologies also exist. Later studies such as Schwan (1971) and Dean & Monod (1985) discuss how these different nomenclatures correlate.

the Amanos Fault, the slip rate on the latter is once again *c.* 3 mm/a. However, we do not understand how displacement 'transfers' between the northern end of the Guzelce Fault and the Amanos Fault (whether by localized slip on an unidentified fault linking the two, in the area west of Aktepe, or by local distributed deformation); further investigation of this point is beyond the scope of this study.

Basalt east of the Guzelce Fault

Çapan *et al.* (1987) first proposed that much of the basalt in the central-southern Karasu Valley originated from the large Büyük Aktepe neck, located *c.* 10 km east of Hassa, just west of the Syrian border. In their interpretation, basalt flowed SSW from this source for >20 km, passing east of Aktepe town and the adjoining neck. The southern limit of exposure of this flow unit is *c.* [BA 750 570], *c.* 8 km south of Aktepe, although it may continue in the subsurface beneath the young alluvium of the Karasu Valley interior.

At a locality that they called Korogha Geri, *c.* 4 km ESE of Aktepe town, Çapan *et al.* (1987) obtained a spiked K–Ar date of 2100 ± 400 ka (±2σ). Basalts from the Karasu Valley were subsequently analyzed geochemically by Polat *et al.* (1997). Although they published no location information, and their full data-set cannot now be traced, enquiries during the preparation of the Yurtmen *et al.* (2002) paper indicated that their 'low SiO_2' set of basalts (the majority that they studied) came from this flow unit. Tatar *et al.* (2004) showed that this flow unit, both in the area near Aktepe (they sampled it at their sites 8 and 9, *c.* 1.5 and *c.* 3 km SSE of Aktepe at Karapınar and Hanobası Fig. 3) and in the area around Büyük

Aktepe neck, is normally magnetized. They also obtained two spiked K–Ar dates for this flow unit, 1420 $\pm$ 240 ka ($\pm 2\sigma$) at site 43 *c.* 4 km west of the Büyük Aktepe neck and 490 $\pm$ 600 ka ($\pm 2\sigma$) at site 42 adjacent to this neck, from what may be a late eruption unrelated to the wider flow unit. The weighted mean of the dates from Çapan *et al.* (1987) and site 43 of Tatar *et al.* (2004) is 1600 $\pm$ 206 ka ($\pm 2\sigma$). This error margin overlaps with the age span of the normal-polarity Olduvai subchron. This flow unit would thus seem to be the oldest known in the study region, predating the reverse-magnetized basalts in the Ceylanlı area (see above). We have not found any sites where this basalt is suitable for dating, but geochemical analysis of sample 01TR48 of it, collected just east of the bridge over the Kargılı River, *c.* 2 km SSE of Aktepe, confirms the mafic and primitive magmatic characteristics observed by Polat *et al.* (1997) (Fig. 8).

We inferred earlier that, south of Aktepe, the left-lateral slip seems to be partitioned with *c.* 0.8 mm/a on the Amanos Fault and *c.* 2.2 mm/a on the Güzelce Fault. Taking a nominal age for the Büyük Aktepe neck flow unit of 1.8 Ma, restoring SW–NE-oriented left-lateral slip on the Güzelce Fault would juxtapose localities east of this fault southwestward by *c.* 4 km, placing the southern end of this basalt outcrop *c.* [BA 722 542]. Such a restoration would roughly align the southern end of this basalt with the northern end of other basalt present east of the Amanos Fault around Karaelma (Fig. 3).

According to Atan (1969), the northern limit of this Karaelma basalt east of the Amanos Fault is *c.* [BA 665 535]. This basalt covers a substantial area east and south of this locality, and may also be present farther north and east in the subsurface, beneath young alluvial-fan deposits shed from the Amanos Mountains. Tatar *et al.* (2004) determined normal geomagnetic polarity in the basalt at Karaelma. Notwithstanding the vague location information that they provided, it appears that they studied the Karaelma basalt, rather than the stratigraphically younger Yassı Tepe basalt that crops out south of Karaelma (see above). Their magnetostratigraphic results from this area may thus again indicate the Olduvai subchron, raising the possibility that the Karaelma basalt forms the offset SW part of the larger Büyük Aktepe flow unit, as is tentatively implied by our above slip restoration. These basalts thus offer scope for future refinement of the Amanos Fault kinematics, but further discussion is beyond the scope of this study.

Vertical crustal motions

As already noted, the principal active faults within the Karasu Valley are not pure left-lateral faults; there are also minor components of normal slip, with downthrow to the east on the Amanos and Guzelce faults and to the west on the East Hatay Fault. Overall, the Karasu Valley therefore resembles a graben, it being thus sometimes (rather loosely) called the 'Karasu Rift' (e.g. by Tatar *et al.* 2004; cf. Mart *et al.* 2005). However, recent kinematic models (e.g. Westaway 2004) indicate that any extension on these faults is more than compensated by the distributed shortening within the surrounding mountain ranges (the Amanos Mountains to the west, and the Kurd Dagh or Jabal al-Akrad to the east) that is occurring to accommodate the transpression along the northern DSFZ.

At several localities where basalt is offset (e.g. Küreci, Fig. 5b; Kısık Tepe, Fig. 6a, and Hacılar, Fig. 10), scarps tens of metres high have developed along the Amanos Fault over hundreds of thousands of years, implying vertical slip rates of *c.* 0.1 mm/a, only a few percent of the *c.* 3 mm/a left-lateral slip rate. As indicated in Figure. 10b and its caption, at Hacılar an estimated *c.* 20 m of heave has developed on this *c.* 50°-dipping fault since the basalt eruption at *c.* 160 ka. The horizontal slip rate can thus be estimated as *c.* 0.13 mm/a, and the vertical slip rate as *c.* 0.13 mm/a $\times$ tan(50°) or *c.* 0.15 mm/a. This is less than the vertical slip rate estimate by Yurtmen *et al.* (2002), who inferred that the bluff along the Amanos Fault at Hacılar developed entirely since the youngest basalt eruption. However, our subsequent fieldwork (summarized above and in the caption to Fig. 10) establishes that part of this bluff already existed at *c.* 160 ka, since the basalt cascaded down it. Moreover, a bluff already existed along the line of this fault in this locality when the older basalt beneath the Hacılar basalt (which we correlated above with the Kesmeli Tepe basalt, with an inferred age of *c.* 400–500 ka), since this basalt likewise has a steep local dip to the ESE. It is thus evident that here, and by analogy elsewhere on the Amanos Fault, the vertical component of slip is constantly 'trying' to add to the local relief, but other local processes, such as erosion, deposition by alluvial fans, and the intermittent eruption of basalts, act to progressively remove this relief.

Geological and topographic maps of the Amanos Mountains have been published by Westaway *et al.* (2006*a*); for cross-sections illustrating their structure see Figure 4. These mountains can be regarded as an asymmetrical anticline oriented SSW–NNE (parallel to the Amanos Fault), with a gentle western limb and a steeper eastern limb. This asymmetry means that, typically, the oldest rocks present (of Late Proterozoic and Palaeozoic ages) crop out near the eastern margin of the range. For instance, inliers of the Late Proterozoic Eğribucak Formation occur near Zeytinoba in the north of the present study region (Fig. 12) and near Ceylanlı in the south (Fig. 4b). However, in detail, this mountain range is more complex, due to the combined effects of folding into many smaller-scale anticlines and localized reverse faulting (Figs 4b & 12).

As already noted, the folded structure of the Amanos Mountains was intensively researched in the 1960s (e.g. Atan 1969; Schwan 1971, 1972); this work led to the conclusion that this folding is ancient, probably related to the Late Cretaceous ophiolite obduction, and thus unrelated to the active crustal deformation. Moreover, at the time when this fieldwork was done, mainly in 1966–1968, the Karasu Valley was thought to be a graben; the component of left-lateral slip across it had not yet been recognized. This component was first recognized across the southern DSFZ by Freund *et al.* (1970); the subsequent history of development of ideas regarding the northern DSFZ (which has included setbacks due to the publication of mistaken hypotheses) has been discussed at length by Yurtmen *et al.* (2002) and Westaway (2003, 2004), and so is not repeated here. Furthermore, when the work on the Amanos Mountains was carried out, the EAFZ had not yet been recognized; its 'discovery' is generally attributed to Arpat & Şaroğlu (1972) and the widespread acceptance of its role as the TR–AR plate boundary to McKenzie (1976). Thus, at the time when the structure of the Amanos Mountains was intensively researched, there was not yet even any suggestion that these mountains are located within the linkage of left-lateral faults forming the AF–AR and TR–AR plate boundaries. Indeed, prior to the Westaway (2003, 2004) studies, the possibility of active transpression associated with this left-lateral faulting had not occurred to anyone. The structure of these mountains was instead considered to be a problem that had already been solved, the solution being of no interest to any consideration of the active tectonics.

Given that the Amanos Mountains can now be seen to form part of a zone of active transpression along the northern DSFZ (cf. Westaway 2003, 2004), it is important to consider the possibility that part (as opposed to none) of their structural development relates to this active deformation. This situation contrasts with that farther south along the Lebanon stepover (Fig. 1). As discussed by Walley (1998), it was formerly thought that the structural development of that area relates entirely to transpression along the DSFZ (cf. Westaway 1995), but more recent analysis has concluded that part (as opposed to all) of the structural development relates to this active deformation, and part of it relates to processes in the Late Cretaceous or at other ancient times (cf. Gomez *et al.* 2006). Thus, the Lebanon and Amanos mountain ranges may have experienced similar histories of deformation, with the most recent phase of both relating to the active transpression. As Westaway (2003, 2004) also noted, the Jebel Nusayriyah or Syrian Coastal Range in between (Fig. 1) can now be seen to be another transpressive stepover along the northern DSFZ, but with less-dramatic deformation than the others because it is less-strongly misaligned to the tangential direction to the DSFZ Euler pole (cf. Gomez *et al.* 2006).

The geomorphology of the Amanos Mountains closely reflects their structure. Along most of their length the drainage divide is near the eastern margin of the range, often only *c.* 2–4 km west of the Karasu Valley. For instance, the few roads that cross these mountains tend to cross cols no more than *c.* 2–4 km from the Karasu Valley. A notable example is the motorway linking Adana to Gaziantep, which crosses this mountain range near Bahçe (Fig. 2). Its summit is in the *c.* 1.2 km long Aslanlı tunnel, from which this motorway emerges overlooking the Karasu Valley. This typically asymmetrical drainage divide suggests that local rates of Late Cenozoic surface uplift have been greatest close to the eastern margin of the mountain range, mimicking their overall structure.

The overall topography of the Amanos Mountains is more symmetrical, with the highest summits typically midway between the western and eastern range fronts. In the south, the mountain range is barely 15 km wide and rises no higher than *c.* 1800–1900 m a.s.l. between İskenderun and Kırıkhan; the highest summit altitude decreases even lower, to 1427 m a.s.l., WSW of Kırıkhan beside the southern end of the Amanos Fault. North of Hassa, the mountain range widens to *c.* 30–40 km, with the highest summits >2200 m a.s.l. The highest topography of all is found in the widest part of the Amanos Mountains in the north, NE of Düziçi and SW of Kahramanmaraş, for instance, the mountains Uluziyaret (2268 m a.s.l.) and Beşikdüldülu (2246 m a.s.l.) (Fig. 2). It can thus be presumed that erosion along the eastern margin of the mountain range has been rapid enough to limit the local growth of topography; hence, the exhumation of the oldest rocks in the region. For comparison, the Kurd Dagh range, east of the East Hatay Fault in NW Syria (Fig. 2), typically rises no higher than *c.* 1100 m a.s.l. (Fig. 5b), although in the north, on the Turkish border, it reaches 1275 m a.s.l.

The only river to transect the Amanos Mountains is the Ceyhan, which enters this range west of Kahramanmaraş and leaves it at Cevdetiye (Fig. 2). Its Berke Gorge passes through the uplands north of Beşikdüldülu mountain and is *c.* 1200-m-deep (river *c.* 300 m a.s.l.; land surface adjoining the gorge *c.* 1500 m a.s.l.), one of the most dramatic river gorges in Turkey. It can thus be inferred that much of the surface uplift that formed the Amanos Mountains occurred at a late stage, after this river had become established (following the Middle Miocene emergence of the region above sea-level); presumably, the Ceyhan had sufficient erosional power to incise in pace with the subsequent surface uplift.

Estimates of the uplift rate of the part of the Amanos Mountains that is transected by the River Ceyhan have recently been obtained by us (and will be published in full elsewhere; see Seyrek *et al.* in press) from Ar/Ar dating of basalt flows that cap fluvial terraces. The basalts in this area form the eastern part of the Ceyhan–Osmaniye volcanic field (cf. Bilgin & Ercan 1981; Arger *et al.* 2000; Yurtmen *et al.* 2000; Gürsoy *et al.* 2003). We have obtained six concordant Ar–Ar dates for basalt abutting the Ceyhan Gorge in the western part of the Amanos Mountains, which yield a weighted mean age of 278 $\pm$ 7 ka ($\pm 2\sigma$), suggesting that a single, brief, episode of volcanism occurred in this area. At Pınarözü, beside the Aslantaş Dam (*c.* 25 km north of Osmaniye; Fig. 2), the basalt reaches no lower than *c.* 195 m a.s.l.; the river was locally *c.* 85 m a.s.l. before the dam flooded it, indicating *c.* 110 m of incision, at a time-averaged rate of 0.39 mm/a. About 10 km farther downstream, at Karagedik, the basalt reaches no lower than *c.* 145 m a.s.l.; the river is locally *c.* 70 m a.s.l., indicating *c.* 80 m of incision, at a time-averaged rate of 0.27 mm/a. Roughly 5–10 km farther downstream, at the western margin of the Amanos Mountains around Sarpınağzı and Cevdetiye (Fig. 2), a well-developed Ceyhan terrace is observed *c.* 35 m above river level. Assuming that these terrace deposits accumulated in MIS 6 (*c.* 140 ka), an uplift rate of 0.25 mm/a is indicated. Upstream extrapolation of this variation in uplift rates would imply a rate of at least *c.* 0.4 mm/a along the Berke Gorge in the core of the Amanos Mountains. If extrapolated back in time, this would predict that the incision of this gorge began around *c.* 3 Ma, roughly when the modern geometry of strike-slip faulting in this region is thought to have developed (cf. Westaway *et al.* 2006*a*).

Quaternary regional surface uplift is widely observed elsewhere in Turkey, away from active fault zones (e.g. Demir *et al.* 2004). However, in the Arabian Platform and along the Mediterranean coastline, its rate seldom exceeds *c.* 0.1 mm/a; the much faster uplift now evident in the Amanos Mountains requires a different explanation. We suggest that this vertical crustal motion, as evidenced by the vertical slip on the Amanos Fault and the fluvial geomorphology (including the dating of Ceyhan terraces, above), is being caused primarily by the component of crustal thickening required to balance the crustal shortening that is required by the transpressive geometry of this fault. Supporting evidence for this view is provided by the orientation of fold axes within the Amanos Mountains, which (as in Fig. 12) are typically subparallel to the Amanos Fault, making them optimally oriented to accommodate fault-normal crustal shortening (cf. Westaway 1995). The asymmetry of the Amanos Mountains, revealed by the asymmetrical drainage and folding (see above) thus implies that the strain rates for distributed crustal shortening and thickening typically increase eastward to maxima close to the Amanos Fault. However, the characteristic eastward downwarping of the structure in close proximity to this fault, evident in Figure 4, means that vertical slip rates on the Amanos Fault will underestimate the peak uplift rates in the interior of the Amanos Mountains. Thus (for instance) the peak uplift rate in the Amanos Mountains west of Hacılar may be several times greater than the local *c.* 0.15 mm/a vertical slip rate on this fault.

Estimation of the total Late Cenozoic surface uplift in the Amanos Mountains is difficult, due to the typical lack of suitable outcrop evidence; as Figs 4 & 12 show, most outcrop is far too old to be used in such a calculation. Probably the best evidence is provided by outcrop, flanking the eastern margin of the southern Amanos Mountains (*c.* 7 km WNW of Kırıkhan; *c.* [BA 565 448]) of Middle Miocene (*c.* 15 Ma) reefal limestone, of the Kepez Formation, which is now *c.* 950 m a.s.l. (cf. Yurtmen *et al.* 2002). This requires a time-averaged uplift rate of *c.* 0.06 mm/a; but, if the local uplift is assumed to have been concentrated after the initiation of the modern geometry of strike-slip faulting (*c.* 3.7 Ma; Westaway *et al.* 2006*a*), then the subsequent local rate of surface uplift may have reached 0.25 mm/a, comparable with the evidence from the Ceyhan terraces.

As Westaway (1995) noted, the spatial average strain rate E_s for the distributed shortening along a transpressive stepover (with deformation partitioned symmetrically on both sides) is:

$$E_s = \frac{V \sin(\theta)}{2H} \qquad (1)$$

where V is the local relative plate velocity and H is the width of the mountain range forming the stepover. Using equation (1), with $V = 8$ mm/a (Fig. 13), $H = 30$ km and $\theta = 52°$, we estimate the spatial average strain rate for crustal shortening in the Amanos Mountains as *c.* 0.1 Ma^{-1}. The shortening factor, if deformation at this rate persists for time t, is $\exp(-ht)$ and is thus *c.* 1.5 if this deformation has persisted since *c.* 3.7 Ma. A predicted shortening strain of this magnitude seems roughly consistent with the evidence (Figs 4 & 12).

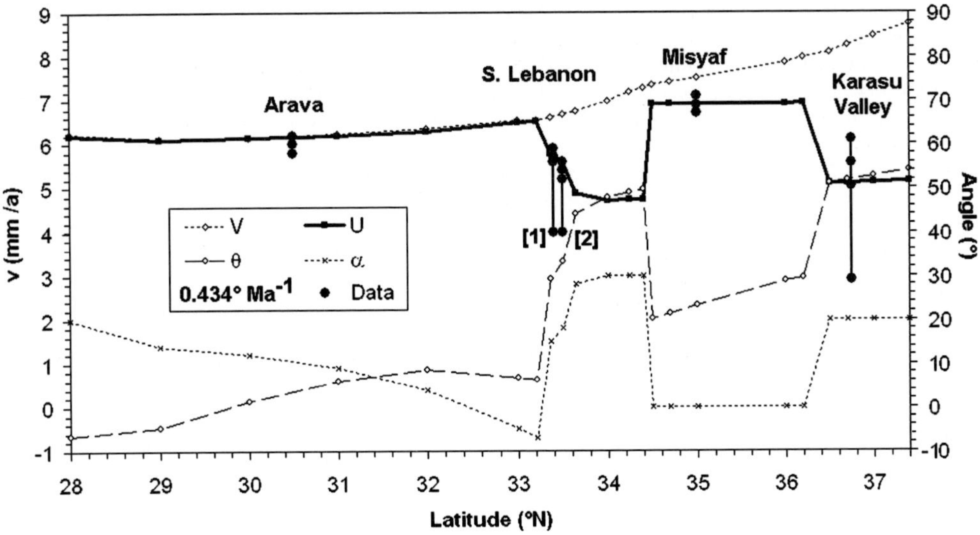

Fig. 13. Summary of the geometry of the Dead Sea fault zone, between the Tiran Strait and Türkoğlu at the northern end of the Amanos Fault. At each point, *V* is the rate of relative motion between the African and Arabian plates, calculated using spherical trigonometry (including the effect of Earth ellipticity) from the position of the point relative to the assumed Euler pole at 31.1°N, 26.7°E, given the preferred 0.434°/Ma rate of relative rotation. α is the observed strike of the principal DSFZ strand, measured clockwise from north. θ is the difference in angle between α and the tangential direction to the Euler pole. *U* is the maximum possible rate of left-lateral slip on this fault strand, calculated as $V \times \cos(\theta)$, where *V* is the local rate of relative plate motion. The data used for comparison with predictions of this kinematic model are discussed in the text. For southern Lebanon, analysis 1 uses the overall time-averaged slip rate for the Serghaya Fault, estimated in the text, whereas analysis 2 uses the Holocene slip rate from Gomez *et al.* (2003). These two estimates have been shown at separate localities, to avoid clutter, although in reality they apply to the same part of the Serghaya Fault.

Neglecting any effects of advection of the thermal boundary at the base of the brittle layer relative to the rock column, or of other thermal effects such as those resulting from erosion, the rate of thickening of the brittle upper crust can be crudely estimated by multiplying the strain rate by the thickness of this brittle layer, which can be estimated as *c.* 15 km. One thus obtains *c.* 1.5 mm/a.

The Amanos mountain range is so narrow that (like the Lebanon Mountains; cf. Khair *et al.* 1993) it is unlikely to be isostatically supported by downward deflection of the underlying mantle lithosphere. The local isostatic balance is probably determined instead by the extent to which local processes (erosion and crustal thickening) can maintain the base of the brittle layer at a different depth beneath this mountain range than beneath its surroundings, given the tendency for such variations to be dynamically removed by diffusion of heat within the crust. Experience of numerical modelling of such effects elsewhere (e.g. Westaway 2002; Westaway *et al.* 2004, 2006*b*) indicates that the surface uplift rate will be only a small proportion of the associated thickening rate of the brittle layer; thus, the observed uplift rates in the Amanos Mountains of a few tenths of 1 mm/a can be tentatively explained. However, formal numerical modelling of this effect is beyond the scope of this study, but will be discussed elsewhere; see Seyrek *et al.*, in press. In the meantime, our preliminary analysis suggests the strong possibility that the observed surface uplift, topography and much of the shortening strain expressed in the structure of the Amanos Mountains and surroundings, have developed during the present phase of strike-slip faulting on the DSFZ, as consequences of the local transpression. We thus see no basis to infer that any left-lateral faulting occurred in the vicinity of the Karasu Valley before *c.* 3.7 Ma; before this time, the northern DSFZ was presumably located farther east (e.g. on the Afrin Fault; Fig. 1), and led into the associated array of faults depicted in the Gaziantep region of SE Turkey in Figure 1.

Discussion

Recurrence of large earthquakes in the Karasu Valley

The magnitude estimate of 7.5 for the 1822 earthquake, from Ambraseys & Jackson (1998), translates to a seismic moment of *c*. 2×10^{20} N m (after Kanamori & Anderson 1975). Taking the length of faulting as 200 km (after Ambraseys & Jackson 1998), the thickness of the brittle upper crust as 15 km, and its shear modulus as 30 GPa, *c*. 2.2 m of coseismic slip can be estimated. This earthquake can be presumed to have de-stressed the whole region, so will not recur until sufficient strain has accumulated between the adjoining plates to match this coseismic slip. Dividing the 2.2 m of coseismic slip by our 5.57 ± 0.54 mm/a ($\pm 2\sigma$) overall slip rate gives an estimated recurrence interval of 395 ± 38 ($\pm 2\sigma$) years. Thus, even though we are now predicting significantly faster overall left-lateral slip across the Karasu Valley than before, no such large earthquake is expected to recur on this part of the DSFZ until the late 22nd century, at the earliest, although (of course) a smaller event may occur sooner.

Comparison with other transpressive DSFZ stepovers

Our analyses of vertical and horizontal crustal motions in and around the Karasu Valley support the suggestion (by Westaway 2003, 2004) that this DSFZ segment is a transpressive stepover. The region thus warrants comparison with the two other significant transpressive stepovers on the DSFZ, formed by the Lebanon/Anti-Lebanon Mountains and the Jabal Nusayriyah (Fig. 1).

Westaway (2003, 2004) determined the DSFZ Euler pole at 31.1°N, 26.7°E. The tangential direction to it, measured at Hacılar (36.75°N, 36.45°E), is N32°W. The Amanos Fault strikes N20°E; thus, it is misaligned by an angle, θ, of 52° in the transpressive sense. Westaway (2004) likewise determined θ for the Lebanon stepover as 48–50°, given the local N30°E fault strike and the N18°W–N20°W tangential direction to the pole. For the Jabal Nusayriyah stepover (between Tell Kalakh and Jisr esh-Shugur; Fig. 1), θ increases northward from 20 to 30°, given the north–south strike of the faulting and the N20°W–N30°W tangential direction to the pole.

The Lebanon and Amanos stepovers thus have similar geometries, with similar values of θ. Both also consist of a central valley flanked by mountain ranges, with the principal active left-lateral faults running along the valley margins. Thus, the Bekaa Valley is analogous to the Karasu Valley, the Lebanon Mountains to the Amanos Mountains, and the Anti-Lebanon Mountains to the Kurd Dagh. The Yammouneh Fault, which typically follows the west side of the Bekaa Valley, is thus analogous to the Amanos Fault; and the Serghaya Fault, which for much of its length is similarly located relative to the eastern margin of this valley, is thus analogous to the East Hatay Fault.

Structural cross-sections through both the Amanos and Lebanon mountain ranges look similar, as can be seen by comparing our Figure 4 with, for instance, Figure 4 of Westaway (2004) or Figure 4 of Gomez *et al.* (2006). In terms of equation (1), H is roughly the same for the Amanos and Lebanon stepovers, and V is less for the latter, because it is nearer the DSFZ Euler pole. One thus expects somewhat higher E_s in the Amanos Mountains than in the Lebanon Mountains. However, the Lebanon Mountains rise much higher than the Amanos Mountains: their highest summit (Qurnat as-Sawda) reaches 3087 m a.s.l.; the highest point in the Anti-Lebanon Mountains (Jabal ash-Shaykh or Mount Hermon) reaches 2814 m a.s.l. These high summits are near the range fronts bounding the major strike-slip faults. They are thus not analogous to the highest summits in the Amanos Mountains; as noted above, erosion has prevented any really high topography from developing along the eastern margin of the Amanos Mountains adjacent to the Amanos Fault. The lesser degree of erosion in the Lebanon Mountains is indicated by the absence of outcrops older than Mesozoic, there being no Palaeozoic or Precambrian inliers (the limited Late Cenozoic erosion of these mountains was also noted by Walley 1998).

In comparison, the Jabal Nusayriyah stepover is a more subdued edifice, reflecting the much lower range of θ. The topography increases northward along it, reflecting the northward increase in θ and E_s. The highest topography is indeed at its northern end, rising to 939 m a.s.l. east of the DSFZ in the Jabal Jubb Sulayman, and to 1562 m west of the DSFZ in the Jabal Nusayriyah (Fig. 1). Like the Amanos stepover, the Jabal Nusayriyah stepover is thus asymmetrical, with higher mountains on its western side. However, only its northern part contains a broad linear valley bounded by left-lateral faults on both sides, the Ghab Basin; in the south, through Misyaf (Fig. 1) the DSFZ consists of only a single active fault, delineating a narrow linear valley. This difference in morphology may relate to the northward increase in E_s that is required by its geometry.

Consistency between local slip rates and regional kinematics

Setting aside localities with multiple en échelon fault strands (e.g. where the Guzelce Fault is present), we have obtained four estimates (all $\pm 2\sigma$) of the slip rate on the Amanos Fault from offset Pleistocene basalt flows: 2.89 ± 0.14 mm/a at Kesmeli Tepe/Hassa; 2.84 $\pm$ 0.41 mm/a at Hacılar; 2.87 $\pm$ 0.05 mm/a at Kısık Tepe/Kocaören; and 3.25 $\pm$ 0.22 mm/a at Küreci. The weighted mean is 2.87 $\pm$ 0.05 mm/a for the first three, which are tightly grouped, and 2.89 $\pm$ 0.05 mm/a ($\pm 2\sigma$) for all four, indicating total slip since 3.73 $\pm$ 0.05 Ma of 10.8 $\pm$ 0.2 km ($\pm 2\sigma$). Westaway (2004) estimated the total slip on the East Hatay Fault as *c.* 10 km from piercing points derived from offset outcrops of the Hatay ophiolite ('K' in Fig. 2). These offsets were measured from maps, since it has proved impossible to carry out fieldwork on the Turkey–Syria border where this fault and its piercing points are located, and so cannot be checked directly. We thus assign a nominal margin of uncertainty of 1 km, making this slip estimate 10 $\pm$ 1 km, giving a time-averaged slip rate since 3.73 $\pm$ 0.05 Ma of 2.68 $\pm$ 0.54 mm/a ($\pm 2\sigma$). Summing these slip rate estimates for the Amanos and East Hatay faults gives 5.57 $\pm$ 0.54 mm/a ($\pm 2\sigma$).

We now combine this data-set with three others from farther south to assess the overall kinematics of the DSFZ (cf. Westaway 2004). First, Ginat *et al.* (1998) estimated *c.* 15 km of left-lateral slip (which we take as 15 $\pm$ 1 km) since *c.* 2.5 Ma in the Arava Valley in southern Israel (*c.* 30.5°N), indicating a time-averaged slip rate of 6.0 $\pm$ 0.2 mm/a. Second, we combine Westaway's (2004) analysis of the offset of the Litani River in southern Lebanon (*c.* 33.4°N) by the Yammouneh Fault, which yielded a *c.* 4.0 mm/a slip rate estimate for the Middle–Late Pleistocene, with the analysis by Gomez *et al.* (2003) of the Serghaya Fault at *c.* 33.6°N that indicated a Holocene slip rate of 1.4 $\pm$ 0.2 mm/a. We thus estimate a 5.4 $\pm$ 0.2 mm/a overall slip rate for both faults at *c.* 33.4°N. Gomez *et al.* (2006) have subsequently reported the total slip on the part of the Serghaya Fault between *c.* 33.5 and *c.* 33.9°N as 6–7 km from drainage offsets in the Anti-Lebanon Mountains. They estimated that this slip post-dates the development of an erosion surface of estimated Messinian age (*c.* 7–5.5 Ma), indicating a time-averaged slip rate of *c.* 1 mm/a. However, if the present geometry of faulting in this area developed at *c.* 3.7 Ma, as is now suggested for the northern DSFZ, this estimate of 6.5 $\pm$ 0.5 km of slip indicates a time-averaged rate of 1.74 $\pm$ 0.14 mm/a. Finally, Meghraoui *et al.* (2003) estimated the slip rate of the DSFZ strand at Misyaf in western Syria (*c.* 35.0°N), from trenching and radiocarbon dating of the slip in great earthquakes over the past *c.* 2 ka, as 6.9 $\pm$ 0.2 mm/a ($\pm 2\sigma$).

Theory by Westaway (1995) shows that the slip rate U for a strike-slip fault within a stepover can be no greater than $V \times \cos(\theta)$. We assume in the following analysis that all faults investigated are slipping at this maximum rate permitted by their geometry, given their misalignment relative to the tangential direction to the AF–AR Euler pole.

As Westaway (2004) noted, the Meghraoui *et al.* (2003) data-set is consistent with a rate of relative rotation of 0.434 $\pm$ 0.012°/Ma about the preferred DSFZ Euler pole. As Figure 13 shows, this Euler vector is also consistent with the data from Israel and Lebanon, and with the lower bound to our estimate for the overall left-lateral slip rate across the Karasu Valley. We thus conclude that there is now no basis to suppose that the northern DSFZ, at the latitude of the Karasu Valley, consists of any other faults with significant slip rates (cf. Westaway 2004); we thus infer (for instance) that the Afrin Fault, farther east (Fig. 1), represents an earlier phase of DSFZ activity, preceding *c.* 3.7 Ma.

However, in detail, our Karasu Valley data-set predicts faster slip than is expected from the regional kinematic model (i.e. the opposite problem to the Westaway 2004 analysis). Greater consistency would result if the total slip on the East Hatay Fault were adjusted to 9 $\pm$ 1 km, making its slip rate 2.41 $\pm$ 0.54 mm/a ($\pm 2\sigma$) and the overall slip rate 5.30 $\pm$ 0.54 mm/a ($\pm 2\sigma$) (Fig. 13). Although further work is thus clearly needed to tie down the precise slip rates, the basis for overall kinematic consistency is now established.

Evidence pertaining to the history and rate of slip on the northern DSFZ is also provided by the left-lateral offset of the Homs basalt in western Syria (cf. Westaway 2003, 2004). Eruption of this basalt began at *c.* 8 Ma but was concentrated during *c.* 6.5-5 Ma (see Westaway 2003, 2004, for detailed maps and syntheses of the dating evidence). Estimates of the left-lateral slip since this basalt eruption have ranged from <10 km to *c.* 20 km (see compilation by Westaway 2003), depending on choices of piercing points. The situation is complicated because, as Westaway (2004) noted, the linear valley along the DSFZ already existed at the time of this volcanism, with basalts flowing into it from both sides, making it impossible to match individual flows across the fault zone. Chorowicz *et al.* (2005) and Gomez *et al.* (2006) have argued that the western end of the main volcanic edifice, east of the DSFZ north of Shin (Fig. 1), can be uniquely matched against the basalt west of the DSFZ in the vicinity of the crusader castle of Crac des Chevaliers (Qalat al-Hisn), giving a left-lateral offset of 20 km. Dividing this by a nominal 6 Ma age for the basalt, Chorowicz *et al.* (2005) estimated a subsequent slip rate of 3.3 mm/a, much less than the 6.9 mm/a prediction for this locality (*c.* 34.7°N) from the present kinematic model (Fig. 13).

Chorowicz *et al.* (2005) did not support their interpretation with proper evidence, such as matching the dates or geochemistry of the basalts at their candidate piercing points; their interpretation instead matched the topography. However, since the transpression in this area is asymmetrical (see above, also Westaway 2003, 2004), being stronger west of the DSFZ, much of the present topography has developed since the basalt eruption, at different rates on opposite sides of the DSFZ; this present topography should thus not be assumed to match that which existed during the basalt eruption. This proposal for piercing points seems unlikely, as available maps (e.g. those by Westaway 2003, 2004) and simple field inspection indicate that the basalt around Crac des Chevaliers is much thinner than that in its supposed counterpart east of the DSFZ.

Westaway (2003) noted that the basalt west of the DSFZ appears thickest near the Syria–Lebanon border, *c.* 15 km south of Crac des Chevaliers. If this point is matched against the Chorowicz *et al.* (2005) piercing point east of the DSFZ, then the left-lateral slip since the basalt eruption has been *c.* 35 km, much higher than any previous estimate. Our kinematic model (Fig. 13) predicts *c.* 26 km of left-lateral slip in this area since the present slip phase began at *c.* 3.7 Ma (assuming a constant slip rate). If the remaining *c.* 9 km of post-basalt slip occurred between *c.* 5 Ma and *c.* 3.7 Ma, then the contemporaneous rate would have been *c.* 6.9 mm/a, as at present. However, if this slip occurred between *c.* 6.5 Ma and *c.* 3.7 Ma, the rate would have been only *c.* 3.2 mm/a, consistent with the view (Westaway 2003, 2004) that prior to the present slip phase much of the AF–AR relative motion was taken up on structures east of the modern northern DSFZ (Fig. 1).

Comparison between local slip rates and predictions from the regional kinematics is also possible farther north, near Jisr esh-Shugur (Fig. 1). The DSFZ locally consists of subparallel active north–south-striking left-lateral faults across a *c.* 20-km-wide zone; this complexity is represented (from west to east in Fig. 1) by three subparallel faults, which Westaway (2003, 2004) called the Qanaya–Babatorun, Salqin, and Armanaz faults (Fig. 2). Our kinematic model (Fig. 13) predicts 6.9 mm/a of left-lateral slip in this area (*c.* 35.75°N), again indicating *c.* 35 km of total slip if this rate has persisted since *c.* 3.7 Ma. As Westaway (2003) noted, sedimentation in the adjacent Ghab Basin (Fig. 1), interpreted as a pull-apart basin at a leftward step between the Misyaf Fault and the Qanaya–Babatorun Fault (Fig. 2), began in the Pliocene, consistent with Pliocene initiation of the present geometry of faulting in this region.

The southern limit of lower Pliocene marine sediment is offset left-laterally across the Qanaya–Babatorun Fault by *c.* 10 km (Westaway 2003, 2004; 'L' in Fig. 2). Taking this as the total slip on this fault, with an age of *c.* 3.7 Ma, gives a time-averaged slip rate of *c.* 2.7 mm/a. East of Jisr esh-Shugur ('M' in Fig. 2), the southern end of the Salqin Fault offsets Early Pleistocene basalt (dated to *c.* 1.1–1.3 Ma; Kopp *et al.* 1999) by *c.* 1.1 km, giving a slip rate of *c.* 1.2–1.4 mm/a (see Mart *et al.* 2005, for a local map). Subtracting these rates from our model prediction (Fig. 13) suggests that *c.* 3 mm/a of left-lateral slip is taken up on the Armanaz Fault and any other subparallel faults in this area. The low left-lateral slip rate on the Salqin Fault was used by Mart *et al.* (2005) to argue against significant left-lateral slip on any part of the DSFZ (these authors have indeed claimed that no more than *c.* 20 km of left-lateral slip has occurred anywhere on the DSFZ). The fact now evident, i.e. that the modern geometry of the northernmost DSFZ has taken up only *c.* 20 km of left-lateral slip, reflects its young (Pliocene) age, and does not invalidate the generally accepted view (e.g. Freund *et al.* 1970; Garfunkel 1981) that >100 km of left-lateral slip has occurred on the southern DSFZ, which became active in the Middle Miocene.

Conclusions

We have reported four new Ar/Ar dates and 18 new geochemical analyses of Pleistocene basalts, from the Karasu Valley of southern Turkey, which have become offset left-laterally by slip on the N20°E-striking Amanos Fault. The geochemical analyses help to correlate some of the less-obvious offset fragments of basalt flows and thus to measure amounts of slip; the dates enable slip rates to be calculated. On the basis of four individual slip-rate determinations, obtained in this manner, we estimate a weighted mean slip rate for this fault of 2.89 ± 0.05 mm/a ($\pm 2\sigma$). We have also obtained a slip rate of 2.68 ± 0.54 mm/a ($\pm 2\sigma$) for the subparallel East Hatay Fault farther east. Summing these values gives 5.57 ± 0.54 mm/a ($\pm 2\sigma$) as the overall left-lateral slip rate across the DSFZ in the Karasu Valley. These slip-rate estimates and other evidence from farther south on the DSFZ are consistent with a preferred Euler vector for the relative rotation of the Arabian and African plates of $0.434 \pm 0.012°$ Ma^{-1} about 31.1°N, 26.7°E. The Amanos Fault is misaligned to the tangential direction to this pole by 52° in the transpressive sense. Its geometry thus requires significant fault-normal distributed crustal shortening, taken up by crustal thickening and folding, in the adjacent

Amanos Mountains. The vertical component of slip on the Amanos Fault is estimated as *c.* 0.15 mm/a. This minor component contributes to the uplift of the Amanos Mountains, which reaches rates of *c.* 0.2–0.4 mm/a. These slip-rate estimates are considered representative of times since 3.73 ± 0.05 Ma, when the modern geometry of strike-slip faulting developed in this region; an estimated 11 km of slip on the Amanos Fault and *c.* 10 km of slip on the East Hatay Fault have occurred since then. It is inferred that both these faults came into being, and the associated deformation in the Amanos Mountains began, at that time; before this the northern part of the Africa–Arabia plate boundary was located farther east.

This work was supported in part by HÜBAK (Harran University Scientific Research Council) grant number 428 (A.S.). This study contributes to IGCP 518 'Fluvial sequences as evidence for landscape and climatic evolution in the Late Cenozoic'. W. Olszewski helped with documentation of the Ar/Ar dating results, which are available in full online at http://www.geolsoc.org.uk/SUP18274. A hard copy can be obtained from the Geological Society Library.

References

AMBRASEYS, N. N. 1989. Temporary seismic quiescence: SE Turkey. *Geophysical Journal*, **96**, 311–331.

AMBRASEYS, N. N. & JACKSON, J. A. 1998. Faulting associated with historical and recent earthquakes in the Eastern Mediterranean region. *Geophysical Journal International*, **133**, 390–406.

ARGER, J., MITCHELL, J. & WESTAWAY, R. 2000. Neogene and Quaternary volcanism of south-eastern Turkey. *In:* BOZKURT, E., WINCHESTER, J. A. & PIPER, J. D. A. (eds) *Tectonics and Magmatism of Turkey and the Surrounding Area.* Geological Society, London, Special Publications, **173**, 459–487.

ARPAT, E. & ŞAROĞLU, F. 1972. Doğu Anadolu Fayı ile ilgili bazı gözlem ve dü şünceler. *MTA Bülteni*, **73**, 1–9.

ATAN, O. 1969. *Eğribucak–Karacören (Hassa)–Ceylânli–Dazevleri (Kırıkhan) Arasındaki Amanos Dağlarının Jeolojisi.* Report **139**, General Directorate of Mineral Research and Exploration, Ankara, 85pp.

BILGIN, A. Z. & ERCAN, T. 1981. Petrology of the Quaternary basalts of the Ceyhan–Osmaniye area. *Bulletin of the Geological Society of Turkey*, **24**, 21–30 (in Turkish with English summary).

BRINKMANN, R. 1976. *Geology of Turkey.* Elsevier, Amsterdam, 158pp.

ÇAPAN, U. Z., VIDAL, P. & CANTAGREL, J. M. 1987. K–Ar, Sr and Pb isotopic study of Quaternary volcanism in Karasu valley (Hatay), N-end of the Dead-Sea rift zone in SE-Turkey. *Yerbilimleri (Bulletin of the Earth Sciences Application and Research Centre of Hacettepe University)*, **14**, 165–178.

CHOROWICZ, J., DHONT, D., AMMAR, O., RUKIEH, M. & BILAL, A. 2005. Tectonics of the Pliocene Homs Basalts (Syria) and implications for the Dead Sea fault zone activity. *Journal of the Geological Society, London*, **162**, 259–271.

DEAN, W. T. & MONOD, O. 1985. A new interpretation of Ordovician stratigraphy in the Bahçe area, northern Amanos Mountains, south central Turkey. *Geological Magazine*, **122**, 15–25.

DEMIR, T., YEŞILNACAR, İ & WESTAWAY, R. 2004. River terrace sequences in Turkey: sources of evidence for lateral variations in regional uplift. *Proceedings of the Geologists' Association*, **115**, 289–311.

DILEK, Y. & DELALOYE, M. 1992. Structure of the Kizildağ ophiolite, a slow-spread Cretaceous ridge segment north of the Arabian Promontory. *Geology*, **20**, 19–22.

DSİ 1975. *Asi havzası hidrojeolojik etüt raporu.* Devlet Su İşleri Genel Müdürlüğü [General Directorate of State Water Works], Adana, 53pp.

FREUND, R., GARFUNKEL, Z., ZAK, I., GOLDBERG, M., WEISSBROD, T. & DERIN, B. 1970. The shear along the Dead Sea rift. *Philosophical Transactions of the Royal Society, London, Series A*, **267**, 107–130.

GARFUNKEL, Z. 1981. Internal structure of the Dead Sea leaky transform (rift) in relation to plate kinematics. *Tectonophysics*, **80**, 81–108.

GINAT, H., ENZEL, Y. & AVNI, Y. 1998. Translocated Plio-Pleistocene drainage systems along the Arava Fault of the Dead Sea Transform. *Tectonophysics*, **284**, 151–160.

GOMEZ, F., KHAWLIE, M., TABET, C., DARKAL, A. N., KHAIR, K. & BARAZANGI, M. 2006. Late Cenozoic uplift along the northern Dead Sea transform in Syria and Lebanon. *Earth and Planetary Science Letters*, **241**, 913–931.

GOMEZ, F., MEGHRAOUI, M. *ET AL.* 2003. Holocene faulting and earthquake recurrence along the Serghaya branch of the Dead Sea fault system in Syria and Lebanon. *Geophysical Journal International*, **153**, 658–674.

GÜRSOY, H., TATAR, O., PIPER, J. D. A., HEIMANN, A. & MESCI, L. 2003. Neotectonic deformation linking the East Anatolian and Karataş–Osmaniye intracontinental transform fault zones in the Gulf of İskenderun, southern Turkey, deduced from paleomagnetic study of the Ceyhan–Osmaniye volcanics. *Tectonics*, **22**, 1067, doi: 10.1029/2003TC001524, 13pp.

HARFORD, C. L., PRINGLE, M. S., SPARKS, R. S. J. & YOUNG, S. R. 2002. The volcanic evolution of Montserrat using $^{40}Ar/^{39}Ar$ geochronology. *In:* DRUITT, T. H. & KOJKELAAR, B. P. (eds) *The Eruption of the Soufrière Hills Volcano, Montserrat, from 1995 to 1999.* Geological Society, London, Memoirs, **21**, 93–113.

KANAMORI, H. & ANDERSON, D. L. 1975. Theoretical basis of some empirical relations in seismology. *Bulletin of the Seismological Society of America*, **65**, 1073–1095.

KHAIR, K., KHAWLIE, M., HADDAD, F., BARAZANGI, M., SEBER, D. & CHAIMOV, T. 1993. Bouguer gravity and crustal structure of the Dead Sea transform fault and adjacent mountain belts in Lebanon. *Geology*, **21**, 739–742.

KOPP, M. P., ADZHAMYAN, Z., IL'YAS, K., FAKIANI, F. & KHAFEZ, A. 1999. Mechanism of formation of the El Ghab wrench graben (Syria) and the Levant transform fault propagation. *Geotectonics*, **33**, 408–422.

LE BAS, M. J., LE MAITRE, R. W., STRECKEISEN, A. & ZANETTIN, B. 1986. A chemical classification of volcanic rocks based on the total alkali–silica diagram. *Journal of Petrology*, **27**, 745–750.

MCCLUSKY, S., BALASSANIAN, S. *ET AL.* 2000. Global Positioning System constraints on plate kinematics and dynamics in the eastern Mediterranean and Caucasus. *Journal of Geophysical Research*, **105**, 5695–5719.

MCKENZIE, D. P. 1976. The East Anatolian Fault: a major structure in eastern Turkey. *Earth and Planetary Science Letters*, **29**, 189–193.

MART, Y., RYAN, W. B. F. & LUNINA, O. V. 2005. Review of the tectonics of the Levant rift system; the structural significance of oblique continental breakup. *Tectonophysics*, **395**, 209–232.

MEGHRAOUI, M., GOMEZ, F. *ET AL.* 2003. Evidence for 830 years of seismic quiescence from palaeoseismology, archaeoseismology and historical seismicity along the Dead Sea fault in Syria. *Earth and Planetary Science Letters*, **210**, 35–52.

MIDDLEMOST, E. A. K. (1989). Iron oxidation ratios, norms, and the classification of volcanic rocks. *Chemical Geology*, **77**, 19–26.

PARLAK, O., KOP, A., ÜNLÜGENÇ, U. C. & DEMIRKOL, C. 1998. Geochronology and geochemistry of basaltic rocks in the Karasu Graben around Kırıkhan (Hatay), S. Turkey. *Turkish Journal of Earth Sciences*, **7**, 53–61.

PEARCE, J. A., BENDER, J. F. *ET AL.* 1990. Genesis of collision volcanism in eastern Anatolia, Turkey. *Journal of Volcanology and Geothermal Research*, **44**, 189–229.

POLAT, A., KERRICH, R. & CASEY, J. F. 1997. Geochemistry of Quaternary basalts erupted along the east Anatolian and Dead Sea fault zones of southern Turkey: implications for mantle sources. *Lithos*, **40**, 55–68.

RENNE, P. R., SWISHER, C. C., DEINO, A. L., KARNER, D. B., OWENS, T. L. & DEPAOLO, D. J. 1998. Intercalibration of standards, absolute ages and uncertainties in $^{40}Ar/^{39}Ar$ dating. *Chemical Geology*, **145**, 117–152.

ROJAY, B., HEIMANN, A. & TOPRAK, V. 2001. Neotectonic and volcanic characteristics of the Karasu fault zone (Anatolia, Turkey): the transition zone between the Dead Sea transform and the East Anatolian fault zone. *Geodinamica Acta*, **14**, 197–212.

SCHWAN, W. 1971. Geology and tectonics of the central Amanos Mountains. *In:* CAMPBELL, A. S. (ed.) *Geology and History of Turkey*. Petroleum Exploration Society of Libya, Tripoli, 283–303.

SCHWAN, W. 1972. Ergebnisse neuer geologischer Forschung en im Amanosgebirge (Süd-Türkei). *Geotektonische Forschungen* [Stuttgart], **42**, 130–160.

SEYREK, A., DEMIR, T., PRINGLE, M., YURTMEN, S., WESTAWAY, R., BRIDGLAND, D., BECK, A. & ROWBOTHAM, G. Late Cenozoic uplift of the Amanos Mountains and incision of the Middle Ceyhan river gorge, southern Turkey; Ar-Ar dating of the Düziçi basalt. *Geomorphology*, in press.

STEIGER, R. H. & JÄGER, E. 1977. Convention on the use of decay constants in geo- and cosmochronology. *Earth and Planetary Science Letters*, **36**, 359–363.

TATAR, O., PIPER, J. D. A., GÜRSOY, H., HEIMANN, A. & KOÇBULUT, F. 2004. Neotectonic deformation in the transition zone between the Dead Sea Transform and the East Anatolian Fault Zone, southern Turkey: a palaeomagnetic study of the Karasu Rift volcanism. *Tectonophysics*, **385**, 17–43.

THOMPSON, R. N., LEAT, P. T., DICKIN, A. P., MORRISON, M. A., HENDRY, G. L. & GIBSON, S. A. 1990. Strongly potassic mafic magmas from lithospheric mantle sources during continental extension and heating; evidence from Miocene minettes of northwest Colorado, U.S.A. *Earth and Planetary Science Letters*, **98**, 139–153.

TOLUN, N. & PAMIR, H. N. 1975. *Explanatory Booklet for the Hatay Sheet of the Geological Map of Turkey, 1:500,000 Scale*. General Directorate of Mineral Research and Exploration, Ankara, 99 pp.

WALLEY, C. D. 1998. Some outstanding issues in the geology of Lebanon and their importance in the tectonic evolution of the Levantine region. *Tectonophysics*, **298**, 37–62.

WESTAWAY, R. 1995. Deformation around stepovers in strike-slip fault zones. *Journal of Structural Geology*, **17**, 831–847.

WESTAWAY, R. 2002. The Quaternary evolution of the Gulf of Corinth, central Greece: coupling between surface processes and flow in the lower continental crust. *Tectonophysics*, **348**, 269–318.

WESTAWAY, R. 2003. Kinematics of the Middle East and Eastern Mediterranean updated. *Turkish Journal of Earth Sciences*, **12**, 5–46.

WESTAWAY, R. 2004. Kinematic consistency between the Dead Sea Fault Zone and the Neogene and Quaternary left-lateral faulting in SE Turkey. *Tectonophysics*, **391**, 203–237.

WESTAWAY, R., DEMIR, T., SEYREK, A. & BECK, A. 2006*a*. Kinematics of active left-lateral faulting in southeast Turkey from offset Pleistocene river gorges: improved constraint on the rate and history of relative motion between the Turkish and Arabian plates. *Journal of the Geological Society, London*, **163**, 149–164.

WESTAWAY, R., GUILLOU, H. *ET AL.* 2006*b*. Late Cenozoic uplift of western Turkey: improved dating of the Kula Quaternary volcanic field and numerical modelling of the Gediz river terrace staircase. *Global and Planetary Change*, **51**, 131–171.

WESTAWAY, R., PRINGLE, M., YURTMEN, S., DEMIR, T., BRIDGLAND, D., ROWBOTHAM, G. & MADDY, D. 2004. Pliocene and Quaternary regional uplift in western Turkey: the Gediz river terrace staircase and the volcanism at Kula. *Tectonophysics*, **391**, 121–169.

YURTMEN, S., GUILLOU, H., WESTAWAY, R., ROWBOTHAM, G. & TATAR, O. 2002. Rate of strike-slip motion on the Amanos Fault (Karasu Valley, southern Turkey) constrained by K–Ar dating and geochemical analysis of Quaternary basalts. *Tectonophysics*, **344**, 207–246.

YURTMEN, S., ROWBOTHAM, G., İŞLER, F. & FLOYD, P. 2000. Petrogenesis of Quaternary alkali volcanics, Ceyhan – Turkey. *In:* BOZKURT, E., WINCHESTER, J. A. & PIPER, J. D. A. (eds) *Tectonics and Magmatism of Turkey and the Surrounding Area*. Geological Society, London, Special Publications, **173**, 489–512.

Strain partitioning of active transpression within the Lebanese restraining bend of the Dead Sea Fault (Lebanon and SW Syria)

F. GOMEZ[1], T. NEMER[1], C. TABET[2], M. KHAWLIE[3], M. MEGHRAOUI[4] & M. BARAZANGI[5]

[1]*Department of Geological Sciences, University of Missouri, Columbia, Missouri 65211, USA (e-mail: fgomez@missouri.edu)*

[2]*Lebanese National Council for Scientific Research, Beirut, Lebanon*

[3]*Lebanese National Center for Remote Sensing, Beirut, Lebanon*

[4]*EOST, Institut de Physique du Globe, UMR 7516, Strasburg, France*

[5]*Institute for the Study of the Continents, Snee Hall, Cornell University, Ithaca, New York 14853, USA*

Abstract: Recent neotectonic, palaeoseismic and GPS results along the central Dead Sea fault system elucidate the spatial distribution of crustal deformation within a large (*c*.180-km-long) restraining bend along this major continental transform. Within the 'Lebanese' restraining bend, the Dead Sea fault system splays into several key branches, and we suggest herein that active deformation is partitioned between NNE–SSW strike-slip faults and WNW–ESE crustal shortening. When plate motion is resolved into strike-slip parallel to the two prominent NNE–SSW strike-slip faults (the Yammouneh and Serghaya faults) and orthogonal motion, their slip rates are sufficient to account for all expected strike-slip motion. Shortening of the Mount Lebanon Range is inferred from the geometry and kinematics of the Roum Fault, as well as preliminary quantification of coastal uplift. The results do not account for all expected crustal shortening, suggesting that some contraction is probably accommodated in the Anti-Lebanon Range. It also seems unlikely that the present kinematic configuration characterizes the entire Cenozoic history of the restraining bend. Present-day strain partitioning contrasts with published observations on finite deformation in Lebanon, demonstrating distributed shear and vertical-axis block rotations. Furthermore, the present-day proportions of strike-slip displacement and crustal shortening are inconsistent with the total strike-slip offset and the lack of a significantly thickened crust. This suggests that the present rate of crustal shortening has not persisted for the longer life of the transform. Hence, we suggest that the Lebanese restraining bend evolved in a polyphase manner, involving an earlier episode of wrench-faulting and block rotation, followed by a later period of strain partitioning.

Restraining bends, by definition, involve plate motions that are oblique to the associated transcurrent faults. An important question pertains to how oblique convergence is accommodated by crustal deformation. In an active tectonic setting, understanding the spatial distribution of crustal strain is important for assessing the earthquake hazard, as well as for understanding the tectonic evolution of the restraining bend.

The Dead Sea fault system (DSFS) is a prominent continental transform in the eastern Mediterranean region (Fig. 1). Along this transform system, a large (*c*.180-km-long) restraining bend along the Dead Sea fault system encompasses present-day Lebanon and SW Syria. Recent palaeoseismic (Gomez *et al.* 2001, 2003; Daeron *et al.* 2004, 2005; Nemer & Meghraoui 2006) and geodetic (e.g. Wdowinski *et al.* 2004; Reilinger *et al.* 2006) studies have shown that the strike-slip faults are unambiguously active within the Lebanese restraining bend, refuting suggestions to the contrary (e.g. Butler *et al.* 1997, 1998).

The 'Lebanese' restraining bend of the DSFS also can serve as an analogue for other large restraining bends along continental transform systems, including the Big Bend of the San Andreas Fault (e.g. Chaimov *et al.* 1990). In contrast to the Big Bend, the Lebanese restraining bend appears to be structurally less complicated, with fewer fault branches. Hence, it may be easier to assess the distribution and controls of crustal deformation, owing to the relative simplicity of the structure.

This paper synthesizes new results and recently published data to assemble a kinematic model for present-day tectonism in the Lebanese restraining

From: CUNNINGHAM, W. D. & MANN, P. (eds) *Tectonics of Strike-Slip Restraining and Releasing Bends.* Geological Society, London, Special Publications, **290**, 285–303.
DOI: 10.1144/290.10 0305-8719/07/$15.00

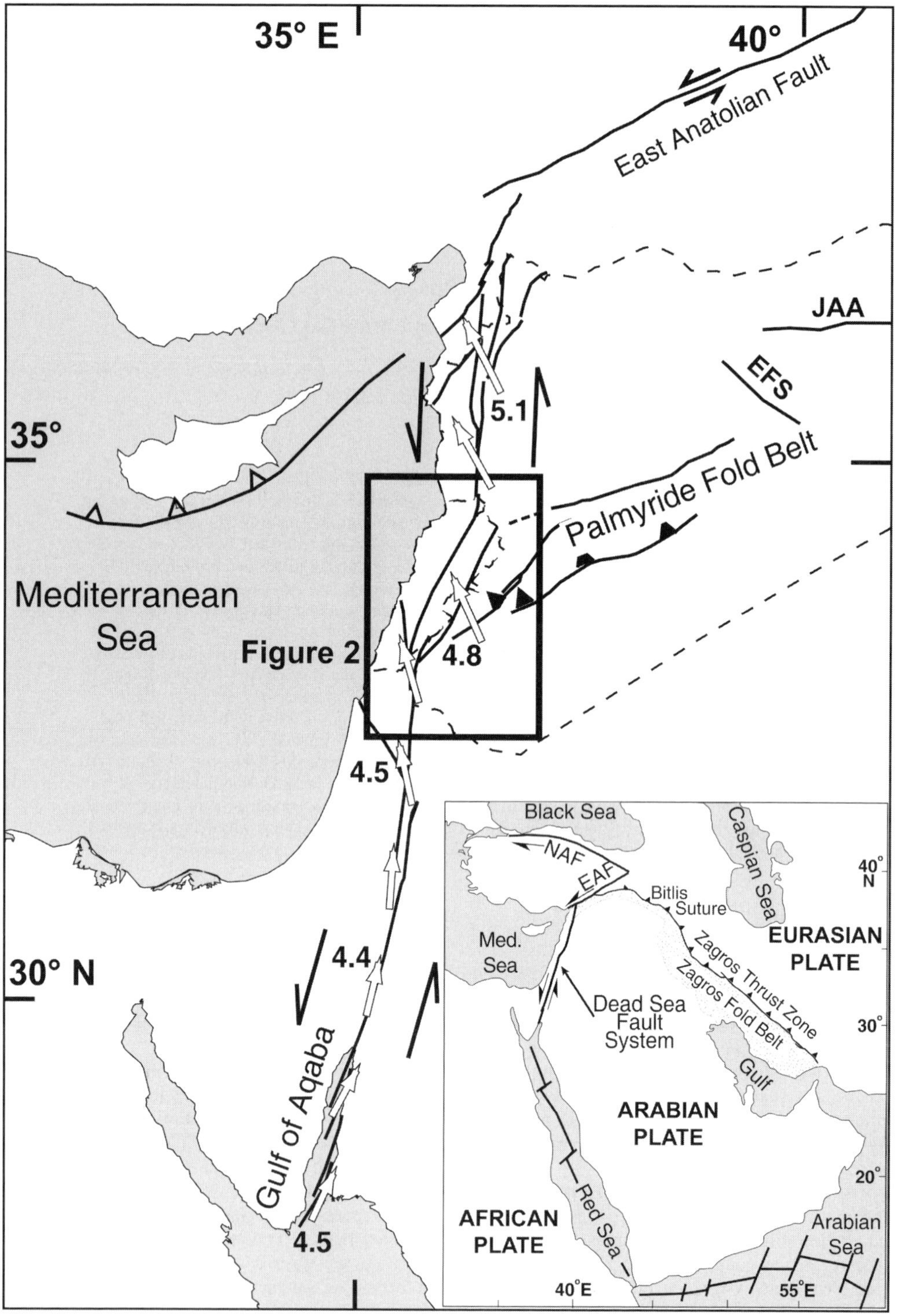

Fig. 1. Regional tectonic map of the Dead Sea fault system. Directions (white arrows) and rates (mm/a) of motion of the Arabian plate relative to the Sinai plate are based on the plate model of Reilinger *et al.* (2006). Abbreviations: JAA, Jebel Abdel Aziz; EFS, Euphrates fault system; NAF, North Anatolian fault; EAF, East Anatolian fault. Inset depicts the plate-tectonic setting of the Arabian–Eurasian collision.

bend. Herein, we suggest that present-day transpression in the Lebanese restraining bend is described in terms of strain partitioning between the NE–SW strike-slip faults and zones of approximately NW–SE folding and thrust faulting. We will explore the implications of this strain partitioning for the long-term development of the restraining bend.

Tectonic setting

Within the tectonic framework of the Eastern Mediterranean, the north–south-striking, left-lateral DSFS spans *c*.850 km from the Gulf of Aqaba to southern Turkey, accommodating differential convergence of the Arabian and Sinai plates relative to Eurasia (Fig. 1). The DSFS comprises three main sections (e.g. Quennell 1984; Garfunkel *et al.* 1981). These are a *c*.400-km-long southern section from the Gulf of Aqaba through the Dead Sea and Jordan River valleys; a *c*.250-km-long, north–south striking section in NW Syria and southern Turkey; and a *c*.180-km-long NE–SW-striking restraining bend through Lebanon and SW Syria that connects the two north–south sections. The restraining bend is the focus of this paper (Fig. 2).

A thorough overview of the history of research along the DSFS is provided by Beydoun (1999), and relevant aspects are summarized here. It is generally agreed that a first episode of tectonism occurred during the Mid- and Late Miocene, with *c*.60 km of left-lateral displacement documented along the southern DSFS (e.g. Freund *et al.* 1970; Quennell 1984). The magnitude of slip along the northern DSFS during this earlier phase remains unknown, owing to a lack of piercing points of sufficient age. A second episode of motion began in the latest Miocene to Early Pliocene, corresponding with initiation of seafloor spreading in the northern part of the Red Sea (Hempton 1987), and continues through the present. The total displacements on the southern and northern sections during this episode are 45 km and 20–25 km, respectively (Freund *et al.* 1970; Quennell 1984) – the difference in displacement may be accommodated by up to 20 km of shortening of the Palmyride fold belt (Chaimov *et al.* 1990).

The present-day relative motion of the Arabian and Sinai plates is well constrained by recent GPS studies (e.g. Wdowinski *et al.* 2004; Mahmoud *et al.* 2005; Reilinger *et al.* 2006). Owing to the position of the Euler pole describing relative plate motion (32.8°N $\pm$ 3.4°, 28.4°E $\pm$ 3.7°; Reilinger *et al.* 2006), the convergent component of plate motion increases northward along the transform (Fig. 1). Consequently, within the Lebanese restraining bend, relative plate motion is oriented approximately 35° to the NNE–SSW strike of the transform (Fig. 1). At the latitude of the restraining bend, the plate model of Reilinger *et al.* (2006) suggests 4.8 ± 0.4 mm/a of relative motion between the Arabian and Sinai plates. It should be noted that, owing to the constraints of the elastic block modelling used, the uncertainties on plate motion are considerably smaller than suggested by the uncertainties in the pole position (Reilinger *et al.* 2006). As shown in Figure 2, velocities of continuous GPS sites spanning the restraining bend clearly demonstrate left-lateral shear across the transform (Reilinger *et al.* 2006).

Physiographically, the Lebanese restraining bend consists of two distinct mountain ranges: the Mount Lebanon and Anti-Lebanon ranges, separated by the intermontane Bekaa Valley – a large synclinorium (Fig. 2). The highest topography in the restraining bend exceeds 3000 m in elevation. By modelling of gravity data, Khair *et al.* (1993) suggested that, despite the topography, the region possesses only a modest crustal root: the Moho is located *c*.27 km beneath Mount Lebanon; *c*.35 km beneath the Anti-Lebanon; and *c*.33 km beneath the Arabian plate farther to the east in western Syria.

Mount Lebanon has the overall structure of a box-anticline (Fig. 3). Although initial folding of the anticlinorium was initiated in the Late Mesozoic/Early Cenozoic (Walley 1988), folding of Miocene and Quaternary strata in the northern Mount Lebanon region indicates that the Mount Lebanon range has experienced geologically recent folding. The Anti-Lebanon is a broad, open anticlinorium. If a probable offshore continuation of the deformation belt associated with Mount Lebanon (Daeron *et al.* 2004) is included, then Lebanese restraining bend is approximately 80–90 km wide. As demonstrated by Hancock & Atiya (1979), the regional shortening is 10–15% (9–16 km), which is consistent with line-length restoration of Upper Cretaceous strata in the regional cross-section in Figure 3. Late Cenozoic deformation has also occurred within the adjacent region of the Arabian plate in the Palmyride fold belt (e.g. Chaimov *et al.* 1990), the Jebel Abdel Aziz of NE Syria (e.g. Brew *et al.* 1999), and the Euphrates fault system (e.g. Litak *et al.* 1997) (Fig. 1). Active internal deformation of the Arabian plate is also expressed by seismicity in the Palmyride fold belt (e.g. Sbeinati *et al.* 2005).

Along the flanks of the Anti-Lebanon, folded Neogene conglomerates indicate a pulse of Late Cenozoic folding (Ponikarov 1964). Remnants of a low-relief palaeosurface are also inferred within the high portions of the Mount Lebanon and Anti-Lebanon ranges (Gomez *et al.* 2006). This

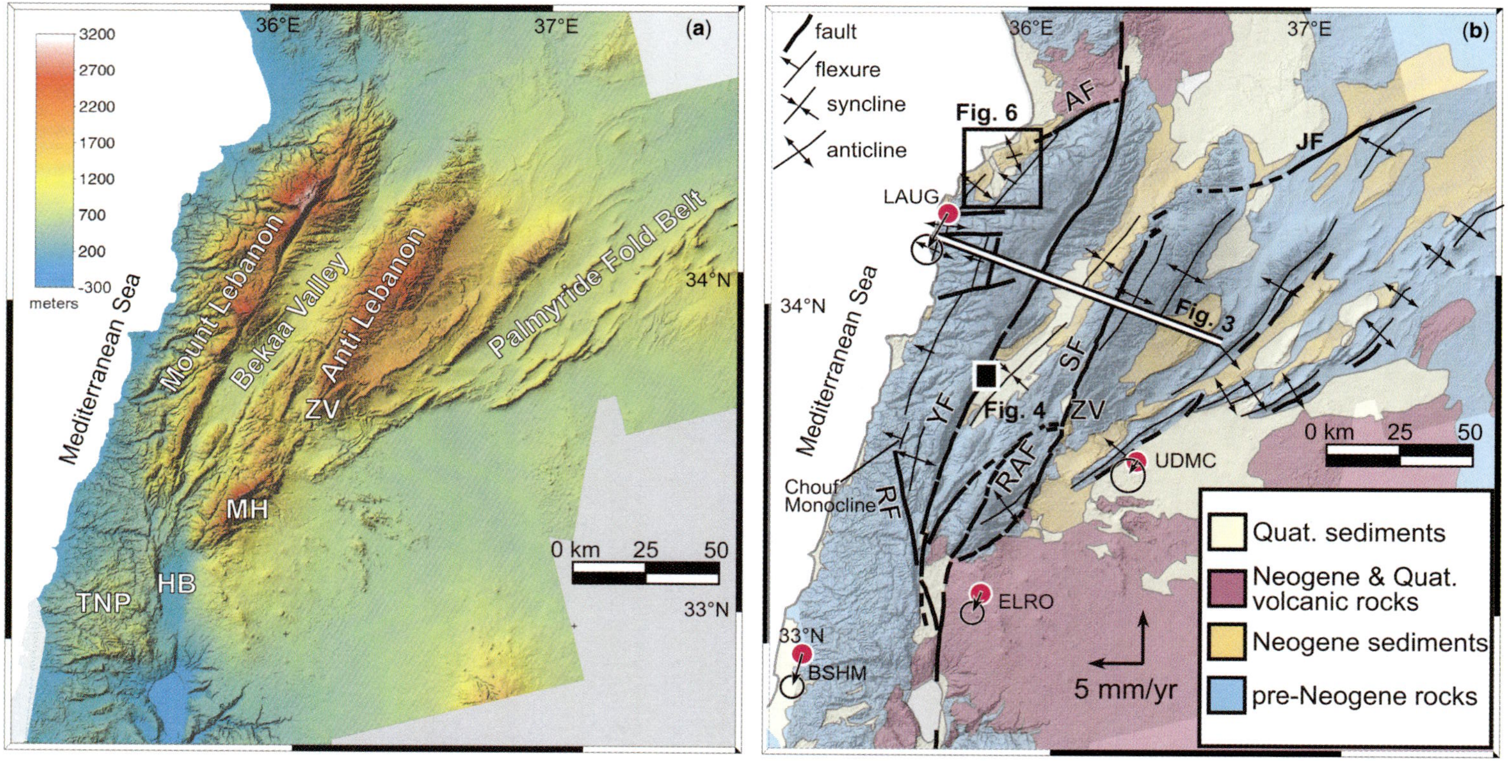

Fig. 2. (**a**) Map showing the topography of the Lebanese restraining bend from a 20-metre-pixel, InSAR-derived DEM (Gomez *et al.* 2006). (**b**) Map showing general geology (simplified from Dubertret 1955) and the structure of the restraining bend (simplified from Gomez *et al.* 2006). Arrows indicate velocities of continuous GPS sites (red circles) in an Arabia-fixed reference frame (Reilinger *et al.* 2006). GPS velocities depict the net left-lateral shear across the Dead Sea fault system in this area. White line = location of the cross-section in Figure 3. Abbreviations: YF, Yammouneh Fault, SF, Serghaya Fault; RF, Roum Fault, RAF, Rachaya Fault; JF, Jhar Fault; HB, Hula Basin; MH = Mount Hermon; TNP, Tyre–Nabatieh Plateau; and ZV, Zebadani Valley.

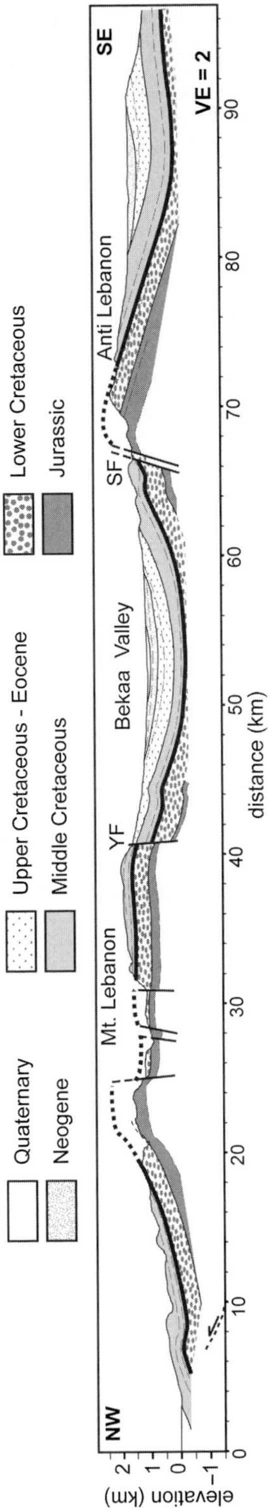

Fig. 3. Geological cross-section across the Lebanese restraining bend (see Fig. 2 for location) simplified from Sabbagh (1962). Line-length restoration of Middle Cretaceous strata (heavy line) suggests 10–14 km of shortening. YF, Yammouneh Fault; SF, Serghaya Fault. Inferred blind thrust faults (heavy dotted lines) are only schematic.

low-relief surface has been interpreted to be a Late Miocene erosional surface, owing to concordance with a regional unconformity of similar age in Syria. Many topographic escarpments parallel to the structural grain of the restraining bend and the Palmyride fold belt are erosional in nature – i.e. cuestaforms, rather than fault scarps (e.g. Dubertret 1955; Ponikarov 1964). Some tectonic maps of the restraining bend erroneously denote these as faults (e.g. Daeron *et al.* 2004) and imply more crustal shortening and thickening than has probably occurred. We emphasize the distinction here, because the crustal shortening is a key aspect of the restraining bend.

Within the restraining bend, the relatively simple north–south trace of the southern DSFS splays into several prominent structures as a 'braided' fault system (Walley 1988). These include the NE–SW-striking Yammouneh, Rachaya and Serghaya faults. Other structures include the north–south-striking Roum Fault and the ENE–WSW striking Akkar Fault – both of which bound the Mount Lebanon uplift. Some of these structures are probably inherited from prior tectonic episodes in the region. For example, the stratigraphic thickness changes across the Roum Fault (e.g. Dubertret 1955; Sabbagh 1962) suggest its activity during a Late Cretaceous rifting episode, and the Jhar Fault in the northern Palmyrides may correspond with a Precambrian suture (Best *et al.* 1990). On the other hand, Mesozoic stratigraphic thicknesses do not change across the Serghaya Fault (e.g. Fig. 3), which suggests that it is not a reactivated normal fault associated with the Late Cretaceous rifting.

Historical records of large, devastating earthquakes also attest to the activity and seismogenic potential of the central and northern DSFS, despite the scarcity of moderate and large events in instrumental records (e.g. Ambraseys & Jackson 1998; Sbeinati *et al.* 2005). For example, it seems likely that a large earthquake in 1759 (Ambraseys & Barazangi 1989) occurred along the Serhgaya branch of the DSFS (e.g. Gomez *et al.* 2003; Daeron *et al.* 2005). By summing estimated moments of large historical earthquakes over the past *c.*2000 years, Ambraseys (2006) estimated a seismic slip rate of 4–5 mm/a for the entire DSFS, which is generally consistent with the total plate motion.

Present-day strike-slip faulting

To date, slip rates for several key strike-slip faults have been reported, including the Serghaya, Roum, and Yammouneh faults (Table 1). These slip rates, along with other structural details, are

Table 1. *Slip rates for major faults in the Lebanese restraining bend*

Fault	Slip rate (mm/a)	Reference
Serghaya	1.4 ± 0.1	Gomez *et al.* (2003)
Roum	0.9 ± 0.2	Nemer & Meghraoui (2006)
Yammouneh	5.1 ± 1.3	Daeron *et al.* (2004)
Yammouneh	5.0 ± 1.1	This study

synthesized in the following sections. We also provide further constraints on the slip rate of the Yammouneh strike-slip fault that have not been previously reported. As will be discussed later, these kinematic constraints may be sufficient to suggest that active transpression is partitioned between strike-slip deformation and perpendicular crustal shortening.

Serghaya/Rachaya faults

The Serghaya Fault splays from the main transform at the NE side of the Hula Basin (Fig. 2). The fault can be followed along the southern flank of Mount Hermon, and it bends NNE to trace obliquely across the Anti-Lebanon range (e.g. Dubertret 1955; Gomez *et al.* 2006). A reverse component of motion along the southern Serghaya Fault adjacent to Mount Hermon is consistent with the vertical juxtaposition of Jurassic strata on top of Cretaceous and Palaeocene strata (Dubertret 1955), as well as the SE vergence of the Mount Hermon anticlinorium (May 1989). The Serghaya Fault appears to terminate in the northern Anti-Lebanon: the distinct, linear geomorphic expression is distinct in the south, but becomes less pronounced to the north where the structure branches into several smaller, discontinuous splays. The termination of the Serghaya Fault suggests a possible linkage with the Jhar Fault as an oblique ramp or tear fault that bounds the Palmyride fold belt (e.g. Walley 1988; Gomez *et al.* 2003).

Walley (1998) suggested 20 km of total displacement along the Serghaya Fault, but the displaced markers are ambiguous. Other evidence of long-term offset can be gleaned from the large river valleys incised into the low-relief erosional surface of the Anti-Lebanon range. Consistent *c.*6 km deflections of the largest river valleys and wind gaps suggest that a similar magnitude of left-lateral offset post-dates the erosional surface (Gomez *et al.* 2006).

A related structure is the Rachaya Fault, which also splays from the NE Hula Basin and traces along the northern flank of Mount Hermon. The linear trace of the Rachaya Fault is clear in the field, and faulted alluvial fans and a palaeoseismic investigation provide compelling evidence for active tectonism (Nemer 2005) – these field investigations find no evidence of reverse faulting along the Rachaya Fault as suggested by Daeron *et al.* (2005). At its northern termination, the Rachaya Fault splays as it bends toward the Serghaya Fault in the Zebadani Valley (e.g. Walley 1988; Gomez *et al.* 2006). The Rachaya Fault is not simply the southern part of the Serhgaya Fault, as suggested by Daeron *et al.* (2004, 2005) – there is a neotectonic fault south of Mount Hermon that connects with the Serghaya Fault. Rather, the geometry of the Serghaya and Rachaya faults, along with *c.*60° block rotations in the Mount Hermon region (Ron 1987) suggest that the Rachaya Fault and the Serghaya Fault may be structurally linked as a strike-slip 'duplex', following the model of Woodcock & Fischer (1986).

A Holocene-averaged slip rate along the Serghaya Fault was determined in the palaeoseismic study of Gomez *et al.* (2003) in the Zebadani Valley. Based on a sequence of three displaced and abandoned channels exposed in the palaeoseismic excavation, a slip rate of 1.4 ± 0.1 mm/a was estimated. A small component of oblique slip on the Serghaya Fault was suggested by fault-plane striations as well as faulted landscape features (Gomez *et al.* 2001). However, the faults are subvertical at the surface, and it is unclear whether the dip-slip component (*c.*0.3 mm/a) reflects regional compression of the Anti-Lebanon range or local extension related to the releasing bend in the Serghaya Fault that corresponds with the Zebadani Valley. Published slip rates for the southern Serghaya Fault and the Rachaya Fault are presently lacking.

Roum Fault

The Roum Fault splays from the Yammouneh Fault in southern Lebanon, north of the Hula Basin. The NNW–SSE strike of the Roum Fault contrasts with the NNE–SSW strike of the other major faults in the restraining bend. The Roum Fault corresponds with a topographic lineament that bounds the topography of the southern Mount Lebanon range with the low relief of the Tyre–Nabatieh Plateau (Fig. 2). The fault lineament can be traced until approximately 33°40′ N latitude. At the northern extent, the trace of the Roum Fault disappears as it bends northward and merges with the hinge of the Chouf monocline – the southern part of the larger Mount Lebanon monoclinorium (Griffiths *et al.* 2000; Khair 2001; Nemer & Meghraoui 2006).

Meso- and micro-structures demonstrate primarily strike-slip displacement along the southern part of the Roum Fault, with increasing oblique slip (east side up) along the northern part of the fault (Griffiths *et al.* 2000; Nemer & Meghraoui 2006). Griffiths *et al.* (2000) and Nemer & Meghraoui (2006) documented a northward decrease in the amount of deflection of major drainages across the Roum Fault. This was inferred to represent a northward decrease in strike-slip offset along the fault.

Evidence of recent movement of the Roum Fault is provided by the misalignment of small streams, as well as sheared alluvium exposed in trench excavations (Nemer & Meghraoui 2006). Along the southern part of the Roum Fault, Nemer & Meghraoui (2006) estimated a slip rate of 0.8–1.1 mm/a, based on the consistent offsets of small streams that incise a caliche horizon. Nemer & Meghraoui observed consistent leftward, stream deflections of 9–11 metres, which were interpreted as left-lateral offsets. A maximum age of the incisions of *c*.6400–8500 BC was provided by radiocarbon dating of detrital charcoal and a bulk soil sample.

Yammouneh Fault

The Yammouneh Fault is the only through-going fault within the Lebanese restraining bend – i.e. it is the only structural link between the southern DSFS in the Jordan River Valley and the northern DSFS in NW Syria. The southern Yammouneh Fault generally bounds the west side of the southern Bekaa Valley, and the northern part of the fault passes through the northern Mount Lebanon Range adjacent to the crest of the range (Fig. 2). The fault's linear map trace suggests a subvertical fault in the upper crust (Fig. 3).

Walley suggested a total displacement of 47 km of left-lateral slip based on correlating Mesozoic geological structures between the Mount Lebanon and Anti-Lebanon ranges. Average, Quaternary slip rates of *c*.5 mm/a have been inferred from the displacement of the Litani River (Walley 1988; Westaway 2004). Displaced landforms, such as faulted alluvial fans and wind gaps, demonstrate that the Yammouneh Fault has experienced only strike-slip movements during the Late Quaternary (e.g. Daeron *et al.* 2004; Gomez *et al.* 2006), i.e. there is no evidence of recent or active dip-slip along the Yammouneh Fault.

Daeron *et al.* (2004) constrained a Late Pleistocene slip rate for the Yammouneh Fault of 3.8–6.4 mm/a, using faulted depositional fans. In the following section, we will provide additional cosmogenic age constraints on the more southerly of the two sites studied by Daeron *et al.* (2004). Despite some differences in the mapping and interpretation of the fault and adjacent landforms, our results complement those of Daeron *et al.* (2004). As discussed, our sampling strategy is different, and we also apply more recently published constraints on cosmogenic nuclide production rates. The integration of our results with those of Daeron *et al.* may improve the uncertainties on the timing of displacement.

Additional constraints on the Yammouneh Fault

Along the western side of the Bekaa Valley, a series of medium-sized alluvial fans are presently abandoned and incised by the modern channel drainage (e.g. Dubertret 1955). Among these, the Zalqa fan is located at approximately 33°50′. Unlike most other fans in the Bekaa Valley, the Zalqa fan is significant, owing to the fact that the Yammouneh Fault truncates the fan very near its apex (Fig. 4a).

In our study, detailed field mapping of the Zalqa fan was supported by high-resolution overhead imagery (2-metre pixel SPIN-2 satellite imagery and 1:20 000 stereo aerial photographs). Field studies of the fan also involved mapping and surveying with kinematic GPS. As in the map made by Daeron *et al.* (2004), we also identified and mapped three alluvial surfaces (Fig. 4a): the oldest fan (Qf1), which comprises the largest volume of sediment; a smaller, but now incised intermediate fan (Qf2); and the active alluvial fan (Qf3). The Zalqa fan consists of sub-angular clasts of Upper Jurassic dolimitic limestone. The clast-supported sediments are poorly sorted, with clast sizes ranging from centimetre-sized gravel to boulders more than 2 m in diameter. Qf1 has a weakly developed soil with small pedogenic structures in the A horizon and very thin carbonate coatings on the clasts in the B horizon.

The present-day Zalqa drainage incises a 5-m canyon exposing sheared alluvium of Qf1 coinciding with the Yammouneh Fault (Fig. 4b) at approximately 1045 m altitude. At this location, the canyon also depicts an abrupt deflection of 5–8 m, which may represent the true offset of the drainage.

The Yammouneh Fault truncates the fan approximately 150 metres from the apex – this permits a confident estimate of the total offset of the fan, rather than the diffuse depositional contacts along its edges. Topographic profiles radiating from the fan's apex down the Qf1 surface were measured using differential GPS, and these profiles depict the steep, but constant, slope of this fan. This suggests a single episode of aggradation has produced the Qf1 surface.

The geometry of the Zalqa fan is well preserved in the topography, and we constrained the location of the head of the fan using available 1:20 000 topographic maps, 2-m pixel Spin-2 satellite photos, and

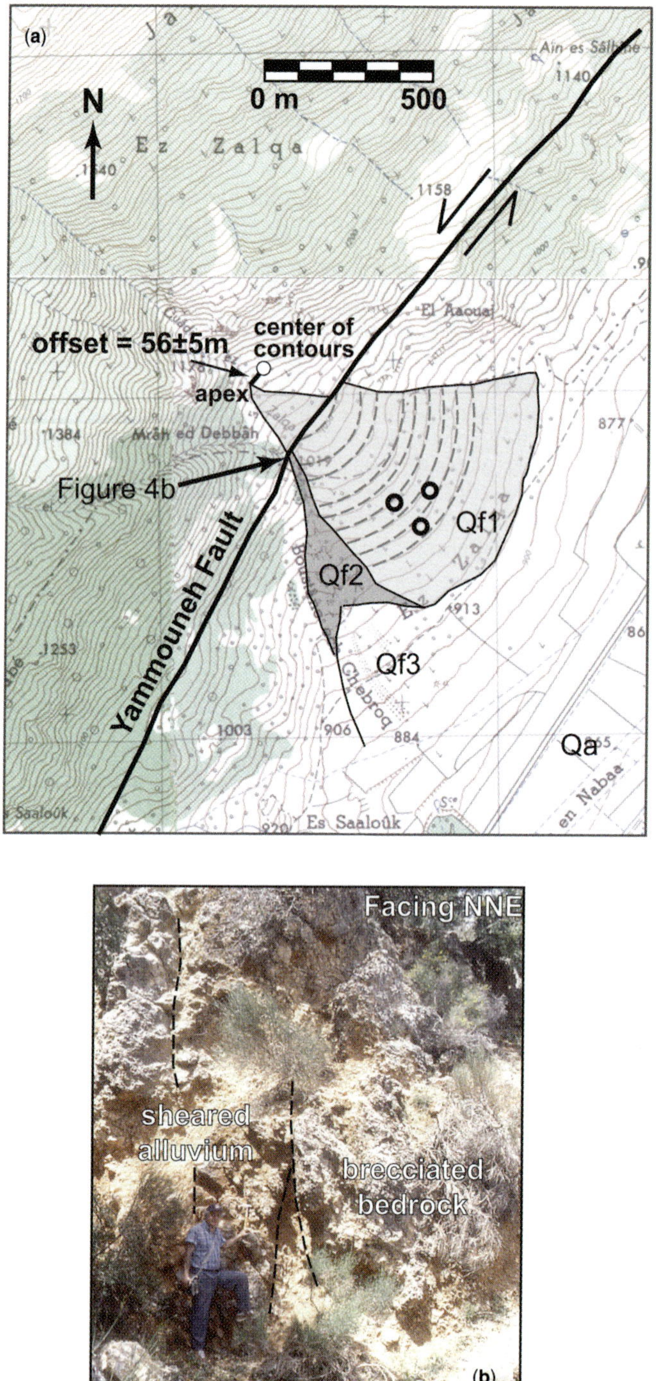

Fig. 4. (**a**) Map of the Zalqa Fan, based on field mapping and stereoscopic aerial photographs. Three fan surfaces are shown: Qf1 (oldest), Qf2 and Qf3 (youngest). Locations of samples for cosmogenic dating are indicated by the black circles. Dashed grey lines indicate the best-fit concentric circles for the topographic contours of the Qf1 surface. The 56-metre displacement of the Qf1 fan is indicated by the offset between the centre of these concentric circles and the present position of the fan's apex. (**b**) Photograph of shear alluvium and brecciated bedrock where the Yammouneh Fault intersects the Zalqa drainage (location shown in Fig. 4a). Dashed line denotes shear zones within the alluvium.

field measurements (kinematic GPS). Offset of the fan was estimated from the displacement between the head of the fan and the fan apex, as defined by the morphology of the lower part of the fan. Alluvial fans typically have a conical geometry, owing to the abrupt deposition at the mouth of a confined mountain drainage (Bull 1977; Pinter *et al.* 1995) – such geometry implies similar slopes radiating from the apex, rather than a single axis. In fact, the topographic contours of the Qf1 surface demonstrate a conical surface (concentric arcs in map view, see Fig. 4). The centre of these concentric arcs should correspond with the relative position of the fan's apex at the time of aggradation, and, following established methods (e.g. Pinter *et al.* 1995; Klinger *et al.* 2000), we believe that our approach correctly uses the properties of an alluvial fan to assess the offset of the Qf1 surface. Topographic contours were digitized and used to determine a best-fit set of concentric circles. The offset between the centre of these circles and the apex provides a measure of the displacement of 56 ± 5 metres.

As with the study of Daeron *et al.* (2004), we used *in situ* cosmogenic ^{36}Cl to date the age of the fan surface. 'Exposure' dating of landforms using cosmogenic nuclide concentrations has become a widely applicable technique (e.g. Gosse & Phillips 2001). *In situ* cosmogenic nuclides result from interactions between cosmic rays and the constituent material in surficial rocks. Hence, concentrations of *in situ* cosmogenic nuclides are directly related to the duration that those particular rocks have been exposed at the surface. Production rates for a particular location can be estimated after considering environmental aspects such as elevation, latitude, topographic slope and shielding. Owing to the predominant carbonate clast lithology in the Zalqa fan, ^{36}Cl was used.

Whereas Daeron *et al.* (2004) dated numerous smaller cobbles from the fan surface, our efforts focused on large boulders of Jurassic limestone embedded within the fan surface. Samples from three distinct boulders embedded in the mid to upper half of the fan's surface were collected for cosmogenic ^{36}Cl dating (see Fig. 4 for locations). As these clasts are embedded in the top of the oldest fan, they provide a maximum age constraint on the timing of abandonment of the fan. All the boulders sampled showed minimal effects of burial or weathering – a 1-m test pit excavated adjacent to one of the boulders demonstrated a relatively intact A horizon, implying negligible soil erosion. The relatively short length of the drainage feeding the fan (*c*.1.5 km) suggests that time in transit after erosion was probably minimal before the boulders were deposited in the fan surface. Following the procedure described by Gosse & Phillips (2001), sample material was collected from the outer two centimetres of the boulders. Boulders showed little evidence of weathering (i.e. solution pits were negligible).

Chemical preparation of the samples was performed by the New Mexico Bureau of Mines, and the ^{36}Cl nuclide concentrations were measured using accelerator mass spectrometry (AMS) at Lawrence Livermore National Laboratory, California, USA. The isotopic concentrations are shown in Table 2. Numerical ages were modelled using the MS Excel-based CHLOE worksheet (Phillips & Plummer 1996) and the production constants provided by Phillips *et al.* (2001). After corrections for environmental conditions, including those described above, the three samples yielded ages of 9.6–12.8 ka BP (Table 2). The production rates of Phillips *et al.* (2001) differ by *c*.25% from those of Stone *et al.* (1998), which were applied by Daeron *et al.* (2005). Consequently, our ages appear younger: applying the production rates of Stone *et al.* yields ages of 12.5–16.6 ka BP.

To improve the precision of the age estimate, we assume that all three boulders aggraded during the same episode – this seems reasonable, owing to the uniform slope of the fan. The total uncertainties include the uncertainty in the production rates and analytical results. Assuming that these uncertainties represent a Gaussian probability density function, the probabilities for each of the dates can be

Table 2. *Results and production rates for ^{36}Cl dating*

Sample	$^{36}Cl/^{35}Cl$ ratio	$^{36}Cl/^{35}Cl$ error	$^{35}Cl/^{37}Cl$ ratio	Age (years BP)
MF-01-1	1.604×10^{12}	2.16×10^{14}	5.21	$12\,800 \pm 1430$
MF-01-2	1.381×10^{12}	3.29×10^{14}	5.56	$11\,410 \pm 1390$
MF-01-3	1.539×10^{12}	2.22×10^{14}	6.07	9810 ± 1210
Production			Mean age	$11\,200 \pm 790$
^{40}Ca spallation	66.8 ± 6.8			
^{39}K spallation	137 ± 60			
Fast neutron	626 ± 105			

Production rates from Phillips *et al.* (2001)

statistically combined to yield a final age of 11.20 ± 0.81 ka BP. This compares favourably with the youngest age group from the results of Daeron *et al.* (2005), especially after accounting for the different production rates used.

The timing of aggradation of the Zalqa Fan (and presumably the other fans of similar size along the western flank of the central Bekaa Valley) are consistent with a shift from cooler and drier conditions to warmer and more wetter conditions approximately 10–11 ka ago in the eastern Mediterranean (this corresponds with the Younger Dryas event in Europe and North America). Rossignol-Strick (1993) interpreted the recent climate history of Lebanon and adjacent regions, based on palynological indicators. Near the end of the Late Pleistocene (approximately 11 ka) an abrupt change from arid to humid conditions was indicated by an abrupt change in vegetation. As described by Bull (2000), such a rapid shift in climate is conducive to alluvial fan aggradation.

These results constrain the maximum ages of abandonment of Qf1. Hence, these ages provided constraints on the minimum rate of slip along the Yammouneh Fault during the Holocene. With 51–61 metres of displacement, the slip rate is 3.9–6.1 mm/a.

Kinematic model

The slip rates for the Yammouneh and Serghaya faults (i.e. the two main NNE–SSW-striking fault systems) provide a basis for assessing how plate motion is accommodated within the restraining bend. Whereas Westaway (1995, 2004) applied a continuum model to finite strain in the Lebanese restraining bend, here we explore a relatively simple geometric model to explain the present-day (i.e. instantaneous) slip.

The recent plate model of Reilinger *et al.* (2006) predicts the movement of the Arabian plate relative to the Sinai plate at the Lebanese restraining bend to be 4.8 ± 0.4 mm/a oriented N10°W. When resolved with the N25°E orientation of the master faults through this restraining bend, this plate motion can be broken down into 3.8 mm/a of NNE–SSW strike-slip displacement on these faults and 3.1 mm/a of perpendicular convergence. Within the uncertainties, this is consistent with the sum of the slip rates on the Serghaya/Rachaya and Yammouneh faults. In fact, the sum of the Late Quaternary strike-slip rates discussed herein (5.2–7.5 mm/a) generally exceeds the total 4.0 mm/a expected from plate motion, which may suggest that geological rates have been overestimated. Since these two faults sufficiently account for all expected strike-slip deformation, there is no need to involve additional distributed simple shear (i.e. wrenching) during the Late Quaternary. Recalling that the Yammouneh Fault demonstrates only strike-slip displacement, the only deformation that remains to be accommodated is the component of shortening perpendicular to the NNE–SSW strike-slip faults. Hence, it appears that oblique plate motion within the restraining bend is partitioned into strike-slip displacements and perpendicular convergence. This is consistent with the suggestion of Griffiths *et al.* (2000) that, during the Late Cenozoic, southern Mount Lebanon folds formed parallel to the NNE–SSW-striking Yammouneh Fault. This suggestion of strain partitioning was supported by the similar orientations of joint-bedding intersections and fold hinges – such a relationship would not be expected if rotation and wrenching were involved.

The newly available slip rates and revised plate-boundary conditions allow the reassessment of our earlier geometric model for the Lebanese restraining bend (Gomez *et al.* 2003). The schematic model shown in Figure 5 also permits some shortening across the Palmyride fold belt, which seems reasonable owing to present-day seismicity in the Palmyride region. This model assumes a fixed fault geometry for Late Quaternary displacements. As in our previous model (Gomez *et al.* 2003), the Serghaya Fault acts as an oblique 'back-stop' to the Palmyride fold belt, and it links with the Jhar Fault in the northern Palmyride region. As illustrated in Figure 5, this geometric model does not involve a simple scalar addition of slip rates, because the Serghaya links directly to the internal deformation of the Arabian plate in the Palmyride fold belt. Considering the geometry of the Serghaya Fault relative to the shortening direction of the Palmyride fold belt, 1.4 mm/a of left-lateral slip corresponds with 1.0 mm/a of shortening across the entire Palmyride region. Hence, internal deformation of the Arabian plate in the Palmyrides may accommodate up to 20% of the total, relative plate motion at this latitude. The remaining 4.0 mm/a of plate motion will be accommodated by the other structures in the restraining bend: *c.*3.2 mm/a of NNE–SSE strike-slip along the Yammouneh Fault, and 2.2 mm/a of orthogonal shortening.

Relatively complete strain partitioning has been documented in other large restraining bends along transform boundaries. For example, the Hispaniola restraining bend along the Caribbean–North American plate boundary appears to involve complete partitioning of slip between two strike-slip faults and an offshore thrust system to the north (Calais *et al.* 2002). Although structurally more complicated than the DSFS and the northern Caribbean plate boundary, strain partitioning is also observed along the Big Bend of the San Andreas Fault system (e.g. Becker *et al.* 2005), as

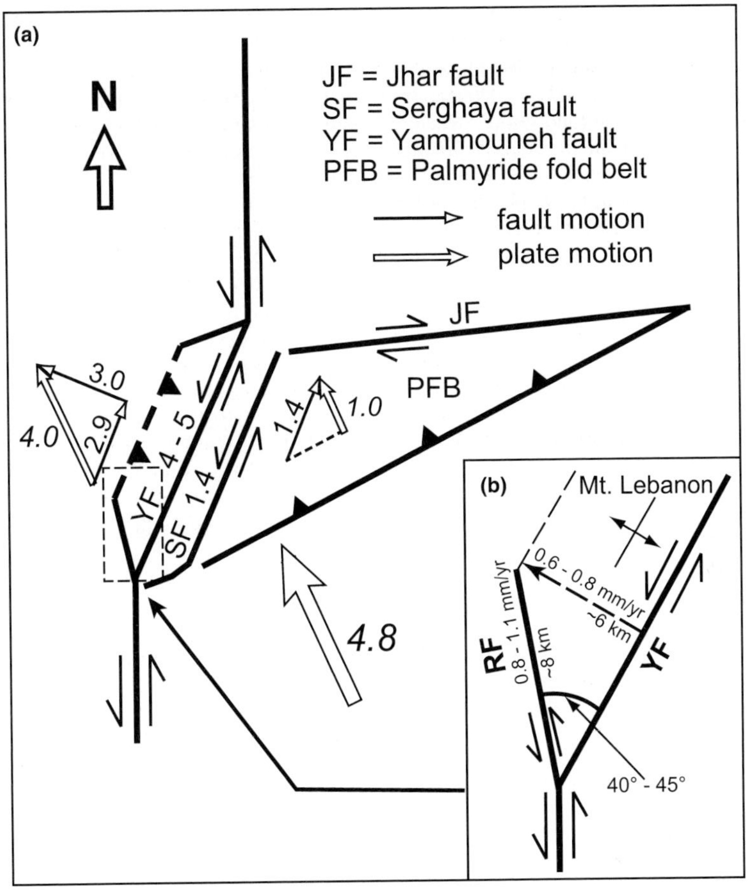

Fig. 5. (**a**) A possible geometric model for the Lebanese restraining bend (revised from Gomez *et al.* 2003). Present-day plate motion is based on Reilinger *et al.* (2006). In this model, the Serghaya Fault serves as a bounding structure of the Palmyride fold belt, and contributes to internal deformation of the Arabian plate. After subtracting that shortening vector, the remaining plate motion is decomposed into strike-slip parallel to the Yammouneh Fault, and orthogonal shortening. Numbers represent rates in mm/a. Heavy dashed line indicates an inferred blind fault corresponding with the Coastal monocline of Lebanon. (**b**) Enlargement of the simplified geometry around the Roum Fault, indicating how the total slip and slip rate can be geometrically related to the total shortening and shortening rate. Total shortening of 6 km across the 30-km-wide region is consistent with the 10–15% total shortening reported by Hancock & Atiya (1979).

well as along the central San Andreas Fault (e.g. Molnar 1992; Page *et al.* 1998), which is also oriented oblique to the relative plate motion. Both cases, as well as the Lebanese restraining bend, involve transform systems with significant total displacement and regionally distributed strain. In such cases, strain partitioning is likely, as it is a more stable kinematic configuration than oblique-slip faulting and wrenching (Molnar 1992).

Contraction and crustal shortening

Although the NE–SW strike-slip kinematics are becoming well constrained, constraints on NE–SW contraction are still quite limited. However, we can explore possible constraints on rates and magnitudes of crustal shortening and thickening in the Mount Lebanon Range from two considerations: (1) the role of the Roum Fault as an accommodation structure, and (2) rates of coastal uplift.

The Roum Fault and Mount Lebanon shortening

Compared with the Yammouneh and Serghaya faults, the Roum Fault is unusual, owing to its N10–15W strike and the displacement gradient along strike. Strain compatibility requires that the inferred northward decrease in displacement along the Roum Fault be accommodated by folding of the adjacent Mount Lebanon Range. Hence, the Roum Fault serves as an 'oblique ramp' structure bounding the Mount Lebanon folds (Griffiths *et al.* 2000).

With this concept in mind, the displacement and slip rate along the Roum Fault can be used to predict the expected WNW–ESE shortening in the southern Mount Lebanon. The Mount Lebanon folds trend *c.*N30°E, implying shortening N60°W. This shortening direction makes a 40–45° angle with the N10–15W strike of the Roum. Based on the geometry of the Roum fault and the WNW–ESE shortening, the *c.*8 km of total displacement (Griffiths *et al.* 2000; Nemer & Meghraoui 2006) suggests that the southern part of Mount Lebanon has been shortened by *c.*6 km (Fig. 5b). Considering the *c.*30 km width of the southern part of Mount Lebanon (between the Yammouneh Fault and the Chouf monocline), this corresponds with 17% horizontal shortening – slightly larger than (but not necessarily inconsistent with) the 10–15% bulk shortening reported by Hancock & Atiya (1979). This similarity between expected and predicted total shortening supports the plausibility of the geometric model in Figure 5.

A similar geometric exercise can be applied to the slip rates. In this case, the 0.8–1.1 mm/a Holocene slip rate would correspond with a shortening rate of 0.6–0.8 mm/a. This shortening rate is significantly less than the 2.2 mm/a predicted from the geometric model in Figure 5. This suggests that WNW–ENE horizontal shortening is presently being accommodated in other parts of the Lebanese restraining bend, such as offshore Lebanon and in the Anti-Lebanon Range.

Coastal uplift

Coastal uplift is another expected manifestation of crustal shortening and thickening within the restraining bend. Late Cenozoic shortening and uplift is particularly well expressed along the coast in northern Lebanon near Tripoli (Fig. 6). South of Tripoli, the Jaouz anticline involves folded Miocene–Pliocene strata (Fig. 7). In the core of the Zagharta syncline to the SW, Lower Quaternary deposits also appear slightly back-tilted, suggesting recent folding. North of Tripoli, Late Cenozoic shortening is also expressed in the deformation of Miocene and Pliocene strata in the Turbol anticline (Fig. 6).

Compelling evidence for recent coastal uplift is suggested by notched shorelines along the Lebanese coast (Fig. 8). The relative sea-level changes represented by notched coastlines may involve both true eustatic variations (which will correlate with global climate variations) and tectonic uplift of the coast. Notched shorelines in northern Lebanon are higher than those along the Syrian coast

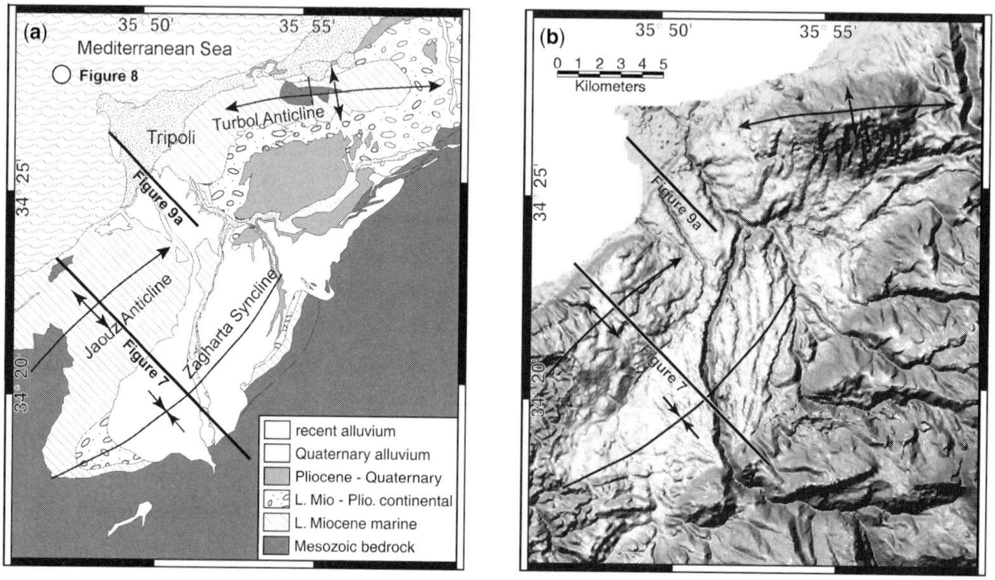

Fig. 6. (**a**) Map showing simplified geology (from Dubertret 1955), and (**b**) map of topography in the region of Tripoli (based on Gomez *et al.* 2006). Locations of the profiles in Figures 7 & 9a are shown, as well as the location of the photograph in Figure 8.

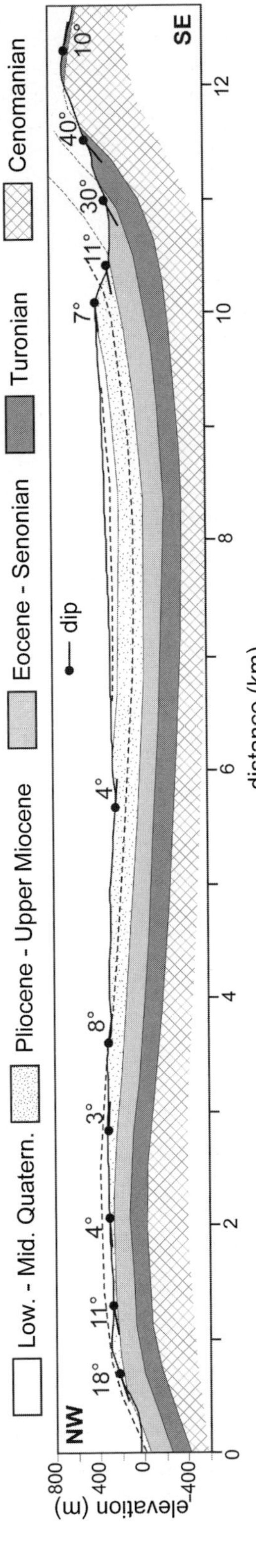

Fig. 7. Geological cross-section south of Tripoli (based on mapping of Dubertret 1955 and structural measurements from our fieldwork). Late Cenozoic deformation is indicated by folded Miocene–Pliocene strata. The cross-section also demonstrates the stratigraphic continuity across the escarpment near the shoreline at the NW end of the profile. See Figure 6 for location.

(Pirazolli *et al.* 1996), which is consistent with the expected increase in crustal shortening within the Lebanese restraining bend. Evidence of Mid- to Late Quaternary uplift is provided by flights of coastal terraces along the Lebanese coast, particularly near Tripoli (e.g. Sanlaville 1974). Along the north Lebanese coast, marine terraces are eroded on to Mesozoic and Tertiary strata (e.g. Sanlaville 1977), and exposures in road cuttings and stream valleys demonstrate these are typically overlain by a thin (2–3 m) veneer of Late Quaternary sedimentary cover. The escarpments corresponding with the risers of these coastal terraces are topographically prominent, involving cliff heights in excess of 50 m (Sanlaville 1977). We emphasize here that these escarpments are erosional in nature, as indicated by the continuity of lithological units across the escarpments (e.g. the cross-section in Fig. 7). Furthermore, our field investigation of these scarps identified no evidence for faulting, such as brecciated bedrock or a gouge zone associated with these escarpments.

Radiocarbon dating of vermetid gastropod shells suggests that notched shorelines near Tripoli with heights of *c.*1 m and *c.*2 m represent two abrupt uplift events that occurred between *c.*770 BC and *c.*550 AD (Sanlaville 1974, 1977; Pirazolli *et al.* 1996 and references cited therein). The most recent may correspond with the large, historically documented earthquake that occurred in Lebanon in 551 AD (Darawcheh *et al.* 2000). Ages of higher terraces and shoreline features are less well constrained. On the basis of Middle Palaeolithic artifacts, Bowen & Jux (1987) suggested that a well-developed terrace at 7–15 metres above sea-level corresponds with the Eem Interglacial (*c.*125 ka BP).

By taking the altitudinal positions of the terraces and their sea-level history into consideration, it is possible to quantify the rates of uplift of the Lebanese coast during the Middle and Late Quaternary (e.g. Bloom *et al.* 1974). Marine terraces are planar surfaces that generally slope seaward up to 3–4°. The 'shoreline angle' where the wave-cut platform meets the steep sea-cliff and is preserved at the top of the terrace indicates sea-level at the high stand. The elevations of these shoreline angles were interpolated from topographic profiles extracted from 1:20 000 topographic maps and spot-checked for several locations in the field. In total, six distinct terrace surfaces were identified in eight profiles north and south of Tripoli – an example is shown in Figure 9a, and mean values are provided in Table 3. We believe that coastal uplift in the northern Mount Lebanon may be representative of the entire Mount Lebanon range, as the most extensive, upper terraces correlate regionally in Lebanon (e.g. Bowen & Jux 1987).

Fig. 8. Photograph from Ramkine Island, offshore of Tripoli, showing a notched shoreline indicating the recent emergence of the coastline. See Figure 6 for location.

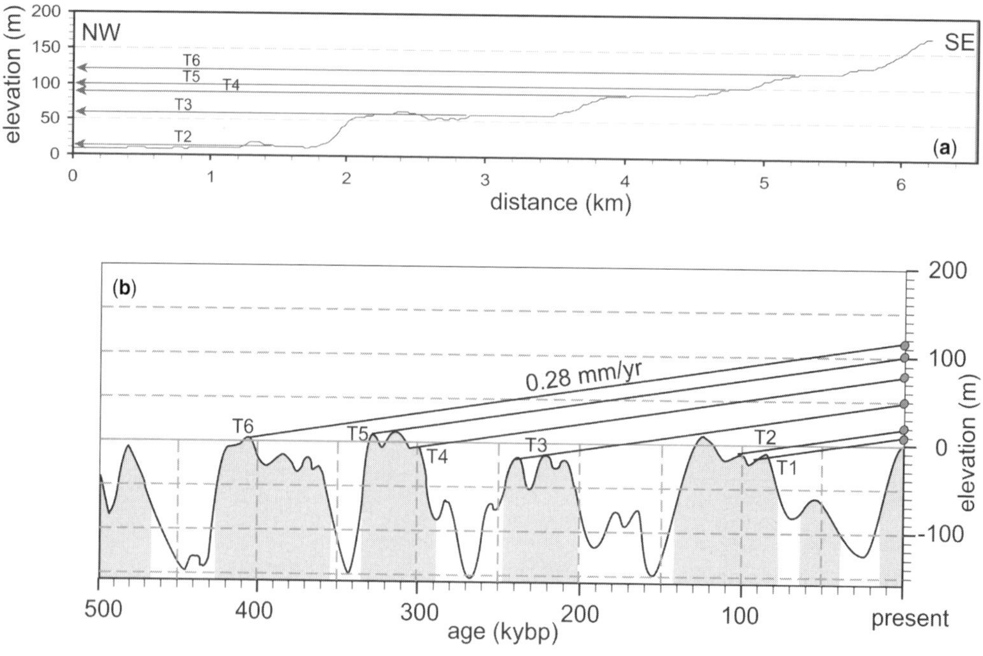

Fig. 9. (**a**) Example topographic profile and shoreline angles in the Tripoli region (see Fig. 6 for location). Six distinct shoreline angles have been identified from eight profiles, and these values are summarized in Table 3. (**b**) Correlation of terrace altitude with sea-level high stands during the past 500 ka (Shackleton & Opdyke 1973; Pinter & Gardner 1989) assuming a constant rate of uplift. An uplift rate of 0.28 mm/a best fits the data.

Table 3. *Mean elevation of coastal terraces in northern Lebanon*

Terrace	Mean elevation (m)
T1	10
T2	22
T3	53
T4	81
T5	104
T6	119

Shoreline angles corresponding with coastal terraces were measured from eight topographic profiles in northern Lebanon.

Following the method outlined by Pinter (1996), we infer rates of coastal long-term uplift by correlating the relative vertical spacing of terrace heights with sea-level variations during the past *c*.500 ka (Shackleton & Opdyke 1973). Marine terraces typically form during sea-level high stands. For this analysis, we assumed that the rate of uplift has been constant during the Late Quaternary. Using these constraints, linear regression resulted in an uplift rate of 0.28 mm/a that best fit the heights of the terraces – the results are graphically depicted in Figure 9b. This rate of uplift is considerably slower than the 0.75–1 mm/a that may be inferred from the notched shorelines discussed by Pirazolli *et al.* (1996). One possible explanation may be that uplift rates can be overestimated over short time periods if earthquakes demonstrate temporal clustering behaviour.

We assume that the uplift deduced from terraces near Tripoli is representative of uplift in the Mount Lebanon region and that this uplift is directly related to crustal thickening and NW–SE shortening. This uplift rate is a minimum estimate on the rate of crustal thickening if there is no isostatic compensation. If the growth in topography is isostatically compensated, thickening is faster. Assuming Airy isostasy, the change in thickening (Δt) and uplift rate are related as:

$$\Delta t = \Delta h \times \{1 + E\rho_c/(\rho_m - \rho_c)]\},$$

where ρ_m is the density of the mantle and ρ_c is the bulk crustal density. Using typical values for these densities $\rho_m = 3.3\ g/cm^3$; $\rho_c = 2.7\ g/cm^3$ yields $\Delta t = 5.5\Delta h$. Hence, 0.28 mm/a of coastal uplift corresponds with a maximum crustal thickening of 1.5 mm/a.

Although there is considerable uncertainty in relating crustal thickening to horizontal shortening, a maximum horizontal shortening rate can be inferred by assuming that all shortening is accommodated by crustal thickening. A first-order estimate will assume a block of crust of uniform thickness. The Mount Lebanon part of the restraining bend is approximately 40–50 km wide and underlain by a 27–30 -km-thick crust (Khair *et al.* 1993). For these dimensions, thickening of 1.5 mm/a will correspond with shortening of 2.1–2.7 mm/a, which is comparable, but slightly less than the 2.6–3.1 mm/a suggested by relative plate motions and the kinematic model in Figure 5. To accommodate the full shortening within the Mount Lebanon Range may require either greater uplift rate (which is not observed) or a wider zone of deformation than the 40 km between the Yammouneh Fault and the limit of the offshore fold belt. As with the discussion of the Roum fault and the southern Mount Lebanon, this suggests that some shortening may be accommodated within the Anti-Lebanon and Palmyride fold belts.

Implications for long-term development of the LRB

Since the present-day slip rates along the main NNE–SSW strike-slip faults account for all expected strike-slip deformation within the restraining bend, there is no need for distributed simple shear away from the faults. Hence, present-day regional kinematics seem well explained in terms of nearly complete regional strain partitioning, which does not involve tectonic wrenching (e.g. Tikoff & Teyssier 1994; Jones & Tanner 1995). On the other hand, this kinematic model does not seem applicable to the long-term, Cenozoic tectonic development of the Lebanese restraining bend: vertical-axis rotations, finite strain expressions of distributed shear, are suggested from palaeomagnetic studies. On Mount Lebanon, counterclockwise rotations of 30° are inferred from Cretaceous rocks (Van Dongen *et al.* 1967; Gregor *et al.* 1974). Similarly, rotations of at least 30° and more have been reported in the southern Anti-Lebanon region around Mount Hermon (Ron 1987). The exact timing of these post-Cretaceous rotations is uncertain: some have attributed the rotations to the present (i.e. neotectonic) activity (e.g. Ron 1987; Westaway 1995), whereas others have suggested that these correspond with earlier deformation (e.g. Walley 1998; Westaway 2004). Although the counterclockwise sense of rotation is consistent with left-lateral shear along the Dead Sea fault system, the Late Quaternary strain partitioning suggests that rotation and wrenching occurred earlier.

To reconcile the present-day kinematics with the finite deformation, we suggest herein that the

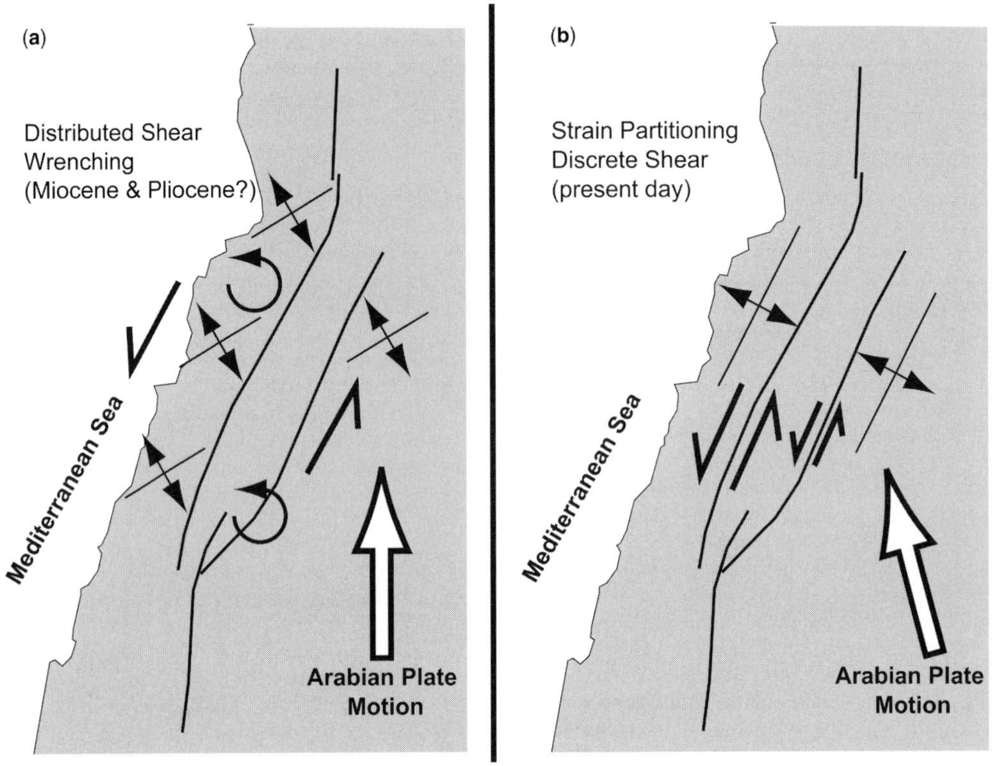

Fig. 10. Proposed two-stage evolution of the Lebanese restraining bend. An earlier phase involves wrenching and rotation (**a**). A later phase involves strain partitioning (**b**). It also seems likely that plate motions associated with these two episodes have changed over time.

Lebanese restraining bend has experienced a polyphase tectonic evolution (Fig. 10):

- An earlier deformation, involving wrenching, distributed shear, block rotation and less crustal shortening (relative plate motion less oblique to the restraining bend) (Fig. 10a).
- A later (and active) phase of tectonism involving strain partitioning and increased convergence (increased obliquity of plate motion relative to the restraining bend) (Fig. 10b).

Wrenching and distributed shear would evolve to regional strain partitioning as wrench-related folds (which originally form oblique to the strike-slip faults) rotate counterclockwise to become oriented subparallel to the strike-slip faults. The result is the present-day configuration of strain partitioning. This proposed kinematic history is similar to that suggested by Miller (1998) for the present-day slip partitioning along the central San Andreas Fault.

Furthermore, the present proportion of WNW–ESE crustal shortening and strike-slip is inconsistent with total estimates of shortening and strike-slip. Today, the obliquity of plate motion relative to the restraining bend resolves fault slip into 4.0 mm/a strike-slip and 2.8 mm/a orthogonal shortening. Assuming that these proportions of strike-slip and shortening have been maintained for the course of the Cenozoic, the total 105 km of strike-slip displacement along the southern DSFS would resolve into 80 km and 67 km of strike-slip and shortening, respectively. Similarly, considering only the 45 km from the second phase of DSFS movement implies 34 km of strike-slip displacement and 29 km of shortening within the restraining bend. Even for the latter second case, the shortening predicted is considerably greater than the 9–16 km of shortening implied by Hancock & Atiya (1979). This suggests either (1) that the present proportions of strike-slip and shortening are characteristic only of the latest Cenozoic, or (2) that some shortening has been accommodated by lateral extrusions (e.g. Jones *et al.* 1997; Gomez *et al.* 1998). Although extensional structures west of the Roum Fault are consistent with some component of lateral extrusion (Griffiths *et al.* 2000), changing plate-boundary conditions are also likely.

Garfunkel (1981) suggested that the Euler pole describing Arabian–Sinai plate motions has evolved through the Late Cenozoic with a progressively eastward position (i.e. closer to the transform). The implication of such a change in plate motion would be increasing obliquity of plate motion resolved along the transform (Fig. 10a). This increased component of convergence along the central and northern DSFS served as the basis for suggesting that the restraining bend, as well as the northern DSFS, was no longer active, owing to the geometrically incompatible orientation of those parts of the transform (Butler *et al.* 1998). Although recent geological and geodetic studies have strongly refuted the suggested inactivity of the central and northern DSFS, the suggestion of an evolving plate motion has merit. Within the context of the two-stage model in Figure 10, we suggest that increased compression within the restraining bend (resulting from increased convergence) may have favoured reactivation of the pre-existing, rotated folds and contractional faults, rather than the development of new structures.

Conclusions

The Late Quaternary kinematics of the Lebanese restraining bend seem adequately reconciled with present-day plate boundary conditions by a model of strain partitioning. This model has implications for the understanding of the regional earthquake hazard, in that multiple, seismogenic structures are active – this has been demonstrated by recent palaeoseismic investigations along key faults (e.g. Gomez *et al.* 2003; Daeron *et al.* 2005; Nemer & Meghraoui 2006). Additionally, the discrete structures that accommodate horizontal shortening may include seismogenic, blind faults, expressed at the surface only as folds such as the Jaouz and Turbol anticlines. These structures warrant further investigation, particularly owing to their proximity to the city of Tripoli.

The present-day strain partitioning model is inconsistent with the long-term (Late Cenozoic) finite deformation in the Lebanese restraining bend. Reconciling the present configuration with these long-term constraints seems to require a change in crustal kinematics during the Late Cenozoic.

This Lebanese restraining bend may serve as an example of the complex tectonic history associated with the development of restraining bends in general. Present-day strain partitioning is observed in large restraining bends along other transform-fault systems (e.g. Hispaniola and the Big Bend of the San Andreas Fault). These restraining bends are long-lived features along fault systems that have hundreds of kilometres of total displacement. Hence, strain partitioning may be characteristic of mature fault systems that have experienced significant regional strain.

We thank F. Phillips and T. Thomas for assistance with the analysis of the ^{36}Cl results. Fieldwork associated with this study benefited from significant logistical support provided by the Lebanese National Center for Remote Sensing. C. Burchfiel and L. Seeber provided helpful reviews of this manuscript. We also thank P. Mann and D. Cunningham for feedback and helpful suggestions on this paper. We also acknowledge the assistance of S. Lepley. Research included in this paper was partially supported by US National Science Foundation grants EAR-0340670 and EAR-0439021. This research was also partially supported by University of Missouri Research Board grant URB04-102.

References

AMBRASEYS, N. N. 2006. Comparison of frequency of occurrence of earthquakes with slip rates from long-term seismicity data: the cases of the Gulf of Corinth, Sea of Marmara and Dead Sea Fault Zone. *Geophysical Journal International*, **165**, 516–526.

AMBRASEYS, N. N. & BARAZANGI, M. 1989. The 1759 earthquake in the Bekaa Valley: implications for earthquake hazard assessment in the Eastern Mediterranean region. *Journal of Geophysical Research*, **94**, 4007–4013.

AMBRASEYS, N. N. & JACKSON, J. A. 1998. Faulting associated with historical and recent earthquakes in the Eastern Mediterranean region. *Geophysical Journal International*, **133**, 390–406.

BECKER, T. W., HARDEBECK, J. L. & ANDERSON, G. 2005. Constraints on fault slip rates of the Southern California plate boundary from GPS velocity and stress inversions. *Geophysical Journal International*, **160**, 634–650.

BEST, J. A., BARAZANGI, M., AL-SAAD, D., GEBRAN, A. & SAWAF, T. 1990. Bouguer gravity trends and crustal structure of the Palmyride Mountain belt and surrounding northern Arabian platform in Syria. *Geology*, **18**, 1235–1239.

BEYDOUN, Z. R. 1999. Evolution and development of the Levant (Dead Sea Rift) Transform System: a historical–chronological review of a structural controversy. *In*: MACNIOCAILL, C. & RYAN, P. D. (eds) *Continental Tectonics*. Geological Society, London, Special Publications, **164**, 239–255.

BLOOM, A. L., BROECKER, W. S., CHAPPELL, J. M. A., MATTHEWS, R. K. & MESOLELLA, K. J. 1974. Quaternary sea level fluctuations on a tectonic coast: new ^{230}Th/^{234}U dates from the Huon Peninsula, New Guinea. *Quaternary Research*, **4**, 185–205.

BOWEN, R. & JUX, U., 1987. The drama of climate change. *In*: *Afro-Arabian Geology: A kinematic view*. Chapman and Hall, New York, 212–232.

BREW, G., LITAK, R., BARAZANGI, M. & SAWAF, T. 1999. Tectonic evolution of northeast Syria: regional implications and hydrocarbon prospects. *GeoArabia*, **4**, 289–318.

BULL, W. B. 1977. The alluvial fan environment. *Progress in Physical Geography*, **1**, 222–270.

BULL, W. B. 2000. Correlation of fluvial aggradation events to times of global climate change. *In*: NOLLER, J. S. & SOWERS, J. M. (eds) *Quaternary Geochronology: Methods and Applications*. American Geophysical Union. AGU Reference Shelf, **14**, 456–464.

BUTLER, R. W. H., SPENCER, S. & GRIFFITHS, H. M. 1997. Transcurrent fault activity on the Dead Sea Transform in Lebanon and its implications for plate tectonics and seismic hazard. *Journal of the Geological Society, London*, **154**, 757–760.

BUTLER, R. W. H., SPENCER, S. & GRIFFITHS, H. M. 1998. The structural response to evolving plate kinematics during transpression: evolution of the Lebanese restraining bend of the Dead Sea Transform. *In*: HOLDSWORTH, R. E., STRAHAN, R. A. & DEWEY, J. F. (eds), *Continental Transpressional and Transtensional Tectonics*. Geological Society, London, Special Publications, **135**, 81–106.

CALAIS, E., MAZABRAUD, Y., MERCIER DE LEPINAY, B., MANN, P., MATTIOLI, G. & JANSMA, P. 2002. Strain partitioning and fault slip rates in the northeastern Caribbean from GPS measurements. *Geophysical Research Letters*, **29**, doi:10.1029/2002GL015397.

CHAIMOV, T. A., BARAZANGI, M., AL-SAAD, D., SAWAF, T. & GEBRAN, A. 1990. Crustal shortening in the Palmyride fold belt, Syria, and implications for movement along the Dead Sea fault system. *Tectonics*, **9**, 1369–1386.

DAERON, M., BENEDETTI, L., TAPPONNIER, P., SURSOCK, A. & FINKEL, R. 2004. Constraints on the post ~25-Ka slip rate of the Yammouneh fault (Lebanon) using in situ cosmogenic ^{36}Cl dating of offset limestone-clast fans. *Earth and Planetary Science Letters*, **227**, 105–119.

DAERON, M., KLINGER, Y., TAPPONNIER, P., ELIAS, A., JACQUES, E. & SURSOCK, A. 2005. Sources of the large A.D. 1202 and 1759 Near East earthquakes. *Geology*, **33**, 529–532.

DARAWCHEH, R., SBEINATI, M. R., MARGOTTINI, C. & PAOLINI, S. 2000. The 9 July 551 AD Beirut earthquake, Eastern Mediterranean region. *Journal of Earthquake Engineering*, **4**, 403–414.

DUBERTRET, L. 1955. *Carte Geologique du Liban*. Lebanese Ministry of Public Works, Beirut.

FREUND, R., GARFUNKEL, Z., ZAK, I., GOLDBERG, M., WEISSBROD, T. & DERIN, B. 1970. The shear along the Dead Sea rift. *Philosophical Transactions of the Royal Society of London, Series A*, **267**, 107–130.

GARFUNKEL, Z. 1981. Internal structure of the Dead Sea leaky transform (rift) in relation to plate kinematics. *Tectonophysics*, **80**, 81–108.

GARFUNKEL, Z., ZAK, I. & FREUND, R. 1981. Active faulting in the Dead Sea Rift. *Tectonophysics*, **80**, 1–26.

GOMEZ, F., KHAWLIE, M., TABET, C., DARKAL, A. N., KHAIR, K. & BARAZANGI, M. 2006. Late Cenozoic uplift along the northern Dead Sea transform in Lebanon and Syria. *Earth and Planetary Science Letters*, **241**, 913–931.

GOMEZ, F., ALLMENDINGER, R., BARAZANGI, M., ER-RAJI, A. & DAHMANI, M. 1998. Crustal shortening and vertical strain partitioning in the Middle Atlas Mountains of Morocco. *Tectonics*, **17**, 520–533.

GOMEZ, F., MEGHRAOUI, M. ET AL. 2001. Coseismic displacements along the Serghaya fault: an active branch of the Dead Sea fault system in Syria and Lebanon. *Journal of the Geological Society, London*, **158**, 405–408.

GOMEZ, F., MEGHRAOUI, M. ET AL. 2003. Holocene faulting and earthquake recurrence along the Serghaya branch of the Dead Sea fault system in Syria and Lebanon. *Geophysical Journal International*, **153**, 658–674.

GOSSE, J. C. & PHILLIPS, F. M. 2001. Terrestrial in situ cosmogenic nuclides: theory and application. *Quaternary Science Reviews*, **20**, 1475–1560.

GREGOR, C. B., MERTZMAN, S., NARIN, A. E. M. & NEGENDANK, J. 1974. Palaeomagnetism and the Alpine tectonics of Eurasia 5: the palaeomagnetism of some Mesozoic and Cenozoic volcanic rocks from the Lebanon. *Tectonophysics*, **21**, 375–395.

GRIFFITHS, H. M., CLARK, R. A., THORP, K. M. & SPENCER, S. 2000. Strain accommodation at the lateral margin of an active transpressive zone: geological and seismological evidence from the Lebanese restraining bend. *Journal of the Geological Society, London*, **157**, 289–302.

HANCOCK, P. L. & ATIYA, M. S. 1979. Tectonic significance of mesofracture systems associated with the Lebanese segment of the Dead Sea transform fault. *Journal of Structural Geology*, **1**, 143–153.

HEMPTON, M. R. 1987. Constraints on Arabian plate motion and extensional history of the Red Sea. *Tectonics*, **6**, 687–705.

JONES, R. R. & TANNER, W. G. 1995. Strain partitioning in transpression zones. *Journal of Structural Geology*, **17**, 793–802.

JONES, R. R., HOLDSWORTH, R. E. & BAILEY, W. 1997. Lateral extrusion in transpression zones: the importance of boundary conditions. *Journal of Structural Geology*, **19**, 1201–1217.

KHAIR, K. 2001. Geomorphology and seismicity of the Roum fault as one of the active branches of the Dead Sea fault system in Lebanon. *Journal of Geophysical Research*, **106**, 4233–4245.

KHAIR, K., KHAWLIE, M., HADDAD, F., BARAZANGI, M., SEBER, D. & CHAIMOV, T. 1993. Bouguer gravity and crustal structure of the Dead Sea transform fault and adjacent mountain belts in Lebanon. *Geology*, **21**, 739–742.

KLINGER, Y., AVOUAC, J. P., ABOU KARAKI, N., DORBATH, L., BOURLES, L. & REYS, J. 2000. Slip rate on the Dead Sea transform fault in the northern Araba Valley (Jordan). *Geophysical Journal International*, **142**, 755–768.

LITAK, R. K., BARAZANGI, M., BEAUCHAMP, W., SEBER, D., BREW, G., SAWAF, T. & AL-YOUSSEF, W. 1997. Mesozoic–Cenozoic evolution of the intraplate Euphrates fault system, Syria: implication for regional tectonics. *Journal of the Geological Society, London*, **154**, 653–666.

MAHMOUD, S., REILINGER, R., MCCLUSKY, S., VERNANT, P. & TEALEB, A. 2005. GPS evidence for northward motion of the Sinai block: implications for E. Mediterranean tectonics. *Earth and Planetary Science Letters*, **238**, 217–227.

MAY, P. R. 1989. *Low-angle Overthrusting of Mount Hermon: A Working Hypothesis*. Annual Meeting—Israel Geological Society, **1989**, 102 pp.

MILLER, D. D. 1998. Distributed shear, rotation, and partitioned strain along the San Andreas fault, central California. *Geology*, **26**, 867–870.

MOLNAR, P. 1992. Brace–Goetze strength profiles, the partitioning of strike-slip and thrust faulting at zones of oblique convergence, and the stress–heat flow paradox of the San Andreas Fault. *In*: EVANS, B. & WONG, T.-F. (eds) *Fault Mechanics and Transport Properties of Rocks*. Academic Press, New York, 435–459.

NEMER, T. 2005. *Sismotectonique et comportement sismique du relais transpressif de la faille du Levant : rôles et effets des branches de failles sur l'aléa sismique au Liban*. Université Louis Pasteur, Strasburg, PhD dissertation, 208 pp.

NEMER, T. & MEGHRAOUI, M. 2006. Evidence of coseismic ruptures along the Roum fault (Lebanon): a possible source for the AD 1837 earthquake. *Journal of Structural Geology*, **28**, 1483–1495.

PAGE, B. M., THOMPSON, G. A. & COLEMAN, R. G. 1998. Late Cenozoic tectonics of the central and southern Coast Ranges of California. *Geological Society of America Bulletin*, **110**, 846–876.

PHILLIPS, F. M. & PLUMMER, M. A. 1996. CHLOE: a program for interpreting in-situ cosmogenic nuclide data for surface exposure dating and erosion studies. *Radiocarbon*, **38**, p. 98.

PHILLIPS, F. M., STONE, W. B. & FABRYKA-MARTIN, J. T. 2001. An improved approach to calculating low-energy cosmic-ray neutron fluxes near the land/atmosphere interface. *Chemical Geology*, **175**, 689–701.

PINTER, N. & GARDNER, T. W. 1989. Construction of a polynomial model of sea level: estimating palaeo-sea levels continuously through time. *Geology*, **17**, 295–298.

PINTER, N. 1996. *Exercises in Active Tectonics*. Prentice-Hall, New York, 166 pp.

PINTER, N. & KELLER, E. A. 1995. Geomorphic analysis of neotectonic deformation, northern Owens Valley, California. *Geologische Rundschau*, **84**, 200–212.

PIRAZOLLI, P. A., LABOREL, J. & STIROS, S. C. 1996. Earthquake clustering in the Eastern Mediterranean during historical times. *Journal of Geophysical Research*, **101**, 6083–6097.

PONIKAROV, V. P. 1964. *Geological Map of Syria*. Ministry of Industry, Damascus, Syria.

QUENNELL, A. M. 1984. The Western Arabia rift system. *In*: DIXON, J. E. & ROBERTSON, A. H. F. (eds) *The Geological Evolution of the Eastern Mediterranean*. Geological Society of London, Special Publications, **17**, 775–788.

REILINGER, R., MCCLUSKY, S. *ET AL*. 2006. GPS constraints on continental deformation in the Africa–Arabia–Eurasia continental collision zone and implications for dynamics of plate interactions. *Journal of Geophysical Research*, **111**, doi:10.1029/2005JB004051.

RON, H. 1987. Deformation along the Yammuneh, the restraining bend of the Dead Sea transform: palaeomagnetic data and kinematic implications. *Tectonics*, **6**, 653–666.

ROSSIGNOL-STRICK, M. 1993. Late Quaternary climate in the Eastern Mediterranean region. *Palaeorient*, **19**, 135–152.

SABBAGH, G. 1962. *Geological Cross Section Across Northern Lebanon*. American University of Beirut Archive, Beirut.

SANLAVILLE, P. 1974. Le role de la mer dans les aplanissesments cotiers du Liban. *Revue de Geographie de Lyon*, **49**, 295–310.

SANLAVILLE, P. 1977. *Etude Géomorphologique de la Région Littorale du Liban*. Lebanese University, Beirut, 859 pp.

SBEINATI, M. R., DARAWCHEH, R. & MOUTY, M. 2005. The historical earthquakes of Syria: an analysis of large and moderate earthquakes from 1365 B.C. to 1900 A.D. *Annals of Geophysics*, **48**, 347–435.

SHACKLETON, N. J. & OPDYKE, N. D. 1973. Oxygen iosotope and palaeomagnetic stratigraphy of equatorial Pacific core V28–238: oxygen isotope temperatures and ice volume on a 10^5 and 10^6 year time scale. *Quaternary Research*, **3**, 39–55.

STONE, J. O. H., EVANS, J. M., FIFIELD, L. K., ALLAN, G. L. & CRESSWELL, R. G. 1998. Cosmogenic chlorine-36 production in calcite by muons. *Geochimica et Cosmochimica Acta*, **63**, 433–454.

TIKOFF, B. & TEYSSIER, C. 1994. Strain modeling of displacement-field partitioning in transpressional orogens. *Journal of Structural Geology*, **16**, 1575–1588.

VAN DONGEN, P. G., VAN DER VOO, R. & RAVEN, T. 1967. Palaeomagnetism and the Alpine tectonics of Eurasia 3: palaeomagnetic research in the central Lebanon mountains and in the Tartous area (Syria). *Tectonophysics*, **4**, 35–53.

WALLEY, C. D. 1988. A braided strike-slip model for the northern continuation of the Dead Sea Fault and its implications for Levantine tectonics. *Tectonophysics*, **145**, 63–72.

WALLEY, C. D. 1998. Some outstanding issues in the geology of Lebanon and their importance in the tectonic evolution of the Levantine region. *Tectonophysics*, **298**, 37–62.

WDOWINSKI, S., BOCK, Y. *ET AL*. 2004. GPS measurements of current crustal movements along the Dead Sea Fault. *Journal of Geophysical Research*, **109**, doi: 10.1029/2003JB002640, 2004.

WESTAWAY, R. 1995. Deformation around stepovers in strike-slip fault zones. *Journal of Structural Geology*, **17**, 831–846.

WESTAWAY, R. 2004. Kinematic consistency between the Dead Sea Fault Zone and the Neogene and Quaternary left-lateral faulting in SE Turkey. *Tectonophysics*, **391**, 203–237.

WOODCOCK, N. H. & FISCHER, M. 1986. Strike-slip duplexes. *Journal of Structural Geology*, **8**, 725–735.

Structural geometry and timing of deformation in the Chainat duplex, Thailand

M. SMITH[1], S. CHANTRAPRASERT[2], C. K. MORLEY[3] & I. CARTWRIGHT[4]

[1]*Department of Petroleum Geosciences, Universiti Brunei Darussalam, Bandar Seri Begawan, Brunei Darussalam*

[2]*Department of Geological Sciences, Chiang Mai University, Thailand*

[3]*PTTEP, 555 Vibhavadi-Rangsit Road, Chatuchak, Bangkok, Thailand*

[4]*School of Geosciences, Monash University, Clayton, Victoria, 3800, Australia*

Abstract: The Chainat duplex is about 100 km in a north–south direction, and was developed along the predominantly sinistral Mae Ping fault zone, which was active during the Cenozoic. The duplex is manifested as eroded, north–south- and NW–SE-striking outliers of Palaeozoic and Mesozoic rocks rising from the surrounding flat plains of the Central Basin (a Pliocene–Recent post-rift basin). Satellite images, geological maps and magnetic maps have been used to reconstruct the structural geometry of the duplex, which is composed of a series of north–south-striking ridges, bounded to the north and south by NW–SE-striking faults. Overall, the duplex has the geometry of analogue restraining-bend models with relatively low displacement. No well-developed duplex-traversing short-cut faults linking the principal displacement zones are apparent. The duplex shows evidence for widespread sinistral motion, as well as some dextral reactivation the latter of which is particularly marked in the eastern part of the duplex. The main sinistral activity ended at about 30 Ma: subsequently, minor, episodic reactivation of the duplex may have occurred. Detailed timing of events cannot be determined from structures within the duplex, but the evolution of adjacent rift basins suggests that stresses developed during episodes of inversion may have also caused reactivation of strike-slip faults (sinistral for NW–SE to north–south striking faults) during the Miocene. Minor episodic dextral motion may also have been of Late Oligocene–Miocene and/or Pliocene–Recent age.

This paper and a companion paper (Morley *et al.* 2007) address the evolution of a large strike-slip duplex developed along the 100-km-long Khlong Lhan restraining bend in the Cenozoic Mae Ping fault zone in Thailand (Fig. 1). The presence of the duplex was largely based upon the regional geometry of outcrops seen on geological maps and satellite data (Morley 2002). This paper presents the results of more detailed geological studies aimed at better documenting the evidence for a strike-slip duplex, its geometry and structural evolution in the Chainat area. The data used to document the Chainat duplex in this paper are: outcrops containing strike-slip-related deformation; cooling ages along the fault zone (Upton *et al.* 1997; Upton 1999; Morley 2004); satellite and outcrop maps (Morley 2002, 2004); magnetic anomaly maps; wells and seismic reflection data from adjacent Cenozoic rift basins (e.g. O'Leary & Hill 1989; Bal *et al.* 1992; Tulyatid 1997). The companion paper focuses on the use of cooling ages to determine the structural evolution of the Chainat Ridge and Mae Ping fault zone, particularly during the Oligocene.

Despite many documented bends in strike-slip fault systems, there are few well-described large strike-slip restraining-bend duplexes in the literature (for examples see reviews in Cunningham *et al.* 1996, 2003; McClay & Bonora 2001; Cowgill *et al.* 2004; Wakabayashi *et al.* 2004). The idealized geometry and evolution of strike-slip duplexes has been addressed by analogue modelling (e.g. Dooley & McClay 1997; McClay & Bonora 2001). Typically, these models consider two main strike-slip fault strands or principal displacement zones (PDZs), which lie en echelon and are connected via an oblique-trending zone of deformation. As this oblique zone opens or closes, a releasing or restraining strike-slip duplex is developed in the overlying material (usually sand). Variations in the structural geometry of the modelled duplexes are primarily a result of changing the angle of the oblique zone to the main strike-slip faults, and the amount of strike-slip motion. Studies of natural large fault systems indicate that there are also significant variations in restraining-bend behaviour, related to whether stresses at the bends are

From: CUNNINGHAM, W. D. & MANN, P. (eds) *Tectonics of Strike-Slip Restraining and Releasing Bends.* Geological Society, London, Special Publications, **290**, 305–323.
DOI: 10.1144/SP290.11 0305-8719/07/$15.00

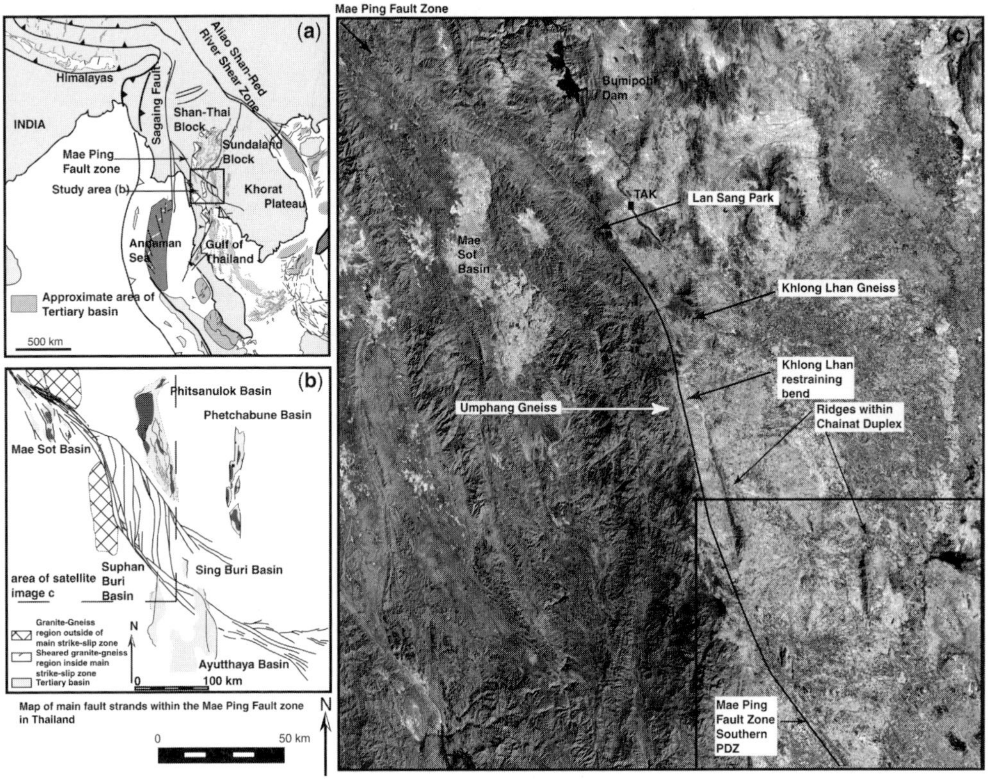

Fig. 1. (**a**) to (**c**) Regional location maps and satellite image showing some of the key features of the Mae Ping fault zone in central Thailand.

strike-slip (i.e. vertical principal stress = σ2), or compressional (i.e. vertical principal stress = σ3) (e.g. Hauksson & Jones 1988; Cowgill *et al.* 2004). More natural examples of large strike-slip duplexes are needed to demonstrate and test the applicability of these analogue and natural models.

In addition to this documentation of examples, the Chainat duplex described in this paper has important regional implications for the way that escape tectonics related to the Himalayan Orogeny have affected central Thailand. In particular, structural relationships within the duplex, and between the duplex and adjacent Cenozoic rift basins, can be used to test whether the simple progression from sinistral to dextral motion with time on major NW–SE-trending fault zones (e.g. Tapponnier *et al.* 1986; Polachan *et al.* 1991; Lacassin *et al.* 1997, 1998) has actually occurred.

Geological background

The Central Plains of Thailand are a north–south-trending, flat-lying region that extends northward about 450 km from the coast around Bangkok. The flatness of the region is due to ongoing post-rift subsidence that commenced sometime during the Late Miocene or Pliocene. The base of the post-rift section is marked by a subhorizontal, generally low-relief unconformity, located several hundred to about 600 m below the land surface. Below this unconformity lie Late Oligocene–Miocene Cenozoic rift basins. The largest are the Suphan Buri, Ayutthaya and the Phitsanulok basins (Flint *et al.* 1988; O'Leary and Hill 1989; Bal *et al.* 1992). The western margin of the Central Plains is marked by a broad, north–south-trending region of high hills (western highlands), which experienced significant uplift and erosion during the Cenozoic (Lacassin *et al.* 1997; Upton 1999; Morley 2004). Traversing these ranges are NW–SE- and north–south-trending Cenozoic strike-slip faults – the two largest of which are called the Three Pagodas fault zone and the Mae Ping (or Wang Chao) fault zone (e.g. Lacassin *et al.* 1997; Morley 2002; Fig. 1). The term 'Chainat Ridge' was applied by O'Leary & Hill (1989) to the region of low-relief ridges of Palaeozoic and

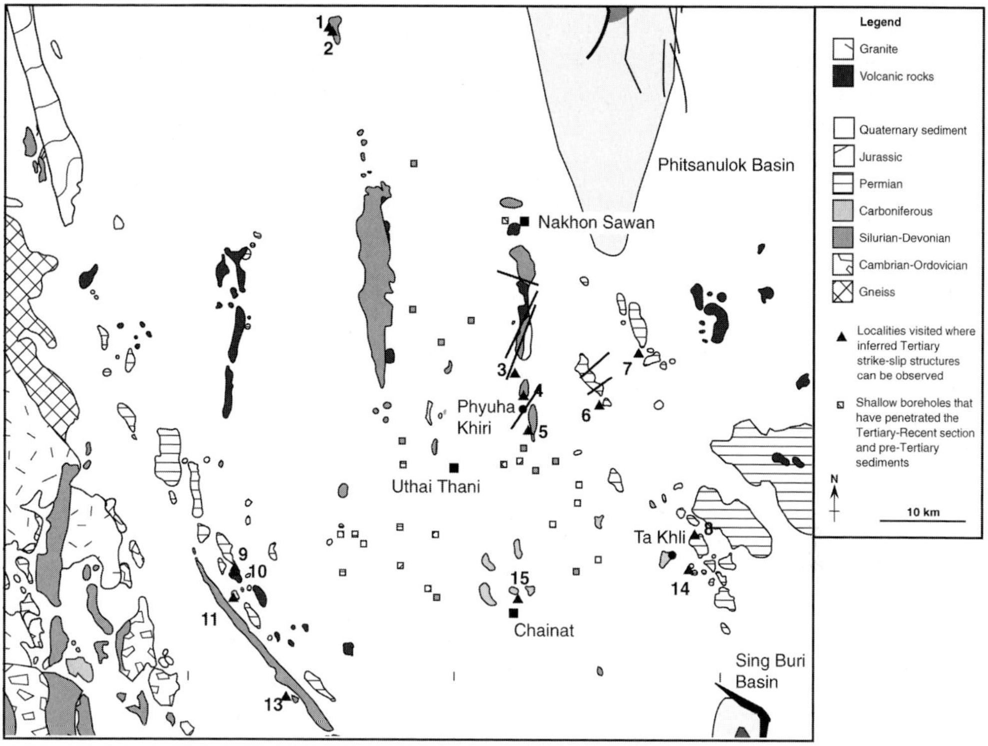

Fig. 2. Geological map of the outlying ridges in the Chainat area. See Figure 1c for location. Small squares are basement penetrations made by shallow boreholes.

Mesozoic rocks that outcrop in the Central Plains (Fig. 2). Satellite images of the Mae Ping fault zone show that where it exits the western highlands to the east, the relatively narrow, discrete fault zone, some 5–10 km wide, broadens out into a region defined by NW–SE- and north–south-striking topographic ridges (Figs 1, 2 & 3a). The proximity of the ridges to the Mae Ping fault zone in the western highlands led Bunopas (1981) and O'Leary & Hill (1989) to interpret them as part of the Mae Ping fault zone. However, a detailed interpretation of the geometry of faults within the Chainat area was not made. A sediment provenance study of the syn-rift section in the Suphan Buri and Ayutthaya basins indicates that the Chainat Ridge was a sediment source area for the Late Oligocene–Miocene basins (O'Leary & Hill 1989).

The mixture of north–south and NW–SE trends is seen elsewhere along the Mae Ping fault zone. The NW–SE-oriented fault zone in the western highlands slices through predominantly north–south-striking, Palaeozoic–Early Mesozoic terrane boundaries (as defined by Barr & Macdonald 1991, Fig. 3). However, several of the north–south splays appear to coincide with the terrane boundaries: the Mae Sariang splay (Fig. 3) coincides with the boundary between the Western Zone and the Inthanon Zone, whilst the major bend in the fault zone that is the focus of this study (here termed the Khlong Lhan bend, Fig. 3), lies between the Sukhothai Zone and the Inthanon Zone. Hence, the influence of major crustal pre-existing fabrics described for examples of large releasing or restraining bends elsewhere in the world (e.g. Corsini *et al.* 1996; Tommasi & Vauchez 1997; Curtis 1998), also appears to have influenced their location along the Mae Ping fault zone.

The relationship between the strike-slip faults and Cenozoic basins have been the subject of two main models: in one model the Cenozoic rift basins are treated as pull-apart basins and owe their origins to strike-slip motion (e.g. Knox & Wakefield 1983; Tapponnier *et al.* 1986; Polachan *et al.* 1991; Bal *et al.* 1992; Lacassin *et al.* 1997; Replumaz & Tapponnier 2003); in the other model most of the Cenozoic rift basins are primarily extensional (Morley *et al.* 2000, 2001). However, the extensional story is complex, due to fluctuations in stress orientation and magnitude

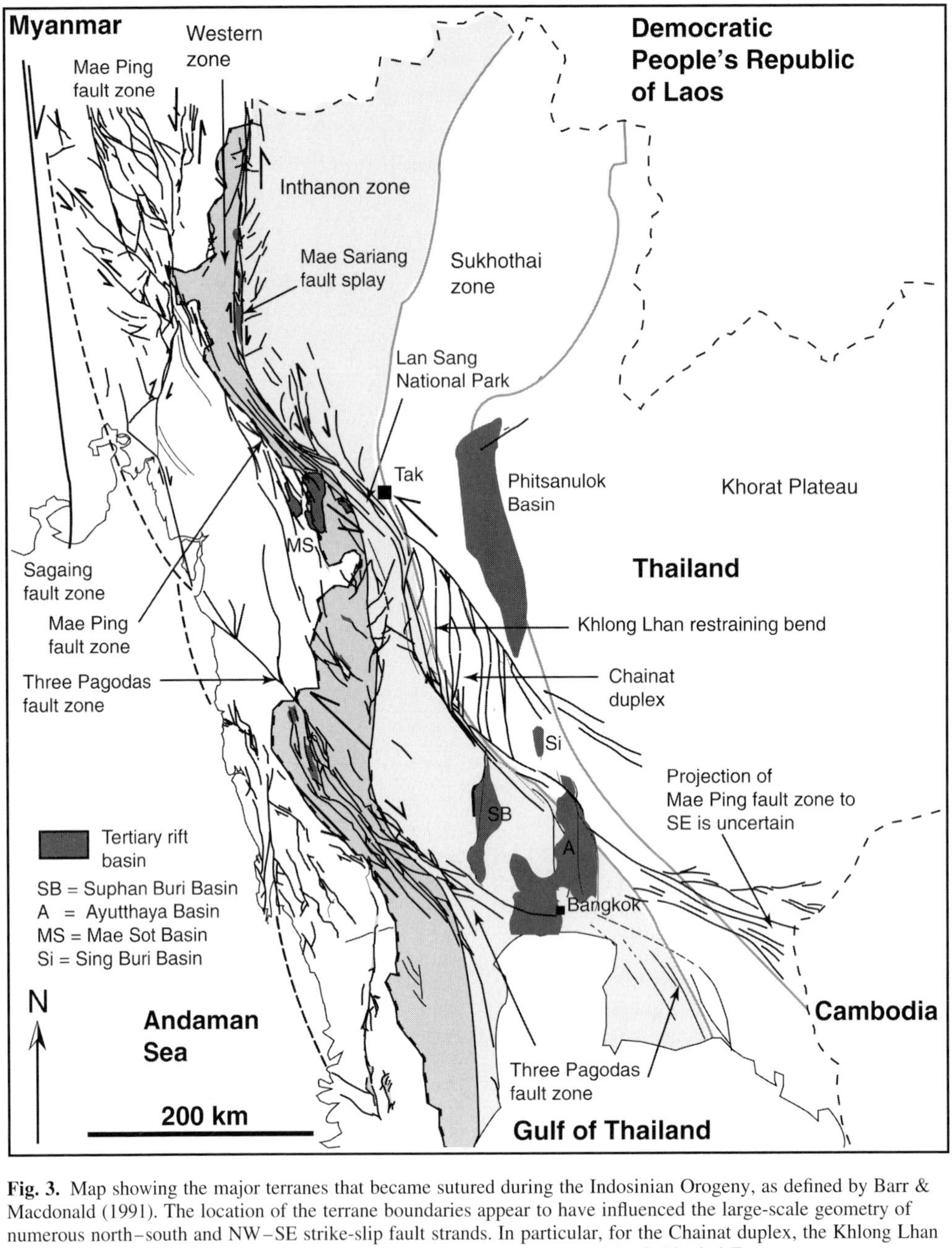

Fig. 3. Map showing the major terranes that became sutured during the Indosinian Orogeny, as defined by Barr & Macdonald (1991). The location of the terrane boundaries appear to have influenced the large-scale geometry of numerous north–south and NW–SE strike-slip fault strands. In particular, for the Chainat duplex, the Khlong Lhan restraining bend is located at the boundary between the Inthanon zone and the Sukhothai Zone.

through time that episodically resulted in periods of basin inversion and strike-slip reactivation (Morley 2002, 2004). Furthermore, there are lateral changes in tectonic stress magnitude, in particular a westward transition from extensional to strike-slip stresses, as indicated by the absence of Late Cenozoic deformation in eastern Thailand; the presence of north–south-trending rift systems in central and offshore Thailand; and the active dextral strike-slip Sagaing Fault in Myanmar (Fig. 3). Thus, within the Three Pagodas and the Mae Ping fault zones in the western ranges, there are genuine pull-apart basins located at north–south-striking releasing bends that formed sometime between the Late Oligocene and the present day (Morley 2002; Fig. 4).

In the strike-slip model, the change from sinistral to dextral displacement on large NW–SE-trending faults is placed around the Late Oligocene (Lacassin *et al.* 1997). The switch to dextral motion then opened up the rift basins along north–south-trending releasing bends (e.g. Polachan *et al.* 1991). This fault evolution fits in with the overall development of escape tectonics related to the Himalayan Orogeny, and fits the pattern of tectonic development identified for the Red River shear zone (e.g. Lacassin *et al.* 1997; Leloup *et al.* 2001). However, Morley (2002) argued that the geometry of the Chainat Ridge failed to match the predictions of the strike-slip model. The heart of the Chainat Ridge is composed of a series of north–south-striking isolated ridges of Palaeozoic–Mesozoic sedimentary, meta-sedimentary, and igneous rocks (Fig. 2). In the south, the north–south trend is replaced by NW–SE-striking ridges. On the basis of satellite images and published geological maps Morley (2002) proposed that the Chainat Ridge was actually a large strike-slip duplex, about 100 km wide (east–west) and 100 km long (north–south). The presence of outcropping pre-Cenozoic rocks between the Phitsanulok Basin to the north, and the Ayutthaya and Suphan Buri Basins to the south (Fig. 1) suggested that the duplex primarily formed by exhuming pre-Cenozoic rocks at a restraining bend during sinistral deformation (Morley 2002). Under significant dextral motion the Chainat duplex would have acted as a large releasing bend. Consequently, the absence of evidence for subsidence and basin formation in the Chainat duplex suggests that dextral motion on the Mae Ping fault zone was minor (Morley 2002). However, in none of the published literature have the detailed relationships between the Cenozoic rift basins and faults forming the Chainat duplex been described. In an unpublished PhD thesis (Tulyatid 1997) an interpretation of the Chainat area was made on the basis of magnetic data, and a number of north–south and NW–SE-striking fault strands were interpreted within the Chainat area. The overall duplex-like fault pattern geometry discussed in this paper was also determined by Tulyatid (1997).

Interpretation of the major fault geometries within the duplex

The traces of the Mae Ping and Three Pagodas faults passing from the western ranges eastward under the Central Plains are difficult to define in detail from geological mapping and satellite data, because surface outcrops are widely spaced apart. The remnant topography has been infilled and onlapped by poorly dated Cenozoic–Recent sediments (Fig. 2). In reality, the location of the major strike-slip faults in eastern Thailand and the Gulf of Thailand are poorly constrained, despite the confidence with which strike-slip fault patterns are commonly deployed on regional maps. The best evidence for the location of the strike-slip faults are the scattered outcrops of the Chainat area (Fig. 2) and magnetic data (Tulyatid 1997; Fig. 4). The change in trend of the northern strand of the Three Pagodas fault zone from NW–SE to east–west under the central plains is particularly well displayed on magnetic data (Nutalaya & Rau 1984; Fig. 4). However, the trace of the Mae Ping fault zone is less clear. The southern PDZ of the Chainat duplex is strongly manifest as outcrop trends and on the magnetic data (NW of Ang Thong; Fig. 4). The curving of ridges and inferred faults from NW–SE to NNW–SSE trends is also a feature of the southern PDZ (Fig. 5a). However, the internal arrangement of north–south-trending ridges does not stand out within the magnetic data. In a few places, outcrop geometries indicate major and secondary faults by their sharp contacts and the juxtaposition of rocks of different ages in adjacent outcropping localities. Along the western ridge of Jurassic rocks (Figs 2 & 5b) and in some other places, the very sharp, linear edge of outcrops also indicates the location of a fault (Fig. 5a). However, more typically, the ridges are linear (Fig. 5c), but outcrop edges are insufficiently well defined to precisely locate the position of the fault traces. It is assumed that the faults controlling the ridges lie on one or both sides of each ridge. Minor faults trending NW–SE and NE–SW do appear to internally break up some of the ridges (Fig. 2). The strongest evidence of slip direction from outcrop data comes from the western ridge of Mesozoic rocks, which displays an anticline–syncline fold pair, that strikes at about 040° oblique to the north–south trend of the ridge. The fold geometry suggests development as en echelon buckle folds during sinistral motion (Fig. 5b).

In the 1980s 81 shallow stratigraphic wells were drilled by BP to the top pre-Cenozoic (Fig. 6). The basement encountered in these wells, plus the outcrop information was used to make the Cenozoic subcrop map in Figure 7. The wells typically drilled 20–50 m of 'young' poorly dated Quaternary to Cenozoic poorly lithified clastic section before reaching basement. Exceptionally, about 100 m of young section was penetrated. However, there was no evidence for any well-developed Cenozoic basins being present between the outcropping pre-Cenozoic ridges,

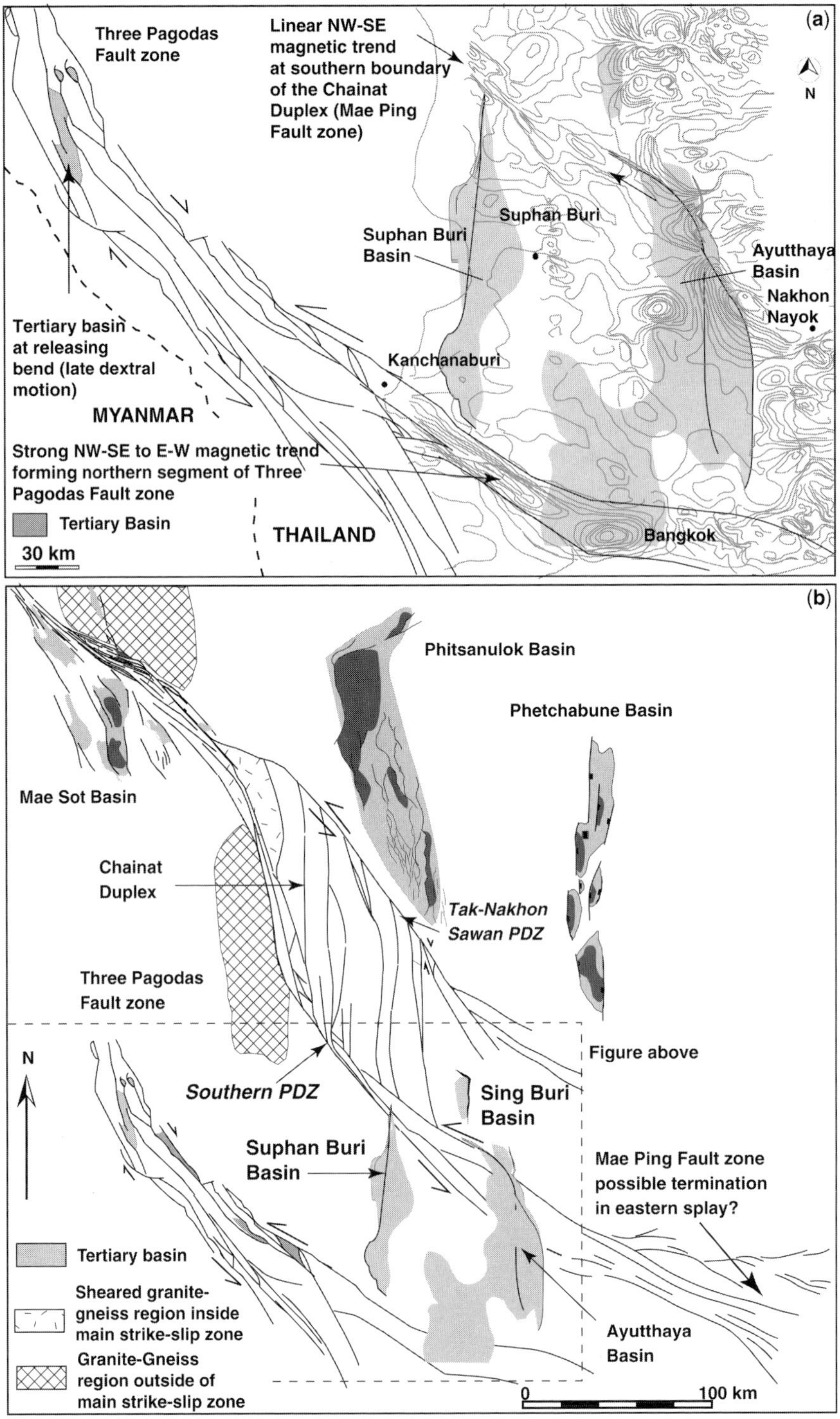

Fig. 4. Maps of fault patterns of the Three Pagodas (**a**) (**b**) and Mae Ping fault zones (b). (a) illustrates the relationship of fault strands interpreted from outcrop, satellite and magnetic anomalies (Nutalaya & Rau 1984). The Three Pagodas fault zone appears to splay into important east–west and NNW–SSE-striking strands. The NW–SE strike of the southern principal displacement zone (PDZ) of the Mae Ping fault zone is a strong feature of the magnetic anomalies.

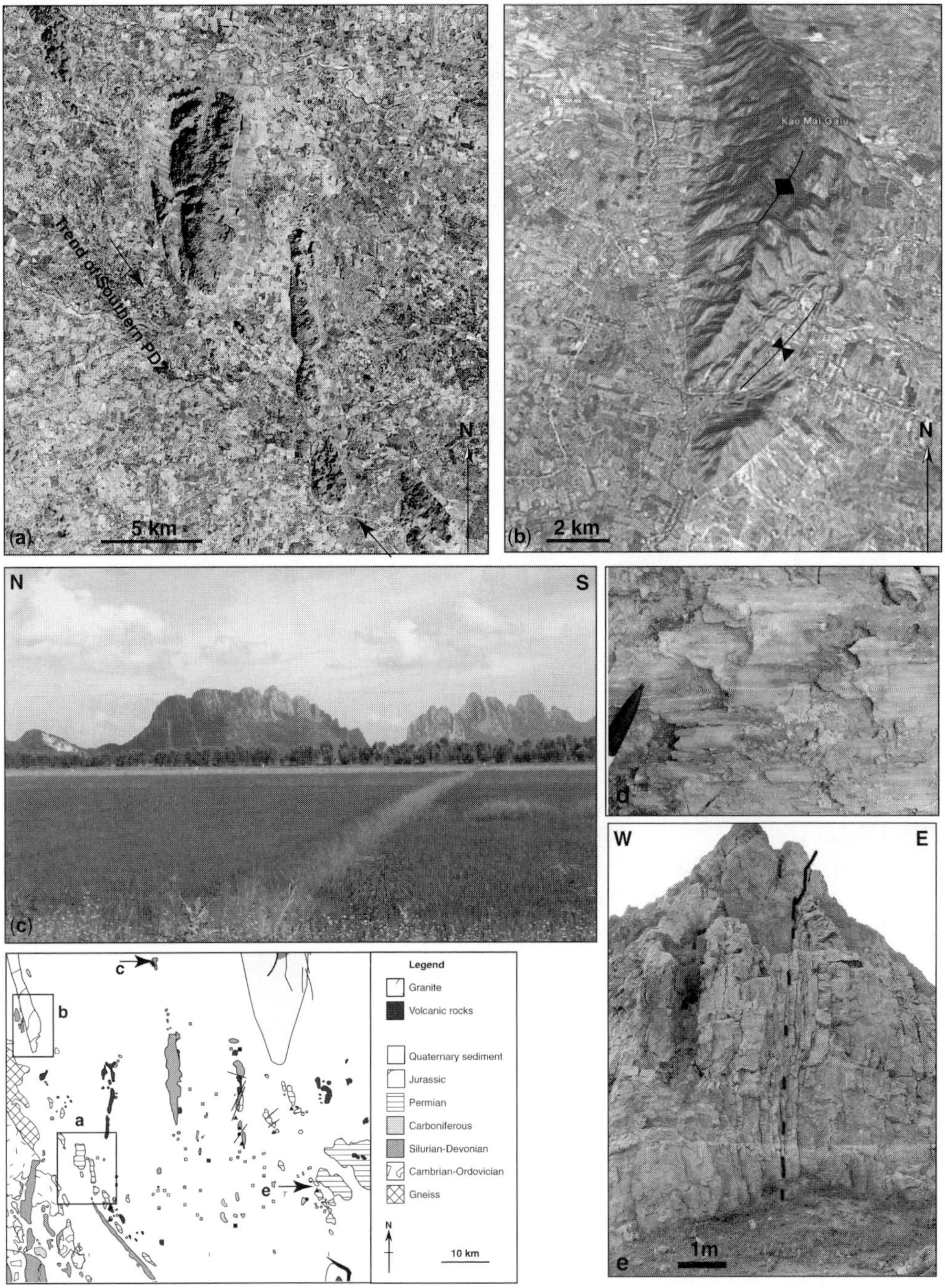

Fig. 5. Images of the geometry and structure of ridges within the Chainat duplex. (**a**) Linear north–south ridges where a number of fault strands splay from the NW–SE-trending Ayutthaya-Ban Klong Pong Fault and curve to a north–south direction. (**b**) The southern end of the Mesozoic ridge east of the Khlong Lhan restraining bend, showing a sharp, linear boundary to the ridge (inferred faulted boundary) and a fold pair consistent with sinistral displacement. (**c**) View to the east of one of the smaller ridges in the Chainat duplex, illustrating the general character of the region: long, narrow, linear ridges rising abruptly from a flat plain. The ridges are of Silurian carbonates. (**d**) and (**e**) Permian carbonate quarry in Takhli, showing the general characteristics of dextral strike-slip faults in the carbonates. They form networks of subvertical fractures, and commonly contain subhorizontal to *c*. 45° plunging slickenfibres. The location map for localities (a)–(e) is the same one as in Figure 2.

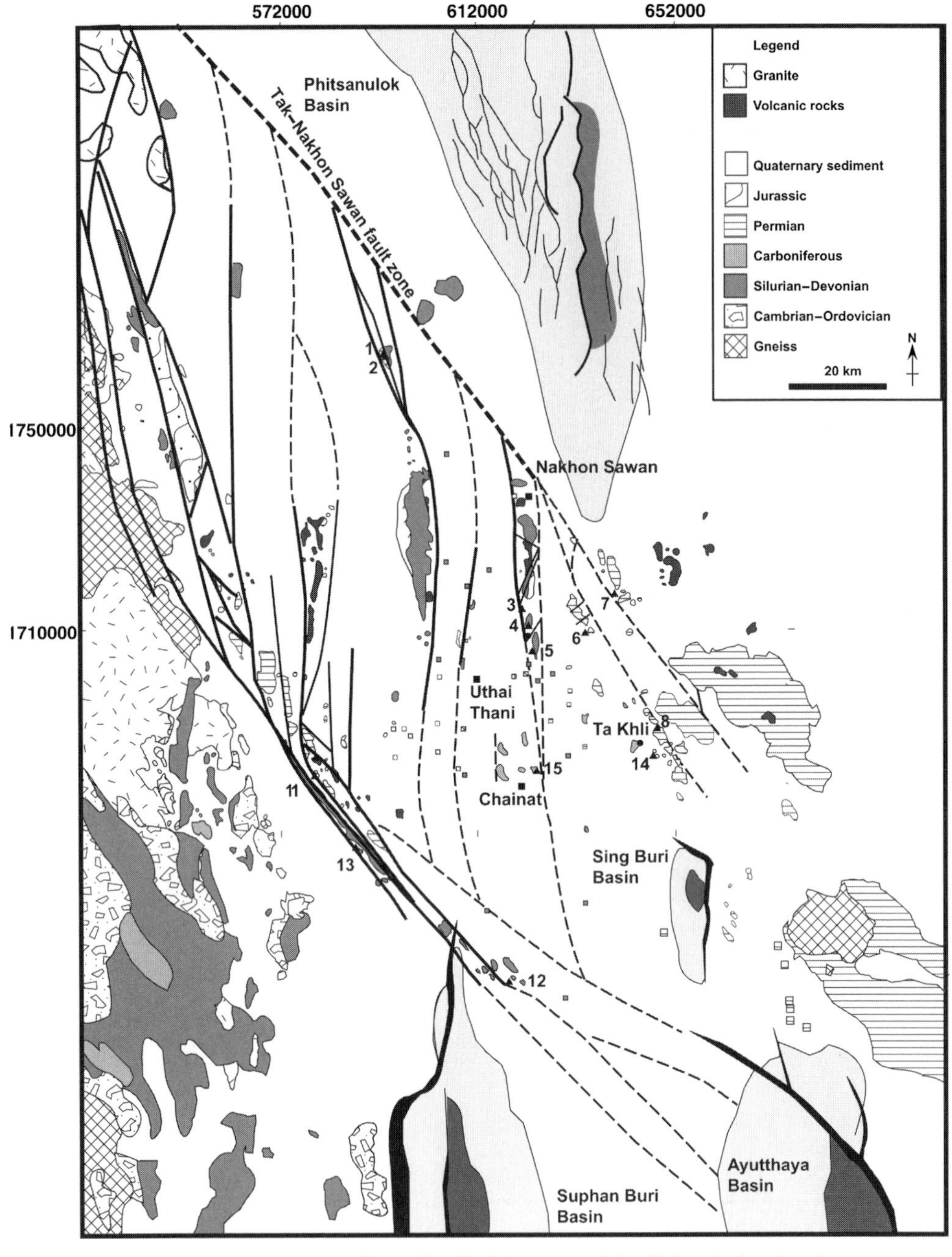

Fig. 6. Figure showing the interpretation of the strike-slip fault pattern of the Chainat duplex, based on outcrop, borehole satellite and magnetic data.

except SE of the duplex, where a simple east-thickening half-graben called the Sing Buri Basin was identified (Fig. 6).

The Permo-Triassic igneous rocks in the eastern region of the duplex give a strong magnetic response (Fig. 4). The areas of Cenozoic sedimentary basins are not very obvious from magnetic data, but are generally regions of long-wavelength, low-amplitude magnetic response. The Cenozoic rift basins associated with the Chainat area are

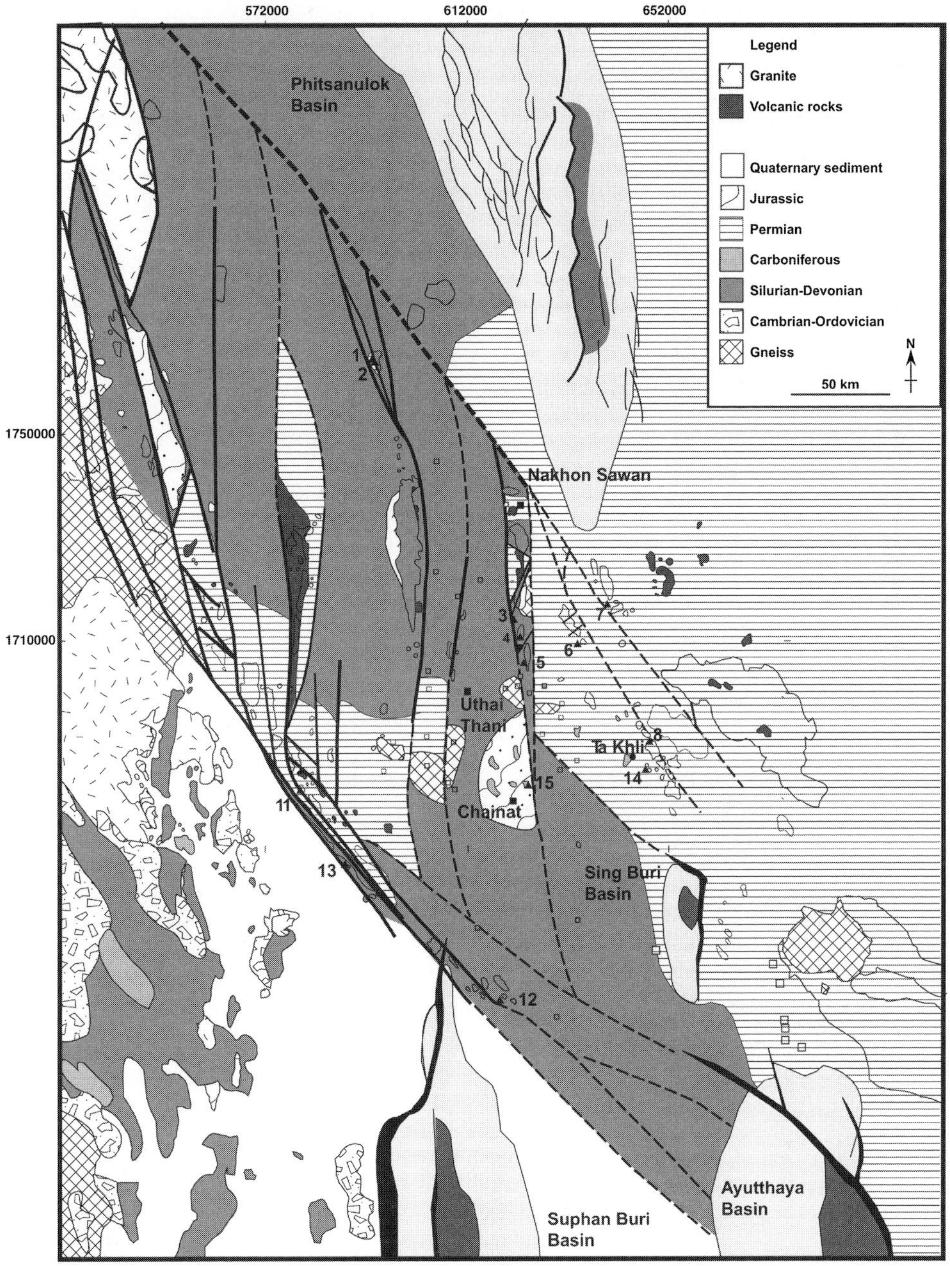

Fig. 7. Interpretation of the pre-Cenozoic subcrop map of the Chainat duplex, based on projecting faults, outcrop data, and borehole pre-Cenozoic penetrations.

known from seismic-reflection data, wells, and some gravity data, and their timing and geometry are well constrained (e.g. Knox & Wakefield 1983; Flint *et al.* 1988; O'Leary & Hill 1989; Bal *et al.* 1992; Ronge & Surarat 2002; Seusuthya & Morley 2004).

Identification of strike-slip deformation in outcrop

The great majority of outcrops within the duplex are Lower to Upper Palaeozoic meta-sediments and meta-volcanics. These rocks have been through at least the Permo-Triassic Indosinian Orogeny, which itself has several deformation episodes (Cooper *et al.* 1989; Charusiri *et al.* 1993); and more if they are of Early Palaeozoic age (Barr & MacDonald 1991). Hence, determining what is 'young' deformation (i.e. Cenozoic–Recent) related to strike-slip motion, and what is older deformation, unrelated to the strike-slip deformation, can be difficult.

In attempting to determine the different ages of deformation we have made the following assumptions:

1. In Triassic and older rocks any thrusts and folds present are assumed to be related to the Indosinian or older orogenies and so cannot be used to help determine the Cenozoic strike-slip kinematics. This assumption is made simply because of difficulties and cost in resolving a younger age for deformation in these rocks.
2. Any steeply dipping strike-slip fault, or normal fault in Triassic and older rocks that has a brittle deformation style and cross-cuts older deformation fabrics (folds, thrust faults, pressure-solution seams, extensive calcite vein arrays), has the possibility of being associated with the late strike-slip deformation.
3. Deformation in Mesozoic clastic sedimentary rocks. The continental Mesozoic section is poorly dated but is present in several areas. Some folding in the region occurred during the Mesozoic, but Late Cretaceous–Early Cenozoic deformation has also occurred and may be significant for the initial development of the Mae Ping fault zone. The nature of the Late Cretaceous–Early Cenozoic deformation in Thailand is poorly described. Morley (2004) suggested that folding and thrusting is associated with a transpressional orogenic belt arising from the collision of the Burma Block with the Shan-Thai Block. This transpressional phase caused the initial development of the Mae Ping and Three Pagodas fault zones. The degree of deformation within the Mesozoic section in Thailand is highly variable. Just to the NE of the Chanait Ridge, east of Phitsanulok, are hills of flat-lying Jurassic clastics, whilst to the SE the western edge of the Khorat Plateau is also flat to gently dipping. Hence, in some areas the Mesozoic section is essentially undeformed. Yet, in other areas, particularly in patchy outcrops in western and central Thailand, including the area NE of Phitsanulok, Mesozoic rocks are deformed by thrusting and folding. In one well in the southern Phitsanulok Basin, the Thai Shell Oil Company obtained a K–Ar date of bedding-oblique cleavage in a cored shale that yielded an age of 74 ± 3 Ma.

The majority of the outcrops within the Chainat Ridge are slaty to phyllitic igneous, volcaniclastic, and metasedimentary rocks assigned a Siluro-Devonian age on geological maps (Fig. 2). These rocks form prominent tree-covered hills, with generally poor, weathered exposures. The dip of foliations in the rocks ranges from horizontal to vertical, and can rapidly vary within an outcrop. Fresh exposures display refolded folds, pervasive foliations, stretching lineations, and shear zones that boudinage veins and create small-scale duplex geometries (Fig. 2, location 5). A wide range of foliation orientations are present: WNW–ESE- to ENE–WSW-striking foliations, oblique to the north–south ridges, are commonly found, but also frequently foliations are subparallel to the ridges.

Unfortunately, the poorly exposed nature of the outcrops means that outside this general observation, no definitive observations of strike-slip-related structures have been found in the Siluro-Devonian rocks. Possibly at location 13 (Fig. 2), shear along the Mae Ping fault zone is recorded in Siluro-Devonian rocks, where a sub-horizontal to up to 20°SE (110°) plunging stretching lineation in quartz rods (up to 2 cm wide, 10 cm long) is present; however, the age of the deformation is unknown.

Carbonates of Siluro-Devonian age and Permo-Triassic age rocks have been mapped within the Chainat duplex. They tend to form spectacular ridges of extensive rocky outcrops (Fig. 5c). However, the natural exposures tend to be weathered, karstified and yield virtually no structural information. Fortunately, the carbonates have local economic value (cement, building stone, roadfill) and many large and small quarries are present. It is these quarries that permit the main structural observations to be made (Localities 1, 2, 6, 7, 8, 10 and 11). This does introduce a sampling bias, since the main quarries occur on the eastern side of the duplex, and in one area north of Nakhon Sawan (Fig. 2). Carbonates display a wide range of characteristics, including: recrystallized marbles with little remaining primary sedimentary fabric; massive carbonates, partially recrystallized and heavily veined; well-bedded massive carbonates and well-bedded carbonates interbedded with shales. Where the carbonates are well bedded, fold-and-thrust structures of probable Indosinian Orogeny age are well displayed (Fig. 2, localities 1 and 8). The strike of the folds and thrusts ranges from north–south to NW–SE.

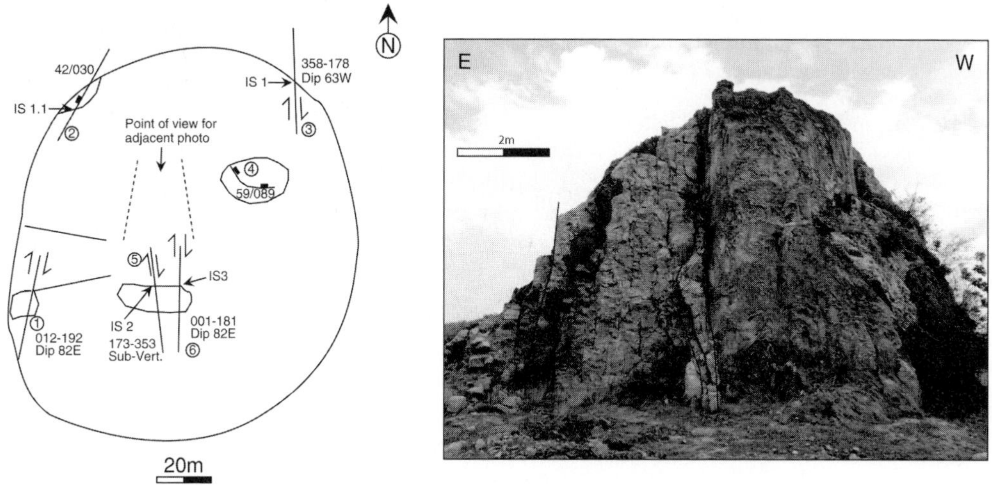

Fig. 8. Sketch map of Permian carbonate quarry in Takhli, and a photograph of a vertical dextral strike-slip fault offsetting a karstified limestone (location 8, Fig. 2). This was the locality where the slickenfibres were analysed for carbon and oxygen stable isotopes.

Within the Permian carbonate quarries it is common to find one or several subvertical brittle fault zones cutting across older fold- and thrust-related structures and vein-arrays (Figs 5d & e & 8). These fault zones are invariably subparallel to the strike of the ridges, and have slickenfibres whose plunge ranges from horizontal to about 45° (Fig. 9). Typically, the fault zones are broken into a number of minor fault-bounded blocks, with the secondary faults displaying a wide range of dips, strikes and slickenside orientations.

At locations 3 and 4 (Fig. 2) is a well-exposed Triassic granite. This granite is intruded by strongly altered mafic dykes, that in places have taken on a subvertical, north–south-striking shear-zone-related foliation. The foliation passes into striated, subvertical fracture zones within the granite that show a sinistral sense of offset. The granite is highly fractured along many different orientations (Fig. 9). However, the dominant sense of offset is sinistral along north–south-striking fractures. Hence, although the timing of motion is uncertain, the deformation is consistent with Cenozoic

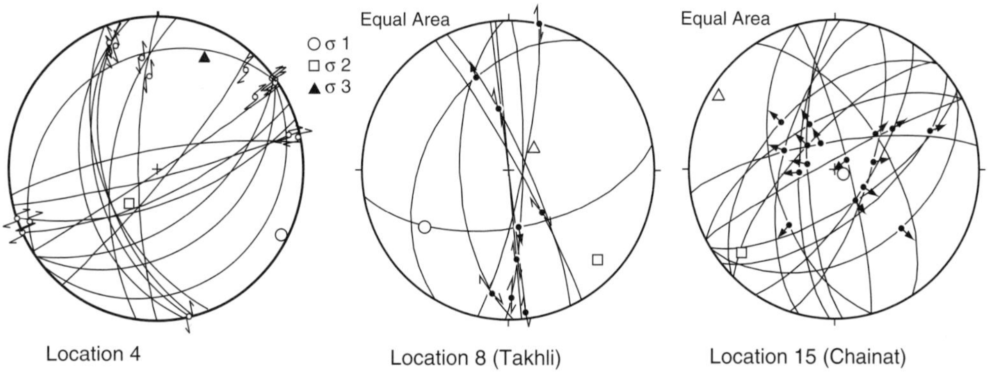

Fig. 9. Example of striations from quarries where a relatively high number of faults with striations were present. Location 4, shows north–south-striking fractures with a sinistral sense of motion, Triassic granite. Location 8 shows north–south to NNW–SSE-striking dextral strike-slip faults, with an oblique extensional component, Permian carbonates, Takhli. Location 12 is in Chainat and shows predominantly extensional faults, some, in particular the north–south striking faults display oblique dextral extension; the faults are within Mesozoic clastics. See Figure 2 for locations.

motion on the Mae Ping fault zone. Apatite fission-track dating of the outcrop shows rapid cooling in the Miocene, with a central age of 18 ± 1 Ma (see Morley *et al.* 2007).

The outcrop data permit the following inferences to be made about the structural evolution. Evidence for sinistral displacement along the north–south to NW–SE-trending faults (usually just a few faults per locality) is found all across the duplex, except in the east in the Permian limestone outcrops. Evidence for dextral displacement along north–south to NW–SE-trending faults is largely confined to the eastern area, particularly the Permian limestone outcrops, and Chainat. Most of the strike-slip structures identified are brittle fault zones, with slickensides.

Temperature of slickenside formation

Figure 8 shows the location of slickenside samples taken for stable oxygen- and carbon-isotope analysis. IS1, 2 and 3 were taken from dextral slickensides of fibrous calcite on N–S-trending strike-slip faults 3, 5 and 6 respectively. IS1.1 was taken on fault 2, which is probably an extensional fault. IS1.1 has both host limestone and the carbonate microbreccia; these are identified as A and B in Table 1. The slickenside fibres are large enough to allow several small areas of each to be analysed. The samples were analysed at Monash University using a Finnigan MAT 252 mass spectrometer, and are expressed in standard δ notation relative to V-SMOW (O) and PDB (C). The CO_2 from carbonate was liberated by acidification using H_3PO_4 in a He atmosphere at 72°C in a Finnigan MAT Gas Bench, and analysed by continuous flow. Results were normalized using IAEA standard CO-1 and internal laboratory standards run at the same time as the samples from this study. Many samples were analysed twice, and the precision (1σ) of both $\delta^{18}O$ and $\delta^{13}C$ values based on replicate analyses is ± 0.1‰. Stable-isotope data are presented in Table 1.

The $\delta^{18}O$ values of the slickensides range from 15.7 to 20.0‰. These $\delta^{18}O$ values may be used to estimate the temperature of their formation and, therefore, the approximate temperature during the majority of fault activity. These temperatures may provide a link between the temperature-dependent uplift dating (AFT) in other parts of the duplex and the timing of the deformation at the sample locality. The fractionation of ^{18}O between two coexisting phases (e.g. water and calcite) is temperature dependent (e.g. Faure 1986; Hoefs 1987; Gill 1997). Therefore, if the $\delta^{18}O$ of the water is known, then this allows the temperature of mineralization to be determined from the $\delta^{18}O$ of calcite veins. For this study, the calcite–water ^{18}O fractionations of Epstein *et al.* (1953) and O'Neil *et al.* (1969) were used to calculate the temperature of slickenside formation. Assuming that the water was ocean water with a $\delta^{18}O$ of 0‰, and that the system is fluid-dominated such that interaction with the rocks did not substantially alter the fluid $\delta^{18}O$ values, the estimated temperatures for slickenside formation are between 77 and 115°C, and the host limestone is 84°C. If the slickenside temperatures are assumed to be correct then faulting was at an approximate depth of 1.5 to 3 km (at a 30°C/km geothermal gradient). The host limestone temperature may seem high, but a detailed study of limestones of similar age in southern Thailand, by Dill *et al.* (2005), indicates that it is consistent with alteration along a burial trend.

The $\delta^{13}C$ values may be used to determine the source of the mineralizing fluid. The $\delta^{13}C$ value of 3.3‰ of the limestone falls within the range of marine carbonates (Emrich *et al.* 1970). However, the $\delta^{13}C$ values of the slickenside values are between −4.5 and −12.3‰. This indicates that it is an open system; however; several factors can

Table 1. *Isotope values and formation temperatures*

Sample	^{18}O SMOW	^{13}C PDB	Sample type	Formation temperature (°C)
IS1	15.7	−4.5	Fibrous calcite slix.	115 ± 20
IS1.1A	19.1	3.3	Host limestone	84.02
IS1.1B	20.0	−12.3	Fibrous calcite slix.	77 ± 20
IS2A	18.2	−12.1	Fibrous calcite slix.	91 ± 20
IS2B	18.1	−12.2	Fibrous calcite slix.	92 ± 20
IS2C	17.7	−12.1	Fibrous calcite slix.	96 ± 20
IS2D	19.2	−10.6	Fibrous calcite slix.	83 ± 20
IS2E	17.1	−10.9	Fibrous calcite slix.	101 ± 20
IS3	17.9	−11.8	Fibrous calcite slix.	94 ± 20

Oxygen and carbon isotopes expressed in standard δ notation relative to SMOW (O) and PDB (O and C). 'Slix' is an abbreviation of slickensides.

make the ^{13}C value more negative. Thus the source of the carbon cannot be determined. The assigned value of 0‰ SMOW for the slickenside formation fluid needs to be evaluated. There are many potential sources of water, but the realistic options are: meteoric water, metamorphic water and pore/connate water from the rock column. Meteoric water (*c.* −4‰ SMOW in tropical regions) may mix more with pore-water (approximate host limestone values) in areas of increased porosity and permeability, e.g. fault zones. The faults may tap into deeper metamorphic water, which can vary between *c.* 3 and 20‰ SMOW. Given the uncertainty of water volume mixing and the lack of hydrothermal minerals in the fault zones to give further temperature indicators, the ^{18}O value of the formation fluid is left at 0‰ SMOW, but with approximately ±2‰ SMOW margins of error.

Relationship of the duplex to adjacent sedimentary basins

It seems reasonably well established that, prior to about 30 Ma, motion on the Mae Ping fault zone was sinistral (e.g. Lacassin *et al.* 1993, 1997; Morley 2004). The Oligocene–Recent history is not well defined, although the assumption has been made that the displacement is dextral (e.g. Polachan *et al.* 1991; Lacassin *et al.* 1997). However, the outcrop data show that both sinistral and, in some places, dextral, brittle strike-slip faults are present along north–south and NW–SE-trending faults. Unfortunately, there are no good superimposed fabrics or cross-cutting relationships that would permit the age relationships to be defined. Hence, the best opportunity for establishing the likely timing and magnitude of these dextral and sinistral events is to investigate the impact of strike-slip faulting on adjacent sedimentary basins. The four key Late Oligocene–Pliocene basins are: the Suphan Buri, Ayutthaya, Phitsanulok and Sing Buri basins (Fig. 4).

Suphan Buri Basin

The Suphan Buri Basin is well imaged on seismic reflection data as a simple rift basin with a west-thickening half-graben geometry. The basin displays maximum displacement in the central part of the basin (O'Leary & Hill 1989; Seusuthya & Morley 2004). Syn-rift sediments are of Late Oligocene–Miocene age (O'Leary and Hill 1989). Displacement dies out northward approaching the southern NW–SE-bounding fault strand of the Chainat duplex (Fig. 6). The basin does not display any large inversion anticlines. There appears to be no displacement transfer or linkage between the Suphan Buri boundary fault, and the Mae Ping fault zone.

Ayutthaya Basin

The Ayutthaya Basin is composed of several east-thickening half-grabens filled with Late Oligocene–Miocene sedimentary rocks (O'Leary & Hill 1989). The northern termination of the basin is marked by a curved boundary fault that changes from a north–south orientation to a NW–SE orientation near the fault termination (Figs 4, 6 & 7). This pattern could be inferred to indicate a pull-apart type geometry. However, the NW–SE fault segment loses displacement towards the northwest, and does not appear to link with any strike-slip fault. Hence, a simple normal fault geometry, influenced by a pre-existing fabric, is the preferred interpretation, rather than linkage with an active strike-slip fault system. More critical still is the relationship of the main NW–SE Mae Ping fault strand with the Ayutthaya Basin (Figs 3, 6 & 7). The basin overlies the Mae Ping fault zone, and both the bounding fault and the basin edge show no offset, indicating very little activity on the southern Mae Ping fault zone from the Late Oligocene to the present. However, some minor reactivation of the Mae Ping fault zone may be indicated by local inversion of two half-graben bounding faults in the vicinity of the underlying Mae Ping fault zone (Fig. 10). If the stresses associated with inversion did induce reactivation of segments of the Mae Ping fault zone, then they would cause minor sinistral motion along the Mae Ping fault zone during the Late Miocene–Pliocene.

Sing Buri Basin

This is a small basin located on the eastern margin of the Chainat duplex (Fig. 6). Like the Auytthaya Basin, it is an eastward-thickening half-graben, with the north–south-bounding fault curving to a NW–SE strike at its northern termination. It shows no evidence for basin inversion. The basin lies south of the termination of the Tak–Nakhon Sawan fault zone which bounds the northern margin of the Chainat duplex.

Southern Phitsanulok Basin

The Phitsanulok Basin lies north of the Chainat duplex and is the largest onshore rift basin in Thailand (Fig. 4). The Phitsanulok Basin shows diminishing basin size (both width and depth) passing to the south, and progressively dies out towards the Chainat duplex (Flint *et al.* 1988; Wongpornchai 1997); a pattern that would not be expected if extension was linked to late dextral

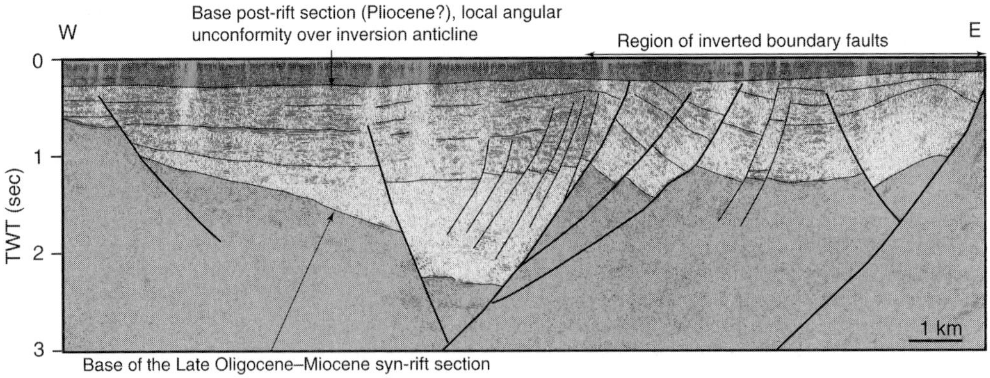

Fig. 10. Inversion geometry of the Ayutthaya Basin, from seismic reflection data (Seusuthya & Morley 2004).

strike-slip displacement (thus making north–south trends releasing bends). The southern Phitsanulok Basin is composed of two sub-basins: the Lahan Graben and the Nong Bua sub-basin (Fig. 11). The Lahan Graben lies closest to the Mae Ping fault zone, and shows evidence for Late Oligocene–Early Miocene extension (Fig. 11). By the end of the Early Miocene extension in the basin had ceased, and this was followed by uplift and widespread erosion. In Particular, in the area between the Lahan and Nong Bua sub-basins, there was pronounced basin inversion, which occurred in the late Early Miocene. Inversion created north–south-trending anticlines and inverted north–south-trending faults. The Nong Bua sub-basin to the east was largely unaffected by the inversion, and shows simple expansion of the section into a half-graben bounding fault from the Late Oligocene to the Late Miocene (Fig. 11). Uplift and inversion of the Lahan sub-basin coincides with the apatite fission-track cooling ages obtained for two granites in the Chainat duplex (22.4 $\pm$ 2.8 Ma and 18.2 $\pm$ 1.6 Ma) (Morley *et al.* 2007). Hence, the uplift and inversion are interpreted to have arisen from an approximately east–west-oriented maximum horizontal compression direction which would have activated Chainat duplex faults either as thrusts or sinistral strike-slip faults. A Late Miocene inversion phase also affected parts of the Phitsanulok Basin (Bal *et al.* 1992; Morley *et al.* 2004).

Summary of tectonic history inferred from Cenozoic basins

For the southern NW–SE Mae Ping fault zone, the Suphan Buri and Ayutthaya basins indicate:

(1) major sinistral motion was prior to the Late Oligocene,
(2) no, or virtually no dextral motion has occurred, and
(3) a Late Miocene phase of minor (a few kilometres maximum displacement) sinistral motion associated with local inversion in the Ayutthaya Basin.

For the Tak–Nakhon Sawan fault zone:

(1) this fault zone shows outcrop evidence for dextral (commonly oblique-normal dextral) motion. Episodes of dextral motion could have occurred during periods of extension in

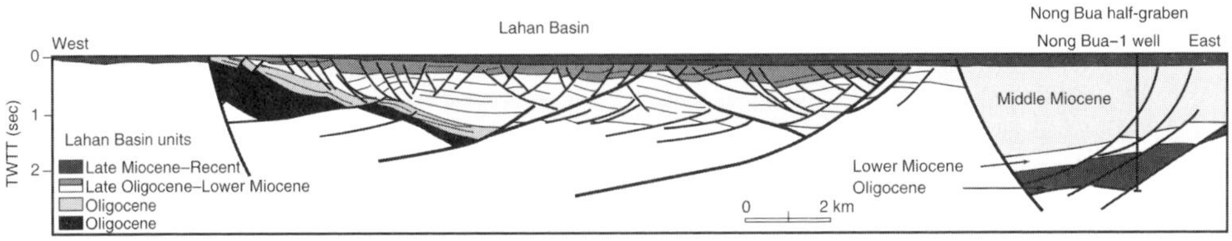

Fig. 11. Cross-section through the Lahan graben, based on seismic-reflection data showing the active early history of the basin, followed by an unconformity in the Early Miocene. This is unusual for the basin, where elsewhere most of the activity was of Miocene age. This suggests that the close proximity of the Lahan graben to the Mae Ping fault zone was a significant factor in the basin's history.

the basins or when compressional or strike-slip stresses with an approximately north–south S_{Hmax} affected the area (like the Recent stress field in northern Thailand, for example). The modern stress field is associated with a north–south S_{Hmax} direction (Morley 2004).

(2) Short episodes of sinistral strike-slip motion were probably associated with phases of basin inversion in the late Early Miocene, and the Late Miocene.

Figure 12 summarizes the likely timing of structural events in the sedimentary basins and corresponding motions within the Chainat duplex. However, although long periods of potential activity are shown in Figure 12, these just indicate the broad time range when parts of the fault zone may have been active. There is no intention to imply sustained fault activity or large displacements during those periods when activity could have occurred.

More regionally, there is also support for the development of dextral motion on the NW–SE-trending faults. The NW–SE-trending Three Pagodas Fault to the south has late Cenozoic basins developed at north–south releasing-bend geometries (Morley 2002; Fig. 4). The low-level earthquake activity that affects northern and western Thailand today is dominated by dextral strike-slip fault-plane mechanisms on NW–SE-striking faults and sinistral focal mechanisms for NE–SW-striking faults: the S_{Hmax} direction is approximately north–south (e.g. Bott *et al.* 1997; Morley 2004). However, much of the dextral strike-slip motion on the Mae Ping fault zone may have been dissipated in the western

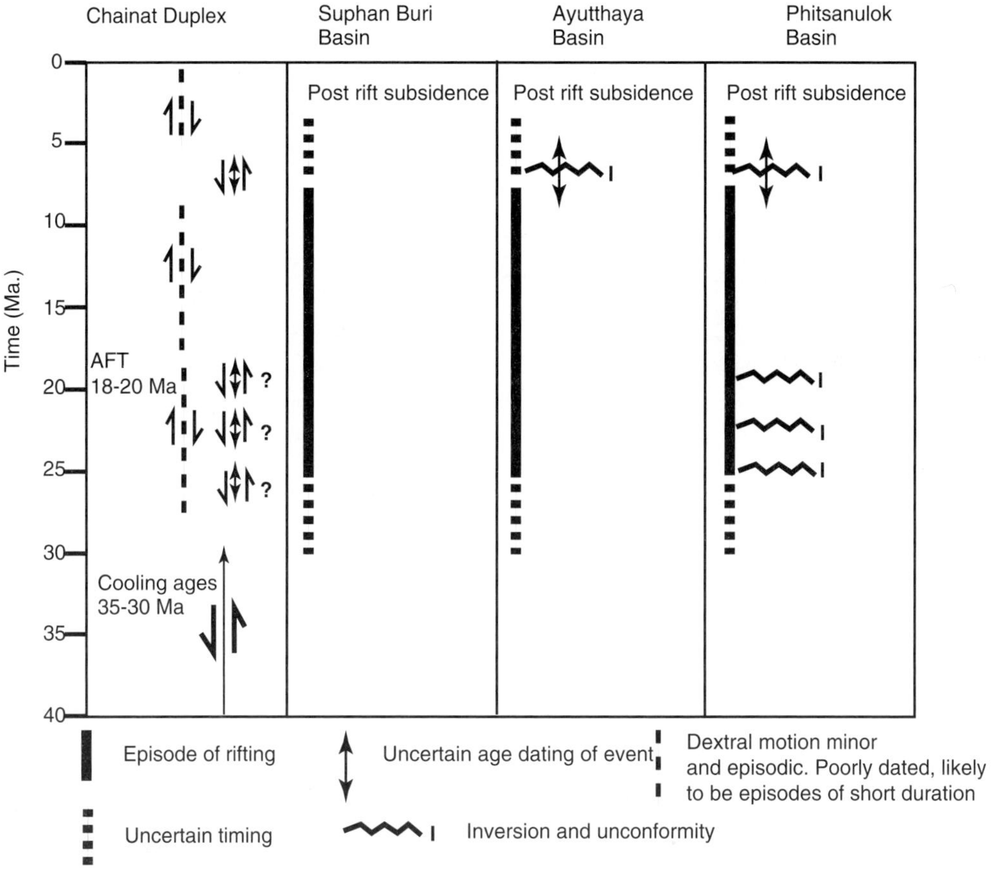

Fig. 12. Summary of the tectonic events in the Cenozoic sedimentary basins around the Chainat duplex, and the likely episodes of deformation in the Chainat duplex. The main sinistral displacement ended around 30 Ma (Lacassin *et al.* 1997). Subsequent deformation, both sinistral and dextral, appears to have been both highly episodic and minor. The chart is not meant to imply that strike-slip deformation drove inversion, it is just that stresses appropriate for inversion may also have resulted in reactivation of parts of the major strike-slip fault zones.

highlands, where the Mae Sot and Mae Lamao basins are present (Fig. 1), and dextral motion may have diminished considerably – passing eastward towards the Chainat duplex area.

Comparision of the Chainat duplex with analogue models

The Chainat duplex appears to have evolved first by sinistral motion, with displacement focused on the southern (Ayutthaya–Ban Klong Pong) fault strand, which is associated with the large Khlong Lhan restraining bend and a prominent long ridge of Jurassic rocks. No such uplifted area is associated with the eastern margin of the duplex where the northern (Tak–Nakhon Sawan) fault zone curves into a restraining-bend geometry, suggesting that much of the sinistral displacement was focused on the Ayutthaya-Ban Klong Pong fault strand. Hence, the duplex geometry is not one of equal displacement transfer between the northern and southern fault strands.

The internal geometry of the duplex is dominated by north–south-trending ridges and associated faults. In the Chainat duplex, the angle made by the restraining bend with respect to the main fault trend is about 35° and the restraining bend is long, hence the 30° stepover model shown in McClay & Bonora (2001, their fig. 3) is the most appropriate (Fig. 13). This model shows that the duplex is dominated by internal faults striking subparallel to the restraining bend, unlike higher stepover angles, where a wider range of fault angles is developed. The model pattern is very reminiscent of the dominant NNW–SSE to north–south strike of ridges within the Chainat duplex, bounded by NW–SE-striking faults to the north and south (Figs 3 & 13). The Chainat duplex shows a higher intensity of imbricate faulting than the analogue model, and a narrower zone of deformation associated with the 'floor' strike-slip fault trend, compared with the analogue model (Fig. 13). However, the greater intensity of deformation in the Southern PDZ compared with the Tak–Nakhon Sawan PDZ in the natural and analogue models suggests that in both most of the displacement occurs along the Southern PDZ, the main restraining bend and then the Tak–Nakhon Sawan PDZ at the exiting bend. Consequently, the Tak–Nakhon Sawan PDZ where it caps the duplex is a relatively low displacement fault zone. This is unlike the classic simple thrust duplex model, where the roof PDZ displays large displacements all the way across the duplex.

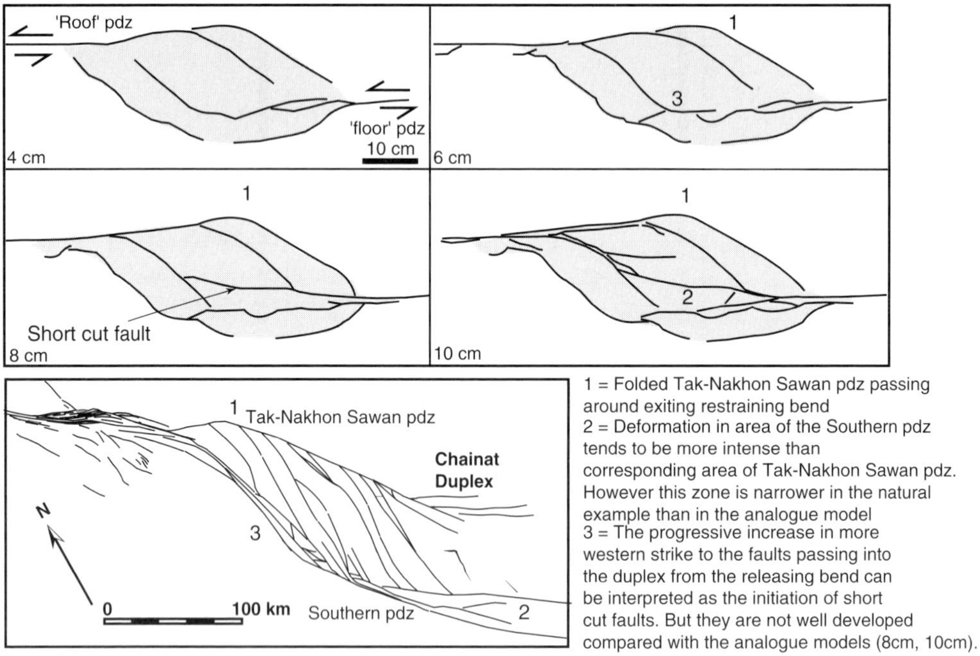

Fig. 13. Sketch of the top surface of the analogue model from McClay and Bonora (2001: their fig. 3), showing the progressive evolution of a restraining-bend duplex, at 30° to the main strike-slip fault principal displacement zones (PDZs), with 4 cm, 6 cm, 8 cm and 10 cm displacement. The analogue model results are compared with the Chainat duplex geometry.

It is quite apparent from analogue models and descriptions of natural examples of strike-slip duplexes (e.g. Laney & Gates 1996; McClay & Bonora 2001; Cunningham *et al.* 2003) that the relatively simple, classic strike-slip duplex geometry becomes complicated by a wide range of fault trends, rotation of faults, and variable fault kinematics, once large displacements become imposed. Hence, the simple geometry of the Chainat duplex suggests a relatively low-displacement feature.

Whilst the comparative simplicity of the Chainat duplex geometry might be misleading, and a function of exposure; the lack of strong uplift (and exposure of higher metamorphic-grade rocks) within the duplex; the long, linear, uninterrupted trend of the Jurassic ridge on the west side of the duplex (Fig. 2); and the 22 Ma to 18 Ma AFT ages, all indicate it is a comparatively young feature, developed late in the history of the Mae Ping fault zone. There is no evidence for a NW–SE-trending short-cut fault zone cutting across these ridges, as is typically seen in analogue models (Fig. 13). As strike-slip motion increases there is the tendency for a short-cut fault to form an oblique link to the two principal displacement zones traversing the centre of the duplex. Consequently, the faults at relatively high angles to the strike-slip fault zones display a comparatively small amount of the total strike-slip displacement, but a high amount of dip slip (e.g. Dooley and McClay 1997; McClay & Bonora 2001). The Dead Sea Basin is a natural example of a short-cut fault linking the PDZs (e.g. Ben Avraham & Zoback 1992; Sagy *et al.* 2003). The analogue models with relatively low displacements (displacements up to about 0.2 × length of the duplex) do not show the short-cut fault. If similar scaling is applied to the Chainat duplex, then, with an NW–SE length of 100 km, the amount of displacement should be in the order of 10–30 km. The comparison of geometries with the analogue models, plus the absence of any unusually deeply buried rocks exhumed within the duplex, indicate that the Chainat duplex has not accommodated much of the estimated 150-km slip on the Mae Ping fault zone (Lacassin *et al.* 1997). Consequently, the Chainat duplex is viewed here as a relatively young feature of the Mae Ping fault zone, developed during the Oligocene, and represents the last stage of major sinistral displacement along the Khlong Lhan restraining bend. The older evolution of this restraining bend is discussed in a companion paper (Morley *et al.* 2007). Following cessation of major sinistral displacement, the Chainat duplex was episodically reactivated by minor sinistral and dextral motions during the Late Oligocene–Recent. Dextral motions appear to have been focused along the northern and eastern part of the duplex, particularly the Tak–Nakhon Sawan fault strand.

Conclusions

This study of the Chainat duplex shows that its structure resembles an early-stage restraining-bend strike-slip duplex geometry when compared with analogue models (McClay & Bonora 2001). The lack of development of an obvious oblique fault zone traversing the duplex linking the PDZs, and the absence of any high-grade metamorphic rocks exposed within the duplex, both point to the relatively low displacement sinistral history. Sinistral displacement is probably in the order of a few tens of kilometres. The Chainat duplex probably began to develop towards the end of the main phase of escape tectonics in Thailand, i.e. around 33–30 Ma (using dates from Lacassin *et al.* 1997). To explain the apparent immaturity of the Chainat duplex at a large restraining bend, and the presence of exhumed middle-crustal rocks nearby at Lan Sang, Morley *et al.* (2007, Paper 12, this volume) suggest that a precursor restraining-bend duplex to the Chainat duplex at the Khlong Lhan restraining bend was shortened and exhumed by passage through the exiting bend. As this region of the crust was removed from the east side of the restraining bend during the main period of sinistral displacement, a new duplex (the Chainat duplex) began to develop in its place during the waning stages.

For the post-30-Ma time period, the dominance of simple right-lateral displacement proposed in earlier regional models (e.g. Polachan *et al.* 1991; Lacassin *et al.* 1997, 1998) does not seem to have occurred. Instead, the stress system exhibited a more complex evolution. Extensional, not pull-apart basins developed during the Late Oligocene and Miocene (*c.* north–south S_{Hmax}); they overlie the older, largely inactive strike-slip fault system. These basins are not offset by the strike-slip faults; however, the major boundary faults do curve into a NW–SE orientation, following older pre-existing weaknesses. The extensional basins were episodically inverted, with the likely S_{Hmax} orientations ranging between ENE–WSW and WNW–ESW. Such stress orientations could give rise to sinistral displacement on NW–SE- and north–south-oriented faults within the Chainat duplex for short episodes in the Early and Late Miocene. Dextral strike-slip motion did occur episodically in parts of the duplex, particularly focused on the Tak–Nakhon Sawan fault strand, but is minor (probably kilometres rather than tens of kilometres). Dextral deformation occurred at relatively shallow depths, with temperatures of slickenside calcite vein fill of between 77° and 115 °C. Provided that the fluids associated with the vein fill were not pumped from a considerable depth, or diluted by meteoric waters, this suggests some 2–3 km depth of burial at the time of dextral motion, assuming a 3 °C/100 m geothermal gradient and a 25° surface temperature. If dextral motion was significant (in the order of tens of kilometres) then the Chainat duplex should form a large releasing-bend pull-apart, but there are no indications of deep basins forming between the ridges. A considerable number of normal faults are associated with the dextral strike-slip faults, and the striations on north–south to NW–SE-striking faults commonly plunge 30–40° with a normal sense of displacement, suggesting that the dextral motion was transtensional. Earthquakes and geomorphology indicate that currently the Mae Ping fault zone is currently undergoing episodic, low-strain-rate dextral motion.

M. Smith would like to acknowledge the Universiti of Brunei Darussalam and the AAPG Foundation Grants in Aid (2003) scheme for providing the funding for fieldwork and associated analytical costs. C. Morley thanks the Universiti of Brunei Darussalam for funding for fieldwork and sample analysis. S. Chantraprasert was funded by the Faculty of Science, Chiang Mai University. Residual magnetic-anomaly data used in fault interpretation were provided by Department of Mineral Resources, Ministry of Natural Resources and Environment, Thailand.

References

BAL, A. A., BURGISSER, H. M., HARRIS, D. K., HERBER, M. A., RIGBY, S. M., THUMPRASERTWONG, S. & WINKLER, F. J. 1992. The Cenozoic Phitsanulok lacustrine basin, Thailand. *Geological Resources of Thailand: Potential for Future development, Department of Mineral Resources, Bangkok, November, 1992*, 247–258.

BARR, S. M. & MACDONALD, A. S. 1991. Towards a late Palaeozoic–early Mesozoic tectonic model for Thailand. *Journal of Thai Geosciences*, **1**, 11–22.

BEN AVRAHAM, Z. & ZOBACK, M. D. 1992. Transform-normal extension and asymmetric basins: an alternative to pull-apart models. *Geology*, **20**, 423–426.

BOTT, J., WONG, I., PRACHUAB, S., WECHBUNTHUNG, B., HINTHONG, C. & SURAPIROME, S. 1997. Contemporary seismicity in northern Thailand and its tectonic implications. *The International Conference on Stratigraphy and Tectonic Evolution of Southeast Asia and the South Pacific*, Department of Mineral Resources, Bangkok, Thailand, 453–464.

BUNOPAS 1981. *Paleogeographic history of western Thailand and adjacent parts of South East Asia – a plate tectonic interpretation*. PhD thesis, Victoria University of Wellington, New Zealand, 810 pp. Reprinted 1982, Geological Survey Papers, **5**, Geological Survey Division, Department of Mineral Resources, Thailand.

CHARUSIRI, P., CLARK, A. H., FARRAR, E., ARCHIBALD, D. & CHARUSIRI, B. 1993. Granite belts in Thailand: evidence from the $^{40}Ar/^{39}Ar$ geochronological and geological synthesis. *Journal of Southeast Asian Earth Sciences*, **8**, 127–136.

COOPER, M. A., HERBERT, R. & HILL, G. S. 1989. The structural evolution of Triassic intermontaine basins in northeastern Thailand. *International Symposium on Intermontane Basins: Geology and Resources*, Chiang Mai, Thailand, 231–242.

CORSINI, M., VAUCHEZ, A. & CABY, R. 1996. Ductile duplexing at a bend of a continental-scale strike-slip shear zone: example from NE Brazil. *Journal of Structural Geology*, **18**, 385–394.

COWGILL, E., YIN, A., ARROWSMITH, J. R., WANG, X. F. & SHUANHONG, Z. 2004. The Akato Tagh bend along the Altyn Tagh fault, northwest Tibet 1: smoothing by vertical-axis rotation and the effect of topographic stresses on bend-flanking faults. *Geological Society of America Bulletin*, **116**, 1423–1442.

CUNNINGHAM, D., DAVIES, S. & BADARCH, G. 2003. Crustal architecture and active growth of the Sutai Range, western Mongolia: a major intracontinental, intraplate restraining bend. *Journal of Geodynamics*, **36**, 169–191.

CUNNINGHAM, D., WINDLEY, B. F., DORJNAMJAA, D., BADAMGAROV, J. & SAANDAR, M. 1996. Late Cenozoic transpression in southwestern Mongolia and the Gobi Altai–Tien Shan connection. *Earth and Planetary Science Letters*, **140**, 67–81.

CURTIS, M. L. 1998. Structural and kinematic evolution of a Miocene–Recent sinistral restraining bend: the Montejunto massif, Portugal. *Journal of Structural Geology*, **21**, 39–53.

DILL, H. G., BOTZ, R., LUPPOLD, F. W. & HENJES-KUNST, F. 2005. Hypogene and supergene alteration of the Late Palaeozoic Ratburi Limestone during the Mesozoic and Cenozoic (Thailand, Surat Thani Province). Implications for the concentration of mineral commodities and hydrocarbons. *International Journal of Earth Science (Geologische Rundschau)* **94**, 24–46.

DOOLEY, T. & MCCLAY, K. 1997. Analog modelling of pull-apart basins. *AAPG Bulletin*, **81**, 1804–1826.

EMRICH, K., EHHALT, D. H. & VOGEL, J. C. 1970. Carbon isotope fractionation during the precipitation of calcium carbonate. *Earth and Planetary Science Letters*, **8**, 363–371.

EPSTEIN, S., BUCHSBAUM, R., LOWENSTAM, H. A. & UREY, H. C. 1953. Revised carbonate–water isotopic temperature scale. *Geological Society of America Bulletin*, **64**, 1315–1326.

FAURE, G. 1986. *Principles of Isotope Geology*, 2nd edn, Wiley, New York, 335 pp.

FLINT, S., STEWART, D. J., HYDE, T., GEVERS, C. A., DUBRULE, O. R. F. & VAN RIESSEN, E. D. 1988. Aspects of reservoir geology and production behaviour of Sirikit Oil Field, Thailand: an integrated study using well and 3-D seismic data. *AAPG Bulletin*, **72**, 1254–1268.

GILL, R. 1997. *Modern Analytical Geochemistry*. Longman, Harlow, Essex, 329 pp.

HAUKSSON, E. & JONES, L. M. 1988. The July 1986 Oceanside (ML = 5.3) earthquake sequence in the continental borderland, southern California. *Bulletin of the Seismological Society of America*, **78**, 1885–1906.

HOEFS, J. 1987. *Stable Isotope Geochemistry*, 3rd edn, Springer-Verlag, Berlin, 241 pp.

KNOX, G. J. & WAKEFIELD, L. L. 1983. An introduction to the geology of the Phitsanulok Basin. *Conference on Geology and Mineral Resources of Thailand, Bangkok*, Department of Mineral Resources, Bangkok, 19–28 November, 9 pp.

LACASSIN, R., LELOUP, P. H. & TAPPONNIER, P. 1993. Bounds on strain in large Cenozoic shear zones of SE Asia from boudinage restoration. *Journal of Structural Geology*, **15**, 677–692.

LACASSIN, R., HINTHONG, C. *ET AL.* 1997. Cenozoic diachronic extrusion and deformation of western Indochina: structure and $^{40}Ar/^{39}Ar$ evidence from NW Thailand. *Journal of Geophysical Research*, **102**, 10 013–10 037.

LACASSIN, R., REPLUMAZ, A. & LELOUP, H. P. 1998, Hairpin river loops and slip-sense inversion on southeast Asian strike-slip faults. *Geology*, **26**, 703–706.

LANEY, S. E. & GATES, A. E. 1996. Extrusional shuffling of horses in strike-slip duplexes: an example from the Lambertville sill, New Jersey, USA. *Tectonophysics*, **258**, 57–70.

LELOUP, P. H., ARNAUD, N. *ET AL.* 2001. New constraints on the structure, thermochronology and timing of the Ailao Shan–Red River shear zone, SE Asia. *Journal of Geophysical Research*, **106**, 6683–6732.

MCCLAY, K. & BONORA, M. 2001. Analog models of restraining stepovers in strike-slip fault systems. *AAPG Bulletin*, **85**, 233–260.

MORLEY, C. K. 2002. A tectonic model for the Cenozoic evolution of strike-slip faults and rift basins in SE Asia. *Tectonophysics*, **347**, 189–215

MORLEY, C. K. 2004. Nested strike-slip duplexes, and other evidence for Late Cretaceous–Palaeogene transpressional tectonics before and during India–Eurasia collision, in Thailand, Myanmar and Malaysia. *Journal of the Geological Society, London*, **161**, 799–812.

MORLEY, C. K., SANGKUMARN, N., HOON, T. B., CHONGLAKMANI, C. & LAMBIASE, J. 2000. Structural evolution of the Li Basin northern Thailand. *Journal of the Geological Society, London*, **157**, 483–492.

MORLEY, C. K., SMITH, M., CARTER, A., CHARUSIRI, P. & CHANTRAPRASERT, S. 2007. Evolution of deformation styles at a major restraining bend, constraints from cooling histories, Mae Ping Fault zone, Western Thailand. *In*: CUNNINGHAM, W. D. & MANN, P. (eds) *Tectonics of Strike-Slip Restraining and Releasing Bends*. Geological Society, London, Special Publications, **290**, 325–349.

MORLEY, C. K., WONGANAN, N., SANKUMARN, N., HOON, T. B., ALIEF, A. & SIMMONS, M. 2001. Late Oligocene–Recent stress evolution in rift basins of northern and central Thailand: implications for escape tectonics. *Tectonophysics*, **334**, 115–150.

MORLEY, C. K., WONGANAN, N., KORNASAWAN, A., PHOOSONGSEE, W., HARANYA, C. & PONGWAPEE, S. 2004. Activation of rift oblique and rift parallel preexisting fabrics during extension and their effect on deformation style: examples from the rifts of Thailand. *Journal of Structural Geology*, **26**, 1803–1829.

NUTALAYA, P. & RAU, J. L. 1984. Structural framework of the Chao Phraya basin, Thailand. *Proceedings of the Symposium on Cenozoic Basins, Thailand: Geology and Resources,* Chiang Mai University, Thailand, 30–36.

O'LEARY, H. & HILL, G. S. 1989. Cenozoic basin development in the Southern Central Plains, Thailand. *Proceedings of the International Conference on Geology and Mineral Resources of Thailand, Bangkok*, Department of Mineral Resources, Bangkok, 1–8.

O'NEIL, J. R., CLAYTON, R. N. & MAYEDA, T. K. 1969. Oxygen isotope fractionation between divalent metal carbonates. *Journal of Chemical Physics*, **51**, 5547–5558.

POLACHAN, S., PRADIDTAN, S., TONGTAOW, C., JANMAHA, S., INTARAWIJITR, K. & SANGSUWAN, C. 1991. Development of Cenozoic basins in Thailand. *Marine and Petroleum Geology*, **8**, 84–97.

REPLUMAZ, A. & TAPPONNIER, P. 2003. Reconstruction of the deformed collision zone between India and Asia by backward motion of lithospheric blocks. *Journal of Geophysical Research*, **108**, doi:1029/2001JB000661.

RONGE, S. & SURARAT, K. 2002. Acoustic impedance interpretation for sand distribution adjacent to a rift boundary fault, Suphan Buri Basin, Thailand. *AAPG Bulletin*, **86**, 1753–1771.

SAGY, A., RECHES, Z. & AGNON, A. 2003. Hierarchic three-dimensional structure and slip partitioning in the western Dead Sea pull-apart. *Tectonics*, **22**. doi:10.1029/2001ITC001323. 4–1 4–17.

SEUSUTHYA, K. & MORLEY, C. K. 2004. Structural style of Suphan Buri Basin related to the regional tectonic setting of Thailand. *Geophysics, Chiang Mai 2004, International Conference Proceedings on Applied Geophysics, 26th–27th November, 2004*, Chiang Mai University, Thailand.

TAPPONNIER, P., PELTZER, G. & ARMIJO, R. 1986. On the mechanism of collison between India and Asia. *In*: COWARD, M. P. & RIES, A. C. (eds) *Collision Tectonics*, Geological Society, London, Special Publications, **19**, 115–157.

TOMMASI, A. & VAUCHEZ, A. 1997. Complex tectonometamorphic patterns in continental collision zones: the role of intraplate rheological heterogeneities. *Tectonophysics*, **279**, 323–350.

TULYATID, J. 1997. *Application of airborne geophysical data to the study of Cenozoic basins in central Thailand.* University of Leeds, unpublished PhD thesis, 325 pp.

UPTON, D. R. 1999. *A regional fission track study of Thailand: implications for thermal history and denudation.* PhD thesis, University of London.

UPTON, D. R., BRISTOW, C. S., HURFORD, C. S. & CARTER, A. 1997. Cenozoic denudation in Northwestern Thailand. Provisional results from apatite fission-track analysis. *Proceedings of the International Conference on Stratigraphy and Tectonic Evolution in Southeast Asia and the South Pacific*, Bangkok, Thailand, 421–431.

WAKABAYASHI, J., HENGESH, J. & SAWYER, T. L. 2004. Four-dimensional transform fault processes: progressive evolution of step-overs and bends. *Tectonophysics*, **392**, 279–301.

WONGPORNCHAI, P. 1977. Origin of formations in the Nong Bua Basin, Central Thailand. *The International Conference on Stratigraphy and Tectonic Evolution of Southeast Asia and the South Pacific*, Department of Mineral Resources, Bangkok, Thailand, 210–217.

Evolution of deformation styles at a major restraining bend, constraints from cooling histories, Mae Ping fault zone, western Thailand

C. K. MORLEY[1], M. SMITH[2], A. CARTER[3], P. CHARUSIRI[4] & S. CHANTRAPRASERT[5]

[1]*PTTEP, 555 Vibhavadi-Rangsit Road, Chatuchak, Bangkok, Thailand (e-mail: chrissmorley@gmail.com)*

[2]*Department of Petroleum Geoscience, Universiti Brunei Darussalam, Bandar Seri Begawan, Brunei Darussalam*

[3]*School of Earth Sciences, Birkbeck College, University of London, UK*

[4]*Department of Geology, Chulalongkorn University, 10330, Bangkok, Thailand*

[5]*Department of Geological Sciences, Chiang Mai University, Thailand*

Abstract: The *c.* 500-km-long Mae Ping fault zone trends NW–SE across Thailand into eastern Myanmar and has probably undergone in excess of 150 km sinistral motion during the Cenozoic. A large, *c.* 150-km-long, restraining bend in this fault zone lies on the western margin of the Chainat duplex. The duplex is a low-lying region dominated by north–south-trending ridges of Mesozoic and Palaeozoic sedimentary, metamorphic and igneous rocks, flanked by flat, post-rift basins of Pliocene–Recent age to the north and south. A review of published cooling-age data, plus new apatite and zircon fission-track results indicates that significant changes in patterns of exhumation occurred along the fault zone with time. Oldest uplift and erosion (Eocene) occurred in the Umphang Gneiss region, west of an inferred thrust-dominated restraining-bend setting. From 36 Ma to 30 Ma, exhumation was strongest north of the duplex, along the NW–SE-trending segment of the fault zone at the (northern) exiting bend of the Chainat duplex. This region of the fault zone is characterized by a mid-crustal level shear zone 5–6 km wide (Lan Sang Gneisses), that passes to the NW into an apparent strike-slip duplex geometry. The deformation is interpreted to have occurred during passage around the northern restraining bend, which resulted in vertical thickening, uplift, erosion and extensional collapse of the northern side of the shear zone. This concentration of deformation at the bends at the ends of the restraining bend is thought to be a characteristic of strike-slip-dominated restraining bends. Following Late Oligocene–Early Miocene extension, there is apatite fission-track evidence for 22–18 Ma exhumation in the Chainat duplex, that coincides with a phase of inversion in the Phitsanulok Basin to the north. The Miocene–Recent history of the Chainat duplex is one of minor sinistral and dextral displacements, related to a rapidly evolving stress field, influenced by the numerous tectonic reorganizations that affected SE Asia during that time.

Restraining bends (cf. Crowell 1974) in strike-slip zones have been identified in many parts of the world (e.g. Anderson 1990; Corsini *et al.* 1996; Laney & Gates 1996; Curtis 1998), but in particular those in California, China and Mongolia have been the subject of numerous and diverse investigations (e.g. reviews in Cunningham *et al.* 1996, 2003; Cowgill *et al.* 2004; Wakabayashi *et al.* 2004). Typically, in these areas, major restraining bends form regions tens of kilometres up to several 100 km long. Models for restraining-bend behaviour have highlighted two end members: thrust dominated, and strike-slip dominated (e.g. Hauksson & Jones 1988; Cowgill *et al.* 2004). Restraining-bend structural evolution displays a range of trends, including synchronous movement on faults within a strike-slip duplex (McClay & Bonora 2001); progressive outward propagation of the active fault system within a duplex away from the original restraining bend (Wakabayashi *et al.* 2004), and complex rotations of faults during simple shear which causes changes in their sense of motion, and degree of activation (Cunningham *et al.* 2003). However, the number of well-documented examples of major restraining bends for developing and testing such models remains low.

Eastern Myanmar and western Thailand display an extensive network of Cenozoic strike-slip faults (e.g. Le Dain *et al.* 1984; Lacassin *et al.* 1997; Morley 2004; Fig. 1). One of the major faults within this system is the Mae Ping fault zone (also known as the Wang Chao fault zone). The

From: CUNNINGHAM, W. D. & MANN, P. (eds) *Tectonics of Strike-Slip Restraining and Releasing Bends.* Geological Society, London, Special Publications, **290**, 325–349.
DOI: 10.1144/SP290.12 0305-8719/07/$15.00

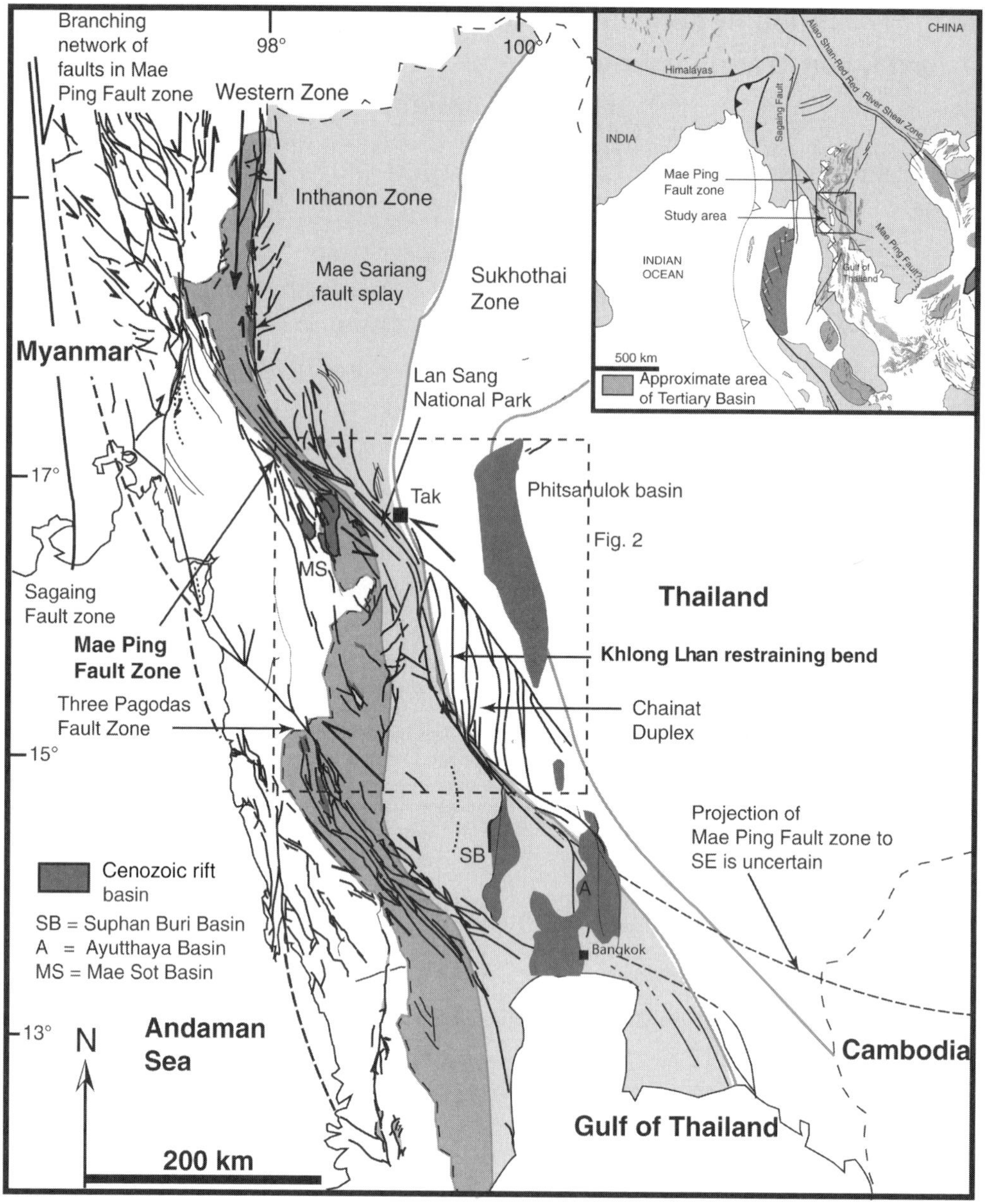

Fig. 1. Regional location map, modified from Morley (2004). Terrane boundaries are from Barr & Macdonald (1991).

Mae Ping fault zone trends predominantly NW–SE, but displays important north–south-trending segments (Morley 2004; Figs 1, 2 & 3). From Myanmar to central Thailand, the Mae Ping fault zone is 500 km long; its continuation to the SE is uncertain. Some interpretations extend the Mae Ping fault zone over 1000 km further to the SE, to reach the Mekong Delta of southern Vietnam (Lacassin *et al.* 1997; Leloup *et al.* 2001). In regional restorations of SE Asian rigid-block motions during the Cenozoic, Replumaz & Tapponier (2003), building of a model published in Leloup *et al.* (2001) required the Mae Ping fault zone to extend to the NW Borneo margin. However, Morley (2002) viewed such an extensive Mae Ping fault zone as unsupported by available data, and contradictory to the known geological history of NW Borneo. There is actually no hard

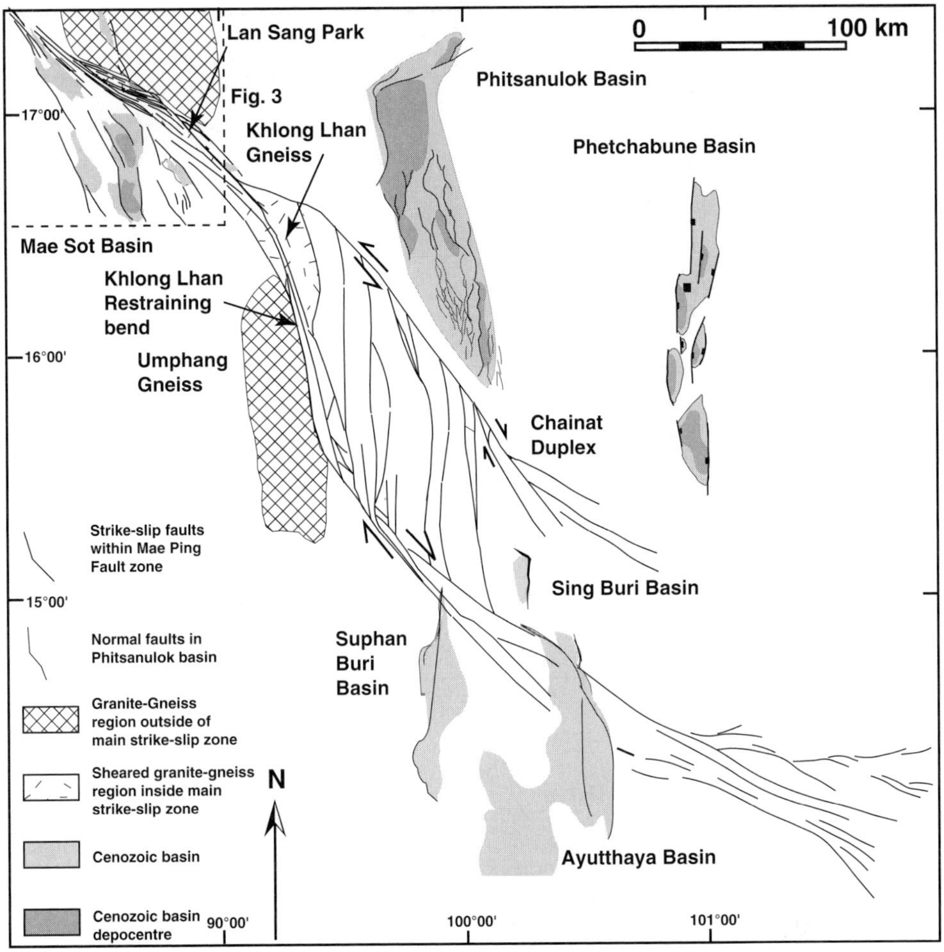

Fig. 2. Regional map of the Mae Ping fault zone (modified from Smith *et al.* 2007, this volume).

evidence for even extending the Mae Ping fault zone through Cambodia and Vietnam to the Mekong Delta, except for the presence of convenient NW–SE-oriented linear features such as the lake, Ton Le Sap. It is even possible that the fault zone splays and ends in eastern Thailand/western Cambodia (Fig. 1).

The Mae Ping fault zone has undergone predominantly sinistral strike-slip motion (Lacassin *et al.* 1993, 1997) where the north–south segments would have acted as restraining bends within the overall NW–SE trend. Horizontal sinistral displacement is estimated at about 150 km (Lacassin *et al.* 1997). Hence, the restraining bends should have experienced considerable strain, uplift and erosion. Strongly entrenched in the literature is the idea of a later, simple reversal to dextral strike-slip motion along the major NW–SE-trending strike-slip faults of SE Asia during the Miocene or Pliocene (e.g. Huchon 1994; Leloup *et al.* 1995; Lacassin *et al.* 1997, 1998). Specifically for the Mae Ping fault zone, Lacassin *et al.* (1997) suggested that the switch occurred post-23 Ma. Smith *et al.* (2007) discuss the evidence for this dextral motion within the Chainat duplex area of the Mae Ping fault zone, and so the subject is not addressed in detail in this volume. However, the conclusions are that evidence for dextral motion can be found, but displacement is minor (a few kilometres of displacement). Dextral displacement increases in magnitude and importance, passing westward into Myanmar. In the Chainat duplex area, dextral motion has alternated with long periods of quiescence, and occasional left-lateral motion during the Miocene, and does not have the history of timing, displacement magnitude or correct detailed structural geometries

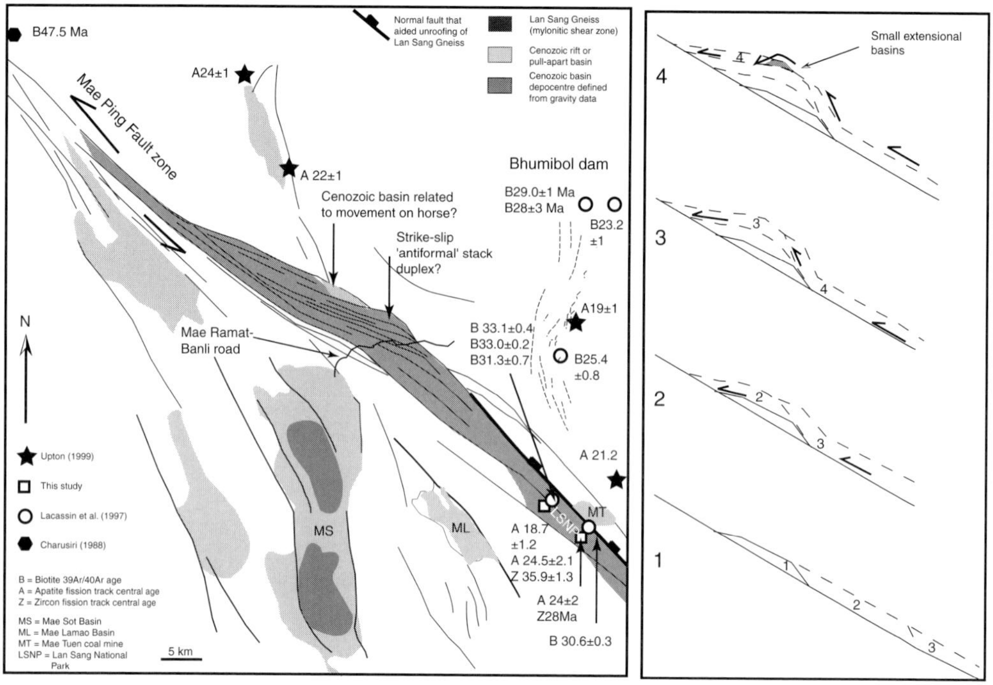

Fig. 3. Map of the Lan Sang to Mae Sot area, showing the antiformal geometry of the Lan Sang Gneisses, based on satellite image interpretation and maps in Lacassin *et al.* (1997).

to be the cause of rift basin development as pull-apart basins (Morley 2002; Smith *et al.* 2007, paper 11, this volume). The absence of large basins within the Chainat duplex area indicates that dextral motion reactivation of the area as a releasing bend must have been minor, and deep basins adjacent to the duplex were formed by extension, not by strike-slip (Morley 2002).

In this paper we show that the 33–30 Ma exhumation documented by Lacassin *et al.* (1997) is only seen locally in the Lan Sang area, and probably represents the results of transpression at the exiting bend of the Khlong Lhan restraining bend, with evolution from a thrust-dominated to strike-slip dominated type restraining bend. The structural evolution of the fault zone during the Cenozoic is defined by summarizing the existing thermochronology data from the region around the Mae Ping fault zone in western and central Thailand, and from new apatite (AFT) and zircon (ZFT) fisson-track ages. This model is compared with existing models for restraining bend evolution.

Regional geological setting

The Mae Ping fault zone in western Thailand is clearly seen on satellite images to form a pronounced NW–SE-trending feature (Figs 1 & 2), with north–south-trending splays branching from it. The NW–SE trend of the fault zone slices through older Palaeozoic–Early Mesozoic terrane boundaries (as defined by Barr & Macdonald 1991), that trend predominantly north–south (Fig. 1). However, several of the splays appear to coincide with the terrane boundaries, hence the Mae Sariang splay (Fig. 1) coincides with the boundary between the Western Zone and the Inthanon Zone, whilst the major bend in the fault zone that is the focus of this study (here termed the Khlong Lhan bend, Fig. 1), lies between the Sukhothai Zone and the Inthanon Zone. Hence, the influence of major crustal pre-existing fabrics described for examples of large releasing or restraining bends elsewhere in the world (e.g. Corsini *et al.* 1996; Tommasi & Vauchez 1997; Curtis 1998), also appears to have influenced their location along the Mae Ping fault zone.

Recently, Morley (2004) has suggested that the Mae Ping fault zone first developed during a Late Cretaceous–Early Cenozoic transpressional event related to collision of the Burma Block with the western margin of Sundaland. This early transpression was a precursor to the main Indian–Eurasian collision when the fault zone underwent further (probably the greatest) sinistral motion during the

Oligocene (Lacassin *et al.* 1997). The best exposure of the mid-crustal levels of deformation associated with the Mae Ping fault zone is the 5–6-km-wide mylonitic to ultramylonitic shear zone in the Lan Sang national park (Lacassin *et al.* 1993, 1997; Fig. 2). North of this area is a north–south-trending region of gneisses and granites which form the highest ranges of hills in Thailand (Fig. 2). These ranges include the hills called Doi Inthanon and Doi Suthep, which have been interpreted as metamorphic core complexes exposed by top-to-the-east shear on low-angle east-dipping detachments (e.g. MacDonald *et al.* 1993; Rhodes *et al.* 1997). However, differences in ages between the dating of the detachment (Eocene) and the timing of exhumation (Early–Middle Miocene) mean that the history of metamorphic core complex is in doubt (e.g. Rhodes 2002).

The Lan Sang Gneisses within the Mae Ping fault zone NW of Lan Sang national park have a distinctive pattern to the trend of their foliation (Fig. 3). Although the main shear zone trends NW–SE, the foliations are curved and lie within a convex stretch of the NE northern boundary (Lacassin *et al.* 1997; Fig. 3). The pattern of foliations and shear zones suggests the strike-slip equivalent of an antiformal duplex geometry (e.g. Woodcock & Fischer 1986). The small Cenozoic sedimentary basin that opened up on the northernmost segment of the duplex suggests that one horse block moved independently from those to the south (Fig. 3). The minor road from Mae Ramat to Banli, which cuts the northern part of the duplex, reveals small outcrops of gneiss and augen gneiss that do not have an imposed sinistral mylonitic fabric, and a few strongly weathered outcrops with a subvertical foliation – a pattern consistent with horses within a duplex. However, the duplex (and the road) lies in remote, jungle-covered, hilly country, and detailed resolution of the structural geometry from outcrops is unlikely to be possible. If the foliation pattern on satellite images does represent an antiformal duplex, then the area would have evolved in a way similar to that illustrated in stages 1–4 of Figure 3. The first horse to move block (1) then ceased motion and became overridden by successive horses that each were transported further to the NW than previous ones. What was possibly the final motion on horse 4 set up a small releasing-bend geometry, resulting in the creation of a minor Cenozoic basin.

Southeast of Lan Sang national park and the western highlands are the broad Central Plains (Fig. 1). This region is a flat-lying area which represents a post-rift basin overlying several Late Oligocene–Miocene rift basins (e.g. Morley *et al.* 2001). The Mae Ping fault zone east of Lan Sang national park broadens and splays into the Central Plains area. In one large region of the Central Plains, some 200 km north–south and 100 km wide, Palaeozoic and Mesozoic sedimentary, metasedimentary and igneous rocks are exposed as isolated hills. These hills tend to trend either north–south or NW–SE. This area between the Cenozoic rift basins was called the Chainat Ridge in O'Leary & Hill (1989). Morley (2002, 2004) interpreted the region as a strike-slip duplex and renamed it the Chainat duplex. A detailed discussion of the evidence for strike-slip deformation in the Chainat Ridge area is provided in a companion paper to this one (Smith *et al.* 2007). Adjacent to the Chainat duplex are the Phitsanulok, Ayutthaya and Suphan Buri rift basins of Late Oligocene–Miocene age (O'Leary & Hill 1989). The timing and structural history of the basins are constrained by well and seismic reflection data gathered for hydrocarbon exploration (e.g. Flint *et al.* 1988; O'Leary & Hill 1989; Wongpornchai 1997; Ronge & Surarat 2002). The history of these basins helps to further constrain the activity of the Mae Ping fault zone. This paper focuses on the exhumation history of the region around the poorly exposed antiformal duplex in the NW illustrated in Figure 3, and the much better-constrained Chainat duplex to the SE (Fig. 1).

Methods and results

This study is based on collating available published and unpublished cooling age data for NW Thailand, and providing additional ZFT and AFT data which infill key areas where there was little published information. The aim of the work is to determine whether the patterns of uplift are consistent with one or more mechanisms of uplift, and specifically to determine patterns of uplift that might be associated with motion along the Mae Ping fault zone.

A number of radiometric dating studies have been conducted in western Thailand, with a range of aims. The locations and cooling ages determined from these studies are shown in Figure 4. Several studies have focused on the uplift and erosion of gneisses in the Doi Suthep and Doi Inthanon areas, with regard to documenting the denudation history of putative metamorphic complexes associated with low-angle extensional detachments (Dunning *et al.* 1995; Rhodes 2002). Ahrendt *et al.* (1993, 1997) have regionally dated granites and gneisses in Thailand, and have related the ages to orogenic events. Charusiri (1989) obtained $^{40}Ar/^{39}Ar$ radiometric age dates from micas and feldspars from parts of the Three Pagodas fault zone and the Mae Ping fault zone, in order to understand the timing and genesis of ore deposits (Fig. 4). Upton *et al.* (1997) and Upton (1999) collected samples for

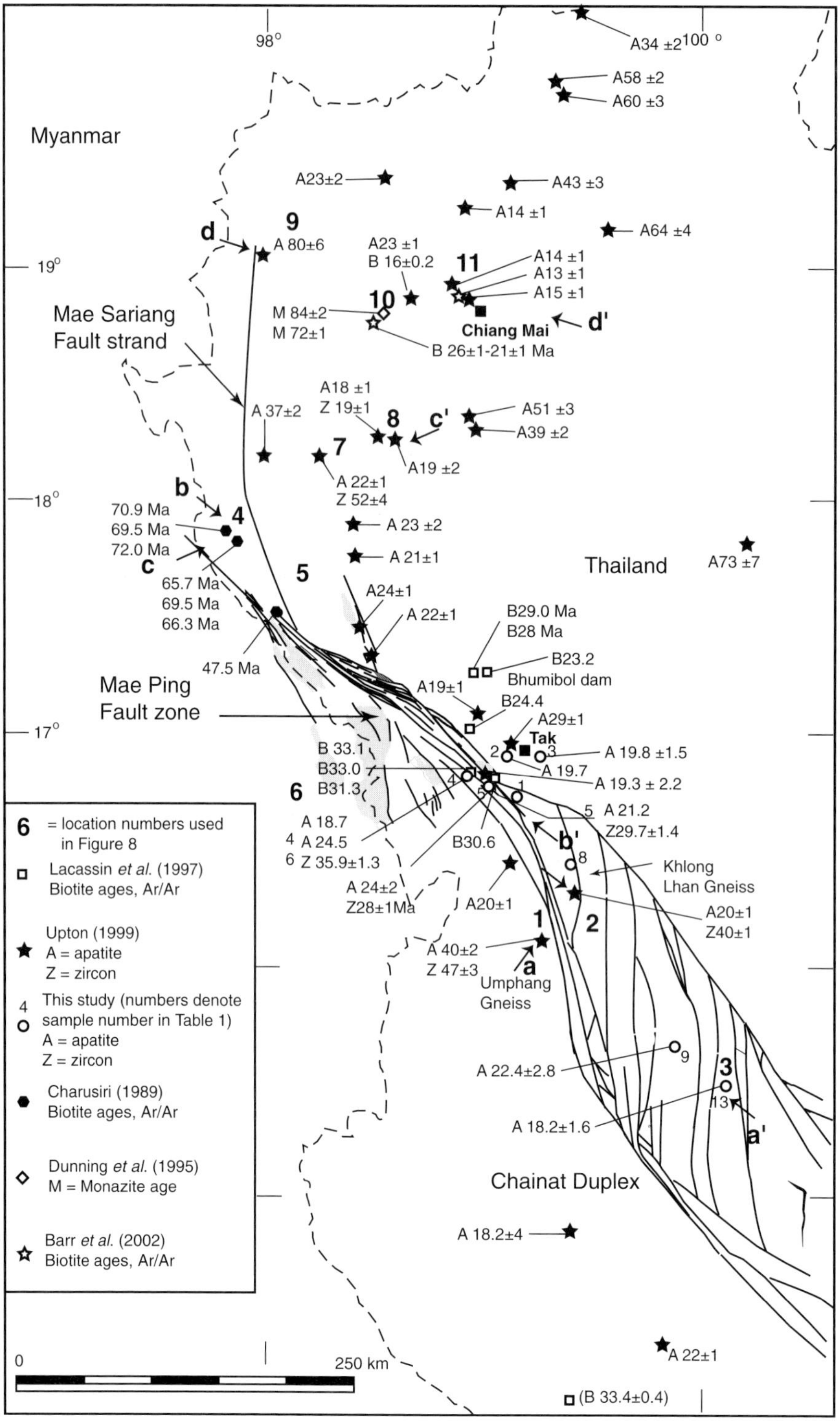

Fig. 4. Regional map of NW Thailand, showing the location of cooling-age data used in this study.

apatite and zircon fission-track analysis in order to build a regional denudation history for Thailand, as well as focusing on more detailed local tectonic and exhumation problems in some areas (such as more concentrated sampling in the regions of the proposed metamorphic core complexes in the western highlands, and around the Mae Ping fault zone). Lacassin *et al.* (1997) specifically sampled the Mae Ping and Three Pagodas fault zones to determine the timing of strike-slip deformation; their results are discussed separately below.

For this study, samples were taken for apatite and zircon fission-track dating from the Lan Sang area into the Chainat Ridge area (Fig. 4) to determine whether any systematic change in ages occurred along the strike, and perpendicular to the Mae Ping fault zone passing away from the Lan Sang area (Table 1). The samples were analysed in the laboratories at the University College of London. The results of this work were partially successful; however, a systematic spread of data could not be obtained, due to unsuitable outcrop lithologies and insufficient apatite or zircon in some samples (UBDA-7, 8, 10, 11 and 12). New dates were obtained for two localities within the Chainat duplex (samples UBDA-9 and UBDA-13, Table 1, Fig. 4). Samples (UBDA-4 , 5, and 6, Table 1, Fig. 4) within the Lan Sang area validated previous results and established similarity between biotite $^{40}Ar/^{39}Ar$ and ZFT cooling ages (Table 2). However, east and NE of the Mae Ping fault zone, around Tak, a cluster of cooling ages (samples UBDA-1, 2 and 3, Table 1, Fig. 4) showed AFT central ages around 19–20 Ma.

Cooling-age studies in western Thailand

The cooling ages available from the studies mentioned above are mostly from $^{40}Ar/^{39}Ar$ biotite ages, zircon and apatite fission-tracks. Complications arising from mineral structure, grain size and previous cooling rates mean that the concept of 'closure temperatures' (Dodson 1979) for many mineral/isotopic systems (e.g. $^{40}Ar/^{39}Ar$) is an oversimplification. For example, chemical composition and the presence of large quantities of fluid inclusions can cause significant changes to standard closure temperatures form micas (e.g. McDougall & Harrison 1999; Dunlap 2003). However, the temperature range below which many of these systems effectively become stable can yield qualitative estimates of cooling rates experienced by a sample. In this context, stability means retention, within the crystal system, of some measurable product of various radioactive decay reactions. As an approximate guide, the temperature range below which the system is effectively stable is as follows (Carter 1999; McDougall & Harrison 1999; Dunlap 2003): $^{40}Ar/^{39}Ar$ for muscovite = 400–250 °C; $^{40}Ar/^{39}Ar$ for biotite 300 ± 50 °C, zircon fission-track 320–200°C, and apatite fission-track 110–60 °C. Consequently, for the high-temperature cooling age map (Fig. 5), dates for biotite $^{40}Ar/^{39}Ar$ and zircon fission-track were combined. Whilst this is clearly a great approximation, where zircon fission-track and biotite $^{40}Ar/^{39}Ar$ ages have been obtained from the same or nearby localities (e.g. Lan Sang, Fig. 4), the resulting cooling ages are very similar (Fig. 4; Tables 1 & 2). The low-temperature cooling-age map (Fig. 6) is entirely based on apatite fission-track ages from Upton (1999) and this study (Table 1).

Determination of the cooling history along the Three Pagodas and Mae Ping fault zones, by Lacassin *et al.* (1997)

Evidence for dating motion on the Mae Ping (Wang Chao) and Three Pagodas fault zones relies considerably upon the work by Lacassin *et al.* (1997) who specifically dated synkinematic micas and feldspars from metasediments and orthogneisses within mylonitic shear zones, using the $^{40}Ar/^{39}Ar$ technique. Biotite cooling ages for the Three Pagodas fault zone suggested that the dates of the onset and end of sinistral motion were ≥36 Ma to 33 Ma, and for the Mae Ping fault zone ≥33 Ma to 30 Ma (Figs 4 & 7). Lacassin *et al.* (1997) modelled the cooling histories of the Lan Sang samples, calibrated by $^{40}Ar/^{39}Ar$ step-heating of a K-feldspar. The results indicate that cooling from 400 °C to 185 °C was rapid between 32.5 Ma and 31 Ma, in order to fit the last 16% of argon release. These authors also identified a second cooling step at about 23.5 Ma, before a final isothermal step (about 75°C). Lacassin *et al.* also stressed that the 33 Ma to 30 Ma dates probably documented the last increments of ductile sinistral deformation. The authors also suggest that late-stage exhumation of the Lan Sang Gneisses might be explained by normal faulting within a transtensional setting. The onset of dextral strike-slip motion was placed at about 23 Ma, but was not constrained by any radiometric dating.

The work by Lacassin *et al.* (1997) also obtained biotite cooling ages between 29 Ma and 23 Ma in some gneisses away from the strike-slip fault zones, including the Bhumipol Dam to the north (Fig. 4). The gneisses at Bhumipol Dam show no evidence for Cenozoic shear, and hence are inferred to represent denudation between 29 Ma and 23 Ma, possibly related to a Cenozoic basin-bounding normal fault (Sam Ngao Fault) to the east (Lacassin *et al.* 1997).

Table 1. *Results of apatite and fission-track dating conducted for this study*

	Mineral	No. of crystals	Dosimeter		Spontaneous		Induced		Age dispersion		Central age (Ma) ±1	Mean track length (μm)	SD	No. of tracks
			ρ_d	N_d	ρ_s	N_s	ρ_i	N_i	$P\chi^2$	RE%				
UBDA-1	Apatite	19	1.060	3122	0.201	100	1.863	927	40	19.5	19.3 ± 2.2	12.55		1
UBDA-2	Apatite	20	1.066	3122	0.476	424	4.294	3821	10	16.0	19.7 ± 1.3	14.27 ± 0.28	1.28	22
UBDA-3	Apatite	20	1.085	3122	0.437	233	4.076	2175	5	12.2	19.8 ± 1.5	14.14 ± 0.19	0.53	9
UBDA-4	Apatite	20	1.091	3122	0.243	401	2.146	3540	<1	28.2	21.2 ± 1.8	13.56 ± 0.15	1.32	82
	Zircon	20	0.409	2921	13.12	4709	9.388	3370	<1	10.1	35.9 ± 1.3			
UBDA-5	Apatite	20	1.098	3122	0.283	303	2.799	2999	15	1.9	18.7 ± 1.2	14.88 ± 0.26	1.38	28
	Zircon	20	0.411	2921	4.336	2560	3.838	2266	<1	13.9	29.7 ± 1.4			
UBDA-6	Apatite	20	1.104	3122	0.280	162	2.135	1234	10	4.2	24.5 ± 2.1	14.51 ± 0.40	0.79	5
UBDA-7	Apatite	5	1.111	3122	0.266	81	1.404	428	<1	60.2	37.2 ± 11.3	None measured		
UBDA-9	Apatite	20	1.123	3122	0.236	144	1.991	1217	<1	34.9	22.4 ± 2.8	15.16 ± 1.55	1.55	2
UBDA-11	Apatite	5	1.129	3122	0.019	5	0.745	187	10	0	5.1 ± 2.3	None measured		
UBDA-12	Apatite	14	1.135	3122	0.028	20	0.049	354	10	24.3	11.4 ± 2.9	None measured		
UBDA-13	Apatite	20	1.142	3122	0.146	215	1.591	2329	5	20.5	18.2 ± 1.6	14.11 ± 0.49	1.19	7

[1] Track densities are ($\times 10^6$ tr cm^2) numbers of tracks counted (N) shown in brackets.
[2] Analyses by external detector method using 0.5 for the $4\pi/2\pi$ geometry correction factor.
[3] Ages calculated using dosimeter glass CN-5; (apatite) ζCN5 = 338 ± 4; CN-2; (zircon) ζCN2 = 127 ± 5, calibrated by multiple analyses of IUGS apatite and zircon age standards (see Hurford 1990).
[4] $P\chi^2$ is the probability of obtaining the χ^2 value for ν degrees of freedom, where ν = number of crystals − 1.
[5] Central age is a modal age, weighted for different precisions of individual crystals.

Table 2. *Comparison of high-temperature cooling ages determined for the Lan Sang Gneisses using zircon fission-track and ^{40}Ar–^{39}Ar of biotite from three separate studies*

Lacassin et al. 1997
TL3 biotite 33.1 ± 0.4
TL7 biotite 33.0 ± 0.2
TL8 biotite 31.3 ± 0.7
TA34 biotite 30.6 ± 0.3
Upton (1999)
THI2264 zircon fission-track 28 ± 1 Ma
This study
UBDA-4 zircon fission-track, 35.9 ± 1.3 Ma
UBDA-5 zircon fission-track, 29.7 ± 1.4 Ma

Patterns of cooling ages in western Thailand

Introduction

The highest density of cooling-age data clusters around the Mae Ping fault zone and the area to the north. The area south of the Mae Ping fault zone, to the Three Pagodas fault zone, is much more sparsely sampled (Fig. 4). Therefore much of this discussion will focus on the northern half of western Thailand. To understand the uplift and erosion history of the Mae Ping fault zone, it is necessary to investigate not only the timing of exhumation around the fault zone, but also the regional pattern.

Upton (1999) sampled outcrops in western Thailand extensively for the purposes of apatite fission-track analysis, and produced composite cooling paths for a subset of those samples, using zircon fission-track and published K–Ar dates. For the apatite fission-track data, Upton *et al.* (1997) and Upton (1999) identified two sample suites that differ in both age and cooling history. The largest subset displayed results that mostly ranged between 24 Ma and 13 Ma. However, three samples with central ages of 40, 34 and 29 Ma were also included. These samples in the first set are characterized by narrow s.d. 1–1.5 μm, unimodal and long (>14 μm) mean track-length distributions, consistent with rapid cooling through the partial annealing zone. Upton (1999) estimated the average cooling rates as between 8.5 °C Ma and 25 °C Ma. The second subset ranged between 80 Ma and 37 Ma, and shows broad track-length distributions, with standard deviations between 2.03–2.49 μm. Mean track-length distributions are relatively short (12–13 μm), typical of samples exposed to short-lived, high temperatures, or prolonged residence in the partial annealing zone to reduce track length. The second suite of samples exhibited much slower average cooling rates, of about 1.85 ± 0.55 °C Ma. The young AFT ages of the first subset form an extensive north–south-trending region along the western highlands of Thailand (Fig. 6). On either side of this north–south trend, in central northern Thailand and along the Thailand–Myanmar border region, older ages of the second subset are found (Figs. 6 & 7).

The variations in cooling-age history in the region are discussed in the context of three different provinces: the Chainat duplex; the Mae Ping fault zone; and the putative metamorphic core complex area between the Mae Sariang–Hot Highway and Chiang Mai. These provinces are exemplified in four cooling-age traverses (Fig. 7), and are discussed below.

Chainat duplex area (Fig. 7 a–a′)

The southern traverse (Figs 4 & 7 a–a′) crosses the Khlong Lhan restraining bend and passes into the Chainat duplex. Location 1 is from the Umphang Gneiss area (Upton 1999; Fig. 4), The deepest crustal rocks exposed in the duplex, called the Umphang Gneiss, lie on the western side of the restraining bend. The U–Pb analysis of the Umphang Gneiss, indicates that the paragneiss underwent high-grade metamorphism during the Late Triassic (Mickein 1997). A ZFT central age of 47 ± 3 was obtained for the Umphang Gneiss by Upton (1999; Fig. 4). An apatite central age of 40 ± 2 Ma for the Umphang Gneiss (Upton 1999) suggests rapid exhumation on the western margin of the Chainat Ridge during the Eocene (Fig. 7 a–a′). At the NW corner of the duplex, where the Mae Ping fault zone splays to the SE, are the Khlong Lhan Gneisses. The U–Pb dating of zircon indicates a slightly younger age for high-grade metamorphism in the Khlong Lhan Gneiss compared with the Umphang Gneiss, of 174 ± 5–6 Ma, whilst a monazite age of 117 ± 3 Ma (Mickein 1997) indicates a subsequent high-temperature metamorphic event during the Cretaceous. ZFT and K–Ar analyses by Upton (1999) from both gneisses show overlap at the 2σ-error level, thus indicating that they had cooled below the 350–260 °C isotherm (Fig. 7) by the end of the Eocene (*c.* 40–43 Ma). However, the Khlong Lhan Gneiss has a 40 ± 1 Ma ZFT central age, and a 20 ± 1 AFT age, indicating either slower exhumation between 40 Ma and 20 Ma or a younger exhumation event imposed on the older Eocene one (Fig. 7 a–a′). Whatever the precise scenario, the cooling history is unlike the Umphang Gneiss, which just shows rapid cooling during the Eocene (Fig. 7).

Within the main part of the Chainat duplex there are only a few Triassic granitic outcrops. Five

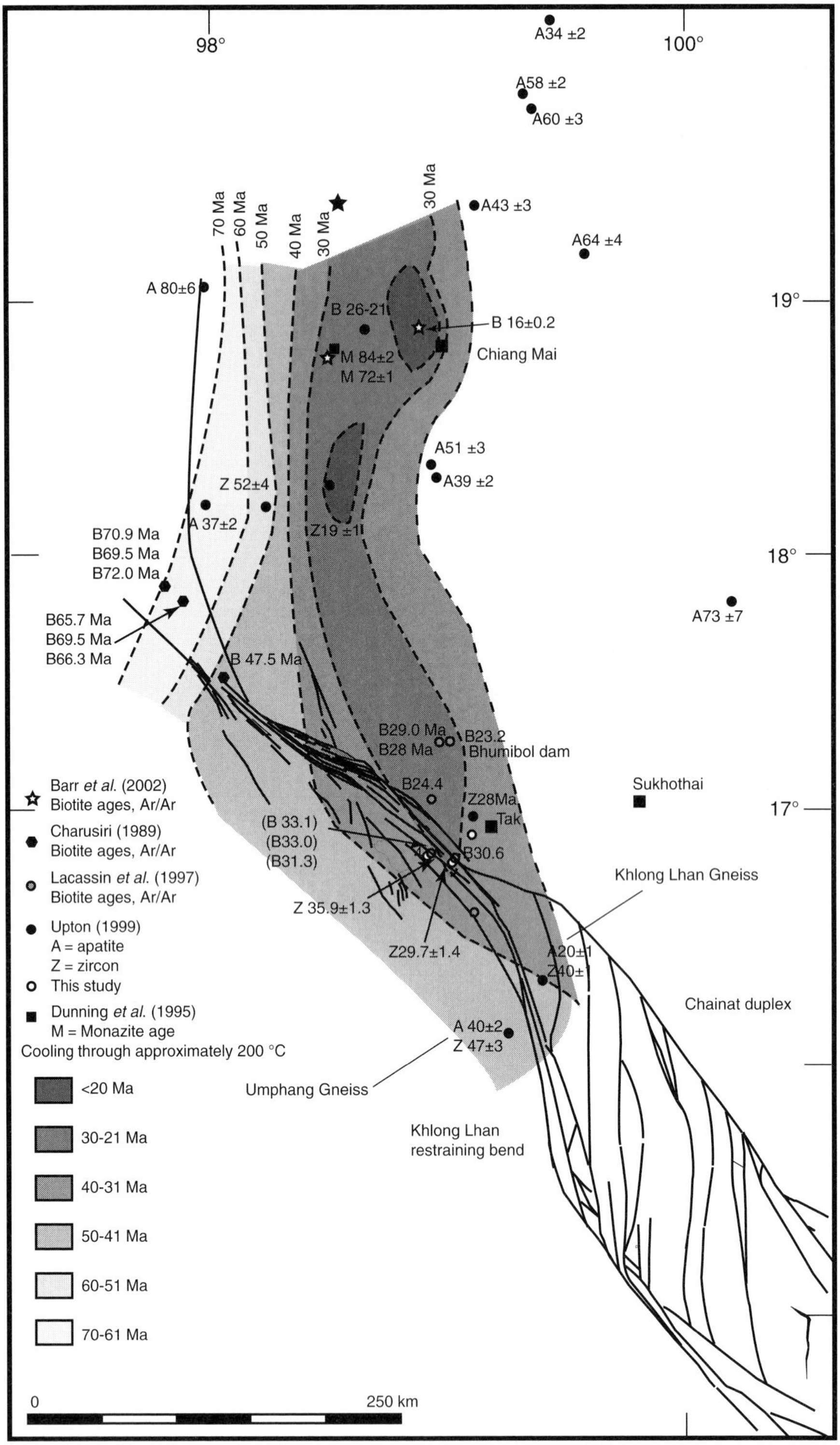

Fig. 5. Map of high-temperature cooling ages for NW Thailand, mostly from zircon fission-track and $^{40}Ar/^{39}Ar$ biotite cooling ages. Some of the older apatite fission-track dates are included because they provide minimum ages for high-temperature cooling.

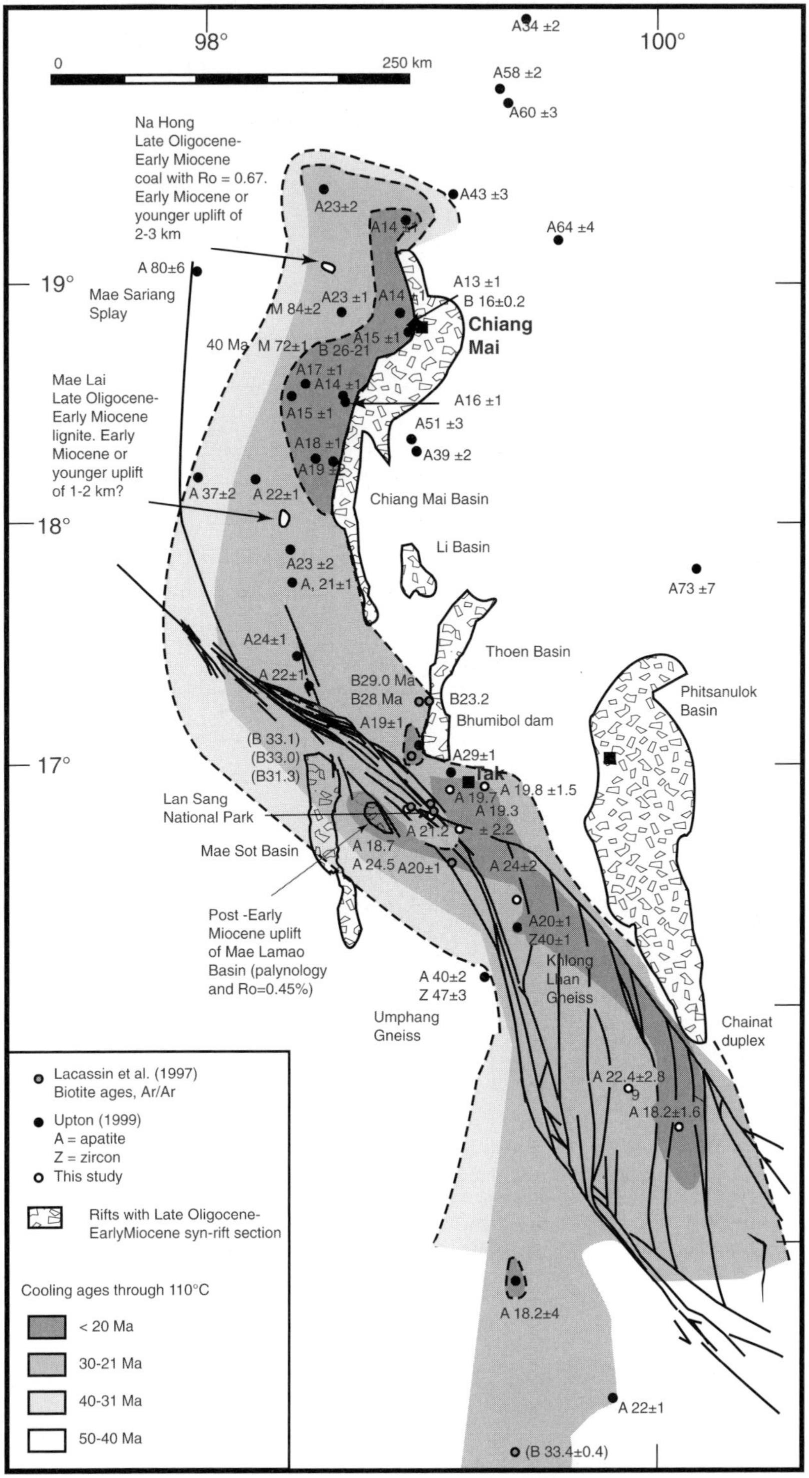

Fig. 6. Map of low-temperature cooling ages for NW Thailand, from apatite fission-track data.

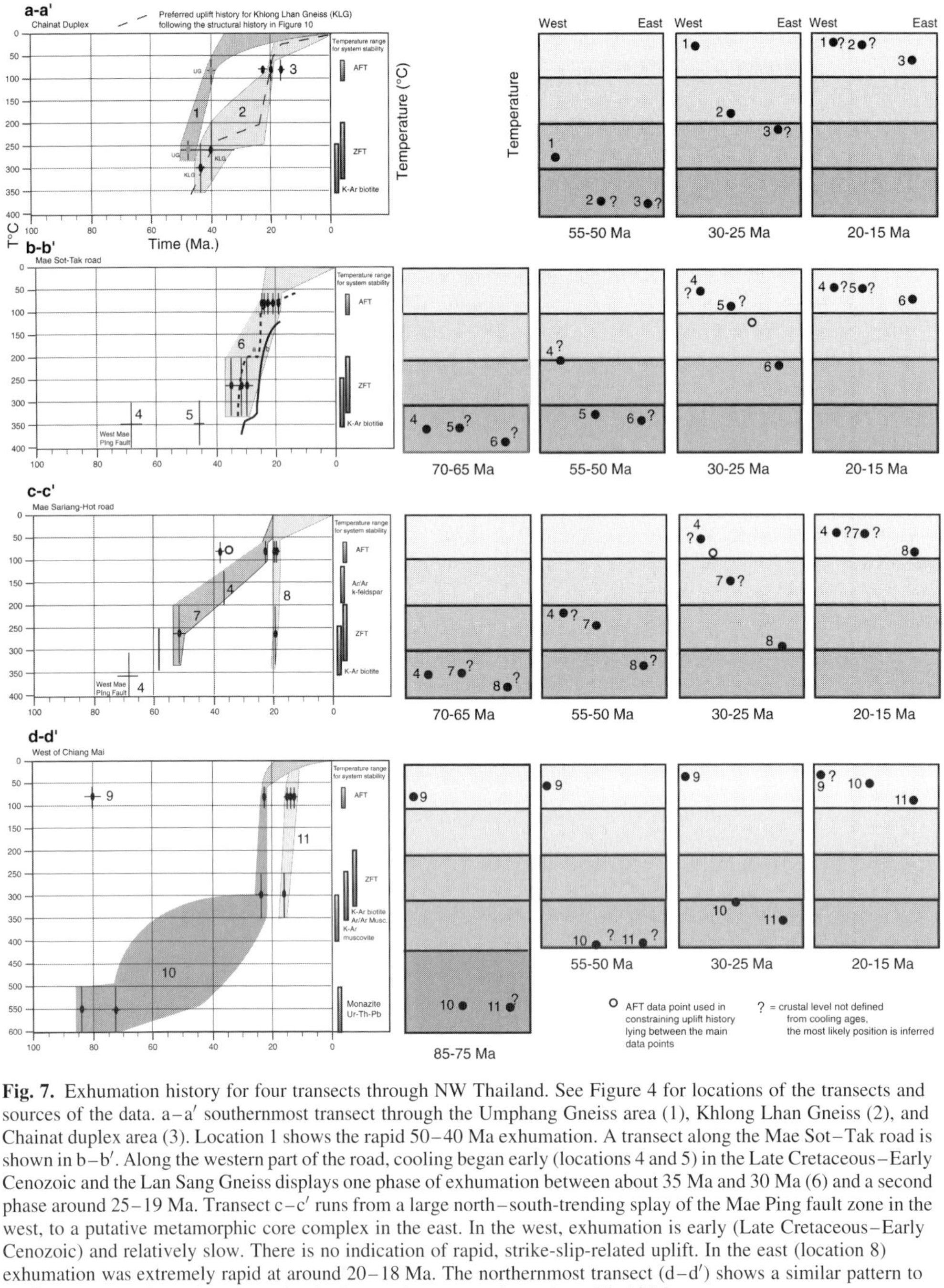

Fig. 7. Exhumation history for four transects through NW Thailand. See Figure 4 for locations of the transects and sources of the data. a–a′ southernmost transect through the Umphang Gneiss area (1), Khlong Lhan Gneiss (2), and Chainat duplex area (3). Location 1 shows the rapid 50–40 Ma exhumation. A transect along the Mae Sot–Tak road is shown in b–b′. Along the western part of the road, cooling began early (locations 4 and 5) in the Late Cretaceous–Early Cenozoic and the Lan Sang Gneiss displays one phase of exhumation between about 35 Ma and 30 Ma (6) and a second phase around 25–19 Ma. Transect c–c′ runs from a large north–south-trending splay of the Mae Ping fault zone in the west, to a putative metamorphic core complex in the east. In the west, exhumation is early (Late Cretaceous–Early Cenozoic) and relatively slow. There is no indication of rapid, strike-slip-related uplift. In the east (location 8) exhumation was extremely rapid at around 20–18 Ma. The northernmost transect (d–d′) shows a similar pattern to c–c′, except that the rapid exhumation in the east is younger, from about 16–13 Ma.

localities were sampled, and two yielded usable AFT data (Table 1). AFT central ages of 22.4 ± 2.8 and 18.2 ± 1.6 were obtained (Fig. 4). It is uncertain from these results alone whether the Early Miocene cooling ages represent regional uplift and erosion, or a specific structural event related to strike-slip deformation within the duplex. Seismic data from the Lahan graben of the Phitsanulok Basin show that the basin ceased to be active in the latest Early Miocene, coincident with the AFT ages (Smith *et al.* 2007, paper 11, this volume). This uplift is not seen in the sedimentary section of the Ayutthaya or Suphan Buri basins on the southern margin of the duplex. Hence, the present available data suggest that uplift occurred in a NW–SE-trending belt, in the northern part of the duplex (Fig. 6).

Mae Ping fault zone (Fig. 7 b–b′)

Passing westward into Myanmar along the Mae Ping fault zone, the muscovite $^{40}Ar/^{39}Ar$ cooling ages increase – a similar trend to that established by the AFT ages further north (Figs 6 & 7 c–c′ and d–d′). Charusiri (1989) obtained $^{40}Ar–^{39}Ar$ cooling ages from micas in granites adjacent to the Mae Ping fault zone in westernmost Thailand (Fig. 4). The oldest ages obtained lie furthest to the west (69.5–72 Ma), and young eastward (69.5–65.7 Ma and 47.5 Ma). A sample of hydrothermal muscovite collected from a wolframite-bearing quartz vein (collected underground) yielded $^{40}Ar–^{39}Ar$ spectra with well-defined plateau ages of *c.* 69.5 ± 0.68 Ma and 70.6 Ma. Muscovite from younger, cross-cutting scheelite–fluorite–calcite–quartz and sphalerite–muscovite–quartz veins yielded fusion dates of *c.* 69.2 Ma and 71.9 Ma. The hydrothermal muscovite probably crystallized at temperatures between 300 and 425 °C, under maximum confining pressures of about 170 to 200 MPa (i.e. depths of 6–7 km).

Further ESE along the trend of the Mae Ping fault zone, Charusiri dated samples from the Mae Suri Mine. Hydrothermal muscovite from a tungsten-rich quartz vein was dated using total fusion and step-heating methods. The total fusion age is *c.* 45.2 Ma, and the integrated age *c.* 47.5 ± 0.51 Ma. The $^{40}Ar–^{39}Ar$ age spectrum displays a well-defined plateau of 46 Ma. The minimum at the first step (21.5 Ma) may be a result of thermal resetting. There was evidence of shearing within the mine that suggested emplacement of the veins during sinistral displacement of the Mae Ping fault zone.

The cooling ages obtained by Charusiri (1989) are not as directly linked to the Mae Ping fault zone as the Lan Sang ages (Lacassin *et al.* 1997). But they are close to the fault zone, and thus show that passing west, close to the fault zone, there is no evidence for large-scale Oligocene regional exhumation related to strike-slip faulting that would have obliterated the older ages. Hence, the exhumation of mid-crustal rocks at Lan Sang between 33 and 30 Ma (Fig. 7 b–b′ location 6) is an atypical and localized feature of the fault zone that requires specific explanation.

The pattern associated with high-temperature cooling (Fig. 6) is very consistent with the AFT results of Upton (1999). The contour patterns (Figs 5 & 6) suggest that, for much of the length of the Mae Ping fault zone, strike-slip motion does not equate with significant Oligocene uplift and erosion, and that if the exhumation is associated with the Mae Ping fault zone, then it is of Late Cretaceous–Palaeogene age (Fig. 7 b–b′). The overall cooling pattern indicates that the strike-slip deformation has not dominated the cooling history. The low-temperature cooling pattern shown in Figure 6 reveals a predominantly north–south-trending exhumation pattern. This pattern may represent both more local tectonic effects such as extensional or inversion related uplift and erosion, and more regional uplift and erosion at least partially related to climate change (Morley & Westaway 2006). For example, the syn-rift basins of northern Thailand show a switch from palynomorphs associated with a temperate climate to tropical forms in the Early Miocene (Songtham 2000; Ratanasthien 2002); this change is also seen in peninsular Malaysia (Morley 1998).

Exhumation of the Lan Sang area can either be interpreted as part of a regional Late Oligocene–Early Miocene north–south-trending event, or as a composite of strike-slip-related deformation superimposed on a north–south striking regional trend. We feel that there is sufficient evidence as presented by Lacassin *et al.* (1997) to justify much of the exhumation at Lan Sang as being related to strike-slip deformation.

Hot–Mae Sariang highway-region west of Chiang Mai (Fig. 7 c–c′ & d–d′)

The two northernmost traverses (Fig. 7 c–c′, d–d′) are through the putative metamorphic core complex area (MacDonald *et al.* 1993; Rhodes *et al.* 1997, 2002). Traverse c–c′ is along the Hot–Mae Sariang highway (Figs 6 & 7). The pattern of cooling was defined by Upton (1999) using AFT, ZFT and K–Ar biotite cooling ages. Passing westward, four AFT ages progressively become older, ranging from 19 ± 2 Ma and 18 ± 1 Ma, through 22 ± 1 Ma in the east to 37 ± 2 Ma in the west (Fig. 7). For the 18 ± 1 Ma sample, Upton (1999) also obtained a 19 ± 1 ZFT age, and nearby a concordant K–Ar biotite cooling age of 20 ± 1 Ma, was reported. For the 22 ± 1 Ma sample, Upton (1999) obtained a ZFT age of 52 ± 4 Ma, and a K–Ar biotite cooling age of 67 ± 2 Ma, indicating a much slower and more prolonged cooling history toward the west. Using isotopic dating, Mickein (1997) also identified a younging-to-the-east pattern along the Hot to Mae Sariang highway.

On the northernmost line (Fig. 7 d–d′) a similar pattern of cooling ages is seen, with slow, prolonged exhumation in the west (AFT central age of 80 ± 6 Ma) and rapid exhumation in the east. The timing of the eastern rapid exhumation becomes younger passing north, along line c–c′ (Fig. 7) with the AFT central ages being Early Miocene. Along line d–d′, the AFT central ages are Middle Miocene, with the nearest biotite $^{40}Ar/^{39}Ar$ age being 16 ± 0.2 Ma. The rapid Early and Middle Miocene cooling occurs in the region identified as the metamorphic-core-complex area. However, one problem with a simple core-complex story is that the rapid exhumation is young, compared with the age of the shearing defined by dating of biotite within the detachement zone, which is of Eocene age (Rhodes 2002).

One possible explanation of the cooling-age pattern lies in the model for basin subsidence in response to sediment loading, a model proposed by Morley & Westaway (2006). There, erosion of the sediment source area and deposition in the sedimentary basin triggers a return flow in the lower crust, from beneath the basin toward the sediment source area. Applied to Thailand, the model predicts lower-crustal flow from beneath the basins of the Gulf of Thailand toward the sediment source areas of the western highlands (Morley & Westaway 2006). If the pattern of erosion and lower-crustal flow shifted northward with time then this may explain the pattern of young, rapid cooling on the eastern side of the highlands.

Before discussing a model for the structural history of the Mae Ping fault zone, other constraints on timing of deformation and exhumation associated with the Mae Ping fault zone from adjacent sedimentary basins and the Chainat duplex area are reviewed.

Significance of Cenozoic basins for exhumation history

Sedimentary basins that were sites of subsidence synchronous with the areas of exhumation provide important constraints for the location and origin of exhumation. Here the basins within the western highlands are discussed.

Mae Lamao

The main exposures of the Mae Lamao basin are coarse conglomerates and sandstones along the main Tak–Mae Sot road, and deeper levels of the basin exposed in a very small coal mine that lies north of the main Tak–Mae Sot road (Fig. 3). The main coal seam is mined from the footwall of a normal fault. This fault strikes 325° and dips 60°. It displays pure dip-slip striations that plunge 60° 238°SW. Bedding dips range between 306°20°SW and 330°24°WSW. The mine lies very close to the Mae Ping fault zone, just a kilometre or two south of one of the main fault strands (Fig. 3). There is little evidence in the outcrop for strike-slip deformation. Instead, the main normal fault and two secondary faults show almost pure dip-slip motions.

In the Mae Lamao Basin, deposits over 500 m thick comprise conglomeratic claystone and sandstone at the base, overlain by shales, oil shales, coal and sandstone. The palynology indicates a Late Oligocene–Early Miocene age (Ratanasthien 1989), which is presumably also the age of the normal faulting. The coal seams pass abruptly laterally into thick conglomeratic sequences, indicating that the sediment from an adjacent uplifted area was dumped into the basin. There is little post-Miocene deposition except for fluvial deposits, indicating Early Miocene or later uplift and erosion. Vitrinite reflectance values from the coals average 0.45 (Ratanasthien 1989), i.e. the rocks experienced maximum temperatures of about 120 °C. If a 30 °C surface temperature is assumed, then, for a geothermal gradient of 3 °C/100 m, burial to 3 km is indicated. If a much higher rift-type geothermal gradient of 6 °C/100 m is assumed, then burial to 1.5 km is indicated. These numbers suggest that a considerably thicker, more extensive basin existed in the past and was removed by Early Miocene or later uplift and erosion. The basin geometry appears to be that of a simple uplifted and eroded rift. Hence, it is uncertain whether strike-slip motion was responsible for basin uplift, or whether it was just part of the more regional uplift and erosion event.

Mae Sot Basin

The Mae Sot Basin is one of the larger rift basins in northern Thailand, and lies just south of the Mae Ping Fault (Figs 2 & 3); hence, its evolution is of great interest for understanding the activity of the Mae Ping Fault. Unfortunately, there is little information in the public domain about the basin. Gibling *et al.* (1985) show a Bouguer gravity map for the Mae Sot Basin which comprises two en échelon NNW–SSE-trending gravity lows, which are 10–20 milligals less in magnitude than the areas of outcropping pre-Cenozoic basement. The largest anomaly indicates a Cenozoic basin centred around Mae Sot about 20 km long and 10 km wide. They also report on a drill-hole (DDH 3-5) made by the Department of Mineral Resources (DMR). The drill-hole penetrated 833 m of Cenozoic strata, dominated by carbonate mudstones and oil shales, without reaching basement. The well was drilled near an outcrop of oil shales reported in Gibling *et al.* (1985). Huminite-reflectance values for the section range between 0.25 and 0.34%, equal to soft brown coal rank (Gibling *et al.* 1985), and vitrinite reflectance values for the outcrops range between 0.25 and 0.4%, suggesting a sedimentary cover 800–1100 m thick that has subsequently been uplifted and eroded. Watanasak (1989) sampled the DMR IMS1 borehole in the Mae Sot Basin from 866–454 m and, on the basis of palynology, determined an early Middle to late Early Miocene age for the section (i.e. probably in the age range of 18–13 Ma). The outcrop described by Gibling *et al.* 1985) is folded into a syncline. This outcrop indicates that compression/transpression affected the basin sometime after the Early Miocene.

A 1997 vintage seismic line across the Mae Sot Basin is presented on website http://www.ccop.or.th/epf/thailand/thailand_petroleum.html. Figure 8 is a line drawing of the seismic line, showing that the basin has a half-graben geometry, expanding to the west. The seismic line also shows folding associated with basin inversion, and confirms the outcrop observations made around Mae Sot as well as the general basin geometry as determined by gravity data. Assuming an interval velocity of 3000 m sec^{-1}, then the maximum depth of the basin on the seismic line is about 2600 m. Hence, it is very likely that a Late Oligocene–Early Miocene section is present in the basin, below the section penetrated by the IMS-1 well.

Mae Tuen coalfield

The Mae Tuen coalfield is located in a small basin that lies along the northern trend of the Mae Ping fault zone, north of Lan Sang national park (Fig. 3). Ratanasthien (1990) describes the Mae Tuen coalfield as having early coals (Late Eocene–Early Oligocene, i.e. probably in the range of 36–32 Ma) unconformably overlain by Late Oligocene–Early Miocene strata. Vitrinite reflectance values from the coals are high (about 0.66% R_o, Ratanasthein, pers. comm., 2005). These values suggest uplift in the order of 2 km if a high (6 °C/100 m) geothermal gradient is assumed. Modern geothermal gradients associated with Thailand rift basins range between about 3 °C and 7 °C (see Morley *et al.* 2001 for a review). The modern values occur at a time when rifting has largely ceased or is very minor, yet they are high, and would seem to indicate a range of gradients appropriate for syn-rift times as well.

The relatively old age of the Mae Tuen Basin is unusual considering that other coal mines in northern Thailand exploit the reserves in basins of Late Oligocene–Miocene age (e.g. Ratanasthien 2002). The Eocene–Early Oligocene age is concomitant with biotite $^{40}Ar/^{39}Ar$ and zircon fission-track cooling ages in the Lan Sang area (Fig. 5). Hence, there is a strong indication that extensional collapse and basin formation occurred on the northern side of the Lan Sang Gneiss region during strike-slip deformation. Widening of the Tak–Mae Sot road just south of the Lan Sang national park has cut

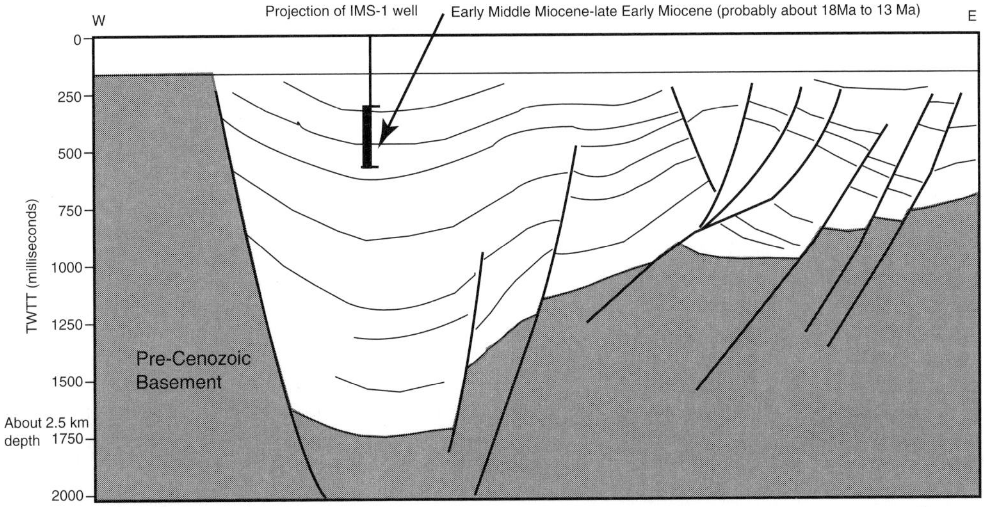

Fig. 8. Line drawing of a seismic line through the Mae Sot Basin, on website http://www.ccop.or.th/epf/thailand/thailand_petroleum.html

into sediments of the Mae Tuen Basin. The road cut ting has revealed a poorly sorted conglomerate, including boulder-sized clasts composed of metamorphic rocks typical of the Lan Sang area, cut by minor normal faults. There are no shales present that could be used for dating. However, the coarse, immature deposits of metamorphic rock clasts are consistent with deposition adjacent to a rapidly uplifted and eroded region.

Structural evolution of the Mae Ping fault zone

The still limited, but more regional, data review undertaken in this study shows that the region of the Lan Sang Gneisses where the Oligocene mica cooling ages have been obtained (Lacassin *et al.* 1997) is very limited geographically (Fig. 5). The Oligocene ages from Lan Sang are bracketed to the NW and SE by Eocene–Late Cretaceous biotite, ZFT and AFT cooling ages (Figs 4, 5 & 7). Hence, the area with the highest number of cooling ages from the fault zone is not really representative of the history of exhumation along the entire fault zone. This is a demonstrably large fault zone in outcrop and on satellite images, a zone which extends hundreds of kilometres into Myanmar (Lacassin *et al.* 1997; Morley 2004). However, accurate quantification of the displacement has not yet been achieved: Lacassin *et al.* (1993) estimated a minimum 40 km of sinistral displacement based on shear-zone geometries. Using the offset of the regional geological markers, Lacassin *et al.* (1997) estimate about 150 km sinistral displacement, whilst the regional rigid-plate reconstructions of Replumaz & Tapponnier (2003) require up to 240 km of 40–30 Ma sinistral motion. The Replumaz & Tapponnier (2003) estimate is model-driven and is not constrained by geological markers, whereas the 150 km estimate is based on a generalized, but reasonable, offset of granitic outcrops, and is the preferred estimate here. However, detailed geochemical typing of offset granites is really necessary to demonstrate offset of the same granite body and to obtain a reasonably constrained offset estimate. Despite the probable large displacements, the exhumation history along the fault zone is highly variable and certainly not consistently in the range of the 33–30 Ma ages determined by Lacassin *et al.* (1997) (Fig. 5). This section discusses how the available cooling ages can be used to explain the structural evolution of the Mae Ping fault zone.

The early history of a Mae Ping fault zone as part of a transpressional orogen spanning the Late Cretaceous–Palaeogene, related to collision of the Burma Block with the Shan Thai Block, has been discussed by Morley (2004), and is not discussed in detail here. The oldest cooling ages in westernmost Thailand (Figs 4 & 5) are part of the evidence for that orogenic event. The starting point for this discussion is the Eocene–Oligocene history of the fault zone. Within the Chainat Ridge area the Umphang Gneiss to the west shows rapid exhumation within the time span of 50 Ma to 40 Ma (Figs 7 & 9). Cooling-age data south of the Umphang Gneisses are sparse (Fig. 4), but regionally appear to fit a north–south trend of Late Oligocene–Early Miocene AFT ages that extend from peninsular Thailand up to northern Thailand (Morley 2004). Hence, the Umphang Gneisses appear to be a patch of locally older exhumation, on the western margin of the Mae Ping fault zone, consistent with exhumation at the restraining bend of a sinistral strike-slip fault system (Fig. 9). However, the subsequent history of the Mae Ping fault zone in the area does not follow such a simple interpretation.

The outcropping geology in the Lan Sang area shows that the deepest crustal levels exposed along the Mae Ping fault zone are not in the restraining-bend area to the SE, but along a NW–SE segment of the fault. This uplift occurred from 36 Ma to 30 Ma (Lacassin *et al.* 1997; Upton 1999; Figs 4, 5, 9 & Table 1), with the ages younging from the NW to the SE. The most intensely deformed part of the fault zone is a belt of gneisses and mylonitic metasediments about 6 km wide, within which are more highly deformed zones of ultramylonite (particularly calc-silicates and marbles) typically about 1 km wide (Lacassin *et al.* 1993). Assuming simple shear, Lacassin *et al.* (1993) estimated lower bounds of 7 to 9 $\pm$ 3 for the shear strain (γ)% within the mylonite zones. The strain estimate implies a minimum of 35–45 km sinistral displacement within a *c.* 5-km-wide shear zone (Lacassin *et al.* 1993). The latest sinistral shear occurred along a retrograde P/T path, and progressed from ductile deformation to below the brittle–ductile transition (Lacassin *et al.* 1993; 1997). Commonly, small-scale conjugate brittle faults cross-cut the ductile shear zone fabrics. They tend to strike east–west (sinistral shear sense) and NNE–SSW (dextral shear sense). Displacements are typically in the order of centimetres to metres, although a few may display tens of metres of displacement. The conjugate brittle faults indicate that the horizontal principal stress was (at least locally) approximately perpendicular to the strike of the gneissic foliation during their formation. In Figure 10b, the sheared mid-crustal rocks seen in Lan Sang are restored to a position east of the Umphang Gneiss. In this position they would have occupied the first Chainat duplex area. During progressive simple shear, the

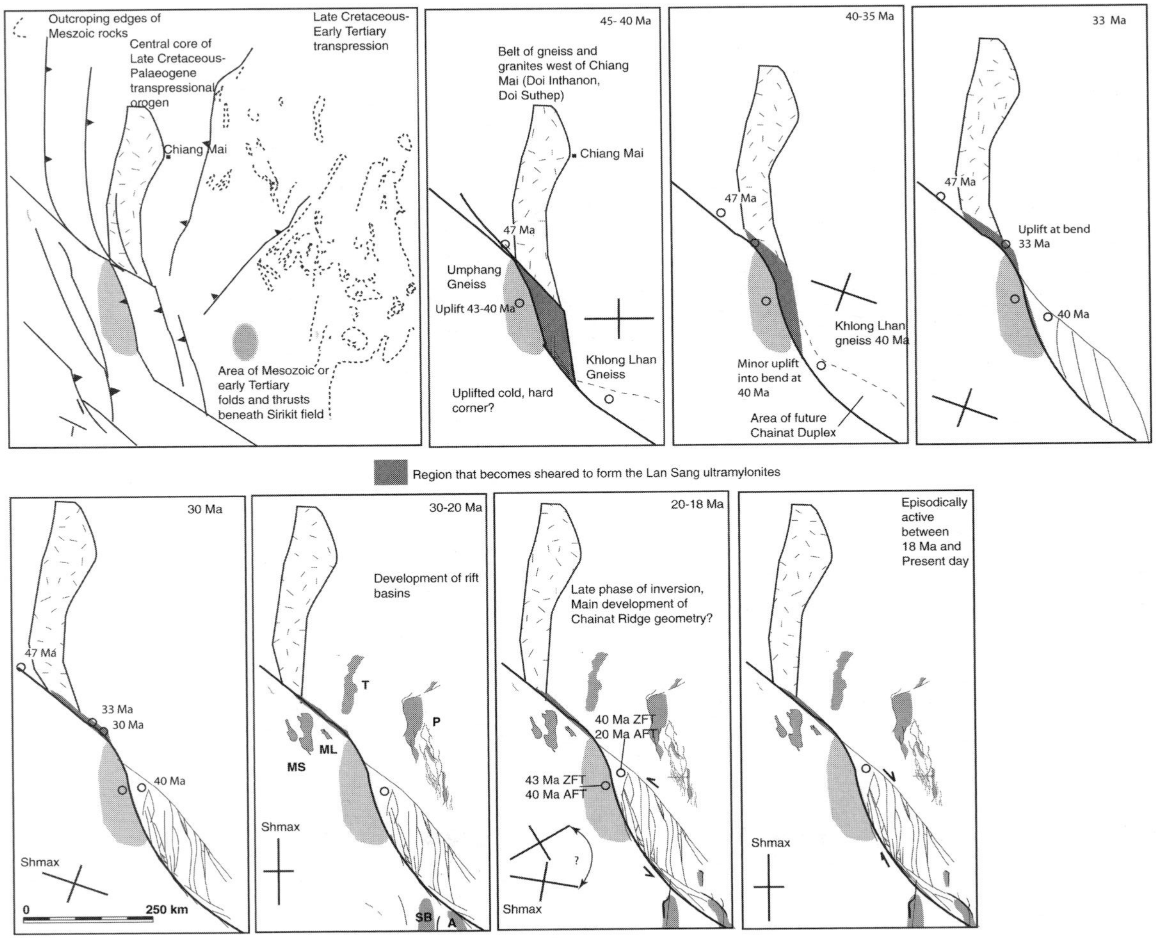

Fig. 9 Proposed evolution of the Mae Ping fault zone to fit the cooling ages and outcrop patterns discussed in this study. Note: motions during the later stages of deformation (30 Ma–present) were probably small (a few kilometres at most), and hence appear insignificant on the maps.

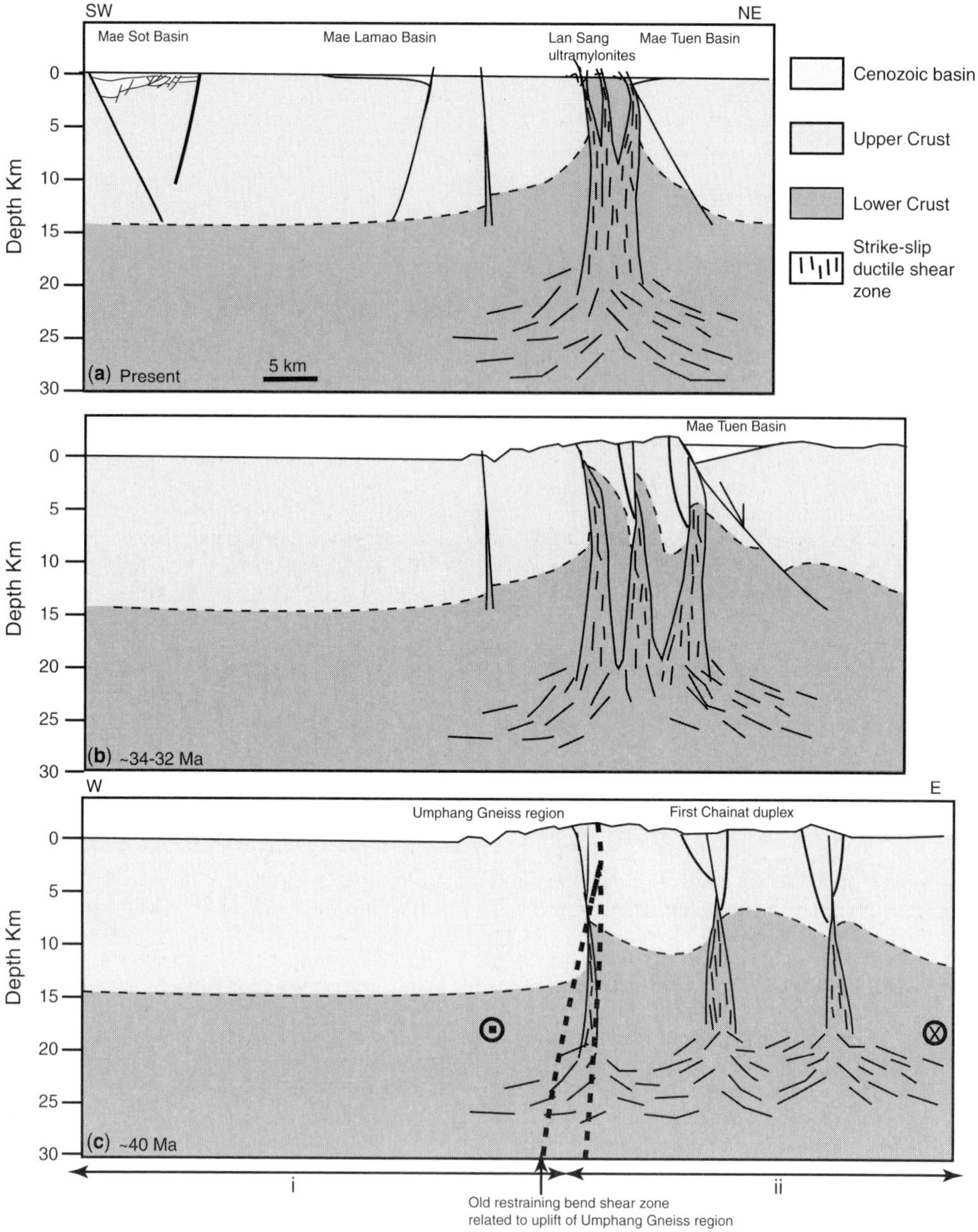

Fig. 10. Schematic cross-section illustrating the structural evolution of the Mae Ping fault zone. The eastern half (**ii**) of the cross-section is kept in a constant location, equivalent to the location of the Lan Sang Gneisses today, whereas the western half (**i**) of the cross-section changes with time as section (**ii**) moves north to NW during sinistral displacement. Hence, for section (**c**), (**i**) is through the Umphang Gneiss area, whilst, for section (**a**), (**i**) is through the Mae Sot and Mae Lamao basins. Section a shows the present-day configuration; however, most of the uplift of the Lan Sang Gneiss was completed by the Late Oligocene, and apart from some erosion and minor strike-slip motion, and development of the Mae Sot and Mae Lamao basins, the geometry of the strike-slip fault zone is likely to have been similar from the Late Oligocene onward. The near-surface Lan Sang Gneiss geometry is based on the cross-section in Lacassin *et al.* (1997). Section (**b**) represents *c.* 34 Ma to 32 Ma ago. Since the Lan Sang Gneiss had to undergo vertical thickening and uplift to be exposed today, restoration of the gneisses through the bend requires that the region of strike-slip deformation becomes broader, with the amalgamated shear zones

duplex was translated, became subject to horizontal simple shear, and became narrower. Vertical thickening is required to produce exhumation of the Palaeozoic–Mesozoic cover and retrograde *P*/*T* conditions within the Lan Sang Gneisses.

Whilst some erosion of the Chainat duplex has occurred, it is noticeable that the area of the duplex is dominated by Palaeozoic–Mesozoic sedimentary, metasedimentary and igneous rocks. Deeper crustal levels are exposed only where the Khlong Lhan Gneisses crop out in the NW corner of the duplex. The Khlong Lhan Gneisses show a cooling history (ZFT = 40 ± 1 Ma, AFT = 20 ± 1 Ma) different from the adjacent Umphang Gneiss (ZFT = 47 ± 3 Ma, AFT= 40 ± 2 Ma) (Fig. 7). In Figure 9, the history of the Khlong Lhan Gneiss is explained as early exhumation occurring during entry into the restraining-bend area in the south of the duplex, and later exhumation where the gneisses entered the northern bend of the duplex. This interpretation of the history of the Khlong Lhan Gneiss implies that the main exhumation of the Lan Sang Gneisses did not occur at the obvious restraining-bend geometry, but as the rocks entered and turned the corner of the bend, passing from a north–south to NNW–SSE-striking fault segment to the NE–SW-striking segment. The cooling-age data are consistent with this interpretation, the oldest (36–33 Ma) ZFT and biotite cooling ages in the Lan Sang Gneisses come from the NW area, whilst the youngest (30 Ma) come from the SE. Whilst the data-set is not sufficient to be definitive, these data fit with rocks entering the bend and then being uplifted and eroded. Probably the exhumed, cooled, and thus relatively strong, region of the Umphang Gneiss acted as a hard anvil or buttress at the bend in the Mae Ping fault trace, and served to focus stresses, as rocks to the NE were translated, uplifted and flattened when passing through the bend.

One of the key questions arising from the model for exhumation of the Lan Sang Gneiss is why was the Umphang Gneiss area exhumed first, then failed to continue reactivating, but instead acted as the hard, resistant buttress against which the Lan Sang Gneisses were flattened and sheared? The answer may lie in the granite intrusions prevalent in western Thailand. Malaysia, Myanmar and Thailand are famous for their extensive suites of granitic rocks, in particular those formed during Triassic and Cretaceous orogenic events (e.g. Beckinsale *et al.* 1979; Charusiri *et al.* 1993). The Umphang Gneiss region is a mixture of para- and ortho-gneisses intruded by granites. Formation of granite melts depletes the lower crust of radiogenic materials and concentrates them in the granites (for example, see the discussion by Sandiford & McLaren 2002). Granite intrusion then transfers those radiogenic materials to higher levels of the crust. Uplift and erosion such as that seen in the Umphang Gneiss region would then remove much of the radiogenic granite to sedimentary basins, and elevate the remaining granite to very high levels in the crust. Consequently, the underlying area of gneiss is likely to be depleted in radiogenic material; have lower geothermal gradients than the surrounding regions; and thus be relatively cold and strong. Hence, the exhumation to the highest levels of the crust of the Umphang Gneiss would have brought the stronger granitic rocks west of the Khlong Lhan restraining bend into contact with weaker sedimentary and metasedimentary rocks east of the Khlong Lhan restraining bend. Deeper in the upper and middle crust, once the depleted, less-radioactive crust cooled, the Umphang Gneiss area would have been colder than the adjacent radioactive granitic Khlong Lhan gneiss region east of the Khlong Lhan restraining bend. Thus, the buttressing effect of the Umphang Gneiss would have developed, due to both mechanical and thermal variations in the upper crust, west and east of the Khlong Lhan restraining bend.

One feature of the northern boundary of the Lan Sang Gneisses is a sharp contact with an adjacent Cenozoic basin, mapped as a northward dipping normal fault by Lacassin *et al.* (1997) (Fig. 3). This Cenozoic basin contains the Mae Tuen coalfield with its Late Eocene–Early Oligocene coals unconformably overlain by a Oligocene–Early Miocene section (Ratanasthien 1990). Lacassin *et al.* (1997) interpreted the normal fault as a Late Oligocene feature. However, the coalfield data indicate that the normal fault probably operated during

Fig. 10. (*Continued*) seen today at Lan Sang becoming more widely separated as they enter the northern bend of the Khlong Lhan restraining bend. Section (**c**) represents *c.* 40 Ma ago, prior to translation of the Lan Sang Gneisses around the northern bend. The Umphang Gneiss area was uplifted, cooled and became inactive. The restraining bend had begun to develop a number of strike-slip fault strands east of the Umphang Gneiss, which would develop into the Lan Sang Gneiss region. The principal strike-slip fault zones are shown as brittle faults in the upper crust passing into narrow, but broader shear zones in the brittle–ductile transition and the upper part of the lower crust. It is uncertain whether the fault zone should be drawn as a narrow, discrete zone all the way through the crust to the Moho (e.g. Leloup *et al.* 2001), or whether the fault zone passes into typical lower-crustal flat-lying shear zones as strike-slip motion becomes accommodated by lower-crustal flow. Since the crust is likely to be hot in this area (Hall & Morley 2004), the lower crust is depicted as deforming along low-angle shear zones.

sinistral displacement as well. Hence, the inference is made here that whilst passing through the bend the sheared and uplifted, hot and thickened area of Lan Sang Gneisses underwent extensional collapse on the northern side, whilst being overthrust to the SW on the southern side of the shear zone (Fig. 10). The evidence from the modelled cooling histories using K-feldspar (Lacassin *et al.* 1997) indicates rapid cooling from 400 to 185 °C between 32.5 Ma and 31 Ma, which is consistent with exhumation occurring in a short burst, and not progressively throughout the strike-slip history of the fault zone. Movement through the bend, uplift, and concomitant extensional unroofing are interpreted here to be the reason for the narrow range of cooling ages.

During the Late Oligocene–Early Miocene (i.e. *c.* 28–22 Ma) there was a period of extensive rift-basin formation, from the Gulf of Thailand, all the way up to northern Thailand (as reviewed by Morley *et al.* 2001). Adjacent to the Mae Ping fault zone, several rift basins developed (the Mae Sot, Mae Lamao, Phitsanulok, Suphan Buri and Ayutthaya basins). The regional extent of these basins suggests a major change in regional stress, probably from an approximately east–west S_{Hmax} direction favourable for sinistral strike-slip deformation, to a north–south S_{Hmax} direction appropriate for east–west extension (e.g. Huchon *et al.* 1994; Morley 2002).

The Chainat duplex area is a region of uplift, but, east of the Umphang Gneiss, deep levels of the crust are not exposed, despite having a restraining-bend geometry under sinistral motion. Relatively young uplift is supported by the 22–18 Ma range of three AFT central ages from the duplex area. As discussed in Smith *et al.* (2007) uplift within the duplex approximately coincides with the cessation of extension in the Lahan graben immediately north of the duplex, and a phase of inversion within the southern Phitsanulok Basin (Bal *et al.* 1992). The interpretation therefore implies a short-lived phase of minor (in the order of kilometres of horizontal displacement) sinistral motion occurred along the Mae Ping fault zone in the Early Miocene and contributed to the present duplex geometry.

The Mae Lamao and Mae Sot basins may have opened under dextral motion on the Mae Ping fault zone, but an oblique extensional origin is also possible. Satellite images show fault strands branching off the Mae Ping fault zone and linking with basin-bounding faults. However, whether the basins are just reactivating older strike-slip trends or are kinematically linked remains uncertain. In the basins the youngest rift fill is of Early Miocene to early Middle Miocene age; there are inversion structures in the Mae Sot Basin; and coal maturity points to removal of somewhere between 1.5 and 3 km of section from the Mae Lamao Basin. These data point to an uplift event of Middle Miocene or younger age. A known Late Miocene inversion event associated with sinistral deformation on NW–SE-trending faults affects the Phitsanulok and Ayutthaya basins (Bal *et al.* 1992; Smith *et al.* 2007, this volume), and hence may also fit with the Mae Sot and Mae Lamao uplift history.

Vertical extent of strike-slip shear zones

There are two main models for the way that the large escape tectonics related shear zones might be behaving in SE Asia. In one model the shear zones penetrate the entire crust and upper mantle, and a broadening – but comparatively narrow and discrete – zone of simple shear (e.g. the Red River fault zone model of Leloup *et al.* 1995). The alternative model considers the fault zones to be essentially upper-crustal features that die out into broadly distributed shear within the middle or lower crust (e.g. England & Houseman 1989). The model in Figure 10 shows the Mae Ping fault zone as dying out within the lower crust. This is not because there is definitive evidence for either model, but because on balance it is the model most favoured by the data at present. First, there are no melts along the Mae Ping fault zone that indicate that magma of mantle origin was being tapped, unlike the model for the Red River fault zone (Leloup *et al.* 1995). Second, in Yunnan, where there are numerous important strike-slip zones, there does appear to be evidence for strike-slip faults dying out in the middle to lower crust – both from magneto-telluric data which indicate the presence of a middle-crustal detachment layer (e.g. Bai & Meju 2003) and from seismic tomography which shows no evidence for any deep perturbation of layers vertically beneath the major strike-slip fault zones (Liu *et al.* 2000). In the case of the Red River fault zone, it may follow a major suture zone at the surface, but tomography indicates that a relict Tethyan subduction zone at lower-crustal and mantle levels lies fifty or more kilometres west of the Red River fault zone (Liu *et al.* 2000). Hence, the upper-crustal zone of weakness does not appear to extend downward vertically throughout the crust to favour the development of a deep-penetrating strike-slip fault zone.

Comparison with other models of restraining-bend development

Cowgill *et al.* (2004) describe large restraining bends in terms of thrust- and strike-slip-dominated types. Thrust-dominated restraining bends display maximum uplift along the main length of the restraining bend, producing restraining-bend 'pop-ups' (e.g. Wakabayashi *et al.* 2004). Thrust-dominated earthquake focal mechanisms from restraining bends in California, such as the Santa Cruz bend, indicate that the vertical principal stress is the minimum principal stress (Hauksson & Jones 1988; Cowgill *et al.* 2004). In strike-slip-dominated restraining bends, the vertical principal stress axis is the intermediate principal stress. Strain and uplift are focused on the areas of changing fault orientation entering and leaving the restraining bend (Cowgill *et al.* 2004). There is also a tendency for the strike-slip fault to undergo vertical-axis rotation to reduce the bend angle (Cowgill *et al.* 2004). The Akato Tagh bend along the Altyn Tagh Fault in China, is cited by Cowgill *et al.* (2004) as such an example.

The Mae Ping fault zone does not appear to show a simple or constant pattern of deformation associated with the restraining-bend geometry of the Chainat duplex area. The oldest documented uplift began in the Umphang Gneiss region on the western margin of the duplex (Fig. 2), and may have spanned the time from about 50 Ma to 40 Ma (Fig. 9). This uplift suggests a thrust-dominated Santa Cruz-type restraining-bend setting (e.g. Hauksson & Jones 1988; Cowgill *et al.* 2004), where uplift of the gneisses occurred along the restraining bend in the hanging wall of a steeply inclined, west-dipping transpressional fault zone.

The next phase of deformation, during the Oligocene, appears to be very different in character, and involved extensive shearing and translation of the Lan Sang Gneiss around the northern bend in the fault zone just west of Tak (Fig. 9). The Khlong Lhan Gneiss underwent uplift moving into the restraining bend at about 40 Ma, and then appears to have been translated with only moderate cooling until a second uplift event occurred at 20 Ma at the exiting bend of the Chainat duplex. Conversely, the Lan Sang Gneisses moving around the exiting bend display rapid Late Eocene–Early Oligocene cooling ages (Figs 5 & 7). This concentration of uplift at the entering and exiting bends is consistent with the strike-slip-dominated restraining-bend model (Cowgill *et al.* 2004), with transpressional deformation just being locally concentrated at the exiting bend. The two styles are also consistent with the regional tectonics, where early fault development occurred within a Late Cretaceous–Palaeogene transpressional orogen (Morley 2004), whilst Oligocene reactivation occurred during Himalayan escape tectonics (Lacassin *et al.* 1997).

McClay & Bonora (2001) presented analogue models for restraining-bend duplex geometries, and they thus generated a range of deformation styles that changed according to: the amount of displacement; the angle between the restraining bend and the main strike-slip trend; and the width of the restraining bend. The last major stage of the Chainat restraining-bend development is the formation of the present-day Chainat duplex, and its geometry appears to be quite appropriate for comparison with the McClay & Bonora (2001) analogue models. In the Chainat duplex, the angle made by the restraining bend with respect to the main fault trend is about 35°, hence the 30° stepover model shown in McClay and Bonora (their fig. 3) is the most appropriate. In this model, the duplex is dominated by internal faults striking subparallel to the restraining bend, unlike higher stepover angles, where a wider range of fault angles is developed. The model pattern is reminiscent of the dominant NNW–SSE to north–south strike of ridges within the Chainat duplex, bounded by NW–SE-striking faults to the north and south (Fig. 2). It is quite apparent from analogue models and descriptions of natural examples of strike-slip duplexes (e.g. Laney & Gates 1996; McClay & Bonora 2001; Cunningham *et al.* 2003) that the relatively simple, classic strike-slip duplex geometry becomes complicated by a wide range of fault trends, rotation of faults, and variable fault kinematics once large displacements become imposed. Whilst the comparative simplicity of the Chainat duplex geometry might be misleading (and a function of exposure), the lack of strong uplift (and exposure of higher metamorphic-grade rocks) within the duplex; the long, linear, uninterrupted trend of the Jurassic ridge on the west side of the duplex (Smith *et al.* 2007, paper 11, this volume); and the 22 Ma to 18 Ma AFT ages, all indicate that it is a comparatively young feature that developed late in the history of the fault zone. It appears to represent the third incarnation of uplift at the restraining bend.

During the Late Oligocene to Pliocene, the rift basins of central and northern Thailand document a series of extensional phases punctuated by periods of inversion (e.g. Morley *et al.* 2000, 2001), and testify to a rapidly evolving stress regime. Two episodes of inversion during the Early Miocene and the latest Miocene to Early Pliocene appear to be quite widespread (Morley *et al.* 2000; 2001), but at least four episodes of inversion have been recorded in some basins (Bal *et al.* 1992; Morley *et al.* 2000). Probably the dominant stress regime was extensional, with S_{Hmax} oriented approximately north–south, as it is today (Bott *et al.* 1997). The orientations of inversion-related folds, inverted normal faults, and episodically active strike-slip faults indicate that during episodes of inversion the stress regime may have ranged from strike-slip to compression, and the S_{Hmax} direction ranged between north–south and east–west (Morley *et al.* 2000, 2001). This brief summary of regional data indicates that the latest history of the Chainat duplex was characterized by short episodes of activity during phases of basin inversion, and there is clear structural evidence for sinistral motion within the duplex, from folded Mesozoic rocks and fault kinematic data (Smith *et al.* 2007).

The NW–SE trending Three Pagodas Fault to the south has Late Cenozoic basins developed at north–south releasing-bend geometries (Morley 2002). The low-level earthquake activity that affects northern and western Thailand today is dominated by dextral strike-slip fault-plane mechanisms on NW–SE-striking faults and sinistral focal mechanisms for NE–SW-striking faults; the S_{Hmax} direction is approximately north–south (e.g. Bott *et al.* 1997; Morley 2004). From these two lines of evidence and the observed dextral slickensides within the duplex, it is concluded that the Chainat duplex was also reactivated episodically under minor dextral motion.

Conclusions

The study by Lacassin *et al.* (1997) remains vitally important to our understanding of the Mae Ping fault zone, but highlights the problem of drawing conclusions from a geographically limited area of the fault zone. Other parts of the fault do not show the same cooling-age histories. Both to the SE (Umphang and Khlong Lhan Gneisses, Fig. 7) and the NW (Fig. 5) of the Lan Sang Gneisses cooling ages become older. The rapid cooling ages of the Lan Sang area do not appear to be representative of the entire fault zone, or even a long segment of it, but instead record an unusual exhumation event, interpreted here to be a passage around the exiting restraining bend. In addition, the regional north–south trend of cooling-age patterns seen for biotite, ZFT and AFT data (Figs 5 & 6) indicates that, at least in part, exposure of the Lan Sang Gneisses is related to more regional exhumation patterns than strike-slip specific uplift and erosion.

Given the available range of major structures in the area (large rift basins, pull-apart basins, low-angle extensional detachments, major strike-slip faults, strike-slip duplexes, the 'extensional collapse' normal fault north of Lan Sang) and the available range of cooling ages (and associated data such as sedimentary-basin history), our ability to construct the structural model remains limited, and numerous questions remain outstanding. Considerably more supplementary data is required to test the models presented in this paper and to develop a good understanding of the relationships between different structural styles. For example, the way that the region of 'metamorphic core complexes' west of Chiang Mai, down to the Mae Sariang–Hot highway (between arrows c–c′, Fig. 4) connects with the Mae Ping fault zone is uncertain. The Umphang Gneiss appears to be an island of Eocene exhumation in the western ranges, surrounded by Oligocene–Miocene cooling ages, but again data south and west of the gneisses are very sparse and additional information is required to fill in the gaps in our knowledge.

Despite the caveats associated with the interpretation of the data and its limitations, a fairly detailed model for the evolution of the fault zone has been proposed in this paper, and can be tested in future studies. The Cenozoic history of the predominantly sinistral Mae Ping strike-slip fault zone shows considerable strain in the vicinity of the Khlong Lhan restraining bend. This deformation can be understood in terms of models proposed for other restraining beds (strike-slip v. thrust-dominated restraining bends Cowgill *et al.* 2004) and analogue modes of early restraining-bend deformation (McClay & Bonora 2001). Initial uplift and erosion on the western side of the restraining bend unroofed the Umphang Gneisses during the Eocene, probably in a thrust-dominated restraining-bend context. Later, as regional deformation evolved from a transpressional orogen related to terrane collision, to escape tectonics associated with the main India–Eurasia collision (Morley 2004) the restraining bend shows strike-slip-dominated characteristics (Cowgill *et al.* 2004). Passing through the northern (exiting) bend in the restraining bend, the northern side of the fault zone was subject to extensive simple shear and vertical thickening, resulting in uplift, erosion and extensional unroofing during passage through the bend. The resulting 5–6-km-wide mid-crustal shear zone exposed at Lan Sang records cooling ages consistent with this passage through the bend. Possibly prior to flattening and simple shear passing through the bend, this zone was originally some 40–50 km wide. The final phase of restraining-bend deformation (Late Oligocene–Recent) occurred under a complexly evolving stress field when episodically relatively small displacements (probably totalling a few kilometres of motion) with both sinistral and dextral sense of motion affected the Chainat duplex area.

M. Smith would like to acknowledge the Universiti of Brunei Darussalam and the AAPG Foundation Grants in Aid (2003) scheme for providing the funding for fieldwork and associated analytical costs. C. Morley thanks the Universiti of Brunei Darussalam for funding for fieldwork and sample analysis. Sarawute Chantraprasert was funded by the Faculty of Science, Chiang Mai University. Residual magnetic-anomaly data used in fault interpretation were provided by the Department of Mineral Resources, Ministry of Natural Resources and Environment, Thailand.

References

AHRENDT, H., CHONGLAKMANI, C., HANSEN, B. T. & HELMCKE, D. 1993. Geochronological cross-section through northern Thailand. *Journal of Southeast Asian Earth Sciences*, **8**, 207–218.

AHRENDT, H., HANSEN, B. T., LUMJUAN, A., MICKEIN, A. & WEMMER, K. 1997. Tectonometamorphic evolution of NW-Thailand deduced from U/Pb–Sm/Nd– and K/Ar-isotope investigations. *The International Conference on Stratigraphy and Tectonic Evolution of Southeast Asia and the South Pacific*, Department of Mineral Resources, Bangkok, Thailand, 314–319.

ANDERSON, R. S. 1990. Evolution of the northern Santa Cruz Mountains by advection of crust past a San Andreas Fault bend. *Science*, **249**, 397–401.

BAI, D. & MEJU, M. A. 2003. Deep structure of the Longling–Ruili fault underneath Ruili basin near the eastern Himalayan syntaxis: insights from magnetotelluric imaging. *Tectonophysics*, **364**, 135–146.

BAL, A. A., BURGISSER, H. M., HARRIS, D. K., HERBER, M. A., RIGBY, S. M., THUMPRASERTWONG, S. & WINKLER, F. J. 1992. The Tertiary Phitsanulok Basin, Thailand. *National Conference on the Geological Resources of Thailand: Potential for Future Development.* Department of Mineral Resources, Bangkok, Thailand, 247–258.

BARR, S. M. & MACDONALD, A. S. 1991. Towards a late Palaeozoic – early Mesozoic tectonic model for Thailand. *Journal of Thai Geosciences*, **1**, 11–22.

BARR, S. M., MACDONALD, A. S., MILLER, B. V., REYNOLDS, P. H., RHODES, B. P. & YOKART, B. 2002. New U–Pb and Ar/Ar ages from the Doi Inthanon and Doi Suthep metamorphic core complexes, Northwestern Thailand. *Symposium on Geology of Thailand, 26–31 August, 2002, Department of Mineral Resources*, Bangkok, Thailand, 284–308.

BECKINSALE, R. D., SUENSILPONG, S., NAKAPADUNGRAT, S. & WALSH, J. N. 1979. Geochronology and geochemistry of granite magmatism in Thailand in relation to a plate tectonic model. *Journal of the Geological Society, London*, **136**, 529–540.

BOTT, J., WONG, I., PRACHAUB, S., WECHBUNTHUNG, B., HINTHONG, C. & SURAPIROME, S. 1997. Contemporary seismicity in northern Thailand and its tectonic implications. *Proceedings of the International Conference on Stratigraphy and Tectonic Evolution of Southeast Asia and the South Pacific*, Department of Mineral Resources, Bangkok, Thailand, 453–464.

CARTER, A. 1999. Present status and future avenues of source region discrimination and characterisation using fission track analysis. *Sedimentary Geology*, **124**, 31–45.

CHARUSIRI, P. 1989. *Lithophile metallogenetic epochs of Thailand: a geological and geochronological investigation.* PhD thesis, Queen's University, Kingston, Canada.

CHARUSIRI, P., CLARK, A. H., FARRAR, E., ARCHIBALD, D. & CHARUSIRI, B. 1993. Granite belts in Thailand: evidence from the $^{40}Ar/^{39}Ar$ geochronological and geological synthesis. *Journal of Southeast Asian Earth Sciences*, **8**, 127–136.

CORSINI, M., VAUCHEZ, A. & CABY, R. 1996. Ductile duplexing at a bend of a continental-scale strike-slip shear zone: example from NE Brazil. *Journal of Structural Geology*, **18**, 385–394.

COWGILL, E., YIN, A., ARROWSMITH, J. R., WANG, X. F. & SHUANHONG, Z. 2004. The Akato Tagh bend along the Altyn Tagh fault, northwest Tibet 1: smoothing by vertical-axis rotation and the effect of topographic stresses on bend-flanking faults. *Geological Society of America Bulletin*, **116**, 1423–1442.

CROWELL, J. C. 1974. Origin of late Cenozoic basins in southern California. *In:* DICKINSON, W. R. (ed.), *Tectonics and Sedimentation.* SEPM Special Publications, **22**, 190–204.

CUNNINGHAM, D., DAVIES, S. & BADARCH, G. 2003. Crustal architecture and active growth of the Sutai Range, western Mongolia: a major intracontinental, intraplate restraining bend. *Journal of Geodynamics*, **36**, 169–191.

CUNNINGHAM, D., WINDLEY, B. F., DORJNAMJAA, D., BADAMGAROV, J. & SAANDAR, M. 1996. Late Cenozoic transpression in southwestern Mongolia and the Gobi Altai–Tien Shan connection. *Earth and Planetary Science Letters*, **140**, 67–81.

CURTIS, M. L. 1998. Structural and kinematic evolution of a Miocene–Recent sinistral restraining bend: the Montejunto massif, Portugal. *Journal of Structural Geology*, **21**, 39–53.

DODSON, M. H. 1979. Theory of cooling ages. *In:* JÄEGER, E. & HUNZIKER, J. C. (eds) *Lectures in Isotope Geology*. Springer-Verlag, Berlin, 194–202.

DUNLAP, W. J. 2003. Crystallisation versus cooling ages of white micas: dramatic effects of K-poor inclusions on $^{40}Ar/^{39}Ar$ age spectra. *Journal of the Virtual Explorer*, **11**.

DUNNING, G. R., MACDONALD, A. S. & BARR, S. M. 1995. Zircon and monazite U–Pb dating of the Doi Inthanon core complex, northern Thailand: implications for extension within the Indosinian Orogen. *Tectonophysics*, **251**, 197–213.

ENGLAND, P. C. & HOUSEMAN, G. 1989. Extension during continental convergence, with application to the Tibetan Plateau. *Journal of Geophysical Research*, **94**, 17 561–17 579.

FLINT, S., STEWART, D. J., HYDE, T., GEVERS, C. A., DUBRULE, O. R. F. & VAN RIESSEN, E. D. 1988. Aspects of reservoir geology and production behaviour of Sirikit Oil Field, Thailand: an integrated study using well and 3-D seismic data. *AAPG Bulletin*, **72**, 1254–1268.

GIBLING, M. R., TANTISUKRIT, C., UTTAMO, W., THANASUTHIPITAK, T. & HARALUCK, M. 1985. Oil shale sedimentology and geochemistry in Cenozoic Mae Sot Basin, Thailand. *American Association of Petroleum Geologists Bulletin*, **69**, 767–780.

HALL, R. & MORLEY, C. K. 2004. Sundaland basins. *AGU Geophysical Monograph*, **149**, 55–85.

HAUKSSON, E. & JONES, L. M. 1988. The July 1986 Oceanside (ML = 5.3) earthquake sequence in the continental borderland, southern California. *Bulletin of the Seismological Society of America*, **78**, 1885–1906.

HUCHON, P., LE PICHON, X. & RANGIN, C. 1994. Indo-China Peninsula and the collision of India and Eurasia. *Geology*, **22**, 27–30.

HURFORD, A. J. 1990. Standardization of fission track dating calibration: recommendation by the Fission Track Working Group of the IUGS Subcommission of Geochronology. *Chemical Geology (Isotope Geoscience Section)*, **80**, 171–178.

LACASSIN, R., LELOUP, P. H. & TAPPONNIER, P. 1993. Bounds on strain in large Tertiary shear zones of SE Asia from boudinage restoration. *Journal of Structural Geology*, **15**, 677–692.

LACASSIN, R., MALUSKI, H., LELOUP, H., TAPPONNIER, P., HINTHONG, C., SIRIBHAKDI, K., CHAUAVIROJ, S. & CHAROENRAVAT, A. 1997. Tertiary diachronic extrusion and deformation of western Indochina: structural and $^{40}Ar/^{39}Ar$ evidence from NW Thailand. *Journal of Geophysical Research*, **102**, 10 013–10 037.

LACASSIN, R., REPLUMAZ, A. & LELOUP, H. P. 1998. Hairpin river loops and slip-sense inversion on southeast Asian strike-slip faults. *Geology*, **26**, 703–706.

LANEY, S. E. & GATES, A. E. 1996. Extrusional shuffling of horses in strike-slip duplexes: an example from the Lambertville Sill, New Jersey. *Tectonophysics*, **258**, 53–70.

LE DAIN, A. Y., TAPPONNIER, P. & MOLNAR, P. 1984. Active faulting and tectonics of Burma and surrounding regions. *Journal of Geophysical Research*, **89**, 453–472.

LELOUP, P. H., LASSASSIN, R. *ET AL.* 1995. The Ailao Shan–Red River shear zone (Yunnan, China), Tertiary transform boundary of Indochina. *Tectonophysics*, **251**, 3–84.

LELOUP, P. H., ARNAUD, N. *ET AL.* 2001. New constraints on the structure, thermochronology and timing of the Ailao Shan–Red River shear zone, SE Asia. *Journal of Geophysical Research*, **106**, 6683–6732.

LIU, F., LIU, J., ZHONG, D., HE, J. & YOU, Q. 2000. The subducted slab of Yangtze continental block beneath the Tethyian orogen in western Yunnan. *Chinese Science Bulletin*, **45**, 466–472.

MCCLAY, K. & BONORA, M. 2001. Analog models of restraining stopovers in strike-slip fault systems. *AAPG Bulletin*, **85**, 233–260.

MACDONALD, A. S., BARR, S. M., DUNNING, G. R. & YAOWANOIYOTHIN, W. 1993. The Doi Inthanon metamorphic core complex in NW Thailand: age and tectonic significance. *Journal of Southeast Asian Earth Sciences*, **8**, 117–126.

MCDOUGALL, I. & HARRISON, T. M. 1999. *Geochronology and Thermochronology by the $^{40}Ar/^{39}Ar$ Method.* 2nd edn, Oxford University Press, New York, 269 pp.

MICKEIN, A. 1997. *U–Pb, Rb–Sr- und K–Ar Untersuchungen zur metamorphen Entwicklung und Altersstellung des 'Präkambriums' in NW-Thailand.* PhD Thesis, Göttinger Arbeiten zur Geologie und Palaeontologie, Göttingen, 1–83.

MORLEY, R. J. 1998. Palynological evidence for Tertiary plant dispersals in the SE Asian region in relation to plate tectonics and climate. *In*: HALL, R. & HOLLOWAY, J. D. (eds) *Biogeography and Geological Evolution of SE Asia.* Backhuyo Publishers, Leiden, The Netherlands, 211–234.

MORLEY, C. K. 2002. A tectonic model for the Tertiary evolution of strike-slip faults and rift basins in SE Asia. *Tectonophysics*, **347**, 189–215.

MORLEY, C. K. 2004. Nested strike-slip duplexes, and other evidence for Late Cretaceous–Palaeogene transpressional tectonics before and during India–Eurasia collision, in Thailand, Myanmar and Malaysia. *Journal of the Geological Society, London*, **161**, 799–812.

MORLEY, C. K. & WESTAWAY, R. 2006. Super-deep Pattani and Malay basins of Southeast Asia: a coupled model incorporating lower crustal flow in response to post-rift sediment loading. *Basin Research*, **18**, 51–84.

MORLEY, C. K., SANGKUMARN, N., HOON, T. B., CHONGLAKMANI, C. & LAMBIASE, J. 2000. Structural evolution of the Li Basin northern Thailand. *Journal of the Geological Society of London*, **157**, 483–492.

MORLEY, C. K., WOGANAN, N., SANKUMARN, N., HOON, T. B., ALIFE, A. & SIMMONS, M. 2001. Late Oligocene–Recent stress evolution in rift basins of northern and central Thailand: implications for escape tectonics. *Tectonophysics*, **334**, 115–150.

O'LEARY, H. & HILL, G. S. 1989. Tertiary basin development in the Southern Central Plains, Thailand. *Proceedings of the International Conference on Geology and Mineral Resources of Thailand*, Chiang Mai University, Chiang, Mai, Thailand, 1–8.

RATANASTHIEN, B. 1989. Depositional environment of Mae Lamao Basin as indicated by palynology and coal petrology. *International Symposium on Intermontane Basins: Geology and Resources*, Chiang Mai, Thailand, 205–215.

RATANASTHIEN, B. 1990. Mae Long Formation of Li Basin, Thailand. *In*: TSUCHI, R. (ed.) *Pacific Neogene Events, their timing and interrelationship.* Tokoyo, Japan, University of Tokyo Press, Proceedings of the Oji International Seminar for ICCP, **246**, 123–128.

RATANASTHIEN, B. 2002. Problems of Neogene biostratigraphic correlation in Thailand and surrounding areas. *Revista Mexicana de Ciencias Geologicas*, **19**, 235–241.

REPLUMAZ, A. & TAPPONNIER, P. 2003. Reconstruction of the deformed collision zone between India and Asia by backward motion of lithospheric blocks. *Journal of Geophysical Research*, **108**, 1–1 to 1–24.

RHODES, B. P. 2002. New U–Pb and Ar/Ar ages from the Doi Inthanon and Doi Suthep metamorphic core complexes, Northwestern Thailand. *Symposium on Geology of Thailand, 26–31 August, 2002*, Department of Mineral Resources, Bangkok, Thailand, 284–308.

RHODES, B. P., BLUM, J. & DEVINE, T. 1997. Geology of the Doi Suthep metamorphic core complex and adjacent

Chiang Mai Basin. *The International Conference on Stratigraphy and Tectonic Evolution of Southeast Asia and the South Pacific*, Department of Mineral Resources, Bangkok, Thailand, 305–323.

RONGE, S. & SURARAT, K. 2002. Acoustic impedance interpretation for sand distribution adjacent to a rift boundary fault, Suphan Buri Basin, Thailand. *AAPG Bulletin*, **86**, 1753–1771.

SANDIFORD, M. & MCLAREN, S. 2002. Tectonic feedback and the ordering of heat producing elements within the continental lithosphere. *Earth and Planetary Science Letters*, **204**, 133–150.

SMITH, M., CHANTRAPRASERT, S., MORLEY, C. K. & CARTWRIGHT, I. 2007. Structural geometry and timing of deformation in the Chainat duplex, Thailand. *In:* CUNNINGHAM, W. D. & MANN, P. (eds) *Tectonics of Strike-Slip Restraining and Releasing Bends*. Geological Society, London, Special Publications, **290**, 305–323.

SONGTHAM, W. 2000. *Palynology of Na Hong Basin Amphoe Mae Chaem Changwat Chiang Mai: Chiang Mai, Thailand*. Chiang Mai University, Graduate School, M.Sc. Thesis, 115 pp.

TOMMASI, A. & VAUCHEZ, A. 1997. Complex tectono-metamorphic patterns in continental collision zones: the role of intraplate rheological heterogeneities. *Tectonophysics*, **279**, 323–350.

UPTON, D. R. 1999. *A regional fission track study of Thailand: implications for thermal history and denudation*. PhD thesis, University of London.

UPTON, D. R., BRISTOW, C. S., HURFORD, C. S. & CARTER, A. 1997. Tertiary denudation in Northwestern Thailand. Provisional results from apatite fission-track analysis. *Proceedings of the International Conference on Stratigraphy and Tectonic Evolution in Southeast Asia and the South Pacific*, Department of Mineral Resources, Bangkok, Thailand, 421–431.

WAKABAYASHI, J., HENGESH, J. & SAWYER, T. L. 2004. Four-dimensional transform fault processes: progressive evolution of step-overs and bends. *Tectonophysics*, **392**, 279–301.

WATANASAK, M. 1989. Palynological zonation of Mid-Tertiary intermontain basins in northern Thailand. *In:* THANASUTHIPITAK, T. (ed.) *Proceedings of the International Symposium on Intermontane Basins: Geology and Resources*, Chiang Mai University Press, Thailand, 216–225.

WONGPORNCHAI, P. 1997. Origin of formations in the Nong Bua Basin, Central Thailand. *The International Conference on Stratigraphy and Tectonic Evolution of Southeast Asia and the South Pacific*, Department of Mineral Resources, Bangkok, Thailand, 210–217.

WOODCOCK, N. & FISCHER, M. 1986. Strike-slip duplexes. *Journal of Structural Geology*, **7**, 725–735.

Evolution of a poly-deformed relay zone between fault segments in the eastern Southern Alps, Italy

D. ZAMPIERI[1,2] & M. MASSIRONI[1]

[1]*Dipartimento di Geoscienze dell' Università di Padova, Via Giotto 1, 35137 Padova, Italy (e-mail: dario.zampieri@unipd.it)*

[2]*CNR, Istituto di Geoscienze e Georisorse, Sezione di Padova, Corso Garibaldi 37, 35137 Padova, Italy*

Abstract: In the eastern Southern Alps (NE Italy), Liassic north–south extensional structures are prominent. The southern Trento Platform also experienced extension during the Palaeogene, when reactivation of some pre-existing faults occurred, coupled with nucleation of new faults. During Neogene shortening, these structures were reactivated once again, but with strike-slip kinematics. In this framework, the Gamonda–Tormeno restraining stepover represents the final result of an overlap zone which evolved through time. In the first stage (Lias to Early Cretaceous) a prominent splay developed at the tip of the Gamonda Fault by lateral propagation and breaching of independent segments. At the same time, there was kinematic interaction between the antithetic Gamonda and Tormeno faults, followed by diachronous motion on crossing faults and the development of a narrow graben. During the second stage of extensional tectonics (Palaeocene to Early Oligocene), the reactivation and propagation of the overlapping faults along with the generation of new faults led to deepening of the graben. In the third stage (Miocene to present), the final structure of a strike-slip restraining stepover was accomplished. Due to the mechanical stratigraphy and complex inherited architecture of the relay zone where stratigraphic sequences with different rheological properties are juxtaposed, the style of shortening is different in the western and eastern sides of the stepover. The Gamonda–Tormeno structure represents a unique example of how a relay zone may change through time.

The existence and geometry of relay zones between fault segments (accommodation and transfer zones) have important implications for seismic hazard evaluation of tectonically active regions, hydrocarbon and groundwater migration and accumulation, and mineral deposition (Faulds & Varga 1998 and references therein). Relay zones are observed in all tectonic settings across a broad range of scales (e.g. Dahlstrom 1969; Boyer & Elliott 1982; Morley *et al.* 1990; Peacock & Sanderson 1991; Peacock & Zhang 1993; Trudgill & Cartwright 1994; Peacock 2003). Due to their effect on topography and sedimentation, relay structures of extensional and strike-slip domains are studied more frequently than those of contractional domains. The best-known transfer structures are extensional relay ramps (Larsen 1988; Childs *et al.* 1995; Crider & Pollard 1998) and strike-slip pull-apart basins (Aydin & Nur 1982; Mann *et al.* 1983).

Whilst studies of relay structures commonly include analysis of features formed in a single tectonic event, except for a few case-histories (e.g. Reijs & McClay 1998), examples of relay structures which evolved through different deformation events are lacking in the literature. However, reworking of inherited structures is quite a common event in nature, and the study of reactivated relay zones can provide unique insights into understanding the evolution of different linkage mechanisms in variable stress fields. The aim of this study is (1) to unravel the evolution of an exceptionally well-exposed kilometre-scale relay structure between conjugate fault segments, and (2) to increase the knowledge of strike-slip contractional relay zones and factors influencing strain distribution and deformational styles inside stepovers.

Regional geology

Stratigraphy

The study area is located within the eastern Southern Alps mountain chain, and during most of the Jurassic it lay within a horst of the passive margin of the Adria plate known as the Trento Platform (Fig. 1). The oldest rocks exposed (Recoaro Phyllites) belong to the pre-Permian Variscan crystalline basement. The overlying sedimentary cover has a thickness of 2.5–3 km. The complex Permian–Middle Triassic stratigraphy includes several siliciclastic and carbonate units intruded and capped by Ladinian volcanics. However, the surface geology is dominated by Upper Triassic to

From: CUNNINGHAM, W. D. & MANN, P. (eds) *Tectonics of Strike-Slip Restraining and Releasing Bends.* Geological Society, London, Special Publications, **290**, 351–366.
DOI: 10.1144/SP290.13 0305-8719/07/$15.00

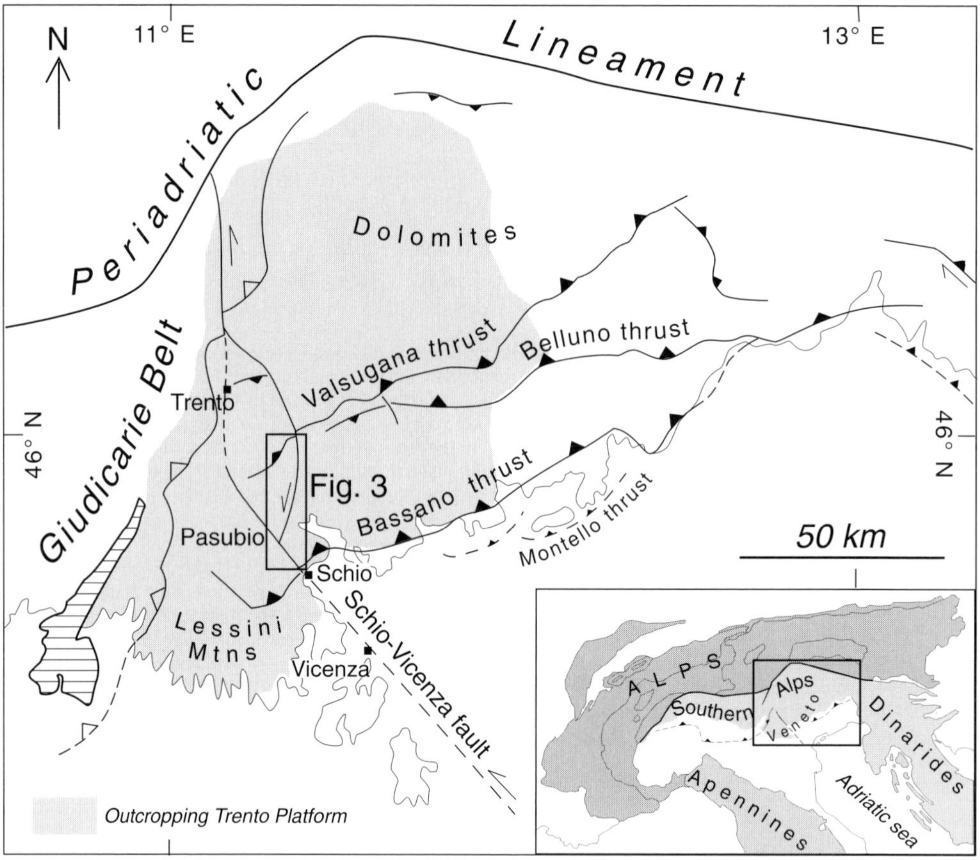

Fig. 1. Simplified structural map of the central-eastern Southern Alps, showing the main faults and the areal extent of the outcropping Trento Platform.

Lower Cretaceous carbonates. Within this interval, the thickest unit is the Upper Triassic Dolomia Principale (*c.* 800 m) (Fig. 2), a peri-tidal succession with a wide regional distribution. In the study area, this formation lies on Middle Triassic volcanics or on the Carnian soft rocks (claystones, conglomerates, evaporites) of the Raibl Group. The overlying Calcari Grigi Group (Early Jurassic) is a non-dolomitized carbonate platform unit. It includes three formations: from the bottom they are the Mount Zugna Formation, a peri-tidal succession; the Loppio oolitic limestone, a massive 20-m-thick bed of oolitic grainstones; and the Rotzo Formation, which is composed of biostromes, micrite beds, marls and black shales (Tobaldo *et al.* 2004). The St. Vigilio oolitic limestone is the youngest shallow-water unit of the Trento Platform. The overlying Middle–Upper Jurassic Ammonitico Rosso unit is composed of a condensed pelagic red nodular limestone. The Lower Cretaceous is made up of pelagic thinly bedded limestones (Maiolica Formation). All rocks are intruded by Palaeogene mafic to ultramafic dykes and explosion necks (De Vecchi 1966).

The mechanical stratigraphy of the outcropping geology is broadly composed of two units: (1) a lower thick competent unit including the Dolomia Principale, Mount Zugna and Loppio limestone Formations, and (2) an upper thin incompetent unit including the Rotzo, St Vigilio oolitic limestone, Ammonitico Rosso and Maiolica formations (Fig. 2).

Kinematic evolution of the eastern Southern Alps

The eastern Southern Alps are a SSE-verging mountain belt (Fig. 1), which mainly developed during the Neogene by contraction and oblique inversion of the Mesozoic passive margin of the Adria plate (e.g. Bertotti *et al.* 1993). The mountains consist of an imbricate fan of thrust sheets

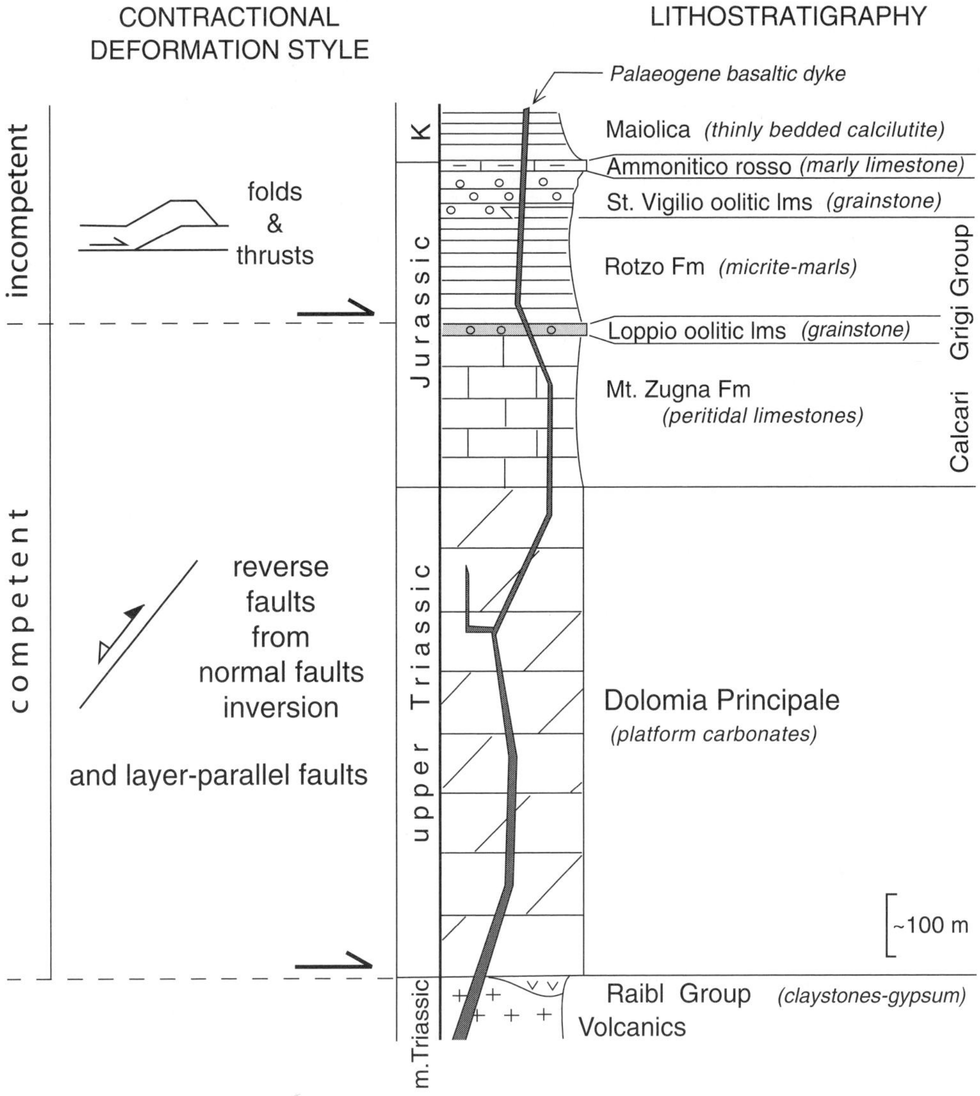

Fig. 2. Schematic column of the stratigraphy cropping out in the study area.

involving the crystalline basement (Doglioni & Bosellini 1987; Doglioni 1990; Schönborn 1999; Transalp Working Group 2002). In plan view, the thrusts show several undulations controlled by inherited features, such as Norian–Early Cretaceous normal faults and, in the western Veneto region, occasional Palaeogene normal faults. The mainly north–south-trending normal faults developed during several extensional phases in the framework of the thinning of the Adria margin (from Norian to Early Cretaceous) and the occurrence of the Early Tertiary magmatic event (De Vecchi *et al.* 1976; Zampieri 1995*a*; Figs 3 & 4).

The subsequent Neogene (Alpine) contraction with a maximum principal stress axis trending NNW (Doglioni & Bosellini 1987; Castellarin *et al.* 1992; Castellarin & Cantelli 2000) reactivated the steep north- to NNE-trending normal faults with sinistral strike-slip kinematics (Zampieri *et al.* 2003; Massironi *et al.* 2006). Recent investigations have recognized that restraining and releasing structures developed at various scales by interaction and linkage of these stepped fault segments (Zampieri 2000; Zampieri *et al.* 2003). The structure described in this study is a complex km-scale restraining stepover, which has deformed a pre-existing extensional

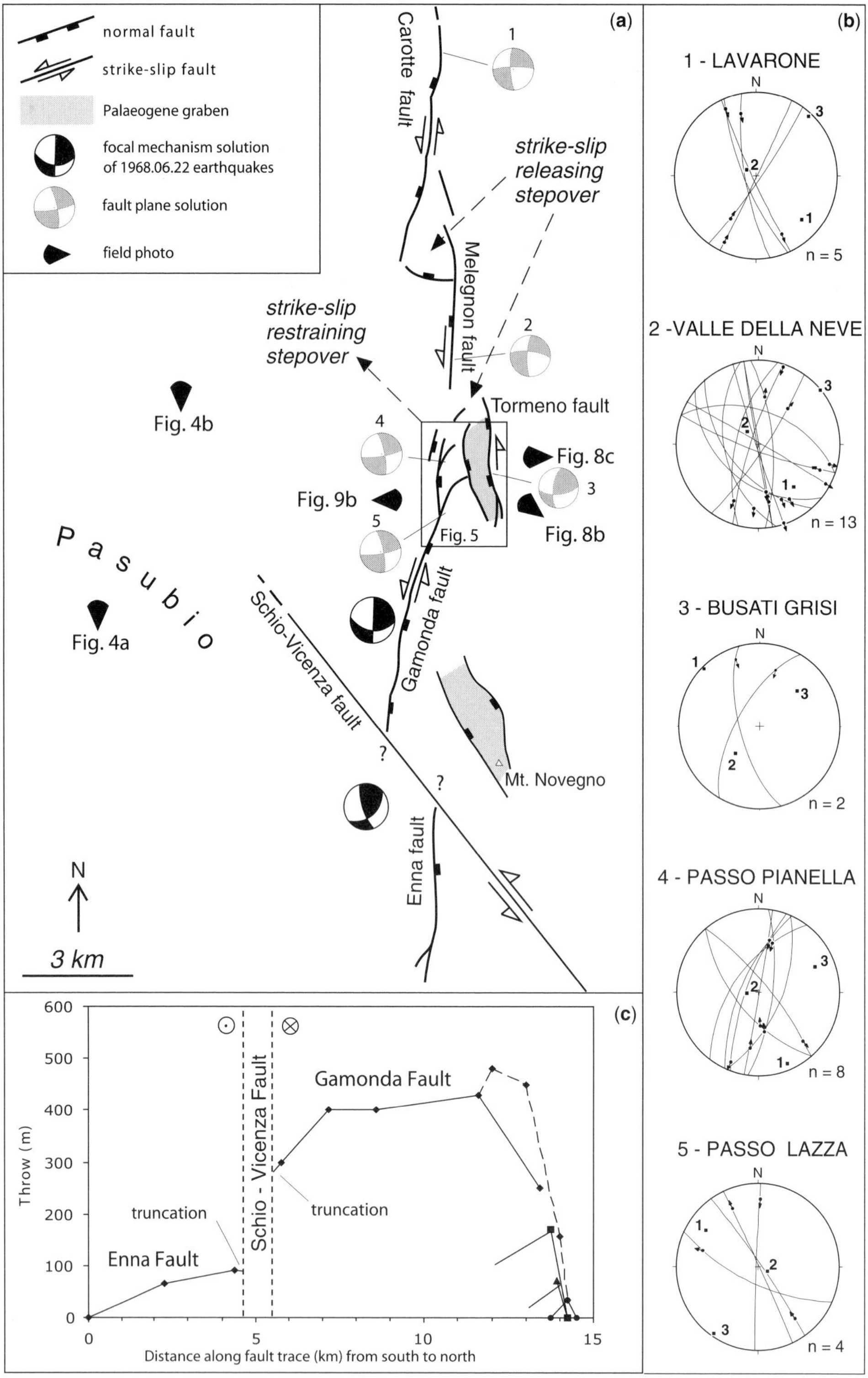
(a)
normal fault
strike-slip fault
Palaeogene graben
focal mechanism solution of 1968.06.22 earthquakes
fault plane solution
field photo
Carotte fault
Melegnon fault
Tormeno fault
strike-slip releasing stepover
strike-slip restraining stepover
Fig. 4b
Fig. 4a
Fig. 8c
Fig. 8b
Fig. 9b
Fig. 5
Pasubio
Schio-Vicenza fault
Gamonda fault
Mt. Novegno
Enna fault
N
3 km
(b)
1 - LAVARONE
n = 5
2 -VALLE DELLA NEVE
n = 13
3 - BUSATI GRISI
n = 2
4 - PASSO PIANELLA
n = 8
5 - PASSO LAZZA
n = 4
(c)
Throw (m)
Distance along fault trace (km) from south to north
Enna Fault
Gamonda Fault
Schio - Vicenza Fault
truncation
truncation

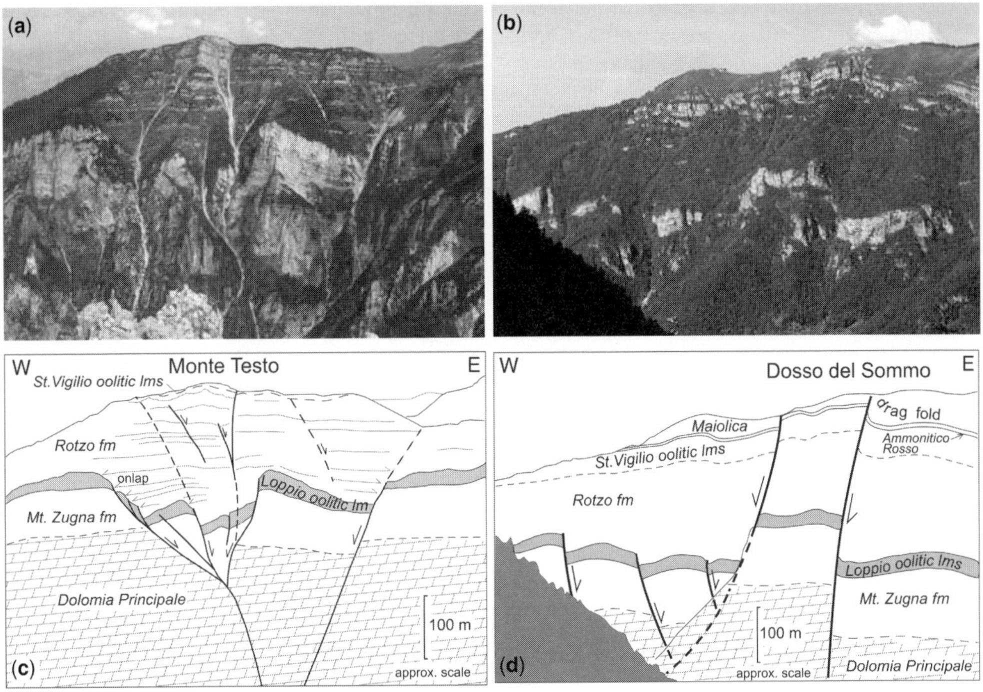

Fig. 4. Liassic extensional structures showing prominent thickness variations of the synkinematic layer, represented by the Rotzo Formation: (**a**) structure exposed on the southern cliff face of Mount Testo in the Pasubio Massif (slightly modified after Tobaldo *et al.* 2004). The location is shown in Figure 3a. (**b**) Structure exposed on the south face of Dosso del Sommo, just east of Pasubio Massif. Note the subsequent reactivation of a fault shown by the drag fold affecting Lower Cretaceous rocks. Location shown in Figure 3a. (**c**) Line drawing of Figure 4a. (**d**) Line drawing of Figure 4b.

structure, at present lying on the hanging-wall block of the SSE-verging Bassano Thrust (Fig. 1).

Present-day activity

The seismotectonics of Italy are the result of the relative motion between Africa and Eurasia. Considerable debate exists about whether the recent motion of the Adria plate occurred in conjunction with or independent of the rotation of Africa (e.g. Muttoni *et al.* 2001; Oldow *et al.* 2002). In the eastern Southern Alps, shallow-crustal seismicity is located on south-verging thrust faults within the Adriatic cover (Chiarabba *et al.* 2005). The southernmost active thrust is located in the foothills, where growth folding occurs (Montello Thrust, Fig. 1; Benedetti *et al.* 2000). The overall north–south contraction causes large thrust-related earthquakes and strike-slip earthquakes on the inherited faults. The seismicity is clustered in the eastern and westernmost parts (Slejko *et al.* 1989; Bressan *et al.* 1998) at the intersection with the Dinaric and Giudicarie belts respectively, whilst the central-western chain front shows an absence of recent seismic activity where future earthquakes are more likely to occur (Chiarabba *et al.* 2005; Galadini *et al.* 2005); the study area lies in the central western chain.

Fig. 3. (**a**) Segmented Mesozoic to Quaternary fault zone, including Tertiary grabens (see Fig. 1 for location) and the location of the Figure 4 synsedimentary extensional structures. The focal mechanisms belong to earthquakes with $M = 4.5$ and 4.1 respectively, occurring nearly contemporaneously on 22 June 1968. Their epicentral location is poorly constrained, but the solutions are consistent respectively with sinistral strike-slip activity along a north-trending fault and possibly a Riedel shear fracture (R of a northward-trending fault or P of the Schio–Vicenza Fault). (**b**) Stereographic lower-hemisphere equal-angle projections of strike-slip fault data (modified after Zampieri *et al.* 2003). The numbers 1, 2 and 3 represent the computed principal palaeostress axes. (**c**) Displacement profile of the Enna and Gamonda normal faults. Throw values have been obtained from published (De Vecchi *et al.* 1986; Braga *et al.* 1968) and unpublished geological maps, and from our field surveys. The dashed line is the summed displacement profile (main fault and splays).

It is interesting to note that in the period 1900–1980 the most important earthquakes were two events of 22 June 1968 ($M = 4.1$ and $M = 4.5$; epicentral area Pasubio), whose focal mechanism solutions are consistent with sinistral transpression along north–south-trending faults (Slejko *et al.* 1989, p. 124, fig. 9; Fig. 3, this paper). In the subsequent 1981–2002 period, once again the three most important events ($4 < M < 5$) plot in the same area (Castello *et al.* 2004). Although focal mechanism solutions are not available, their position is consistent with that of the two previous events and two of them define a north–south alignment east of Pasubio. In addition, data from the seismic network of the Provincia Autonoma di Trento for the 1982–1997 time span show activity along the north–south fault system (Galadini & Galli 1999, p. 172, Fig. 1). Therefore, all the available seismic data, along with geological studies (Zampieri *et al.* 2003; Massironi *et al.* 2006), suggest that the north–south fault array lying between the Lessini Mountains and the Valsugana Valley (Fig. 1) is presently active.

The Gamonda–Tormeno relay structure

Some fault segments (Enna, Gamonda, Tormeno, Melegnon and Carotte) compose a prominent north–south-trending fault array cutting the region between the Lessini Mountains and the Valsugana Valley for about 30 km (Braga *et al.* 1968; Figs 1 & 3). This fault array is inferred to be a synsedimentary Liassic structure, because in this area north–south-trending extensional synsedimentary structures can be demonstrated from well-documented thickness variations of Liassic units and also in outstanding exposures on east–west-trending cliff faces (Zampieri 1995*b*; Tobaldo *et al.* 2004; Fig. 4). These faults control differences in thickness mainly within the Rotzo Formation, pointing to a Sinemurian–Pliensbachian climax of activity.

The Gamonda Fault is an extensional east-dipping segment whose trace length is *c.* 8.5 km. The maximum throw is *c.* 400 m. Considering the fault length, the thickness of the sedimentary cover (2.5–3 km), and that outcropping Anisian units are clearly cross-cut, the fault is presumed to penetrate the metamorphic basement. Towards the south, the fault intersects the NW-trending strike-slip Schio–Vicenza Line, which displays recent sinistral kinematics (Castellarin & Cantelli 2000; Zampieri *et al.* 2003; Massironi *et al.* 2006). On the other side of the Schio–Vicenza Fault, an east-dipping normal-fault segment (the Enna Fault) appears to be the offset continuation of the Gamonda Fault.

The Tormeno Fault is an extensional westward-dipping segment, whose trace length is *c.* 4.8 km. The maximum throw is *c.* 340 m. The fault partially overlaps with the northern termination of the Gamonda Fault, creating a right stepover. The fault spacing is *c.* 2 km. The stepover is composed of two parts, separated by the eastward-dipping Malga Zolle Fault, which is synthetic to the Gamonda Fault and antithetic to the Tormeno Fault. Therefore, the structure composed by the Tormeno and the Malga Zolle faults is a narrow graben, where Lower Cretaceous rocks (Maiolica) are still preserved (Figs 3 & 5).

Within the overstepping zone, basaltic dykes belonging to the Palaeogene mafic to ultramafic magmatism of the southern Trento Platform are widespread (Fig. 5; De Vecchi *et al.* 1976), whereas in the surrounding area they are rare. This magmatism was coupled with ENE extension affecting the region from the Palaeocene to the Early Oligocene (Zampieri 1995*a*).

Exposed fault planes generally show indications of the last phase of activity; the latest transcurrent movements obliterated the former dip-slip slickenlines. Fault-slip data relative to the strike-slip activity (Fig. 3*b*) show that the plunges of the slip vectors are nearly equally distributed between northerly and southerly quadrants. Therefore, the resulting overall slip of the Gamonda Fault can be assumed to be subhorizontal, with a negligible vertical component.

Evolution of the polyphase Gamonda–Tormeno relay structure

Extension and evolution of extensional splays, overlaps and crossing faults

The extensional origin of the north–south-trending fault array is readily seen in the Schio sheet of the Carta Geologica d'Italia (Braga *et al.* 1968), where it is also possible to recognize the westward dip of the Carotte, Melegnon and Tormeno Faults and the eastward dip of the Gamonda and Enna faults.

A link between the Gamonda and Enna Faults was first proposed by De Boer (1963), and later reported by De Vecchi *et al.* (1986). A throw versus distance profile was constructed (Fig. 3c) using original and published data and taking into account the base of the Dolomia Principale as a reference level. This diagram is a confident representation of the cumulative throw of the extensional stages, since the vertical component of strike-slip activity is negligible (see the stereoplots in Fig. 3b). The prominent difference of the throw of the Gamonda and Enna faults at the Schio–Vicenza intersection can be explained by a transfer fault connecting the two fault segments. In this case, the observed offset of the Gamonda and

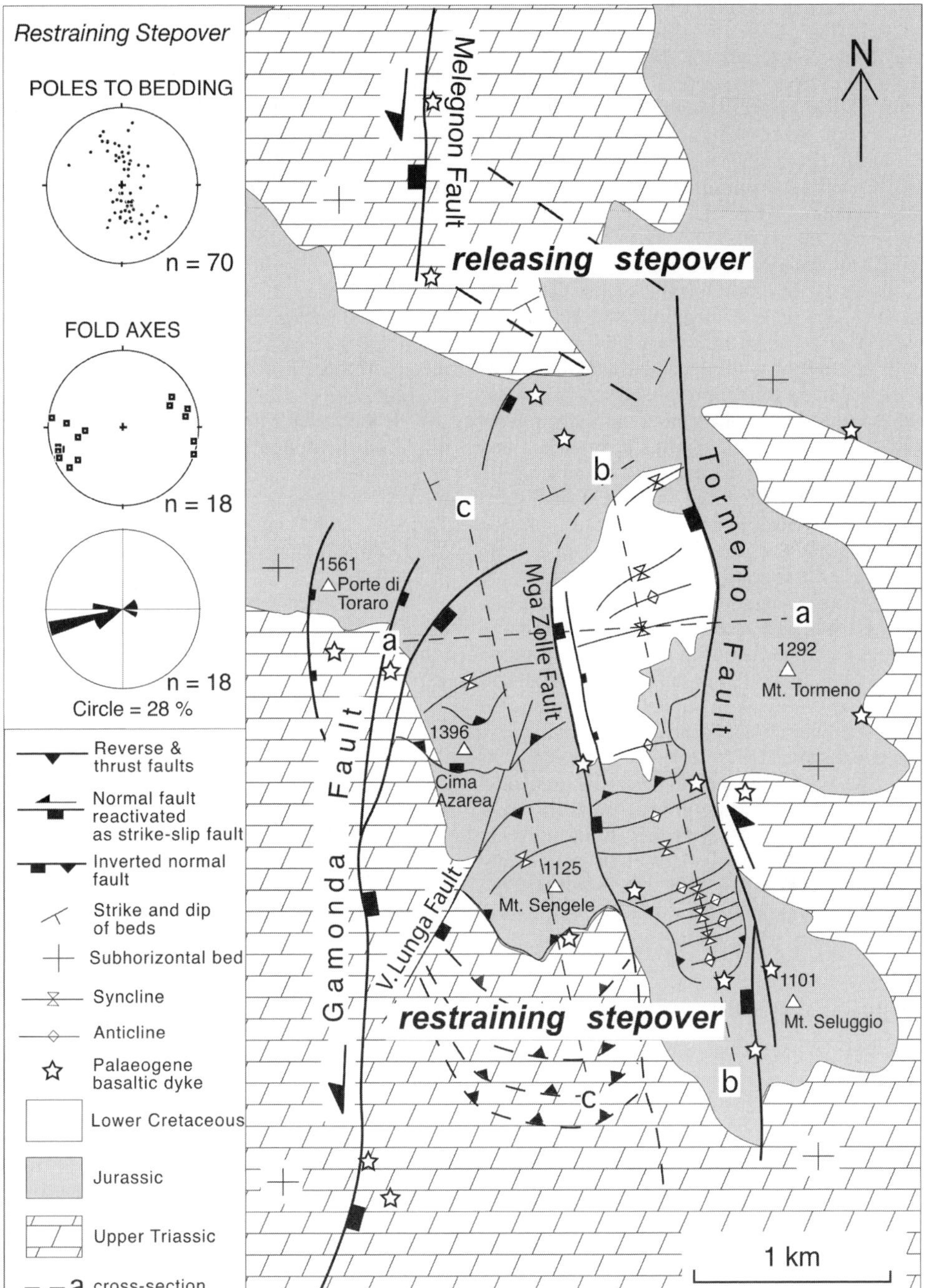

Fig. 5. Structural map of the Gamonda–Tormeno restraining stepover (see Figs 1 & 3 for location).

Enna faults does not indicate the strike-slip displacement of the Schio–Vicenza Fault.

The northern termination of the Gamonda Fault presents a horsetail splay typical of the tip damage zones of normal faults (McGrath & Davison 1995). The splay is truncated by the synthetic Malga Zolle Fault, which with its antithetic conjugate (the Tormeno Fault), forms a narrow graben where

Lower Cretaceous sediments (Maiolica) are still preserved (Figs 3 & 5). The involvement of Lower Cretaceous rocks in the graben demands extensional Upper Cretaceous and/or Tertiary activity of the Malga Zolle and Tormeno faults (see also Fig. 4d). Alternatively, the same result may have been obtained by synsedimentary Lower Cretaceous extensional activity of the two graben-bounding faults. It is likely that the high throws are the result of long-lived activity, ranging from the Lias to the Early Tertiary.

Looking to the throw profile of the Gamonda Fault (Fig. 3c), it is evident that the cumulative curve has an asymmetrical shape characterized at its northern end by a displacement maximum followed by a steep gradient termination. This effect could be explained with phases of synchronous activity of the overlapping Gamonda and Tormeno, and/or Gamonda and Malga Zolle faults. Therefore, the throw maximum and steep gradient at the northern termination of the Gamonda Fault suggest development of a true relay zone at the overlapping area during Mesozoic to Tertiary extension. A further consideration that should be taken into account when interpreting the throw profile of the Gamonda Fault is the evolution of its synthetic splays. Some authors have modelled fault growth processes by lateral propagation and breaching of initially independent overlapping segments in extensional relay zones (e.g. Childs *et al.* 1995). These studies and numerically computed experiments of intersecting normal faults (Maerten *et al.* 1999) show high slip gradients within relay zones and at fault intersections. It is very likely that similar processes also occurred at the Gamonda fault tip, inducing a further contribution to the asymmetry of the profile.

Balancing the extensional Gamonda–Tormeno cross-section

The Gamonda–Tormeno overlap zone is a narrow region of crustal extension typically composed of conjugate normal faults (Fig. 6a). Crossing conjugate normal faults are quite common in nature (e.g. Odonne & Massonnat 1992), and the surface geometry of the studied structure is consistent with an evolution by mutual offsetting of the conjugate faults. It has been demonstrated that simultaneous slip on crossing conjugate normal faults requires loss, gain, or localized redistribution of the cross-sectional area (Odonne & Massonat 1992; Ferrill *et al.* 2000). In the case-study, a possible area gain could have occurred by intrusion of Palaeogene magma, but the existence of large magmatic bodies is not proved, and the cumulative thickness of the basaltic dykes across any section allows only a very limited contribution to balancing of the section if simultaneous slip has occurred. Other mechanisms, such as salt injection, salt removal, pressure solution and sediment compaction must be excluded. In contrast, the balancing of the cross-section of Figure 6a without invoking additional hypotheses can be achieved through: (1) the model of alternating sequential movement on crossing conjugate faults (Ferrill *et al.* 2000); (2) the flattening of faults along a detachment located at the base of Dolomia Principale. No consistent evidence of extensional detachments has been found in the study area, therefore we have opted for the cross-cutting conjugate planar fault model. The balanced cross-section of Figure 6a is therefore one viable solution to the field data-set.

The long-term extensional activity of the region may have permitted alternating periods of simultaneous motions of the main faults (the development of relay zones between the Gamonda and Tormeno faults and between the Gamonda and Malga Zolle faults) and of diachronous reactivation of the faults.

The cross-section in Figure 6a includes four primary fault segments. Assuming a planar geometry for these faults, we interpret the following sequence depicted in Figure 7 as follows:

1. The east-dipping Gamonda Fault formed (stage 2), interacted (stage 3) with, and was subsequently cut by, the west-dipping Tormeno Fault (stage 4).
2. The displaced deep section of the Gamonda Fault reactivated and propagated upward (stage 5). The shallow fault resulting from propagation corresponds with the Malga Zolle Fault, whilst the resulting east-dipping fault offsetting the Tormeno Fault is composed of the shallow Malga Zolle and the deep Gamonda faults.
3. Reactivation and downward propagation of the shallow Gamonda Fault along with formation of a new synthetic fault occurred (stage 6).

The northern part of the Gamonda Fault may have been active during the last stages of extension (stages 6–7), as suggested by the injection of Palaeogene basaltic dykes. This sequence may well explain the extensional Gamonda–Tormeno overlap structure, which is composed of prominent normal faults even in the presence of successive contractional features (Fig. 5).

Contraction and evolution of the restraining stepover

After extension, the relay zone between the Gamonda and Tormeno faults experienced contraction shown by compressional structures like reverse

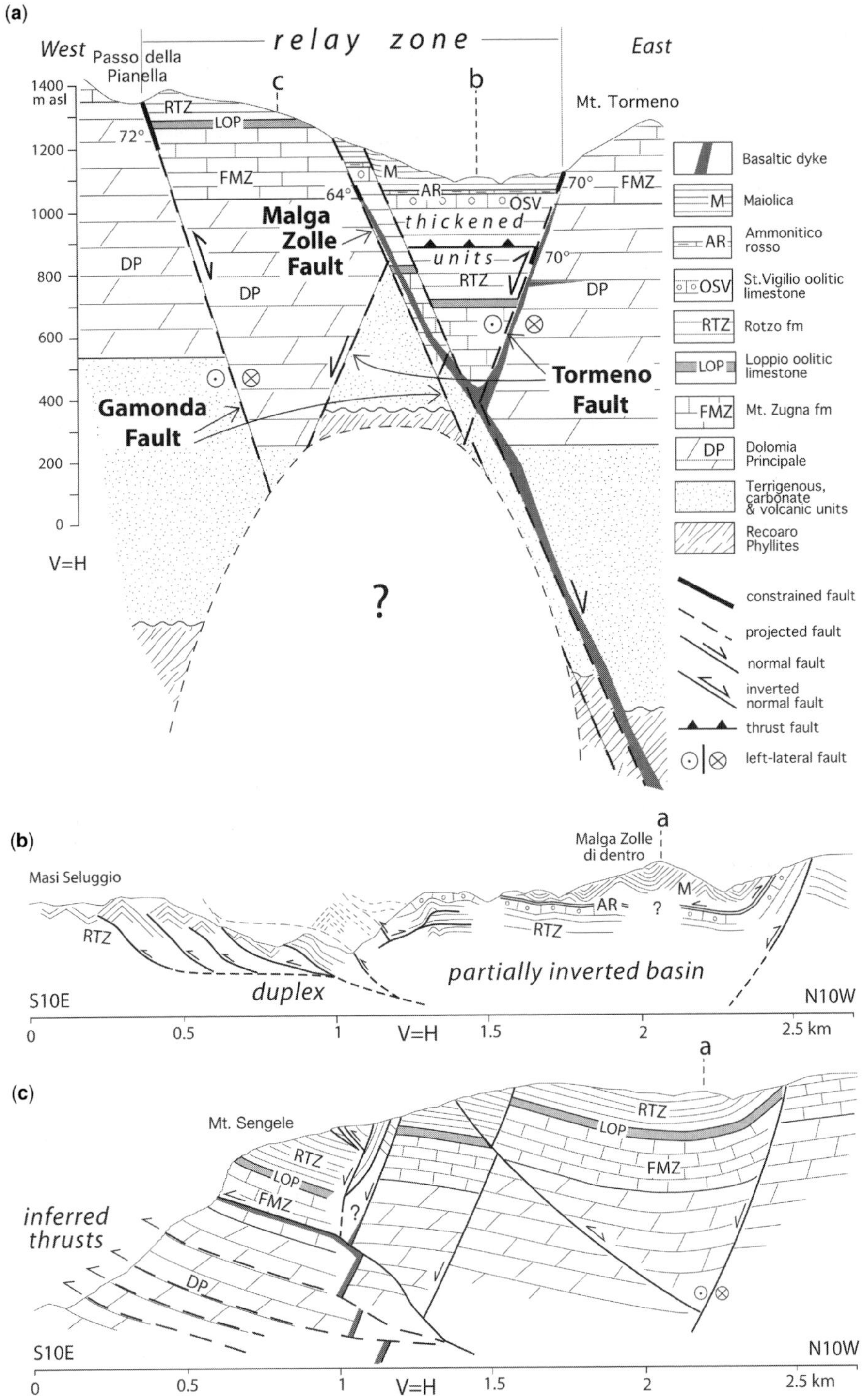

Fig. 6. (**a**) Schematic transverse cross-section of the Gamonda–Tormeno stepover. (**b**) Along-fault strike cross-section of the eastern sector of the stepover. (**c**) Along-fault strike cross-section of the western sector of the stepover. Locations shown in Figure 5.

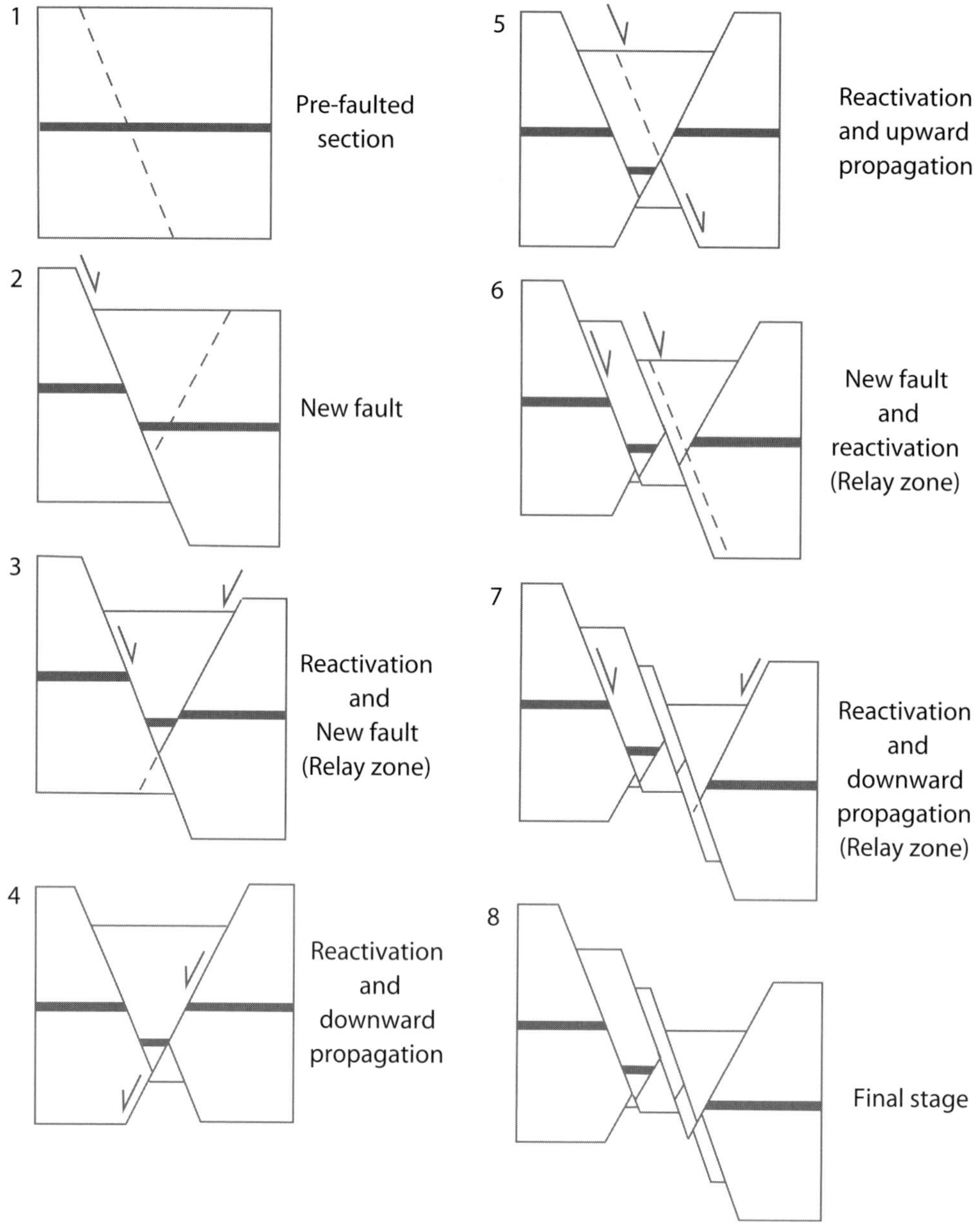

Fig. 7. Model of sequential faulting for the cross-section of Figure 6a. The synsedimentary extensional activity started in Mesozoic times (stages 1–5), and ended in Palaeogene times (stages 6–7). It is likely that ephemeral relay zones between conjugate and synthetic faults developed during some stages (stages 3, 6 and 7).

Fig. 8. Structures exposed on the eastern subzone of the Gamonda Tormeno stepover. (**a**) Digital elevation model of the Gamonda–Tormeno stepover from the SE. (**b**) Oblique view of the Mount Tormeno Fault, juxtaposing the Lower Cretaceous rocks (K) of the hanging wall (HW) against the Jurassic rocks (J) of the footwall (FW). (**c**) Large box fold in the HW of the Mount Tormeno Fault. (**d**) Detail of Figure 8c. (**e**) Line drawing of Figure 8d.

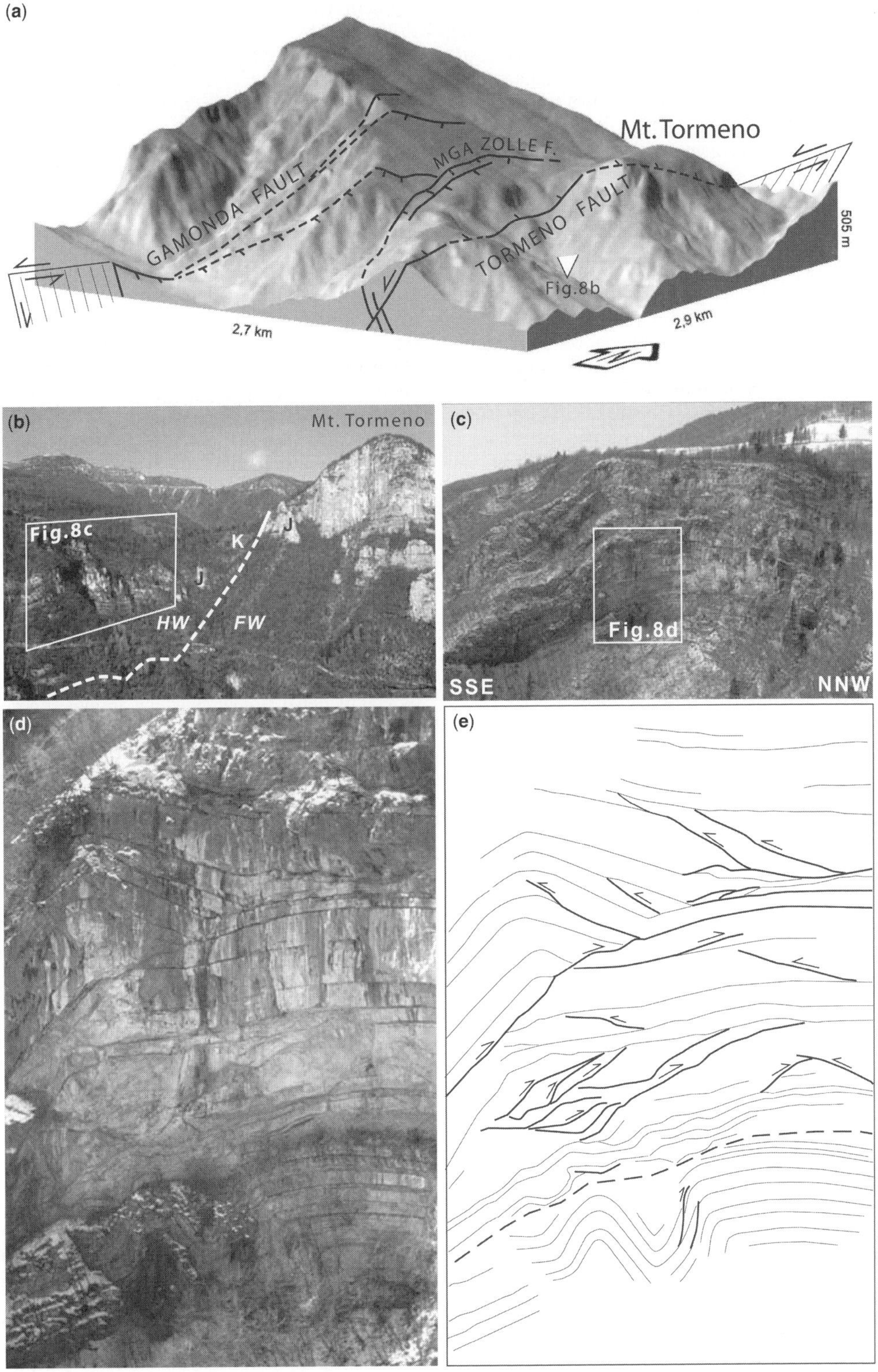
(a)
Mt.Tormeno
MGA ZOLLE F.
GAMONDA FAULT
TORMENO FAULT
Fig.8b
505 m
2,7 km
2,9 km
(b)
Mt. Tormeno
Fig.8c
K
J
J
HW
FW
(c)
Fig.8d
SSE
NNW
(d)
(e)

and thrust faults and folds. In contrast, in the surrounding area, the sedimentary rocks are flat-lying and nearly undeformed (Fig. 5). This is easily explained by reactivation of the bounding faults under strike-slip activity, which produced a restraining stepover. The geometry of the Gamonda and Tormeno faults interference is that of a right stepover. Therefore, local compression needs sinistral strike-slip activity of the faults, as proved by the kinematic indicators on fault planes (Zampieri *et al.* 2003) and Miocene–Quaternary orientation of the stress field with the maximum principal axis trending NNW to NW (Doglioni & Bosellini 1987; Castellarin *et al.* 1992; Castellarin & Cantelli 2000). In the southern Trento Platform, two sets of slickenlines on north–south to NNE fault planes are very common. The first generation of slickenside lineations is systematically dip-slip or oblique-slip extensional. The second generation is always sinistral strike-slip or shallow dipping, and overprints the first one (Zampieri 2000; Zampieri *et al.* 2003; Fig. 3b). Only in the westernmost part, close to the Giudicarie front belt, are dip-slip reverse movements found on north- to NNE fault planes (Prosser 1998). Therefore, the Gamonda–Tormeno strike-slip stepover is consistent with the regional kinematics. Its reactivation has been strongly controlled by inherited architecture and dramatically different deformation styles developed at the eastern and western blocks of the relay zone.

Eastern subzone. The eastern part of the restraining stepover is bound to the east by the Tormeno Fault and to the west by the Malga Zolle Fault (Figs 5, 6 & 7). These convergent antithetic normal faults bound a relatively small rock volume defining the narrow graben composed of the Rotzo, Ammonitico Rosso and the lower part of the Maiolica formations. These units form a weak mechanical body, because they are thin bedded and contain clay interlayers where slip localizes.

The local NNW–SSE compression induced by the development of a restraining stepover structure has produced a number of ramp–flat structures and attendant folds at various scales. The NNW- and SSE-verging thrust faults are associated with ENE–WSW-trending folds. Major folds have a chevron geometry, but box folds are also present (Figs 6 & 8). Due to intense flexural slip, duplexes, domino structures and pressure-solution cleavage occur on limbs, whilst in hinge zones saddle reefs are common. A rough estimation of the amount of the shortening along the cross-section of Figure 6b is 60%, corresponding with 1.5 km. This value corresponds with the minimum horizontal slip along the Gamonda and Tormeno faults.

Western subzone. The western part of the restraining stepover is bound to the west by the northern termination of the Gamonda Fault and to the east by the Malga Zolle Fault. These synthetic normal faults dip to the east. The rock volume in between is from the upper part of the Dolomia Principale and the Calcari Grigi Group. The Rotzo Formation is preserved at the core of two wide synclines with the axis subparallel to NE-trending steep faults (Fig. 5c). These folds are interpreted to be Mesozoic hanging-wall drag folds adjacent to normal-fault segments belonging to the splay termination of the Gamonda Fault. During subsequent contraction, these folds were ultimately reworked and normal faults were locally inverted. This is clearly documented on the western face of Cima Azarea, where a NNW-dipping fault shows both extensional and contractional features (Fig. 9). The western subzone shows less contraction than the eastern subzone. A rough estimate of the amount of the shortening accommodated by the structures exposed along the cross-section of Figure 6c is $<10\%$, corresponding to *c.* 200 m. Because the stepover structure transfers strike-slip displacement from the Gamonda Fault to the Tormeno Fault, the shortening on any north–south profile must be more or less the same and equal to the transferred displacement. Therefore, we infer that the corresponding difference in shortening of the eastern and western parts of the stepover is accommodated by contractional structures which are not exposed. The only zone where such structures may exist is the lower slope of Mount Sengele (Fig. 5), composed of the Dolomia Principale Formation and covered by vegetation. In this slope, the beds dip 20° towards NNW. In the upper part of the slope, beds belonging to the base of the Mount Zugna Formation show prominent layer-parallel faults. It is very likely that pervasive layer-parallel faulting also occurred in the rocks below. In this way, slip partitioned on to a large number of bedding surfaces may account for the apparent difference in shortening between the eastern and the western subzones of the stepover. The different structures of the eastern and western parts of the stepover are highlighted in the 3D sketch of Figure 10.

Conclusions

The poly-deformed Gamonda–Tormeno relay zone records two different tectonic regimes (Fig. 11). Prolonged extensional activity started in the Lias and ended in the Palaeogene, followed by contractional activity in the Neogene.

During the first extensional stage (Lias–Early Cretaceous), a prominent splay at the northern termination of the Gamonda normal fault developed

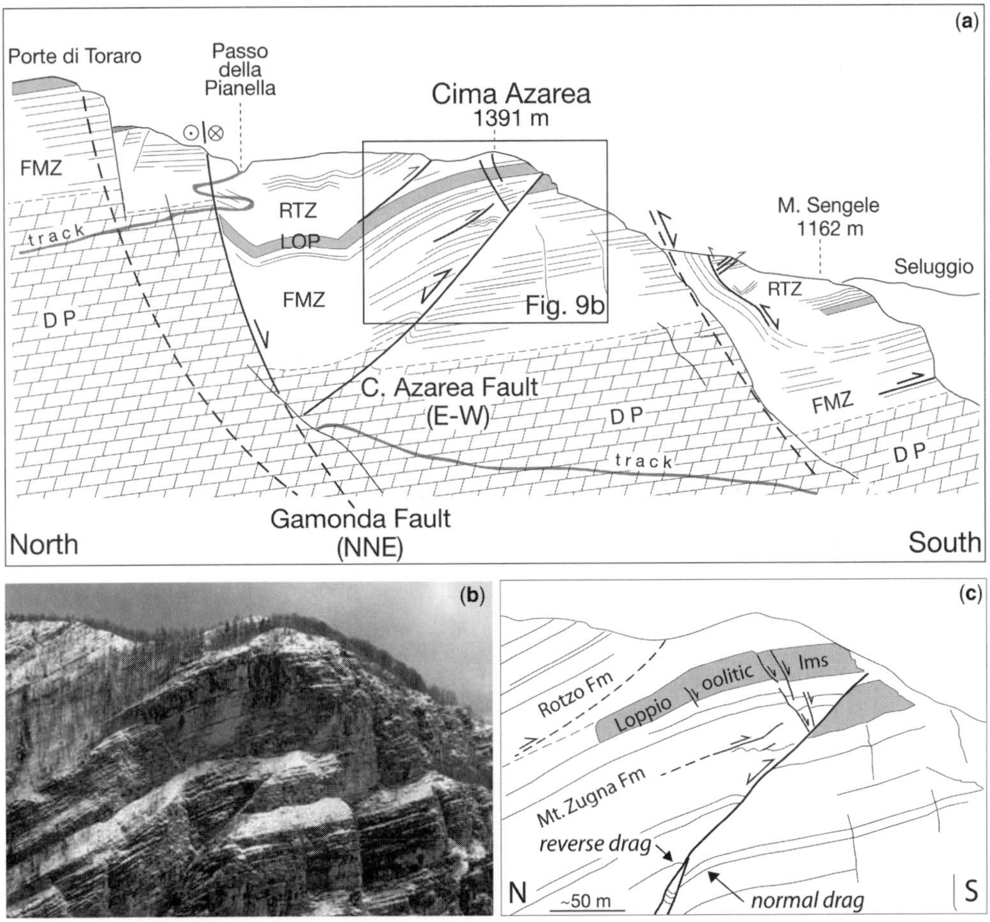

Fig. 9. (**a**) Line drawing of the western subzone structure of the Gamonda–Tormeno stepover looking east. DP: Dolomia Principale; FMZ: Mount Zugna Formation; LOP: Loppio oolitic limestone; and RTZ: Rotzo Formation. (**b**) The Cima Azarea western face. (**c**) Line drawing of the Cima Azarea structures.

in conjunction with an ephemeral transfer zone with the antithetic Tormeno Fault.

During the second extensional stage (Palaeogene), the narrow graben bound by the Malga Zolle and Tormeno faults deepened by reactivation and propagation of normal faults along with formation of new faults. The graben partly overlaps with the Gamonda fault splay juxtaposing low-competence younger rocks against the older competent rock pile.

In the third stage of evolution (Neogene to Present), the two overlapping Gamonda and Tormeno faults were reactivated with sinistral strike-slip kinematics, and the complex overlap zone became a restraining stepover. The eastern and western subzones of the stepover, composed respectively of incompetent and competent rocks, experienced deformation with varying structural styles. The eastern subzone was shortened by development of thrusts and folds, while the western subzone was shortened by inversion of pre-existing normal faults and layer-parallel faulting.

The Tormeno–Gamonda restraining stepover is an example of a relay structure which has experienced a multi-stage history. Structural relationships within the stepover demonstrate that early-formed overlap zones between normal faults are weakness zones susceptible to subsequent strike-slip reactivation in a different tectonic regime. In addition, this example shows the strong influence of inherited fault geometries and different rheological properties of rocks inside the stepover on the style of deformation of different subzones.

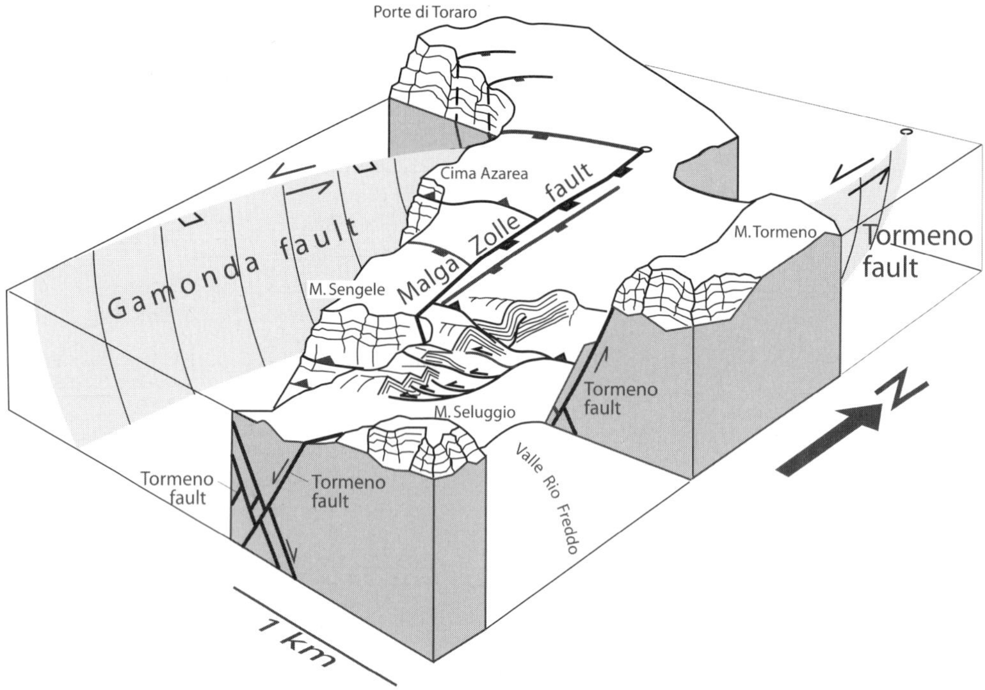

Fig. 10. Three-dimensional reconstruction of the Gamonda–Tormeno stepover.

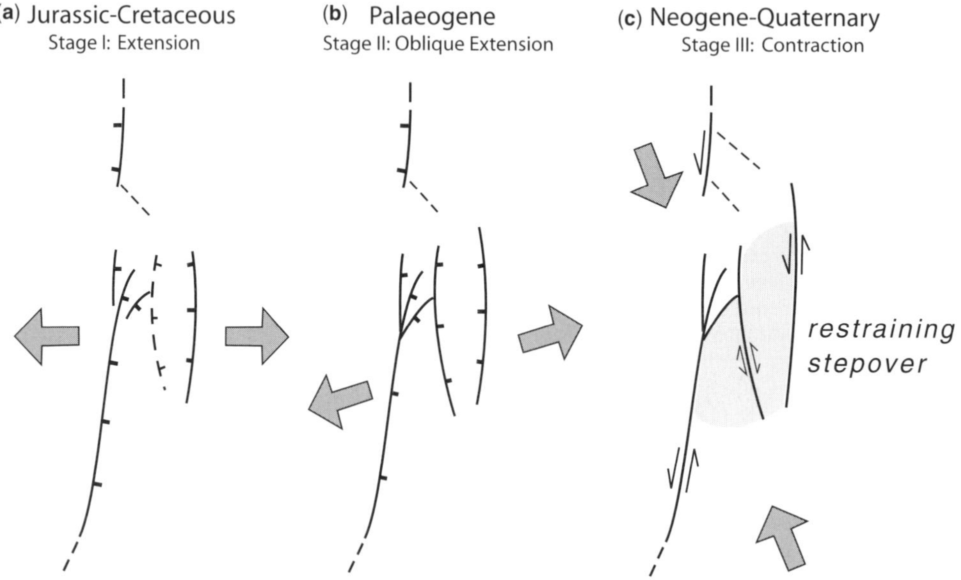

Fig. 11. Schematic summary of the structural evolution of the Mount Tormeno poly-history relay zone from the Lias to the present.

This research was supported by COFIN 2002 project (2002043912_002, national leader G. Cello, sponsored by MIUR) and 'Faults and fluids' 2004-05 project (Istituto di Geoscienze e Georisorse, sponsored by CNR). The paper has benefited considerably from thoughtful reviews and comments by S. Mazzoli and C. Childs.

References

AYDIN, A. & NUR, A. 1982. Evolution of pull-apart basins and their scale independence. *Tectonics*, **1**, 91–105.

BENEDETTI, L., TAPPONNIER, P., KING, G. C. P., MEYER, B. & MANIGHETTI, I. 2000. Growth folding and active thrusting in the Montello region, Veneto, northern Italy. *Journal of Geophysical Research*, **105**, 739–766.

BERTOTTI, G., PICOTTI, V., BERNOULLI, D. & CASTELLARIN, A. 1993. From rifting to drifting: tectonic evolution in the South-Alpine upper crust from the Triassic to the Early Cretaceous. *Sedimentary Geology*, **86**, 53–76.

BOYER, S. E. & ELLIOTT, D. 1982. Thrust systems. *AAPG Bulletin*, **66**, 1196–1230.

BRAGA, G. P., CASTELLARIN, A., CORSI, M., DE VECCHI, G. P., GATTO, G. O., LARGAIOLLI, T., MONESE, A., MOZZI, G., RUI, A., SASSI, F. P. & ZIRPOLI, G. 1968. F° 36 (Schio) alla scala 1:100.000. Servizio Geologico d'Italia, Roma.

BRESSAN, G., SNIDARCIG, A. & VENTURINI, C. 1998. Present state of tectonic stress of the Friuli area (Eastern Southern Alps). *Tectonophysics*, **292**, 211–227.

CASTELLARIN, A. & CANTELLI, L. 2000. Neo-Alpine evolution of the Southern Eastern Alps. *Journal of Geodynamics*, **30**, 251–274.

CASTELLARIN, A., CANTELLI, L. *ET AL.* 1992. Alpine compressional tectonics in the Southern Alps. Relationships with the N-Apennines. *Annales Tectonicae*, **6**, 62–94.

CASTELLO, B., MORO, M., CHIARABBA, C., DI BONA, M., DOUMAZ, F., SELVAGGI, G. & AMATO, A. 2004. *Seismicity Map of Italy*. Istituto Nazionale di Geofisica e Vulcanologia, World Wide Web Address: http://www.ingv.it/CSI/

CHIARABBA, C., JOVANE, L. & DI STEFANO, R. 2005. A new view of Italian seismicity using 20 years of instrumental recordings. *Tectonophysics*, **395**, 251–268.

CHILDS, C., WATTERSON, J. & WALSH, J. J. 1995. Fault overlap zones within developing normal fault systems. *Journal of the Geological Society, London*, **152**, 535–549.

CRIDER, J. G. & POLLARD, D. D. 1998. Fault linkage: three-dimensional mechanical interaction between échelon normal faults. *Journal of Geophysical Research*, **103**, 373–391.

DAHLSTROM, C. D. A. 1969. Balanced cross-sections. *Canadian Journal of Earth Sciences*, **6**, 743–757.

DE BOER, J. 1963. The geology of the Vicentinian Alps (NE Italy). *Geologica Ultraiectina*, **11**, 1–178.

DE VECCHI, G. P. 1966. I filoni basici ed ultrabasici dell' Altipiano di Tonezza (Alto Vicentino). *Memorie degli Istituti di Geologia e Mineralogia dell'Università di Padova*, **25**, 1–58.

DE VECCHI, G. P., DI LALLO, E. & SEDEA, R. 1986. Note illustrative della Carta geologica dell' area di Valli del Pasubio–Posina–Laghi, alla scala 1:20.000. *Memorie di Scienze Geologiche, Padova*, **38**, 187–205.

DE VECCHI, G. P., GREGNANIN, A. & PICCIRILLO, E. M. 1976. Tertiary volcanism in the Veneto: magmatology, petrogenesis and geodynamic implications. *Geologische Rundschau*, **65**, 701–710.

DOGLIONI, C. 1990. The Venetian Alps Thrust Belt. *In:* MCCLAY, K. (ed.), *Thrust Tectonics*, Chapman & Hall, London, 319–324.

DOGLIONI, C. & BOSELLINI, A. 1987. Eoalpine and Mesoalpine tectonics in the Southern Alps. *Geologische Rundschau*, **76**, 735–754.

FAULDS, E. J. & VARGA, R. J. 1998. The role of accommodation zones and transfer zones in the regional segmentation of extended terranes. *In*: FAULDS, J. E. & STEWARD, J. H. (eds) *Accommodation Zones and Transfer Zones: the Regional Segmentation of the Basin and Range Province*, Geological Society of America Special Papers, **323**, 1–45.

FERRILL, A., MORRIS, A. P., STAMATAKOS, J. A. & SIMS, D. W. 2000. Crossing conjugate normal faults. *AAPG Bulletin*, **84**, 1543–1559.

GALADINI, F. & GALLI, P. 1999. Palaeoseismology related to the displaced Roman archaeological remains at Egna (Adige Valley, northern Italy). *Tectonophysics*, **308**, 171–191.

GALADINI, F., POLI, M. E. & ZANFERRARI, A. 2005. Seismogenic sources potentially responsible for earthquakes with $M > 6$ in the eastern Southern Alps (Thiene–Udine sector, NE Italy). *Geophysical Journal International*, **161**, 739–762.

LARSEN, P. H. 1988. Relay structures in a Lower Permian basement-involved extension system, East Greenland. *Journal of Structural Geology*, **17**, 1011–1024.

MCGRATH, A. & DAVISON, I. 1995. Damage zone geometry around fault tips. *Journal of Structural Geology*, **10**, 3–8.

MAERTEN, L., WILLEMSE, E. J., POLLARD, D. D. & RAWNSLEY, K. 1999. Slip distribution on intersecting normal faults. *Journal of Structural Geology*, **21**, 259–271.

MANN, P., HEMPTON, M. R., BRADLEY, D. C. & BURKE, K. 1983. Development of pull-apart basins. *Journal of Geology*, **91**, 529–554.

MASSIRONI, M., ZAMPIERI, D. & CAPORALI, A. 2006. Miocene to Present major fault linkages through the Adriatic indenter and the Austroalpine–Penninic collisional wedge (Alps of NE Italy). *In*: MORATTI, G. & CHALOUAN, A. (eds) *Active Tectonics of the Western Mediterranean Region and North Africa*. Geological Society, London, Special Publications, **262**, 245–258.

MORLEY, C. K., NELSON, R. A., PATTON, T. L. & MUNN, S. G. 1990. Transfer zones in the East African Rift System and their relevance to hydrocarbon exploration in rifts. *AAPG Bulletin*, **74**, 1234–1253.

MUTTONI, G., GARZANTI, E., ALFONSI, L., BIRILLI, S., GERMANI, D. & LOWRIE, W. 2001. Motion of Africa and Adria since the Permian: paleomagnetic and paleoclimatic constraints from northern Libya. *Earth and Planetary Science Letters*, **192**, 159–174.

ODONNE, F. & MASSONNAT, G. 1992. Volume loss and deformation around conjugate fractures: comparison between natural example and analogue experiments. *Journal of Structural Geology*, **14**, 963–972.

OLDOW, J. S., FERRANTI, L. ET AL. 2002. Active fragmentation of Adria, the north African promontory, central Mediterranean orogen. *Geology*, **30**, 779–782.

PEACOCK, D. C. P. 2003. Scaling of transfer zones in the British Isles. *Journal of Structural Geology*, **25**, 1561–1567.

PEACOCK, D. C. P. & SANDERSON, D. J. 1991. Displacement, segment linkage and relay ramps in normal fault zones. *Journal of Structural Geology*, **13**, 721–733.

PEACOCK, D. C. P. & ZHANG, X. 1993. Field examples and numerical modelling of oversteps and bends along normal faults in cross-section. *Tectonophysics*, **234**, 147–167.

PROSSER, G. 1998. Strike-slip movements and thrusting along a transpressive fault zone: the North Giudicarie line (Insubric line, northern Italy). *Tectonics*, **17**, 921–937.

REIJS, J. & MCCLAY, K. 1998. Salar Grande pull-apart basin, Atacama Fault System, northern Chile. *In:* HOLDSWORTH, R. E., STRACHAN, R. A. & DEWEY, J. F. (eds) *Continental Transpressional and Transtensional Tectonics*. Geological Society, London, Special Publications, **135**, 127–141.

SCHÖNBORN, G. 1999. Balancing cross sections with kinematic constraints: the Dolomites (northern Italy). *Tectonics*, **18**, 527–545.

SLEJKO, D., CARULLI, G. B. ET AL. 1989. Seismotectonics of the eastern Southern Alps. *Bollettino di Geofisica Teorica ed Applicata*, **31**, 109–136.

TOBALDO, M., ZANDONAI, F., AVANZINI, M., MIORANDI, M. & ZAMPIERI, D. 2004. Note illustrative della carta geologica del settore nord occidentale del Monte Pasubio (Trentino, Italia). *Studi Trentini di Scienze Naturali–Acta Geologica*, **79**, 161–180.

TRANSALP WORKING GROUP 2002. First deep seismic images of the Eastern Alps reveal giant crustal wedges. *Geophysical Research Letters*, **29**, 10 1029–10 1032.

TRUDGILL, B. & CARTWRIGHT, J. 1994. Relay-ramp forms and normal-fault linkages, Canyonlands National Park, Utah. *Geological Society of America Bulletin*, **106**, 1143–1157.

ZAMPIERI, D. 1995*a*. Tertiary extension in the southern Trento Platform, Southern Alps, Italy. *Tectonics*, **14**, 645–657.

ZAMPIERI, D. 1995*b*. L'anticlinale di roll-over liassica dei Sogli Bianchi nel Monte Pasubio (Vicenza). *Atti Ticinensi di Scienze della Terra (Serie speciale)*, **3**, 3–9.

ZAMPIERI, D. 2000. Segmentation and linkage of the Lessini Mountains normal faults, Southern Alps, Italy. *Tectonophysics*, **319**, 19–31.

ZAMPIERI, D., MASSIRONI, M., SEDEA, R. & SPARACINO, V. 2003. Strike-slip contractional stepovers in the Southern Alps (northeastern Italy). *Eclogae Geologicae Helvetiae*, **96**, 115–123.

Transpressional structures on a Late Palaeozoic intracontinental transform fault, Canadian Appalachians

J. W. F. WALDRON[1], C. ROSELLI[2] & S. K. JOHNSTON[3]

[1]*Department of Earth and Atmospheric Sciences, University of Alberta, Edmonton, Alberta, T6G 2E3, Canada (e-mail: john.waldron@ualberta.ca)*

[2]*Department of Earth and Atmospheric Sciences, University of Alberta, Edmonton, Alberta, T6G 2E3, Canada*

[3]*Saint Mary's University, Halifax, Nova Scotia, B3H 3C3, Canada*

Abstract: The east–west Minas fault zone, separating the Early Palaeozoic Meguma and Avalon terranes of the Appalachians, experienced dextral strike-slip motion during the Carboniferous. Abundant oblique contractional structures indicate localized dextral transpression, immediately south of the zone, probably associated with a restraining bend. Subsurface data indicate that the deformed Horton Group clastic rocks are thrust above younger Windsor Group evaporites.

Excellent exposures on wave-cut platforms of the Bay of Fundy show structures developed in transpression, including NE-trending upright and inclined folds; south-verging thrust and reverse faults; and NW-striking normal faults. Northwest-trending boudins, which are perpendicular and slightly rotated in a clockwise sense relative to fold hinges, provide a field indicator for dextral transpression. The earliest folds (F_1) are curvilinear and may have formed by deformation of wet sediment. F_2 tectonic folds show weak axial-planar cleavage. Locally, these have been rotated into reclined orientations; spectacular downward-facing folds are probably due to refolding by more east–west F_3 folds. The structures observed are consistent with pure-shear-dominated transpression, with the local angle of convergence α increasing over time. This strain history is compatible with progressive strain partitioning, probably associated with the spreading of topography developed at the restraining bend.

In the Northern Appalachians of Atlantic Canada, the Meguma and Avalon terranes represent peri-Gondwanan crust that was amalgamated with Laurentia (ancestral North America) in Early Palaeozoic times. The boundary between these terranes is marked by the Minas fault zone that strikes east–west through Nova Scotia, Canada (Fig. 1). The zone, also known as the Cobequid–Chedabucto fault zone (Webb 1969), Minas geofracture (Keppie 1982) or Minas fault system (Gibbons *et al.* 1996), was tectonically active at intervals throughout the Late Palaeozoic. Its lateral extent, linear trace, and a variety of kinematic indicators suggest that it was a dextral intracontinental transform fault during the late stages of the development of the Appalachians.

South of the terrane boundary, on the macrotidal wave-cut platforms of the Minas Basin of the Bay of Fundy (Figs 2 & 3), exceptional exposures of the Mississippian Horton and Windsor Groups display spectacular systems of folds, including overturned and downward-facing structures indicating at least several kilometres of shortening. Faults are also abundant, and include examples with extensional, contractional, and strike-slip offsets. Seismic and drilling data from petroleum exploration imply duplication of stratigraphy above a zone of Mississippian evaporites. An environment of dextral transpression at a restraining bend on the Minas fault zone is suggested by the predominance and orientations of contractional and strike-slip structures, and by the restricted lateral extent of this intensely deformed zone along the relatively linear former terrane boundary.

Following the work of Sanderson and Marchini (1984), numerous theoretical studies have outlined the relationship between incremental and finite strains in belts of transpression and transtension (e.g. Fossen & Tikoff 1993, 1998; Tikoff & Teyssier 1994; Jones & Tanner 1995; Teyssier *et al.* 1995; Dewey *et al.* 1998). The area of this study provides a unique opportunity to document and test interpretations of the outcrop-scale kinematic development of a transpressional zone in sedimentary rocks at low metamorphic grade.

Regional geological setting

South of the Minas fault zone, the Meguma terrane is characterized by a thick (>10 km) succession of Cambrian–Ordovician metasedimentary rocks, the Meguma Group, overlain by a Late Ordovician to

From: CUNNINGHAM, W. D. & MANN, P. (eds) *Tectonics of Strike-Slip Restraining and Releasing Bends.* Geological Society, London, Special Publications, **290**, 367–385.
DOI: 10.1144/SP290.14 0305-8719/07/$15.00

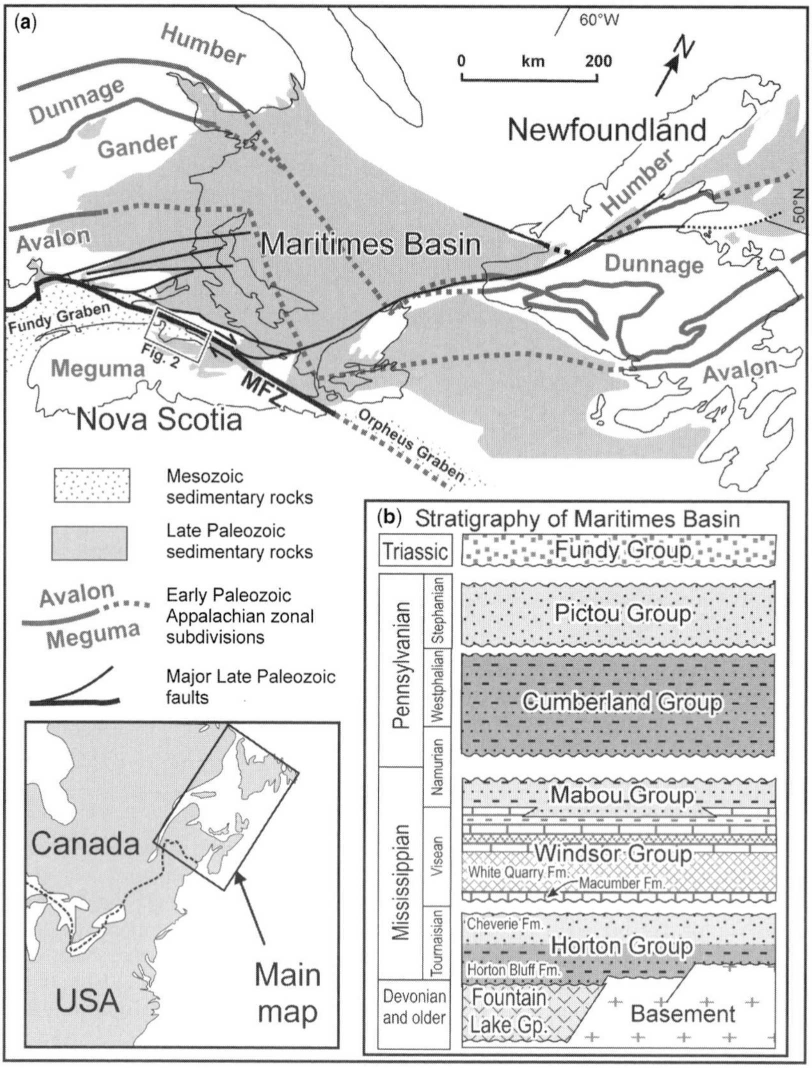

Fig. 1. (**a**) Regional map of the Appalachians of Atlantic Canada, showing the major tectonic units and boundaries, modified from Gibling *et al.* (1995). MFZ, Minas fault zone. Inset shows location in North America. (**b**) Simplified stratigraphic relationships in the fill of the Maritimes Basin. Stipple patterns: clastic sediments. Brick patterns: carbonates. Hatching: evaporities. V-patterns: volcanics.

Early Devonian succession of volcanics and shelf sedimentary rocks (Fig. 1). The entire package was tightly folded in mid-Devonian Neo-Acadian deformation, and intruded by Devonian granitoid plutons. North of the boundary, Proterozoic to Silurian basement rocks of the Avalon terrane are exposed in the Cobequid and Antigonish Highlands, together with extensive Devono-Carboniferous intrusive and volcanic rocks, and mainly fault-bounded slivers of Carboniferous sedimentary units. S–C protomylonites and other structural features in Devonian granitoid rocks close to the Minas fault zone indicate predominantly dextral strike-slip motion during the later stages of Palaeozoic terrane accretion (Eisbacher 1969, 1970; Mawer & White 1987; Culshaw & Liesa 1997).

The timing of deformation along the Minas fault zone is in part constrained by stratigraphic relationships in the mainly Carboniferous fill of the Maritimes sedimentary basin (Fig. 1), which developed across all the former terranes of the Atlantic Canadian Appalachians. At least some of the deformation along the Minas fault zone predated deposition of the Tournaisian Horton Group

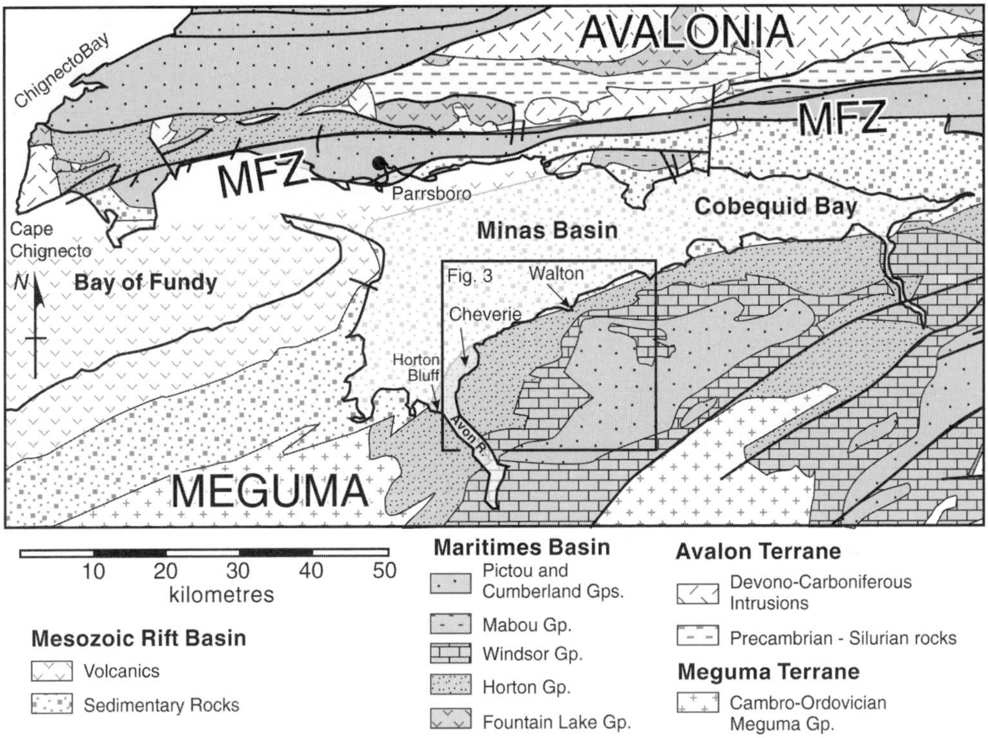

Fig. 2. Geologic map of the area around the Minas Basin, Nova Scotia, after Keppie (2000). A box encloses the study area, shown in Figure 3. MFZ, Minas fault zone.

(Murphy *et al.* 1994). In the area of this study, Horton Group rocks are intensely deformed, but a mafic intrusion that cross-cuts fabrics is dated by Kontak (2000) at 315 ± 4 Ma, indicating that significant deformation occurred by Early Westphalian times. To the east, relationships at Stellarton (Fig. 1; Chandler *et al.* 1998; Waldron 2004) also indicate major deformation close to the Namurian–Westphalian boundary, with continuing dextral strike-slip motion; localized episodes of both transtension and transpression, probably associated with releasing and restraining bends, continued until at least the Late Westphalian (Fralick & Schenk 1981; Yeo & Gao 1987; Waldron 2004).

Subsequently, during the Early Mesozoic opening of the Atlantic Ocean, the former terrane boundary was reactivated as an oblique-slip, sinistral transtensional feature, bounding the Fundy half-graben (Fig. 1), in which were deposited clastics and volcanics of the Fundy Group (Newark Supergroup; Olsen & Schlische 1990; Wade *et al.* 1996). The thin, up-dip, southern erosional margin of this succession rests unconformably on Carboniferous rocks overlying the Meguma Terrane in the area of this study (Fig. 4a).

Field relationships

Stratigraphic units

Figure 3 represents a summary of the principal features of the geology, based on previous mapping by the Geological Survey of Canada (Weeks 1948; Boyle 1957; Stevenson 1959; Bell 1960; Crosby 1962) and the Nova Scotia Department of Natural Resources (Giles & Boehner 1982; Ferguson 1983; Moore 1993*a*, *b*, 1994; Moore 1996; Moore *et al.* 2000).

Much of the coastal exposure is of the Tournaisian Horton Group, which in this area is divided into a lower, mainly grey lacustrine unit named the Horton Bluff Formation, overlain by predominantly fluvial red and grey beds of the Cheverie Formation (Fig. 3). Relatively undeformed strata characterize the type sections of the two formations, respectively west and east of the Avon River estuary (Fig. 2). These sections are interpreted to rest unconformably on Meguma basement.

Stratigraphically above the Cheverie Formation, the Viséan Windsor Group includes a thin (2–10 m) basal unit of laminated limestone (Macumber

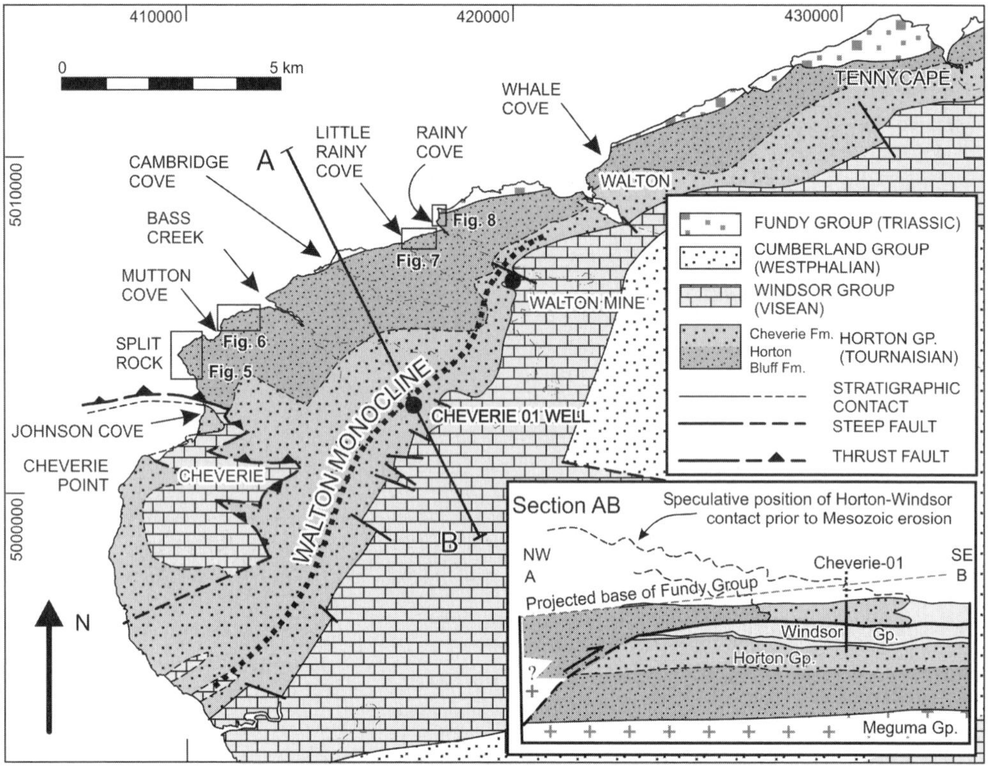

Fig. 3. (**a**) Geological map of Cheverie–Walton area, areas of detailed maps, line of cross-section, and other geographical features mentioned the in text (with information from maps by Weeks 1948; Boyle 1957; Stevenson 1959; Crosby 1962; Ferguson 1983; Moore *et al.* 2000). Coordinates show a Universal Transverse Mercator (UTM) grid. (**b**) Schematic cross-section, showing an inferred unit of the allochthonous Horton Group above a blind thrust at depth.

Formation). The Macumber Formation was the site of significant mineralization at Walton (Fig. 3), which was a major producer of baryte in the mid-20th century. Up-section, the Macumber Formation invariably passes transitionally into complex brecciated carbonates, informally termed the Pembroke Breccia. Lavoie *et al.* (1995) have argued that the Pembroke Breccia has a composite origin, resulting partly from synsedimentary brecciation and debris flow, and partly from later evaporite solution and collapse. We argue that tectonic movement in the overlying Windsor evaporites has also played a role in the development of the Pembroke Breccia. These basal carbonate units are overlain by thick anhydrite and halite successions, represented in outcrop by gypsum and anhydrite that contain complex structures suggesting folding, diapiric flow, fracturing, and brecciation.

The youngest lithified strata in the area are those of the Triassic Fundy Group (Newark Supergroup), which dip gently northward under the Bay of Fundy, and rest with profound angular unconformity on the Horton Group (Fig. 4a). Most faults and folds in the Horton Group are truncated at this surface. However, the unconformity is offset by widely spaced (0.5–5 km), north- to NE-striking steep faults of inferred Mesozoic age, presumed to be associated with sinistral reactivation of the Minas fault zone in Mesozoic times.

Map-scale structure

Relatively undeformed strata characterize the type sections of the Horton Bluff and Cheverie formations, respectively west and east of the Avon Estuary (Figs 2 & 3). To the NE, at Johnson Cove (Fig. 3), a gentle anticline in the Cheverie Formation plunges beneath the Windsor Group (the Macumber Formation overlain by the Pembroke Breccia). Scattered outcrops of highly deformed Windsor gypsum and anhydrite occur in adjacent coastal cliffs. However, immediately north of Johnson Cove, where a Windsor succession would

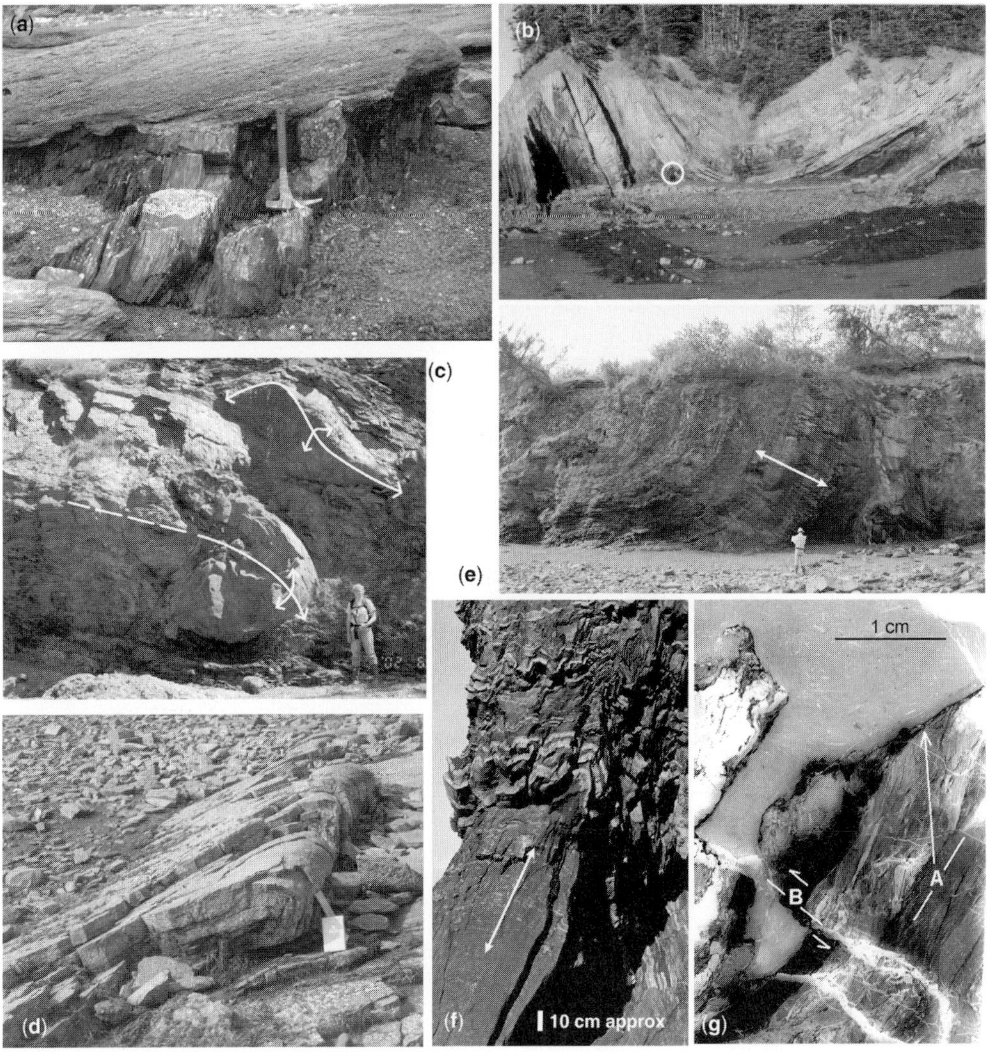

Fig. 4. (**a**) Unconformity of the subhorizontal Fundy Group (Triassic) on the subvertical Horton Group (Carboniferous) at Whale Cove, Walton (Fig. 3). (**b**) View of a syncline at Split Rock (Fig. 5) (human figure for scale). (**c**) Curvilinear folds interpreted as F_1, Little Rainy Cove (Fig. 7) (human figure for scale). (**d**) Fault propagation fold, 1 km south of Split Rock (Fig. 5). The notebook used for scale is 18 cm high. (**e**) F_2 folds in a cliff at Little Rainy Cove (Fig. 7), showing F_2 fold and axial-planar spaced cleavage (arrow) in sandy palaeosol unit. Figure for scale. (**f**) Slaty cleavage at Split Rock (Fig. 5), axial-planar to F_2 folds. (**g**) Thin section of the contact of a diabase intrusive (upper part of slide) cross-cutting the fabric in a slate at A, but cut by a later fault at B.

be expected, is a region of coast within which Windsor and Horton lithologies are intermixed and have highly variable bedding orientations, interpreted as a megabreccia. Moore (1996) interpreted a north-dipping thrust fault through this zone (Fig. 3). This inferred thrust is overlain by the Horton Bluff Formation exposed in a succession of ENE- and WSW-trending folds with axial-planar cleavage increasing in intensity northwards, as far as Split Rock (Figs 4b & 5). The Horton Bluff Formation is locally cut by mafic intrusions which cross-cut the fabric in the slates, although the intrusions are themselves cut by brittle faults (Fig. 4g). An intrusion from this suite was dated by Kontak (2000) at 315 ± 4 Ma, close to the Mississippian–Pennsylvanian boundary.

East of Cheverie, Horton Bluff Formation strata continue to be exposed in coastal cliffs, between

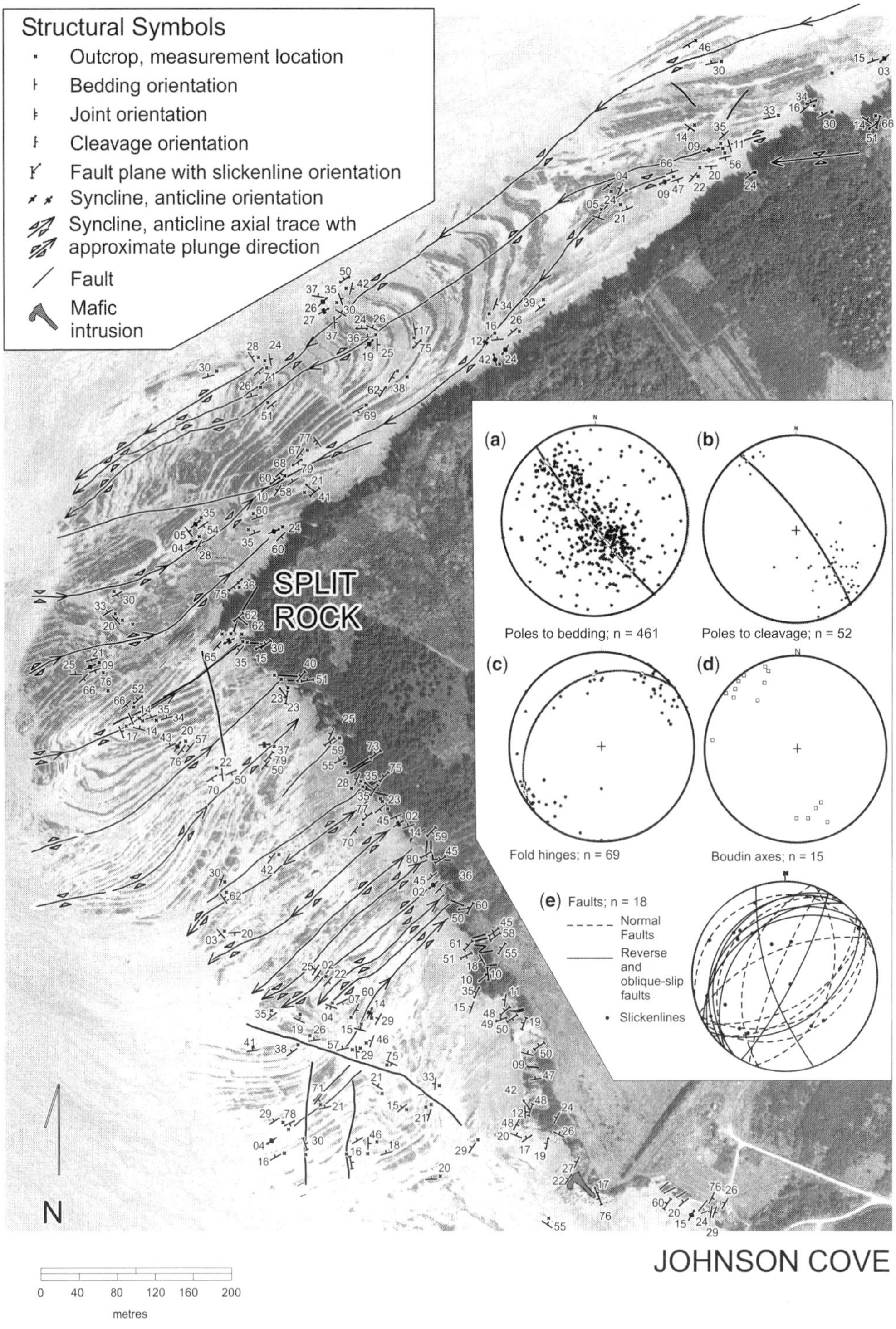

Fig. 5. Map of the area between Johnson Cove and Split Rock, NE of Cheverie (location shown in Fig. 3), superimposed on an aerial photograph of the shoreline exposed at low tide. (**a–d**) Equal-area lower-hemisphere projections, showing the orientations of the principal structures. Location shown in Figure 3.

regions of unconformable Triassic cover. Typically, the formation is folded by ENE-trending folds, and cut by steep faults that strike NW, marked by stream valleys. Some of the structures on the shore were described and interpreted by Fyson (1964), who proposed regional SE movements on thrusts to account for the largest folds, and later more southerly transport to rotate a first generation of folds to recumbent and then downward-facing orientations. In a later work (Fyson 1967), the main folds were interpreted as inclined away from Meguma basement highs; this pattern was inferred to support gravity sliding as a possible mechanism responsible for folding.

Inland, in areas of poor exposure, the Horton Bluff Formation appears to pass stratigraphically upward into the steeply dipping, generally south-younging Cheverie Formation. A sinuous belt of lower Windsor Group rocks, including the Macumber Formation, marks the SE edge of the belt of Horton rocks. It is marked by sinkholes and steeply dipping carbonate outcrops, indicating that this belt represents the steep limb of a somewhat sinuous fold, here termed the Walton monocline (Fig. 3). Extensive mineralization of the basal Windsor Group at Walton, and elsewhere on this zone, has led to a history of drilling and mining operations.

Subsurface structure

Cheverie-01 well (Fig. 3) was drilled approximately 6 km inland of the coastal section, immediately NW of the Walton monocline (McDonald 2001). The uppermost 404 m of the well intersected a succession of grey to reddish-grey sandstones and shales, lithologies similar to the surface exposure of the Horton Group (Cheverie Formation) at the wellsite. From 404 m to 890 m, the well penetrated Windsor Group carbonates and evaporites. From 890 m to 1410 m, the well intersected the Horton Group again: red to grey, shales and fine-to-coarse, locally quartzitic sandstones of the Cheverie Formation (McDonald 2001). Thus, we infer that the Cheverie-01 well intersected a sheet of deformed Horton Group rocks at the surface, thrust over a décollement in younger Windsor Group rocks, which in turn rest on a much less deformed, autochthonous Horton succession (Fig. 3 inset).

Additional subsurface information is available from mining exploration boreholes around the Walton baryte mine and elsewhere. The Walton deposit is located at the intersection between the Walton monocline and a prominent NW–SE fault zone that dips steeply to the NE (O'Sullivan & Macfie 1988). To the south of this fault, in numerous boreholes, the Horton Group is intersected twice, above and below an interval of the Windsor Group, suggesting that this fault cuts a recumbent SE-facing syncline or a south-verging thrust.

We interpret these relationships as indicating that the relatively little-deformed type sections of the Horton Bluff and Cheverie formations around the Avon Estuary are autochthonous, but above them a décollement in Windsor evaporites has allowed transport of a much more deformed, allochthonous Horton Bluff succession. The Pembroke Breccia, and highly deformed evaporites on the coast of Cheverie Harbour, probably represent the thick décollement zone. This interpretation is similar to that of Boehner (1991), who interpreted Chevron Canada seismic lines as showing a low-angle, north-dipping thrust fault, the Kennetcook Thrust, but with a mapped trace SE of the Walton monocline. In our interpretation, the décollement does not crop out at the surface to the SE, and therefore is categorized as a 'blind' thrust. Seismic lines and mapping to the east of Tennycape (Fig. 3) show no indication of large-scale duplication of the stratigraphy. We infer that the Kennetcook Thrust loses displacement to the east as well as to the south. The restricted area of the deformed, allochthonous unit NW of the Walton monocline suggests that it was expelled from a localized zone of compression at a restraining bend on the Minas fault zone, resulting in outward spreading of highly deformed units above mobile Windsor evaporites (Fig. 3 inset).

Outcrop-scale structures

Folds

Cheverie to Cambridge Cove. Folds are the most conspicuous features of many coastal outcrops between Cheverie and Walton. In the western part of Figure 3, from Cheverie to Cambridge Cove, the pattern of folding is relatively simple and consistent: folds are subhorizontal to gently ENE- or WSW-plunging. The majority of axial surfaces are upright to steeply inclined, with a predominance of steep dips to the NW, especially in areas where cleavage is developed. Profile views abound in cliffs (Fig. 4b), whereas wave-cut platforms show oblique sections, which are particularly clear in aerial photographs (Figs 5 & 6). The largest folds have wavelengths from 50 to 200 m, with rounded hinge regions 10–20 m across. Most are open to close, but they range from gentle to tight (according to the classification of Fleuty 1964). Sandstone and dolostone layers show Class 1 geometry (according to the classification of Ramsay 1967) and extensional fractures are found on outer surfaces at fold hinges, indicating tangential strain and suggesting that the sandstones were relatively

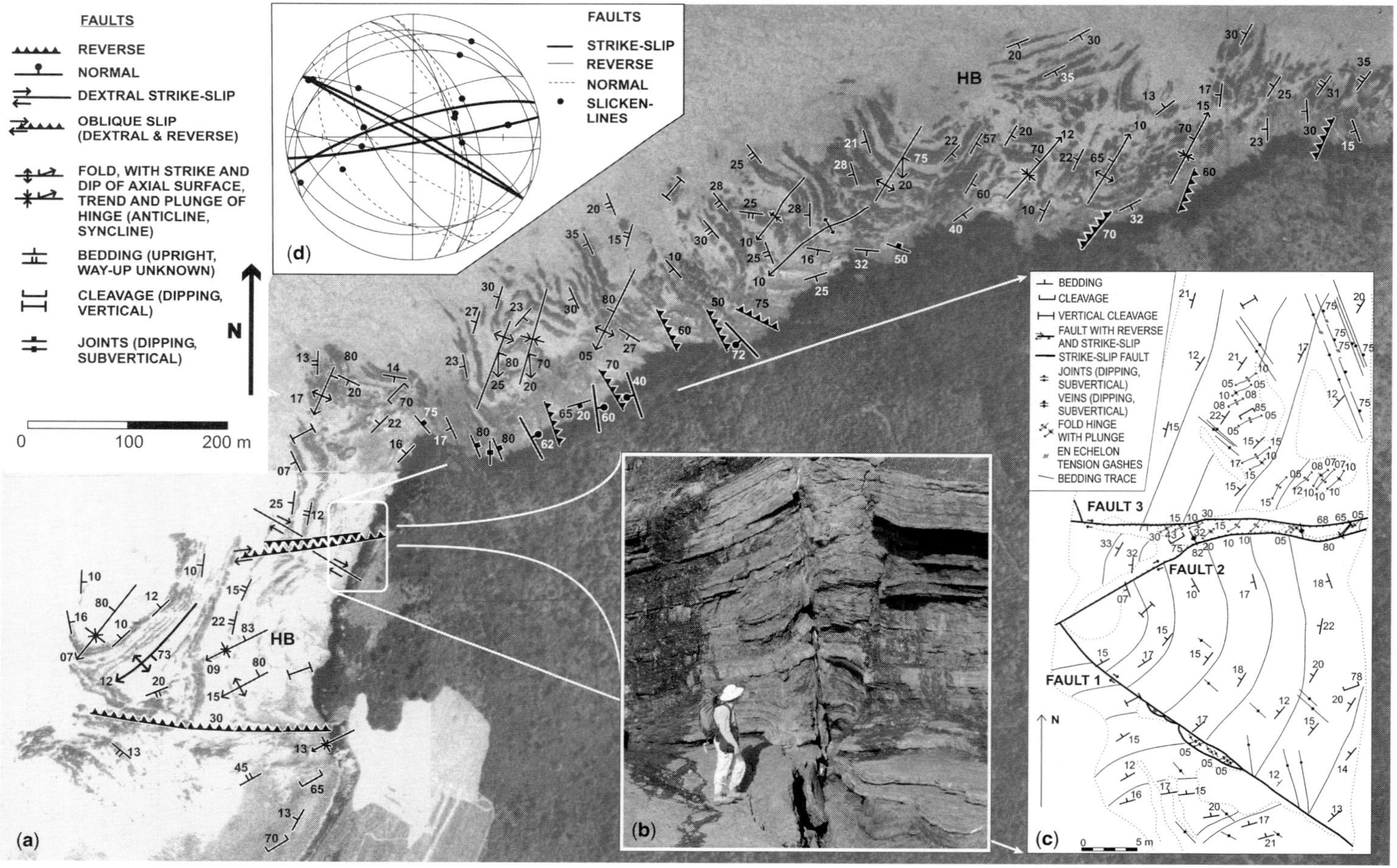

Fig. 6. (**a**) Map of the area east of Mutton Cove (location shown in Fig. 3) superimposed on an aerial photograph. (**b**) View of cliff (looking east) showing positive flower structure in outcrop (human figure for scale). (**c**) Enlarged map of wave-cut platform below (b), showing the orientations of the structures. (**d**) Equal-area projection showing fault orientations.

competent during folding. Intervening shale layers are thickened at the fold hinges (Class 3).

Smaller-scale folds (wavelengths of 1–10 m) are of several types. Some are obviously parasitic folds related to the larger-scale features, with senses of vergence consistent with their position on the limbs of larger folds. However, intra-folial fold pairs, lacking a consistent relationship with the large-scale folds, also occur. Locally, these show strongly curved hinges (Fig. 4c), suggesting either that they formed under conditions of very inhomogeneous bulk strain, or that they were refolded, or both. The occurrence and geometry of these folds show little relation to their position on the limbs of the larger folds. We infer that they formed before the larger-scale main folds, possibly whilst the sediment was not fully lithified. We designate these folds F_1, and the larger-scale folds and their smaller parasitic counterparts are F_2.

Little Rainy Cove. At Little Rainy Cove the pattern of folding becomes more complex. Figure 7 shows that map-scale fold axial traces strike NW–SE, almost perpendicular to the strike further west. However, fold hinges trend south to SW, down the dip of the axial surfaces (Fig. 7e), making these folds reclined to near-reclined in their orientation. Despite this dramatic change in orientation, these folds resemble the F_2 folds in the Cheverie area: their wavelengths, open to close shape, and the presence of weak axial-planar cleavage (Fig. 4e), are all identical. These folds overprint smaller, intra-folial folds that have very variable orientations (Fig. 7c), and strongly curvilinear hinges (Fig. 4c), suggesting deformation whilst the sediments were still soft. These small-scale folds are again designated F_1, and the cliff-scale, cleavage-related folds are F_2. The difference in orientation of F_2 between this section and those in the Cheverie area, and the slight girdle distribution shown by S_2 cleavage poles (Fig. 7d), suggest the existence of a third generation of folds F_3, as indicated in Figure 7f, although these cannot be observed directly at this location.

Rainy Cove. On the east side of Rainy Cove (Fig. 8), there are still more startling changes in geometry. The cliff profile (Fig. 9), a favourite stop for student field trips, displays three large, downward-facing folds (folds A, B and C). B is an antiformal syncline, whilst A and C are synformal anticlines. All three folds plunge moderately west. In the southern part of the cliff section, folds D and E face upward. Fold D has a box-fold style; two hinge regions, D1 and D2, can be distinguished. D1 appears to fold a duplex of small slices in a competent bed of sandstone approximately 10 cm thick. Fold E is a south-facing structure, showing detachment-fold characteristics, with an isoclinal pinched closure in sandstone that became detached from underlying shale. Smaller-scale upward-facing parasitic folds with wavelengths from 0.5 to 5 m and similar orientations are abundant between folds E and D. Local development of axial-planar cleavage in these parasitic folds suggests that they are equivalent to the F_2 folds seen elsewhere. Parasitic folds are also present between folds B and C, but display a mixture of upward- and downward-facing directions.

The presence of cliff-high downward-facing folds requires the presence of at least two generations of large-scale folds (>10 m wavelength). Figure 10 shows two overprinting hypotheses that are consistent with the observed geometry. In hypothesis 1 (Fig. 10a), folds A, B, C, and probably E, are F_2 folds, equivalent to those seen farther west, but refolded by large-scale later folds (F_3). A broad F_3 hinge zone D in the southern part of the cliff explains the variation in F_2 orientation. In hypothesis 2 (Fig. 10b), synform A and antiform E are the only major F_2 folds, whereas folds B, C, and D are later (F_3).

Cleavage is only developed at scattered locations in this cliff profile, but the available cleavage measurements can be used to discriminate between the two hypotheses. For example, cleavage between folds B and C dips to the north, whereas in hypothesis 1 the axial planes of F_2 folds would be inferred to dip south in this area. Therefore, we prefer hypothesis 2, in which fold A is the hinge region of a large, recumbent to reclined F_2 fold, refolded about several F_3 folds (B, C) that strike east–west. This hypothesis is compatible with the scenario for F_3 folding suggested at Little Rainy Cove (Fig. 7f).

Walton. Farther east, around Walton (Fig. 3), the pattern of folding is at first sight similar to that between Cheverie and Cambridge Cove. Upright folds strike and trend ENE–WSW. However, in outcrops at Whale Cove, these refold smaller-scale, tight folds in an interference pattern as shown in Figure 11a.

Axial-planar cleavage is not clearly developed at Walton, and hence the correlation of fold generations with Rainy Cove is difficult. Based on the relatively constant hinge orientations observed, we correlate the earlier folds in Figure 11a with the F_2 folds seen farther west; the more upright structures are probably related to F_3 at Rainy Cove.

Furthermore, we stress that the overprinting relationships observed may not represent distinct episodes of deformation in any regional sense. In a zone of strike-slip motion, it is likely that deformation is diachronous; therefore folds identified

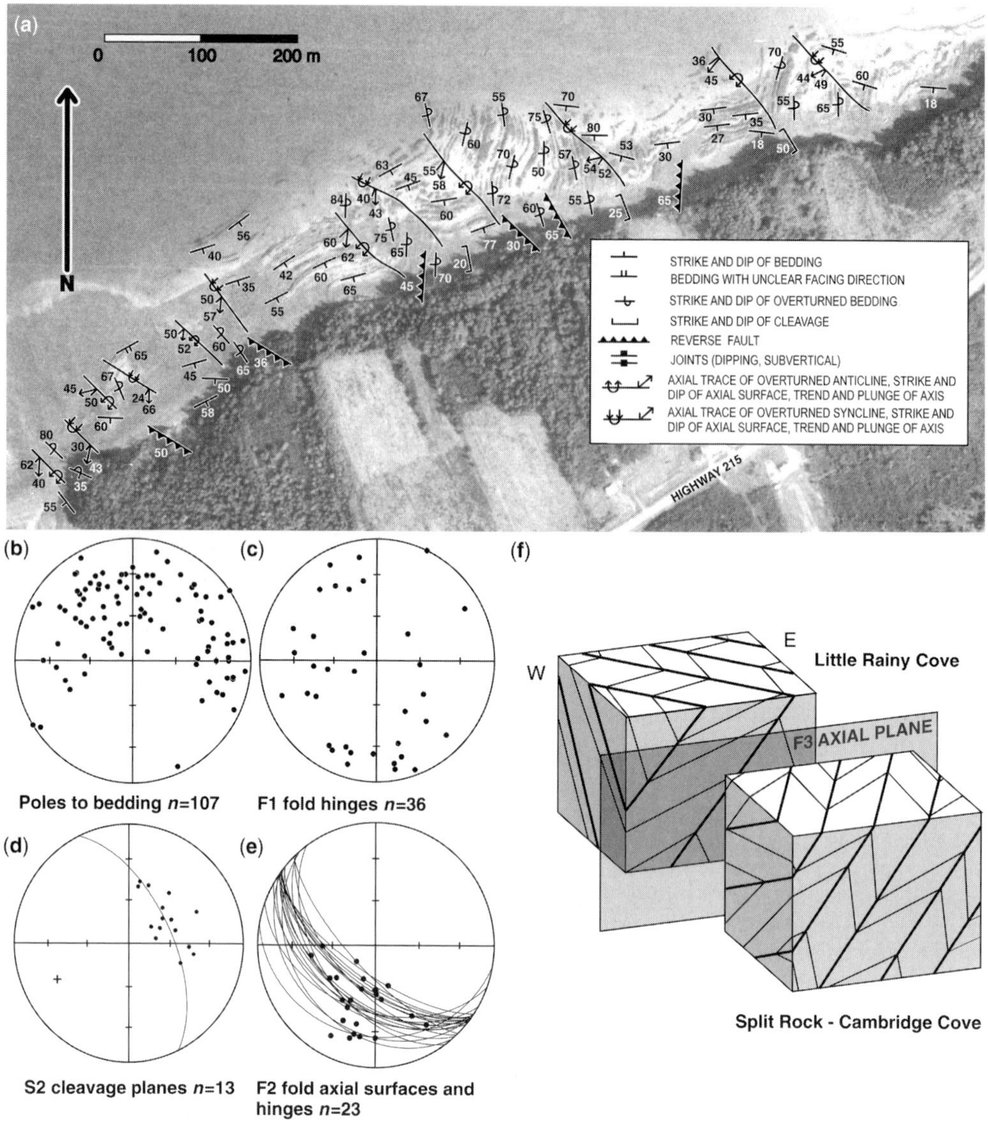

Fig. 7. (**a**) Map of the area east of Little Rainy Cove (location shown in Fig. 3) superimposed on an aerial photograph. (**b–e**) Lower-hemisphere equal-area projections showing the orientations of structures. (**f**) Block diagram showing interpreted relationship of F_2 folds with NW-striking axial surfaces at Little Rainy Cove, compared with NE-striking F_2 folds farther west, interpreted to result from overprinting by F_3.

as F_1 in one location may be simultaneous with F_2 or F_3 elsewhere.

Boudinage

Boudinage structures are common in thinly interbedded and interlaminated sediments of varying lithology. Boudins are typically 5 to 50 mm wide, producing a distinctive ribbed pattern on bedding surfaces (Fig. 12). Boudinage was clearly strongly controlled by lithological contrasts. Strongly heterolithic intervals of interlaminated sandstone and shale may show boudinage of sandstone layers; intervening more homogeneous layers may show no trace of boudinage.

Boudin axes are typically oriented at high angles to both fold hinges and bedding–cleavage intersection lineations (Fig. 12a), indicating extension in

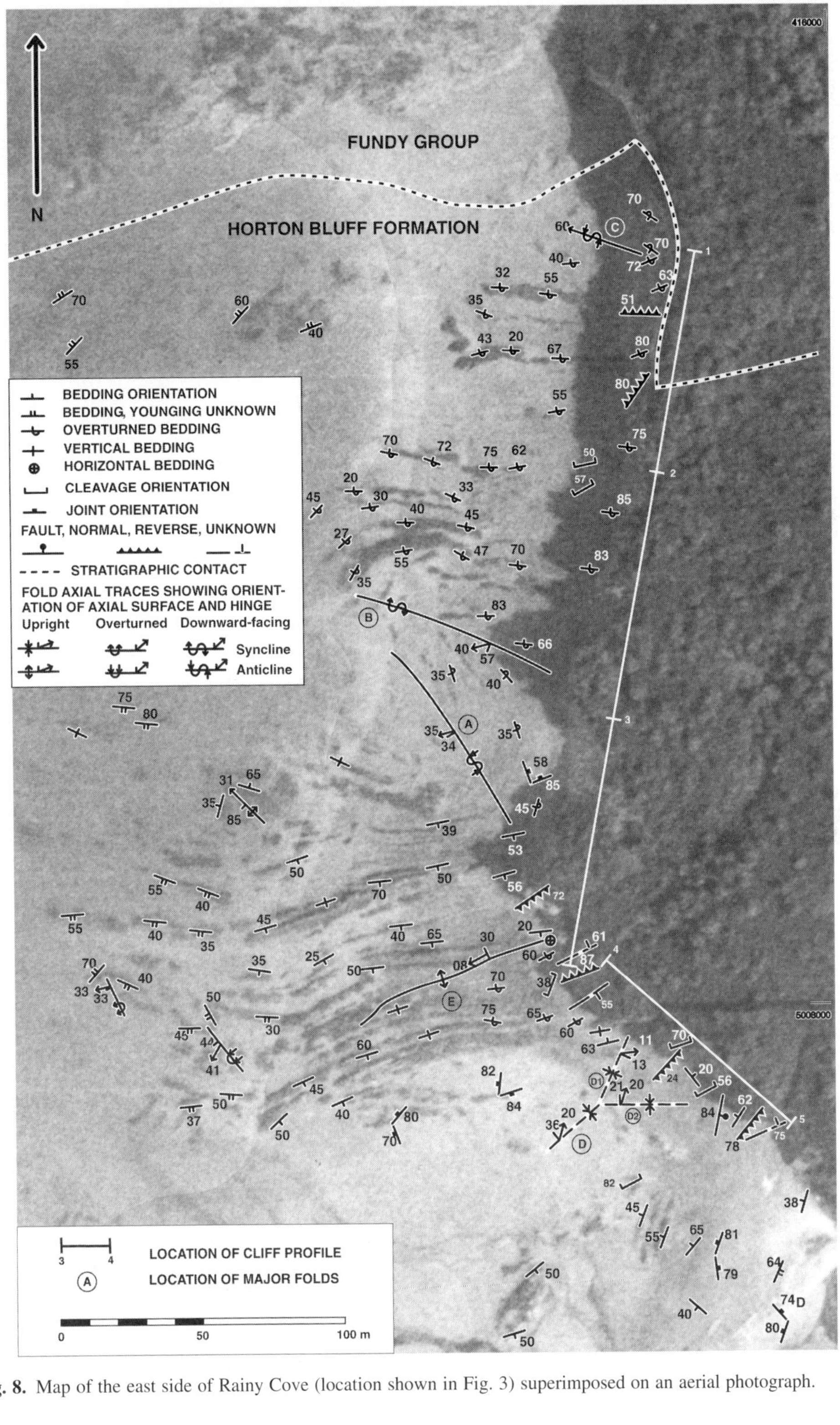

Fig. 8. Map of the east side of Rainy Cove (location shown in Fig. 3) superimposed on an aerial photograph.

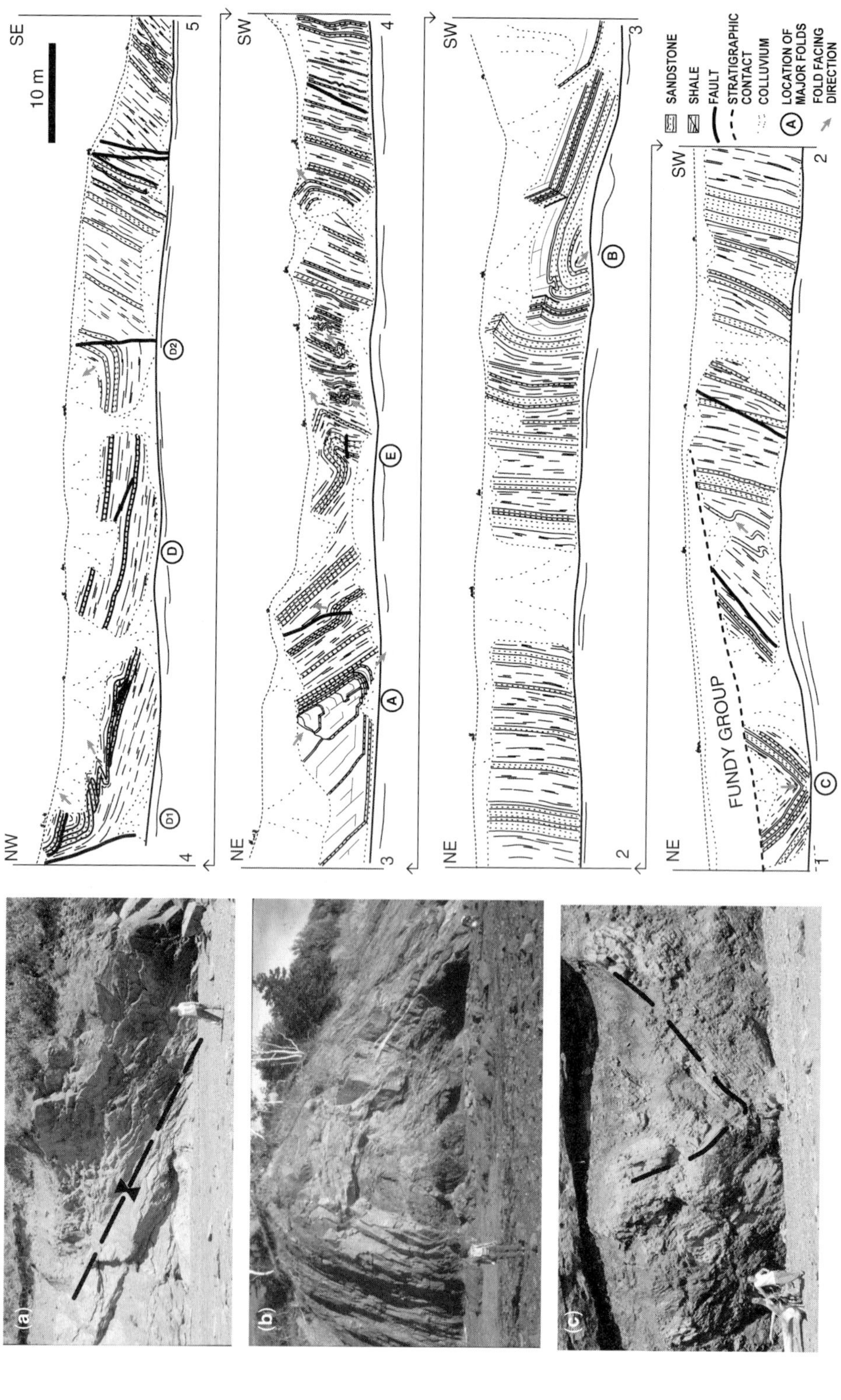

Fig. 9. **(a–c)** Cliff views, looking east, of downward-facing folds A, B, and C, shown in Figure 8 (human figures for scale). **(d)** Composite cliff profile at Rainy Cove (location shown in Fig. 8) showing the principal structures.

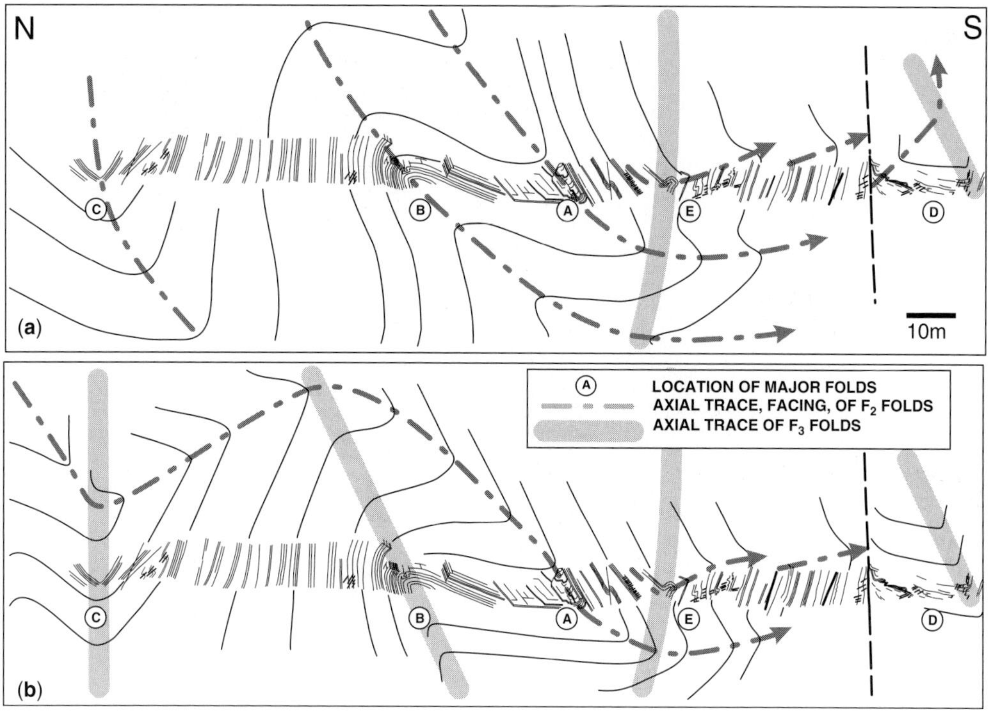

Fig. 10. Contrasting interpretations of downward-facing folds in a cliff profile at Rainy Cove (Fig. 9). (**a**) Interpretation with multiple F_2 folds refolded by a single large F_3 fold. (**b**) Preferred interpretation with a single large F_2 fold refolded by multiple F_3 folds.

Fig. 11. (**a**) Fold interference pattern at Walton; upright F_3 folds are superimposed on recumbent F_2. Clipboard for scale 30 cm long. (**b**) Close-up of the base of the cliff showing en echelon tension gashes indicative of dextral shear. Hammer head in the foreground is *c*. 20 cm long.

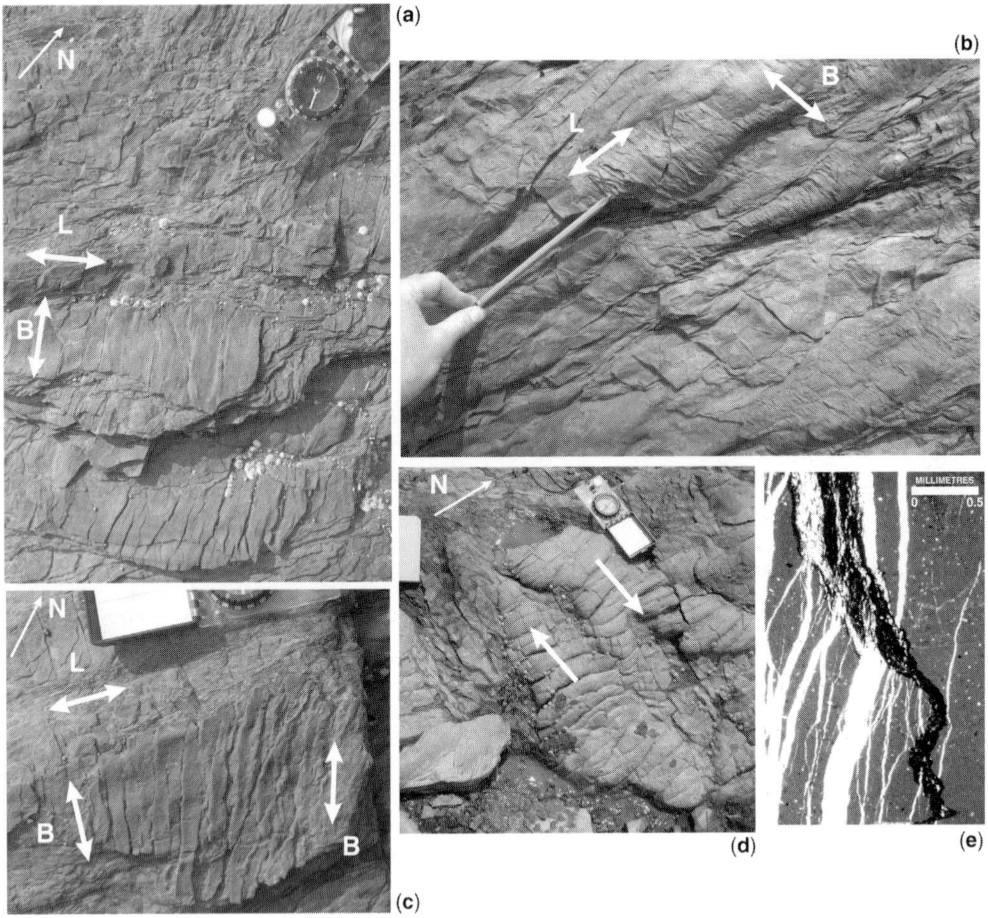

Fig. 12. Boudinage structures. Body of compass in (a), (c) and (d) is 10 cm long. (**a**) Top surface of bed, showing bedding-cleavage intersection lineation (L) and boudin long-axes (B). Johnson Cove–Split Rock section (Fig. 5). (**b**) Top surface of a bed, showing the variable orientation of the boudins (B). Johnson Cove–Split Rock section (Fig. 5). (**c**) Bottom surface of a bed, showing intersection lineation (L) and boudin long axis (B), Johnson Cove–Split Rock section (Fig. 5). (**d**) Top surface of a limestone bed at Bass Creek (Fig. 3), showing the rotation of boudins by dextral shear (arrows). (**e**) Thin-section view of bed in (d), cut perpendicular to bedding and boudins, showing carbonate-filled veins cut (white) cut by a stylolite (black).

directions close to the fold axis. On some folded surfaces, there is significant variation in the orientation of boudins, from precisely perpendicular to the fold axis, to significantly rotated in a clockwise sense (Figs 12c & d). On other surfaces, all the boudins are significantly oblique to fold axes, showing the same clockwise sense of rotation when viewed from above (Fig. 12b shows an undersurface, with an apparent anticlockwise sense).

At a microscopic scale, the spaces between boudins are typically occupied by shale, which has flowed during boudinage. However, in some carbonate-rich beds, inter-boudin spaces are veins filled by carbonate material. In some cases, these veins are the locus of later stylolites, indicating that extension was followed by shortening (Fig. 12e).

We interpret both the formation of boudins by extension parallel to fold axes, and the clockwise rotation of already-formed boudins, as products of dextral shear during transpression. Overprinting of boudins by stylolites is inferred to represent rotation of boudins so that the former extension axis comes to lie in the shortening field of the strain ellipse in bedding. Figure 13 shows the inferred kinematics of boudin formation in transpression.

Cleavage

Cleavage is only sporadically developed in the Horton Group. Through most of the area, cleavage is manifested as poorly defined, slightly recessive bands, particularly apparent in rooted dolomitic

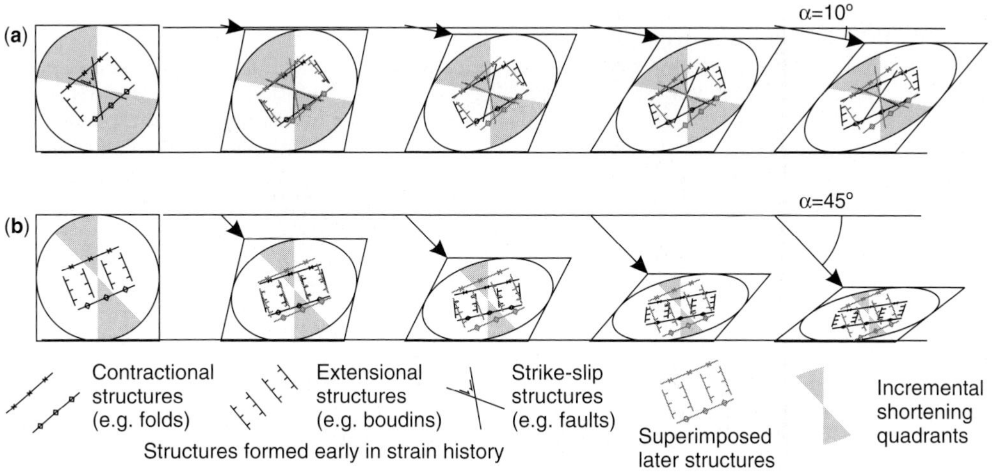

Fig. 13. (**a**) Diagrams showing the orientation of structures predicted to develop in wrench-dominated transpression at low convergence ($\alpha = 10°$), with progressive strain. In later increments, the effect of progressive strain on early-formed structures is shown in black; orientation of new superimposed structures is shown grey. Instantaneous extension and shortening directions in (**b**) and (**c**) are as determined by Tikoff & Teyssier (1994); finite deformation was simulated by transformations in the program CorelDraw^{TM}. Lines of no incremental extension were determined according to the formulas of Ramsay and Huber (Ramsay & Huber 1983) for finite strain, using a shear-strain increment of 0.001. (b) Diagrams showing the orientation of structures predicted to develop in pure-shear-dominated transpression at high convergence ($\alpha = 45°$). Symbols and methods are the same as in (a).

siltstones identified as palaeosols (Fig. 4e). Interbedded, finer-grained shales representing subaqueous deposition typically display only a strong bed-parallel fissility. Locally, distorted in situ fossil-tree impressions imply a strain ellipse in the plane of bedding, with its long axis oriented NE–SW, generally parallel to the regional strike of cleavage. In the area around Split Rock (Fig. 5), cleavage increases rapidly in intensity to the NW, so that at Split Rock all the mudrocks show a strong slaty cleavage (Fig. 4f).

In several areas, cleavage is clearly oblique to fold axial surfaces (e.g. Fig. 6). In all cases observed, the transection is clockwise, suggesting a component of dextral strike-slip (Borradaile 1978; Treagus & Treagus 1992). However, both cleavage and folds show too much variability from place to place for transection to be clearly demonstrable from simple plots of fold and cleavage orientations.

Faults and fractures

Faults and fractures are common in coastal exposures of the Horton Bluff Formation between Cheverie and Walton. Most are marked by white quartz and/or calcite veins; in many cases fibres indicate the direction and sometimes sense of slip. The history of faulting is complex. In many locations we have observed faults that are folded; elsewhere, faults separate blocks having contrasting bedding orientations, suggesting that the faults had a strong primary curvature. Hence, we have not attempted to apply commonly used statistical approaches to the kinematics or dynamics of fault populations, because we believe that high strains involving significant post-faulting rotations invalidate the assumptions of these methods.

Figure 5e shows the orientations of faults at Split Rock. The most abundant faults strike generally NE–SW, parallel to the fold hinges. Many show NW dips and contractional offsets of stratigraphy; at a number of locations, these faults are clearly closely related to folds, and are interpreted to have developed synchronously with F_2 folding. For example, in Figure 4d in the Split Rock section, a tight fold, interpreted as a fault propagation fold, is present in the hanging wall of an outcrop-scale south-verging thrust fault. Where mineral fibres are present on contractional fault planes of this group, they typically show orientations varying between 'pure' thrust-sense dip-slip motion and strike-slip, with a predominance of dextral oblique slip.

NE–SW-striking normal faults are also present in the section, typically in conjugate pairs. At one location, faults with this orientation bound a small outlier of the Triassic Fundy Group, indicating that at least some are of Mesozoic age.

A second prominent population of faults strikes generally NW–SE. These are particularly abundant in the coastal section east of Mutton Cove (Fig. 6d). In many cases the presence of these faults is marked by erosional gullies on the shoreline, with poor exposure. Mapped offsets of stratigraphy are recorded inland by Moore *et al.* (2000) at many of these faults. Exposed examples display both normal and reverse offsets of stratigraphy, with reverse faults oriented slightly clockwise of the predominant orientation of normal faults.

A third, smaller population of faults is marked by steep surfaces striking east–west, ENE–WSW, and NW–SE. Particularly clear examples are seen immediately east of Mutton Cove (Figs 3 & 6), where straight fault traces can be followed from the cliff to the beach. Figure 6c shows the pattern of anastomosing faults, joints and veins at this location. Faults 1, 2 and 3 are subvertical; subhorizontal slickenfibres indicate dextral displacement. Fault 1 strikes NW–SE; anastomosing curved fault surfaces branch from the main strand, bounding regions that display small-scale en echelon folds and veins, with orientations consistent with dextral strike-slip. Subvertical faults 2 and 3 have a general east–west strike but show changes in orientation on a scale of metres. Where these two faults intersect the cliff, they are seen to diverge upward (Fig. 6b), in a configuration resembling a positive flower structure (Harding 1974). The relative uplift of the central, fault-bounded block is a combined effect of folding and small dip-slip components of motion on the bounding faults.

En echelon sets of tension gashes are common in some parts of the coastal outcrop. Most common are east–west zones indicative of dextral-sense motion as shown in Figure 11b. Subordinate north–south zones display sinistral-sense motion.

Discussion

Overall tectonic environment

The array of structures exposed on the Cheverie–Walton coast strongly indicates finite NW–SE shortening, finite NE–SW extension, and a strong clockwise (dextral) sense of rotation (vorticity) in the deformation. Structures indicative of major NW–SE tectonic shortening include: abundant folds; cleavage striking generally NE–SW; a finite strain ellipse in the plane of bedding with its long axis NE–SW; and abundant thrust and reverse faults verging SE. Structures indicative of minor NE–SW extension include normal faults, boudins, and tension gashes. Where contractional and extensional structures occur together in the same outcrops, NW–SE shortening appears much greater than NE–SW extension. Structures indicating clockwise vorticity include boudins rotated clockwise from their inferred initiation direction perpendicular to folds; folds transected in a counterclockwise sense by cleavage; strike-slip faults with dextral kinematic indicators; and en echelon vein sets indicative of dextral shear. Taken together, these structures imply an environment of bulk dextral transpression.

The overall vergence of folds and thrust faults is toward the SE, suggesting that the core of the transpressional zone lies to the north of the coastal sections, beneath the Minas Basin or along its north shore. Highly strained rocks of Early Carboniferous age are in fact widespread along the north shore in the Parrsboro area, where MacInnes & White (2004) have described phyllonites and mylonites with steeply dipping foliations and dextral kinematic indicators. Lineations within that zone show variations in orientation from subhorizontal in the outer parts of the zone to steeply plunging in the centre. We infer that vertical extrusion of material from this zone built up a substantial topographic ridge during deformation, providing the gravitational potential to drive outward thrusting in the area of this study, especially where ductile Windsor evaporites provided a convenient décollement horizon. The overall deformation zone thus probably had the form of a positive flower structure (Fig. 14).

Development of minor structures in transpression

Transpression encompasses a range of deformation styles, from almost ideal strike-slip motion to almost pure convergence. The style of transpression in a general shear zone can be measured by the angle of convergence α, the angle between the

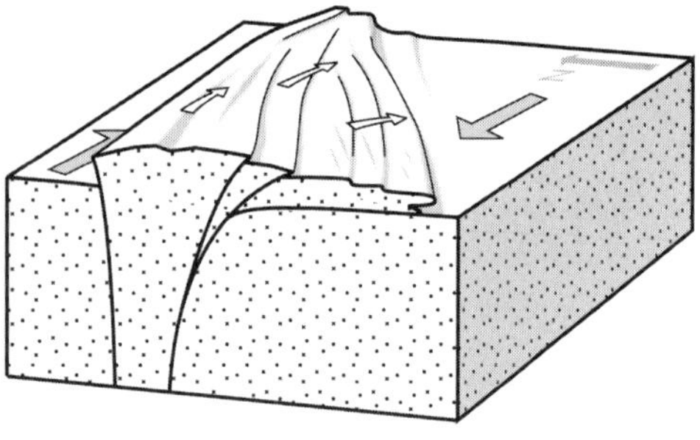

Fig. 14. Block diagram showing the inferred large-scale geometry of the deformation zone in the Cheverie–Walton area, interpreted as a positive flower structure at a restraining bend. Arrows show the variation in transport direction across the zone.

direction of transport and the shear zone boundary, which ranges from 0° for pure strike-slip motion to 90° for pure convergence (e.g. Sanderson & Marchini 1984). Fossen & Tikoff (1993) and Tikoff & Teyssier (1994) were able to distinguish two styles of transpression: wrench-dominated at $\alpha < 20°$, and pure-shear dominated at higher values of α. Wrench-dominated zones show horizontal incremental stretching directions but a finite stretching direction which switches from horizontal to vertical with increasing strain. In pure-shear-dominated zones the direction of maximum extension is always vertical. Further modelling of transpressive zones (Tikoff & Teyssier 1994; Jones & Tanner 1995; Teyssier *et al.* 1995; Holdsworth *et al.* 2002; De Paola *et al.* 2005) has focused on the partitioning of strain, particularly between zones dominated by simple shear and pure shear – a likely consequence of the development of through-going strike-slip faults early in the deformation history.

Figure 13 shows the predicted development, progressive rotation and overprinting of minor structures in horizontal strata during dextral transpression at low to moderate strains, at low and high local values of α respectively. Constant-volume deformation is assumed (reduction in area is compensated by vertical stretching). At low angles of convergence (wrench-dominated transpression; Fig. 13a), the intermediate incremental strain axis is vertical; in brittle rocks, deformation by vertical strike-slip faults would initially be predicted. However, finite vertical stretching cannot be achieved purely by vertical faults, so other contractional structures such as folds or reverse faults must form (Tikoff & Teyssier 1994). If subhorizontal strong layers are present (as in a bedded sedimentary succession) boudins may form perpendicular to the horizontal incremental stretching direction. All structures, except those parallel to the zone boundary, undergo clockwise rotation; however, extensional structures form in orientations subject to most rapid rotation, and therefore show clockwise rotations relative to contemporary shortening structures such as cleavage and folds. With progressive rotation, extensional structures would be predicted to rotate into the field of shortening, potentially leading to inversion of original normal faults and overprinting of boudins by stylolites.

At high α (pure-shear-dominated transpression; Fig. 13b) the incremental and finite extension axes are vertical, and structures such as reverse faults and folds are expected to dominate deformation, although strike-slip faults may develop to accommodate the simple shear component (Tikoff & Teyssier 1994). However, initially horizontal layers are still subject to extension in a direction oblique to the shear-zone boundary and extensional structures are predicted to form in orientations shown by the grey quadrants in Figure 13b. Inversion is initiated at an earlier stage in the deformation history, because structures rotate more rapidly out of the narrow field of extension.

Structures in the area between Cheverie and Cambridge Cove (Fig. 3) are consistent with a transpressional strain history. The asymmetrical relationships between folds and boudins shown in Figure 12 are particularly distinctive, and are a useful new field indicator for deformation involving a component of strike-slip. North- and NW-striking faults show combinations of normal and reverse offsets, with reverse faults typically slightly clockwise of normal faults, suggesting that initially extensional faults may have been inverted in compression, consistent with the progressive strain histories seen in Figure 13. Overall, the zone is dominated by shortening and extension structures; strike-slip faults, although spectacularly exposed (Fig. 6), are relatively rare. Thus, it is likely that the environment of deformation was pure-shear-dominated transpression, with local angles of convergence α greater than 20° – similar to that portrayed in Figure 13b.

The orientation relationships of folds observed at Rainy Cove, where F_2 folds with NE–SW-trending hinges are overprinted by F_3 folds with more east–west trends (Fig. 7f), are not consistent with either of the strain histories portrayed in Figure 13. In both models, later folds are predicted to show orientations counterclockwise of rotated earlier-formed folds. The simplest explanation for the observed geometries is a progressive increase in convergence over time. This might have been achieved by a change in the overall direction of plate movement at the Minas fault zone; by the more localized effects of a restraining bend; or by an increase in strain partitioning between through-going strike-slip faults and intervening panels dominated by pure shear (Tikoff & Teyssier 1994; Jones & Tanner 1995; Teyssier *et al.* 1995). Increasing convergence may have been encouraged by the geometry of the south margin of the Minas fault zone, at which the deformed Horton Group rocks have been thrust southward above Windsor evaporites. This suggests that the Horton Group was initially deformed in the transpressional zone at relatively low α, but that as the core of the deformation zone developed topographic expression, outward spreading of the elevated central region was facilitated by the presence of a mobile evaporite layer in the adjoining Meguma terrane (Fig. 14). Horton Group rocks may thus have experienced a progressive increase in the pure shear component of deformation as they moved from the central zone of tectonically driven transpression to a region of gravitationally driven spreading in the superficial parts of a positive flower structure.

Conclusions

The shores of the Bay of Fundy provide a natural laboratory for the study of structures developed in transpression. Deformation occurred in mid-Carboniferous times, probably as a result of the passage of a restraining bend on the east–west Minas fault zone – the boundary between the Avalon and Meguma Terranes of the Appalachians. A zone of deformed Early Carboniferous Horton Group was subject to folding, thrusting, and cleavage development, with a shortening direction oriented roughly NW–SE. Extensional structures, including boudins and faults, were developed in response to NE–SW stretching, and underwent clockwise rotation relative to contemporary contractional structures. Upright to recumbent F_2 folds with NE trends were overprinted by more east–west F_3 folds, locally producing spectacular downward-facing structures. These relationships can be explained by a change from low to high angles of convergence (α), associated with a change from deformation in a steep transpressional shear zone to gravity-driven spreading of the resulting topographic ridge, promoted by detachment in mobile evaporites.

The authors thank Devon Canada for access to seismic profiles and subsurface data from the Cheverie-01 well. Field research was supported by the Natural Sciences and Engineering Research Council (Canada) through Discovery Grant A8508, and by a start-up research grant from the University of Alberta. C. Roselli was supported by University of Alberta teaching and graduate research assistantships. D. Hagman of Summerville, Nova Scotia, and K. Branchard of Rainy Cove, Nova Scotia, helped us in the field. W. Loogman, journal reviewer M. Swanson, and an anonymous referee contributed comments which improved the final version. We are also grateful to J. White and E. MacInnes (University of New Brunswick); to R. Boehner, D. Kontak and J. McMullen (Nova Scotia Department of Natural Resources); and to R. Moore (Acadia University, Nova Scotia) for discussions and access to data.

References

BELL, W. A. 1960. *Mississippian Horton Group of Type Windsor–Horton District, Nova Scotia*. Geological Survey of Canada, Memoirs, **314**.

BOEHNER, R. C. 1991. Seismic interpretation, potential overthrust geology and mineral deposits in the Kennetcook Basin, Nova Scotia. *Nova Scotia Department of Natural Resources, Mineral Resources Branch Report of Activities*, **1990**, 37–47.

BORRADAILE, A. J. 1978. Transected folds: a study illustrated with examples from Canada and Scotland. *Geological Society of America Bulletin*, **89**, 481–493.

BOYLE, R. W. 1957. *Geology, Walton–Cheverie Area, Nova Scotia*. Geological Survey of Canada, Bulletin, **166**.

CHANDLER, F. W., WALDRON, J. W. F., GILES, P. & GALL, Q. 1998. *Geology, Stellarton Gap, Nova Scotia (11E10, Parts of 11E7, 8, 9, 15, 16)*. Geological Survey of Canada Open Files, **3525**, scale 1:50 000.

CROSBY, D. G. 1962. *Wolfville Map Area, Nova Scotia (21H/1)*. Geological Survey of Canada Memoirs, **325**.

CULSHAW, N. & LIESA, M. 1997. Alleghenian reactivation of the Acadian fold belt, Meguma zone, southwest Nova Scotia. *Canadian Journal of Earth Sciences*, **34**, 833–847.

DE PAOLA, N., HOLDSWORTH, R. E., MCCAFFREY, K. J. W. & BARCHI, M. R. 2005. Partitioned transtension: an alternative to basin inversion models. *Journal of Structural Geology*, **27**, 607–625.

DEWEY, J. F., HOLDSWORTH, R. E. & STRACHAN, R. A. 1998. Transpression and transtension zones *In*: DEWEY, J. F., HOLDSWORTH, R. E. & STRACHAN, R. A. (eds) *Continental Transpressional and Transtensional Tectonics*. Geological Society, London, Special Publications, **135**, 1–14.

EISBACHER, G. H. 1969. Displacement and stress field along part of the Cobequid Fault, Nova Scotia. *Canadian Journal of Earth Sciences*, **6**, 1095–1104.

EISBACHER, G. H. 1970. Deformation mechanisms of mylonitic rocks and fractured granites in the Cobequid Mountains, Nova Scotia, Canada. *Geological Society of America Bulletin*, **81**, 2009–2020.

FERGUSON, S. A. 1983. *Geological Map of the Hantsport Area*. Nova Scotia Department of Natural Resources Map, **83–1**.

FLEUTY, M. J. 1964. The description of folds. *Proceedings of the Geologists' Association*, **75**, 461–492.

FOSSEN, H. & TIKOFF, B. 1993. The deformation matrix for simultaneous pure shear, simple shear, and volume change, and its application to transpression/transtension tectonics. *Journal of Structural Geology*, **15**, 413–422.

FOSSEN, H. & TIKOFF, B. 1998. Extended models of transpression and transtension, and application to tectonic settings. *In*: HOLDSWORTH, R. E., STRACHAN, R. A. & DEWEY, J. F. (eds) *Continental Transpressional and Transtensional Tectonics*. Geological Society, London, Special Publications, **135**, 15–33.

FRALICK, P. W. & SCHENK, P. E. 1981. Molasse deposition and basin evolution in a wrench tectonic setting: the Late Paleozoic, Eastern Cumberland Basin, Maritime Canada. *In*: MIALL, A. D. (ed.) *Sedimentation and Tectonics in Alluvial Basins*. Geological Association of Canada Special Papers, **23**, 77–97.

FYSON, W. K. 1964. Repeated trends of folds and cross-folds in Palaeozoic rocks, Parrsboro, Nova Scotia. *Canadian Journal of Earth Sciences*, **1**, 167–183.

FYSON, W. K. 1967. Gravity sliding and cross-folding in Carboniferous rocks, Nova Scotia. *American Journal of Science*, **265**, 1–11.

GIBBONS, W., DOIG, R., GORDON, T., MURPHY, B., REYNOLDS, P. & WHITE, J. C. 1996. Mylonite to megabreccia: tracking fault events within a transcurrent terrane boundary in Nova Scotia, Canada. *Geology*, **24**, 411–414.

GIBLING, M. R. 1995. Upper Paleozoic rocks, Nova Scotia. *In*: WILLIAMS, H. (ed.) *Geology of the Appalachian–Caledonian Orogen in Canada and Greenland*, Geological Society of America, The Geology of North America, **F-1**, 493–523.

GILES, P. S. & BOEHNER, R. C. 1982. *Geological Map of the Shubenacadie and Musquodoboit (1:50,000)*. Nova Scotia Department of Natural Resources, Mineral Resources Branch, Map **ME 1982-4**.

HARDING, T. P. 1974. Petroleum traps associated with wrench faults. *AAPG Bulletin*, **58**, 1290–1304.

HOLDSWORTH, R. E., TAVERNELLI, E., CLEGG, P., PINHEIRO, R. V. L., JONES, R. & MCCAFFREY, K. J. W. 2002. Domainal deformation patterns and strain partitioning during transpression: an example from the Southern Uplands terrane, Scotland. *Journal of the Geological Society, London*, **159**, 401–415.

JONES, R. & TANNER, P. 1995. Strain partitioning in transpression zones. *Journal of Structural Geology*, **17**, 793–802.

KEPPIE, J. D. 1982. The Minas geofracture. *In*: ST. JULIEN, P. & BÉLAND, J. (eds) *Major Structural Zones and Faults of the Northern Appalachians*. GAC Special Papers, **24**, 263–280.

KEPPIE, J. D. 2000. *Geological Map of the Province of Nova Scotia*. Nova Scotia Department of Natural Resources, Map, **ME 2000-1**, scale 1:500 000.

KONTAK, D., ANSDELL, K. & DOUGLAS, A. 2000. Zn–Pb mineralization associated with a mafic dyke at Cheverie, Hants County, Nova Scotia: implications for Carboniferous metallogeny. *Atlantic Geology*, **36**, 7–26.

LAVOIE, D., SANGSTER, D. F., SAVARD, M. M. & FALLARA, F. 1995. Multiple breccia events in the lower part of the Carboniferous Windsor Group, Nova Scotia. *Atlantic Geology*, **31**, 197–207.

MCDONALD, E. 2001. Final Well Report for Cheverie-01 Stratigraphic Test. Capstain Holdings Ltd, Trenton, Nova Scotia, Report to Devon Canada Corporation (unpublished).

MACINNES, E. & WHITE, J. C. 2004. Geometric and kinematic analysis of a transpression terrane boundary: Minas fault system, Nova Scotia, Canada. *In*: ALSOP, G. I., HOLDSWORTH, R. E., MCCAFFREY, K. J. W. & HAND, M. (eds) *Flow Processes in Faults and Shear Zones*. Geological Society, London, Special Publications, **224**, 201–214.

MAWER, C. K. & WHITE, J. C. 1987. Sense of displacement on the Cobequid–Chedabucto fault system, Nova Scotia, Canada. *Canadian Journal of Earth Sciences*, **24**, 217–223.

MOORE, R. G. 1993*a*. *Geological Map of Cheverie–Lower Burlington Quadrangle, [NTS 21H01] [1:10 000]*. Nova Scotia Department of Natural Resources, Mineral Resources Branch, Open File Map **ME 1993-1**.

MOORE, R. G. 1993*b*. *Geological Map of Cogmagun River–Goshen Quadrangle, [NTS 21H01] [1:10 000]*. Nova Scotia Department of Natural Resources, Mineral Resources Branch, Open File Map **ME 1993-2**.

MOORE, R. G. 1994. *Geology of the Walton–Rainy Cove Brook map area [NTS 21H01-Z2 and Z4], Hants County, Nova Scotia*.Nova Scotia Department of Natural Resources, Mineral Resources Branch, Open File Reports, **1994-2**.

MOORE, R. G. 1996. *Geology of the Cambridge Cove/Bramber and Red Head Map Areas 21H/01-Z1, Z3 and 21H/01-Y2*. Nova Scotia Department of Natural Resources Open File Reports, **96-002**.

MOORE, R. G., FERGUSON, S. A., BOEHNER, R. C. & KENNEDY, C. M. 2000. *Preliminary Geological Map of the Wolfville/Windsor Area, Hants and Kings Counties, Nova Scotia [21H/01 and Parts of 21A/16C and D]*. Nova Scotia Department of Natural Resources, Open File Map **ME 2000-3**.

MURPHY, J. B., STOKES, T. R., MEAGHER, C. & MOSHER, S.-J. 1994. Geology of the Eastern St. Mary's Basin, central mainland Nova Scotia. *Geological Survey of Canada Current Research*, **1994-D**, 95–102.

OLSEN, P. E. & SCHLISCHE, R. W. 1990. Transtensional arm of the early Mesozoic Fundy rift basin: penecontemporaneous faulting and sedimentation. *Geology*, **18**, 695–698.

O'SULLIVAN, J. & MACFIE, R. 1988. *Base Metals, Silver, Barite, Walton, Hants County, Nova Scotia. Report on Mining History, Geology, Magnetic, VLF-EM and IP Surveys, Drilling and Drill Core Assays*. Nova Scotia Department of Natural Resources, Mineral Resources Branch, Assessment Reports, **ME 1988-200**.

RAMSAY, J. G. 1967. *Folding and Fracturing of Rocks*. San Francisco, McGraw-Hill.

RAMSAY, J. G. & HUBER, M. I. 1983. *The Techniques of Modern Structural Geology. Volume 1: Strain Analysis*. Academic Press, London.

SANDERSON, D. J. & MARCHINI, W. 1984. Transpression. *Journal of Structural Geology*, **6**, 111–117.

STEVENSON, I. M. 1959. *Shubenacadie and Kennetcook Map Areas, Colchester, Hants and Halifax Counties, Nova Scotia*. Geological Survey of Canada Memoirs, **302**.

TEYSSIER, C., TIKOFF, B. & MARKLEY, M. 1995. Oblique plate motion and continental tectonics. *Geology*, **23**, 447–450.

TIKOFF, B. & TEYSSIER, C. 1994. Strain modeling of displacement-field partitioning in transpressional orogens. *Journal of Structural Geology*, **16**, 1575–1588.

TREAGUS, S. H. & TREAGUS, J. E. 1992. Transected folds and transpression: how are they associated? *Journal of Structural Geology*, **14**, 361–367.

WADE, J., BROWN, D., TRAVERSE, A. & FENSOME, R. A. 1996. The Triassic–Jurassic Fundy Basin, Eastern Canada; regional setting, stratigraphy, and hydrocarbon potential. *Atlantic Geology*, **32**, 189–231.

WALDRON, J. W. F. 2004. Anatomy and evolution of a pull-apart basin, Stellarton, Nova Scotia. *Geological Society of America Bulletin*, **116**, 109–127.

WEBB, G. W. 1969. Paleozoic wrench faults in Canadian Appalachians. *In*: KAY, M. (ed.) *North Atlantic Geology and Continental Drift*. AAPG, Memoirs, **12**, 745–786.

WEEKS, L. J. 1948. *Londonderry and Bass River Map Areas, Colchester and Hants Counties, Nova Scotia*. Geological Survey of Canada, Memoirs, **245**.

YEO, G. M. & GAO, R.-X. 1987. Stellarton graben: an upper Carboniferous pull-apart basin in northern Nova Scotia. *In*: BEAUMONT, C. & TANKARD, A. J. (eds) *Sedimentary Basins and Basin-forming Mechanisms*. Canadian Society of Petroleum Geologists Memoirs, **12**, 299–309.

Terminations of large strike-slip faults: an alternative model from New Zealand

V. MOUSLOPOULOU[1*], A. NICOL[2], T. A. LITTLE[1] & J. J. WALSH[3]

[1]*School of Earth Sciences, PO Box 600, Victoria University of Wellington, Wellington, New Zealand (e-mail: vmouslopoulou@gns.cri.nz)*

[2]*GNS Science, PO Box 30368, Lower Hutt, New Zealand*

[3]*Fault Analysis Group, University College Dublin, Dublin 4, Ireland*

[*]*Present address: Fault Analysis Group, Department of Geological Sciences, University College Dublin, Dublin 4, Ireland (e-mail: vasso@fag.ucd.ie)*

Abstract: The 500-km-long strike-slip North Island Fault System (NIFS) intersects and terminates against the Taupo Rift. Both fault systems are active, with strike-slip displacement transferred into the rift without displacing normal faults along the rift margin. Data from displaced landforms, fault-trenching, gravity and seismic-reflection profiles, and aerial photograph analysis suggest that within 150 km of the northern termination of the NIFS, the main faults in the strike-slip fault system bend through 25°, splay into five principal strands and decrease their mean dip. These changes in fault geometry are accompanied by a gradual steepening of the pitch of the slip vectors, and by an anticlockwise swing (up to 50°) in the azimuth of slip on the faults in the NIFS. As a consequence of the bending of the strike-slip faults and the changes in their slip vectors, near their intersection, the slip vectors on the two component fault systems become subparallel to each other and to their mutual line of intersection. This subparallelism facilitates the transfer of displacement from one fault system to the other, accounting for a significant amount of the NE increase of extension along the rift, whilst maintaining the overall coherence of the strike-slip termination. Changes in the slip vectors of the strike-slip faults arise from the superimposition of rift-orthogonal differential extension outside the rift margin, resulting in differential motion of the footwall and hanging-wall blocks of each fault in the NIFS. The combination of rift-orthogonal heterogeneous extension (dip-slip) and strike-slip, results in a steepening of the pitch of the slip vectors on the terminating fault system. Slip vectors on each splay close to their terminations are, therefore, the sum of strike-slip and dip-slip components, with the total angle through which the pitch of the slip vectors steepens being dependent on the relative values of both these two component vectors. In circumstances where interaction of the velocity fields for the intersecting fault systems cannot resolve to a slip vector that is boundary-coherent, either rotation about vertical axes of the terminating fault relative to the through-going fault system may take place to accommodate the termination of the strike-slip fault system, or the rift may be offset by the strike-slip fault system rather than terminating into it. At the termination of the NIFS, an earlier phase of such rotations may have produced the 25° anticlockwise bend in fault strike and contributed up to about one-third of the anticlockwise deflection in slip azimuth. On the terminating strike-slip NIFS, therefore, rotational and non-rotational termination mechanisms have both played a role, but at different times in its evolution, as the thermal structure, the rheology and the thickness of the crust in the rift intersection region have changed.

Continental strike-slip faults commonly traverse the Earth's upper crust for tens to hundreds of kilometres. They often operate as transform faults, linking and transferring displacement between other plate-boundary segments (Joffe & Garfunkel 1987; Freymueller *et al.* 1999; Norris & Cooper 2000), as large strike-slip faults within continental plate-boundary zones (Davis & Burchfiel 1973; Bellier & Sébrier 1995; Quebral *et al.* 1996; Tsutsumi & Okada 1996; Mouslopoulou *et al.* 2007) or as isolated structures within otherwise rigid plates (Molnar & Lyon-Caen 1989; Taymaz *et al.* 1991; Jackson *et al.* 1995; Bayasgalan *et al.* 1999). Each of these different forms of strike-slip faults must terminate, yet the mechanisms by which they terminate are often poorly resolved by the available data.

Terminations of strike-slip faults are of three basic types (Fig. 1).

(1) A free tip termination in which the strike-slip fault dies out within a rock volume that appears unfaulted at a regional scale of observation (Fig. 1a). The eastern end of the *c.* E–W-striking left-lateral Kunlun Fault in

From: CUNNINGHAM, W. D. & MANN, P. (eds) *Tectonics of Strike-Slip Restraining and Releasing Bends.* Geological Society, London, Special Publications, **290**, 387–415.
DOI: 10.1144/SP290.15 0305-8719/07/$15.00

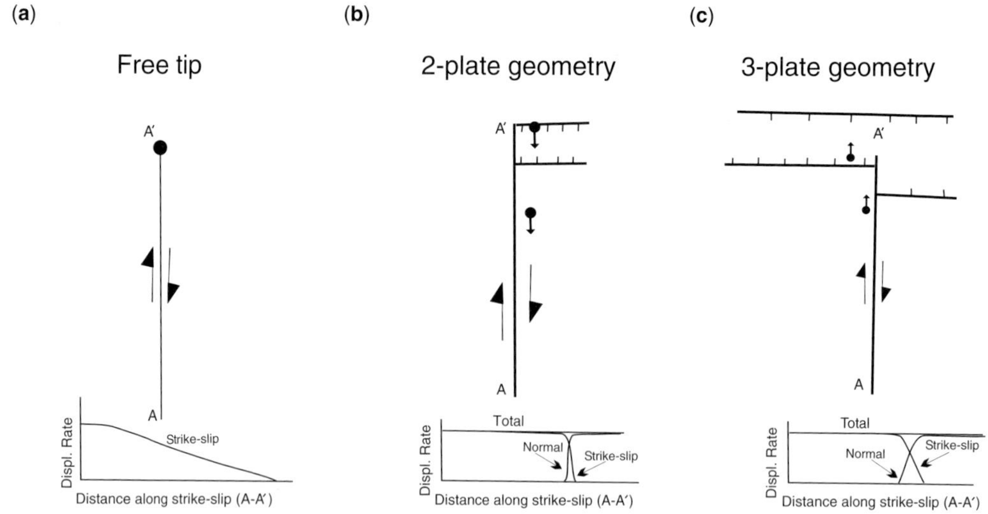

Fig. 1. Schematic diagram illustrating the three basic types of strike-slip fault termination: (**a**) strike-slip fault terminates within undeformed rock (free tip); (**b**) strike-slip terminations produced by normal or reverse faults formed at their tips (two-plate geometry); and (**c**) strike-slip fault terminates against another through-going fault or fault system (three-plate geometry). Arrows attached to filled circles represent slip vector azimuths. Schematic displacement profiles along each strike-slip fault system are indicated.

Tibet is an example of this type of termination (Kirby 2006).

(2) A termination produced by normal or reverse faults that intersect the tip of the strike-slip fault (i.e. horsetail splay). Slip azimuths on the normal or reverse faults ideally will be parallel, and of equal magnitude, to those of the strike-slip fault (Fig. 1b). This type of termination is a two-plate geometry in classical plate-tectonic terminology (Le Pichon & Francheteau 1976) and can be observed where the southern tip of the strike-slip Mogod Fault in central Mongolia ends on a reverse fault (Bayasgalan *et al.* 1999), where the Sumatra Fault in Indonesia terminates in the Sunda Strait Rift (Lelgemann *et al.* 2000), where the

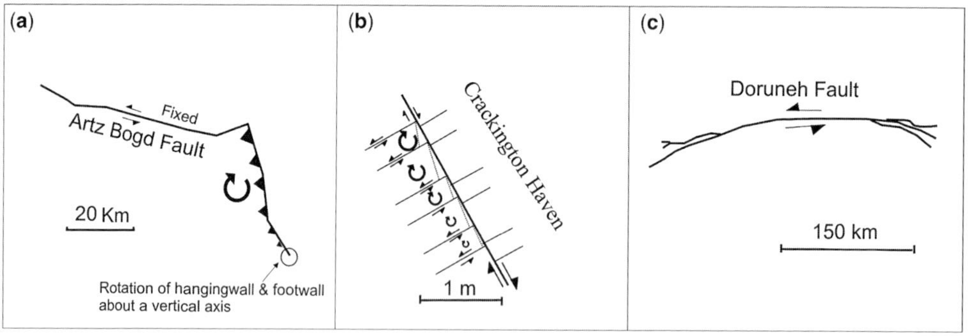

Fig. 2. (**a**) The eastern termination of the Artz Bogd Fault in Mongolia is achieved through the development of a rotating thrust at its tip. The displacement on the thrust decreases with distance from the strike-slip fault as a result of the relative rotation of the footwall and hanging-wall blocks of the thrust about a vertical axis (map modified after Bayasgalan *et al.* 1999). (**b**) The termination of the small-scale strike-slip fault at Crackington Haven, SW England. Strike-slip displacement gradually decreases towards the tip of the fault, in association with the rotational deformation of small blocks in the wall-rock that are bounded by bedding discontinuities. Dotted lines indicate extensional cracks formed by rotation (map modified after Kim *et al.* 2000). (**c**) The Doruneh Fault, Iran, terminates at both tips through bifurcation of the principal slip surface (map modified after Freund 1974).

Median Tectonic Line in Japan ends in the Yatsushiro graben (Kamata & Kodama 1994) and at the NE tip of the Hope Fault in New Zealand (Van Dissen & Yeats 1991).

(3) A termination produced where strike-slip fault intersects and terminates against (often in an abutting relation) another through-going reverse, normal- or strike-slip fault system (Fig. 1c). This three-plate setting can be observed where the Dead Sea Transform intersects the Red Sea Rift (Joffe & Garfunkel 1987), at the western end of the Garlock Fault where it intersects the San Andreas Fault (Davis & Burchfiel 1973) and at the SW tip of the Hope Fault where it intersects the Alpine Fault of New Zealand (Freund 1974).

Each of the three types of strike-slip fault termination may be associated with vertical-axis rotations of the rock close to the tip, as has been suggested by Bayasgalan *et al.* (1999) and Kim *et al.* (2000) for the tips of the Artz Bogd Fault in Mongolia (e.g. Fig. 2a), and the small scale strike-slip faults at Crackington Haven in SW England (e.g. Fig 2b), respectively, and/or with branching of the principal slip surface into multiple splays, as occurs at both tips of the Doruneh Fault in Iran (Freund 1974; e.g. Fig. 2c). These processes decrease the displacement on the principal fault surface by redistributing the slip into the rock volume around the fault.

Both vertical-axis rotations and fault bifurcation appear to contribute to the northern termination of the strike-slip North Island Fault System (NIFS), New Zealand, against the Taupo Rift. In this New Zealand example, however, neither vertical-axis rotations nor fault bifurcation by itself decrease strike-slip displacement to zero at the rift margin or permit transfer of strike-slip into the rift without displacing the rift margin. In this paper we show that, in addition to an earlier phase of vertical-axis rotations and fault bifurcation, termination of the NIFS and transfer of strike-slip into the Taupo Rift is achieved today because rift-related heterogeneous extension is distributed into the fault blocks outside the rift and is differentially accommodated in footwall and hanging-wall blocks of each of the main faults in the NIFS. As a result of this heterogeneous extension, slip vectors on the northern end of the strike-slip faults become gradually subparallel to slip vectors in the rift and to the line of intersection between the two fault systems. This subparallelism of fault slip vectors reduces potential space problems at the intersection of the two fault systems, allowing the kinematics and geometry of faulting to change efficiently whilst accommodating the overall deformation imposed by the regional kinematics and without the requirement of vertical-axis rotations.

Tectonic setting and fault data

The North Island Fault System (NIFS) is the principal strike-slip fault system in the upper Australian Plate of the Hikurangi margin, which is forming in response to oblique (*c.* 40–70°) subduction of the Pacific Plate (DeMets *et al.* 1994) (Fig. 3, inset). Oblique relative plate motion produces a margin-parallel component of motion of 26–33 mm/a, which is mainly accommodated by regional vertical-axis rotations of the deforming Hikurangi margin with respect to the relatively stable Australian Plate and by strike-slip faulting in the upper plate, with little or no strike-slip on the subduction thrust (Wallace *et al.* 2004; Nicol & Wallace 2007). The NIFS accommodates between 20 and 70% of the margin-parallel relative plate motion (Beanland 1995; Wallace *et al.* 2004).

The NIFS comprises two main strands, here referred to as the eastern and the western strand (Beanland 1995; Fig. 3). Each strand traverses the rugged terrain of Mesozoic greywacke basement along much of its length, both exploiting pre-existing terrane boundaries and mélange zones and cutting across these structures to define their own path (Mortimer 1994; Begg, pers. comm., 2006). The eastern strand extends for approximately 250 km, from the Wellington region in the south to southern Hawke's Bay in the north, where it is mainly a reverse-dextral fault (Beanland 1995; Fig. 3). The western strand is approximately 500 km long and traverses the North Island from Wellington to the Bay of Plenty coastline (Fig. 3). In this paper we focus on the northern termination of the western strand, and hereafter refer to this strand as the NIFS.

The NIFS ends where it intersects at a *c.* 45° strike angle the active Taupo Rift, a 200-km-long and up to 40-km-wide extensional basin associated with intense volcanic activity (Fig. 3; Wilson *et al.* 1995; Rowland & Sibson 2001). At its northern termination the NIFS comprises five main splays, spaced at a distance of 5–10 km, that distribute the displacement of the fault system across a zone that is 40–50 km wide. The intersection of the two fault systems is in part exposed onshore, providing an excellent opportunity to examine the geometry and kinematics of the termination of the strike-slip faulting.

The active NIFS has accommodated up to 5–8 km of strike-slip during the last 1–2 Ma (Beanland 1995; Kelsey *et al.* 1995; Nicol *et al.* 2007). Aggregated Late Quaternary dextral strike-slip rates, across the two main strands of the NIFS (i.e. eastern

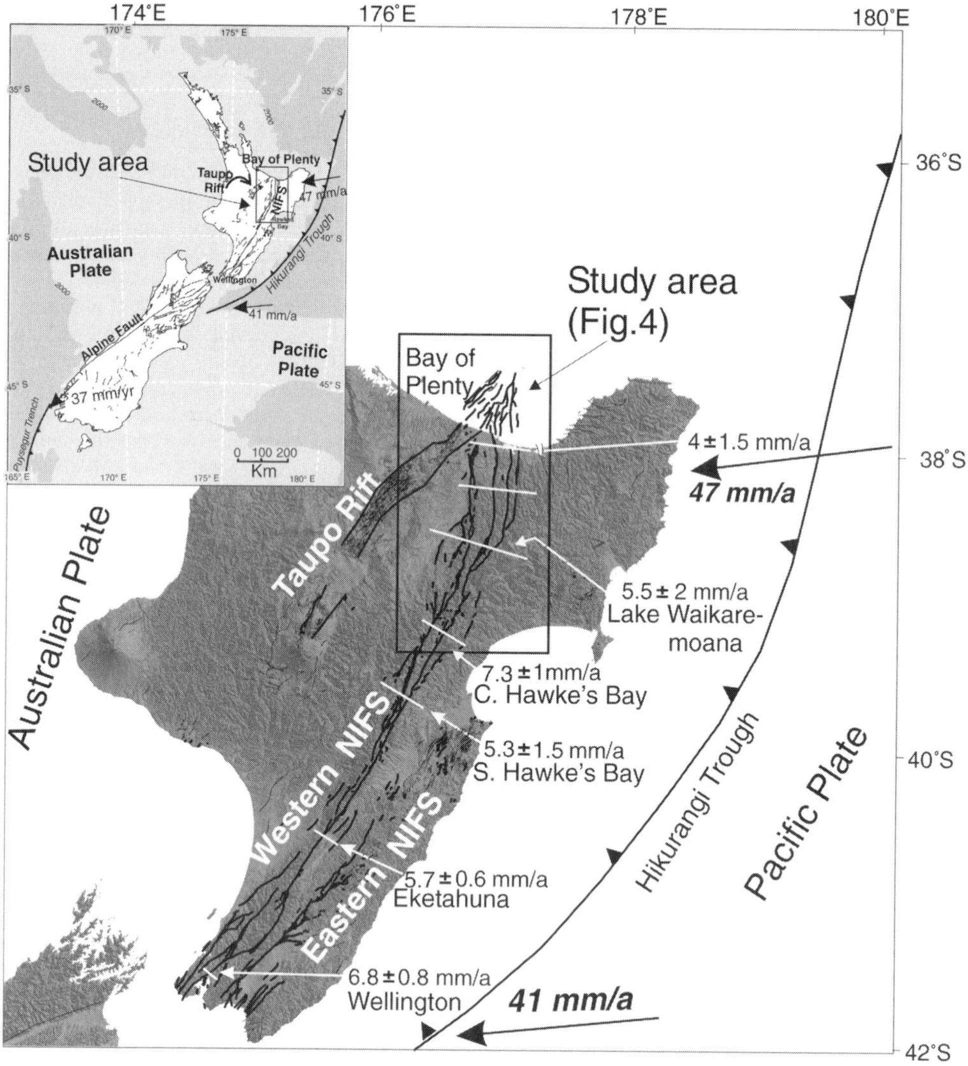

Fig. 3. Digital elevation model illustrating the large strike-slip North Island Fault System (NIFS) traversing the North Island of New Zealand and terminating against the Taupo Rift. The eastern strand of the NIFS extends from Wellington to Hawke's Bay, whilst the western strand traverses the North Island from Wellington to the Bay of Plenty. The Taupo Rift is also indicated. The black lines that bound the rift indicate the geographical extent of the rift. The along-strike variation of the strike-slip rate for the western strand of the NIFS is shown for six transects. Slip-rate data are from: Beanland & Berryman (1987); Cutten *et al.* (1988); Beanland (1995); Van Dissen & Berryman (1996); Kelsey *et al.* (1998); Schermer *et al.* (2004); Langridge *et al.* (2005); this study. Arrows with associated velocities indicate relative plate motion between the Pacific and Australian plates (DeMets *et al.* 1994). Inset: the New Zealand plate-boundary setting; the study area is located on the upper plate of the Hikurangi subduction margin.

and western) range from *c.* 18 mm/a in the south (Beanland 1995; Van Dissen & Berryman 1996; Langridge *et al.* 2005), near Wellington, to *c.* 4 mm/a in the north (this study), near the Bay of Plenty (see Fig. 3 for further details). During the Late Quaternary, the southernmost *c.* 400 km of the NIFS has chiefly carried dextral strike-slip and only minor reverse- or normal-slip (typical horizontal:vertical slip ratios are 10:1; Beanland 1995). Along the northern *c.* 100 km of the fault system, however, strike-slip and normal dip-slip decrease and increase northwards, respectively, resulting in a change of the

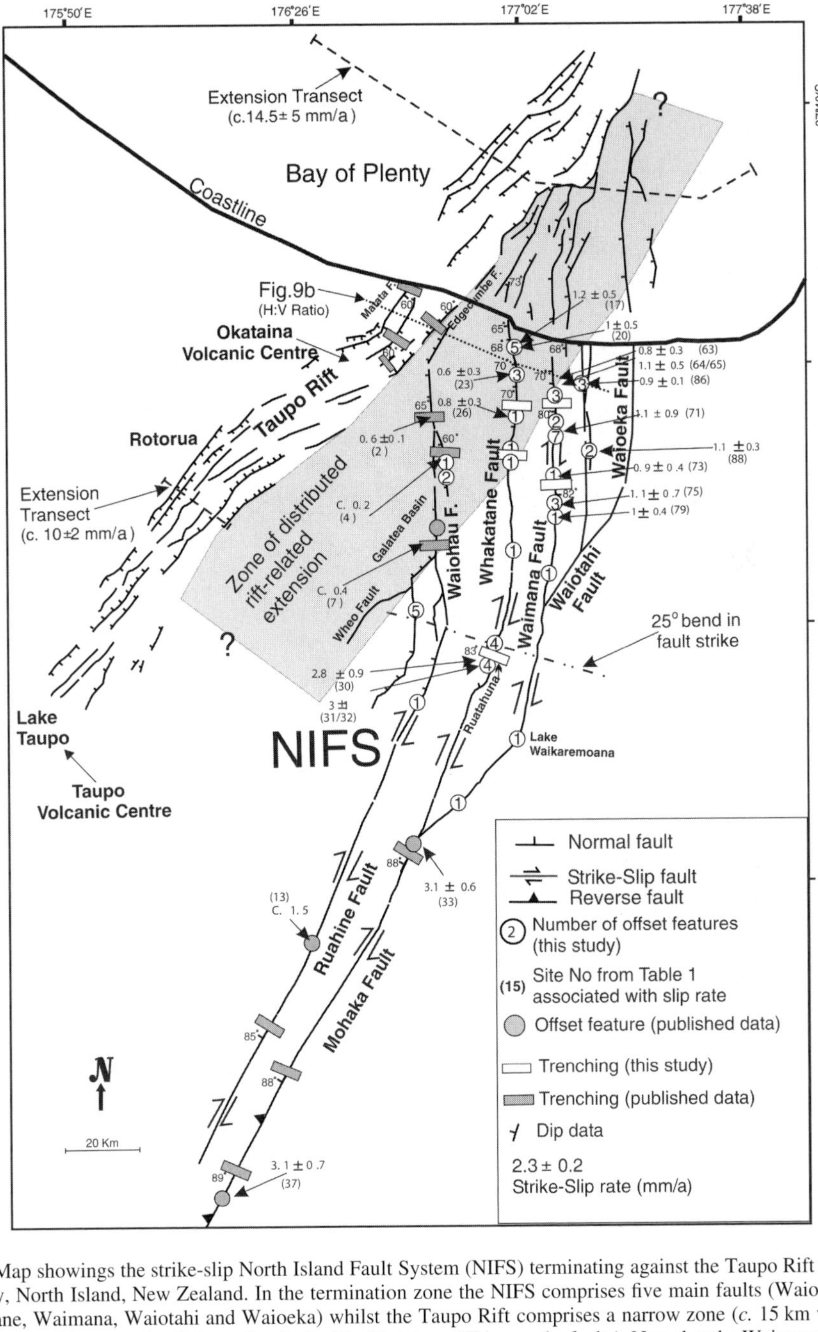

Fig. 4. Map showings the strike-slip North Island Fault System (NIFS) terminating against the Taupo Rift in the Bay of Plenty, North Island, New Zealand. In the termination zone the NIFS comprises five main faults (Waiohau, Whakatane, Waimana, Waiotahi and Waioeka) whilst the Taupo Rift comprises a narrow zone (*c.* 15 km wide) of normal faults with two main rift-bounding faults (the Matata and Edgecumbe faults). Note that the Waimana, Waiotahi and Waioeka faults merge southward with the Whakatane Fault, which becomes the Mohaka Fault still further south. The sites studied and the associated strike-slip rates (where available) are indicated (see also Table 1). Dashed transects indicate the locations of extension profiles, one across the onshore and one across the offshore Taupo Rift. The approximate location of the *H*:*V* slip-ratio transect across the intersection of the two fault systems in Fig. 9b, is indicated by the dotted line. Regional distributed extension outside the rift is indicated by grey shading. Published data are from: Hull (1983); Raub (1985); Ota *et al.* (1988); Beanland (1989*b*); Beanland *et al.* (1989); Berryman *et al.* (1998); Hanson (1998); Woodward-Clyde (1998); Taylor *et al.* (2004). Offshore data are from Davey *et al.* (1995), Lamarche *et al.* (2000) and Taylor *et al.* (2004).

Table 1. *Summary of the geometric and kinematic attributes of the faults at the sites studied*

Site No.	Site name	Grid ref. (NZMS 260)	Strike (°)	Dip (°)	Distance along fault (km)[1]	Dextral separation (m)	Vertical separation (m)	*H*:*V* ratio
				Waiohau–Ruahine				
1	Matahina gr	V16/453360	360	65 ± 5	23	–	700 ± 100	–
2	Tasman 1	V16/453298	350	65 ± 5	29	–	*c.* 7 (min.)	0.4
3	Cornes tr	V16/472252	360	65 ± 10	34	–	*c.* 7	–
4	Cornes	V16/472251	360	65 ± 10	34.2	3 ± 1	8 ± 1	0.4
5	Waiohau1	V16/471246	008	–	34.7	3 ± 2	4 ± 2	0.75
6	Waiohau2	V16/475213	040	–	38.2	–	5 ± 3	–
7	Troutebeck	V17/446057	010	*c.* 70	55	–	*c.* 9	0.5
8	Whirinaki	V17/401916	340	–	79.1	*c.* 24	–	–
9	Whirinaki1	V17/401913	340	–	79.4	38	–	–
10	Whirinaki2	V17/402914	340	–	79.3	*c.* 24	–	–
11	Whirinaki3	V17/397928	340	–	78	*c.* 24	–	–
12	Skips Hut	V18/404692	–	–	92.8	–	–	–
13	Gorge Stream	U20/083061	020	–	130	*c.* 10	*c.* 5	2.0
14	Davis	U20/092077	015	85	163	55	10	5.5
				Whakatane–Mohaka				
15	F1	W14/787761	360	60 ± 10	*c.* –40	–	*c.* 610	–
16	Poroporo	W15/599532	–	65 ± 5	6	–	660 ± 100	–
17	Pukehoko	W15/614459	345	–	9	26 ± 5	32 ± 2	0.81
18	Anemos	W15/614455	345	–	9.2	23 ± 5	–	–
19	Galini	W15/615452	345	–	9.4	5 ± 4	5 ± 1	1.0
20	Tophouse	W15/617450	345	68 ± 10	9.5	22 ± 5	30 ± 2	0.73
21	Karenza	W15/618444	345	–	9.8	25 ± 5	–	–
22	Awahou1	W16/622389	360	–	15.3	5 ± 3	–	–
23	Awahou2	W16/621388	355	70 ± 10	15.4	14 ± 6	13 ± 2	1.07
24	Awahou grav	W16/622346	–	70 ± 5	18	–	500 ± 100	–
25	Te Whetu	W16/619327	010	68 ± 2	19	–	10 ± 1	–
26	Noti	W16/615298	010	–	24.5	15 ± 7	15.5 (min.)	0.97
27	Te Marama	W16/603193	350	55 ± 5	37	–	1.7 ± 0.3	–
28	Matangi-Mariri	W16/603177	342	–	38	4 ± 0.5	2.7 ± 0.5	1.5
29	Waikare	W17/608984	009	65 ± 2	57	–	–	–
30	Ru-1/Armyra	W17/556802	018	81 ± 5	75.1	20 ±2	3 ± 1	6.67
31	Ru-2	W17/555802	020	–	75.3	22 ± 2	2 ± 0.5	11.0
32	Thalassa/Helios	W17/555800	020	85 ± 5	75.3	22 ± 2	2 ± 0.5	11.0
33	Ru-3	W17/555800	020	–	75.6	8.5 ± 6	–	–
34	Ru-4	W17/555799	017	–	75.9	9.5 ± 5	–	–

Down side	Offset feature	Data source	Age (min–max) (ka)[4]	Net slip rate (mm/a)	Pitch (°)	Reference
			Waiohau–Ruahine			
W	Basement	Gravity profile	1000 ± 200[3]	0.7 ± 0.2 (throw rate)	–	This study
E&W	Scarp	Trench	17.8[6,7]	0.7 ± 0.2	70	Woodward-Clyde (1998)[2]
W	Scarp	Trench	25[6,7]	–	–	Woodward-Clyde (1998)[2]
W	Stream	RTK[5]	13.8[6]	0.7 ± 0.2	69	This study
W	Stream	Tape measure	–	–	–	This study
W	Scarp	Tape measure	–	–	–	This study
W	Scarp	Trench	17.6[6]	0.7 ± 0.2	63	Beanland (1989*b*)
W	Stream	Aerial photo	–	–	–	This study
W	Spur	Aerial photo	–	–	–	This study
W	Spur	Aerial photo	–	–	–	This study
W	Spur	Aerial photo	–	–	–	This study
W	Basement	Mapping	–	–	–	This study
W	River margin	Tape measure	*c.* 3.5[6]	1.5 ± 0.5	–	Beanland & Berryman (1987)
W	Scarp	Trench	*c.* 13.8[6]	–	10	Hanson (1998)
			Whakatane–Mohaka			
W	Basement	Seismic–reflection profiles	600 ± 30[3]	*c.* 1.2 (dip-slip)	–	Davey *et al.* (1995)[2]
W	Basement	Gravity profile	600 ± 30[3]	1.1 ± 0.2 (throw rate)	–	This study
W	Spur	Tape measure	17.6–30[6]	2.2 ± 1	51	This study
W	Spur	Tape measure	17.6–30[6]	–	–	This study
W	Spur	RTK	–	–	45	This study
W	Spur	RTK	17.6–30[6]	2.2 ± 0.9	54	This study
W	Spur	Tape measure	17.6–30[6]	–	–	This study
W	Stream	RTK	17.6–30[6]	–	–	This study
W	Spur	RTK	17.6–30[6]	1 ± 0.5 (min.)	43	This study
W	Basement	Gravity profile	600 ± 30[3]	0.8 ± 0.4 (throw rate)	–	This study
W	Scarp	Trench	*c.* 30[6,7]	0.4 (min.)	–	This study
W	Spur	RTK	17.6–25[6]	1.5 (min.)	46	This study
W	Terrace margin	Trench	Post *c.* 0.8 [7]	–	–	This study
W	Stream on river terrace	Tape measure	<3.5[6]	–	34	This study
W	Basement	Striations	–	–	70 & 50 NW	This study
E	River margin	RTK/ trench	*c.* 6–9.5[6,7]	3 ± 1.1	9	This study
E	Stream	RTK	*c.* 6–9.5[6,7]	3.1 ± 1	5	This study
E	Stream	Trench	*c.* 6–9.5[6,7]	3.1 ± 1	–	This study
–	Stream	RTK	–	–	–	This study
–	Stream	RTK	–	–	–	This study

(*Continued*)

Table 1. Continued

Site No.	Site name	Grid ref. (NZMS 260)	Strike (°)	Dip (°)	Distance along fault (km)[1]	Dextral separation (m)	Vertical separation (m)	*H*:*V* ratio
35	Ru-5	W17/555799	020	–	76	17 ± 7	2 ± 0.5	14.0
36	Ru-7	W17/559811	020	–	74	15 ± 2	2 ± 0.5	7.5
37	Ru-8	W17/559810	020	–	74.3	12 ± 5	–	–
38	Ru-9	W17/558809	020	–	74.4	15 ± 5	–	–
39	Ru-10	W17/558808	020	–	74.6	5.5 ± 1.5	1.2 ± 0.3	4.5
40	Ru-11	W17/557805	022	–	74.9	14 ± 1	–	–
41	Ru-13	W17/561817	020	–	73.6	11 ± 5	–	–
42	Ru-15	W17/565828	028	–	73	15 ± 8	2 ± 1	7.5
43	Ru-16	W17/566829	028	–	72.8	–	2 ± 0.5	–
44	Ru-17	W17/566829	028	–	72.75	4.5 ± 2.5	–	–
45	Ru-18	W17/567832	022	–	72.5	15 ± 5	3 ± 1	5.0
46	Opuhou	W17/562821	020	–	73.3	–	0.5 ± 0.2	–
47	Opuhou1	W17/562821	020	–	73.2	90 ± 10	–	–
48	Kakanui	W17/551795	048	–	76.5	–	2 ± 1	–
49	Te Hoe	V19/379395	020	87	132	–	2.9	–
50	N. Te Hoe	V19/373400	020	88	131.7	*c.* 40	3 ± 1	11.70
51	Syme	V20/137986	030	85	170	60	6	10.0
52	McCool 1	U21/010746	020	89	196	66	4	16.5
53	C. Mohaka	U22/876485	024	90	225	*c.* 33	4.5	7.5
54	Hughes 1	T24/377761	030	87	314	60 ± 5	1.5 ± 1	40.0
55	Bennet	T24/372753	037	86	315	50 ± 6	–	–
56	Emerald Hill	R27/873102	062	–	*c.* 400	104–940	10–90	*c.* 10.0
				Waimana				
57	F2	W14/817760	360	60 ± 5	*c.* –40	–	390 ± 100	–
58	Ohope	W15/769485	–	68 ± 3	0.1	–	500 ± 100	–
59	Ohiwa1	W15/748458	–	–	3.1	–	2	–
60	Ohiwa2	W15/743438	–	–	5.2	–	2	–
61	Ohiwa3	W15/737426	–	–	6.5	–	2	–
62	Don1	W16/725377	350	–	11.3	24 ± 15	–	–
63	Don2	W16/725379	350	–	11.2	13.5 ± 2	2 ± 0.5	6.75
64	Moana-Iti	W16/725379	360	88 ± 2	11.2	20 ± 3	2 ± 1	10.0
65	Moana	W16/725379	357	47 ± 2	11.25	20 ± 3	2 ± 1	10.0
66	Don3	W16/727371	345	–	11.6	19 ± 2	–	–
67	Stevens2	W16/728342	350	–	15	20 ± 2.5	–	–
68	Nukuhou-gr	W16/724328	–	>80	15.3	–	<50 m	–
69	SH2	W16/716329	010	–	15.5	–	4 ± 1	–
70	Tom1	W16/707293	018	–	18.8	20 ± 5	–	–
71	Timoti 2	W16/703278	014	–	20.3	13 ± 7	2 ± 1	6.5

Down side	Offset feature	Data source	Age (min–max) (ka)[4]	Net slip rate (mm/a)	Pitch (°)	Reference
E	Spur	RTK	–	–	7	This study
W	Spur	Tape measure	–	–	8	This study
–	Stream	RTK	–	–	–	This study
E	Spur	RTK	–	–	–	This study
W	Spur	RTK	–	–	14	This study
–	Stream	Tape measure	–	–	–	This study
–	Stream	RTK	–	–	–	This study
E	Spur	RTK	–	–	10	This study
E	Scarp	RTK	–	–	–	This study
W	Stream	RTK	–	–	–	This study
E	Spur	RTK	–	–	11	This study
W	Scarp	Tape measure	–	–	–	This study
–	Stream	Tape measure	–	–	–	This study
SE	Scarp	Tape measure	–	–	–	This study
–	Scarp	Trench	9.5[6]	3.5 ± 0.8	–	Hull (1983)
W	Stream	Tape measure	9–13[6,7]	3.7 ± 0.3	5	Berryman *et al.* (1988)
E	Scarp	Trench	10.1[6,7]	–	–	Hanson (1998)
W	Stream	Trench	16.8 ± 0.5[6,7]	*c.* 4.2	–	Hanson (1998)
W	River margin /scarp	Tape measure	10–13[6]	3.2 ± 0.6	8	Raub (1985)
SE	Stream	Tape measure	*c.* 10.8[7]	4.9–6.2(max.) (dextral)	–	Langridge *et al.*(2005)
NW	Stream	RTK	*c.* 8.5 [7]	5.1–6.7 (min.) (dextral)	–	Langridge *et al.*(2005)
SE	River margins	Tape measure	14–140 [6]	6–7.6 (dextral)	–	Van Dissen & Berryman (1996)
			Waimana			
W	Basement	Seismic reflection profiles	600 ± 30[3]	*c.* 0.8 (dip-slip)	–	Davey *et al.* (1995)[2]
W	Basement	Gravity profile	600 ± 30[3]	0.8 ± 0.4 (throw rate)	–	This study
–	Holocene sediments	Drill-hole	2.7–8[6]	–	–	Hayward *et al.* (2003)
–	Holocene sediments	Drill-hole	2.7–8[6]	–	–	Hayward *et al.* (2003)
–	Holocene sediments	Drill-hole	2.7–8[6]	–	–	Hayward *et al.* (2003)
–	Stream	RTK	–	–	–	This study
W	River margin	RTK	15–23[6,7]	0.8 ± 0.4	–	This study
W	Scarp	Trench–RTK	15–23[6,7]	1.2 ± 0.5 (max.)	–	This study
W	Stream	Trench–RTK	15–23[6,7]	1.2 ± 0.5 (max.)	0 and 20	This study
–	Stream	Tape measure	–	–	–	This study
W	Stream	Tape measure	–	–	–	This study
W	Basement	Gravity profile	600 ± 30	–	–	This study
W	Scarp	Tape measure	–	–	–	This study
W	Spur	Tape measure	–	–	–	This study
E	Spur	RTK	9.5–30[6]	1.2 ± 0.9	9	This study

(*Continued*)

Table 1. Continued

Site No.	Site name	Grid ref. (NZMS 260)	Strike (°)	Dip (°)	Distance along fault (km)[1]	Dextral separation (m)	Vertical separation (m)	*H*:*V* ratio
72	Timoti 1	W16/702276	010	–	20.4	12 ± 6	2 ± 1	6.0
73	Sonny Riser	W16/704170	360	–	31	13 ± 6	2.5 ± 0.5	5.2
74	Sonny scarp	W16/704180	360	–	30.5	–	5 ± 1	–
75	Ahirau 4	W16/705158	007	–	32.3	10 ± 5	2 ± 1	5.0
76	Ahirau 1	W16/705161	008	82	31.6	–	2 ± 1	–
77	Ahirau 2	W16/705161	352	70	31.7	–	2 ± 1	–
78	Tuhora 2	W16/705160	352	–	32.1	18 ± 2	–	–
79	Waiiti	W16/704145	010	–	33.8	5.5 ± 2	2.5 ± 0.5	2.2
80	Panoanoa	W18/672902	020	–	60	330 ± 50	–	–
81	Hopuruahine	W18/632701	020	–	80.5	–	–	–
82	Mangatoatoa	W18/515537	020	–	100	–	5 ± 2	–
				Waiotahi				
83	F3	W14/898756	360	60 ± 10	*c.* −40	–	300 ± 70	–
84	F4	W14/912755	360	60 ± 10	*c.* −40	–	330 ± 70	–
85	Waiotahi Beach	W15/775485	–	68 ± 3	0.1	–	300 ± 50	–
86	Waiotahi N	W16/766397	350	–	9	16 ± 1	2 ± 0.5	8.0
87	Waiotahi n	W16/767396	340	–	9.1	–	5 ± 1	–
88	Waiotahi S	W16/761308	340	–	18	17 ± 3	2 ± 0.5	8.5
				Whakatane Graben				
89	Edgecumbe 1	V15/477502	055	55 ± 5	9.2	–	4.7 ± 0.8 (dip-slip)	–
90	Edgecumbe 2	V15/477502(?)	055	55 ± 5	6–12	–	2.5 ± 0.2	–
91	Rotoitipakau	V15/345436	055	60	20.5	–	11.5 ± 0.7 (dip-slip)	–
92	Braemar	V15/358480	045	60	16	–	5.5 ± 2.2 (dip-slip)	–
93	Matata	V15/414602	045	60	1.5	–	9.2	–
94	White Island Fault	*c.* W15/ 670676	*c.* 040	67 ± 2	*c.* −20	–	–	–
95	R1	*c.* W15/ 650708	*c.* 057	61 ± 2	*c.* −20	–	830 ± 130 (max.)	–
96	R2	*c.* W15/ 648719	*c.* 057	61 ± 2	*c.* −20	–	830 ± 130 (max.)	–
97	R3	*c.* W15/ 642727	*c.* 057	61 ± 5	*c.* −20	–	830 ± 130 (max.)	–

[1]Distance from Bay of Plenty coast.
[2]Reinterpreted data.
[3]Fleming (1955) and Beu (2004).
[4]Calibrated ages from Zachariasen & Van Dissen (2001).
[5]Real-Time Kinematics.
[6]Tephrochronology.
[7] ^{14}C dating.

Down side	Offset feature	Data source	Age (min–max) (ka)[4]	Net slip rate (mm/a)	Pitch (°)	Reference
E	Spur	RTK	9.5–30[6]	1.1 ± 0.9	9	This study
E	River terrace riser	RTK	13.8[6](max.)	1 ± 0.5(min.)	11	This study
E	Scarp	RTK				This study
W	Stream	RTK	8–13.8[6]	1.2 ± 0.8	11	This study
E	Scarp	Trench	9.5[6,7]	–	–	This study
E	Scarp	Trench	9.5[6]	–	–	This study
E	Spur	Tape M.	–	–	–	This study
W	River margin	Tape measure	5.6[6] (max.)	1.2 ± 0.5(min)	24	This study
–	Stream	Aerial photo	–	–	–	This study
–	Basement	Fault gouge	–	–	–	This study
W	Scarp	Mapping	–	–	–	This study
			Waiotahi			
W	Basement	Seismic-reflection profiles	600 ± 30[3]	0.6 ± 0.2 (dip-slip)	–	Davey *et al.* (1995)[2]
W	Basement	Seismic-reflection profiles	600 ± 30[3]	0.7 ± 0.2 (dip-slip)	–	Davey *et al.* (1995)[2]
W	Basement	Gravity profile	600 ± 30[3]	0.5 ± 0.1 (throw rate)	–	This study
W	Stream	Tape measure	17.6[6]	1 ± 0.1	7	This study
W	Scarp	Tape measure	–	–	–	This study
W	Stream	Tape measure	13.8–17.6[6]	1. 2 ± 0.4	7	This study
			Whakatane Graben			
W	Scarp	Trench	Today-1.8[6,7]	2.6 ± 1.4	–	Beanland *et al.* (1989)
W	Cultural features	Topographic survey	March 1987	–	55(plunge)	Beanland *et al.* (1989)
W	Scarp	Trench	5.6[6,7]	2.2 ± 0.25	–	Berryman *et al.* (1998)
E	Scarp	Trench	8.7[6,7]	0.7 ± 0.3	–	Beanland (1989*b*)
E	Scarp	Trench	5.6[6,7]	1.8 ± 0.2	–	Ota *et al.* (1988)
W	Scarp	Seismic-reflection profiles	–	–	–	Taylor *et al.* (2004)
W	Scarp	Seismic-reflection profiles	1340 ± 510	1.4 ± 0.3	–	Taylor *et al.* (2004)
W	Scarp	Seismic-reflection profiles	1340 ± 510	1.4 ± 0.3	–	Taylor *et al.* (2004)
W	Scarp	Seismic-reflection profiles	1340 ± 510	1.4 ± 0.3	–	Taylor *et al.* (2004)

Table 2. *Representative electron microscope analyses of Late Quaternary tephras*

Grid ref. (NZMS 260)	Site(s)	SIO_2	TiO_2	Al_2O_3	FeO	MnO	MgO
W15/617450	17–21	78.02435064	0.150309057	12.48245932	0.916175309	0.098091675	0.147202849
W15/614455	17–21	78.20429093	0.125243206	12.33958067	0.897010204	0.09401229	0.128103381
W16/615298	26	76.48350851	0.163634339	13.14231646	1.151104501	0.132208109	0.239496842
W16/621388	23	77.24903693	0.163957707	13.10296928	1.066033391	0.082107918	0.189004327
W16/703278	71 and 72	77.23769278	0.175606495	13.04701472	1.209107098	0.066909542	0.172188881
W17/555800	30–32	78.28981531	0.138429804	12.25884391	1.178333151	0.085028924	0.119598668
W16/615298	26	78.04501236	0.142036469	12.50896867	1.028611105	0.052291719	0.142744509
W16/615298	26	77.36781658	0.134989157	12.52832752	0.862295766	0.083827794	0.18263948
W15/617450	17–21	78.13548128	0.139410968	12.34420868	1.023404642	0.032036565	0.150776724
W15/617450	17–21	78.26190126	0.132460846	12.45292186	1.071220364	0.056924951	0.124949203
W16/705158	75	78.0802471	0.129365758	12.66561124	0.813489481	0.094105154	0.119002363
W16/704170	73 and 74	78.34149993	0.125452757	12.58958724	0.858707208	0.077521273	0.111601113
W16/725379	63–65	78.15790731	0.133026627	12.78216948	0.930848824	0.061047388	0.140006083
W17/555800	30–32	77.91066293	0.145856682	12.37275675	0.953738156	0.085464393	0.126876874
W16/705161	76 and 77	78.59251256	0.137313974	12.51771047	0.931526968	0.049695927	0.120368041
W16/725379	63–65	77.7384353	0.121395709	12.97947964	0.909595006	0.095025578	0.127070073
W16/704145	79	77.83847525	0.120459518	12.39247615	0.789705751	0.085504277	0.109238061

For details of each site, refer to Table 1. Glass chemistry determined using an electron microprobe (10 μm beam diameter). Analyses recalculated to 100% (volatile-free).Tephra correlations from: 1, Froggatt & Solloway (1986); 2, Lowe & Hogg (1986); 3, Lowe (1988); 4, Eden *et al.* (1993); 5, Manning (and references therein) (1995); 6, Newnham *et al.* (1995); 7, Lowe *et al.* (1999); 8, Wilmhurst *et al* (1999); 9, Shane & Hoverd (2002); 10, Pillans & Wright (1992); 11, Nairn *et al.* (2004); n = number of analyses per sample.

horizontal to vertical slip ratio (H:V) from 10:1 to 1:1 (Beanland 1995; this study).

In this paper we use displaced landforms ≤50 000 years old, fault-trench data, gravity profiles and seismic-reflection lines to track changes in the kinematics and geometries of the NIFS approaching its intersection with the Taupo Rift. Eighty-eight sites were studied along the northernmost 250 km of the four main strands in the NIFS (Waiohau–Ruahine, Whakatane–Mohaka, Waimana and Waiotahi faults; Fig. 4, Table 1). Fault geometries were determined from outcrop exposures, trenching, aerial-photo interpretation, gravity modelling and seismic-reflection profiles. Slip-directions were derived from the analysis of offset geomorphic linear markers (e.g. spurs, abandoned stream axes, river margins) and slickenside striations, and slip-rates were estimated from offset landforms, 3D-trenching (i.e. perpendicular and parallel to the fault) and dating of key geomorphic stratigraphic markers by tephrochronology (Table 2) and ^{14}C dating (Fig. 4, Table 1).

Most of the slip-rate estimates derive from fault traces that traverse and displace young geomorphic surfaces (≤50 000 years old) which are often mantled by alternating layers of loess, palaeosols and volcanic tephra. Datable volcanic tephra layers were sourced from two major volcanic centres located adjacent to the NIFS (Taupo and Okataina volcanic centres) and have erupted repeatedly throughout the Late Quaternary (Fig. 4, Table 2) (Nairn & Beanland 1989). Individual tephra layers of known ages have been identified by comparing their distinctive glass chemistry to a New Zealand database of results (see caption of Table 2 for references). Wherever possible, these ages were confirmed by radiocarbon dating of associated sediments (Table 1). Most of the displaced landforms (i.e. abandoned stream channels, river-terrace risers and ridges) occur in proximity to fluvial aggradational terrace surfaces that are today elevated above modern river beds. In the study area, these river terraces have typical ages of *c.* 5.6 ka, 13.8 ka, 17.6 ka, 30 ka and 50 ka (Table 2 and references therein).

Fault displacements and slip vectors have been measured using topographic surveys (GPS–Real-Time-Kinematics) of linear landforms that are offset by the fault to produce piercing points (e.g. Fig. 5; Table 1). These piercing points were identified and correlated across faults by constructing detailed topographic maps, from which profiles were made parallel (and adjacent) to the fault in both the footwall and the hanging wall (e.g. Fig. 5). Where the displaced lineaments have been eroded or modified by deformation close to the fault, they have been projected as an inclined

CaO	Na_2O	K_2O	Cl	Total	*n*	Tephra ID	Cal. Age (ka)	Sample code	References
0.824311561	3.845583447	3.291318757	0.217084683	100	10	Rotoehu	*c.* 50	Q3	5,10
0.789958935	3.856770419	3.403530531	0.164397333	100	10	Rotoehu	*c.* 50	An	5,10
1.089616989	4.528127589	2.883478131	0.188039011	100	10	Mangaone	*c.* 30	NF	5
1.014611831	4.041584503	2.915147446	0.173451148	100	10	Mangaone	*c.* 30	Awa Manga	5
1.031165677	4.003858275	2.879559627	0.18267911	100	10	Mangaone	*c.* 30	TimotiD	5
1.094131287	3.579481201	3.076558184	0.176254372	100	10	Kawakawa	26	T(0)	1,10
0.880936193	3.603643074	3.413256722	0.178353591	100	10	Te Rere	25	Noti B	1,9
0.830687157	3.969037141	3.882644786	0.157763347	100	10	Rerewhakaaitu	17,6	Ng	2,3,4,7,9
0.906501784	3.608334206	3.517315703	0.145591665	100	10	Rerewhakaaitu	17,6	TH/c	2,3,4,7,9
0.971485785	3.470119464	3.274041168	0.179901569	100	10	Rerewhakaaitu	17,6	GB	2,3,4,7,9
0.815461965	3.78036434	3.347047862	0.156738584	100	10	Waiohau	13,8	Ah4-C	2,4,7,9
0.801367917	3.691214224	3.254299427	0.135902441	100	10	Waiohau	13,8	Auger4	2,4,7,9
0.841391423	3.613392058	3.17301722	0.167185641	100	10	Waiohau	13,8	T1-Moana	2,4,7,9
0.873409594	3.978884498	3.372082016	0.181488297	100	10	Rotoma	9,5	T1	2,3,7,9
0.815984733	3.317571194	3.36958933	0.144329966	100	10	Rotoma	9,5	TuhD	2,3,7,9
0.770644313	3.666433001	3.437026468	0.153705947	100	10	Rotoma	9,5	M29	2,3,7,9
0.720401993	3.955755884	3.82794325	0.157464513	100	10	Whakatane	5,6	Sonny1	2,7

lineation to intersect the fault surface from both sides of the fault. Errors in slip estimates are chiefly derived from projection uncertainties rather than measurement precision (i.e. Fig. 5a). Kinematic attributes such as fault strike, dip, displacement components and Late Quaternary slip-rates at 97 localities in the northern NIFS ($N = 88$) and Taupo Rift ($N = 9$) are summarized (with estimated errors) in Table 1.

Northern termination of the NIFS

Fault geometries

Strike-slip faults in the NIFS strike NNE–SSW along the southernmost 350 km of their length (Fig. 3). To the north of Lake Waikaremoana, however, this consistent NNE strike swings *c.* 25° anticlockwise to a N–S direction (Figs 3 & 4). This change in strike, the possible origin of which is discussed in the 'fault termination' section, is accompanied by some bifurcation of the main faults. Immediately south of the bend, the NIFS consists of two main faults, the Mohaka and the Ruahine faults, whilst north of the bend, the Mohaka Fault splays into the Whakatane, Waimana, Waiotahi and Waioeka faults, and the Ruahine Fault becomes the Waiohau Fault which is associated with other secondary splays (e.g. Wheo Fault; Fig. 4). These five main splays extend northward for a distance of *c.* 100 km on land and up to 50 km offshore, and form a 40–50-km-wide zone at their northernmost onshore extent. Still further north, however, the north–south striking faults deflect clockwise in strike by *c.* 20–35° to the NE within *c.* 10–15 km of the rift's SE margin (Fig. 4). For the Waiohau Fault, this change of strike close to the intersection takes place onshore, whereas for the Whakatane, Waimana and Waiotahi faults it occurs offshore (Fig. 4; Lamarche *et al.* 2000; Taylor *et al.* 2004; Mouslopoulou *et al.* 2007). Similarly, the faults of the Taupo Rift, south of their intersection with the NIFS, strike ENE, whilst northward, as the intersection is approached, they swing anticlockwise by *c.* 20° to a NE strike (Fig. 4).

Along the southern 350 km of the NIFS, the faults often dip steeply at the surface, *c.* 80–90°, either to the east or west (Beanland 1995). Close to the northern termination of the NIFS, however, the dips of the faults gradually become shallower, at 60–70°, and are consistently west-dipping and downthrown to the west (Fig. 4) (e.g. sites 1, 15, 16, 57, 58, 83 in Table 1). The lower dips at the northern end of the NIFS are comparable with those measured on normal faults in the Taupo Rift itself from Beanland *et al.* (1989), Lamarche *et al.* (2000) and Taylor *et al.* (2004) (for details see fig. 4 from Mouslopoulou *et al.* 2007).

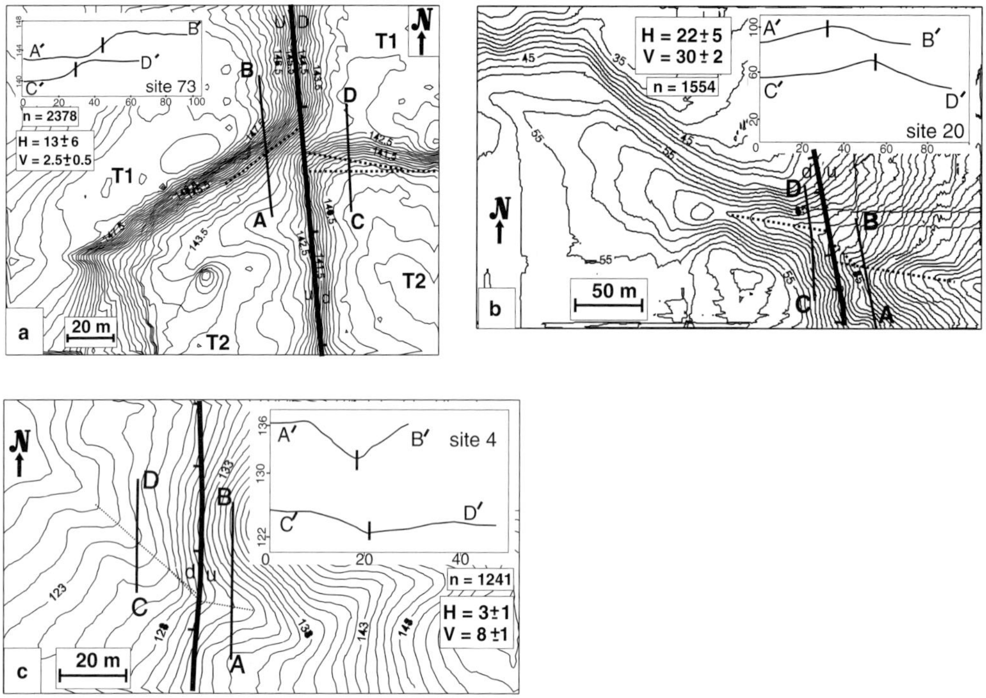

Fig. 5. (**a**) to (**c**) Detailed contour maps of three representative offset features from faults within the northern NIFS, constructed by GPS–Real-Time Kinematics. The topographic data were gridded, contoured and 'sliced' in order to estimate slip-vectors. Offset profiles (e.g. the A–B line) were derived by sampling a data-line approximately perpendicular to the offset marker and parallel to the fault plane. These profiles were projected on to the fault plane on each side of the fault (e.g. A′–B′) and matched with one another to produce a series of piercing points (the actual piercing points utilized are indicated by the thick vertical lines on each profile). Dashed lines represent offset piercing points, whilst their divergence represents errors in the projection. (a) Displaced river margin on the Waimana Fault (T1 = higher river terrace, T2 = younger river terrace); (b) displaced ridge spur in northern Whakatane Fault; (c) displaced stream channel on the Waiohau Fault (see Table 1 for more details on each of these sites). The downthrown side is indicated. *n* = number of topographic data points/site. The associated displaced profiles are indicated as: A–B profile parallel to the fault plane. A′–B′ projection of A–B on the fault plane.

Late Quaternary displacement rates and slip vectors

The relatively uniform Late Quaternary strike-slip rate recorded along the southernmost *c.* 350 km of the Wellington–Mohaka Fault (*c.* 6.8 mm/a in the Wellington region, *c.* 5.7 mm/a in Eketahuna, *c.* 5.3 mm/a in south Hawke's Bay, *c.* 7.3 in central Hawke's Bay) begins to decrease northwards within *c.* 100 km of its intersection with the Taupo Rift (Figs 3, 4 & 8). In order to assess how the termination of the northern NIFS is accomplished, we examine changes in the Late Quaternary rates of strike-slip, dip-slip, and net-slip on four of the main strands of the fault system towards its northern tip. These changes are reflected in corresponding changes of the slip vectors along each of the strands of the NIFS as they approach the Taupo Rift. In this section, we document the northward decrease in the H:V slip ratio from 10 to 1, signifying a steepening of the pitch of the slip vectors.

Waiohau–Ruahine Fault. The Waiohau Fault, which is the westernmost strand of the NIFS, has been the least active fault within the NIFS during the last *c.* 17 ka (Fig. 6). In the south, where this strand is referred to as the Ruahine Fault, it is predominantly strike-slip (H:V = 5.5, site 14 in Table 1), has a Late Quaternary dextral slip-rate of *c.* 1.5 mm/a (site 13 in Table 1 & Fig. 4) and slip vectors plunging *c.* 10°N on a steeply dipping (i.e. 85°) fault (Hanson 1998). The geomorphic signature of the Waiohau Fault along its northernmost 60–70 km, however, differs significantly from that of the Ruahine Fault. Steep scarps greater than 100 m high and triangular facets occur along the

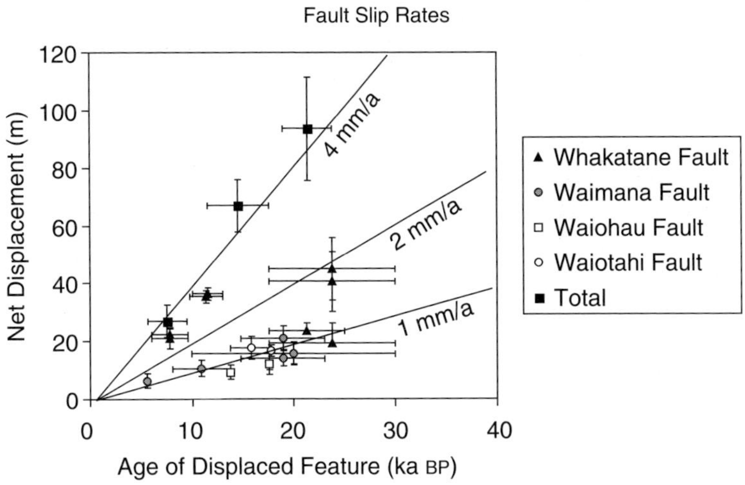

Fig. 6. Displacement rates for the faults within the northern NIFS. Data suggest that the Whakatane Fault is presently the most active fault within the northern NIFS. The total displacement rate is *c.* 4 mm/a, and has remained approximately uniform for the last *c.* 30 ka.

eastern margin of the Galatea Basin, where they record a significant component of normal displacement on this section of the fault. Estimates of Late Quaternary slip-rates, on the northern Waiohau Fault (Beanland 1989*b*; Woodward-Clyde Ltd. 1998; this study) (sites 2–5 & 7 in Table 1, Fig. 5c) indicate a strike-slip rate of <0.5 mm/a, significantly less than the slip-rate of 1.5 mm/a documented on the Ruahine Fault in the south. This northward decrease in the strike-slip rate is associated with a steepening in the pitch of the slip vectors by up to 60° (i.e. from 10° to 69°W) and a 20° shallowing in the mean dip of the fault (i.e. from 80–90° to 60°W). These changes in slip vector pitch are associated with a *c.* 50° anticlockwise rotation of the slip azimuth, and also a 50% decrease in the net-slip rate from 1.5 ± 0.5 mm/a in the south, to 0.7 ± 0.2 mm/a in the north (site 4 in Table 1).

Whakatane–Mohaka Fault. The Whakatane Fault is presently the fastest moving of the splays in the northern NIFS and, like the Waiohau Fault, it experiences a change in its kinematics towards its northern end (Figs 4 & 6). In the south, at Ruatahuna (see Fig. 4 for location), the Whakatane Fault strikes 020°, has a relatively straight trace and is a strike-slip-dominated fault with an average *H*:*V* ratio of *c.* 6.5. Dextral offsets of *c.* 4.5 to 90 m were observed with respect to eight channels, five spurs and one river terrace riser (sites 30–48 in Table 1). At site 32, for example, 3D-trenching across an abandoned channel incised into mainly pre-Holocene deposits (Fig. 7), indicates a dextral offset of 22 ± 2 m. Based on trenching of the channel, we suggest that the youngest deposits that predate incision of the channel are about 10 ka old (Rotoma tephra, calibrated 9.5 ka BP, Table 2), whilst a radiocarbon date of 5728–5586 calibrated years BP from a peat layer close to the base of channel infill deposits indicates that the channel was abandoned about 6 ka ago (Fig. 7). From these ages, we estimate the age of the offset channel to be 8 ± 2 ka which, in combination with strike-slip of 22 ± 2 m, indicates a dextral slip-rate of *c.* 3 ± 1 mm/a.

At the northern end of the fault, within 10 km of the coastline, the fault consistently dips to the west and has a significant component of normal dip-slip. Normal displacements are indicated by numerous prominent scarps that are down to the west (e.g. Fig. 5b); by a normal slip recorded in a fault trench (site 25 in Table 1); and by a westward throw of at least 660 m on the top of the basement inferred from modelling of a gravity survey (site 16 in Table 1; this study). Accordingly, the average dip-slip rate on this part of the fault is 1.5 ± 0.5 mm/a (sites 17–21 in Table 1). Dip-slip rates at these localities are greater than associated strike-slip rates of *c.* 1.1 ± 0.5 mm/a during the last *c.* 30 ka. In summary, the strike-slip rate along the Whakatane Fault decreases from *c.* 3 ± 1 mm/a in the south, to *c.* 1 mm/a some 65 km to the north. This decrease in strike-slip rate is accompanied by an increasing component of dip-slip rate from *c.* 0.2 in the south to *c.* 1.5 mm/a near the coast. Despite this change in the individual components of slip, the magnitude

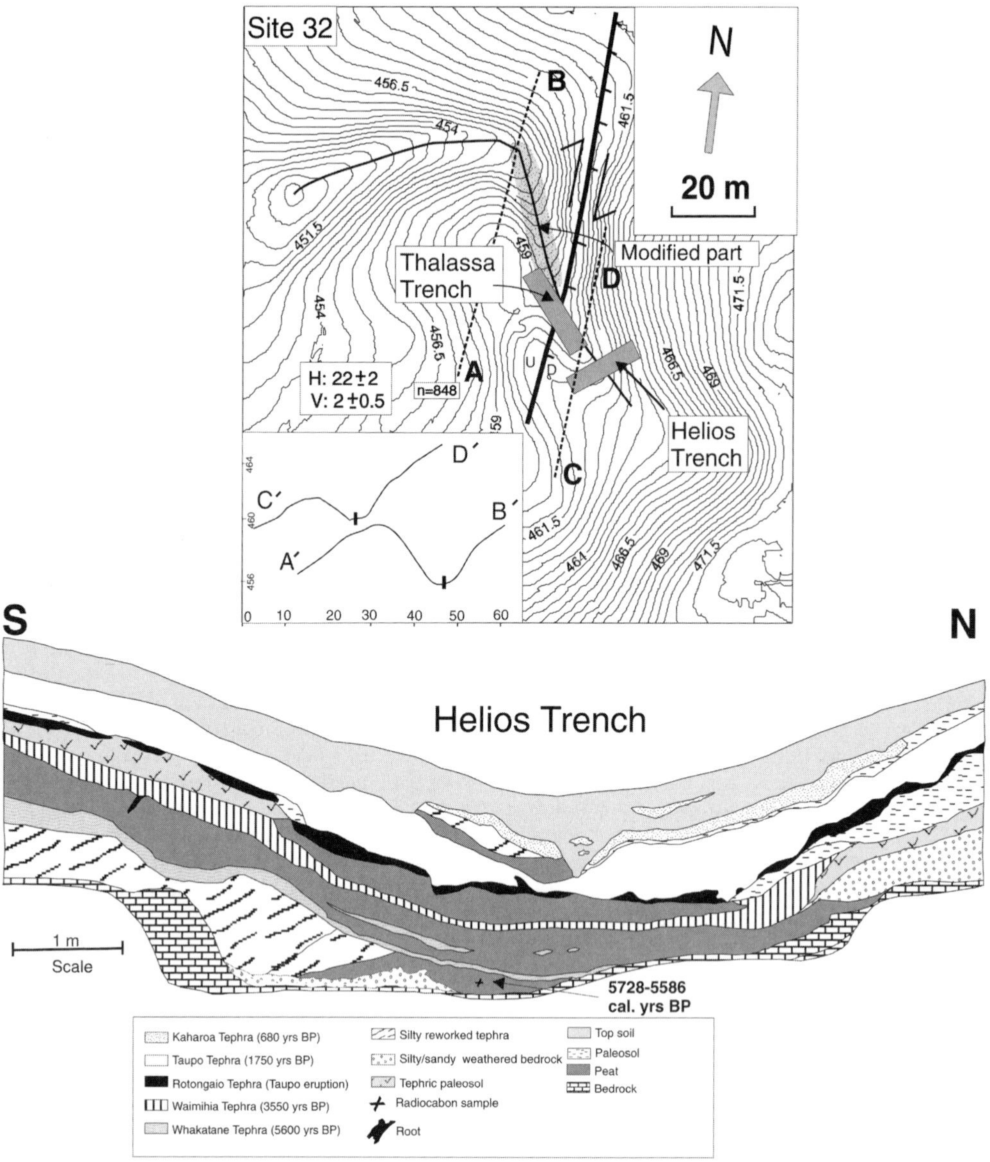

Fig. 7. Detailed contour map of site 32, Ruatahuna, where the Whakatane Fault dextrally displaces an abandoned stream channel by 22 ± 2 m. Tephra stratigraphy revealed by trenching across and parallel to the fault helped to determine a slip rate of 3 ± 1 mm/a for this section of the fault (see the text for further discussion). The log of the west wall of the stream crossing trench 'Helios', excavated normal to the stream channel (see site map for location), shows the record of the volcanic and sedimentary layers deposited in the abandoned channel.

of the net-slip rate remains approximately constant along the northernmost 100 km of the Whakatane Fault (Table 1).

As expected, the changing kinematics along the Whakatane Fault are expressed by a northward steepening in the pitch of measured slip vectors (Table 1). In the south these are subhorizontal (i.e. 0–10°; sites 50 and 53 in Table 1), whereas further north, close to the coast, they steepen to a *c.* 55° NW pitch (sites 17 and 20). This northward transition occurs gradationally over a strike distance of *c.* 60 km (see also Fig. 6a from Mouslopoulou

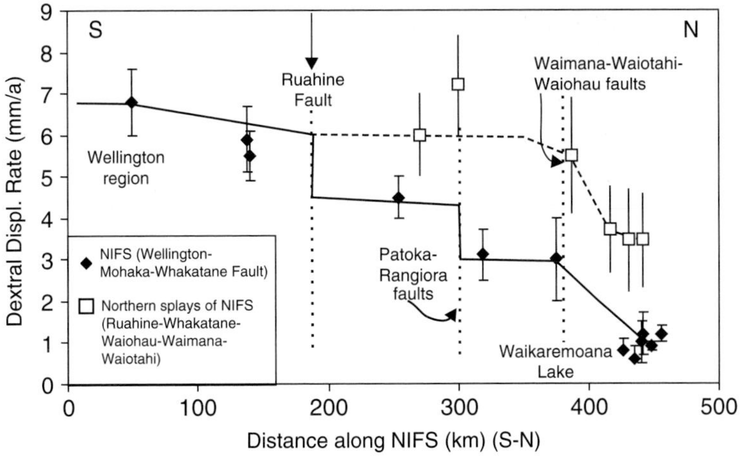

Fig. 8. Displacement profile along the Wellington–Mohaka–Whakatane Fault (filled rhombs) and its splays to the north (open squares). Each of the inferred steps on the profile coincides with a branch point and the transfer of some displacement onto additional fault splays. Note that the total strike-slip across the NIFS decreases northward. Fault-slip rate data from: Hull (1983); Raub (1985); Beanland & Berryman (1987); Cutten *et al.* (1988); Ota *et al.* (1988); Beanland (1989*b*); Beanland *et al.* (1989); Beanland (1995); Van Dissen & Berryman (1996); Berryman *et al.* (1998); Hanson (1998); Kelsey *et al.* (1998); Villamor & Berryman (2001); Schermer *et al.* (2004); Langridge *et al.* (2005).

et al. 2007) in conjunction with a northward shallowing in the dip of the faults (from 80–90° to *c.* 60° down to the west) and is reflected as a 50° anticlockwise deflection in the slip azimuth.

Waimana Fault. The Waimana Fault, which splays eastward from the Whakatane Fault, is the second-most active fault within the northern NIFS (Figs 4 & 6). Within the onshore study region, the Waimana Fault is predominantly strike-slip. Near the coastline and offshore, however, it appears to become increasingly dip-slip (Davey *et al.* 1995; this study).

Along the northernmost 35 km of its onshore trace, the Waimana Fault strikes approximately N–S and has fault scarps that face either to the east or west (Fig. 4). The fault traverses a diachronous landscape with displaced surface features ranging in age from <5.6 ka to 23 ka (Table 1). We observed dextral displacements of five channels, three spurs and three river terraces by offsets of 5.5–24 m (Figs 4 & 5a, sites 62–79 in Table 1). The average *H*:*V* ratio along this section of the fault is *c.* 6.5, with slip vectors plunging at *c.* 10°N on a fault that dips >80° (e.g. sites 64, 68, 76 in Table 1). The fault has a uniform Late Quaternary strike-slip rate of *c.* 1 mm/a over an onshore strike-distance of *c.* 35 km. Dip-slip rates are typically only *c.* 0.1–0.2 mm/a (Table 1).

Along its northernmost 50 km, the Waimana Fault accumulates increasing normal displacement as its dip decreases (sites 57 and 58 in Table 1). Gravity and seismic-reflection profiles acquired at the coast (site 58 in Table 1) and some 40 km offshore (site 57 in Table 1) respectively, suggest that the kinematics and geometries of the fault in the north, near the rift, differ from those observed along its onshore trace further south (Davey *et al.* 1995; this study). For example, the *c.* 500 m cumulative throw on the top of the greywacke basement during the past *c.* 600 ka (sites 57 and 58 in Table 1), suggests an increase in the dip-slip rate of the Waimana Fault from 0.1–0.2 mm/a onshore, to *c.* 0.8 mm/a offshore. This increase in the normal displacement occurs in conjunction with a decrease of the fault-dip from >80° in the south to *c.* 60°W offshore (Table 1).

The changing kinematics, from onshore to offshore along the Waimana Fault, are manifest as a northward-steepening in the pitch of its Late Quaternary slip vectors (Table 1). The pitch of the slip vectors onshore, in the south, is subhorizontal (i.e. 10°N) (sites 65, 72 and 73 in Table 1), and, based on the documented increase of the normal dip-slip rate, we infer that the pitch of the slip vectors increases significantly into the offshore region, in conjunction with an anticlockwise deflection in the azimuth of these vectors.

Waiotahi Fault. The Waiotahi Fault splays eastward from the Waimana Fault and, similar to the Waimana Fault, does not change kinematics in the

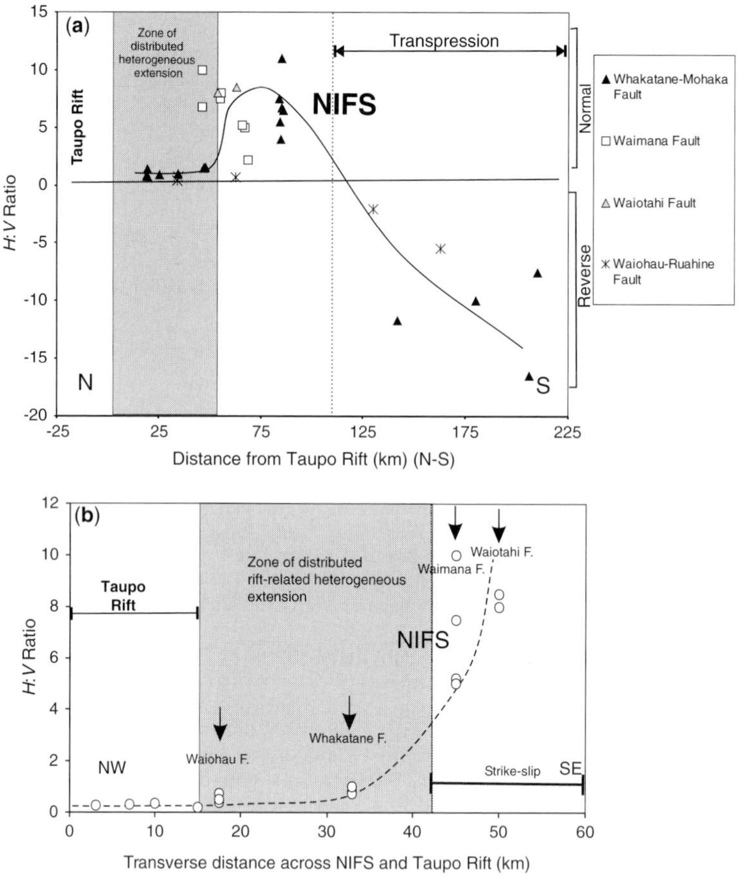

Fig. 9. Horizontal to vertical (*H*:*V*) slip ratios for each of the faults in the NIFS determined using displaced landforms. The data suggest that the ratio of strike-slip to dip-slip decreases, both, along (**a**) and across (**b**) the NIFS towards its intersection with the Taupo Rift. Regional distributed extension outside the rift is indicated by grey shading. See Figure 4 for the location of the *H*:*V* transect across the two fault systems.

onshore study area, although it does appear to become more dip-slip in the offshore region proximal to its northern termination. Onshore and some 10–20 km south of the coastline, the Waiotahi Fault strikes N–S, and is chiefly a strike-slip fault with an *H*:*V* slip ratio of *c.* 8 and dextral offsets of up to 17 m. These offsets are associated with horizontal and vertical slip-rates of *c.* 1 mm/a and <0.2 mm/a, respectively (sites 86 and 88 in Table 1 and Figs 4 & 6).

In the north, however, gravity modelling and seismic surveys suggest a gradual increase in the normal (down-to-the-west) displacement on a relatively shallow (i.e. 60°) fault plane near the coast and offshore (Davey *et al.* 1995; this study). As expected, these changes are reflected in the gradual northward increase of the dip-slip rate on the Waiotahi Fault from <0.2 mm/a some 20 km south of the coastline, to *c.* 0.5 mm/a at the coast and up to 0.7 mm/a about 40 km north of the coastline and close to its northern end (sites 83–85 in Table 1; Davey *et al.* 1995; this study).

In a similar manner to the Waimana Fault, the northward change in kinematics on the Waiotahi Fault is associated with corresponding deflections in the pitch of the Late Quaternary slip vectors. The pitch of the slip vectors in the south is sub-horizontal (i.e. 7°N) (sites 86 and 88 in Table 1), and we infer that the increase of the normal dip-slip rate offshore to the north is accompanied by a steepening in the pitch of the slip vectors on the Waiotahi Fault, together with an anticlockwise deflection of its slip azimuth.

Extension rates

Strike-slip rates aggregated across all splays of the NIFS decrease from *c.* 5.5 mm/a in the south, near Lake Waikaremoana, to *c.* 4 mm/a close to the Bay of Plenty coastline (Figs 3 & 8). This decrease in the horizontal slip-rate is accompanied by an increase in the aggregated dip-slip rate across all strands of the NIFS from $\leq$0.5 mm/a in the south to *c.* 2.5 mm/a at the coast (Table 1). These changes in cumulative strike-slip and dip-slip rates produce a net-slip rate aggregated across all fault strands which, within the uncertainties, is approximately constant for the northern 150 km of the NIFS. For example, the 5.5 $\pm$ 1.8 mm/a of aggregated net-slip rate across the coastal section is comparable with the 6 $\pm$ 2 and 7.3 $\pm$ 1 mm/a estimated for the Waikaremoana and central Hawke's Bay slip-rate transects, respectively (Table 1 & Fig. 3).

Figure 9 illustrates how the *H*:*V* slip ratio and the component of rift-orthogonal NW–SE extension decreases and increases, respectively, along and across the NIFS towards the Taupo Rift. Figure 9a indicates a progressive kinematic transition on the Whakatane–Mohaka Fault (black triangles) from minor transpression (strike-slip with a minor reverse component) in the south, to minor transtension (strike-slip with a minor normal component) and to oblique-normal slip in the north, over a strike distance of *c.* 150 km. Similarly, Figure 9b suggests that oblique-normal slip is accommodated on the Waiohau and Whakatane faults outside the rift, whilst the onshore Waimana and Waiotahi faults are principally strike-slip. These relationships indicate that the component of dip-slip within the NIFS increases northwestwards towards the Taupo Rift. The SE transition from oblique-normal slip to strike-slip across the NIFS, and the corresponding increase in *H*:*V* slip ratio (Fig. 9b), has led us to map a kinematic boundary which trends subparallel to the margin of the rift and at *c.* 45° to the NIFS (shaded zone in Fig. 4), and separates a region of oblique extension to the NW from one dominantly strike-slip to the SE.

The presence of normal oblique-slip in the northern NIFS indicates that some rift-orthogonal extension is distributed outside the main rift. This extension, which here is inferred to have formed in association with NW–SE rifting across the crust of the Taupo Rift, increases towards the rift (Fig. 9a & b) and covers an area that is at least 150 km long (i.e. parallel to the rift) and up to 50 km wide (i.e. perpendicular to the SE margin of the rift) (Fig. 4). Rift-parallel normal faulting in the region to the SE of the topographically defined rift margin is supported by offshore seismic-reflection data acquired across the Taupo Rift (Davey *et al.* 1995; Lamarche *et al.* 2006; see northern extension transect in Fig. 4), which show that zones of minor normal faults occur beyond the main rift. The rate of NW–SE extension to the SE of the main rift margin is *c.* 1 mm/a (Davey *et al.* 1995; this study) and appears to be comparable with rates of extension to the NW of the main rift (Davey *et al.* 1995), accounting for between *c.* 6 and 15% of the estimated total 7–15 mm/a extension across the greater Taupo Rift (Davey *et al.* 1995; Darby *et al.* 2000; Villamor & Berryman 2001; Wallace *et al.* 2004) (see extension transects in Fig. 4).

Fault termination

At the termination of the NIFS, these strike-slip faults are not orthogonal to the Taupo Rift, and therefore the far-field slip azimuths on the component faults are non-parallel. As the margins of the rift do not appear to have been displaced by the strike-slip faults, the termination of these faults cannot be attained by rigid block translations. Instead, at their northern termination, the faults in the NIFS undergo a change in their slip vectors, such that these become parallel to the line of their intersection with the Taupo Rift. This parallelism allows the intersecting fault systems to be kinematically coherent and the displacement to be transferred into the rift without any offset of the margin of the rift. The present quasi-stability on this triple junction may have been aided by an earlier phase of anticlockwise vertical-axis rotations of the northern tip of the faults in the NIFS relative to the southern part of the fault system and the Taupo Rift (e.g. Fig. 10a), and currently it is conditioned by an additional swing in the slip vectors along each fault that takes place in proximity to the rift margin (Fig. 10c). The inferred vertical-axis fault rotations (25° change in fault strike near Lake Waikaremoana) and changes of the slip vectors due to heterogeneous distributed extension may account for about one-third and two-thirds of the total 75° required swing in slip azimuths, respectively. In the following sections, we document how the termination of the northern NIFS has included aspects of fault bending and bifurcation, as well as the currently operating swing in their slip vectors that is permitted by distributed and heterogeneous off-fault extension orthogonal to the rift.

Vertical-axis fault-block rotations

Vertical-axis rotations of rocks proximal to the tips of large strike-slip faults have been widely discussed as a mechanism for their termination

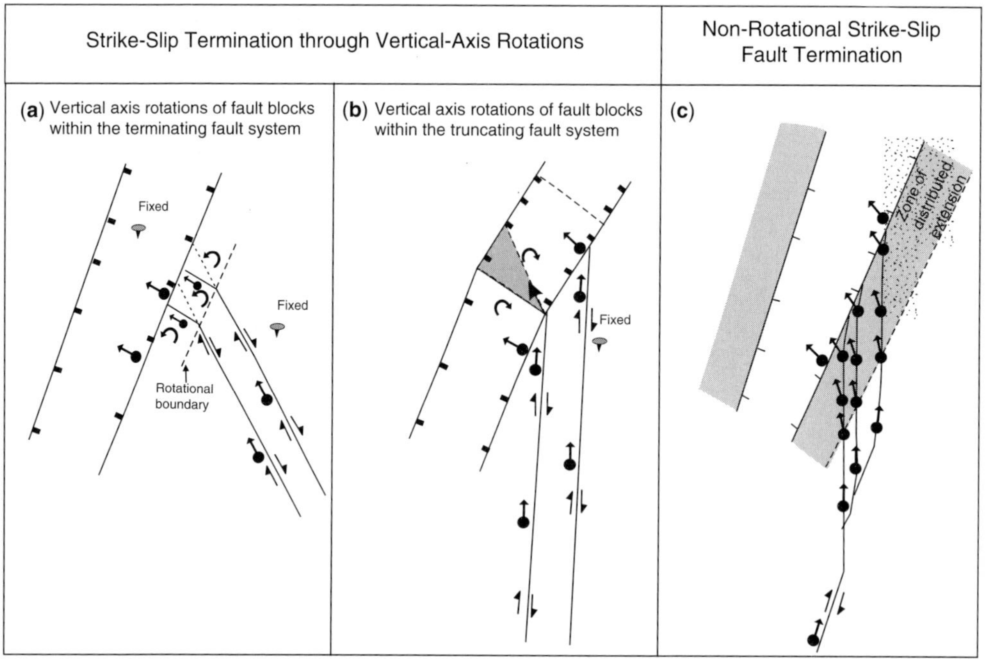

Fig. 10. Schematic end-member termination models for regional strike-slip faults that terminate against another fault or fault system. Termination can be achieved through either: (**a**) rotations about vertical axis of the crust enclosed between the tip of the fault and the truncating fault or fault system (i.e. Little & Roberts 1997); (**b**) rotations about vertical axis of the crust enclosed within the truncating fault or fault system (i.e. Taymaz *et al.* 1991); and/or (**c**) superimposition of heterogeneous extension onto strike-slip, resulting in changes in the slip vectors of the strike-slip fault (this study). Regional distributed extension outside the rift is indicated by grey shading. The dotted pattern indicates higher strain rates. Arrows attached to filled circles represent slip vector azimuths.

(Jackson & McKenzie 1984; McKenzie & Jackson 1986; Taymaz *et al.* 1991; Little & Roberts 1997). Figures 10a & b illustrate termination of a strike-slip fault by vertical-axis rotations of the truncated fault system relative to the through-going one. A fault termination mechanism similar to the end-member illustrated in Figure 10a is documented on the NE tip of the strike-slip Clarence Fault in New Zealand (Little & Roberts 1997), whereas a mechanism similar to the one illustrated in Figure 10b is proposed for central Greece, where clockwise rotations of the NW-trending normal faults accommodate the oblique right-lateral motion documented on the NE-trending faults of the North Anatolian Fault and allow abrupt changes in the type of faulting (Taymaz *et al.* 1991).

Anticlockwise vertical-axis rotations of the northern NIFS fault blocks relative to the southern part of the fault system and the Taupo Rift may have produced the abrupt 25° anticlockwise bend (i.e. from NNE–SSW to N–S) in the strike of the NIFS *c.* 150 km south of its intersection with the Taupo Rift (Figs 3 & 4). Palaeomagnetic data from sites to the east of the bend in the NIFS suggest that the change in fault strike may have resulted from differential vertical-axis rotation of the faults on either side of the bend (Walcott 1989; Thornley 1996; Nicol *et al.* 2007). Such a change in strike of the faults might produce a comparable deviation in the strike-slip azimuths without any change in the pitch of those vectors. This appears to be the case in the NIFS. In addition, fault rotation is facilitated by the bifurcation of the strike-slip fault. Bifurcation reduces displacement gradients on individual faults, and transfers displacements from faults into the adjacent blocks which subsequently are not rigid. A consequence of such a vertical-axis rotation, therefore, would have been to decrease the divergence in the slip azimuths between the two intersecting fault systems from *c.* 75° to *c.* 50°, and thus to increase the compatibility of fault motions between the rift and the NIFS by bringing the strike of the northern end of the NIFS more parallel with its line of intersection with the rift. Global Positioning System (GPS) data provide no evidence that the differential rotation required to produce the observed 25° anticlockwise bend in the NIFS is still operating today

(Wallace *et al.* 2004; Nicol & Wallace 2007). As this relict anticlockwise bend in the fault strike occurred in the opposite sense to the one expected for a dextral strike-slip fault that terminates through 'horsetail type' splaying, here we argue that this rotation about vertical axis might have taken place during an initial stage of the development of the NIFS–Taupo Rift intersection, when the crust was thicker and colder, in order to increase the compatibility between the fault motions. As the rift evolved, the crust in proximity to the Taupo Rift–NIFS junction became thinner through stretching and associated normal faulting, and the termination of the strike-slip fault occurred in conjunction with heterogeneous extension distributed outside the main margin of the Taupo Rift.

Slip vector changes due to distributed heterogeneous extension

Data from the NIFS suggest that the azimuth and pitch of slip vectors on each of the main faults in this system change by up to 50° from south to north, approaching the fault-system termination (i.e. Fig. 10c). Here we suggest that these adjustments in slip vectors on the NIFS are produced by heterogeneous NW–SE-directed extension distributed across the off-fault regions to the SE of the rift margin. How these changes in extension produce the observed changes in fault slip vectors can be visualized with the aid of Figure 11, in which the termination of the NIFS and the velocities of the three blocks (AUST, RAUK and AXIR) adjacent to the fault system tip are shown. The boundaries between these three blocks are the NIFS (between the RAUK and AXIR blocks) and Taupo Rift (between the AUST and RAUK and the AUST and AXIR blocks). The Australian Block is assumed to be fixed. Within the rift, extension increases to the NE and this increase is primarily taking place across the main faults in the NIFS (Wallace *et al.* 2004; Mouslopoulou *et al.* 2007). Within the RAUK Block, the highest rates of extension outside the rift occur close to the main rift-bounding fault (black arrows within the stippled region in Fig. 11), whereas the rates of extension to the SE of the Taupo Rift are distributed more evenly across the AXIR Block (white arrows in Fig. 11). The velocity vectors (with respect to a fixed Australian block) across the rift and within the RAUK & AXIR blocks are shown schematically in Figure 11. In proximity to the SE margin of the Taupo Rift, in the area of localized high strain-rate on the RAUK block, the differential extension between the RAUK and the AXIR blocks requires a local component of extension on the transcurrent fault separating these blocks. This extension

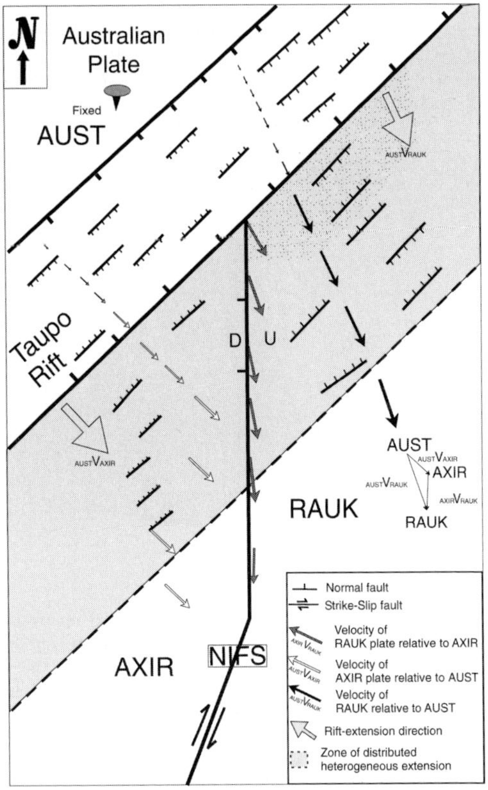

Fig. 11. Schematic cartoon illustrating deformation and velocity vectors for the three blocks (i.e. Australian (AUST), Raukumara (RAUK) and Axial Ranges (AXIR)) enclosing the NIFS termination. Regional distributed differential extension outside the rift is indicated by grey shading. The velocity field of each block relative to the fixed Australian Plate is indicated. Note that the rates of extension are different, and unevenly distributed, in the footwall and hanging-wall blocks (RAUK and AXIR) of the NIFS (the dotted pattern indicates higher strain rates). Also note that all faults of the NIFS are collapsed onto a single slip surface. The far-field velocity circuit is indicated. The differential extension between the RAUK and AXIR blocks reflects the oblique-slip documented in the northern NIFS.

introduces a net oblique slip on the faults of the NIFS (Fig. 12). Indeed, when each of the faults of the NIFS enters this zone of distributed differential extension, a component of normal dip-slip is added onto its strike-slip (Fig. 12). At each location along the NIFS and within this zone of distributed off-fault deformation, therefore, slip vectors are the sum of strike-slip and dip-slip, with the total angle through which the pitch of these vectors deflect being dependent on the relative displacements of the interacting fault systems outside the rift and

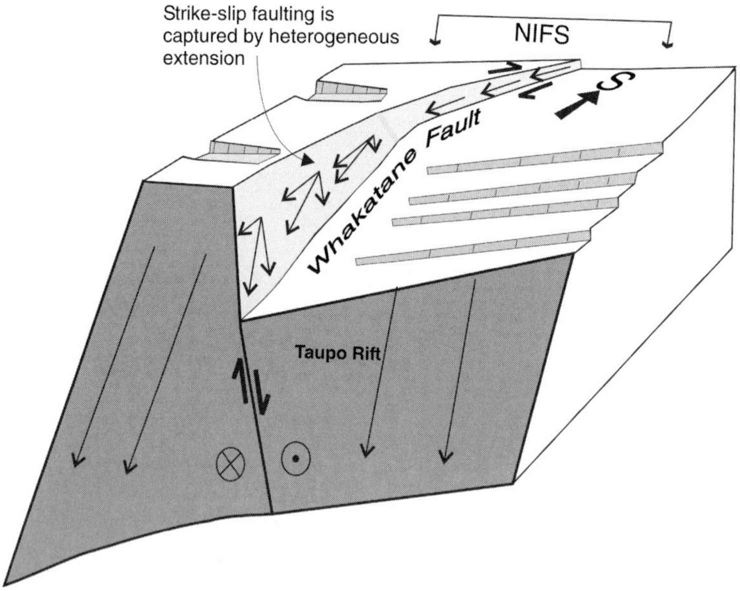

Fig. 12. Block diagram illustrating the gradual superimposition of rift-related dip-slip onto strike-slip at the northern termination of the NIFS. Given that extension and strike-slip rates increase and decrease towards the Taupo Rift, respectively, the pitch of the slip vectors steepens as the northern end of the fault is approached. The subparallel slip at the junction of the two fault systems results in the kinematic coherence of the termination.

close to the termination of the strike-slip fault system (Fig. 12). Because the strike-slip on these faults decreases to the north, and the normal dip-slip increases in this direction, the pitch of slip on the fault surfaces steepens towards the NIFS termination. The angular change in the pitch of slip vectors of the terminating fault system will depend on a number of factors, including the difference in strike of the two intersecting fault systems, and the relative values of strike-slip and extension accommodated by the terminating fault. If, for example, no differential extension were accommodated across the northern NIFS outside the rift, then the slip vectors on the strike-slip faults would not change in magnitude or orientation. In such cases the strike-slip fault would most likely offset the margin of the rift. Therefore, a key element of our model is that significant NW–SE extension is distributed in a heterogeneous way outside the rift. This heterogeneous extension, which is associated with the formation of the weak Taupo Rift, therefore captures the pre-existing strike-slip faults of the NIFS.

For the NIFS, each of the splays accommodates 1–2 mm/a of normal dip-slip rate and *c.* 1 mm/a of strike-slip rate close to their northern tips (Table 1, Fig. 4). About 50 km south of the termination of the NIFS, the rate of normal dip-slip is approximately zero, whilst the strike-slip rate is 1–3 mm/a (Table 1, Fig. 4). The change in the rates of strike-slip and dip-slip along the main faults of the NIFS, results in significant northward steepening (by up to *c.* 50°) of the pitch of the slip vectors on the fault plane and in a northward 50° anticlockwise swing in the slip azimuth (Fig. 13). These changes of the slip vectors take place onshore for the Waiohau and Whakatane faults, and offshore for the Waimana and Waiotahi faults. The change in slip vector orientations in the NIFS results in the subparallelism of the slip azimuths on the intersecting fault systems (Fig. 13). Slip vectors are also oriented subparallel to the line of intersection of the planes of the two fault systems, permitting strike-slip transfer into the rift without lateral offset of the rift's marginal faults.

Termination of the NIFS may be assisted by fault bifurcation. Within 150 km of their northern end, the faults in the NIFS bifurcate, distributing fault slip across a wider area, and reducing the strike-slip on individual faults within the system. The Mohaka and Ruahine faults splay northwards into five subparallel slip surfaces (Fig. 4), resulting in the redistribution of the total *c.* 7 mm/a of strike-slip rate onto multiple slip surfaces, each accommodating 1–1.5 mm/a strike-slip. Decreases in strike-slip along the faults of the NIFS arising from fault

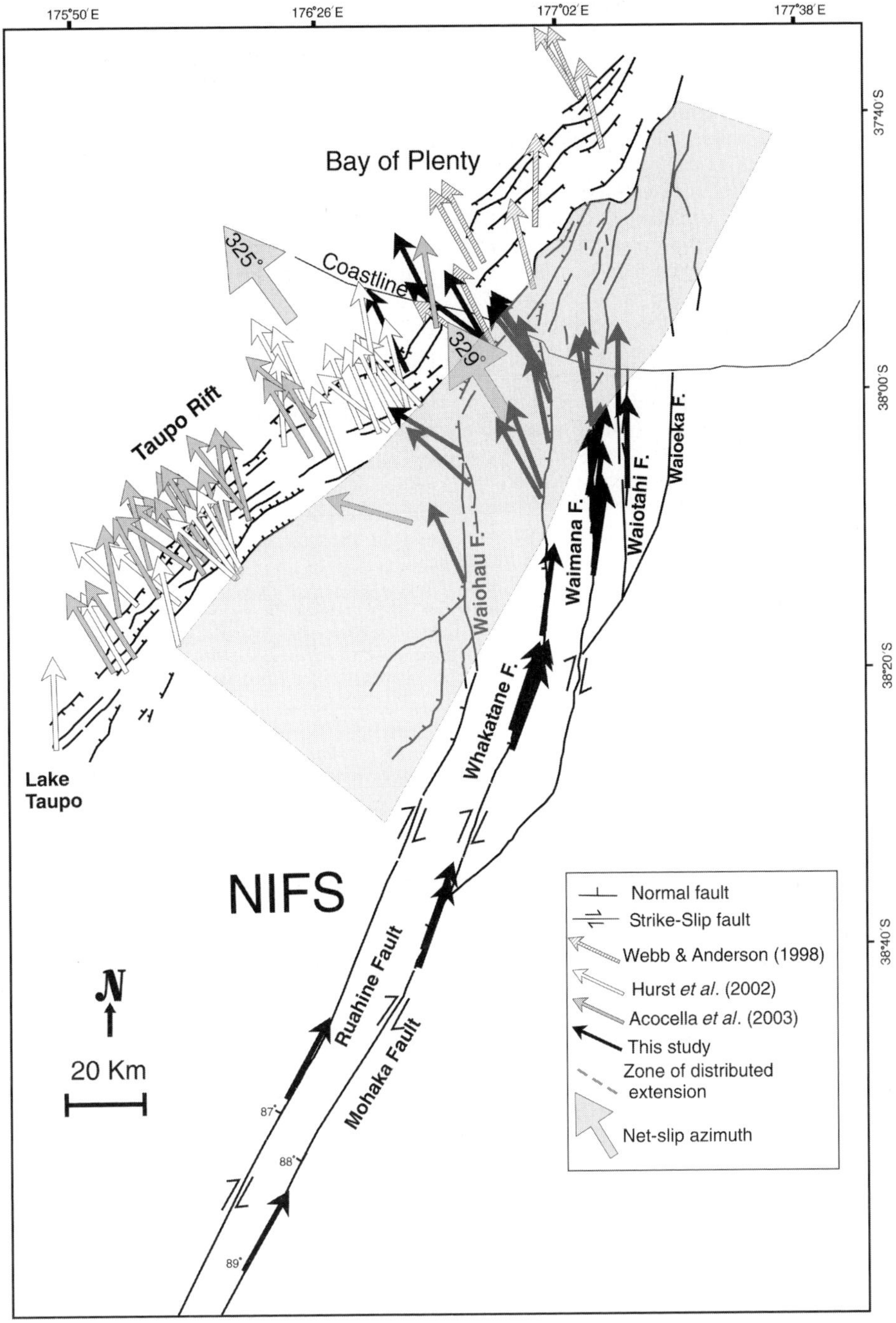

Fig. 13. Map showing the azimuths of the net-slip vectors for the NIFS (Hull 1983; Raub 1985; Hanson 1998; this study) and the Taupo Rift (Webb & Anderson 1998; Hurst *et al.* 2002; Acocella *et al.* 2003), which have been derived from a combination of outcrop geology and focal mechanism solutions. Note the anticlockwise deflection in the azimuth of net-slip on the faults in the northern NIFS, such that these become subparallel to the slip azimuth documented on the faults within the rift. Regional distributed differential extension outside the rift is indicated by grey shading. Large grey arrows represent average values of net-slip azimuths for the onshore Taupo Rift and northern NIFS. Note that the two arrows are approximately parallel.

bifurcation may influence the degree to which the slip vector orientations and magnitudes change approaching the rift, by reducing the required intra-fault block strain rates. The potential influence of fault bifurcation can be rationalized if we consider that for each fault within the NIFS the rates of normal dip-slip close to the rift is approximately the same (*c.* 1–2 mm/a) and (within the uncertainties) appear not to vary with changes in the rates of strike-slip. Therefore, northward steepening in slip vector pitch is dependent on a reduction in strike-slip rate. Where the strike-slip rate is higher than the dip-slip, the change in the pitch of the slip vector will be less, and additional off-fault strain will be required. If, for example, the northern tip of the NIFS had not bifurcated and all 7 mm/a strike-slip was carried on a single displacement surface into the rift, then the *H*:*V* ratio would have remained high (e.g. 3.5–7), and the pitch of the slip vectors on the fault would not have steepened into subparallelism with the line of fault intersection. Thus, a single strike-slip fault would be more likely than a bifurcating fault to displace the margin of the rift. Moreover, the bifurcation of the fault results in the distribution of the strike-slip-related deformation over a larger rock volume, and reduces the requirements for strain accommodation on each of the displacement surfaces whilst the total displacement budget across the fault system is preserved. Lower displacement and displacement gradients on each splay would, therefore, decrease the volumetric problems that may arise at the intersection of two fault systems with non-parallel slip vectors.

Discussion

One question associated with the evolution of the NIFS is why it terminates to the north. To answer this question, we need to consider the regional kinematics of the North Island and their impact on the kinematics of the NIFS. Along the Hikurangi margin, the margin-parallel motion of the oblique subduction of the oceanic Pacific Plate beneath the continental Australian Plate is principally

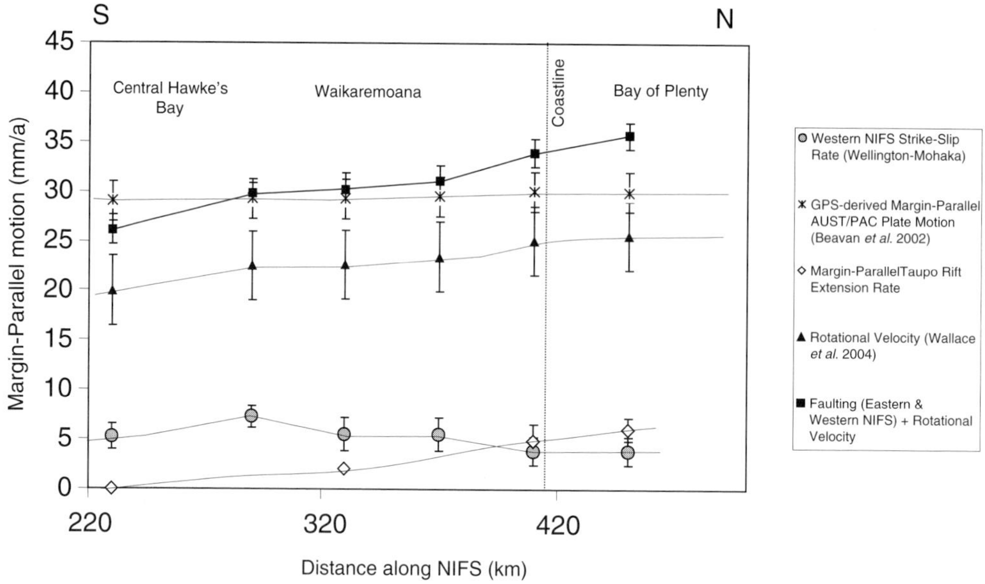

Fig. 14. Plot summarizing the horizontal kinematics of the northern North Island of New Zealand. Strike-slip rates for the NIFS are summed for the Wellington–Mohaka Fault and its northern splays (Ruahine, Waiohau, Whakatane, Waimana, Waiotahi). The calculated extension rate within the Taupo Rift is estimated for a margin-parallel direction. Rotational velocities show the margin-parallel component of relative plate motion accommodated by clockwise vertical-axis rotations (Wallace *et al.* 2004). Please refer to Wallace *et al.* (2004) for further discussion of conversion of vertical-axis rotation to margin-parallel motion in mm/a. The margin-parallel component of the relative Australian/Pacific plate motion is taken from Beavan *et al.* (2002). Onshore transects are shown in Figure 3, whilst offshore transects are given in Figure 4. Fault-slip rate data are from: Hull (1983); Raub (1985); Beanland & Berryman (1987); Cutten *et al.* (1988); Ota *et al.* (1988); Beanland (1989*b*); Beanland *et al.* (1989); Beanland (1995); Van Dissen & Berryman (1996); Berryman *et al.* (1998); Hanson (1998); Kelsey *et al.* (1998); Villamor & Berryman (2001); Schermer *et al.* (2004); Langridge *et al.* (2005).

accommodated by clockwise vertical-axis rotations (of the NE North Island relative to the stable Australian Plate) and strike-slip faulting in the overriding Australian Plate (Fig. 3; Walcott 1989; Wallace *et al.* 2004; Nicol *et al.* 2007). Figure 14 illustrates this interrelation between faulting and rotations, as derived by geological and GPS data respectively, by comparing the margin-parallel rate of the relative motion between Australian and Pacific plates with the strike-slip rate on the eastern and western strands of the NIFS, the rate of the extension in the rift which is parallel to the margin and the rotational velocity of the NE North Island relative to stable Australia at six margin-normal transects located at different latitudes (see Figs 3 & 4 for the location of the transects). Collectively, these data suggest that there is a gradual northward increase of the clockwise rotational velocity of the NE North Island relative to the stable Australian Plate, from *c.* 20 to *c.* 26 mm/a (Fig. 14). At the latitude of the NIFS termination, vertical-axis rotation of the NE North Island relative to the stable Australian Plate accounts for more than 80% of the total Pacific–Australian margin-parallel velocity accommodated across the entire actively deforming plate-boundary zone. These high rotational velocities documented in the NE North Island (Wallace *et al.* 2004) significantly decrease the requirement for strike-slip faulting in the upper plate. The northward increase in clockwise vertical-axis rotations of the NE North Island is accompanied by a decrease in strike-slip on the NIFS in the same direction, from *c.* 7 mm/a to *c.* 4 mm/a (Fig. 8). Despite the northward changes in the fault-slip rates and rotational velocities, the summation of those two components (rotational velocity and fault-slip rates) on each of the transects, compares favourably with the total margin-parallel motion between the Australian–Pacific plates. The slightly higher values of the summed slip and rotational velocities (as compared to the total margin-parallel relative plate motion) may be due to uncertainties in fault-slip and vertical-axis rotational velocities (Fig. 14).

Although the strike-slip fault system terminates at the edge of the rift, the *c.* 4 mm/a strike-slip does not, and is transferred into the rift. Because the strike-slip faults in the NIFS strike at 45° to the rift, the slip transferred from the NIFS is expected to produce about 3 mm/a strike-slip and extension within the rift. The additional extension contributed by the NIFS into the Taupo Rift, therefore, accounts for a significant component of the observed NE increase in extension rates within the rift, and it is illustrated in Figure 15 by the difference between the AUST/AXIR and AUST/

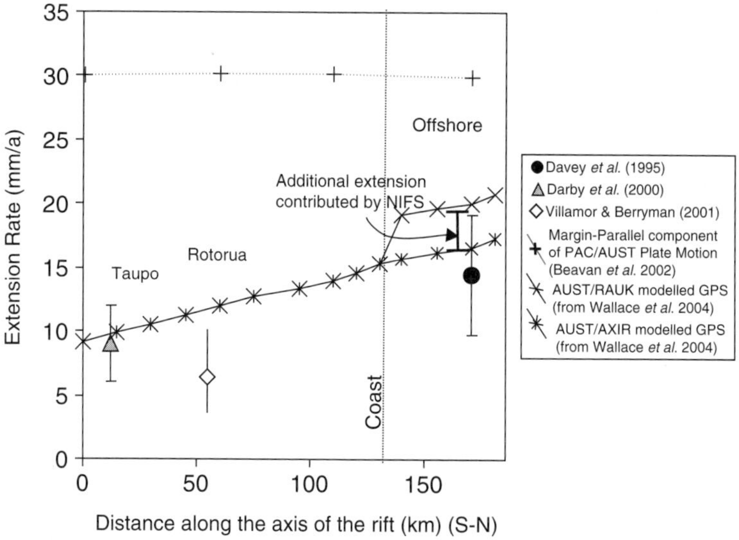

Fig. 15. Plot illustrating extension rates plotted against distance along the axis of the rift. Extension transects across the Taupo Rift and the NIFS are indicated in Figure 4. Both GPS and geological data indicate a northward increase of the extension rate (see the text for further discussion). The margin-parallel vector of the relative Pacific–Australian plate motion is plotted. Note that the extension rate associated with the NIFS accounts for a significant amount of the difference between the AUST/AXIR and AUST/RAUK relative plate motion (rates estimated from poles published in Wallace *et al.* 2004). The AUST/AXIR extension rates are calculated to be *c.* 30 km to the north of the northern end of AUST/AXIR block boundary in Wallace *et al.* (2004).

RAUK relative plate motion (rates estimated for the Euler poles published in Wallace *et al.* 2004). An additional effect of the increase of the extension along the rift and across its junction with the strike-slip faults of the NIFS, is the transition from subaerial to submarine rifting across the intersection zone in the Bay of Plenty.

The strike-slip fault termination accommodated by rift-orthogonal heterogeneous extension may be a generally applicable mechanism for strike-slip terminations globally. For this reason, it is worthwhile considering under what circumstances this non-rotational termination mechanism that involves slip vector changes may take place instead of fault-block rotations about vertical axis. The New Zealand example shows that the presence of some distributed heterogeneous off-fault deformation associated with the truncating fault (i.e. the rift) and the relativity in the rates of the two interacting velocity fields (i.e. rates of the truncating fault v. rates of the terminating strike-slip fault) may play important roles in the functionality of our model. If, for example, all deformation on the truncating fault was confined to a single slip surface, then the displacement fields of the intersecting faults would not overlap and the slip vector on the terminating fault would be constant up to the intersection. Therefore, the superimposition of the velocity fields helps to produce subparallel slip vectors at the fault intersection.

For superimposition of the velocity fields to significantly impact on the slip vector, however, the displacement rate of the strike-slip fault should not significantly exceed the rate associated with the distributed deformation on the truncating fault. The ratio between strike-slip and dip-slip rates controls the total angle through which the pitch of the slip vectors on the strike-slip faults deflects from the horizontal. In circumstances where the strike-slip rates are far larger than the rates of the truncating fault system, slip vector changes will be small and may not bring slip on the terminating fault into subparallelism with the line of fault intersection. In circumstances where slip vectors do not swing into subparallelism with the line of intersection, vertical-axis rotations of the terminating strike-slip fault relative to the truncating one may become important. The western termination of the strike-slip North Anatolian Fault against rotating blocks that are bounded by normal faults in central Greece (Taymaz *et al.* 1991; Fig. 10b) is such an example.

Moreover, the rheology of the crust may control whether fault rotations about a vertical axis would be the dominant termination mechanism as compared with non-rotational. Relative vertical-axis rotations between rigid fault blocks and abutting strike-slip fault systems are most likely to occur where the former consists of a cold and thick continental crust (McKenzie & Jackson 1986; Little & Roberts 1997; Fig. 10a & b). On the contrary, a strike-slip fault system transected by a rift that has hotter and thinner continental crust could terminate by distributed extension outside the rift, allowing the rift to ‘capture’ the strike-slip fault system by means of its differential extension. We anticipate that the potential for temporal transition from continental to oceanic crust during rifting will produce localization of the rift-related deformation and, therefore, narrower transition zones from strike-slip to dip-slip faulting (Mouslopoulou *et al.* 2007). The NIFS in New Zealand, however, may have employed both mechanisms, at different times in its evolution, in order to meet the requirements for subparallel slip at the intersection of the two fault systems.

Conclusions

The strike-slip North Island Fault System in New Zealand intersects the Taupo Rift at 45°, and terminates without displacing the margin of the rift. Examination of the northern end of the strike-slip fault system suggests that its termination is accomplished by bending and splaying of the main faults into numerous strands, and by changes in the trend and pitch of the slip vectors by up to 50°. Slip vectors progressively change on faults within the NIFS as they approach the rift. These changes arise because rift-orthogonal extension distributed outside the margins of the rift is accommodated differentially in the footwall and hanging-wall blocks of each fault in the NIFS. The combination of rift-related differential extension (dip-slip) and strike-slip results in a steepening of the pitch in the slip vectors on the terminating fault system. Interaction of the velocity fields of the two intersecting fault systems produces a northward gradual subparallelism of the slip vectors on the terminating fault system with the line of fault intersection. This subparallelism facilitates the transfer of strike-slip displacement into the rift whilst maintaining the overall coherence of the strike-slip termination. In cases where the superimposition of the two velocity fields does not provide subparallelism of the slip vectors of the two intersecting fault systems, fault rotations about vertical axes may take place. In our case, such rotations, that have taken place in the past, account for up to one-third of the termination mechanism, whilst today the rift-orthogonal extension, distributed into the fault blocks outside the rift and differentially accommodated across the northern NIFS, accounts for the primary termination mechanism. Although the NIFS terminates, its displacement is transferred into the Taupo Rift, resulting in a northeastward increase in rift-related extension.

This work is the result of a PhD study undertaken at the Victoria University of Wellington, New Zealand, funded by the Earthquake Commission of New Zealand and supported by the Foundation of Research Science and Technology of New Zealand (Enterprise Scholarship) and GNS Science. J. Begg provided a constructive review of an earlier version of this manuscript, which improved it significantly. We are grateful to F. Davey for providing us with raw seismic-reflection data. J. Rowland and P. Barnes are thanked for constructive reviews. Special thanks go to the QMAP team of GNS Science and H. Seebeck for assisting in the fieldwork, and to the Maori tribe of Tuhoe for welcoming us into their home: Te Urewera forest.

References

ACOCELLA, V., SPINKS, K., COLE, J. & NICOL, A. 2003. Oblique back-arc rifting of Taupo Volcanic Zone, New Zealand. *Tectonics*, **22**, 1045, doi:10.1029/2002TC001447.

BAYASGALAN, A., JACKSON, J., RITZ, J. F. & CARRETIER, S. 1999. Field examples of strike-slip fault terminations in Mongolia and their tectonic significance. *Tectonics*, **18**, 394–411.

BEANLAND, S. 1989*b*. Detailed mapping in the Matahina Dam region, New Zealand. *New Zealand Geological Survey Client Reports*, **89/8**, 45 pp.

BEANLAND, S. 1995. *The North Island dextral fault belt, Hikurangi subduction margin, New Zealand*. PhD thesis, Victoria University of Wellington, New Zealand.

BEANLAND, S. & BERRYMAN, K. R. 1987. Ruahine Fault reconnaissance. *New Zealand Geological Survey Reports*, **EDS 109**, 15 pp.

BEANLAND, S., BERRYMAN, K. R. & BLICK, G. H. 1989. Geological investigations of the 1987 Edgecumbe earthquake. *New Zealand Journal of Geology and Geophysics*, **32**, 73–91.

BEAVAN, J., TREGONING, P., BEVIS, M., KATO, T. & MEERTENS, C. 2002. Motion and rigidity of the Pacific Plate and implications for plate boundary deformation. *Journal of Geophysical Research*, **107**, 2261, doi:10.1029/2001JB000282.

BELLIER, O. & SÉBRIER, M. 1995. Is the slip rate variation on the Great Sumatran Fault accommodated by fore-arc stretching? *Geophysical Research Letters*, **22**, 1969–1972.

BERRYMAN, K. R., BEANLAND, S., CUTTEN, H. N. C., DARBY, D. J., HANCOX, G. T., HULL, A. G. & READ, S. A. L. 1988. Seismotectonic hazard evaluation for the Mohaka River Power Development. *New Zealand Geological Survey Contract Report*, 91 pp.

BERRYMAN, K. R., BEANLAND, S. & WESNOUSKY, S. 1998. Paleoseismicity of the Rotoitipakau Fault Zone, a complex normal fault in the Taupo Volcanic Zone, New Zealand. *New Zealand Journal of Geology and Geophysics*, **41**, 449–465.

BEU, A. G. 2004. Marine mollusca of oxygen isotope stages of the last 2 million years in New Zealand. Part 1: revised generic positions and recognition of warm-water and cool-water migrants. *Journal of the Royal Society of New Zealand*, **32**, 111–265.

CUTTEN, H. N. C., BEANLAND, S. & BERRYMAN, K. 1988. The Rangiora Fault, an active structure in Hawkes Bay. *New Zealand Geological Survey Record*, **35**, 65–72.

DARBY, D. J., HODGKINSON, K. M. & BLICK, G. H. 2000. Geodetic measurement of deformation in the Taupo Volcanic Zone, New Zealand: the north Taupo network revisited. *New Zealand Journal of Geology and Geophysics*, **43**, 157–170.

DAVEY, F. J., HENRYS, S. & LODOLO, E. 1995. Asymmetric rifting in a continental back-arc environment, North Island, New Zealand. *Journal of Volcanology and Geothermal Research*, **68**, 209–238.

DAVIS, G. A. & BURCHFIEL, C. B. 1973. Garlock fault: an intracontinental transform structure, southern California. *Geological Society of America Bulletin*, **84**, 1407–1422.

DEMETS, C. R., GORDON, R. G., ARGUS, D. & STEIN, S. 1994. Effect of recent revisions to the geomagnetic reversal time scale on estimates of current plate motions. *Geophysical Research Letters*, **21**, 2191–2194.

EDEN, N. D., PALMER, A. S., FROGGATT, P. C., TRUSTRUM, N. A. & PAGE, M. J. 1993. A multiple-source Holocene tephra sequence from Lake Tutira, Hawke's Bay, New Zealand. *New Zealand Journal of Geology and Geophysics*, **36**, 233–242.

FLEMING, C. A. 1955. Castlecliffian fossils from Ohope Beach, Whakatane (N69). *New Zealand Journal of Science and Technology Section*, **B36**, 511–522.

FREUND, R. 1974. Kinematics of transform and transcurrent faults. *Tectonophysics*, **21**, 93–134.

FREYMUELLER, J. T., MURRAY, M. H., SEGALL, P. & CASTILLO, D. 1999. Kinematics of the Pacific–North America plate boundary zone, Northern California. *Journal of Geophysical Research, B, Solid Earth and Planets*, **104**, 7419–7441.

FROGGATT, P. C. & SOLLOWAY, G. J. 1986. Correlation of Papanetu Tephra to Karapiti Tephra, central North Island, New Zealand. *New Zealand Journal of Geology and Geophysics*, **29**, 303–314.

HANSON, J. A. 1998. *The neotectonics of the Wellington and Ruahine faults between the Manawatu Gorge and Puketitiri, North Island, New Zealand*. PhD thesis, Massey University, New Zealand.

HAYWARD, B., COCHRAN, U. *ET AL*. 2003. Micropaleontological evidence for the Holocene earthquake history of the eastern Bay of Plenty, New Zealand. *Quaternary Science Reviews*, **23**, 1651–1667.

HULL, A. G. 1983. Trenching of the Mohaka Fault near Hautapu River, Hawkes Bay. *New Zealand Geological Survey Reports*, **File 831/26**.

HURST, W. A., BIBBY, H. M. & ROBINSON, R. R. 2002. Earthquake focal mechanism in the Central Volcanic Zone and their relation to faulting and deformation. *New Zealand Journal of Geology and Geophysics*, **45**, 527–536.

JACKSON, J. A. & MCKENZIE, D. P. 1984. Active tectonics of the Alpine–Himalayan Belt between western Turkey and Pakistan. *Geophysical Journal of the Royal Astronomical Society*, **77**, 185–264.

JACKSON, J., HAINES, J. & HOLT, W. 1995. The accommodation of Arabia–Eurasia plate convergence in Iran. *Journal of Geophysical Research*, **100**, 15 205–15 219.

JOFFE, S. & GARFUNKEL, Z. 1987. Plate kinematics of the circum Red Sea – a re-evaluation. *Tectonophysics*, **141**, 5–22.

KAMATA, H. & KODAMA, K. 1994. Tectonics of an arc–arc junction: an example from Kyushu Island at the junction of the Southwest Japan Arc and the Ryukyu Arc. *Tectonophysics*, **233**, 69–81.

KELSEY, H. M., CASHMAN, S. M., BEANLAND, S. & BERRYMAN, K. R. 1995. Structural evolution along the inner forearc of the obliquely convergent Hikurangi margin, New Zealand. *Tectonics*, **14**, 1–18.

KELSEY, H. M., HULL, A. G., CASHMAN, S. M., BERRYMAN, K. R., CASHMAN, P. H., TREXLER, J. H. & BEGG, J. G. 1998. Paleoseismology of an active reverse fault in a forearc setting: the Poukawa fault zone, Hikurangi forearc, New Zealand. *Geological Society of America Bulletin*, **110**, 1123–1148.

KIM, Y. S., ANDREWS, J. R. & SANDERSON, D. J. 2000. Damage zones around strike-slip fault systems and strike-slip fault evolution, Crackington Haven, southwest England. *Geosciences Journal*, **4**, 53–72.

LAMARCHE, G., BARNES, P. M. & BULL, J. M. 2006. Faulting and extension rate over the last 20,000 years in the offshore Whakatane Graben, New Zealand continental shelf, *Tectonics*, **25**, TC4005, doi: 10.1029/2005TC001886.

LAMARCHE, G., BULL, J. M., BARNES, P. M., TAYLOR, S. K. & HORGAN, H. 2000. Constraining fault growth rates and fault evolution in New Zealand. *EOS*, **81**, 485–486.

LANGRIDGE, R. M., BERRYMAN, K. R. & VAN DISSEN, R. J. 2005. Defining the geometric segmentation and Holocene slip rate of the Wellington Fault, New Zealand: the Pahiatua section. *New Zealand Journal of Geology and Geophysics*, **48**, 591–607.

LELGEMANN, H., GUTSCHER, M.-A., BIALAS, J., FLUEH, E., WEINREBE, W. & REICHERT, C. 2000. Transtensional basins in the western Sunda Strait. *Geophysical Research Letters*, **27**, 3545–3548.

LE PICHON, X. & FRANCHETEAU, J. 1976. Plate Tectonics. Elsevier, New York.

LITTLE, T. A. & ROBERTS, A. P. 1997. Distribution and magnitude of Neogene to present-day vertical-axis rotations, Pacific–Australia plate boundary zone, South Island, New Zealand. *Journal of Geophysical Research*, **102**, 20 447–20 468.

LOWE, J. D. 1988. Stratigraphy, age, composition, and correlation of late Quaternary tephras interbedded with organic sediments in Waikato Lakes, North Island, New Zealand. *New Zealand Journal of Geology and Geophysics*, **31**, 125–165.

LOWE, J. D. & HOGG, A. G. 1986. Tephrostratigraphy and chronology of the Kaipo Lagoon, an 11,500 year-old montane peat bog in Urewera National Park, New Zealand. *Journal of the Royal Society of New Zealand*, **16**, 25–41.

LOWE, J. D., NEWNHAM, M. R. & WARD, M. C. 1999. Stratigraphy and chronology of a 15 Ka sequence of multi-sourced silicic tephras in a montane peat bog, eastern North Island, New Zealand. *New Zealand Journal of Geology and Geophysics*, **42**, 566–579.

MANNING, D. A. 1995. *Late Pleistocene tephrostratigraphy of the eastern Bay of Plenty, North Island, New Zealand.* PhD thesis, Victoria University of Wellington, Wellington, New Zealand.

MCKENZIE, D. & JACKSON, J. A. 1986. A block model of distributed deformation by faulting. *Journal of the Geological Society, London*, **143**, 349–353.

MOLNAR, P. & LYON-CAEN, H. 1989. Fault plane solutions of earthquakes and active tectonics of the Tibetan plateau and its margins. *Geophysical Research International*, **99**, 123–153.

MORTIMER, N. 1994. Origin of the Torlesse Terrane and coeval rocks, North Island, New Zealand. *International Geology Review*, **36**, 891–910.

MOUSLOPOULOU, V., NICOL, A., LITTLE, T. A. & WALSH, J. J. 2007. Displacement transfer between intersecting strike-slip and extensional fault systems. *Journal of Structural Geology*, **29**, 100–116.

NAIRN, I. A. & BEANLAND, S. 1989. Geological setting of the 1987 Edgecumbe earthquake, New Zealand. *New Zealand Journal of Geology and Geophysics*, **32**, 1–13.

NAIRN, I. A., SHANE, P. R., COLE, J. W., LEONARD, G. J., SELF, S. & PEARSON, N. 2004. Rhyolite magma process of the AD 1315 Kaharoa eruption episode, Tarawera volcano, New Zealand. *Journal of Volcanology and Geothermal Research*, **131**, 265–294.

NEWNHAM, R. M., LOWE, D. J. & WIGLEY, G. N. A. 1995. Late Holocene palynology and paleovegetation of tephra-bearing mires at Papamoa and Waihi Beach, western Bay of Plenty, North Island, New Zealand. *Journal of the Royal Society of New Zealand*, **25**, 283–300.

NICOL, A. & WALLACE, L. M. 2007. Temporal stability of deformation rates: comparison of geological and geodetic observations, Hikurangi subduction margin, New Zealand. *Earth and Planetary Science Letters*, doi: 10.1016/j.epsl.2007.03.039.

NICOL, A., MAZENGARB, C., CHANIER, F., RAIT, G., URUSKI, C. & WALLACE, L. 2007. Tectonic evolution of the active Hikurangi subduction margin, New Zealand, since the Oligocene. *Tectonics*, **26**, TC4002, doi:10.1029/2006TC002090.

NORRIS, R. J. & COOPER, A. F. 2000. Late Quaternary slip rates and their significance for slip partitioning on the Alpine Fault, New Zealand. *Journal of Structural Geology*, **23**, 507–520.

OTA, Y., BEANLAND, S., BERRYMAN, K. R. & NAIRN, I. A. 1988. The Matata Fault: active faulting at the north-western margin of the Whakatane graben, eastern Bay of Plenty. *New Zealand Geological Survey Record*, **35**, 6–13.

PILLANS, B. & WRIGHT, I. 1992. Late Quaternary tephrostratigraphy from the southern Havre Trough–Bay of Plenty, northeast New Zealand. *New Zealand Journal of Geology and Geophysics*, **35**, 129–143.

QUEBRAL, R. D., PUBELLIER, M. & RANGIN, C. 1996. The onset of movement on the Philippine Fault in eastern Mindano: a transition from a collision to a strike-slip environment. *Tectonics*, **15**, 713–726.

RAUB, M. L. 1985. *The neotectonic evolution of the Wakarara area, Hawke's Bay*. MSc thesis, Geology Department, University of Auckland, New Zealand.

ROWLAND, J. V. & SIBSON, R. H. 2001. Extensional fault kinematics within the Taupo Volcanic Zone, New Zealand: soft-linked segmentation of a continental rift system. *New Zealand Journal of Geology and Geophysics*, **44**, 271–283.

SCHERMER, E. R., VAN DISSEN, R. J., BERRYMAN, K. R., KELSEY, H. M. & CASHMAN, S. M. 2004. Active faults, paleoseismology, and historical fault rupture in northern Wairarapa, North Island, New Zealand. *New Zealand Journal of Geology and Geophysics*, **47**, 101–122.

SHANE, P. & HOVERD, J. 2002. Distal record of multi-sourced tephra in Onepoto basin, Auckland, New Zealand: implications for volcanic chronology, frequency and hazards. *Bulletin of Volcanology*, **64**, 441–454.

TAYLOR, S. K., BULL, J. M., LAMARCHE, G. & BARNES, P. M. 2004. Normal fault growth and linkage in the Whakatane Graben, New Zealand, during the last 1.3 Myr. *Journal of Geophysical Research*, **109**, B02408.

TAYMAZ, T., JACKSON, J. & MCKENZIE, D. 1991. Active tectonics of the north and central Aegean Sea. *Geophysical Journal International*, **106**, 433–490.

THORNLEY, R. S. W. 1996. *Tectonics of the Raukumara Peninsula, New Zealand*. PhD thesis, Victoria University of Wellington, New Zealand.

TSUTSUMI, H. & OKADA, A. 1996. Segmentation and Holocene surface rupture faulting on the Median Tectonic Line, southwest Japan. *Journal of Geophysical Research*, **101**, 5855–5871.

VAN DISSEN, R. J. & BERRYMAN, K. R. 1996. Surface rupture earthquakes over the last ~1000 years in the Wellington region, New Zealand, and implications for ground shaking hazard. *Journal of Geophysical Research. Solid Earth*, **101**, 5999–6019.

VAN DISSEN, R. & YEATS, R. S. 1991. Hope Fault, Jordan Thrust, and uplift of the seaward Kaikoura Range, New Zealand. *Geology*, **19**, 393–396.

VILLAMOR, P. & BERRYMAN, K. 2001. A late Quaternary extension rate in the Taupo Volcanic Zone, New Zealand, derived from fault slip data. *New Zealand Journal of Geology and Geophysics*, **44**, 243–269.

WALCOTT, R. I. 1989. Paleomagnetically observed rotations along the Hikurangi margin of New Zealand. *In:* KISSEL, C. & LAJ, C. (eds) *Paleomagnetic Rotations and Continental Deformation*. Kluwer Academic Publishers, Dordrecht, 459–471.

WALLACE, L. M., BEAVEN, J., MCCAFFREY, R. & DARBY, D. 2004. Subduction zone coupling and tectonic block rotations in the North Island, New Zealand. *Journal of Geophysical Research*, **109**, 2406, doi:10.1029/2004JB003241.

WEBB, T. & ANDERSON, H. 1998. Focal mechanisms of large earthquakes in the North Island of New Zealand: slip partitioning at an oblique active margin. *Geophysical Journal International*, **134**, 40–86.

WILMHURST, J. M., EDEN, D. N. & FROGGATT, P. C. 1999. Late Holocene forest disturbance in Gisborne, New Zealand: a comparison of terrestrial and marine pollen records. *New Zealand Journal of Botany*, **37**, 523–540.

WILSON, C. J. N., HOUGHTON, B. F., MCWILLIAMS, M. O., LAMPHERE, M. A., WEAVER, S. D. & BRIGGS, R. M. 1995. Volcanic and structural evolution of Taupo Volcanic Zone, New Zealand: a review. *Journal of Volcanology and Geothermal Research*, **68**, 1–28.

Woodward-Clyde Ltd 1998. Matahina Dam strengthening project. Geological completion report. *GNS Science Report to Electricity Corporation of New Zealand*, NH 95/80.

ZACHARIASEN, J. & VAN DISSEN, R. 2001. Paleoseismicity of the northern Horohoro Fault, Taupo Volcanic Zone, New Zealand. *New Zealand Journal of Geology and Geophysics*, **44**, 391–401.

Segment linkage and the state of stress in transtensional transfer zones: field examples from the Pannonian Basin

L. I. FODOR

Geological Institute of Hungary Budapest, Stefánia út 14, H-1143, Hungary (e-mail: fodor@mafi.hu)

Abstract: Metre- to kilometre-scale en échelon strike-slip faults were mapped at the 1:10 000 scale in the Gánt mining area of the Vértes Hills in central Hungary. Good exposures allow detailed observation of brittle structures within the transfer zones of the overstepping fault segments. The strike-slip segments are subvertical to steeply dipping, with a rake of 20–30° accommodating a noticeable dip-slip. Displacement transfer between strike-slip faults was achieved through transtensional relay ramps, which represent a specific type of transfer zone. The breaching faults are oblique-normal or pure normal types. Their strike, dip and the obliquity of their striae change systematically. This occurred in order to accommodate extension across the relay ramp. Low-angle connecting faults occur at a high angle to the main fault, and form preferentially when the striae approach the orientation of the strike of the main fault. Slip vectors on the main and secondary connecting fault planes are subparallel, so defining a coherently moving hanging wall. A stress-inversion method was applied for the main fault segments, which characterize a state of stress of regional significance. Calculations were performed for the whole data-sets, including data for the guided connecting faults of the transtensional relay ramps, even though, in theory, stress inversion is not realistic for faults with a guided slip. Indeed, calculations gave a state of stress that is differs from that of the regional stress state. The results should be a warning to geologists that fault geometry has to be clarified in overstepping fault arrays before using the data-set for fault-slip inversion.

Brittle fault zones are composed of numerous fault segments, which may have slightly different strikes and dip angles, as demonstrated both in natural examples and in laboratory experiments. Fault segments commonly interact during their evolution; the development of one fault segment in one or more directions of fault strike and dip will be influenced by another fault segment. Whether segmented faults initiate or grow as a 'coherent fault array' (Walsh *et al.* 2003), or whether fault interaction develops later during 'individual fault propagation' (e.g. Morley *et al.* 1990; Trudgill & Cartwright 1994), is an ongoing debate.

Walsh & Watterson (1991) and Trudgill & Cartwright (1994) classified fault connectivity or linkage on the basis of the geometry and continuity of the overstepping fault segments. By definition, there is no linkage of any form between unlinked faults. Between soft-linked fault segments, geometric continuity is achieved by continuous deformation, e.g. the reorientation of bedding, fault-related folds or various types of small-scale fractures, which are too small to be individually represented at the scale of observation. All of these structures occur between the overstepping fault segments in a transfer zone (Morley *et al.* 1990), a stepover (Woodcock & Schubert 1994), an overlap zone (Childs *et al.* 1995), a relay ramp (Larsen 1988), a relay zone (Huggins *et al.* 1995), or a linking damage zone (Kim *et al.* 2004); see the more extensive summary of definitions by Peacock *et al.* (2000). Among these terms, that of relay ramp has a particular meaning, referring to an area between overstepping faults, where reorientation (tilting) of strata occurs (Larsen 1988; Peacock & Sanderson 1991). The step from soft- to hard-linked faults corresponds with the appearance of breached relay ramps (Childs *et al.* 1993), which is the third step in the evolutionary scheme of relay ramps proposed by Peacock & Sanderson (1994). This step seems to be controlled by the ratio of relay displacement (fault displacement within the relay zone) and fault spacing (Soliva & Benedicto 2004; Acocella & Neri 2005). Finally, the development of discrete connecting faults across the transfer zones results in the formation of a composite hard-linked fault with a zigzag trace.

A classical type of strike-slip fault linkage is the development of a releasing double bend or pull-apart basin (Aydin & Nur 1982; Mann *et al.* 1983). Strike-slip relay ramps were documented by Peacock & Sanderson (1995) and modelled by McClay & Bonora (2001) from a transpressional setting, which occurs at restraining stepovers. In these cases the breaching strike-slip faults, if they occur, have a reverse-slip component. An overview of the structural characteristics of strike-slip linking damage zones is given by Kim *et al.* (2004). In these

From: CUNNINGHAM, W. D. & MANN, P. (eds) *Tectonics of Strike-Slip Restraining and Releasing Bends.* Geological Society, London, Special Publications, **290**, 417–431.
DOI: 10.1144/SP290.16 0305-8719/07/$15.00

cases, the overlapping faults are either subvertical or dip toward each other. However, a few studies describe stepovers in a transtensional setting, where strike-slip faults dip in the same direction and have a dominant strike-slip and a smaller normal-slip component. These transfer zones can be similar to certain normal-fault relay ramps (Peacock & Sanderson 1994) or oblique-normal relay ramps (Crider 2001), because of their identical fault dip direction, but they differ from these by the way in strike-slip is dominant.

This paper describes a set of well-exposed faults that show segmentation and linkage and medium-scale stepovers in a transtensional setting. Faults of metre to kilometre length are well exposed in the bauxite-mining area of Gánt village in the Vértes Hills of central Hungary. Almost completely three-dimensional views of the structures are possible, because of the mining activity that has removed soft sediments from the hanging wall. The transtensional strike-slip faults are hard-linked by moderately to gently dipping oblique-normal connecting faults. Kinematic indicators (slickenlines) were measured along the main fault segments and within the linking damage zones. Fault slip inversion was carried out using different groups of faults in order to demonstrate how the smaller connecting faults of the transfer zones modify the result of the palaeostress calculations.

Geological settings and brief structural evolution

The Vértes Hills of central Hungary are part of the Transdanubian Range (TR), an elevated ridge of Permian–Palaeogene rocks that stands above the surrounding Miocene sediments of the extensional Pannonian Basin (Fig. 1). The study area of *c.* 20 km^2 is located in the SE part of the Vértes Hills, on the Eastern Vértes Ridge (Fig. 2.) This NNE-striking range is built of Middle to Upper Triassic platform carbonates deposited on the passive margin of the Tethys (Taeger 1909; Budai *et al.* 2005). The Mesozoic sequence was deformed (folded, faulted) during several Mid-Cretaceous shortening phases (Tari 1994). During the period from the Late Cretaceous to the Early Eocene, the tilted Triassic carbonates underwent terrestrial denudation, subtropical peneplanation and karstification. The resulting subhorizontal, gently undulating surface represents the reference plane for structural analysis. Depressions of this low-relief surface were infilled with bauxite, considered to be Late Palaeocene to Early Eocene in age (Mindszenty *et al.* 1989). The bauxite is covered by a Middle Eocene brackish to marine transgressive series (Bignot *et al.* 1985). Younger formations are missing from the Eastern Vértes Ridge. Late Miocene sediments occur both in the Gánt depression to the west and in the flatlands to the east.

The Pannonian Basin (including the elevated Vértes Hills) underwent several phases of Tertiary faulting, resulting in a dense network of faults of variable orientation. Stress axes gradually rotating in a clockwise sense governed the fault pattern (Bergerat 1989; Csontos *et al.* 1991; Fodor *et al.* 1999). Major episodes are exemplified by the typical stereograms of Figure 2b, and include: (1a) a Mid- to Late Eocene transpressional phase with WNW–ESE compression; (1b) reactivation of strike-slip faults in a similar stress regime during the Early Miocene; (2) a late Early to early Mid-Miocene main rifting phase; (3) a second phase of rifting in the late Mid-Miocene; and (4) Late Miocene–Pliocene transtension of the so-called 'post-rift phase'.

Strike-slip faults were formed in each phase, although their importance has varied. Phase 1 was dominantly marked by the formation of a conjugate set of strike-slip faults. The most important of them is the Mid-Hungarian shear zone (MHZ on Fig. 1), which accommodated *c.* 300 km dextral displacement (Tari 1994). Within the Transdanubian Range, sinistral faults strike NW–SE to NNW–SSE, whilst dextral ones are E–W to ENE–WSW-oriented. Within the Vértes Hills, the two major dextral strike-slip faults show 14 and 1.2 km of displacement respectively (Nagykovácsi and Gesztes faults, Fig. 1a, Balla & Dudko 1989 and Gyalog 1992, respectively) whilst sinistral faults show less than 1 km of slip. Strike-slip faults of the rifting and post-rift phases (2, 3 and 4) mainly accommodated displacement transfer between oppositely oriented grabens or differently extended blocks. Due to the changing palaeostress field, the Mid-Hungarian shear zone became a sinistral strike-slip zone during the second rifting and the post-rift phase (phases 3 and 4 on Fig. 2b) (Csontos *et al.* 1991).

Structures in the Gánt mining area

General characteristics

The faults of the Gánt mining area preserve signs of all of the regional deformation phases (Fig. 2). Superimposed or curved slickenlines document the kinematic changes from phase 1 to phases 3 or 4 (Almási 1993; Márton & Fodor 2003). Displacements during phase 1a reached only a few metres, but the faults provided pathways for fluids depositing Fe-crusts along fractures and the basal Eocene unconformity (Germán-Heins 1994; Fodor *et al.* 2005). However, some faults have not experienced kinematic changes; thus, presumably they were active during only one phase.

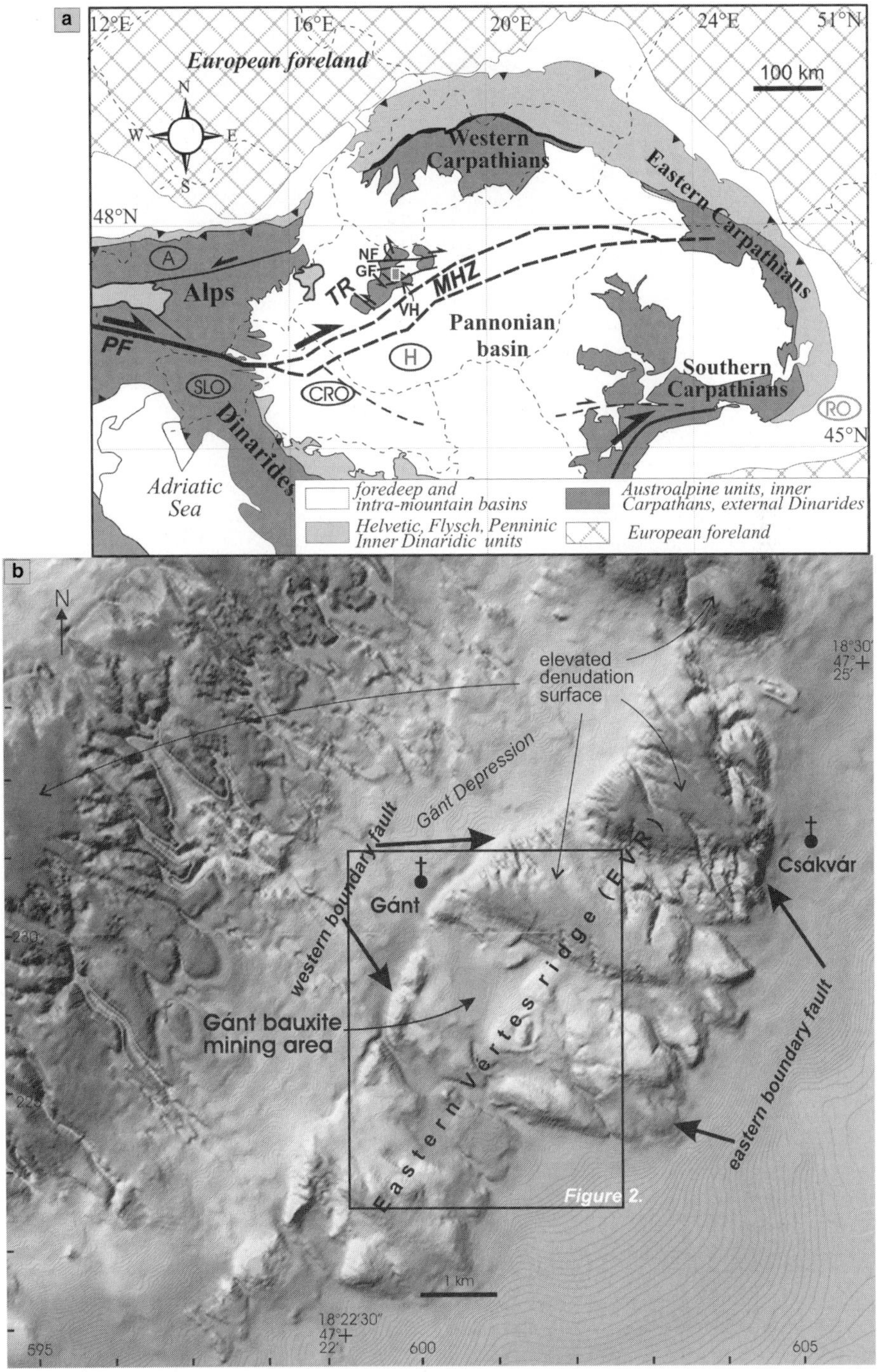

Fig. 1. (**a**) Location of the Vértes Hills (VH; central Hungary) in the context of the Alpine–Carpathian Orogen and the Pannonian Basin. The major strike-slip zones of Eocene to Early Miocene age include the Periadriatic Fault (PF) and the Mid-Hungarian shear zone (MHZ) (dashed when covered by younger sediments). Smaller faults within the Transdanubian Range (TR) are also shown, such as the Nagykovácsi and Gesztes faults (NF, GF). The white box (VH) indicates the location of Figure 1b. (**b**) Digital terrain model of the SE Vértes Hills, using 50 × 50 m grid (courtesy of the Hungarian Military Mapping Office). Numbers correspond with kilometres in the Hungarian Unified Grid.

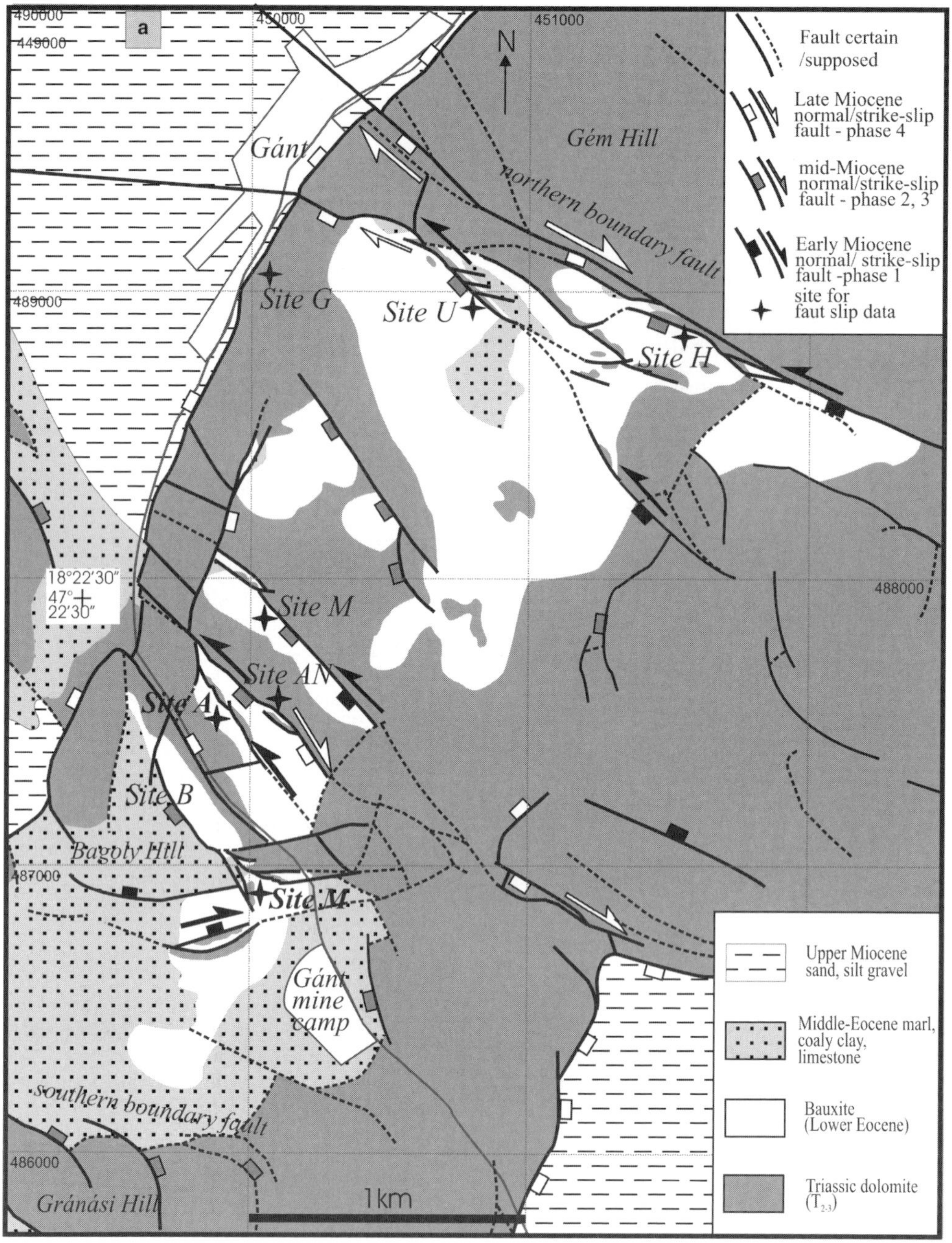

Fig. 2. Geological setting of the study area. (**a**) Geological map of the Gánt mining area. Quaternary cover and mining waste are omitted for clarity. Note that the formation boundaries are partly affected by the present shape of the mines. Numbers indicate the Hungarian Unified Grid in metres. (**b**) Tertiary evolution of the stress field shown by typical stereograms of calculated stress axes and faults measured in the study area. The box under the stereograms shows the main symbols used in Figures 2 & 5. A lower-hemisphere projection of the Schmid net is used for all the stereograms. $\Phi = \sigma_2 - \sigma_3/\sigma_1 - \sigma_3$; ANG indicates the average and standard deviation of misfit angles between the measured data and ideal striae. At the lower left-hand corner, the numbers show the slickensided faults used in the calculations, as wall as the other fractures used.

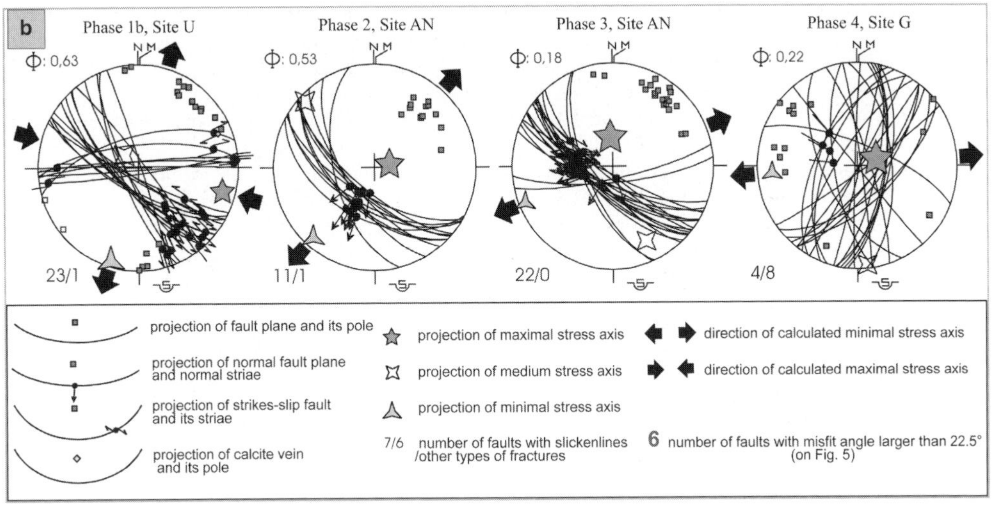

Fig. 2. (*Continued*)

Fourteen NW- to W-striking major faults, whose exposed length is greater than 0.5 km, dissect the Gánt bauxite mining area. Two of them represent the northern and southern boundary faults, the footwall of which has no Eocene cover but has an elevated denudation surface (see the Gém and Gránási hills in Fig. 2). Both boundary fault zones can be followed across the whole Eastern Vértes Ridge, whilst the traces of the shorter faults disappear after less than 1 km, where both the hanging wall and footwall have the same lithology and the faults lack morphological expression, or are covered by dumped mine waste. The mining area is limited by a NNE-striking fault to the west, whilst the eastern boundary is more segmented. NNE- and NW-striking faults are mutually cross-cutting; the northern boundary fault displaces the NNE-striking fault in the village of Gánt, while the NW-trending fault at site M (Meleges) terminates against a NNE-trending fault (Fig. 2). This observation suggests that NNE- and NW-striking faults were active at the same time during the youngest Late Miocene to Pliocene faulting phase 4.

Fault planes are almost completely exposed along 0.2–2.2 km length within the open-cast mines. Due to the excellent outcrop conditions, fine details of the fault planes can be seen. The faults mostly dip to the SW, and the intervening blocks are slightly tilted to the NE, demonstrated by the gentle dip of the strata and the locally preserved Eocene marine cover in the hanging wall immediately to the SW.

The fault pattern is characterized by overstepping fault segments with a wide range of lengths from metres to kilometres. Branch faults frequently occur near the fault tips, representing horsetail splays (e.g. near the village of Gánt). The largest fault array is located between sites U (Újfeltárás) and H (Harasztos, Fig. 2). The three main left-stepping segments bound 130–200-m-wide stepovers. Although the linking damage zones between the main segments are not perfectly exposed, faults seem to be hard-linked with dominantly E–W-striking, curved connecting faults. The basal Tertiary surface (the 'horizontal reference frame') is SE dipping at about 5–10° within the stepovers, which is a typical feature in relay ramps (Larsen 1988). The main faults at sites H and U have three slip events, belonging to phases 1, 2 and 4 respectively (see Fig. 2). Because most of the displacement was accumulated during phase 2 (normal slip on the main fault planes) the majority of the deformation within the stepovers is attributed to phase 2. The presence of connecting fault splays, the dominant dip-slip character and the tilt all suggest that the stepovers are hard-linked relay ramps (*sensu* Walsh & Watterson 1991) or belong to stage 3 (breached relay ramps) in the evolutionary scheme of Peacock & Sanderson (1994) and of Peacock (2002). However, slickenlines demonstrate that the main fault segments were initiated as sinistral strike-slip faults (site U in Fig. 2b), and that the stepovers might have already started to form during the early strike-slip phase 1. In this case, the stepovers might have been initiated as a strike-slip relay ramp (in the sense of Peacock & Sanderson 1995) and have changed later to a normal relay ramp. Nevertheless, the initial strike-slip relay ramp was different from that described by Peacock & Sanderson (1995), because their examples come from contractional stopovers, whilst the Gánt example is in a transtensional stepover.

In order to avoid such complexity, two fault arrays (sites M and A) were selected for detailed description, because they were not reactivated by a different stress regime after their formation in phase 1 (Eocene to Early Miocene). We will concentrate on those examples where the two overstepping faults are oblique-slip faults having a dominant strike-slip character, and the connecting fault splay is completely exposed.

Site A (Anger-rét Mine)

The fault zone of the Anger-rét Mine (site A) is composed of NW-trending en échelon segments. The dominant dip direction is to the SW, but locally changes to the NE, resulting in undulating fault planes in vertical cross-section (Fig. 3). However, the slickenside rake is constant, and ranges from 10° to 35° to the SE. Individual segments are 10–40 m in length, with a spacing of 1–8 m, whilst the complete fault array is exposed along *c.* 200 m (Fig. 3a & b); further away, the fault is covered by a mine-waste dump. Several fault-sense criteria (Petit 1987) indicate sinistral slip, such as well-developed Riedel fractures (Fig. 3c); displaced fragments at the end of the grooves; and shaded ridges behind the clasts. A few conjugate dextral faults have an east–west strike, with up to 10 cm displacement. Where observed, they post-date the main sinistral faults.

Overlap of fault segments may reach 1–10 m, but this value could be larger if one assumes the continuation of faults below the talus (see the dotted lines on Fig. 3a). In some cases, more than two fault segments show oversteps – a feature that can be called ‘multiple overstep’. Fault terminations in the hanging-wall sides of the oversteps are difficult to see because of a non-exploited thin basal layer of bauxite and/or scree cover, so exact measurement of the overlap length is prevented. In most cases, this cover also prevented the exact characterization of the stepovers. The good-quality hanging-wall bauxite was removed up to the fault contact, up to the thin basal poor-quality bauxite layer, or up to the Triassic dolomite. This may suggest that the lack of any observed connecting fault splays in most of the covered stepovers may actually represent soft-linked or partially breached strike-slip relay ramps.

Few exposed examples of segment linkage occur at the SE end of the fault array (Fig. 3d). Connecting fault splays are curved and gradually curve into the main fault planes, forming a continuous sinuous or slightly zigzag fault geometry. Stepovers have widths of 1–3 m. The most southeasterly fault block between segments is dissected by several connecting faults, which has resulted in brecciation (Fig. 3d). Breccia clasts and fault planes are coated by Fe-minerals, and slip has subsequently occurred on the surfaces. The brecciation, the dense network of fractures and the connecting faults all led to the tilt of the formerly subhorizontal basal surface of the bauxite.

In contrast to subvertical ‘main segments’, the connecting faults show moderate dips of 65–45° and their strike differs by 15–40° from that of the main strands. The change in strike is anticlockwise with respect to the main sinistral faults (Fig. 5). As prescribed for the model of Muller & Aydin (2004), this deviation promotes sinistral oblique-normal slip on the connecting fault planes. The striae on these fault planes are closely parallel to the intersection line of the main and the connecting faults. However, the striae on the main strike-slip faults have a tendency to be less inclined than the intersection line (Fig. 5).

Because the overstepping segments dominantly dip in the same direction (SW), the oversteps can be considered as synthetic transfer zones, using the terminology of Morley *et al.* (1990). On the other hand, the dominantly strike-slip character of the faults and the tilt of a formerly horizontal reference layer would suggest the presence of a strike-slip relay ramp. To distinguish it from the contractional strike-slip relay ramp of Peacock & Sanderson (1995), the structure can be termed a transtensional relay ramp. The strike-slip segments are hard-linked and are connected by an upper-ramp breach (Crider 2001).

Site M (Museum Fault)

The site M (or Museum Fault) is an E–W-striking fault array, which is exposed for 150 m below the Mining Museum. Farther to the east, the fault zone can be detected for another 400 m, whilst to the west it merges with the NW–SE-striking main fault of site B (Fig. 2). Individual fault segments are 15–100 m long, and are quite linear, with only local along-strike bends of <5 m length (Fig. 4a & b). Faults are subvertical to steeply south-dipping, although local undulations with a northerly dip exist.

The main faults have slickenlines with rakes of 20° and 30°. Kinematic indicators demonstrate a normal-dextral sense of slip (Almási 1993). A considerable dip-slip component of displacement resulted in staircase-lowering positions of the basal Tertiary discordance surface. Several overstepping faults occur in the hanging wall of the northern fault segment. One fault seems to be connected to the northern segment by oblique branch faults, and is part of a ‘transtensional strike-slip duplex’ around a fault-bound block. Other segments

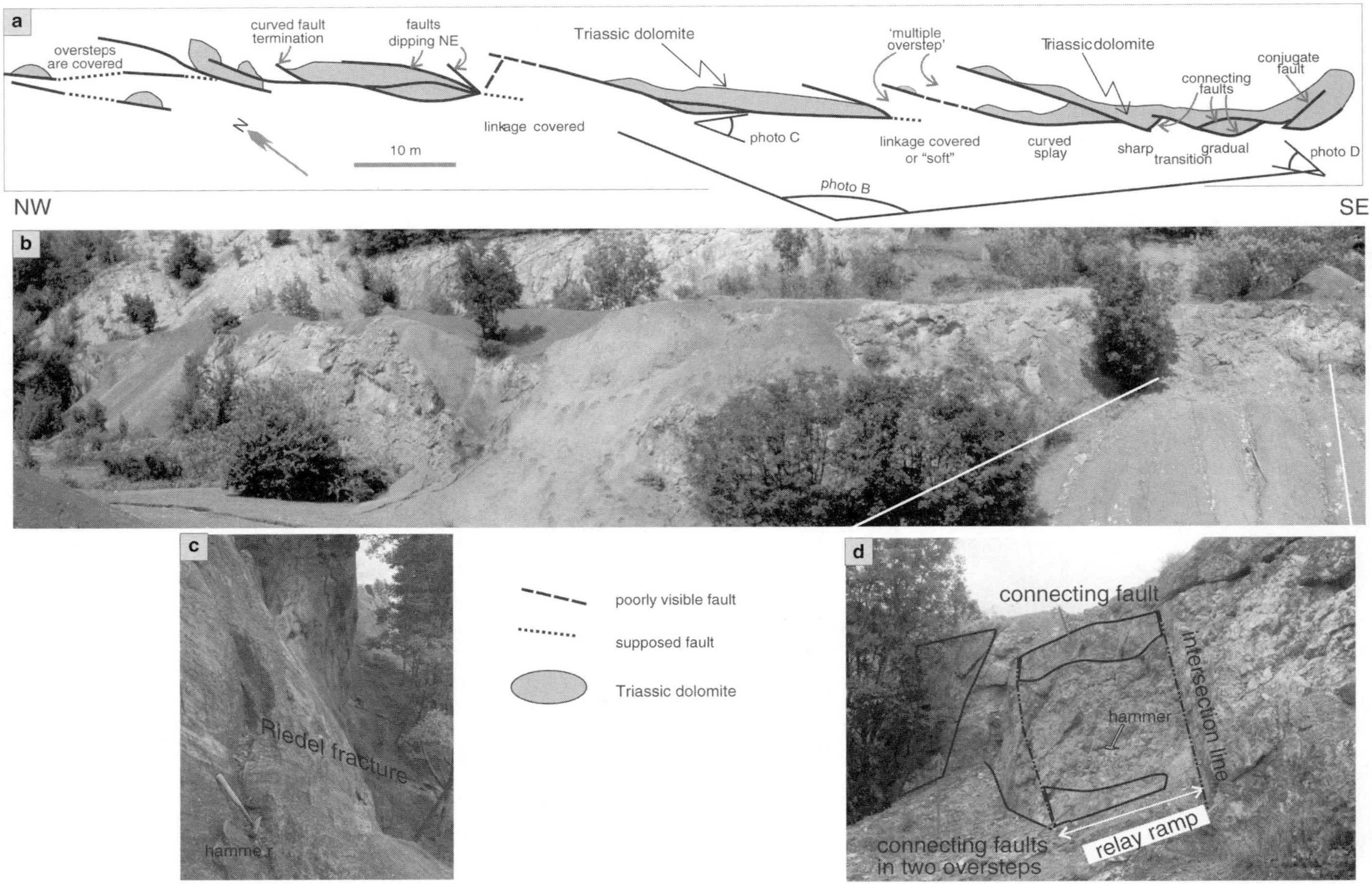

Fig. 3. Diagrams of the fault array at site A (Angerrét Mine). (**a**) Sketch map of the site A fault array. Note the segmented nature of the faults, overlaps and connecting structures. (**b**) A panoramic view of the fault array, viewed from the south. (**c**) Oblique sinistral Riedel shears and striae, looking down (SE) along the slip direction. (**d**) Strike-slip relay ramp at the SE end of the fault array, looking from SE to NW, along strike. Note the curved connecting faults; the deformed nature of the relay ramp; and the limonite encrustation. Hammer for scale.

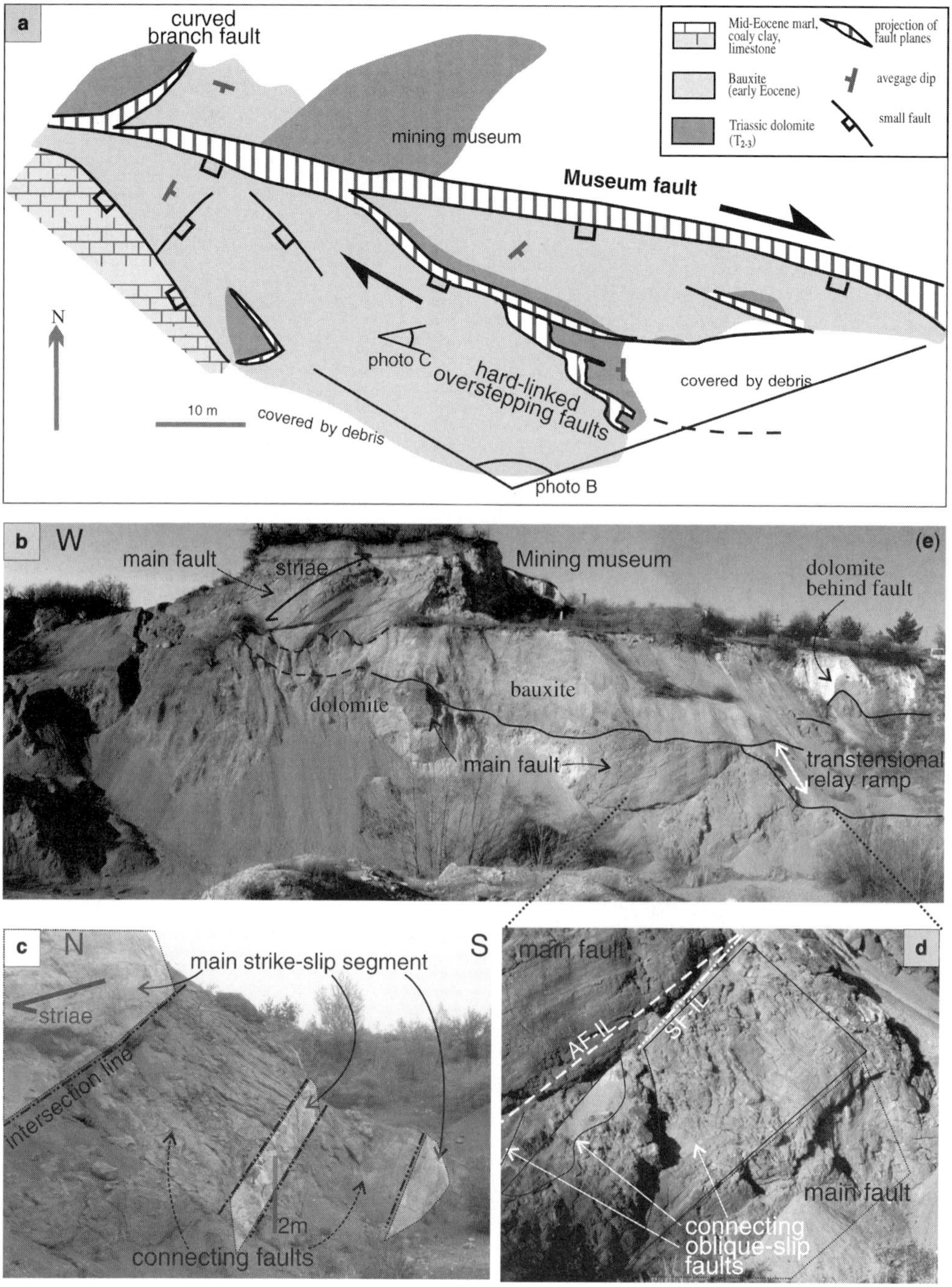

Fig. 4. Map and photographs of the fault array at site M (the Museum Fault). (**a**) Sketch map of the site M fault array. Note the segmented nature of faults, overlaps and connecting structures. (**b**) A panoramic view of the fault array, viewed from the south. (**c**) Relay ramp at the eastern end of the fault array, looking from west to east, along strike. Note the continuous geometry of the main and the connecting faults, and the deformed nature of the relay ramp. The grey bar near the geologist is for scale (2 m). (**d**) Strongly fractured rocks in the transtensional relay ramp along the connecting fault. CF–IL is the intersection line of the main and the connecting fault, and AF–IL is the intersection line of the main fault and the smoothed surface above all the fault-block edges of the connecting splays.

near the quarry level are overstepping, with overlaps of 4–8 m (Fig. 4).

These subvertical faults bound 3–6-m-wide transtensional relay ramps, which are strongly deformed (Fig. 4c). Relay ramps are cut through by one major connecting fault splay, which forms a hard linkage between the fault segments (Almási & Fodor 1995). In the hanging wall of this connecting fault, the transtensional relay ramps are pervasively dissected by smaller faults, and are strongly brecciated (Fig. 4d). Connecting splays curve into the main planes, forming a continuous but zigzag-shaped fault plane.

Connecting faults have strikes of 30–70° (up to 90°) clockwise from the main strike-slip segments. Dip angle varies from 60 to 40°, but locally can be 30°. Dip and strike vary systematically with respect to the main faults, with the dip shallowing for larger strike deviations (Fig. 5).

Connecting faults have oblique dextral-normal or pure normal kinematics. The ratio of dip-slip and strike-slip components varies with the connecting fault strike, with larger strike deviation associated with smaller rake. The intersection lines of the main fault with individual connecting splays (CF–IL) are close to the orientations of striations on secondary faults. However, a small angular difference (of up to 15°) can be observed between the fault intersection lines and the striae of the main faults (Fig. 4d). This geometry will be analysed further in the Discussion section.

State of stress from fault-slip inversion

Inversion of fault-slip data was applied to approximate the stress states that characterized the faulting. Calculations were carried out separately for the 'main faults' and for all of the 'secondary fault segments', and then for the whole data-set. The 'main faults' represent the principal structural elements, which are planar, with lengths larger than 10 m, and

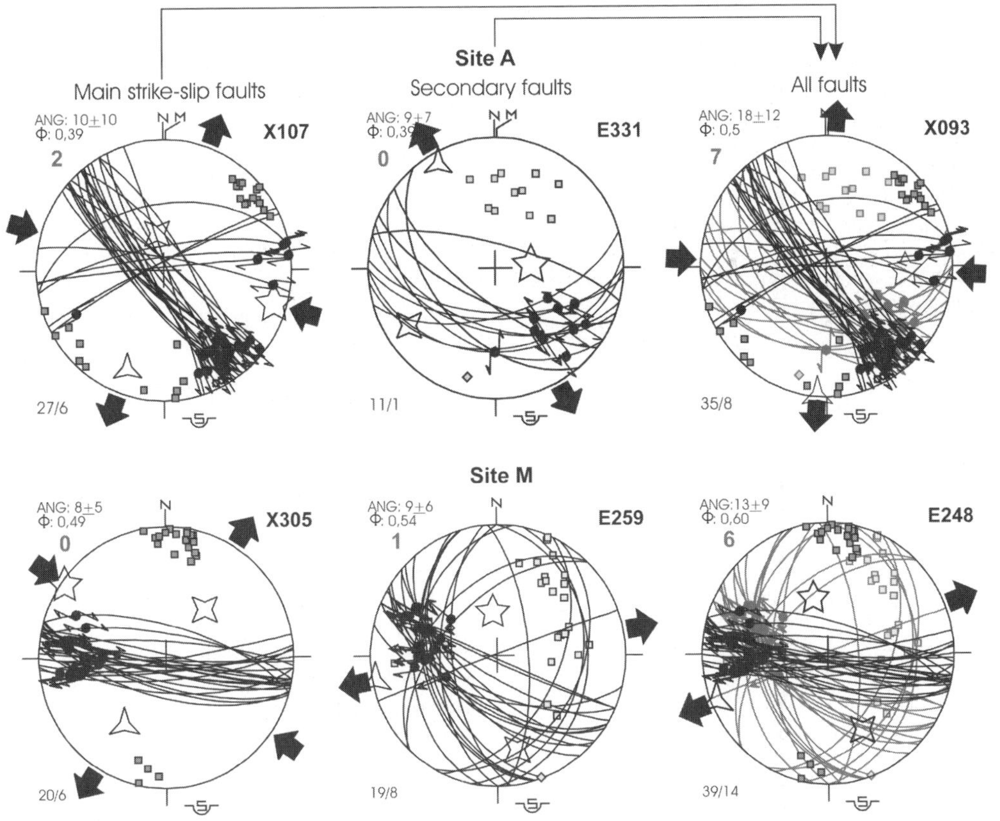

Fig. 5. Stereograms and calculated stress axes for the fault-slip data from the overstepping transtensional faults. Inversion of fault-slip data was carried out separately for the 'main faults', 'secondary faults' and for the combined data-set (all faults); in this latter data-set, the 'secondary faults' are show in grey. Symbols on stereograms are the same as in Figure 2b. Numbers in the upper left-hand corner give the numbers of the data misfit angles larger than 22.5°. Numbers in the upper right-hand corner show the directions of tension (E) and of compression (X).

show an overstepping geometry. The 'secondary fault' class contains all of the connecting faults of the linking damage zones and the fault segments that show a bend in strike or in the angle of dip. This classification is derived from field observation and not from automatic classification according to the strike or rake of the faults.

The software package of Angelier (1984, 1990*a*) was used for stress determination. Three criteria were taken into account to check the consistency of the data: the misfit angle for individual faults between the observed striation and the computed slip direction; the average misfit angle; and the number of data with a misfit angle larger than 22.5°. The value of $\Phi = \sigma_2 - \sigma_3 / \sigma_1 - \sigma_3$ was also considered. When it is close to 0 or 1, the direction of the stress axes is not considered as well constrained, because of a possible flip of σ_2 versus σ_3, and σ_2 versus σ_1 axes, respectively.

The 'main faults' are characterized by a strike-slip stress regime with a horizontal σ_1 and σ_3 (Fig. 5). There are no data with a misfit angle larger than 22.5°, and the average misfit angles are small (10 ± 10 and 8 ± 5° in sites A and M, respectively). The stress axes obtained are similar to those of the faulting phase 1 determined in several sites within the Transdanubian Range (Márton & Fodor 2003), and are considered to characterize a state of stress which has a regional significance.

One of the basic assumptions of the method of Angelier (1990*a*, *b*) is that fault-slip inversion can be executed effectively if the slips on the different fault planes are independent from each other. This assumption is difficult to validate at any field measurement sites, thus fault inversion always carries a certain ambiguity. On the other hand, if the field relationships can predict the presence of guided slip, than those fault planes should be excluded from the stress calculations. However, such a geometric relationship is difficult to establish in outcrops where the measured faults are not closely spaced. In the case of the Gánt sites, the slip on the 'secondary faults' seems to be influenced or completely guided by the 'main faults'. Thus, this case represents a 'controlled test' of stress calculations, showing to what extent the incorporation of guided faults into the data-set modifies the results of the stress calculation.

The 'secondary faults' of sites A and M seem to fit a tensional stress regime, which has σ_3 and σ_2 horizontal. The internal consistency of the data is also good. However, the σ_3 axis deviates from those calculated for the strike-slip stress regime. Deviation is counterclockwise for sinistral (site A) and clockwise for dextral (site M) faults (Fig. 5).

The calculation for the total fault-slip data-set shows noticeable misfit. The average misfit angle and its standard deviation increase from 10 ± 10 to 18 ± 12, and from 8 ± 5 to 13 ± 9, at sites A and M, respectively, between the computations for the separated 'main fault group' and the total data-set (Fig. 5). Also, the number of data with a larger than 22.5° misfit angle increased from zero to six or seven. The discrepancy is not very large, but a precise calculation would pose the question of whether all faults can be considered to slip at the same stress state, and should exclude data with a larger than 22.5° misfit angle from the final stress calculation. However, the lack of overprinting striae on any fault planes, and the observed geometric coherency of the faults suggest that all the faults were formed in the same stress state, thus theoretically they should fit one stress state.

On the other hand, the stress regime for the whole data-set is strike-slip type in site A, and tensional in site M (Fig. 5), so in the latter case a change from a strike-slip to a tensional state of stress occurs. In site M, the minimal stress axis σ_3 for all faults deviates considerably from the one calculated for the main strike-slip fault segments, and is more similar to the tensional stress axis σ_3 computed for the separated 'secondary faults'. Although the 'main faults' and the 'secondary faults' are equal in quantity, the latter 'dominate' the stress calculation. The stress state computed for all of the faults would be more similar to the regional stress field of phase 2 or 3 (see Fig. 2b), placing the ages of the 'main strike-slip faults' several million years younger than the faults of phase 1.

Discussion

Geometry of faults in a transtensional relay ramp

Relay ramps in normal and strike-slip faults play a similar role in transferring the displacement between the two overlapping fault segments. In the case of normal-fault relay ramps, transfer of displacement requires tilting of rocks toward the hanging wall. The tilting will accommodate the vertical component of strain between points A and B in Figure 6a. Deformation can be accommodated by continuous deformation in the early stage of evolution and by discrete fault(s) later (Peacock & Sanderson 1994). The beaching faults will probably occur when tilting, or another type of continuous deformation will not be able to transfer the displacement. On the other hand, the distance between points located in the hanging-wall and footwall parts of the normal relay ramp will not change horizontally parallel to the fault trace (in map view) because of the lack of an along-strike component of displacement (Fig. 6).

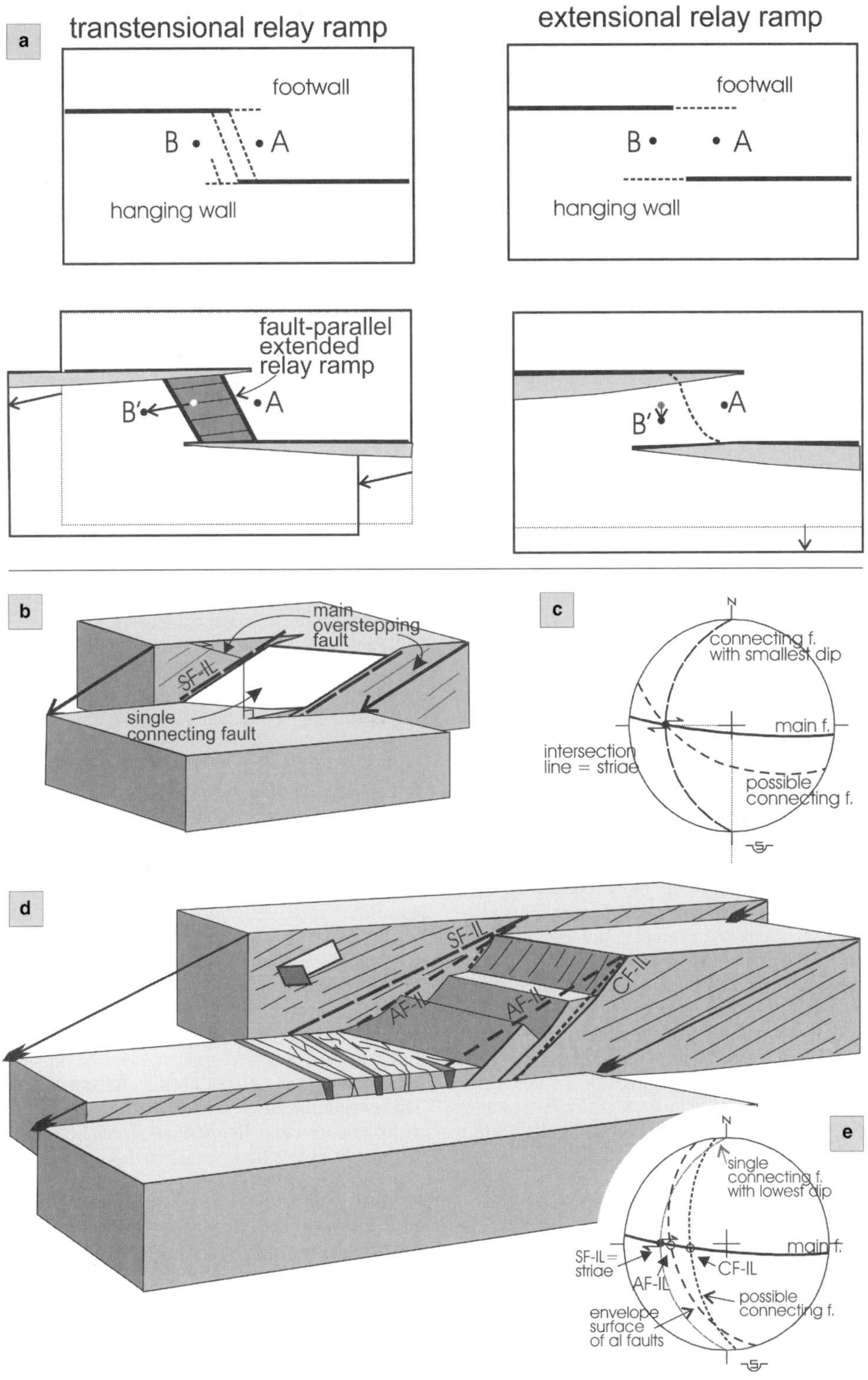

Fig. 6. *For caption see p. 428.*

In the case of pure strike-slip faults, tilting may not occur, thus relay ramps will not form (Peacock & Sanderson 1995). In transtensional relay ramps, the degree of reorientation of the bedding depends on the rake of the main fault, and decreases toward the end-member pure strike-slip fault, because of the smaller magnitude of vertical strain to be accommodated. On the other hand, the horizontal distance of two points near the ramp will increase as much as the strike-slip component of the main faults (Fig. 6). This requires considerable extension across the ramp, subparallel to the overstepping faults. This can be accommodated by ductile strain; by connecting normal faults; or by other deformation mechanisms. Thus, the need for accommodation of the horizontal component of strain governs the geometry of the transtensional relay ramp and the development of the connecting fault.

Assuming rigid footwall and hanging-wall blocks, an ideal single connecting fault would intersect the main fault parallel to its striae (slip direction; Fig. 6b & c). Slip on the combined zigzag-shaped fault prevents further propagation of the main fault, and thus further increase in the overlap; the structure resembles a double-bended fault more than a true relay ramp. The steepness of the connecting faults will depend on the rake of the main fault and on the difference in strike between the main and connecting faults; the larger this strike difference, the lower the angle of the connecting fault (Fig. 6c). The rake of slickenlines on the connecting faults will change in a similar way; the lowest-angle fault shows pure dip-slip striae, and this fault will show the great deviation in strike as compared with the main fault. The tendency toward moderate- to low-angle dip of the connecting faults can contribute to the fault-parallel extension.

The well-exposed examples presented here are used to suggest a slightly more complex three-dimensional model for transtensional relay ramps (Fig. 6d & e). The connecting splays have an oblique-normal slip sense and 30–80° deviation in strike from the main faults; they are moderately to gently dipping. The slickenlines of the connecting faults are subparallel to those of the main strike-slip faults, so the main and connecting faults seem to define a geometrically coherent fault system, which has an almost rigid hanging-wall block.

When viewed more closely, however, the intersection lines of the main fault and most of the connecting faults (CF–IL in Fig. 6d & e) are steeper than the slip vector (for the striae) on the main fault. The upper edges of the blocks bounded by the connecting faults define a smoothed envelope surface, which approximates the surface of the relay ramp. This envelope surface provides another intersection line (AF–IL on Fig. 6d & e). Assuming a rigid hanging-wall block, this line would be parallel to the slip on the main fault planes (SF–IL on Fig. 6d & e).

The model shows a small angular difference between the intersection line AF–IL and the main fault slip vector SF–IL, which involves a smaller horizontal component of slip with respect to an ideal single connecting fault, meaning a ‘deficit in extension’ across the relay ramp. The necessary additional extensional deformation can be accommodated by brecciation and veins (Fig. 4d & 6d). Another feature is that the main faults show small displacements between the fault tip and the connecting faults. This may be inherited from the former phase, when the relay ramp was not breached.

The angular difference of the CF–IL and AF–IL lines (or SF–IL in an ideal case) depends on the geometry, spacing and displacement of the connecting faults, and on the rake of the slickenlines on the main fault. The lower dip angle of a connecting fault provides a CF–IL line closer to the SF–IL line than a steeper one (Fig. 6). Closely spaced high-angle normal faults would define the same AF–IL as a few gently dipping faults. Small-displacement connecting faults would need to be more numerous in order to make the AF–IL line parallel to the slickenlines.

Probably the most important factor influencing the geometry of the connecting faults is the rake of the lineation on the main fault. Gently dipping slickenlines promote shallower connecting faults, which tend to be at a high angle to the main fault. Alternatively, a larger number of high-angle faults may also have an AF–IL close to the slickenside lineation, despite the fact that the individual intersection line (CF–IL) will be different.

Fig. 6. (*Continued*) Simplified geometric models for transtensional and extensional relay ramps. (**a**) Plan view of relay ramps related to pure-normal- and oblique-slip faults, which are compared with regard to displacement across the relay ramps. Arrows show the displacement, and the dashed lines show the future ruptures. (**b**) Block diagram of the overstepping main faults and a single ideal connecting fault in a transtensional setting; SF–IL corresponds with their ideal intersection line. (**c**) Stereogram showing the projection of the main fault, a possible connecting fault, and that with the lowest dip angle. (**d**) Block diagram of a transtensional relay ramp, based on examples near Gánt. CF–IL: intersection line of a connecting fault and the main fault; AF–IL: intersection line of the main fault and a smoothed surface of all the connecting fault blocks. (**e**) Stereographic projections of the main fault, and the three possible types of intersection lines.

The low-angle connecting faults can effectively accommodate horizontal strain across the transtensional relay ramps. However, it is questionable whether the high-angle connecting faults could form a network dense enough to accommodate this strain. It is thus suggested, although not proven, that the lower rake of the striae on the main fault will promote the formation of low-angle connecting faults, which will be at a high angle to the main faults. As seen on Figure 6c, the low dip of the connecting faults will imply a larger rake for the striae, producing characteristics similar to those of dip-slip normal faults.

Aydin & Schultz (1990) concluded from examples worldwide that the overlap increases proportionally with the separation of segmented (en échelon) strike-slip faults. An (1997) suggested that the linkage of overlapping strike-slip segments occurs at a certain ratio (threshold value) of fault length and relay ramp width. For normal faults, Soliva & Benedicto (2004) suggested a correlation between the spacing of overstepping faults and the displacement occurring on faults at the mid-relay ramp position. Acocella *et al.* (2005) found that the development of linkage faults and their exact character depend on the extension rates. Although all of these aspects influence linkage, all of these studies considered the cases of kinematic end members. The considerations presented suggest that the threshold values for fault linkage *also* depend on the exact kinematics of the faults in question. The obliquity of slip defines the amount of fault-parallel extension to be accommodated across the linking damage zones, thus markedly influencing the geometry and temporal evolution of the linking structures.

Stress field and mechanical interaction of faults

Inversion methods for fault-slip data (Carey & Brunier 1974; Angelier 1984) assume that all faults considered for calculation should move independently in a homogeneous stress field. The probable mechanical interaction between the overstepping and connecting faults, and the presence of ‘guided slip’ on ‘secondary faults’ at Gánt do not seem to satisfy this assumption. The calculations for the complete data set do indeed indicate deviations from the strike-slip-type stress state that are considered to be regionally significant. The difference between calculations can be large enough to misleadingly attribute faults to a younger or older stress field, particularly if the stress-field evolution was more complex than in the Pannonian Basin.

Alternatively, the difference between stress states of regional significance and of local meaning can be considered as the sign of an inhomogeneous (perturbed) stress field around the relay ramps. In fact, numerical models show a perturbation in the magnitude and in the direction of resolved stress for both strike-slip (Aydin & Schultz 1990; Muller & Aydin 2004) and for normal faults (Maerten *et al.* 1999; Maerten 2000). The deviation in stress is caused by mechanical interaction between intersecting faults.

The numerical models of Maerten (2000) and Muller & Aydin (2004) produce rake deviation on both mechanically interacting fault planes. In the numerical models, the slip vectors will change, but will not turn parallel to the intersection line on any of the fault planes, whilst this is the case in the field examples presented. This suggests that either the mechanical interaction is larger between the fault segments (because of their close spacing), or that the slip vectors were mainly guided by geometric constraints caused by the movement of the rigid footwall and hanging-wall blocks. Field observations and the suggested geometric model would favour this latter solution.

The observations presented here reaffirm the conclusion of Angelier (1984) and Maerten (2000) that stress inversion procedures should not include data from within the zones of fault interaction. This result warns geologists that field observations should clarify the geometry of fault linkage and should, as much as possible, exclude the connecting faults of transfer zones from the stress calculations.

Conclusions

Detailed observations on well-exposed overlapping strike-slip faults in the Gánt mining area of central Hungary permit the characterization of transfer zones (relay ramps) in a transtensional setting. Slip vectors on all faults are subparallel to each other and to the intersection line of the main and connecting faults, thus defining a geometrically coherent fault system and relatively rigid fault blocks. The hard-linked relay ramps are breached through a set of oblique-normal and normal faults striking obliquely or at a high angle to the main fault segments. The strike, dip angle and density of the connecting faults change systematically in a manner such that the surface of the relay ramp would intersect the main fault subparallel to its striae. The system of connecting faults accommodates horizontal strain across the relay ramp. Because this strain will increase with the decreasing rake of the slickenlines on the main fault, it is proposed that the small rake promotes the formation of low-angle connecting faults, which have a pure dip-slip rather than an oblique-slip character.

Inversion of fault-slip data reveals noticeable variation in the stress axes calculated for the main and connecting secondary fault segments. The stress deviations and uncertainty in stress-field determination can be explained by the fact that the connecting faults within the relay ramps have a guided slip, which was not governed by the regional stress field but was imposed by the geometry of the fault system and by the mechanical interaction of the overlapping fault segments. This result warns researchers that inversion of fault-slip data must be considered carefully in cases where fault segmentation and linkage may occur in the studied fault array.

A. Mindszenty of Eötvös University, Budapest, introduced the author to the impressive geology of the Gánt mining area. I. Almási, a former MSc student at Eötvös University, contributed to the results of this paper. The Vértes Mapping Project of the Geological Institute of Hungary and the Hungarian Science Foundation (OTKA No. 42799, Publication P16) supported the research. The author received funding from a Bolyai János Scholarship awarded by the Hungarian Academy of Sciences. G. Csillag, G. Paulheim and L. Sásdi (Geological Institute of Hungary, Budapest) assisted in the preparation of a few of the figures. The digital terrain model was prepared by the Hungarian Military Mapping Office. The constructive reviews of D. Peacock and J.-C. Hippolyte considerably improved the earlier version of this paper, whilst the editorial comments of D. Cunningham were also helpful and encouraging.

References

Acocella, V. & Neri, M. 2005. Structural features of an active strike-slip fault on the sliding flank of Mt. Etna (Italy). *Journal of Structural Geology*, **27**, 343–355.

Acocella, V., Morvillo, P. & Funiciello, R. 2005. What controls relay ramps and transfer faults within rift zones? Insights from analogue models. *Journal of Structural Geology*, **27**, 397–408.

Almási, I. 1993. *Structural geological research of the Gánt bauxite mining area*. MSc thesis, Eötvös University, Hungary (in Hungarian).

Almási, I. & Fodor, L. 1995. Relay ramp structures and related stress field variations. Abstracts of the 57th EAGE Conference and Technical Exhibition, Glasgow, Scotland, p. 532.

An, L. J. 1997. Maximum link distance between strike-slip faults: observations and constraints. *Pure and Applied Geophysics*, **150**, 19–36.

Angelier, J. 1984. Tectonic analysis of fault slip data sets. *Journal of Geophysical Research*, **89**, 5835–5848.

Angelier, J. 1990*a*. Inversion of field data in fault tectonics to obtain the regional stress – III. A new rapid direct inversion method by analytical means. *Geophysical Journal International*, **103**, 363–373.

Angelier, J. 1990*b*. Tectonique cassante et neotectonique. *Annales de la Société Géologique de Belgique*, **112**, 283–307.

Aydin, A. & Nur, A. 1982. Evolution of pull-apart basins and their scale independence. *Tectonics*, **1**, 91–105.

Aydin, A. & Schultz, R. A. 1990. Effect of mechanical interaction on the development of strike-slip faults with echelon patterns. *Journal of Structural Geology*, **12**, 123–129.

Balla, Z. & Dudko, A. 1989. Large-scale Tertiary strike-slip displacements recorded in the structure of the Transdanubian Range. *Geophysical Transactions*, **35**, 3–65.

Bergerat, F. 1989. From pull-apart to the rifting process: the formation of the Pannonian Basin. *Tectonophysics*, **157**, 271–280.

Bignot, G., Blondeau, A. *et al.* 1985. Age and characteristics of the Eocene transgression at Gánt (Vértes Mts., Transdanubia, Hungary). *Acta Geologica Hungarica*, **28**, 29–48.

Budai, T., Fodor, L., Csillag, G. & Piros, O. 2005. Stratigraphy and structure of the southeastern part of the Vértes Mountain (Transdanubian Range, Hungary). *Annual Report of the Geological Institute of Hungary, 2004*, 189–203. Geological Institute of Hungary, Budapest, Hungary (in Hungarian with English abstract).

Carey, E. & Brunier, B. 1974. Analyse théorique et numerique d'un modéle mécanique élémentaire appliqué á l'étude d'une population de failles. *Compte Rendues de l'Académie de Sciences de France*, **279**, 891–894.

Childs, C., Easton, S. J., Vendeville, B. C., Jackson, M. P. A., Lin, S. T., Walsh, J. J. & Watterson, J. 1993. Kinematic analysis of faults in a physical model of growth faulting above a viscous salt analog. *Tectonophysics*, **228**, 313–329.

Childs, C., Watterson, J. & Walsh, J. J. 1995. Fault overlap zones within developing normal fault systems. *Journal of Geology*, **152**, 535–549.

Crider, J. G. 2001. Oblique slip and the geometry of normal-fault linkage: mechanics and a case study from the Basin and Range in Oregon. *Journal of Structural Geology*, **23**, 1997–2009.

Csontos, L., Tari, G., Bergerat, F. & Fodor, L. 1991. Evolution of the stress fields in the Carpatho-Pannonian area during the Neogene. *Tectonophysics*, **199**, 73–91.

Fodor, L. *et al.* 2005. Tectonic development, morphotectonics and volcanism of the Transdanubian Range: a field guide. *In:* Fodor, L. & Brezsnyánszky, K. (eds) *Proceedings of the Workshop on Application of GPS in Plate Tectonics, in Research on Fossil Energy Resources and in Earthquake Hazard Assessment*. Occasional Papers of the Geological Institute of Hungary, **204**, 68–86. Geological Institute of Hungary, Budapest, Hungary.

Fodor, L., Csontos, L., Bada, G., Györfi, I. & Benkovics, L. 1999. Tertiary tectonic evolution of the Pannonian basin system and neighbouring orogens: a new synthesis of paleostress data. *In:* Durand, B., Jolivet, L., Horváth, F. & Séranne, M. (eds) *The Mediterranean Basins: Tertiary Extension within the Alpine Orogen*. Geological Society, London, Special Publications, **156**, 295–334.

GERMÁN-HEINS, J. 1994. Iron-rich encrustation on the footwall of the Gánt bauxite (Vértes Hills, Hungary) – evidence for preservation of organic matter under exceptional conditions? *Sedimentary Geology*, **94**, 73–83.

GYALOG, L. 1992. Contribution to the structure-geological knowledge of the Várgesztes area. *Annual Report of the Hungarian Geological Institute from 1990*, 69–74. Hungarian Geological Institute, Budapest, Hungary.

HUGGINS, P., WATTERSON, J., WALSH, J. J. & CHILDS, C. 1995. Relay zone geometry and displacement transfer between normal faults recorded in coal-mine plans. *Journal of Structural Geology*, **17**, 1741–1755.

KIM, Y. S., PEACOCK, D. C. P. & SANDERSON, D. J. 2004. Fault damage zones. *Journal of Structural Geology*, **26**, 503–517.

LARSEN, P.-H. 1988. Relay structures in a Lower Permian basement-involved extensional system, East Greenland. *Journal of Structural Geology*, **10**, 3–8.

MCCLAY, K. & BONORA, M. 2001. Analog models of restraining stepovers in strike-slip fault systems. *AAPG Bulletin*, **85**, 233–260.

MAERTEN, L. 2000. Variation in slip on intersecting normal faults: implications for paleostress inversion. *Journal of Geophysical Research*, **105**, 25 553–25 565.

MAERTEN, L., WILLEMSE, E. J. M., POLLARD, D. D. & RAWNSLEY, K. 1999. Slip distributions on intersecting normal faults. *Journal of Structural Geology*, **21**, 259–271.

MANN, P., HEMPTON, M. R., BRADLEY, D. C. & BURKE, K. 1983. Development of pull-apart basins. *Journal of Structural Geology*, **91**, 529–554.

MÁRTON, E. & FODOR, L. 2003. Tertiary paleomagnetic results and structural analysis from the Transdanubian Range (Hungary); rotational disintegration of the Alcapa unit. *Tectonophysics*, **363**, 201–224.

MINDSZENTY, A., SZÖTS, A. & HORVÁTH, A. 1989. Karstbauxites in the Transdanubian Midmountains. *In:* CSÁSZÁR, G. (ed.) *10th Regional Meeting, International Association of Sedimentologists, Excursion Guidebook*, 11–48. Hungarian Geological Institute, Budapest, Hungary.

MORLEY, C. K., NELSON, R. A., PATTON, T. L. & MUN, S. G. 1990. Transfer zones in the East African rift system and their relevance to hydrocarbon exploration in rifts. *AAPG Bulletin*, **74**, 1234–1253.

MULLER, J. R. & AYDIN, A. 2004. Rupture progression along discontinuous oblique fault sets: implications for the Karadere rupture segment of the 1999 Izmit earthquake, and future rupture in the Sea of Marmara. *Tectonophysics*, **391**, 283–302.

PEACOCK, D. C. P. 2002. Propagation, interaction and linkage in normal fault system. *Earth-Science Reviews*, **58**, 121–142.

PEACOCK, D. C. P. & SANDERSON, D. J. 1991. Displacements, segment linkage and relay ramps in normal fault zones. *Journal of Structural Geology*, **13**, 721–733.

PEACOCK, D. C. P. & SANDERSON, D. J. 1994. Geometry and development of relay ramps in normal fault systems. *AAPG Bulletin*, **78**, 147–165.

PEACOCK, D. C. P. & SANDERSON, D. J. 1995. Strike-slip relay ramps. *Journal of Structural Geology*, **17**, 1351–1360.

PEACOCK, D. C. P., KNIPE, R. J. & SANDERSON, D. J. 2000. Glossary of normal faults. *Journal of Structural Geology*, **22**, 291–305.

PETIT, J. P. 1987. Criteria for sense of movement on fault surfaces in brittle rocks. *Journal of Structural Geology*, **9**, 597–608.

SOLIVA, R. & BENEDICTO, A. 2004. A linkage criterion for segmented normal faults. *Journal of Structural Geology*, **26**, 2251–2267.

TAEGER, H. 1909. Die Geologishen Verhältnisse des Vértesgebirges. *Annals of the Geological Institute of the Hungarian Kingdom*, **1**, 1–256.

TARI, G. 1994. *Alpine tectonics of the Pannonian basin.* PhD Thesis, Rice University, Texas, USA.

TRUDGILL, B. & CARTWRIGHT, J. 1994. Relay-ramp forms and normal fault linkages, Canyonland National Park, Utah. *Geological Society of America Bulletin*, **106**, 1143–1157.

WALSH, J. J. & WATTERSON, J. 1991. Geometric and kinematic coherence and scale effects in normal fault systems. *In:* ROBERTS, A. M., YIELDING, G. & FREEMAN, B. (eds) *The Geometry of Normal Faults.* Geological Society, London, Special Publications, **99**, 193–203.

WALSH, J. J., BAILEY, W. R., CHILDS, C., NICOL, A. & BONSON, C. G. 2003. Formation of segmented normal faults: a 3-D perspective. *Journal of Structural Geology*, **25**, 1251–1262.

WOODCOCK, T. L. & SCHUBERT, C. 1994. Continental strike-slip tectonics. *In:* HANCOCK, P. L. (ed.) *Continental Tectonics.* Pergamon Press, Oxford, 251–263.

The structural evolution of dilational stepovers in regional transtensional zones

N. DE PAOLA[1,2], R. E. HOLDSWORTH[2], C. COLLETTINI[1], K. J. W. MCCAFFREY[2] & M. R. BARCHI[1]

[1]*Gruppo di Geologia Strutturale e Geofisica (GSG), Department of Earth Sciences, University of Perugia, Piazza Università 1, 06100 Perugia, Italy (e-mail: nicola.de-paola@durham.ac.uk)*

[2]*Reactivation Research Group (RRG), Department of Earth Sciences, University of Durham, DH1 3LE, UK*

Abstract: We propose a theoretical model, supported by a field study, to describe the patterns of fault/fracture meshes formed within dilational stepovers developed along faults accommodating regional scale wrench-dominated transtension. The geometry and kinematics of the faulting in the dilational stepovers is related to the angle of divergence (α), and differs from the patterns traditionally predicted in dilation zones associated with boundary faults accommodating strike-slip displacements (where $\alpha = 0°$). For low values of oblique divergence ($\alpha < 30°$) and low strain, the fault–fracture mesh comprises interlinked tensile fractures and shear-extensional planes, consistent with wrench-dominated transtension. At higher values of strain, a switch occurs from wrench- to extension-dominated transtension, leading to the reactivation and/or disruption of the early-formed structures. These structural processes lead to the development of a geometrically complex and kinematically heterogeneous fault pattern, which may affect and/or perturb the development of a through-going fault linking and facilitating the slip transfer between the two overlapping fault segments. As a result, dilational stepover zones will tend to form long-lived sites of localized extension and subsidence in regional transtensional tectonic settings. Cyclic increases/decreases of structural permeability will be related to slip on the major boundary faults that control the distribution of fluid-flow paths and, consequently, the long- and short-term structural evolution of these sites. Our model also predicts complex and more realistic subsurface fluid migration pathways relevant to our current understanding of hydrothermal ore deposits and hydrocarbon migration and storage.

Strike-slip fault systems are often schematically represented in map view as a series of spatially juxtaposed segments that lie subparallel to regional slip vectors, hereafter referred to as straights (Fig. 1, Aydin & Nur 1982; Mann *et al.* 1983; Woodcock & Fischer 1986). Regions of offset along straights are commonly referred to as bends and stepovers, depending on whether or not a through-going fault linking the offset fault segments is present (Fig. 1). Thus, bends are continuous linking segments oblique to regional slip vectors, whilst stepovers are zones of slip transfer between overstepping discontinuous faults. The kinematic behaviour of straights is directly determined by regional strain fields, whilst the kinematic and structural evolution of bends and stepovers is additionally influenced by secondary local strains determined by the local offset fault geometry.

Stepovers are commonly sites of localized deformation along strike-slip fault systems, where displacements are accommodated by structurally complex and kinematically heterogeneous fault–fracture meshes (Segall & Pollard 1980; Sibson 1985, 1996). The structural processes occurring at these particular sites and their dynamic evolution exert a strong control on:

1. seismic rupture nucleation and propagation (King & Nabelek 1985; Sibson 1985;
2. the dynamic character of permeability influencing coseismic and post-seismic fluid redistribution within deformation zone volumes (Nur & Booker 1972; Sibson 1985, 1996, 2000; Peltzer *et al.* 1996; Connolly & Cosgrove 1999*a*; Bosl & Nur 2002);
3. hydrothermal fluid paths (Cox 1999; Cox *et al.* 2001; Sibson 2001, 2004; Micklethwaite & Cox 2004; Cox & Ruming 2004);
4. the geometry and architecture of deformation patterns influencing fracture interconnectivity and fluid flow in potential hydrocarbon reservoirs (Connolly & Cosgrove 1999*b*; Sibson 2000; Kim *et al.* 2004).

The structural character, mechanics and evolution of stepovers associated with strike-slip faults have been extensively investigated in many field-based (Aydin & Nur 1982, 1985; Peacock & Sanderson 1995; Kim *et al.* 2001, 2003; Waldron

From: CUNNINGHAM, W. D. & MANN, P. (eds) *Tectonics of Strike-Slip Restraining and Releasing Bends.* Geological Society, London, Special Publications, **290**, 433–445.
DOI: 10.1144/SP190.17 0305-8719/07/$15.00

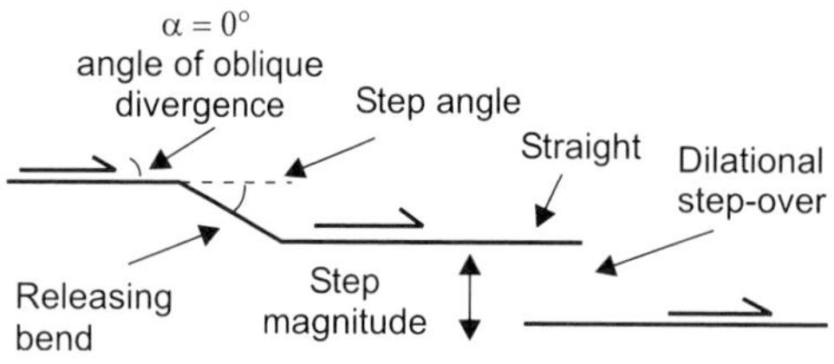

Fig. 1. Schematic representation of the geometry and terminology of overstepping strike-slip faults. The angle of divergence (α) is the angle between the regional displacement direction and the main fault trend. In this case, α = 0°, e.g. these are true strike-slip faults.

2004, 2005), theoretical (Segall & Pollard 1980; Woodcock & Fischer 1986; Du & Aydin 1995) and analogue studies (Dooley & McClay 1997; Rahe *et al.* 1998; Simms *et al.* 1999; Connolly & Cosgrove 1999*b*). However, despite the widespread recognition of transtension on regional scales (e.g. Dewey 1975; Woodcock 1986; Tikoff & Teyssier 1994; Teyssier *et al.* 1995; Dewey 2002; McClusky *et al.* 2003; Oldow 2003; De Paola *et al.* 2005*a*, *b*), a comprehensive study of structural processes and dynamic evolution of stepovers developed along faults active within regional transtensional deformation zones ($\alpha > 0°$, where α is the angle of oblique divergence) is still generally lacking (although see Peacock & Sanderson 1995; Dooley *et al.* 2004). Hereafter, we refer to such cases as 'transtensional dilational stepovers'.

In this paper we will develop a model to explain the structural character and evolution of transtensional dilational stepovers and apply it to well-exposed natural examples from the Carboniferous Northumberland Basin, UK. The implications of this study will then be discussed, highlighting some of the key differences between stepovers associated with regional wrench and transtensional deformation.

Strike-slip dilational stepovers: an overview

Dilational stepover zones develop between two discontinuous subparallel strike-slip faults where the sense of offset is the same as the sense of displacement (Fig. 1). Such discontinuous faults are typically composed of interacting segments, between which deformation tends to localize (Segall & Pollard 1980; Peacock 1991). Overstepping zones are potential sites where complex fault–fracture meshes can develop with a symmetrical pattern relative to the local stress directions (Fig. 2; Sibson 1985, 1996). In general, within such dilational overstepping zones, compressive normal stress values between the interacting segments are reduced compared with far-field values (Segall & Pollard 1980). This results in the development of an area of enhanced tensile and shear failure which may potentially bridge the segments (Fig. 2a). Importantly, the principal stress/infinitesimal strain axes are significantly rotated in the overstepping zone relative to the far-field orientations (Fig. 2).

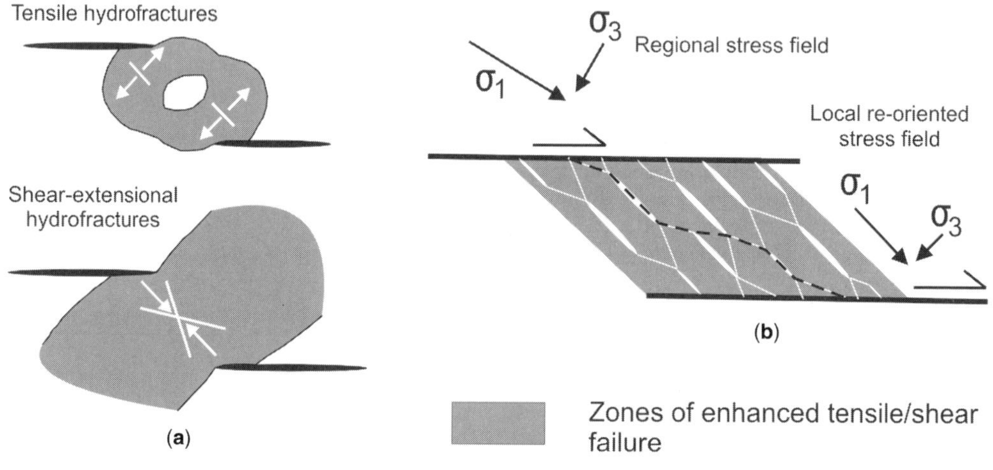

Fig. 2. (**a**) Zones of enhanced tensile (top) and shear failure (bottom) highlighted between overstepping faults. The orientation of hydrofractures and shear-extensional hydrofractures is also shown (after Segall & Pollard 1980). (**b**) 'Hill-type' mixed extensional/shear-extensional fracture mesh (Hill 1977) developed within a dilational stepover between two overlapping faults (slightly modified after Sibson 1985). The staircase trajectory of a hypothetical through-going fault is also shown by the black dashed line. Note the local stress reorientation (clockwise rotation) within the dilational stepover compared with the far field reference state. Note than the amount of local stress reorientation also depends on the rock'elastic properties.

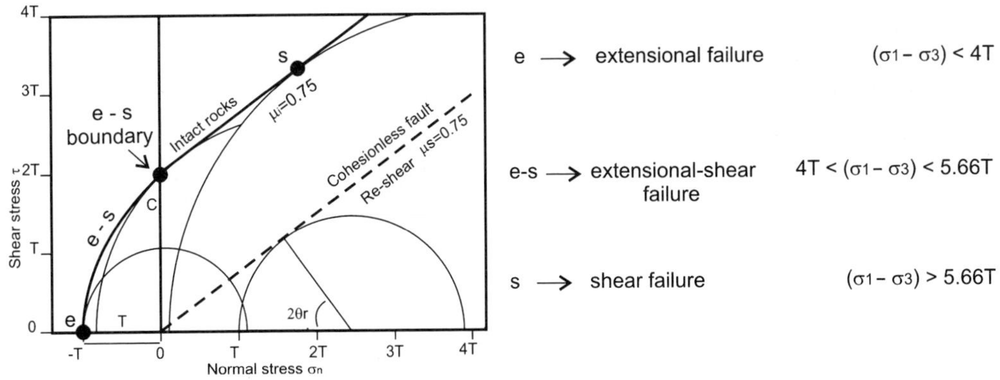

Fig. 3. Generic Mohr diagram showing a composite Griffith Coulomb failure envelope for intact rocks, with $\mu_i = 0.75$, where μ_i is the coefficient of internal friction. The re-shear criterion for cohesionless faults is also shown (dashed line) for $\mu_s = 0.75$, where μ_s is the coefficient of sliding friction (after Sibson 1996). Critical stress circles are shown, representing different failure modes (*e* = extensional failure; e–s = shear-extensional failure; s = shear failure) and the re-shear stress state for an optimally oriented existing fault ($\mu_s = 0.75$, $2\theta_r = 53°$).

The sense of rotation depends on the sense of shear (clockwise for dextral and anticlockwise for sinistral faults).

The conditions required for the development of a so-called 'Hill-type' extensional/shear fracture mesh (Hill 1977), made up of interlinked tensile hydrofractures and conjugate strike-slip shear fractures (Fig. 2b), depend on tensile strength (T), differential stress values ($\sigma_1 - \sigma_3$) and pore-fluid pressure (P_f) (Fig. 3, Sibson 1996, 2000). Dilational stepovers are sites of dynamic permeability, where tensile fluid-pressure conditions ($P_f > \sigma_3$) promote the activation of the fault–fracture mesh. Hydrofracturing and shear failure processes tend to increase permeability, allowing fluid discharge, whilst cementation processes tend to restore lower values of permeability.

With further deformation, a through-going fault, linking the two overstepping segments, is thought to form, initially with a stair-step and/or curvilinear trajectory due to the linking of adjacent shear and tensile fracture segments in the fault–fracture mesh (Fig. 2b). The formation of a through-going fault inhibits the local attainment of the tensile fluid-pressure condition ($P_f > \sigma_3$) and therefore leads to the cessation of further fracture mesh development (Fig. 3). However, as fluid migration can localize in stepovers, cementation and alteration processes may allow rapid recovery of shear strength (e.g. Tenthorey *et al.* 2003).

Transtensional deformation: theory

The 3D character of transtensional strain

Published geodetic data-sets show that crustal deformation in zones of oblique divergence is typically diffuse, either at regional or local scales (e.g. McClusky *et al.* 2003; Oldow 2003). Obliquely divergent relative plate motions will inevitably lead to 3D (non-plane) strains, even in the simplest case where the deformation is homogeneously distributed (Dewey *et al.* 1998; Dewey 2002). A 3D transtensional deformation can be modelled as the simultaneous and combined action of margin-parallel wrench simple shear and a margin-orthogonal extensional pure shear, resulting from a far-field direction of extension (α) oriented obliquely to the deformation zone boundaries (Fig. 4a, Sanderson & Marchini 1984). The combination of these two plane-strain end-member deformations results in 3D transtensional non-coaxial strain (Fig. 4; Withjack & Jamison 1986; Fossen & Tikoff 1993). Predictable geometric relationships exist between the orientations of the deformation zone boundary and both the axis of maximum infinitesimal strain ($x_i \simeq \sigma_3$), expressed by the angle β, and the far-field transport direction expressed by the angle α (Fig. 4). There are two distinct states of infinitesimal transtensional strain: wrench- and extension-dominated, whose development depends on the value of the divergence angle α (Fig. 4b). Infinitesimal elastic strain modelling shows that wrench-dominated transtension is characterized by quadrimodal sets, with orthorhombic symmetry, of strike-slip-oblique faults (i.e. $\sigma_1 \simeq z_i$ and $\sigma_3 \simeq x_i$ both horizontal, e.g. 3D strain deformations in Reches 1978, 1983; Krantz 1988). This contrasts strongly with extension-dominated transtension, which is typically characterized by quadrimodal sets of normal-oblique faults trending at low angles to the deformation zone boundaries (i.e. $\sigma_1 \equiv z_i$ vertical and $\sigma_3 \equiv x_i$ horizontal) (Withjack & Jamison 1986; De Paola *et al.* 2005*a*, *b*).

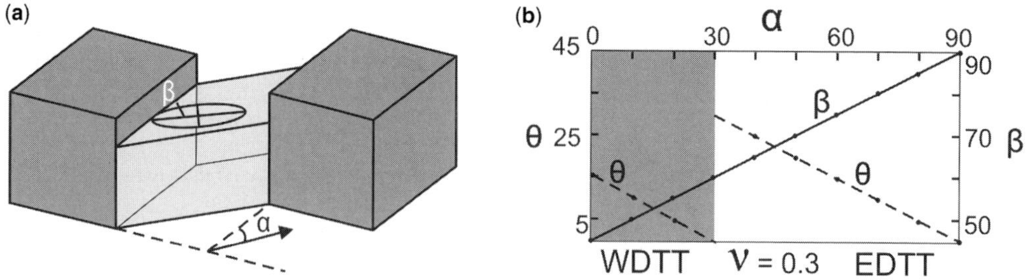

Fig. 4. (**a**) The geometry of homogeneous transtension, where α is the angle between the main boundary fault and the imposed displacement direction and β is the angle between the boundary fault and the horizontal infinitesimal maximum strain axis $x_i \cong \sigma_3$. (**b**) α vs. β–θ diagram displaying the predictable relationship the exist between the angle α and the angles β and θ (which is the angle between the boundary fault and the synthetic (R) fracture planes). The fields for wrench- (WDTT) and extension-dominated transtension (EDTT) are shown for a Poisson's ratio value of $\nu = 0.3$ (after De Paola *et al.* 2005a).

During transtensional deformations (i.e. combined pure shear and simple shear) equal increments of extensional pure shear accumulate multiplicatively, leading to a more rapid accumulation of finite deformation compared with the wrench simple shear (Tikoff & Teyssier 1994; Tikoff & Greene 1997). As a result, with increasing finite strain intensity, an initially wrench-dominated transtension ($\alpha < 30°$) will eventually develop a finite strain ellipsoid similar to one associated with an extension-dominated transtension where the axis of maximum extension (x_f) is always horizontal, whilst the axis of maximum shortening (z_f), initially horizontal, becomes vertical (Fig. 5; Tikoff & Teyssier 1994; Tikoff & Greene 1997). The stability field of wrench-dominated transtension during deformation may be increased by positive volume change associated, for example, with syntectonic magmatic intrusion (Fig. 5a; Teyssier & Tikoff 1999). In general, transtension zones will develop kinematically complex and heterogeneous fault patterns. Basement-controlled strain partitioning can lead to the development of 'wrench-' and 'extension-dominated' transtension domains. In addition, in areas where α values are low ($<30°$), strain-magnitude-controlled switches in finite strain axes can also develop. Simultaneously active strike-slip-oblique and normal-oblique faults during transtensional deformations have been observed and described in the field (Dewey 2002; De Paola *et al.* 2005*a*, *b*).

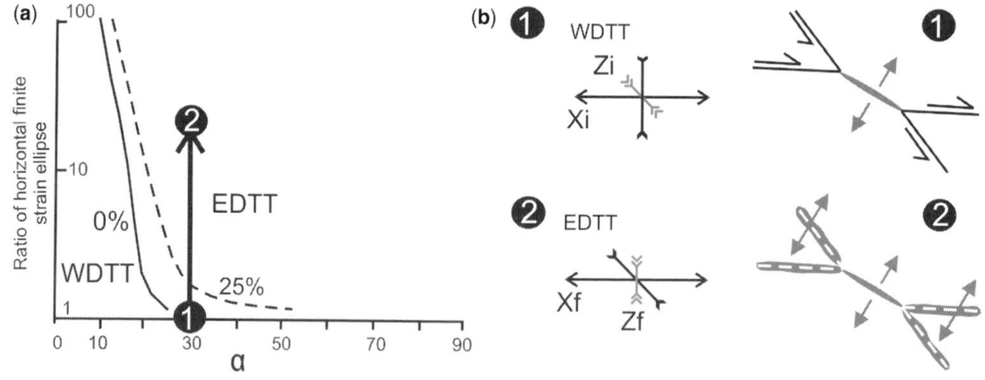

Fig. 5. (**a**) The fields of WDTT (horizontal $z_{i\text{-}f}$) and EDTT (vertical z_f) plotted on an α vs. stain magnitude (ratio of finite horizontal strain axes) diagram for a constant-volume deformation ($\Delta V = 0\%$). The boundary between these fields (solid curved line) represents the point at which the two strain axes are the same length (a prolate strain) and across which an increase in finite stain will lead to strain axis switching (see Dewey 2002). The dashed curve (from Teyssier & Tikoff 1999) represents the upward shift of the WDTT–EDTT boundary when there is a 25% volume increase during transtensional deformation. (**b**) Diagram showing the switch from wrench- (stage 1, top right of the figure) to extension-dominated transtension (stage 2, bottom right of the figure) for increasing finite strain (see text for details).

Geometry, structural processes and evolution of transtensional dilational stepovers

In this section, we apply the theoretical concepts of transtension to dilational stepover zones developed between faults accommodating a regional, obliquely divergent deformation. In the case of a transtensional dilational stepover, the orientation of the infinitesimal strain/stress field within the dilational zone is a function of the divergence angle α (Fig. 6a). In particular, the infinitesimal strain field is three-dimensional and rotated compared with the strike-slip case (Fig. 6a–c). In the case of wrench-dominated transtension, the angle β will be higher (up to 60°) to the deformation zone boundary, compared with the strike-slip case (Fig. 6a, c). Under favourable conditions for the development of fault–fracture meshes (Fig. 3), the deformation pattern is consequently rotated relative to the strike-slip-oblique faults (Fig. 6). The deformation pattern is inevitably three-dimensional and cannot be accommodated by simple fault patterns, but rather by complex polymodal fault–fracture patterns with mainly oblique-slip kinematics (Reches 1978, 1983; Krantz 1988; Healy *et al.* 2006).

As finite strain increases within the wrench-dominated transtensional dilational stepover zone, the local strain field switches from wrench- to extension-dominated transtension, i.e. the maximum shortening axis becomes vertical (Fig. 5). This new local strain field will favour the development of new normal-oblique faults and/or the reactivation of the earlier strike-slip-oblique faults (Fig. 5).

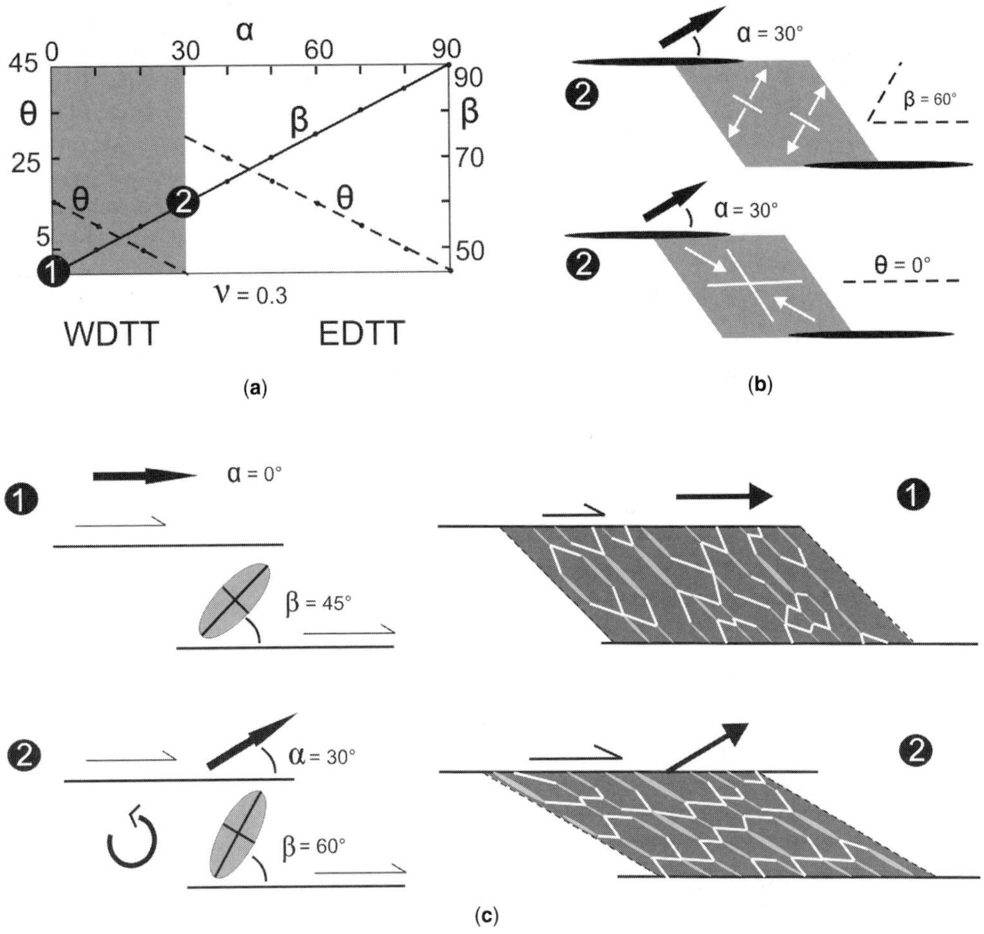

Fig. 6. (**a**) Conditions for wrench (1, α = 0°) and wrench-dominated transtension (2, α = 30°) at dilational stepover plotted on an α vs. β–θ diagram. (**b**) Infinitesimal strain field and orientation of associated extensional (top) and shear-extensional (bottom) fractures within dilational stepovers developed in a bulk wrench-dominated transtensional strain (α = 30°). (**c**) Strain field and geometry of associated fault/fracture meshes developed within a strike-slip (1, α = 0°) and a transtensional (2, α = 30°) dilational stepover, respectively.

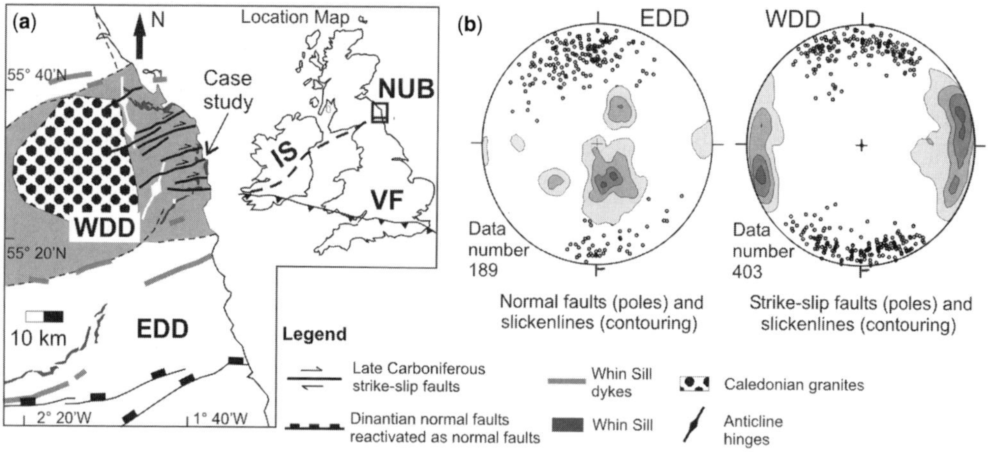

Fig. 7. (**a**) Schematic structural map of the Northumberland Basin (UK) showing location the of the Late Carboniferous extension-dominated (EDD) and wrench-dominated (WDD) domains. Top right inset shows location map of the Northumberland Basin (NUB). IS, Iapetus Suture; VE, Variscan Front (for details see De Paola *et al.* 2005). (**b**) Summary equal-area stereoplots of structures measured within the extension-dominated domain (EDD, Late Carboniferous normal faults) and the wrench-dominated domain (WDD, Strike-slip faults) in the Northumberland Basin, England, UK.

The development of a geometrically complex, strain-intensity-dependent and kinematically unstable deformation pattern at wrench-dominated transtensional dilational stepovers potentially has important implications for the dynamic evolution of these sites, which differs significantly from the strike-slip case (see below).

Case study: transtensional dilational stepovers in the Northumberland Basin

Geological setting

The Northumberland Basin, NE England, is a Carboniferous basin located north of the Variscan Orogen (Fig. 7a, location map). Syndepositional extensional faulting in the Dinantian (Early Carboniferous) caused marked changes in sediment thickness, with up to 4 km of Dinantian sediments recognized close to the fault system bounding the southern margin of the basin. By contrast, only 1.5 km of the same sediments occur in adjacent areas to the north (Kimbell *et al.* 1989). The basin has long been considered to have experienced compressional inversion related to the far-field effects of Variscan collision, leading to the development of reverse faulting and closely associated north–south-trending folds (e.g. Collier 1989; Leeder *et al.* 1989). However, new geological mapping and detailed strain analyses have led to the new interpretation that the Late Carboniferous deformation patterns of the Northumberland Basin are the result of a single phase of dextral transtensional deformation (De Paola *et al.* 2005*a*). Importantly, the distribution of this strain is highly heterogeneous, due to kinematic partitioning of the regional deformation into structurally distinct, NE–SW-trending dextral wrench-dominated domains (WDD) and (pure shear) extension-dominated domains (EDD) (see Figs 7b & 8), whose trend is most likely controlled by pre-existing structures in the basement at depth. A conjugate set of ENE–WSW- to east–west-trending normal faults are the dominant structures that accommodate Late Carboniferous deformation within the EDD (Figs 7b & 8). In sharp contrast, the dominant regional-scale structures in the WDD are ENE–WSW-trending dextral shear zones in which quadrimodal ENE (dextral) and ESE (sinistral) strike-slip-oblique faults are developed in association with minor structures such as folds, thrusts and stylolites (Figs 7b & 8). Integration of the partitioned deformation patterns in the NUB suggests that regional extension was NNE–SSW-directed (Fig. 8; De Paola *et al.* 2005*a*). This rifting was also associated with the emplacement of tholeiitic intrusions (known locally as the Whin Sill suite, Fig. 7a) that make up part of a regional igneous assemblage (recognized across NW Europe) formed during Late Carboniferous–Early Permian times (e.g. see Timmerman 2004 and references therein). Field observations and new palaeomagnetic data (Liss *et al.* 2004) independently confirm that the deformation in both domains is broadly synchronous

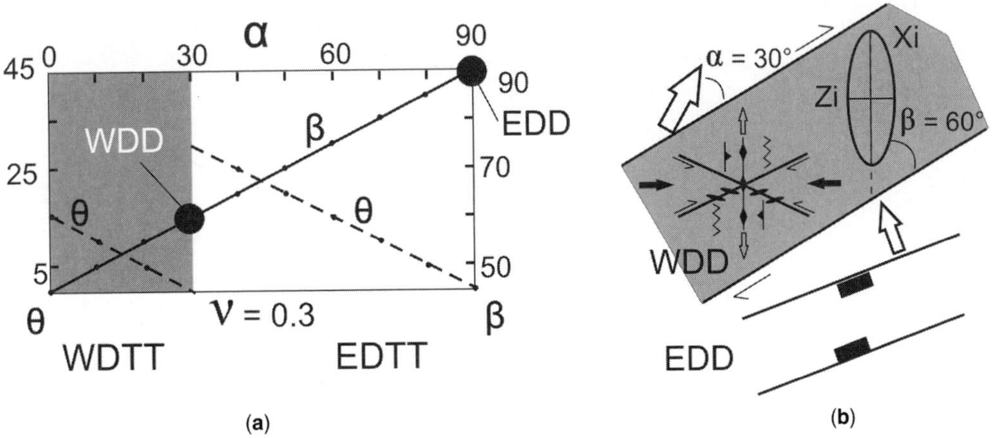

Fig. 8. (**a**) α vs. β–θ diagram displaying the reconstructed strain fields and geometry of the associated structures for the wrench-dominated domain (WDD α = 30°) and extension-dominated domain (EDD α = 90°). See text for details. (**b**) Structural plan-view model of the Northumberland Basin during the Late Carboniferous–early Permain deformation. The structural assemblage of minor structure in the WDD is also shown.

with emplacement of the Whin Sill suite. Hence, the structures in the WDD and EDD are contemporaneous (De Paola *et al.* 2005*a*).

We now describe and discuss in detail the nature and development of a dilational stepover zone associated with a dextral strike-slip zone developed within the WDD.

Transtensional deformation and fracture mesh development

A dilational stepover zone has formed between two major dextral faults (displacements up to few tens of metres) which were active within a bulk wrench-dominated transtensional deformation zone (Fig. 9a, see location in Fig. 7a). The host rocks belong to the Middle Limestone Group (Viséan) and the age of their recorded deformation is Late Carboniferous (Collier 1989; Leeder *et al.* 1989; De Paola *et al.* 2005*a*). The depth at which the observed deformation developed is up to 1.5 km based on the preserved thickness of the Carboniferous stratigraphic infill at this time (Kimbell *et al.* 1989). The bulk wrench-dominated transtensional strain accommodated by the major structures is well documented by evidence for contemporaneous dextral shear and fault-normal extension and by the geometry and patterns of the associated mesoscale deformation (see De Paola *et al.* 2005*a*).

The overlapping area between the two main dextral faults represents a dilational stepover characterized by intense deformation and fracturing compared with the surrounding areas, where the strain is very low and characterized by widely spaced joints (Fig. 9). The boundary between the high- and low-deformation zones is abrupt and defined by sinistral strike-slip-oblique antithetic shear planes (Fig. 9a). Brittle deformation within the dilational stepover is accommodated by a complex and well-developed fault–fracture mesh (Fig. 9), whilst ductile deformation is accommodated by upright folds in the subhorizontal limestone beds (Fig. 9). The formation of folds has been interpreted as being due to the horizontal shortening component of wrench-dominated transtensional strain. Fold orientations are consistent with other mesoscale shortening structures (i.e. north–south-trending reverse faults and stylolites) recorded in the WDD (Figs 7–8, see also De Paola *et al.* 2005*a*). Thin (up to 5 cm), coarse-grained doleritic dykes (belonging to the Whin Sill suite) are also present within the dilational stepover associated with the fracture mesh (Fig. 9a). The dykes are interpreted to have been intruded during the Late Carboniferous transtensional deformation event.

The fault–fracture mesh is composed of a complex pattern of strike-slip/oblique-slip shears, dykes, normal/oblique-slip faults and veins (Figs 10–11). ENE-dextral and ESE-sinistral trending shear planes are associated with gently plunging slickenlines and display a quadrimodal geometry with an acute angle between the pseudo-conjugate fault sets of about 40° (Fig. 10a, b & d). These planes are commonly connected and intruded by thin (up to 5 cm) approximately east–west-trending subvertical doleritic dykes (Fig. 10a, c & e). Normal-oblique ENE-dextral and ESE-sinistral shear planes and calcite-infilled extensional veins (Fig. 11a and stereoplots) are commonly observed to reactivate and/or disrupt the dominantly wrench faults of the

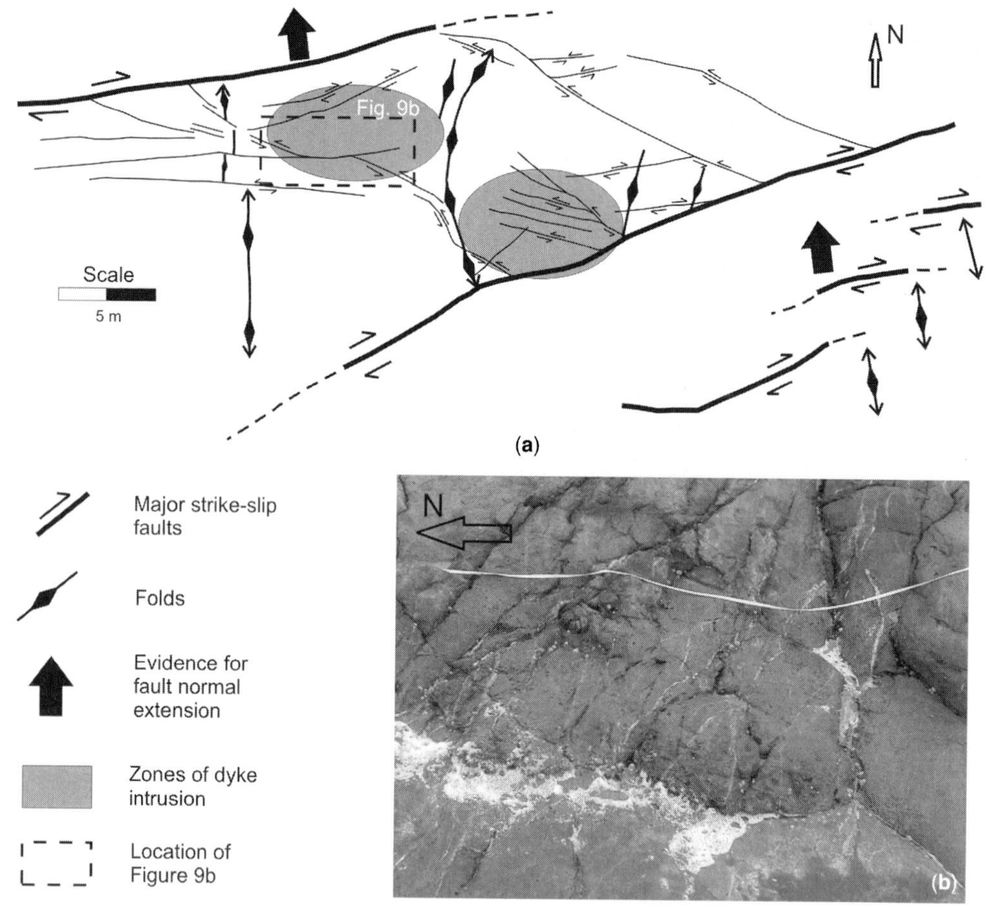

Fig. 9. (**a**) Detailed structural map of Lower Carboniferous rocks exposed at Dunstanburgh Castle, Northumberland Basin (for location see Fig. 7a). Major and minor shear planes within the dilational stepover zone have been mapped together with folds and zones of dyke intrusion (shaded areas). (**b**) Photograph of the typical fault–fracture mesh viewed in plan within a bedding surface.

structural assemblage (Fig. 11c & d). Wall-rock breccias within dilational jogs, interpreted to have formed by hydraulic implosion of the host rock, are abundant along the normal/oblique-slip shear planes (Fig. 11a & b). Several episodes of brecciation and cementation are indicated by incremental opening for the extension veins, and the recemented breccias have a low clast/matrix ratio and angular wall-rock clasts, confirming that they have not originated by frictional attrition.

The intimate association of strike-slip shear planes and dykes confirms a Late Carboniferous age for the development of the observed deformation pattern in accord with the age of the Whin Sill suite of intrusions. The fault/fracture meshes of the earlier strike-slip/oblique-slip and later normal/oblique-slip shear planes display geometries and kinematics consistent with 3D wrench-dominated and extension-dominated transtension, respectively (Figs 10–11). Both patterns are symmetrical about a vertical Y_i–Z_i (i.e. σ_2–σ_1) plane with the maximum shortening axis oriented in a horizontal and vertical position, respectively. The overall geometry of the fracture patterns is quadrimodal in accord with the three-dimensional nature of the strain field (stereoplots in Figs 10–11).

Discussion

Fracture mesh evolution

On the basis of the high symmetry of the strain fields associated with the strike-slip/oblique-slip and normal/oblique-slip structures (both patterns are symmetrical about a vertical Y_i–Z_i,

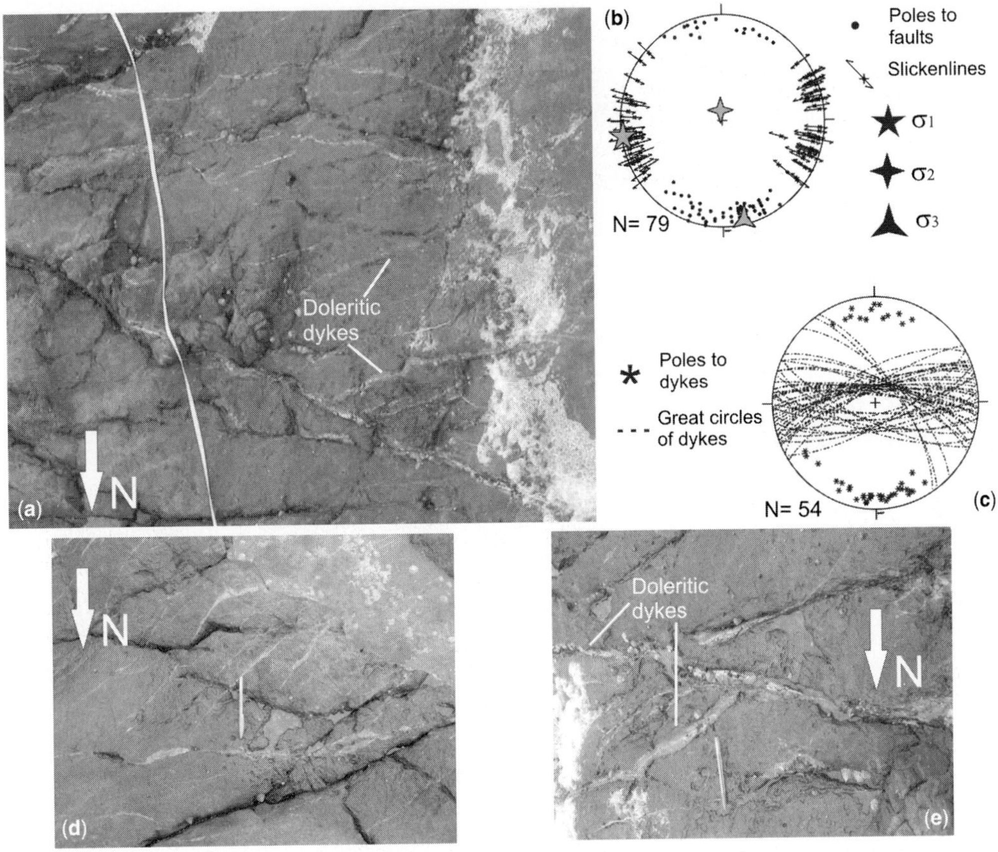

Fig. 10. (**a**) Detail of the fault/fracture mesh with coarse grained doleritic dykes intruded within the otherwise carbonate-infilled fracture mesh. (**b**) Equal-area stereoplots of strike-slip, slightly extensional, faults and shallowly plunging slickenlines measured within the dilational stepover zone. Stress inversion yields a stress field with a vertical σ_2 and horizontal East–West trending σ_1 and North–South trending σ_3. (**c**) Equal-area stereoplot of sibvertical, East–West trending, doleritic dykes intruded within the fracture mesh in the dilational stepover zone. (**d–e**) Detail of the fault/fracture mesh (**d**) with conjugate dextral and sinistral shear fractures and doleritic dykes (**e**). All photo views are in plan.

i.e. σ_2–σ_1 plane), we interpret the observed structural pattern to result from a progressive 3D transtensional deformation rather than two distinct tectonic events. This symmetry is unlikely to occur in different tectonic events, whereas it fits in well with a progressive transtensional deformation where the axis of maximum extension was always in the horizontal plane (cf. an east–west trend for dykes and tensile veins in the stereoplots of Figs 10–11) and the intermediate and maximum shortening axes switched position in the vertical plane when threshold finite strain intensities were reached (Fig. 5). This is in accord with the observation that the sense of movement of the reactivated structures is maintained during the deformation, i.e. dextral and sinistral strike-slip/oblique-slip planes are reactivated as dextral and sinistral normal/oblique-slip planes (Fig. 11c) and dykes are reactivated as tensile veins (Fig. 11d).

Wrench-dominated transtensional strains ($\alpha = 30°$) within the dilational stepover zone have been initially accommodated by a fracture mesh of strike-slip/oblique-slip faults, doleritic dykes and tensile fractures with carbonate infills (Fig. 12a). The deformation zone was characterized by volume increase (dyke intrusion, carbonate mineralization) which expanded and stabilized the field of wrench-dominated transtension (Figs 5a & 12a, stage 1). The structural permeability within the dilational stepover zone is expected to have been very high during the fracture mesh development. At this stage, pronounced pore-pressure gradients between the dilational stepover zone and surrounding areas would probably drive intense fluid flux

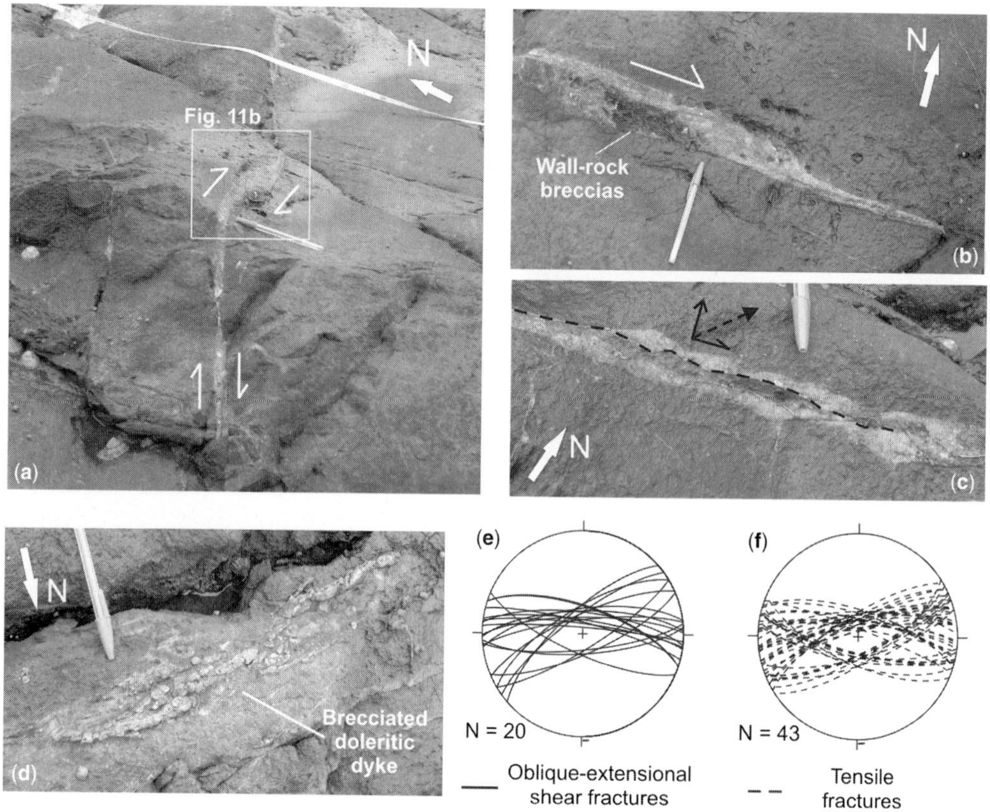

Fig. 11. (**a**) Three-dimensional (plan and cross-sectional) view of a dextral strike-slip fault reactivated as a dextral oblique-extensional fault. (**b**) Dilational jog developed along a dextral oblique-extensional fault. Wall-rock breccias show evidence for rock implosion and suction-pump behaviour. Calcite crack and seal textures indicate multiple episodes of opening. (**c**) Dextral strike-slip plane (dashed black line) reactivated as a dextral oblique-extensional shear fracture. The dashed black arrow displays the oblique component of displacement which can be resolved into components orthogonal and parallel to the fault, respectively (thick black arrows) (**d**) Brecciated doleritic dyke. (**e**) Equal-area stereoplots of oblique extensional shear fractures measured within the dilational stepover zone. These structures reactivated previously formed strike-slip shear fractures (**f**) Equal-area stereoplots of tensile fractures measured within the dilational stepover zone.

within the dilational stepover zone (Sibson 1985, 1996; Cox 2005) when high-permeability zones connect to fluid reservoirs (e.g. doleritic magmas and associated hydrothermal fluids).

For increasing values of strain within the dilational stepover zone a strain-magnitude-controlled switch will occur and extension-dominated transtension conditions will locally develop (Figs 5a & 12b, stage 2). The accommodation of such extensional strain field occurs by the development of normal/oblique-slip faults and/or by the reactivation of the pre-existing structures of the fracture mesh (Fig. 12b). Mechanically, the switch from wrench- to extension-dominated transtension is represented by the transition from strike-slip to extensional conditions. Early-formed wrench structures, having a mean dip of about 70° are favourably oriented for reactivation within the new, locally developed extensional strain field. Field evidence suggests that large fluid flux has occurred cyclically within the dilational stepover zone at this stage, as shown by crack–seal textures observed in thick veins and dilational jogs (up to 10 cm) along oblique-slip and normal faults (Fig. 11b–d). Suction-pump behaviour (Sibson 1985, 2000) may have been particularly active during this stage of deformation, as evidenced by widespread sings of implosion brecciation at dilational jogs along reactivated fault planes (Fig. 11a & b).

Implications of a 3D structural model

The three-dimensional, time-dependent, structural model that we present here accounts for the structural

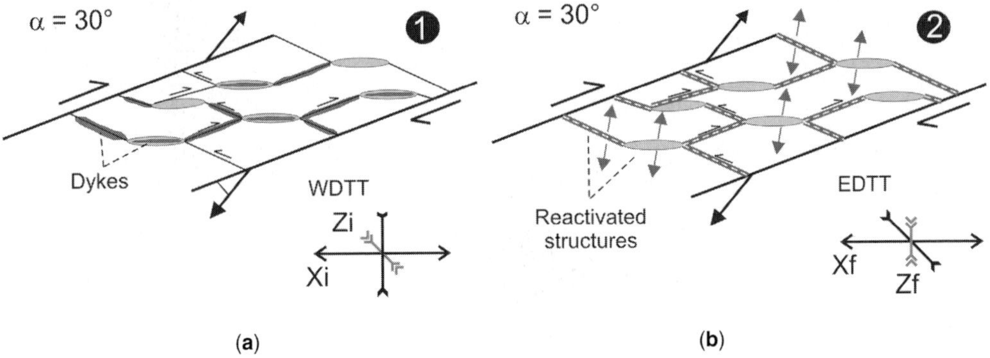

Fig. 12. (**a–b**) Schematic representation of the fault–fracture mesh developed within the studied transtensional dilational stepover, respectively active during (a) the early phase of wrench-dominated transtension (stage 1 in Fig. 5a) and (**b**) the later phase of extension-dominated transtension (stage 2 in Fig. 5a) following a strain-magnitude-controlled switch in finite strain axes.

character, mechanical behaviour and dynamic evolution of the deformation pattern formed at dilational stepovers developed within a regional transtensional deformation zone. Since fault fracture meshes developed at these particular sites have been long recognized to play a critical role in controlling the dynamics of structural permeability and fluid redistribution, we now discuss some of the more relevant implications and applications of our findings.

The structural processes occurring at transtensional dilational stepovers, as described by theoretical models and recognized in the field, predict that fracture mesh geometries and kinematic evolution will be strongly controlled by the angle of divergence α (Figs 4–5). At low angles of divergence ($\alpha < 30°$), the kinematic development will be complex, heterogeneous and unstable as it evolves from a wrench- to an extension-dominated transtension with increasing magnitudes of finite strain (Figs 5 & 12). We have seen that the predicted switching of straining styles causes reactivation and/or disruption of early-formed structures (Figs 9–11).

The fabric disruption processes, documented within the studied dilational stepover, are likely to inhibit the development of a through-going fault formed by linking up of extensional and shear structures within the fracture mesh (Fig. 2b; e.g. the staircase trajectory faults in Sibson 1985, 2000). We speculate that this can lead to localized and long-lived subsidence in transtensional dilational stepovers. The time-partitioned progressive deformation model that we present may be also a valid alternative to polyphase deformation models previously proposed in other stepover regions (e.g. Wakabayashi *et al.* 2004).

As pointed out by Sibson, the development of a fault/fracture mesh will create a significant proportion of gaping tensile fractures and shear-extensional planes, strongly increasing the local structural permeability and enhancing the hydraulic gradients capable of driving fluids through the dilational zone. These patterns can act as preferential pathways for hydrothermal fluid and magma emplacement (e.g. the carbonate veins and associated Whin Sill intrusions in the study area). Localized conditions of sudden fluid-pressure drop (e.g. dilational jogs along irregular faults in Figs 10–11) can favour mineral precipitation processes leading to mineralized veins and shear-extensional planes (e.g. Sibson 1985, 2000). However, the extreme complexity of the fault pattern developed means that the permeability distribution and the resultant preferential fluid paths may be difficult to predict in detail.

N. De Paola gratefully thanks the universities of Durham (UK) and Perugia (Italy) for financial support during fieldwork. S. Cox and J. Waldron are thanked for their perceptive and constructive reviews, which have significantly improved the paper.

References

AYDIN, A. & NUR, A. 1982. Evolution of stepover basins and their scale independence. *Tectonics*, **1**, 91–105.

AYDIN, A. & NUR, A. 1985. The types and role of stepovers in strike-slip tectonics. *In*: BIDDLE, K. T. & CHRISTIE-BLICK, N. (eds) *Strike-Slip Deformation, Basin Formation, and Sedimentation*. SEPM Special Publications, **37**, 35–44.

BOSL, W. J. & NUR, A. 2002. Aftershocks and pore fluid diffusion following the 1992 Landers earthquake. *Journal of Geophysical Research*, **107**, 2366, doi:10.1029/2001JB000155.

COLLIER, R. E. 1989. Tectonic evolution of the Northumberland Basin: the effects of renewed extension upon an inverted extensional basin. *Journal of the Geological Society, London*, **146**, 981–989.

CONNOLLY, P. T. & COSGROVE, J. W. 1999*a*. Prediction of induced permeability and fluid flow in the crust from experimental stress data. *AAPG Bulletin*, **83**, 4.

CONNOLLY, P. T. & COSGROVE, J. W. 1999*b*. Prediction of static and dynamic fluid pathways within and around dilational jogs. *In*: MCCAFFREY, K. J. W., LONERGAN, L. & WILKINSON, J. J. (eds) *Fractures, Fluid Flow and Mineralization*. Geological Society, London, Special Publications, **155**, 105–121.

COX, S. F. 1999. Deformational controls on the dynamics of fluid flow in mesothermal gold systems. *In*: MCCAFFREY, K. J. W., LONERGAN, L. & WILKINSON, J. J. (eds) *Fractures, Fluid Flow and Mineralization*. Geological Society, London, Special Publications, **155**, 123–140.

COX, S. F. 2005. Coupling between deformation, fluid pressures, and fluid flow in ore-producing hydrothermal systems at depth in the crust. *Economic Geology*, 100th Anniversary Volume, 39–76.

COX, S. F. & RUMING, K. 2004. The St Ives mesothermal gold system, Western Australia–a case of golden aftershocks? *Journal of Structural Geology*, **26**, 1109–1125.

COX, S. F., KNACKSTEDT, M. A. & BRAUN, J. 2001. Principles of structural control on permeability and fluid flow in hydrothermal systems. *Reviews in Economic Geology*, **14**, 1–24.

DE PAOLA, N., HOLDSWORTH, R. E., MCCAFFREY, K. J. W. & BARCHI, M. R. 2005*a*. Partitioned transtension in the late Carboniferous of Northern Britain – a radical alternative to basin inversion models. *Journal of Structural Geology*, **27**, 607–625.

DE PAOLA, N., HOLDSWORTH, R. E. & MCCAFFREY, K. J. W. 2005*b*. The influence of lithology and pre-existing structures on reservoir-scale faulting patterns in transtensional rift zones. *Journal of the Geological Society, London*, **162**, 471–480.

DEWEY, J. F. 1975. Finite plate evolution: some implications for the evolution of rock masses at plate margins. *American Journal of Science*, **275-A**, 260–284.

DEWEY, J. F. 2002. Transtension in arcs and orogens. *International Geology Review*, **44**, 402–439.

DEWEY, J. F., HOLDSWORTH, R. E. & STRACHAN, R. A. 1998. Transpression and transtension zones. *In*: HOLDSWORTH, R. E., STRACHAN, R. A. & DEWEY, J. F. (eds) *Continental Transpressional and Transtensional Tectonics*. Geological Society, London, Special Publications, **135**, 1–14.

DOOLEY, T. & MCCLAY, K. 1997. Analog modeling of pull-apart basins. *AAPG Bulletin*, **81**, 1804–1826.

DOOLEY, T., MONASTERO, F., HALL, B., MCCLAY, K. & WHITEHOUSE, P. 2004. Scaled sandbox modelling of transtensional pull-apart basins; applications to the Coso geothermal system. *Geothermal Research Council Transactions*, **28**, 637–641.

DU, Y. & AYDIN, A. 1995. Shear fracture patterns and connectivity at geometric complexities along strike-slip faults. *Journal of Geophysical Research*, **100**, 18 093–18 102.

FOSSEN, H. & TIKOFF, B. 1993. The deformation matrix for simultaneous simple shearing, pure shearing and volume change and its application to transpression/transtension tectonics. *Journal of Structural Geology*, **15**, 413–422.

HEALY, D., JONES, R. R. & HOLDSWORTH, R. E. 2006. Three-dimensional brittle shear fracturing by tensile crack interaction. *Nature*, **439**, 64–67.

HILL, D. P. 1977. A model for earthquake swarms. *Journal of Geophysical Research*, **82**, 1347–1357.

KIM, Y. S., ANDREWS, J. R. & SANDERSON, D. J. 2001. Reactivated strike-slip faults: examples from north Cornwall, UK. *Tectonophysics*, **340**, 173–194.

KIM, Y. S., PEACOCK, D. C. P. & SANDERSON, D. J. 2003. Strike-slip faults and damage zones at Marsalform, Gozo Island, Malta. *Journal of Structural Geology*, **25**, 793–812.

KIM, Y. S., PEACOCK, D. C. P. & SANDERSON, D. J. 2004. Fault damage zones. *Journal of Structural Geology*, **26**, 503–517.

KIMBELL, G. S., CHADWICK, R. A., HOLLIDAY, D. W. & WERNGREN, O. C. 1989. The structure and evolution of the Northumberland Trough from new seismic reflection data and its bearing on modes of continental extension. *Journal of the Geological Society, London*, **146**, 775–787.

KING, G. & NABELEK, J. 1985. Role of fault bends in the initiation and termination of earthquake rupture. *Science*, **228**, 984–987.

KRANTZ, R. W. 1988. Multiple fault sets and three-dimensional strain: theory and application. *Journal of Structural Geology*, **10**, 225–237.

LEEDER, M. R., FAIRHEAD, D., LEE, A., STUART, G., CLEMMEY, H., AL-HADDEH, B. & GREEN, B. 1989. Sedimentary and tectonic evolution of the Northumberland Basin. *Proceedings of the Yorkshire Geological Society*, **47**, 207–223.

LISS, D., OWENS, W. H. & HUTTON, D. H. W. 2004. New palaeomagnetic results from the Whin Sill complex: evidence for a multiple intrusion event and revised virtual geomagnetic poles for the late Carboniferous for the British Isles. *Journal of the Geological Society, London*, **161**, 1–12.

MCCLUSKY, S., REILINGER, R., MAHMOUD, S., BEN SARI, D. & TEALEB, A. 2003. GPS constraints on Africa (Nubia) and Arabia plate motion. *Geophysical Journal International*, **155**, 126–138.

MANN, P., HEMPTON, M. R., BRADLEY, D. C. & BURKE, K. 1983. Development of pull-apart basins. *Journal of Geology*, **91**, 529–554.

MICKLETHWAITE, S. & COX, S. F. 2004. Fault-segment rupture, aftershock-zone fluid flow, and mineralization, *Geology*, **32**, 813–816.

NUR, A. & BOOKER, J. R. 1972. Aftershocks controlled by pore fluid flow? *Science*, **175**, 885–887.

OLDOW, J. S. 2003. Active transtensional boundary zone between the western Great Basin and Sierra Nevada block, western U.S. Cordillera. *Geology*, **31**, 1033–1036.

PEACOCK, D. C. P. 1991. Displacements and segment linkage in strike-slip fault zones. *Journal of Structural Geology*, **13**, 1025–1035.

PEACOCK, D. C. P. & SANDERSON, D. J. 1995. Pull-aparts, shear fractures and pressure solution. *Tectonophysics*, **241**, 1–13.

PELTZER, G., ROSEN, P., ROGEZ, F. & HUDNUT, K. 1996. Postseismic rebound in fault step-overs caused by pore fluid flow. *Science*, **273**, 1202–1204.

RAHE, B., FERRIL, D. A. & MORRIS, A. P. 1998. Physical analog modeling of pull-apart basin evolution. *Tectonophysics*, **285**, 21–40.

RECHES, Z. 1978. Analysis of faulting in three-dimensional strain fields. *Tectonophysics*, **47**, 109–129.

RECHES, Z. 1983. Faulting of rocks in three-dimensional strain fields II. Theoretical analysis. *Tectonophysics*, **95**, 133–156.

SANDERSON, D. J. & MARCHINI, W. R. D. 1984. Transpression. *Journal of Structural Geology*, **6**, 449–458.

SEGALL, P. & POLLARD, D. D. 1980. Mechanics of discontinuous faults. *Journal of Geophysical Research*, **85**, 4337–4350.

SIBSON, R. H. 1985. Stopping of earthquake ruptures at dilational fault jogs. *Nature*, **316**, 248–251.

SIBSON, R. H. 1996. Structural permeability of fluid-driven fault–fracture meshes. *Journal of Structural Geology*, **18**, 1031–1042.

SIBSON, R. H. 2000. Fluid involvement in normal faulting. *Journal of Geodynamics*, **29**, 469–499.

SIBSON, R. H. 2001. Seismogenic framework for hydrothermal transport and ore deposition. *Reviews in Economic Geology*, **14**, 25–50.

SIBSON, R. H. 2004. Controls on maximum fluid overpressure defining conditions for mesozonal mineralisation. *Journal of Structural Geology*, **26**, 1127–1136.

SIMMS, D., FERRIL, D. A. & STAMATAKOS, J. A. 1999. Role of a ductile décollement in the development of pull-apart basins: experimental results and natural examples. *Journal of Structural Geology*, **21**, 533–554.

TENTHOREY, E., COX, S. F. & TODD, H. F. 2003. Evolution of strength recovery and permeability during fluid–rock interaction in experimental fault zones. *Earth and Planetary Science Letters*, **206**, 161–172.

TEYSSIER, C. & TIKOFF, B. 1999. Fabric stability in oblique convergence and divergence. *Journal of Structural Geology*, **21**, 969–974.

TEYSSIER, C., TIKOFF, B. & MARKLEY, M. 1995. Oblique plate motion and continental tectonics. *Geology*, **23**, 447–450.

TIKOFF, B. & GREENE, D. 1997. Stretching lineations in transpressional shear zones: an example from the Sierra Nevada Batholith, California. *Journal of Structural Geology*, **19**, 29–39.

TIKOFF, B. & TEYSSIER, C. 1994. Strain modelling of displacement field partitioning in transpressional orogens. *Journal of Structural Geology*, **16**, 1575–1588.

TIMMERMAN, M. J. 2004. Timing, geodynamic setting and character of Permo-Carboniferous magmatism in the foreland of the Variscan Orogen, NW Europe. *In*: WILSON, M., NEUMANN, E.-R., DAVIS, G. R., TIMMERMAN, M. J., HEEREMANS, M. & LARSEN, B. T. (eds), *Permo-Carboniferous Magmatism and Rifting in Europe*. Geological Society, London, Special Publications, **223**, 41–74.

WAKABAYASHI, J., HENGESH, J. V. & SAWYER, T. L. 2004. Four-dimensional transform fault processes: progressive evolution of step-overs and bends. *Tectonophysics*, **392**, 279–301.

WALDRON, J. W. F. 2004. Anatomy and evolution of a pull-apart basin, Stellarton, Nova Scotia. *Geological Society of America Bulletin*, **116**, 109–127.

WALDRON, J. W. F. 2005. Extensional fault arrays in strike-slip and transtension. *Journal of Structural Geology*, **27**, 23–34.

WITHJACK, M. O. & JAMISON, W. R. 1986. Deformation produced by oblique rifting. *Tectonophysics*, **126**, 99–124.

WOODCOCK, N. H. 1986. The role of strike-slip fault systems at plate boundaries. *Philosophical Transactions of the Royal Society of London*, **A317**, 13–29.

WOODCOCK, N. H. & FISCHER, M. 1986. Strike-slip duplexes. *Journal of Structural Geology*, **8**, 725–735.

The 3D fault and vein architecture of strike-slip releasing- and restraining bends: evidence from volcanic-centre-related mineral deposits

B. R. BERGER

US Geological Survey, Federal Center MS 964, Denver, CO 80225, USA
(e-mail: bberger@usgs.gov)

Abstract: High-temperature, volcanic-centre-related hydrothermal systems involve large fluid-flow volumes and are observed to have high discharge rates in the order of 100–400 kg/s. The flows and discharge occur predominantly on networks of critically stressed fractures. The coupling of hydrothermal fluid flow with deformation produces the volumes of veins found in epithermal mineral deposits. Owing to this coupling, veins provide information on the fault–fracture architecture in existence at the time of mineralization. They therefore provide information on the nature of deformation within fault zones, and the relations between different fault sets. The Virginia City and Goldfield mining districts, Nevada, were localized in zones of strike-slip transtension in an Early to Mid-Miocene volcanic belt along the western margin of North America. The Camp Douglas mining area occurs within the same belt, but is localized in a zone of strike-slip transpression. The vein systems in these districts record the spatial evolution of strike-slip extensional and contractional stepovers, as well as geometry of faulting in and adjacent to points along strike-slip faults where displacement has been interrupted and transferred into releasing and restraining stepovers.

The mass flux of hydrothermal fluids in the discharge regime of shallow, high-temperature volcanic-rock-related hydrothermal systems is typically 100–400 kg/s (Berger & Henley 1989). Sustaining such flow rates and high temperatures for long periods of time requires that fluid flow be predominantly through networks of interconnected fractures (e.g. Björnsson & Bodvarsson 1990). Stress-dependent fracture permeability forms when fractures open and close depending upon orientation with respect to loading and applied stresses. At increased depth, stress-induced dilation occurs on favourably oriented fractures near their point of failure (Zhang & Sanderson 1996; Min *et al.* 2004), thus producing anisotropic fluid pathways. Near the point of failure, permeability can increase by orders of magnitude. Thus, hydrothermal veins originating from this fluid flow provide valuable evidence about stress-field systematics and hydraulically conductive faults at the time of fluid flow. In some instances, vein systems provide insights into fault-system evolution (e.g. McKinstry 1948).

Epithermal, volcanic-centre-related hydrothermal systems produce economically recoverable concentrations of metals (i.e. orebodies) that are discrete entities along longer-strike-length, sub-economic veins. Most commonly, individual orebodies, which can consist of single or multiple veins, have a greater vertical than horizontal dimension, and are located preferentially in meshes of tensile and shear fractures within their controlling fault systems (Henley & Berger 2000; Cox *et al.* 2001). These attributes of vein arrays and the orebodies within them imply that ore formation, in contrast to sub-economic vein formation, requires the focusing of fluid flow into vertically interconnected fracture networks and the inhibition of extensive lateral fluid flow. A ramification is that vein systems, together with their enclosed orebodies, can provide insights into the spatial evolution and architecture of fault systems and serve as guides to reconstructing fault systematics.

This paper presents evidence from two economically important precious-metal mining districts (Virginia City, Goldfield) and one economically less-important area (Camp Douglas) related to Lower to Middle Miocene volcanic centres in western Nevada (Fig. 1). These districts provide insights into how strike-slip fault-fracture systems evolve and, through this evolution, provide insights into the structural controls on epithermal ore formation and the conditions most favourable to this formation. The geometry of the veins in the districts under discussion delineates two extensional and one contractional strike-slip stepovers, as well as the geometry of faults at and adjacent to points of interruption of strike-slip faulting and the onset of extensional or contractional stepping. In addition, geological and geophysical evidence is provided for each district, in support of the likely role of misoriented basement shear zones in the interruption of strike-slip faulting and consequent nucleation of stepovers. Although a spatial relation between epithermal deposits and strike-slip fault systems is

From: CUNNINGHAM, W. D. & MANN, P. (eds) *Tectonics of Strike-Slip Restraining and Releasing Bends.* Geological Society, London, Special Publications, **290**, 447–471.
DOI: 10.1144/SP290.18 0305-8719/07/$15.00

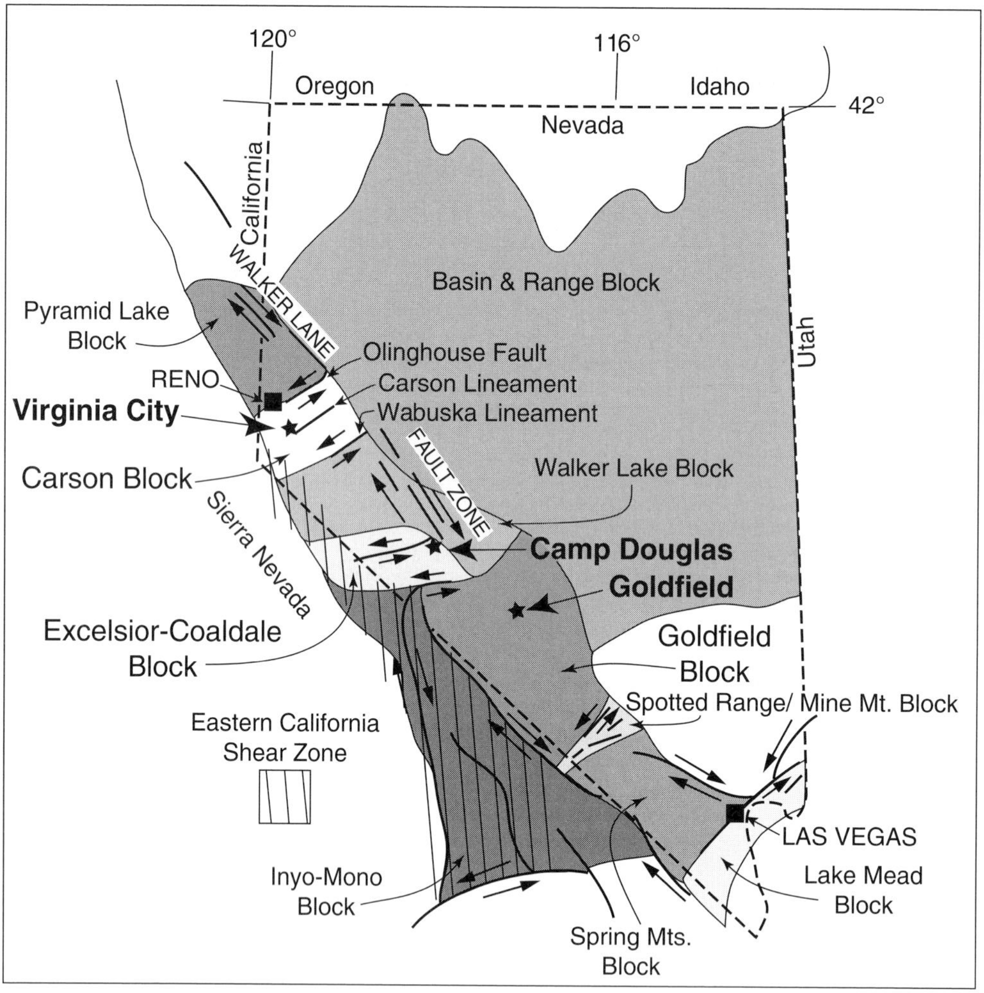

Fig. 1. Locations of the Virginia City, Camp Douglas, and Goldfield mining districts, Nevada, in the context of the structural blocks of Stewart (1988).

broadly recognized (e.g. Sillitoe 1993), little has been published detailing how veins provide insights into fault-system evolution and architecture and the structural localization of orebodies. This paper is a complement to works by Berger *et al.* (2003) that described the structural evolution of fault-related permeability at Virginia City, Nevada, and by Berger *et al.* (2005) that described the likely role of orogen-transverse faults in the localization of hydrothermal systems in western Nevada.

Coupled fluid flow and deformation

The high hydrothermal discharge rates (kg/s) observed in active high-temperature geothermal systems (cf. Elder 1966; Henley 1985) require the coupling of deformation with fluid flow to achieve the requisite fracture permeability (cf. Zhang & Sanderson 1996). Zhang & Sanderson (1996) show that the required fracture permeability may be obtained if potentially conductive fractures are stressed to near failure. The optimal orientation for fracture failure is determined by the strike of the fracture relative to the orientation of the maximum principal stress (σ_1) acting on the fracture surface (Fig. 2). The angle (α) between shear failure and σ_1 must be acute and generally $\leq 55^\circ$ to accommodate slip (e.g. Sibson 1977, 1998) and, therefore, become hydraulically conductive. For $\alpha > 55^\circ$, fractures are not optimally oriented and are said to be misoriented. Nevertheless, field

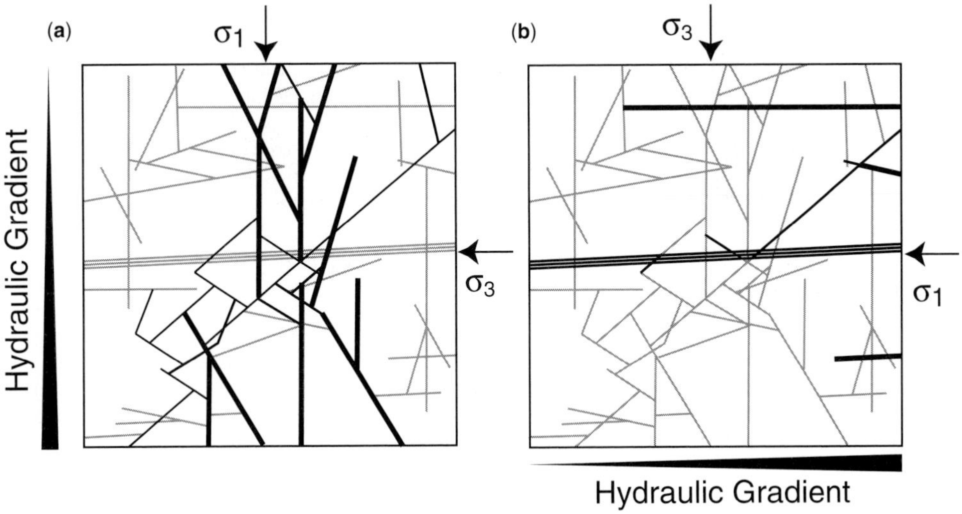

Fig. 2. Hypothetical model of the response of fracture-controlled hydraulic conductivity to a change in the orientation of the principal stresses acting on a set of fractures in a hydraulic gradient. In (**a**) the maximum principal stress is vertical, whereas in (**b**) the maximum principal stress is horizontal.

data provide evidence that misoriented faults can perturb the local strain field, affect the interaction of nearby faults and, consequently, influence the permeability structure of systems of linked faults.

Permeability (k) is a macroscopic quantity measured by the rate of flow through a rock, and is commonly expressed as the hydraulic conductivity (K):

$$K = k\rho_f g/\mu \quad (1)$$

where ρ_f is the density of the fluid, g is the gravitational constant, and μ is the dynamic viscosity of the fluid (e.g. de Marsily 1986). K varies with fluid properties (e.g. ρ/μ) and properties of the medium (e.g. pore shape, pore connectedness, distribution of pore sizes; Furbish 1997).

The geometries of veins and the orebodies contained within them set constraints on the permeability structure of mineralized fault–fracture systems. Typically, individual veins or sets of veins in an epithermal deposit have lateral continuities from metres to kilometres, and the lateral dimensions of veins commonly exceed their known vertical dimensions. In contrast, orebodies within longer strike-length veins typically vary in length from metres to tens of metres; are most commonly longer in their vertical than lateral dimension; and may be modelled as upright, cylindrical zones of fractures. Snow (1968) modelled the hydraulic conductivity (K_f) for a set of continuous planar fractures contained within such a pipe-like geometry. K_f was found to vary as the cube of the fracture aperture and fracture density. Therefore, under a given hydraulic gradient, aperture and fracture density and connectivity are of most importance to fracture-controlled fluid flow. Blenkinsop (2002) argues that circular, pipe-like geometries may be as much as three orders of magnitude more conductive than long-strike-length, planar fractures not contained within pipe-like geometries. The Goldfield mining district, described below, provides some empirical evidence for this experimental observation.

Fault-system evolution and architecture

The accommodation of strains along plate-margin boundaries is heterogeneous, and the margins are commonly segmented into structural blocks (e.g. Stein & Sella 2002). Stresses applied along block boundaries influence the style of deformation within them. The strike-slip fault systems discussed herein may be reflective of the segmented character of the western North American plate boundary as proposed by Stewart (1988), and they are not necessarily parts of a continuous, regional strike-slip fault system along the margin in the sense of the regionally extensive present-day Walker Lane fault zone (Fig. 1).

Volcanism is commonly localized along strike-slip fault zones in subduction-boundary zones. Within volcanic centres, related hydrothermal systems can deposit epithermal-style veins along the fault-fracture systems, the veins being typified by tabular, banded quartz and adularia, or tabular siliceous masses consisting of quartz–alunite–kaolinite ± pyrophyllite alteration. Owing to the coupling of stress and fluid flow in hydrothermal systems, the geometry of veins and relations between veins provide evidence regarding how strains were accommodated within the vein-controlling fault–fracture zones; how the controlling faults interacted; how the fault–fracture systems evolved spatially; and the structural geometries and conditions most favourable to ore formation.

Spatial migration of extensional stepovers

Bends and stepovers having either extensional or contractional character are common features along strike-slip fault zones (e.g. Crowell 1974; Christie-Blick & Biddle 1985). Wakabayashi *et al.* (2004) described the tectonic inversion of previously extending areas; inversions that cannot be explained by regional tectonic stresses; and the associated progressive spatial migration of strike-slip stepovers. Geological relations in the Virginia City (e.g. Berger *et al.* 2003) and Goldfield mining districts (e.g. Berger *et al.* 2005) may be interpreted to show relations similar to those discussed by Wakabayashi (2004).

Stewart (1988) divided western Nevada into nine structural blocks (Fig. 1). Each block was interpreted to have acted independently of adjacent blocks. His map reflects basin-and-range tectonics as well as the partitioning of strike-slip and related faulting east of the San Andreas transform fault in California. Nevertheless, many of Stewart's blocks show a history that predates basin-and-range extension that began in the Virginia City and Camp Douglas areas sometime after approximately 13–12 Ma (e.g. Stewart & Perkins 1999; Henry & Perkins 2001; Petronis *et al.* 2002) and possibly as old as 17–14 Ma in the Goldfield area (e.g. Bonham & Garside 1979). Tectonically, the Virginia City district is in Stewart's (1988) Carson structural block, and Goldfield is in the Goldfield structural block. Strike-slip faulting occurred within some structural blocks at various times during the Cenozoic, for example during the Oligocene (e.g. Shawe & Byers 1999) and Miocene (e.g. Rowley *et al.* 1992). Stewart (1988) noted that strike-slip faults typically do not extend from one structural block into another.

Figures 3 & 4 are geological maps of the Virginia City and Goldfield mining districts, respectively, and Figure 5 shows geological cross-sections of each district. Both districts were sites of Early Miocene andesitic to dacitic volcanism localized in transtensional zones along NW-striking right-lateral strike-slip fault systems (Berger *et al.* 2003; Berger *et al.* 2005). The earliest volcanism was andesitic, and was overlapped through time and then succeeded by predominantly dacitic to rhyodacitic volcanism. In the Virginia City area, volcanism extended into the Mid-Miocene (Castor *et al.* 2005), whereas, in Goldfield, volcanism was short-lived and ended during the late Early Miocene (Ashley 1974). In both districts, andesitic and dacitic volcanism spatially overlap only partially. The locus of strike-slip faulting migrated in time, with an apparent concurrent change in magma chemistry. The sequencing of points of interruption of strike-slip faulting and release into stepovers reflect spatial and temporal changes in the locus of strike-slip and related extensional faulting. The sequencing may be inferred from the location and orientation of fold axes; changes in the sense of displacement on normal faults; the age and distribution of the volcanic units; and the spatial migration of loci of hydrothermal activity. Striation data from slip surfaces, in conjunction with geological and fault interrelations, are used to support the interpretation of fault types and senses of displacement (Fig. 6). For both the Virginia City and Goldfield districts, these data are interpreted in conjunction with geological data to show that NW-striking faults are, in general, strike-slip faults with a significant extensional component.

In the Virginia City district, there were two periods of early andesitic volcanism, one between 18.3 and 17.1 Ma (Castor *et al.* 2005) and a second between 15.8 and 15.3 Ma (Castor *et al.* 2005). West of Virginia City, near Jumbo (Fig. 3), the older andesites are in fault contact with pre-Miocene rocks along a NW-striking fault zone (Hudson *et al.* 2003), interpreted herein to be predominantly strike-slip. A NW-striking zone of silicification, with associated sericite in the fault zone, has been dated at 18.11 ± 0.54 (Castor *et al.* 2005), and provides a minimum age for deformation along this fault trend. Opposing dips in Oligocene ash-flow tuffs across the trend that are not observed in the Miocene volcanic rocks suggest that NW-striking faults may be pre-Miocene. Northeast-striking normal faults clearly link into the NW-striking fault zone and make up a series of right bends in the vicinity of Jumbo (Fig. 3). A prominent right-bend in the strike-slip zone and associated normal faults at Jumbo are interpreted to be a zone of extensional release along the NW-striking fault zone. Therefore, the sense of displacement on the NW-striking fault was right-lateral, consistent with other NW-striking faults in the district (Fig. 6a). Lacustrine sedimentary rocks occur at the top of the older andesite sequence, and crop out only to the NE of the strike-slip fault zone. These sedimentary rocks indicate that, at least by approximately 17.1 Ma, a shallow, strike-slip

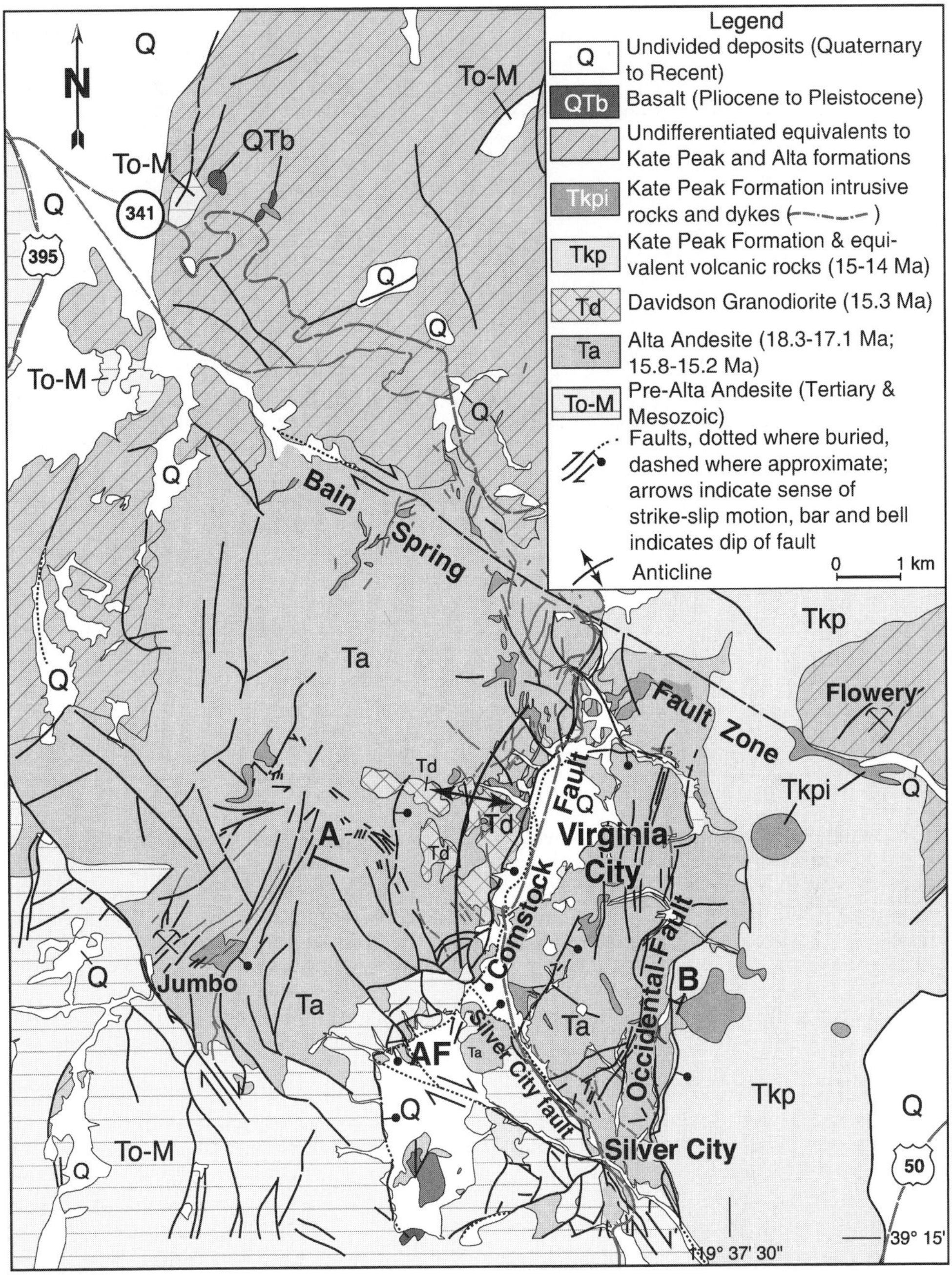

Fig. 3. Generalized geological map of the Virginia City mining district and vicinity. AF indicates the location of American Flat, and A–B indicates the location of the cross-section shown in Figure 5a (after Calkins & Thayer 1945; Thompson 1956; Hudson *et al.* 2003).

fault-bounded basin had formed across the heart of the Virginia City district.

During the second period of early Virginia City andesitic volcanism, adularia-bearing quartz veins (15.49 $\pm$ 0.04 Ma, Castor *et al.* 2005) formed along NE-striking normal faults at Jumbo (Fig. 3). The veins set a minimum age for extensional releasing along the NW strike-slip fault zone at this location.

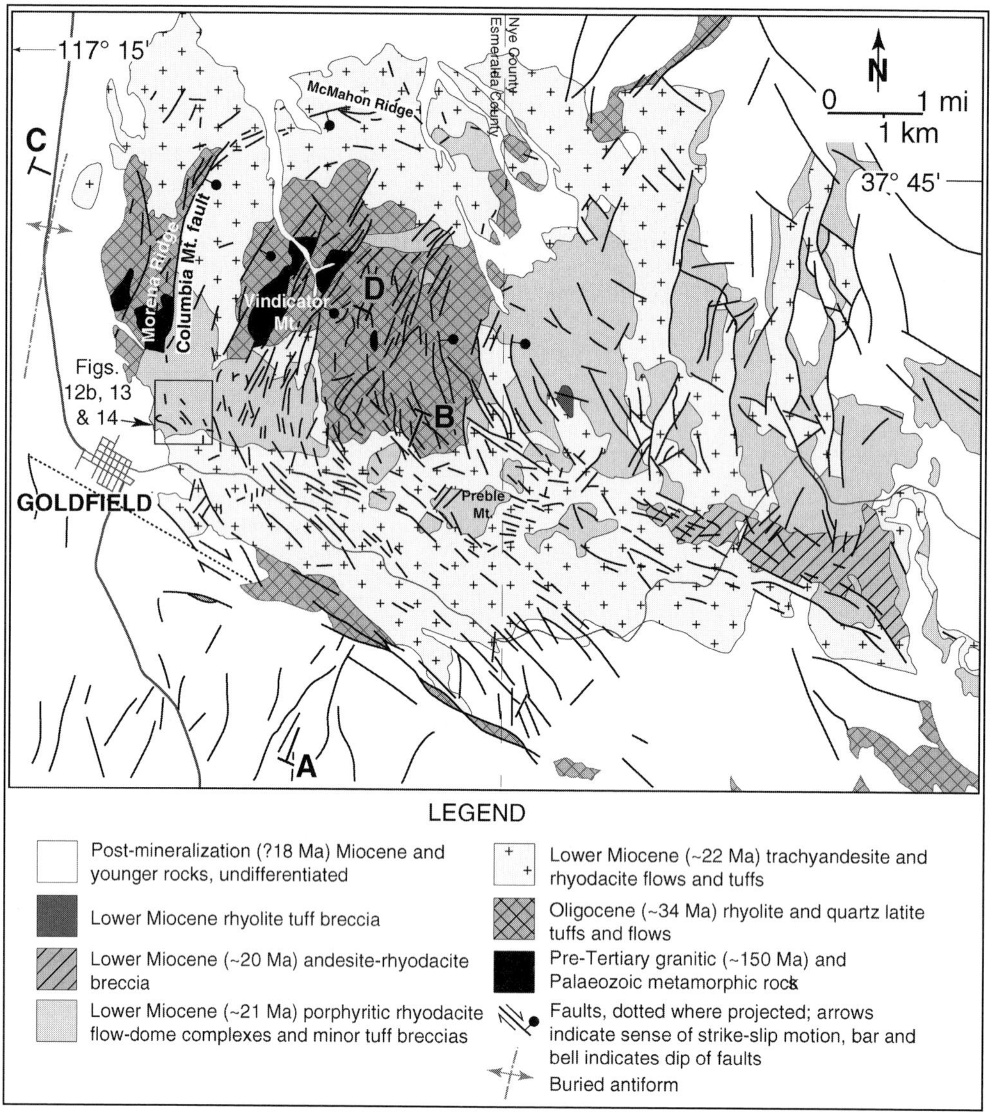

Fig. 4. Generalized geological map of the Goldfield mining district and vicinity. A–B and C–D indicate the locations of cross-sections shown in Figure 5b (after Ashley 1975; Ashley & Keith 1975).

The time that strike-slip faulting began to migrate NE from the vicinity of Jumbo to the Silver City fault zone (Fig. 3) can be no older than the lacustrine sedimentary rocks and, since the 15.3 Ma Davidson Granodiorite is the footwall of the Comstock Fault, complete capture of the locus of strike-slip faulting and associated extension could not be older than the stock. Quartz–alunite masses in the footwall of the Comstock Fault are approximately 15.4 Ma (Castor *et al.* 2005), and some are along NW-striking fractures that parallel a linear SW boundary of the 15.3 Ma Davidson Granodiorite (see Fig. 7) which is, in itself, collinear with the Silver City Fault zone. The dates indicate that fractures along the trend of the Silver City Fault were critically stressed concurrent with hydrothermal activity at Jumbo. All of these relations taken together are suggestive that the migration of the locus of strike-slip faulting to the NE was progressive and not abrupt. Figure 8

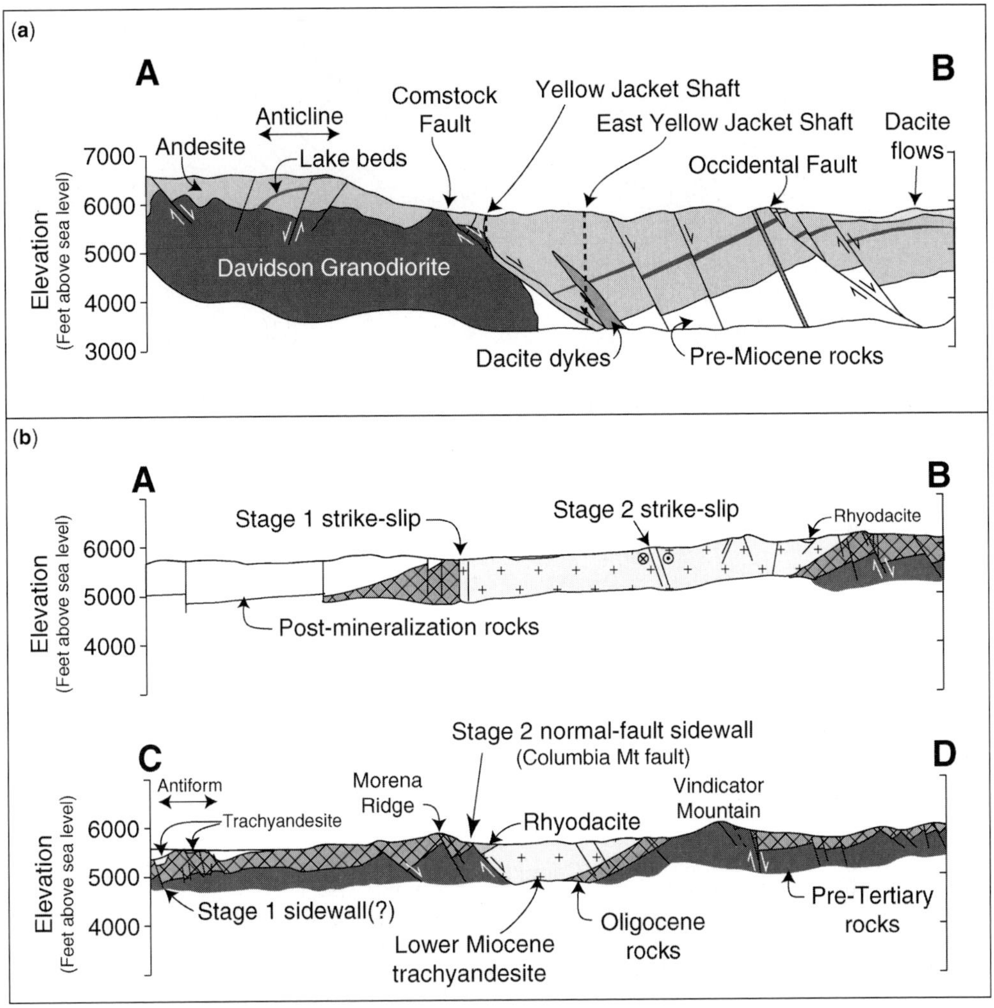

Fig. 5. Geological cross-sections showing fault relations in the Virginia City and Goldfield mining districts, Nevada. (**a**) East–west section (A–B) through the Comstock Fault at Virginia City, showing the spatial relation of the footwall anticline to extension in the hanging wall of the Comstock Fault. The location of the cross-section is shown in Figure 3 (cross-section after Calkins & Thayer 1945; Thompson 1956; Hudson 2003). (**b**) Cross-sections in the Goldfield district. Section A–B shows the relation of the locus of stage 1 strike-slip faulting to the locus of stage 2 strike-slip faulting on the SW margin of the stepover. Section C–D is through the stage 2 Columbia Mountain Fault bounding Morena Ridge on its east flank, showing the relation of the footwall antiform to extension in the hanging wall of the Columbia Mountain fault. Locations of the cross-sections are shown in Figure 4. (Cross-sections after Ashley 1975; Metallic Ventures Gold Ltd. 2007).

summarizes the time and spatial shift of faulting and volcanism in the district.

At approximately 15 Ma, a new stage of andesitic volcanism began in the Virginia City area. The duration of the volcanism was from 14.9 to 14.2 Ma (Castor *et al.* 2005). It consists of dykes and small plugs within the heart of the district and flow–dome complexes to the north and east of Virginia City proper. Hydrothermal activity and the deposition of banded quartz–adularia veins ensued at 14.1 Ma (Castor *et al.* 2005) along the Comstock and Silver City fault zones. Striations, mullions, the geometry of hanging-wall faults in the Comstock fault zone, and geological relations all indicate that it initially had normal-sense displacement (Berger *et al.* 2003). However, striations on

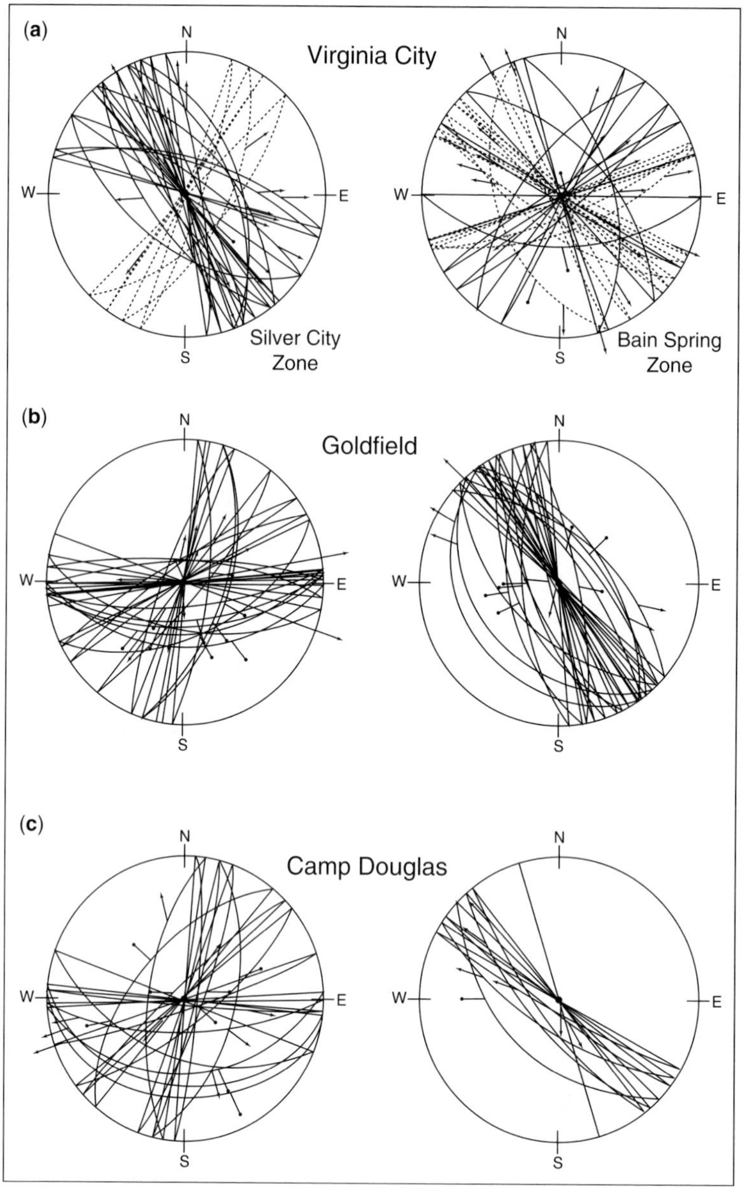

Fig. 6. The plunges of striations on fault surfaces in the (**a**) Virginia City, (**b**) Goldfield, and (**c**) Camp Douglas mining districts, Nevada. Plunges $\leq 45°$ are indicated by a pointed arrow symbol. Plunges $>45°$ are indicated by a filled circle symbol.

the hanging-wall side of veins and along fractures obliquely cross-cutting the footwall veins, together with unlikely vein dips and geometries; vein-parallel folds and stretching; and the pre-ore hydraulic crushing of veins, indicate that strains in the fault zone changed from transtensional to transpressional. The relation of ore minerals to vein textures indicate that this inversion took place during the later stages of hydrothermal activity (Berger *et al.* 2003). The inversion does not affect the Comstock Fault trend south of its intersection with the Silver City Fault or north of its intersection with the Bain Spring Fault and, therefore, must reflect a local, not regional, response to changing stress conditions. The inversion stresses are interpreted to result from the migration of the locus of extension east of the Comstock Fault to the Occidental Fault. An anticlinal fold in the footwall of the Comstock Fault may have formed in response to the inversion, although Thompson (1956) suggests that it reflects doming above the Davidson Granodiorite. Since the fold axis is situated east of the bulk of the pluton; is a generally tight fold; and there are transverse kinks in it behind a kink in the Comstock Fault footwall, it is considered herein to be most likely post-intrusion in age and related to the timing of fault-slip inversion along the Comstock Fault.

A spatially evolving pattern of volcanism and faulting at Goldfield analogous to Virginia City is suggested in the distribution of strike-slip faults, early-stage andesite lavas (22.4–21.8 Ma; Vikre *et al.* 2005) and succeeding rhyodacite volcanic rocks (21.8–21.5 Ma; Vikre *et al.* 2005), and a structural antiform or horst block along the western edge of the Goldfield district (Figs 4 & 9). The normal fault that bounded the stage 1 stepover on its NW margin (Fig. 9a), herein defined as a 'stepover sidewall fault' (cf. Dooley & McClay 1997), is not exposed. Its location is inferred from the location of the antiform (Figs 4 & 9b) that has been delineated by exploration drilling (Metallic Ventures Gold 2007). As in Virginia City, at Goldfield the shift in the locus of NW-striking, strike-slip faulting was to the NE, and appears to have been concurrent with a change in magma chemistry.

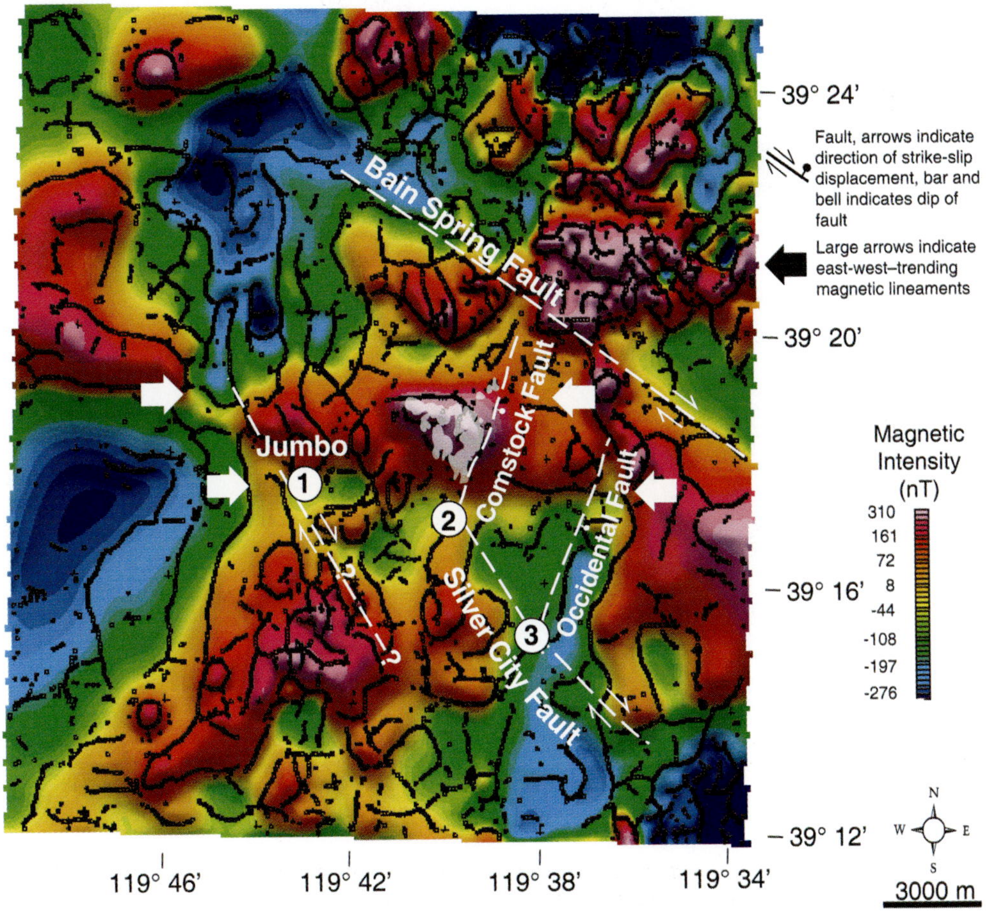

Fig. 7. Reduced-to-pole aeromagnetic map of the Virginia City mining district and vicinity, Nevada. The small circles are points of inflections in the magnetic frequency data, and continuous lines of circles are interpreted to be boundaries between domains of differing magnetic susceptibility. White, dashed lines indicate approximate locations of principal faults; white arrows indicate trends of east–west shear zones manifested as prominent magnetic boundaries; numbered circles indicate loci of principal points of interruption of strike-slip faulting and nucleation of extensional stepping; and light-grey areas indicate outcrops of a lower Middle Miocene granodiorite intrusion.

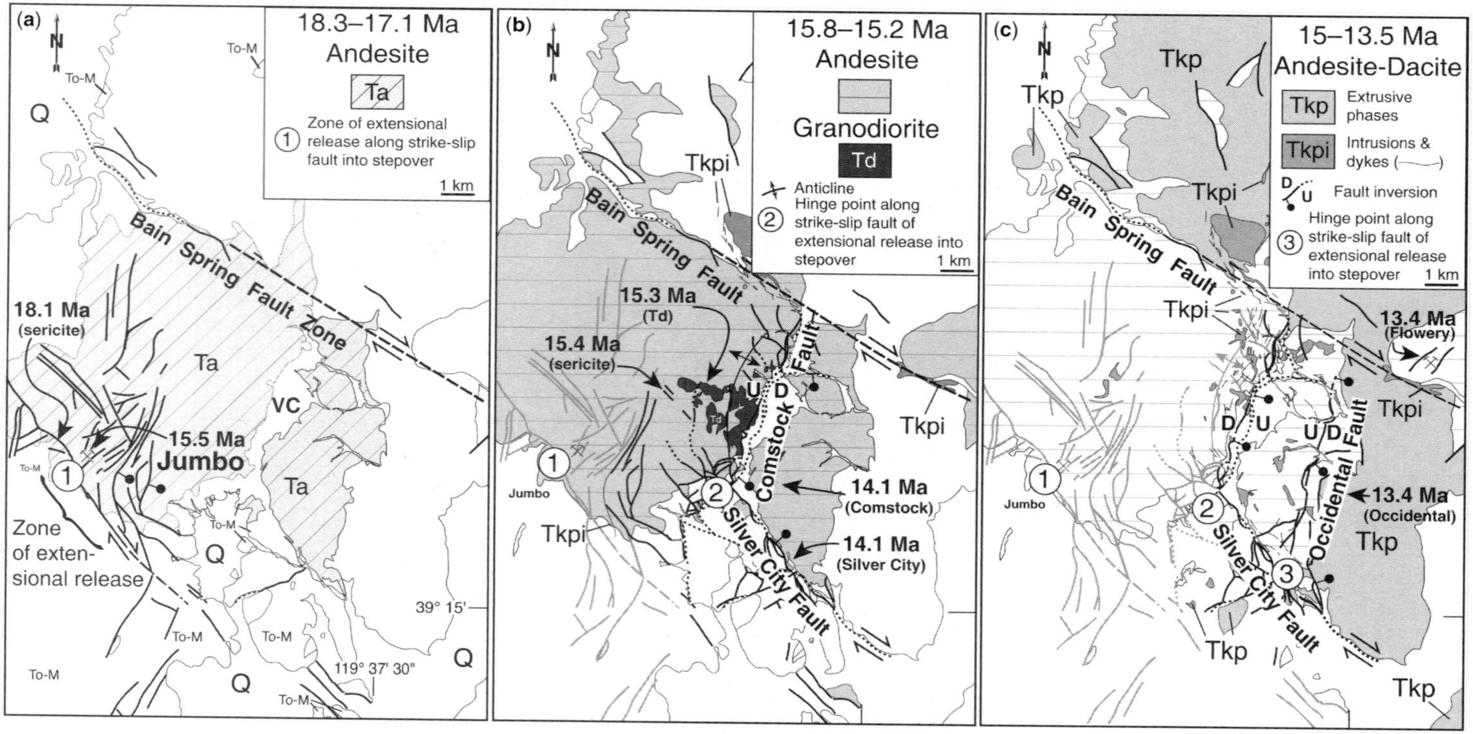

Fig. 8. Distribution of three stages of Upper to Middle Miocene volcanic rocks in outcrop in the Virginia City mining district, Nevada. Circled numbers mark the loci of principal points of interruption of strike-slip faulting and nucleation of extensional stepping and their interpreted sequencing. The $^{40}Ar/^{39}Ar$ thermochronological dates are from Castor *et al.* (2005). (**a**) Stage 1 lava flows of predominantly andesite composition. A sequence of lacustrine rocks is locally conformable on Stage 1 andesites. (**b**) Stage 2 lava flows and flow breccias of predominantly andesite composition. (**c**) Stage 3 lava flows, domes, and dykes of predominantly dacite composition.

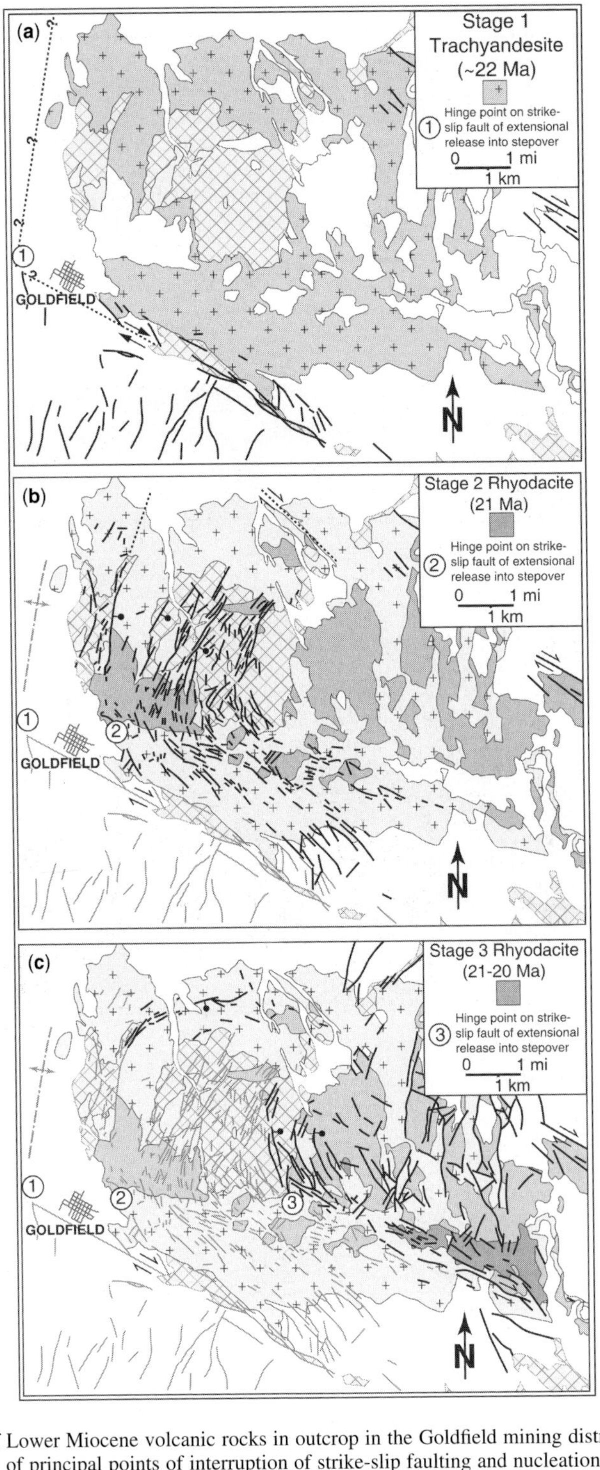

Fig. 9. Distribution of Lower Miocene volcanic rocks in outcrop in the Goldfield mining district, Nevada. Circled numbers mark the loci of principal points of interruption of strike-slip faulting and nucleation of extensional stepping, and their interpreted sequencing. (**a**) Stage 1 lava flows of predominantly trachyandesite composition. Some lacustrine beds occur in the section. (**b**) Stage 2 rhyodacite domes and related flows. (**c**) Heterolithic breccia units and continued emplacement of rhyodacite domes.

Striation data from faults in the Goldfield district are shown in Figure 6b. Most of the data from NW-striking faults was collected within the zone of extension and is reflective of the transtensional nature of the faulting. Cross-section A–B in Figure 5b is across the SW strike-slip boundary of the Goldfield stepover, and shows the lateral dimension of the northerly migration from the stage 1 strike-slip locus to its stage 2 locus.

Figures 7 & 10 are aeromagnetic maps of the Virginia City and Goldfield districts, respectively. For the Virginia City area, the principal strike-slip faults and related normal faults are shown in thin, white lines, whereas on the Goldfield map all faults from Ashley (1975) are shown, and the black masses are fault–fracture-controlled quartz–alunite–pyrophyllite alteration. Large, white arrows indicate the trends of interpreted

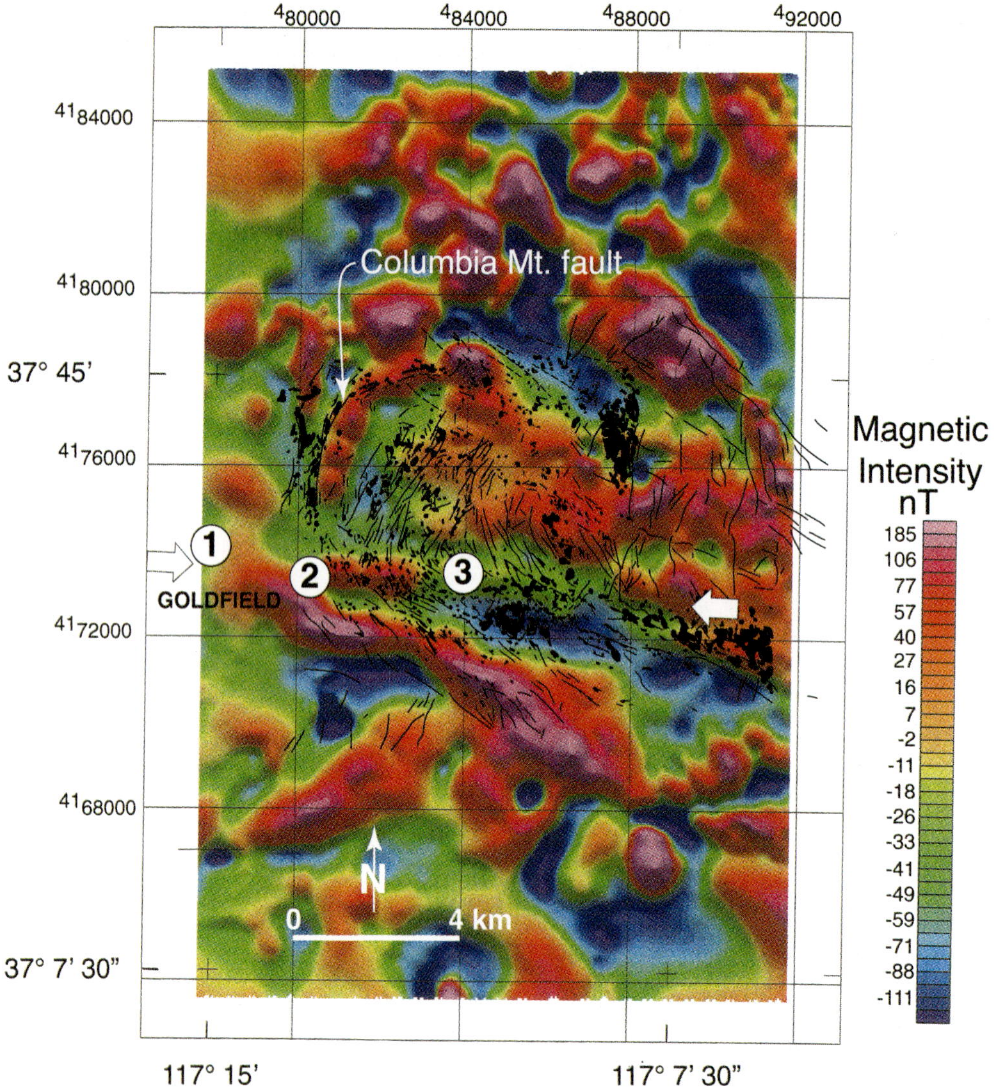

Fig. 10. Reduced-to-pole aeromagnetic map of the Goldfield mining district and vicinity, Nevada. Thin black lines are faults (after Ashley 1975); black masses are outcrops of quartz–alunite ± pyrophyllite alteration (after Ashley & Keith 1975); white arrows indicate the trend of a WNW–ESE shear zone manifested as a prominent magnetic lineament; and the numbered circles indicate the loci of principal points of interruption of strike-slip faulting and nucleation of extensional stepping.

transverse, basement shear zones indicated by magnetic troughs in both high- and low-frequency data. For both districts, the circled numbers indicate interpreted locations of the principal points at which strike-slip faulting was interrupted and strains transferred into an extensional stepover. The interpreted basement transverse shear zones are misoriented with respect to the inferred Miocene 18–15 Ma stress field in order to accommodate displacement, and they primarily only disrupt the local stress field. There is a spatial correlation between the transverse zones and points of interruption of strike-slip displacement (Figs 7 & 10). Field evidence for the transverse magnetic lineaments reflecting basement shear zones includes east–west-elongated zones of hydrothermal alteration; the en échelon stepping of NW-striking veins across transverse faults; and the local accommodation of normal slip, for example along the McMahon Ridge (Fig. 4) at Goldfield, where the local stress field was strongly perturbed.

Transpressive fault deformation in the Virginia City district

As noted above, fault geometries, together with vein textures and geometries; rotated veins; and the development of an anticline in the Comstock footwall, provide evidence that displacement was inverted along the Comstock fault zone (Berger *et al.* 2003). The effects of inversion vary along the fault zone – an apparent manifestation of the relative mechanical strengths of the hanging wall and footwall. From the point where strike-slip displacement on the Silver City Fault was interrupted at its intersection with the Comstock Fault (Fig. 8) and thence north along the Comstock Fault for approximately 1 km, the footwall consists of weakly to moderately altered andesite on weakly altered Davidson Granodiorite and pre-Tertiary rocks. Along this reach, the hanging wall is argillized andesite. Quartz veins against the footwall were hydraulically crushed or sheared out altogether during inversion, with a hanging-wall vein opposite the sheared-out portion of the footwall vein reversing dip from east to the west and being truncated against a 15–20° east-dipping flat on the footwall fault. This geometry is shown in a cross-section through the Yellow Jacket Mine (Fig. 11a, below). North of the Yellow Jacket Mine along the lode (Fig. 11b), the footwall is propylitically altered granodiorite with no andesite. In the Hale & Norcross Mine, north–south shortening of the footwall fault suggests that it has accommodated little reverse-sense motion, in contrast to along the footwall in the Yellow Jacket Mine. However, some right-lateral displacement and horizontal stretching of the footwall vein in the Hale & Norcross hanging wall are evident. Hanging-wall veins have ramps, flats, and inverted dips similar to the hanging wall in the Yellow Jacket Mine. The striations found on quartz-vein band surfaces and hanging-wall faults plunge approximately 45° south (e.g. King 1870) implying a normal-oblique sense of displacement, and, in outcrop, oblique-slip lines superimposed on vertical-slip lines imply that the lateral component of slip was right-sense – consistent with evidence on the footwall.

The geometry of faulting at points of strike-slip release into stepovers

Owing to the coupling of fluid flow and stress, veins and vein-form alteration, together with the geometry of mapped mine stopes, provide images of deformation within stepovers. At Virginia City and Goldfield, the geometric relations of mineralized rocks suggest that at the point of termination of strike-slip faulting and consequent transfer of displacement into an extensional stepover, the footwall of the nucleated normal-fault stepover sidewall migrated out and around the tip of the strike-slip zone.

At Virginia City, the mapping of underground workings by Becker (1882) shows the NW-striking, NE-dipping Silver City Fault to terminate against an arcuate, concave-east bend of the Comstock Fault footwall (Fig. 12a). The bend is a short-wavelength, low-amplitude synform that plunges east parallel to displacement on the Comstock Fault. To the north, there is a complementary antiformal fault bend of the Comstock hanging wall. On the hanging-wall side of the Silver City fault zone, a series of fault veins are arcuate and link the Silver City Fault to the Comstock hanging wall. These arcuate faults are inferred to be a series of horsetail faults (Berger *et al.* 2003).

In the Goldfield mining district, mine stopes at different depths along faults in the January and Combination mines outline the 3D geometry at a fault tip along a NW-striking, NE-dipping strike-slip fault zone (Fig. 12b). Overall, the geometry of the terminated fault and linked normal stepover faults is similar to the Silver City–Comstock intersection (Fig. 12a). Displacement along the strike-slip fault is truncated against a short-wavelength, concave-east bend in the NE-striking normal faults that arch out and around the strike-slip zone. In contrast to Virginia City, a complementary antiform further into the stepover is not evident from the available data.

The geometry of faulting within extensional stepovers

In the Virginia City and Goldfield mining districts, the stepover-bounding normal faults discussed are

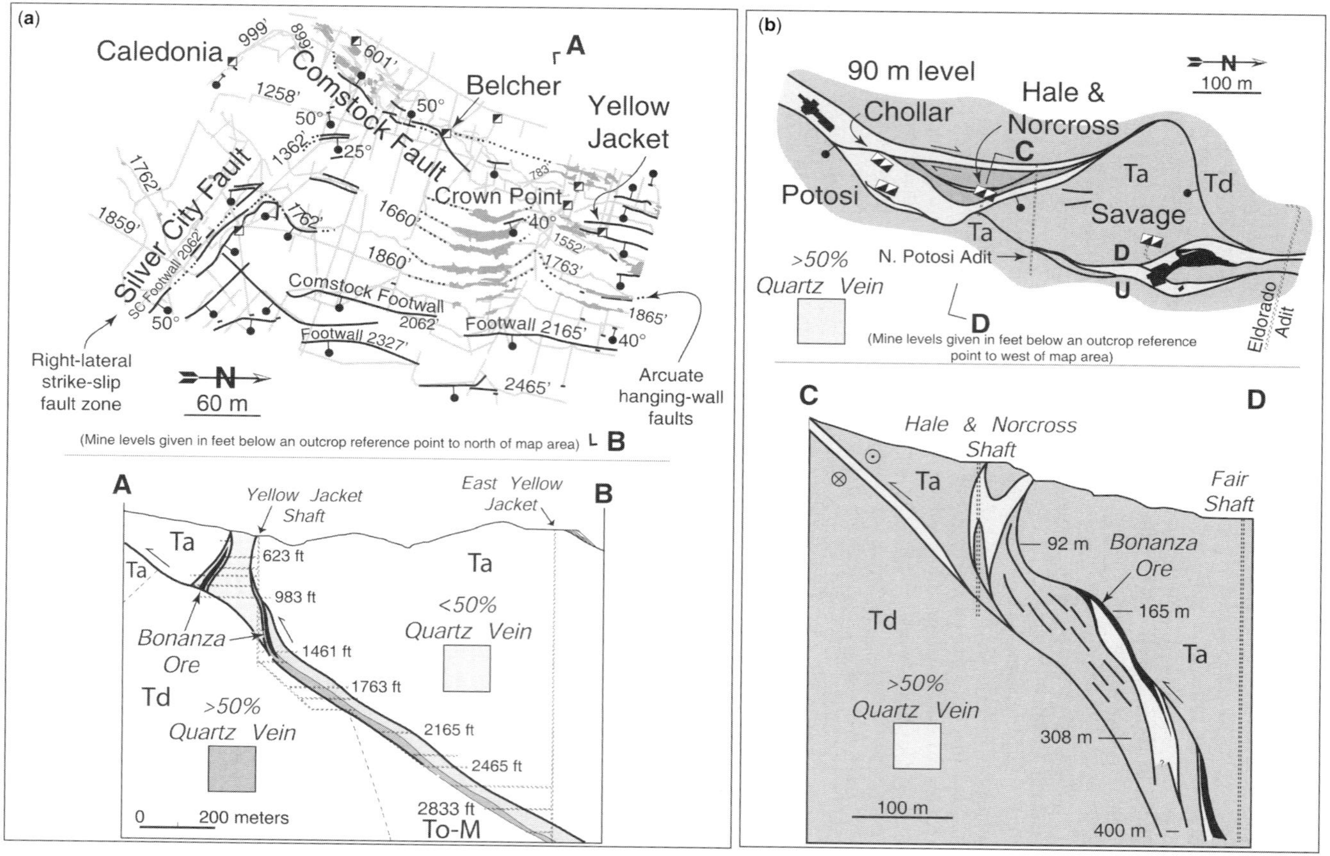

Fig. 11. Plans and cross-sections illustrate the transpressional deformation along the Comstock fault zone, Virginia City, Nevada. Faults are dotted where projected and bar and bell symbols indicate the dip direction of faults. (**a**) Plan view of mine workings and faults in the southern part of the Comstock fault zone (after F. Calkins, unpublished data, 1946). The medium-grey-shaded masses are stoped areas within the Comstock hanging-wall veins (after King 1870). SC stands for Silver City, and A–B denotes location of the cross-section through the Yellow Jacket Mine. Cross-section A–B illustrates the inversion of displacement along the Comstock fault zone, following a migration of the locus of extension to the east. Geological unit symbols are the same as in Figure 3 (Berger *et al.* 2003 after King 1870 and Becker 1882). (**b**) Plan view of the Comstock Fault on the 90 m level in the Hale & Norcross Mine in the central part of the Comstock fault zone. The black masses are stoped areas within the Comstock fault zone (Berger *et al.* 2003, after Becker 1882). C–D denotes the location of the cross-section through the Hale & Norcross Mine. Cross-section C–D illustrates the right-lateral transpressional deformation of the Comstock fault zone, following a migration of the locus of extension to the east (Berger *et al.* 2003, after Becker 1882). Geological unit symbols are the same as in Figure 3.

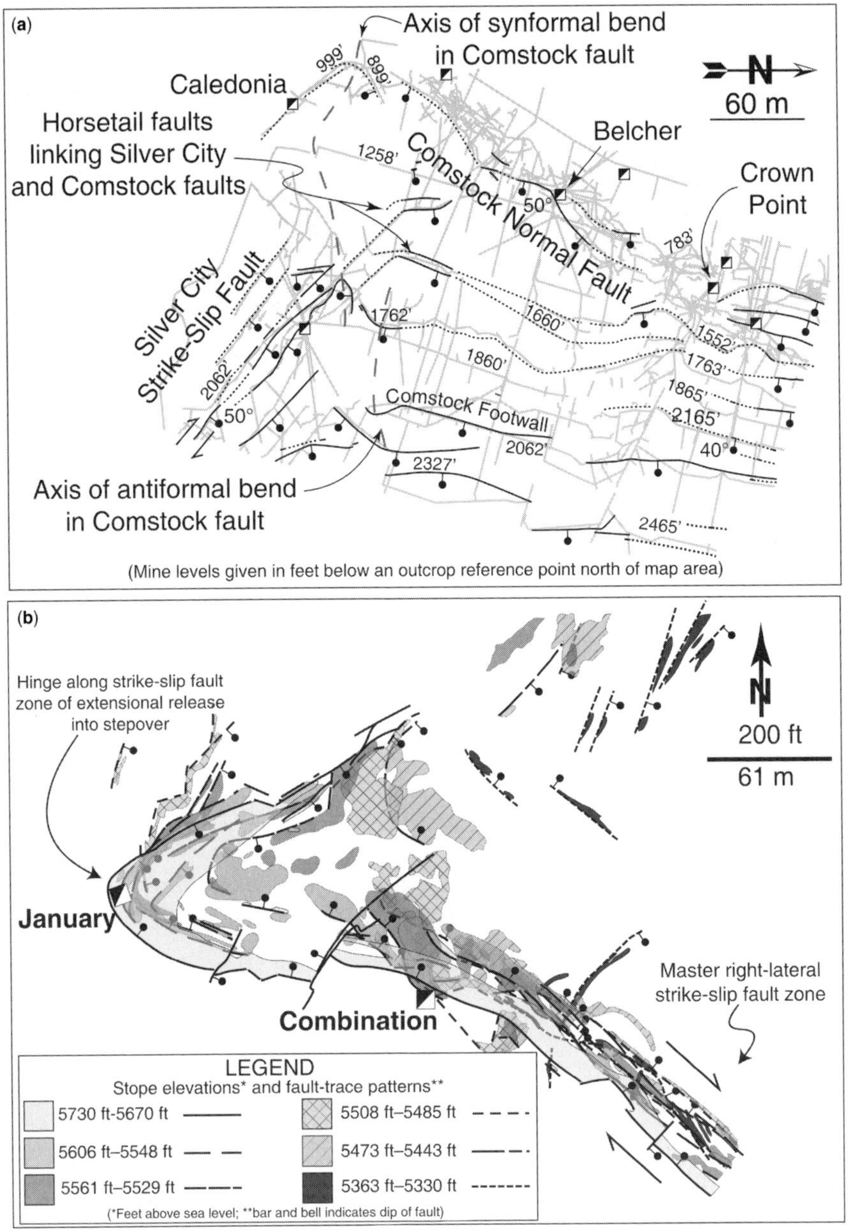

Fig. 12. Mine workings along faults provide images of the geometry of faults in zones of strike-slip fault interruption at points of extensional release. Faults are dotted where projected, and bar and bell symbols indicate the dip direction of the faults. (**a**) In the Virginia City district, the mine workings along the vein-controlling Silver City and Comstock faults, together with some mapped slip surfaces, make up 3D outlines of the fault surfaces, thereby elucidating the fault geometries at their point of intersection. The intersection consists of a synformal bend in the Comstock Fault out and around the tip of the Silver City Fault. The synform is complemented by an antiformal bend in the Comstock Fault. Horsetail faults in the hanging wall of the Silver City fault zone link into the hanging wall of the Comstock Fault (modified from Becker 1882; Berger *et al.* 2003). (**b**) Mine stopes at six elevations in the January and Combination mines in fault-controlled mineralization in the Goldfield district, Nevada, provide a 3D image of the fault geometries at the intersection of a strike-slip fault and a linked normal-oblique fault. The intersection consists of a synformal bend of a normal-oblique fault out and around the tip of the NW-striking right-lateral fault zone (Berger *et al.* 2005, after Ashley & Keith 1975).

the Comstock and Columbia Mountain faults, respectively (Figs 3 & 4). These normal sidewall faults are relatively linear and continuous in plan; accommodated the greatest measurable displacement in each district (e.g. Hudson 2003; Ashley 1975); and have listric geometry. Banded veins are only present on the Comstock Fault. The bands parallel the walls of the fault, except where NW- or NE-striking faults intersect the Comstock Fault from the hanging-wall or footwall sides. In these instances, the bands also strike NW or NE, although they are wholly contained within the NNE-striking walls of the Comstock Fault. Thus, bands preserve evidence of the state-of-stress on the hanging wall and footwall of normal faults.

In both districts, faults complexly deform the hanging wall of the stepover normal-fault boundary. Along the length of the Comstock fault zone at Virginia City, a series of steeply dipping, long-wavelength and slightly overlapping, convex-east faults deform and structurally segment the hanging wall. Orebodies in the Crown Point and Yellow Jacket mines near the Silver City Fault (shaded light-grey areas in Fig. 11a) illustrate the geometry of two such segments. Each segment along the Comstock Zone differs in amplitude. A lack of underground access precludes observing the details of segment intersections and the temporal sequencing of segmentation. Nevertheless, banded quartz–adularia veins provide information about the mode of fault opening and displacement gradients along fault segments. The banded veins thicken in the axial part of the arcuate faults; thin into linking NW- and north-striking zones with adjacent segments; and have bands that generally parallel vein margins. These relations indicate that predominantly normal-sense displacement was accommodated in the hanging wall during much of the duration of vein formation.

In the Goldfield district, pre-ore-stage quartz–alunite $\pm$ pyrophyllite hydrothermal alteration delineates a complex network of faults in the hanging wall of the Columbia Mountain Fault (Figs 4 & 13). Hanging-wall strains were accommodated in a complex manner, especially as the master strike-slip fault at the southern termination of the Columbia Mountain Fault is approached. North of the intersection with the strike-slip fault in the January Mine (Fig. 13a), hanging-wall displacement appears to have been partitioned on honeycomb-like meshes of extensional and shear faults at various spatial scales. Farther into the stepover in the Red Top Mine (Fig. 13a), the along-strike continuity of normal faulting is greater and is less strongly disrupted by shear. In cross-section (Fig. 13b), the fault-controlled quartz–alunite $\pm$ pyrophyllite alteration shows the hanging-wall faults to generally parallel the Columbia Mountain Fault, but flatten somewhat when stratigraphic contacts are approached (e.g. Searles 1948).

Figure 14 illustrates the complexity of the Goldfield Fault meshes. The zigzag pattern of the mine stopes at different depths in the Mohawk Mine (Fig. 14a) provides a 3D representation of how heterogeneously and asymmetrically shear and tensile strains were partitioned between faults of different orientations. The more generally linear stopes in the Red Top Mine (Fig. 14a) suggest that, further into the stepover, there was a decrease in strain partitioning. The Clermont Mine developed ores on faults further into the hanging wall of the Columbia Mountain Fault than the Mohawk Mine. In this mine (Fig. 14b), a detailed plan view of the fault surface is preserved in the geometry of the quartz–alunite $\pm$ pyrophyllite alteration zone. Short-wavelength rolls in the alteration zone along its southern arm probably reflect a different, perhaps local, response to tensile and shear stresses than larger-wavelength zigzags further north that probably reflect larger-scale stresses. Although the overall geometry of the fault-controlled alteration in the Clermont and Mohawk mines is similar, the vertical, pipe-shaped ore stopes indicate that the permeability structure in the Clermont Mine differed in significant ways from permeability in the Mohawk Mine at the time of ore formation which, in both cases, post-dates the quartz–alunite $\pm$ pyrophyllite alteration.

The geometry of faulting in contractional stepovers

In the western United States, epithermal mineralization is less common in contractional strike-slip fault stepovers than in extensional stepovers. An example of mineralization in a contractional stepover occurs the Camp Douglas area (Fig. 1) in the eastern Excelsior Mountains, Nevada. In this region, an east–west zone of transverse faults has affected the localization of intrusions and hydrothermal activity intermittently since the Jurassic. Figure 15 shows a general geological map of the eastern Excelsior Mountains region, and Figure 16 is a reduced-to-pole aeromagnetic map of this same region, showing the major faults irrespective of age. The eastern Excelsior Mountains are seismically active today, owing to the partitioning of a portion of the displacement along the San Andreas transform plate margin from California east into Nevada and on to NW-striking right-lateral faults in what is known as the central Walker Lane fault zone (location in Figs 1 & 16). Mineralization in the district long preceded the timing of transform-related faulting in this region.

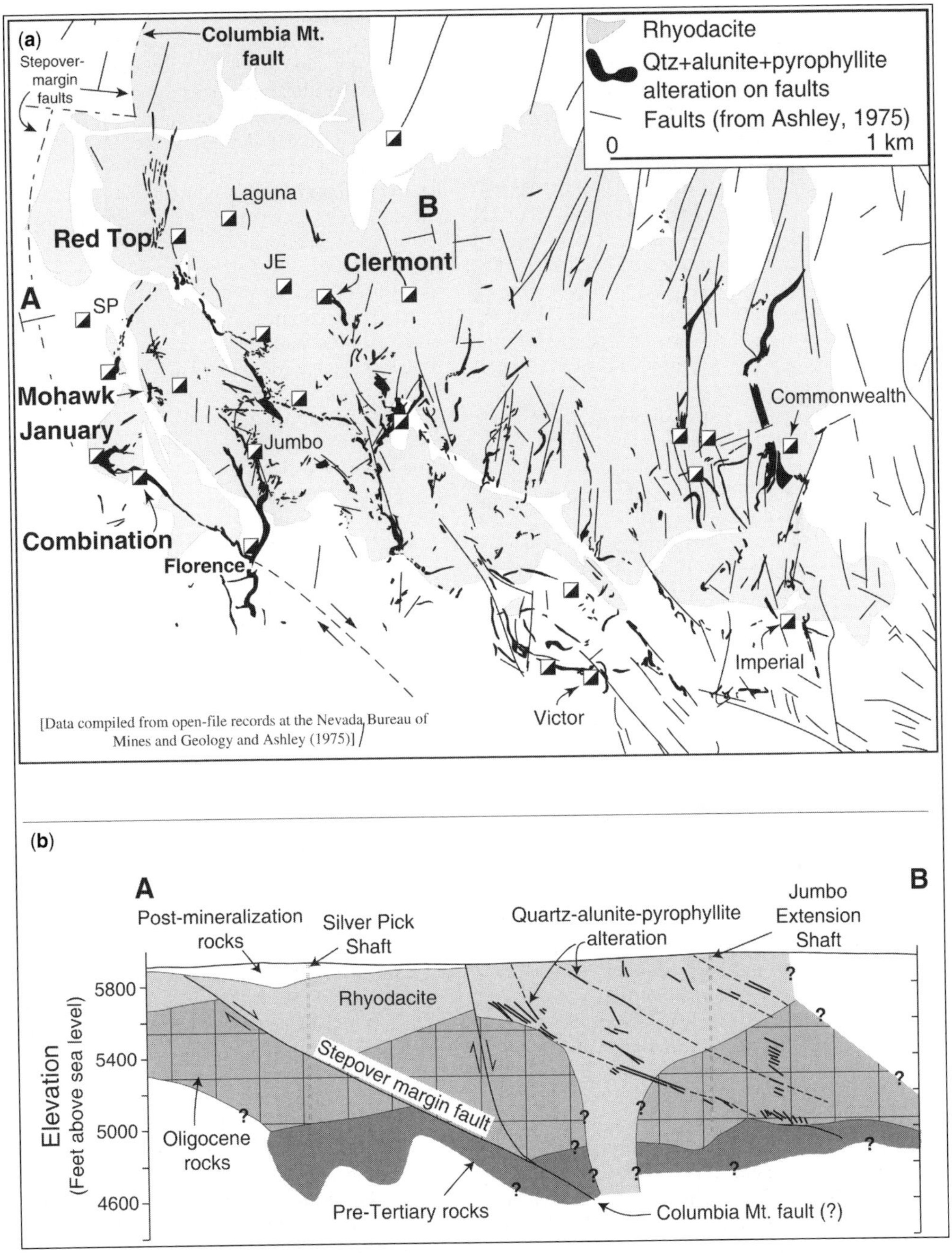

Fig. 13. Fault-controlled quartz–alunite–pyrophyllite alteration in the Goldfield mining district, Nevada, outlines the complex geometry of faulting within an extensional stepover near the point of interruption of strike-slip faulting and release into the stepover. (**a**) Plan view of the outcrop pattern of quartz–alunite–pyrophyllite alteration delineating a zigzag pattern characteristic of localization within a mesh of extension and shear fractures. Selected mines are shown for reference; SP is the Silver Pick shaft and JE is the Jumbo Extension shaft (data compiled from open-file records, Nevada Bureau of Mines and Geology). (**b**) Generalized cross-section (A–B) through the Silver Pick and Jumbo Extension shafts showing the structural relation of faults in the hanging wall of the stepover boundary fault to the stepover normal-fault boundary and stratigraphic contacts (after Searles (1948) and data in open files of the Nevada Bureau of Mines and Geology).

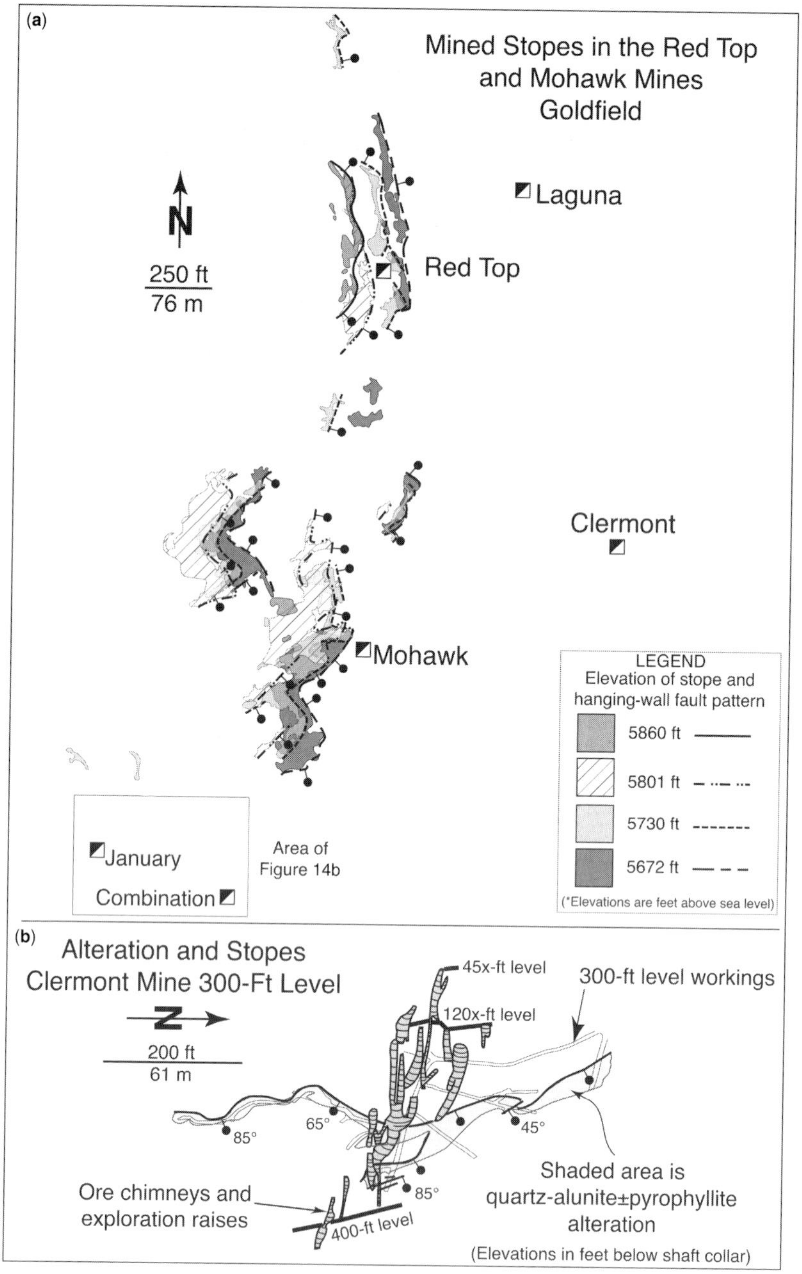

Fig. 14. Maps of mine stopes in mineralized faults in the Goldfield mining district, Nevada, illustrate the complex way in which strains can be accommodated adjacent to a point of interruption in strike-slip faulting and extensional release into a stepover. The bar and bell symbols indicate the direction of dip of the faults. (**a**) Map of stopes on four mined levels down-dip along a fault in the Mohawk and Red Top mines, Goldfield, provides a 3D image of the fault geometry. The trace of the fault on its hanging wall on the different mine levels is shown, the zigzag pattern of which is interpreted to represent the partitioning of strains in a mesh of extension and shear fractures. Strong strain partitioning is more evident in the Mohawk Mine than in the Red Top mine farther north into the stepover. (**b**) Map of fault-controlled quartz–alunite–pyrophyllite alteration and the trace of the fault footwall in the Clermont Mine (location shown on Fig. 14a) illustrates the undulatory and zigzag architecture of an unmined alteration zone, in particular the heterogeneity and differences in scale and magnitude of strain partitioning along the fault zone. (Data compiled from open-file records, Nevada Bureau of Mines and Geology.)

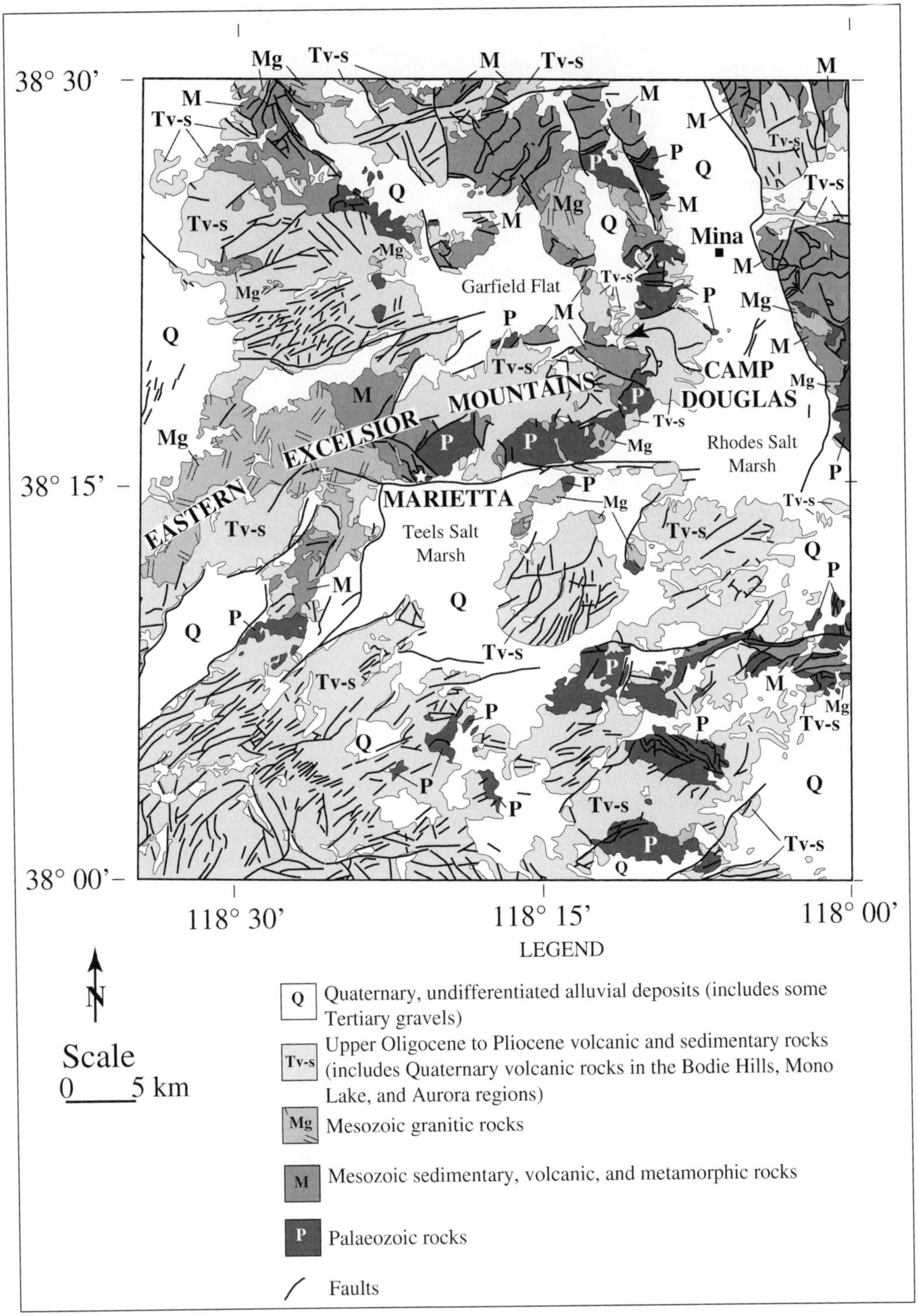

Fig. 15. Regional geological map of the eastern Excelsior Mountains, Nevada, and vicinity (after Stewart 1982).

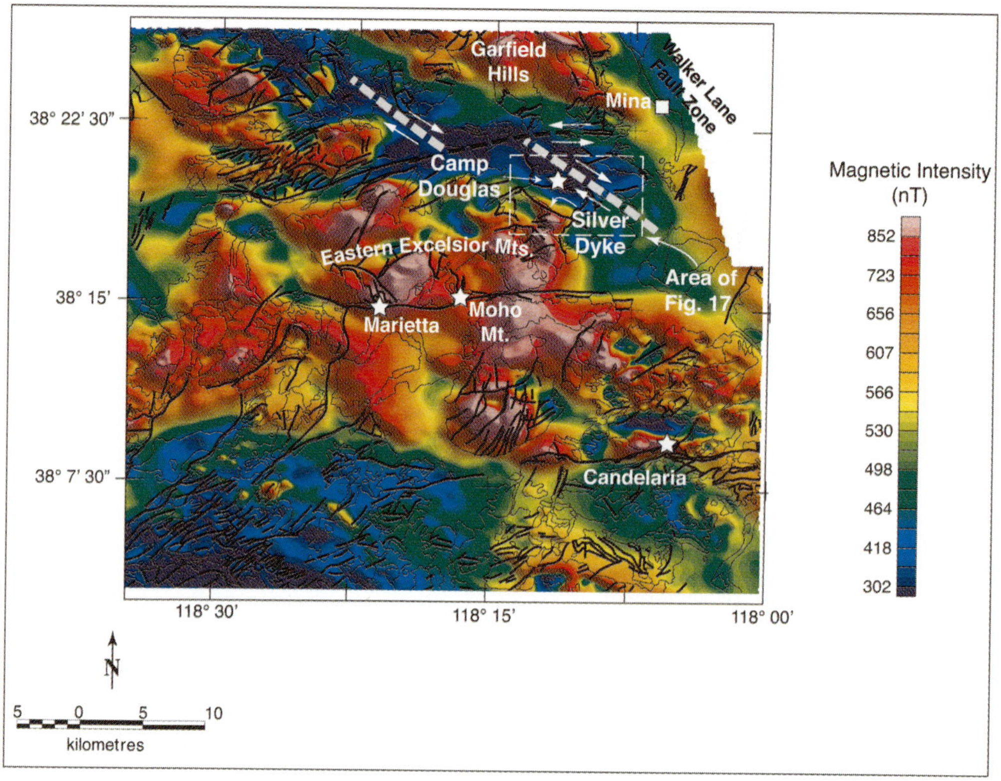

Fig. 16. Reduced-to-pole aeromagnetic map of the eastern Excelsior Mountains, Nevada, and the Camp Douglas mining district. Grey dashed line indicates array of magnetic lows known at the most southeasterly and most northwesterly ends to be coincident with NW-striking, strike-slip faults. Solid black lines are faults; stars indicate mineralized areas. The white box outlines the shown in area Figure 17.

During Early Miocene volcanism in the Camp Douglas area, lavas extruded from local vents on to a basement of Oligocene ash-flow tuffs and Mesozoic and Palaeozoic metavolcanic, metasedimentary and igneous rocks. There were at least three distinct episodes of hydrothermal activity associated with the Miocene volcanism. Potassium–argon dates (Garside & Silberman 1978) show that the first episode of quartz–adularia vein formation occurred along the Silver Dyke fault zone (Figs 16 & 17) at about 18–17 Ma, followed by quartz–adularia veins along an east–west fault zone at Camp Douglas (Figs 16 & 17) at about 16–15 Ma. Immediately following the Camp Douglas mineralization, silicification and advanced-argillic to argillic alteration occurred to the east of Camp Douglas and NE of the Silver Dyke fault zone (areas with diagonal lines in Fig. 17).

A generalized tectonic map of the eastern Excelsior Mountains is shown in Figure 17. Striation data show that NW-striking faults accommodated strike-slip displacement (Fig. 6c). The generally east–west faults at Camp Douglas (star symbol, Figs 15–16) and the Silver Dyke Fault are reverse faults that at some localities place Cretaceous rocks above Tertiary volcanic rocks. The structural arch (Fig. 17) between the two reverse-fault zones is referred to as the Thunder Mountain Block. The reverse faults link along strike into NW-striking right-lateral faults, for example at the Kernick Mine (Fig. 17). A strike-slip fault zone at the easternmost end of the range, indicated by the dashed white line in Figure 16, corresponds with a NW-striking zone of low magnetic intensity. Northwest-striking faults are exposed in outcrop primarily to the SE of Camp Douglas, and are well delineated by silicification and alunite–clay alteration (diagonal pattern in Fig. 17). From the available geological data, it is interpreted that the low-magnetic-intensity lineament corresponds with a right-lateral fault zone, a portion of which has been obfuscated by present-day tectonism. The left bend is interpreted to have resulted in the

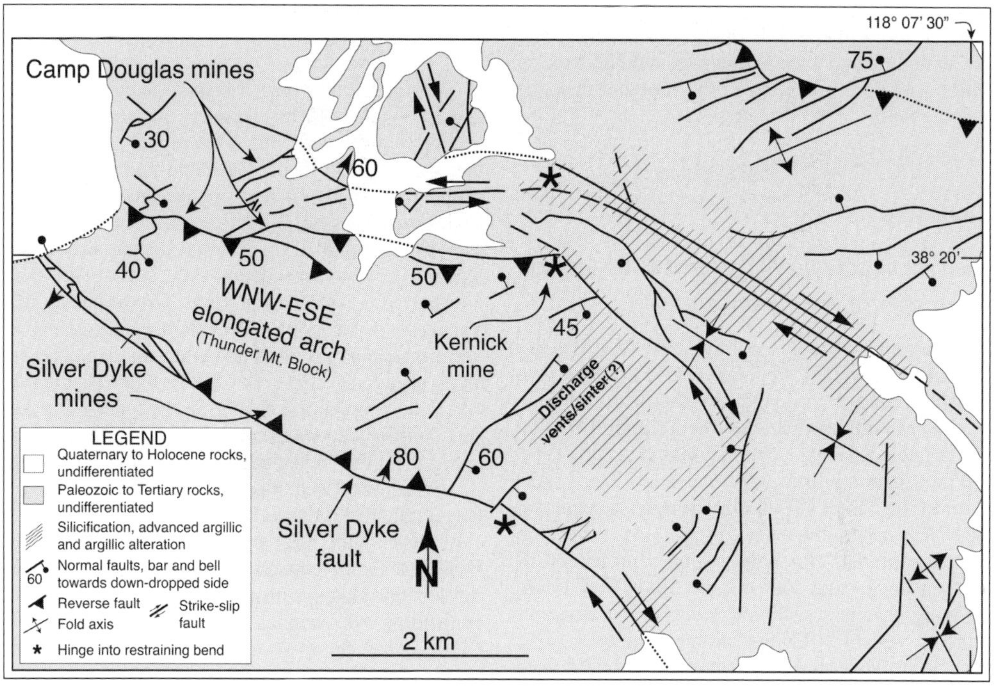

Fig. 17. Tectonic map of the Camp Douglas mining district, illustrating the relationship of NW-striking, right-lateral faults to east–west reverse faults. The location of the map area is shown in Figure 16 (from R. E. Anderson & B. R. Berger, unpublished mapping, 2006, after Garside 1979).

formation of a restraining stepover in the Camp Douglas area. (Fig. 17).

The Thunder Mountain contractional uplift is asymmetrical, with the reverse faults on the north margin dipping less steeply (*c.* 50–60°) than those on the south margin (*c.* 70–90°). Palaeomagnetic data from the Mesozoic intrusion along the Silver Dyke Fault (Fig. 17) indicate that the uplifted mountain block may have been rotated up to 20° clockwise (J. Geissman, pers. comm., 2005).

The Silver Dyke reverse-fault zone (Fig. 17) forms a generally unsegmented 5-km-long south boundary to the block, although parallel faults and their penetration back into the hanging wall vary along strike. Northeast- to NNE-striking faults also occur in the hanging wall and increase in density toward the eastern terminus of the Silver Dyke Zone, where it links (asterisk symbol in Fig. 17) with a right-lateral strike-slip fault. In the Thunder Mountain Block within a kilometre of the point of linkage, NE-striking veins in the hanging wall increase in density and become nearly 100% of the rock. Where the Silver Dyke Fault intersects the NW-striking right-lateral fault, it is truncated by a NE-striking, SE-dipping fault. In addition, as the intersection is approached, east–west faults in the Silver Dyke hanging-wall bend to a more NE orientation and merge into the footwall of the block-bounding NE-striking fault.

Fault and vein relations along the Silver Dyke Fault imply that contraction was occurring perpendicular to its strike whilst the hanging and footwalls were extending at high angles to its strike. North-northeast-striking banded quartz veins are consistent with strains related to regional stresses. The NE-striking banded veins probably reflect extension related to the probable clockwise rotation of the Thunder Mountain Block, and the increase in vein density approaching the link with the strike-slip fault discussed above probably reflects the accommodation of rotation-related extension.

Northeast-striking faults occur in the Thunder Mountain block at its east and west ends (Fig. 17). At the west end, these faults dip shallowly into the centre of the block (irregularly curved lines in Fig. 17), whereas at the east end they dip vertically to steeply towards the centre of the block. Both sets of faults have normal-sense displacement and probably accommodated extensional strains related to the domal arching of the block.

The north margin of the Thunder Mountain Block is more segmented than its south-side complement. A single fault does not delimit the footwall along the length of the block; each fault has an undulatory trace in outcrop; and NE-striking faulting is more prevalent in the footwall than the hanging wall.

Discussion

Were changes in the loci of strike-slip faulting progressive or temporally abrupt?

To determine whether the migration of strike-slip faulting and related stepping was abrupt or progressive requires more precise thermochronology than is presently available. In the Virginia City district, a later-stage dyke SE of Jumbo (Fig. 8b, Tkpi) and fault patterns from American Flat (Fig. 3, AF) north to the Silver City Fault strongly suggest that the NE migration was progressive. Further, quartz–adularia veins occur along a projection of the Comstock Fault south of the Silver City Fault, implying that it accommodated some strain, presumably during Comstock-age mineralization (14.1 Ma). Nevertheless, it may be inferred from the intrusion of the Davidson Granodiorite and eruption of second-stage andesitic lavas around it; the structural localization of Davidson-related hydrothermal alteration; and the general restriction of second-stage andesite to the area NE of the Silver City Fault (Fig. 8b) that, even if progressive, the migration of the locus of strike-slip faulting took place over a short period of time and the earlier strike-slip fault zone was orphaned.

What drove the migration of fault loci?

It may be inferred from the temporal correlation between the migration of strike-slip loci and the diminishing volume of andesitic volcanism and increase in dacitic volcanism in the Virginia City and Goldfield districts, that migration was not driven wholly by very local perturbations in the strain fields. However, because both districts provide evidence of migrating fronts of strike-slip and linked extensional faulting, they did so over different time spans and, therefore, at different rates. This implies something more fundamental in the evolution of strike-slip stepovers than calling on some change in plate-boundary orientations to drive stepover migration. The similarity between the districts in the ostensibly progressive migration of magmatism from SW and west to NE and east through time is intriguing, but, given the small sample size, the similarity in direction may be merely fortuitous.

What roles do misoriented basement fault zones play in stepover formation?

Misoriented, transverse basement shear zones that accommodated little Tertiary displacement are spatially correlated with extensional stepovers in the Virginia City and Goldfield mining districts (Figs 7 & 10). When a change in the locus of initial stage strike-slip faulting occurred in each district, a new point of extensional release was established where a subsequent stage strike-slip fault intersected the transverse zone. This suggests that transverse shear zones can interact sufficiently to perturb local strain fields and interrupt displacement along strike-slip zones. In the Virginia City district, when there was not a change in the locus of strike-slip faulting, the interruption of strike-slip faulting along the Silver City Fault at its intersection with the Comstock Fault migrated SE along the Silver City Fault to a new location that is not obviously correlated with the transverse zone (Fig. 7). However, the new location of strike-slip fault interruption correlates with the intersection of a vein-controlling NE-striking fault zone and the Silver City Fault at the town of Silver City (Fig. 3). The implication is that the interruption of strike-slip faulting and nucleation of stepovers is facilitated at major fault intersections whether or not the intersecting faults are misoriented.

Compression and transpression behind migrating loci of extension

Compressional deformation followed the spatial migration of the locus of strike-slip fault interruption in the Virginia City and Goldfield districts. In both districts, antiforms parallel the footwalls of migrated stepover normal-fault boundaries. Fold axes are not explained by either far-field stresses or some later tectonic event (e.g. basin–range extension). Mineralization localized along fractures parallel to fold axes link these antiforms in time to stepover evolution. Further, some ore textures at Virginia City can only be explained by syn-inversion hydrothermal activity. In both districts, it is permissive that the intrusion of dykes and small domes into the hanging walls of active normal faults resulted in shifts in the locus of extension not accompanied by a shift in the locus of strike-slip faulting (e.g. Berger *et al.* 2003).

Effects of hydrothermal alteration on the style of deformation along stepover-bounding faults

The different styles of deformation in the hanging walls of stepover-bounding normal faults in the Virginia City and Goldfield districts prior to fault inversion are interpreted to reflect differences in the stratigraphy and effects of hydrothermal alteration assemblages in each district. Within the Comstock stepover, the hanging wall is predominantly andesite (Fig. 3) with little textural variation. A stratovolcanic edifice above the core Davidson Granodiorite intrusion was variably argillized owing to the condensation of acidic magmatic vapours that emanated from the intrusion through the edifice into cool groundwaters from 15.4–15.2 Ma. By the time of the 14 Ma Comstock mineralization, this alteration was primarily preserved in the hanging wall of the Comstock normal-fault zone. Thus, the 14 Ma hydrothermal fluid flow was through a mechanically weak, strongly altered hanging wall against a relatively strong, weakly altered footwall. The hanging wall deformed by bowing outward away from the footwall along a series of convex fault segments. During fault inversion along the Comstock Zone, the 14 Ma hydrothermal activity altered pre-existing argillic alteration along the edges of the Comstock quartz–adularia veins to massive kaolinite seams up to 0.5 m thick (King 1870). The clays probably facilitated fault-slip inversion along the Comstock fault zone and helped to compartmentalize hydrothermal fluid flow.

Although the alteration in the Goldfield district is similar to the 15.4–15.2 advanced-argillic/argillic alteration at Virginia City, magmatic-vapour-stage alteration at Goldfield was imposed on unaltered, rigid rocks. Within the stepover, displacement was accommodated on discrete, linear fault planes that became more segmented by synthetic and antithetic shears as the master strike-slip fault was approached. The regularity of the style of fault deformation resulted in the localization of hard, dense masses of quartz–alunite $\pm$ pyrophyllite alteration on the faults, with broad, intervening zones of magmatic–steam quartz–alunite alteration in the matrices of older ash-flow tuffs (Fig. 4) and a restriction of argillization to the andesites and, then, primarily as selvages within a few metres of the flow-controlling fractures. From the pipe-like geometry of many orebodies (e.g. Clermont Mine, Fig. 14b), it may be inferred that, were there any vertical-axis rotational strains in the hanging wall of the Columbia Mountain Fault (Fig. 4) resulting from transpressional stresses, the strains were apparently accommodated in the junctions of the extension and shear faults that made up the fault meshes near the point of interruption of strike-slip faulting and release into the stepover.

Restraining bends in a region dominated by transtension

The development of a restraining bend in the eastern Excelsior Mountains is not the consequence of contractional deformation marginal to a regional zone of transtension. Rather, this style of deformation in the Walker Lane strike-slip fault zone persists in an area that is 40–50 km from north to south (Fig. 15) and extends east of the Excelsior Mountains across the whole of the Walker Lane structural belt at this latitude. Further, the east–west transverse fault zone that bounds the present-day eastern Excelsior Mountains clearly has accommodated significant amounts of displacement in the past (e.g. Mesozoic), as well during the period following Early Miocene Camp Douglas mineralization. Whatever its origin, this transverse zone has been a segmentation boundary in the western North American plate margin for a considerable period of time.

Implications for minerals exploration

Both Virginia City and Goldfield are considered 'world-class' (Singer 1995) epithermal mining districts, based on the amount of gold and silver produced from them. *What* made them world-class is an important problem in economic geology. Although the styles of epithermal mineralization in each district differ, they have a characteristic in common. In both districts a migration in the locus of strike-slip displacement spatially correlates with a change from predominantly andesitic magmatism to predominantly dacitic magmatism. Further, hydrothermal activity concurrent with the dacitic magmatism overlapped a period during which the locus of extension migrated. High-grade, bonanza ores were deposited in the orphaned, formerly normal-fault zones, following the spatial migration. Since ores deposited prior to the migration in each district are sub-economic to low-grade and/or smaller in volume than ore following the migration, it may be inferred that the change in permeability structure of the fault zone in the wake of the migrated extension played some role in the formation of major volumes of bonanza-grade ores.

Another implication is that anticlines formed behind migrated extensional fronts are favourable sites for mineral deposition. At Goldfield, exploration drilling has discovered high-angle, fracture-controlled mineralization parallel to the axis of an antiform, as well as stratiform mineralization in permeable layers between the fractures.

Conclusions

Because magmatic–hydrothermal fluid flow is coupled to deformation, vein-form mineralization provides useful insights into the nature and evolution of the fault systems within which the mineralization is localized. In the Lower to Middle Miocene magmatic arc in Nevada, right-lateral strike-slip fault stepovers played a primary role in the localization of volcanic centres at Virginia City, Goldfield, and Camp Douglas, as well as related hydrothermal systems.

Strike-slip transtension dominated in the Virginia City and Goldfield districts, and the loci of strike-slip faulting and related points of interruption of strike-slip faulting and nucleation of extensional stepovers migrated through time. Contractional deformation of sites of earlier extension, including folding and fault-slip inversion, occurred behind the new loci of extension. Volcanism in Virginia City lasted longer than at Goldfield, yet both districts show analogous patterns of stepover evolution. Whatever the driving forces, the migration of the loci of faulting as well as volcanism is independent of the duration of magmatism in each district.

In the Camp Douglas district, a left bend along a right-lateral fault zone crossing through a transverse basement shear zone resulted in an asymmetrical, reverse-fault-bound uplift, a restraining stepover. Longitudinal stretching and vertical-axis rotational strains within the uplifting block were accommodated on normal- to normal-oblique faults.

When hydrothermal activity occurs on faults that were abandoned during stepover migration, there is a possibility that the resulting permeability structure of the orphaned faults is favourable to high-grade ore formation. This has important implications for how vein systems are explored in strike-slip stepover settings.

The Mineral Resources Program of the US Geological Survey supports this research as part of an on-going investigation into the coupling of deformation and fluid flow in epizonal hydrothermal systems. The goal is to reduce spatial uncertainties in estimating numbers of undiscovered mineral deposits in mineral-resource assessments, and to provide insights into the formation of economically valuable, world-class ore deposits. The work has benefited from numerous interactions with colleagues, most especially R. E. Anderson and R. W. Henley.

References

ASHLEY, R. P. 1974. Goldfield mining district. *In*: *Guidebook to the Geology of Four Tertiary Volcanic Centers in Central Nevada.* Nevada Bureau of Mines and Geology Reports, **19**, 49–66.

ASHLEY, R. P. 1975. *Preliminary Geologic Map of the Goldfield Mining District, Nevada.* US Geological Survey Miscellaneous Field Investigations Map **MF-681**.

ASHLEY, R. P. & KEITH, W. J. 1975. *Distribution of Gold and Other Ore-related Elements in Silicified Rocks of the Goldfield Mining District, Nevada.* US Geological Survey Professional Papers, **843-B**, 17 pp.

BECKER, G. F. 1882. *Geology of the Comstock Lode and the Washoe District.* US Geological Survey Monographs, **3**, 422 pp.

BERGER, B. R. & HENLEY, R. W. 1989. Advances in the understanding of epithermal gold–silver deposits, with special reference to the western United States. *Economic Geology Monograph*, **6**, 405–423.

BERGER, B. R., ANDERSON, R. E., PHILLIPS, J. D. & TINGLEY, J. V. 2005. Plate-boundary transverse deformation zones and their structural roles in localizing mineralization in the Virginia City, Goldfield, and Silver Star mining districts, Nevada. *In*: RHODEN, H. N., STEININGER, R. C. & VIKRE, P. G. (eds) *Geological Society of Nevada Symposium 2005: Window to the World*, 269–281. Geological Society of Nevada, Reno, Nevada.

BERGER, B. R., TINGLEY, J. V. & DREW, L. J. 2003. Structural localization and origin of compartmentalized fluid flow, Comstock Lode, Virginia City, Nevada. *Economic Geology*, **98**, 387–408.

BJÖRNSSON, G. & BODVARSSON, G. 1990. A survey of geothermal reservoir properties. *Geothermics*, **19**, 17–27.

BLENKINSOP, T. G. 2002. Ore deposit geometry in deformation zone-hosted gold deposits. *In*: MCLELLAN, J. G. & BROWN, M. C. (eds) *Deformation, Fluid Flow and Mineralisation.* James Cook University, Economic Geology Research Unit Contributions, **60**, 4–10.

BONHAM, H. F., JR. & GARSIDE, L. J. 1979. *Geology of the Tonopah, Lone Mountain, Klondike, and Northern Mud Lake quadrangles, Nevada.* Nevada Bureau of Mines and Geology Bulletins, **92**, 142 pp.

CALKINS, F. C. & THAYER, T. P. 1945. *Preliminary Geologic Map of the Comstock Lode District.* US Geological Survey Open-File Map, scale 1:24,000.

CASTOR, S. B., GARSIDE, L. J., HENRY, C. D., HUDSON, D. M. & MCINTOSH, W. C. 2005. Epithermal mineralization and intermediate volcanism in the Virginia City area, Nevada. *In*: RHODEN, H. N., STEININGER, R. C. & VIKRE, P. G. (eds) *Geological Society of Nevada Symposium 2005: Window to the World*, 125–134. Geological Society of Nevada, Reno, Nevada.

CHRISTIE-BLICK, N. & BIDDLE, K. T. 1985. Deformation and basin formation along strike-slip faults. *In*: BIDDLE, K. T. & CHRISTIE-BLICK, N. (eds) *Strike-Slip Deformation, Basin Formation, and Sedimentation.* SEPM Special Publications, **37**, 1–34.

COX, S. F., KNACKSTEDT, M. A. & BRAUN, J. 2001. Principles of structural control on permeability and fluid flow in hydrothermal systems. Society of Economic Geologists, Reviews in Economic Geology, **14**, 1–24.

CROWELL, J. C. 1974. Origin of late Cenozoic basins of southern California. *In*: DICKINSON, W. R. (ed.) *Tectonics and Sedimentation.* SEPM Special Publications, **22**, 190–204.

DE MARSILY, G. 1986. *Quantitative Hydrogeology.* Academic Press, New York, 440 pp.

DOOLEY, T. & MCCLAY, K. 1997. Analog modeling of pull-apart basins. *AAPG Bulletin*, **81**, 1804–1826.

ELDER, J. W. 1966. *Heat and Mass Transfer in the Earth: Hydrothermal Systems.* New Zealand Department of Scientific and Industrial Research Bulletins, **169**, 115 pp.

FURBISH, D. J. 1997. *Fluid Physics in Geology.* Oxford University Press, New York, 476 pp.

GARSIDE, L. J. 1979. *Geologic Map of the Camp Douglas Quadrangle, Mineral County, Nevada*. Nevada Bureau of Mines and Geology Map **63**.

GARSIDE, L. J. & SILBERMAN, M. L. 1978. New K–Ar ages of volcanic and plutonic rocks from the Camp Douglas quadrangle, Mineral County, Nevada. *Isochron/West*, **22**, 29–32.

HENLEY, R. W. 1985. The geothermal framework of epithermal deposits. *Society of Economic Geologists, Reviews in Economic Geology*, **2**, 1–24.

HENLEY, R. W. & BERGER, B. R. 2000. Self-ordering and complexity in epizonal mineral deposits. *Annual Review of Earth and Planetary Sciences*, **28**, 669–719.

HENRY, C. D. & PERKINS, M. E. 2001. Sierra Nevada–Basin and Range transition near Reno, Nevada; two-stage development at 12 and 3 Ma. *Geology*, **29**, 719–722.

HUDSON, D. M. 2003. Epithermal alteration and mineralization in the Comstock district, Nevada. *Economic Geology*, **98**, 367–385.

HUDSON, D. M., CASTOR, S. B. & GARSIDE, L. J. 2003. *Preliminary Geologic Map of the Virginia City Quadrangle, Nevada*. Nevada Bureau of Mines and Geology Open-File Reports, **03-15**.

KING, C. 1870. *The Comstock Lode*. US Geological Survey of the 40th Parallel, **3**, 11–96.

MCKINSTRY, H. E. 1948. *Mining Geology*. Prentice-Hall, Englewood Cliffs, New Jersey, 680 pp.

Metallic Ventures Gold 2007. *Goldfield Project. World Wide Web Address*: http://www.metallicventuresgold.com.

MIN, K., RUTQVIST, J., TSANG, C. & JING, L. 2004. Stress-dependent permeability of fractured rock masses: a numerical study. *International Journal of Rock Mechanics and Mining Science*, **42**, 1191–1210.

PETRONIS, M. S., GEISSMAN, J. W., OLDOW, J. S. & MCINTOSH, W. C. 2002. Paleomagnetic and $^{40}Ar/^{39}Ar$ geochronologic data bearing on the structural evolution of the Silver Peak extensional complex, west-central Nevada. *Geological Society of America Bulletin*, **114**, 1108–1130.

ROWLEY, P. D., SNEE, L. W. *ET AL*. 1992. *Structural Setting of the Chief Mining District, Eastern Chief Range, Lincoln County, Nevada*. US Geological Survey Bulletins, **2012**, H1–H17.

SEARLES, F. S., JR. 1948. *A Contribution to the Published Information on the Geology and Ore Deposits of Goldfield, Nevada*. Geology and Mining Series No. 48. University of Nevada Bulletins, **42**, 21 pp.

SHAWE, D. R. & BYERS, F. M. 1999. *Geologic Map of the Belmont East Quadrangle, Nye County, Nevada*. US Geological Survey Geologic Investigations Series, **I-2675**.

SIBSON, R. H. 1977. Fault rocks and fault mechanisms. *Journal of the Geological Society of London*, **133**, 191–213.

SIBSON, R. H. 1998. Conditions for rapid large-volume flow. *In*: AREHART, G. B. & HULSTON, J. R. (eds) *Water–Rock Interaction*. A.A. Balkema, Rotterdam, 35–38.

SILLITOE, R. H. 1993. Epithermal models: genetic types, geometrical controls and shallow features. *In*: KIRKHAM, R. V., SINCLAIR, W. D., THORPE, R. I. & DUKE, J. M. (eds) *Mineral Deposit Modeling*. Geological Association of Canada Special Papers, **40**, 403–417.

SINGER, D. A. 1995. World class base and precious metal deposits – A quantitative analysis. *Economic Geology*, **90**, 88–104.

SNOW, D. T. 1968. Rock fracture spacings, openings, and porosities. *Journal of Soil Mechanics*, **94**, 73–91.

STEIN, S. & SELLA, G. F. 2002. Plate boundary zones: concept and approaches. *In*: STEIN, S. & FREYMUELLER, J. T. (eds) *Plate Boundary Zones*. American Geophysical Union Geodynamic Series, **30**, 1–26.

STEWART, J. H. 1982. *Geologic Map of the Walker Lake 1° by 2° Quadrangle, California and Nevada*. US Geological Survey Miscellaneous Field Studies Map **MF-1382A**.

STEWART, J. H. 1988. Tectonics of the Walker Lane belt, western Great Basin: Mesozoic and Cenozoic deformation in a zone of shear. *In*: ERNST, W. G. (ed.), *Metamorphism and Crustal Evolution of the Western United States*. Prentice-Hall, Englewood Cliffs, New Jersey, 668–713.

STEWART, J. H. & PERKINS, M. E. 1999. *Stratigraphy, tephrochronology, and structure of part of the Miocene Truckee Formation in the Trinity–Hot Springs Mountains area, Churchill County, west-central Nevada*. US Geological Survey Open-File Reports, **99–330**, 23 pp.

THOMPSON, G. A. 1956. *Geology of the Virginia City Quadrangle, Nevada*. US Geological Survey Bulletins, **1042-C**, 45–77.

VIKRE, P., FLECK, R. & RYE, R. 2005. *Ages and Geochemistry of Magmatic Hydrothermal Alunites in the Goldfield District, Esmeralda County, Nevada*. US Geological Survey Open-File Reports, **2005–1258**.

WAKABAYASHI, J., HENGESH, J. V. & SAWYER, T. L. 2004. Four-dimensional transform fault processes: progressive evolution of step-overs and bends. *Tectonophysics*, **392**, 279–301.

ZHANG, X. & SANDERSON, D. J. 1996. Effects of stress on the 2-D permeability tensor of natural fracture networks. *Journal of Geophysics International*, **125**, 912–924.

Index

Note: Page numbers relating to figures are denoted in *italics*, page number relating to tables are in **bold** type.